Coastal Waters Monitoring Using Remote Sensing Technology

Coastal Waters Monitoring Using Remote Sensing Technology

Editors

Stefano Vignudelli
Jérôme Benveniste

MDPI • Basel • Beijing • Wuhan • Barcelona • Belgrade • Manchester • Tokyo • Cluj • Tianjin

Editors
Stefano Vignudelli
Area della Ricerca CNR S. Cataldo
Italy

Jérôme Benveniste
European Space Agency (ESA-ESRIN)
Italy

Editorial Office
MDPI
St. Alban-Anlage 66
4052 Basel, Switzerland

This is a reprint of articles from the Special Issue published online in the open access journal *Remote Sensing* (ISSN 2072-4292) (available at: https://www.mdpi.com/journal/remotesensing/special_issues/Coastal_RS).

For citation purposes, cite each article independently as indicated on the article page online and as indicated below:

LastName, A.A.; LastName, B.B.; LastName, C.C. Article Title. *Journal Name* **Year**, *Volume Number*, Page Range.

ISBN 978-3-0365-1232-7 (Hbk)
ISBN 978-3-0365-1233-4 (PDF)

Cover image courtesy of ESA.

Contents

About the Editors

Stefano Vignudelli is a senior scientist employed at the Consiglio Nazionale delle Ricerche (National Research Council) in Pisa, Italy. He has over 25 years of scientific experience in the area of satellite remote sensing (radar altimetry, in particular) for studying coastal and inland environments (water level variability, in particular). Major interests include processing methods for data analysis, validation with local field observations, multi-sensor synergy, exploitation. His most significant accomplishment has been leading the development of satellite radar altimetry in the coastal zone to provide improved measurements for sea level research and applications. He is one of the organizers of a regular series of coastal altimetry workshops (Silver Spring 2008, Pisa, 2008, Frascati 2009, Porto 2010, San Diego 2011, Riva del Garda 2012, Boulder 2013, Lake Constance 2014, Reston 2015, Florence 2017, Frascati 2018 and 2020) and co-editor of the Springer Book *Coastal Altimetry* (20 chapters, 70 people involved, and top 25% Springer books) and subsequent reviews of the topic (Survey Geophysics 2019, Elsevier 2019). He is actively involved in international cooperation through joint projects, exchange visits, attendance at workshops and capacity building. He is also co-author of five chapters of books and more than 100 publications (half in peer-reviewed journals). He is coordinator/partner/evaluator of scientific projects. He is a member of organizing/scientific committees and international societies. He is actually associate editor for Elsevier's Advances in Space Research journal in the area of satellite oceanography, vice-president elect of the Pan Ocean Remote Sensing Conference (PORSEC) association and chair of the COSPAR Sub-Commission A2 Ocean Dynamics, Productivity and the Cryosphere for the period 2021–2024.

Jérôme Benveniste received his PhD in Oceanography from Space from the University of Toulouse, France, in 1989. After a post-doc in space data assimilation in ocean models at MIT (Massachussets Institute of Technology, Boston, USA), he was recruited by the European Space Agency. He has been at the ESA Earth Observation data centre (ESRIN) near Rome, Italy, since 1992, where he is in charge of the ERS-1, ERS-2, ENVISAT, CryoSat, Sentinel-3 and Sentinel-6 radar altimetry and GOCE gravimetry data exploitation. He interacts with ESA Principal Investigators, fosters the scientific community by organizing scientific symposia and workshops and regularly initiates research and development projects. He was promoted "Senior Advisor" at ESA in 2008. He is co-editor of a Springer book on Coastal Zone Radar Altimetry, published in 2011. He is editor of a peer-reviewed scientific journal and guest editor of five journal Special Issues. He is author or co-author of over 100 peer-reviewed articles. Jérôme Benveniste launches and steers ESA-funded R&D activities, such as the Climate Change Initiative Sea Level Project (2009–) and its current sequel focused on the Coastal Zone, projects on Sea Level Budget Closure, Inland Water Altimetry, Runoff, River Discharge, Ocean Heat Content, Eastern Boundary Upwelling Systems, Coastal Oceanography and Coastal Hazards. Jérôme Benveniste was elected chair of the Oceanography and Cryosphere Scientific Sub-Commission (A2) of the Committee on Space Research (COSPAR), for a 2008–2012 mandate, and re-elected for a second mandate (2012–2016). He has since been elected vice-chair of the COSPAR Earth Observation Commission (A) for a 2016–2024 mandate (re-elected in 2020). He is a Solicited Member of CEOS-COAST as ambassador for the ESA. He is a Solicited Member of the COAST-Predict Advisory Committee. He is the designated ESA coordinator for the UN Decade of Ocean Science and Sustainable Development.

Preface to "Coastal Waters Monitoring Using Remote Sensing Technology"

At present, about 10% of the global population lives in the world's coastal zones, mostly concentrated in the world's largest megacities. In many regions, the population is exposed to a variety of natural hazards (e.g., extreme weather, such as damaging cyclones and storm surges), to consequences of global climate change (e.g., sea level rise), and to the direct impacts of human activities. In low-lying coastal areas, some factors combine negatively, thus increasing risks for coastal dwellers. For example, climate-related sea level rise increases the risk of flooding and coastal erosion during extreme events and can also cause salt water intrusion into rivers and coastal aquifers on which people depend. Land subsidence, caused by groundwater, oil, and gas extraction in coastal megacities, is another example of an amplifier of the impacts of climate-related sea level rise. In addition, because of strong anthropogenic pressures, coastal zones are already suffering ecological and biological stresses, for example, poor water quality, pollution, and destruction of marine ecosystems. Space-based observations, complemented by in situ networks, have demonstrated their capability to provide precise and systematic information about processes acting in the coastal zones worldwide, among them extreme events and phenomena related to climate change and variability, as well as evolving anthropogenic conditions. This volume is a collection of papers that originated as a Special Issue, focused on some recent advances related to the usage of remote sensing observations alone or in synergy with in situ measurements and modeling tools in order to monitor ocean processes or exploit them in applications in the coastal zone. Examples include coastal sea level changes; land–sea interaction (river flow and river plumes); water quality (phytoplankton and sediment load); small-scale shelf currents; ocean tides; upwelling and sea surface temperature variability; wave climate; bathymetry estimation. Applications refer to renewable energy; aquaculture; extreme events (storm surges and hurricanes). The editors of this book are grateful to all the contributing authors, reviewers, journal editors, and the production team.

Stefano Vignudelli, Jérôme Benveniste

Editors

Article

The Spatial-Temporal Distribution of GOCI-Derived Suspended Sediment in Taiwan Coastal Water Induced by Typhoon Soudelor

Pham Minh Chau [1,2], Chi-Kuei Wang [1,*] and An-Te Huang [1]

[1] Department of Geomatics, National Cheng Kung University, No. 1 University Road, East District, Tainan City 701, Taiwan; chaupm.ctt@vimaru.edu.vn (P.M.C.); end7879520@gmail.com (A.-T.H.)

[2] Department of Civil Engineering, Vietnam Maritime University, 484 Lach Tray Street, Haiphong City 180000, Vietnam

* Correspondence: chikuei@ncku.edu.tw; Tel.: +886-6-275-7575 (ext. 63825)

Abstract: This paper discusses the use of a Geostationary Ocean Color Imager (GOCI) to monitor the spatial–temporal distribution of suspended sediment (SS) along the coastal waters of northern Taiwan which was affected by Typhoon Soudelor from 8 to 10 August 2015. High temporal resolution satellite images derived from GOCI were processed to generate four-day average images of SS for pre- and post-typhoon periods. By using these four-day average images, characteristics of SS along the north of Taiwan coastal water can be tracked. The results show that SS concentration increased in the four-day average image immediately after the typhoon (11–14 August), and then decreased in the four-day average image 9 to 12 days after the typhoon (19–22 August). The mouths of the Dajia River and Tamsui River were hotspots of SS, ranging from 9 to 15 g/m^3 during the two post-typhoon periods. Moreover, the maximum suspended sediment (SS_{max}) and its corresponding time (t_{max}) can be computed using GOCI hourly images for the post-typhoon period from 08:30 on 11 August to 08:30 on 22 August. The results show that SS_{max} occurred in the west coastal water within 4 days post-typhoon, and SS_{max} occurred in the east coastal water 9 to 12 days post-typhoon. Furthermore, an exponential decay model was used to compute the time when 90% of typhoon-induced SS was dissipated after Typhoon Soudelor (t_{90}). It was found that t_{90} in the mouths of the Tamsui River and Heping River was the longest among all coastal waters of our study area, with a range of 360–480 h. River discharge and ocean currents with suspended sediment concentration are discussed.

Keywords: GOCI; suspended sediment; Typhoon Soudelor; spatial–temporal distribution

1. Introduction

Suspended sediment (SS) is a key part of studying shallow waters, such as coastal regions, because of its influence on the marine environment and ecosystems [1]. Therefore, monitoring the characteristics of SS can aid in better understanding the bio-geomorphological processes and validate spatially distributed hydrodynamic and transport models in coastal water regions [2]. There are many monitoring methods, such as in situ measurements with a cruise, station observations, numerical models, remote sensing, etc. In situ measurements with a cruise, numerical models, and station observations are costly and time-consuming [3,4]. Remote sensing provides a viable solution for monitoring SS in coastal waters because it can cover large areas at the same time. Moreover, compared with other methods, satellite images also offer richer spatial information and can overcome operational cost issues due to state-of-the-art technologies. For example, the first geostationary ocean color observation satellite has been used for coastal water turbidity and Sentinel-3 missions for scientific observations of the ocean [5,6].

Many regions around the world are affected by tropical storms, including Taiwan. Typhoon-induced suspended sediment (SS) in the coastal water region has an impact on

Citation: Chau, P.M.; Wang, C.-K.; Huang, A.-T. The Spatial-Temporal Distribution of GOCI-Derived Suspended Sediment in Taiwan Coastal Water Induced by Typhoon Soudelor. *Remote Sens.* **2021**, *13*, 194. https://doi.org/10.3390/rs13020194

Received: 6 November 2020
Accepted: 5 January 2021
Published: 8 January 2021

Publisher's Note: MDPI stays neutral with regard to jurisdictional claims in published maps and institutional affiliations.

the marine environment. For instance, Typhoon Morakot had an influence on the marine environment in the East China Sea inner shelf and Okinawa Trough [7,8]. The heavy rains and episodic cyclones associated with typhoons increase the total suspended sediment, sea surface temperature, and phytoplankton. In addition, the high waves and strong wind speeds of typhoons, inducing re-suspension of bottom sediments, have been discussed [9]. The ocean surface current related to Hurricane Sandy, Typhoon Morakot, and Typhoon Saola caused the spreading of suspended sediments from the coast to the open sea [10–12]. No previous studies have used remote sensing to investigate the spatiotemporal distribution of suspended sediment in Taiwan coastal waters, induced by a typhoon. We used satellite images to do this in the north of Taiwan.

Several international studies have used satellite images to assess sediment in coastal water regions. For example, remote sensing has been used to assess typhoon-induced SS concentrations. In Apalachicola Bay, Florida, USA, observations of typhoon-induced SS were conducted by using 250 m Terra MODIS (Moderate Resolution Imaging Spectroradiometer) images during Hurricane Frances [13]. The impact of Typhoon Saomai on SS concentration in the East China Sea was calculated using Aqua and Terra MODIS images [14]. Combinations of multi-satellite images (including MODIS, MERIS (the Medium Resolution Imaging Spectrometer), and GOCI) were used to show the dynamics of suspended sediment associated with Typhoon Tembin in the East China Sea [15]. The sediment transport in the Taiwan Strait induced by Typhoons Soulik and Morakot has been monitored by using Aqua MODIS images [16,17]. The spatial–temporal distribution of SS has not yet been considered because of the limitation of quality data under typhoons. Therefore, this paper tries to bridge the gap between the spatial–temporal distribution of SS induced by a typhoon and data limitations.

The Geostationary Ocean Color Imager (GOCI), a satellite sensor, can overcome the limitation of quality data under typhoon weather conditions due to its temporal resolution [18]. The GOCI is operated by the Korea Ocean Satellite Center (KOSC) at the Korea Institute of Ocean Science and Technology (KIOST). It is the first ocean color satellite placed in geostationary orbit to provide eight hourly images during the daytime (from 08:30 to 15:30 local time at one-hour intervals) with a spatial resolution of 500 m. GOCI covers about 2500 km × 2500 km centering on the Korean Peninsula (at the center of 130° E, 36° N), including the north of Taiwan. It has six visible bands from 412 to 680 nm and two near-infrared bands at 745 and 865 nm. The bands at wavelengths of 555 and 660 nm are used for suspended sediment extraction [19–22]. All of the existing studies related to GOCI-derived suspended sediment focused on monitoring the temporal variation of water turbidity and the diurnal dynamics of suspended sediment in coastal water. For instance, GOCI hourly images have been used to monitor the diurnal and seasonal variability of suspended sediment concentration in a macro-tidal estuary [23]. GOCI images have been used to monitor the suspended sediment in Taihu Lake [24] and the coastal waters of Zhejiang, China [25], as well as Gyeonggi Bay on the west coast of Korea [26]. Moreover, GOCI also monitors long-term suspended sediment concentration and estimates ocean surface currents hourly [27]. However, using GOCI to monitor the spatiotemporal distribution of typhoon-induced SS in coastal waters has not been considered in previous case studies.

This study used GOCI to monitor the spatial and temporal distribution of SS pre- and post-Typhoon Soudelor, which made landfall in Taiwan in August 2015. Furthermore, by taking advantage of GOCI with time-series hourly images of SS after the typhoon, the temporal decay of the SS pattern can be computed by an exponential regression, and the time SS recovered to its pre-typhoon value can be estimated. This approach, which is the core of the study, quantifies the typhoon-induced spatial and temporal distribution of SS along Taiwan coastal water. Finally, factors such as river discharge and ocean currents could have affected the discussed spatiotemporal distribution of SS.

2. Materials and Methods

2.1. Study Area

The study area is located on the northern coast of Taiwan (Figure 1). There are 9 rivers administered by the Taiwan central government (ATCG) [28]. Seven of the rivers are on the west side of Taiwan Island, and two are on the east side.

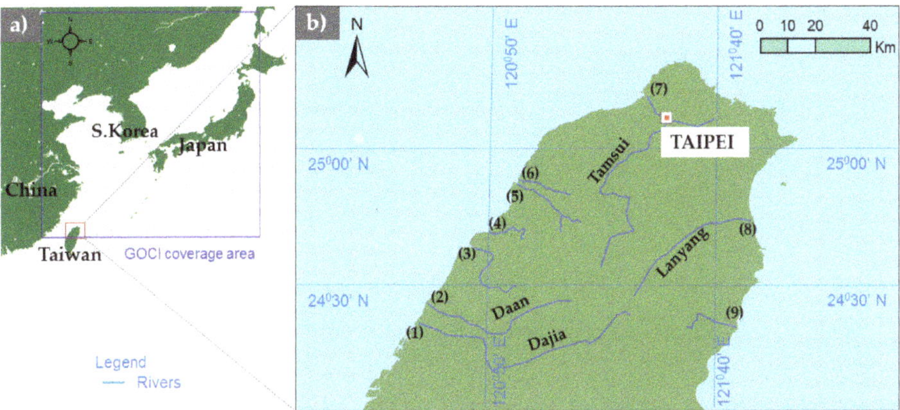

Figure 1. (**a**) Visualization of Geostationary Ocean Color Imager (GOCI) coverage area; (**b**) study area.

The statistics of the annual discharge of the 9 rivers were provided by the Taiwan River Restoration Network [28]. Tamsui River has the largest annual discharge of 7443 m^3/s, followed by Lanyang and Dajia with 2773 m^3/s and 2569 m^3/s, respectively. Other rivers have a lower discharge of less than 2000 m^3/s, and the Fengshan River has the lowest discharge of 376 m^3/s.

2.2. Typhoon Soudelor

Typhoon Soudelor formed in the middle of the Pacific Ocean on 20 July 2015, and became a super typhoon (category 5 on the Saffir–Simpson hurricane wind scale) on 29 July [29]. Typhoon Soudelor made landfall in the east of Taiwan at 04:40 local time on 8 August 2015 and brought torrential rain. The typhoon then moved north-westwards through eastern China and degraded to a tropical depression on 9 August 2015 [30–33].

2.3. GOCI Satellite Images

In this study, GOCI level-2 images were downloaded from the NASA Ocean Color website (https://oceancolor.gsfc.nasa.gov/). Then, SS was extracted from GOCI hourly images with a spatial resolution of 500 m. These hourly images were binned using the arithmetic mean algorithm, implemented in SeaDAS [34,35], to create daily SS images with a spatial resolution of 500 m (Figure 2).

During the period of Typhon Soudelor, cloud cover caused data voids in many of the hourly GOCI images. The problem was more severe on 8–10 and 15–18 August, when GOCI images were not available for our study area (Figure 2). To better visualize the spatial distribution of typhoon-induced SS, i.e., minimize the data voids, 4 daily GOCI images were further binned to generate a 4-day average image with a spatial resolution of 500 m.

A 4-day average image immediately before the typhoon ($\overline{SS}_{4-7}$), binned from the daily SS images of 4–7 August 2015, was generated. Similarly, a 4-day average image immediately after the typhoon ($\overline{SS}_{11-14}$), binned from 11–14 August, was generated. Moreover, to monitor the decrease in suspended sediment, a 4-day average image 9 to 12 days after the typhoon ($\overline{SS}_{19-22}$), binned from the daily SS images of 19–22 August, was generated. Figure 2 shows the data processing of GOCI images in this study.

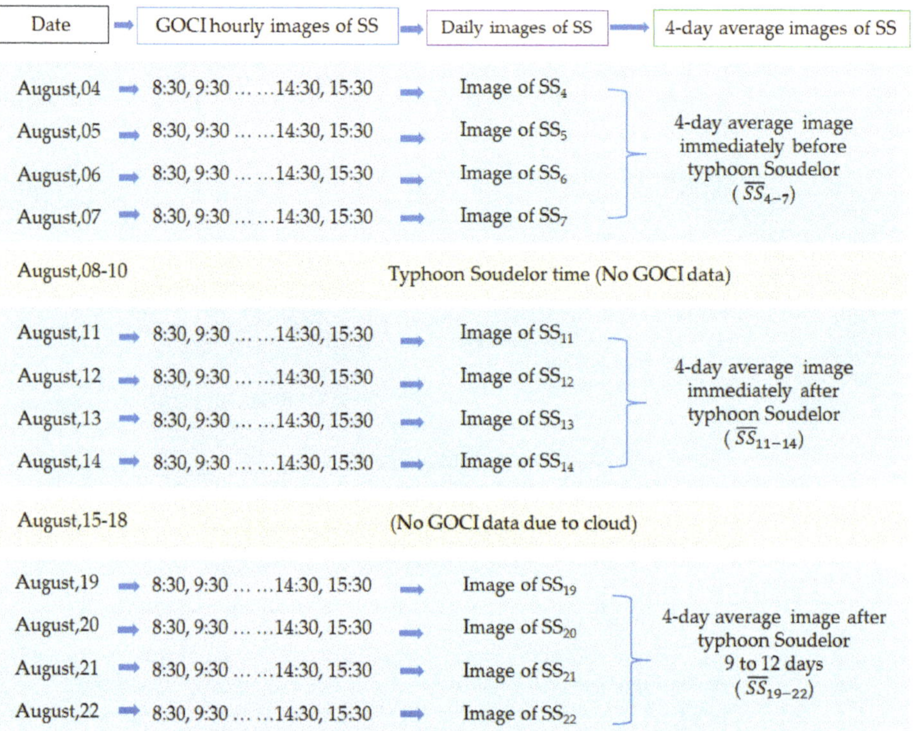

Figure 2. GOCI data processing for generating $\overline{SS}_{4-7}$, $\overline{SS}_{11-14}$, and $\overline{SS}_{19-22}$. SS: suspended sediment.

2.4. Quantitative Retrieval Algorithm of SS

In this paper, SS images were derived from level-2 GOCI images by using the algorithm developed by Moon et al. (2010) based on in-situ SS samples (Equation (1)) [20]:

$$SS = 945.07 \times (Rrs(555))^{1.137} \tag{1}$$

where Rrs (555) is remote sensing reflectance at a wavelength of 555 nm, and SS is reported in g/m^3. The algorithm was implemented in the GOCI Data Processing System (GDPS) by KOSC [36].

2.5. Temporal Decay of SS after Typhoon

When showing the temporal history of hourly SS of a GOCI pixel (Figure 3), the maximum SS value (SS_{max}) was reached in a few days, varying with pixel locations, after Typhoon Soudelor made landfall on Taiwan. In this research, SS_{max} of each pixel was determined as the largest hourly SS value of that pixel from 11–22 August. The time when the pixel reaches SS_{max} is denoted as t_{max}.

An interesting feature in the temporal history of SS (Figure 3) is its decaying pattern, where the SS value decreases after t_{max}. We proposed using a decaying model to quantify the pattern by fitting an exponential curve to hourly SS data for each GOCI pixel via regression. It was found (with internal trials) that a robust result of the regression was obtained for GOCI pixels that had more than 5 hourly SS values available after their t_{max}. Figure 3 shows an example of the regression result depicting the decaying exponential model after t_{max}, despite the GOCI pixel not having data from 15 to 18 August due to cloud cover.

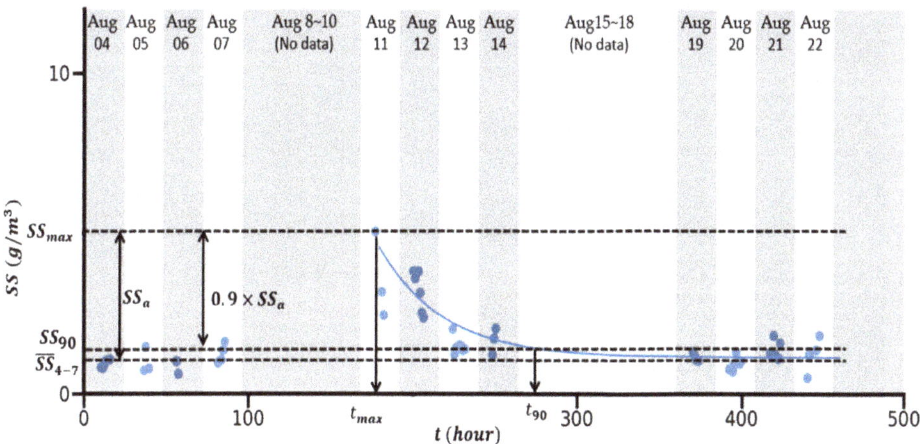

Figure 3. Example of temporal decay of suspended sediment (*SS*). Corresponding SS_{90} with t_{90} are denoted.

Ideally, taking advantage of this decaying model, the time for the pixel to return to its pre-typhoon state could be estimated using the regression result. However, it was found that the decaying exponential model of many GOCI pixels only approaches the pre-typhoon state asymptotically, i.e., the regression line shown in Figure 3 does not intersect with $\overline{SS}_{4-7}$. Instead, we demonstrate the use of this model by computing dissipation of 90% increased SS. This would fit for the application of coastal ecology management, as the ecosystem is resilient to a certain increase in SS for a short period of time.

The amount of increased SS, denoted as SS_a, for each GOCI pixel, can be computed as

$$SS_a = SS_{max} - \overline{SS}_{4-7} \tag{2}$$

and SS_{90}, which represents 90% of SS_a, is dissipated from its maximum value (Figure 3).

$$SS_{90} = SS_{max} - 0.9 \times SS_a \tag{3}$$

Furthermore, the time corresponding to SS_{90} can be identified via the regressed exponential decay and is denoted as t_{90} (Figure 3).

3. Results

3.1. Spatial–Temporal Analysis of SS Pre- and Post-Typhoon Soudelor

During the pre-typhoon period ($\overline{SS}_{4-7}$ in Figure 4a), SS with a concentration of 3–6 g/m³ was mainly distributed along the west coastal water. The mouths of both the Dajia River and Tamsui River had an SS greater than 6 g/m³. The SS concentration in the east coastal water was less than 3 g/m³.

During the post-typhoon period of 11–14 August ($\overline{SS}_{11-14}$ in Figure 4b), the SS of the Taiwan coastal water generally increased. The SS along the west and east coastal waters increased to values greater than 6 and 3 g/m³, respectively. Meanwhile, the mouths of the Dajia River and Tamsui River (west coastal water) had SS values greater than 9 g/m³ and the Lanyang River mouth (east coastal water) had an SS value greater than 6 g/m³.

During the post-typhoon period of 19–22 August ($\overline{SS}_{19-22}$ in Figure 4c), the general distribution of SS along the west coastal water was similar to that of the pre-typhoon period (Figure 5a), except for the mouths of the Dajia River and Tamsui River having an SS value greater than 9 g/m³. A belt with relatively high SS values (greater than 3 g/m³) was found along the east coastal water with a width of approximately 3 km. In addition, the mouths of both the Lanyang River and Heping River had SS values greater than 6 g/m³.

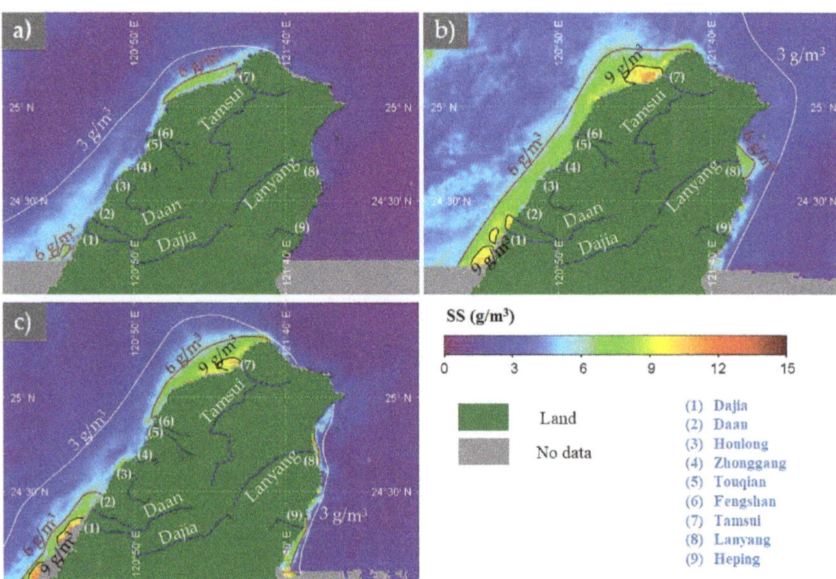

Figure 4. Spatial distribution of SS pre- and post-typhoon: (**a**) $\overline{SS}_{4-7}$; (**b**) $\overline{SS}_{11-14}$; (**c**) $\overline{SS}_{19-22}$.

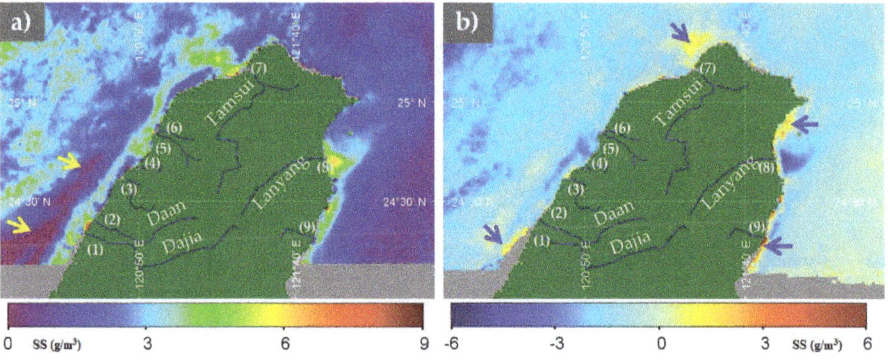

Figure 5. Temporal difference of SS: (**a**) $\overline{SS}_{11-14} - \overline{SS}_{4-7}$; (**b**) $\overline{SS}_{19-22} - \overline{SS}_{11-14}$.

Comparing the SS between the post-typhoon period of 11–14 August ($\overline{SS}_{11-14}$) and pre-typhoon ($\overline{SS}_{4-7}$), it was found that all pixels increased more than 1 g/m³ (Figure 5a), except for a tongue-shaped area (denoted by arrows in Figure 5a) located 10–20 km off the west coastal water of Taiwan. In addition, the distribution of increased SS showed prominent heterogeneity along the coastal water of Taiwan. A high increase in SS (i.e., greater than 6.0 g/m³) was found at the mouths of the Dajia, Daan, Tamsui, and Lanyang Rivers. Three regions with a low SS increase (less than 2 g/m³) along the coast, from the Daan River mouth to Zhonggan River mouth, the Fengshan River mouth to Tamsui River mouth, and the Tamsui River to Lanyang River mouth, were also identified (Figure 5a).

Comparing the SS between two post-typhoon periods of 19–12 August ($\overline{SS}_{19-22}$) and 11–14 August ($\overline{SS}_{11-14}$), it was found that most of the pixels decreased while some remained with increased SS values (Figure 5b). Most of the pixels decreased to less than 3 g/m³. The Lanyang River mouth showed the most significant decrease of 6 g/m³, followed by the Zhonggang and Tauqian River mouths, with a reduction of 3 g/m³. In contrast, the Dajia River mouth showed an increased SS value greater than 3 g/m³ and the Heping

River mouth showed an increased SS value of 6 g/m^3. It is also interesting to note that neighboring regions with increased SS (greater than 3 g/m^3) were found northeast of the Tamsui River mouth and north of the Lanyang River mouth. The regions with increased SS are indicated by arrows in Figure 5b.

With the GOCI average SS data, it was observed that SS in the west coastal water was consistently greater than that on the east coastal water regardless of the effect of Typhoon Soudelor. In addition, hotspots of high SS value (greater than 9 g/m^3) were found at the Dajia and Tamsui River mouth in the two post-typhoon periods ($\overline{SS}_{11-14}$ and $\overline{SS}_{19-22}$). With the GOCI data for pre-typhoon ($\overline{SS}_{4-7}$) and post-typhoon ($\overline{SS}_{11-14}$), it was observed that the Taiwan coastal water showed a prominent increase in SS induced by Typhoon Soudelor.

3.2. SS_{max} and t_{max}

Figure 6a shows a visualization of the maximum suspended sediment (SS_{max}) for each GOCI pixel of Taiwan coastal water during the post-typhoon period (from 08:30 on 11 August to 15:30 on 22 August). Figure 6b shows its corresponding time (t_{max}) derived from GOCI hourly data.

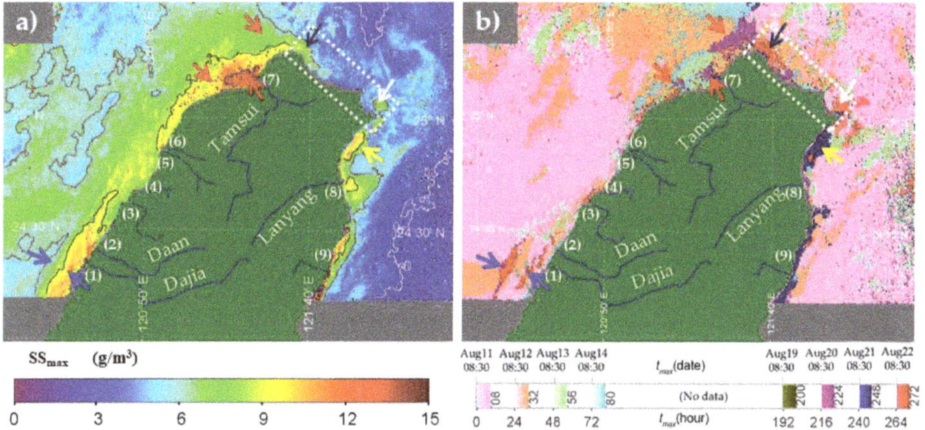

Figure 6. (**a**) Maximum suspended sediment (SS_{max}); (**b**) its corresponding time (t_{max}).

Generally, the west coastal water from the Dajia River mouth to the Tamsui River mouth showed SS_{max} in the range of 9–15 g/m^3 with a corresponding t_{max} of 0–80 h (Figure 6b). This indicates that most of the west coastal water reached SS_{max} during the post-typhoon period of 11–14 August (within four days after Typhoon Soudelor). The exceptions were two strips with t_{max} greater than 264 h located 5 and 15 km away from the Dajia River mouth (denoted by blue arrows in Figure 6b), and three regions with t_{max} greater than 216 h located near the Tamsui River mouth (denoted by red arrows in Figure 6b). Further small regions with t_{max} greater than 264 h sporadically occurred in the coastal water from the Houlong River mouth to Zhonggang River mouth and the Feshang River mouth to Tamsui River mouth.

It is interesting to note that the general appearance of the SS_{max} of the northeast coastal water (indicated by the dotted white rectangle) was lower than the west and east coastal waters, with a range of less than 6 g/m^3, while its t_{max} value showed a large variation of 0–272 h. Most of the coastal water of the northern coast had SS_{max} values in the range of 3–6 g/m^3 with t_{max} of 24–80 h (within four days after Typhoon Soudelor). At the west and east ends (indicated by black and white arrows, respectively, in Figure 6) of the northeast coast, SS_{max} was in the range of 6–9 g/m^3 with a corresponding t_{max} greater than 264 h. It also indicates that both SS_{max} and t_{max} of the northern coastal waters are continuous data from the west and east coastal water. Even though there are no ATCG rivers in this region, SS_{max} and t_{max} have been linked to river-derived suspended sediment.

The east coastal water generally took a long time to reach SS_{max}, with a range of 9–15 g/m^3 compared to the west and northeast coastal waters, with t_{max} of 192–272 h (9–12 days after Typhoon Soudelor). The exception is the coastal water near the Lanyang River mouth, which reached an SS_{max} with t_{max} of 24–80 h (within 4 days after the typhoon). The coastal water near Heping River mouth showed SS_{max} with a range of 6–15 g/m^3 corresponding to t_{max} greater than 240 h (11–12 days after the typhoon). Interestingly, a region north of the Lanyang River mouth (indicated by yellow arrows) also showed a local high SS_{max} of 9–15 g/m^3 with a corresponding t_{max} value similar to the Heping River mouth.

3.3. SS_{90} and t_{90}

Figure 7a shows a visualization of SS_{90}, which means 90% of increasing SS dissipated from its maximum value for each GOCI pixel of Taiwan coastal water. Figure 7b shows the corresponding time (t_{90}) via regressed exponential decay.

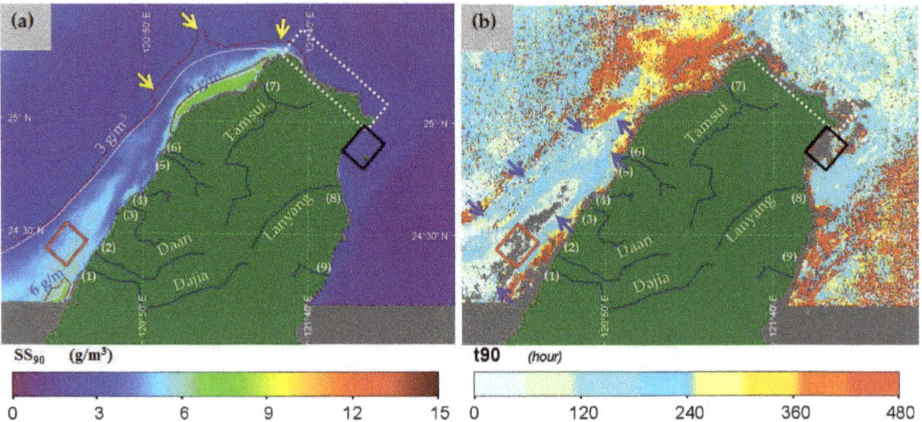

Figure 7. (**a**) SS concentration of each pixel reduced to 90% after typhoon-induced impact; (**b**) its corresponding time t_{90}. White contour lines indicate $\overline{SS}_{4-7}$; red contour lines indicate SS_{90}.

The comparison between SS_{90} (Figure 7a) and $\overline{SS}_{4-7}$ (Figure 4a) indicates that SS_{90} was similar to the pre-typhoon state ($\overline{SS}_{4-7}$), except for the SS of the Tamsui River mouth (within 3–6 g/m^3; indicated by arrows in Figure 7a). Particularly, SS within 3–6 g/m^3 was observed mainly in the west coastal water, while it was less than 3 g/m^3 in the east coastal waters; SS greater than 6 g/m^3 also only appeared at the Tamsui and Dajia River mouths (Figures 7a and 4a). There was a slight difference in SS located 40 km off the Tamsui River mouth. SS_{90} showed that the SS of 3 g/m^3 extended farther into the sea than in $\overline{SS}_{4-7}$.

The west coastal waters from the Dajia River mouth to the Tamsui River mouth showed a t_{90} of more than 240 h, except for a tongue-shaped area with a range of 10–40 km off the west coastal water (indicated by arrows in Figure 7b), showing a t_{90} of less than 240 h. The Tamsui River mouth extending within 20 km of the coastline indicated a t_{90} with a range of 240–480 h.

The northeast coastal water to the Lanyang River mouth (Figure 7a) showed an SS_{90} associated with a t_{90} of less than 240 h, except for small scattered regions that appeared closest to the coastline with a t_{90} of around 480 h (Figure 7b). Lastly, the east coastal waters from the Lanyang to Heping River mouth showed a t_{90} of 360–480 h (Figure 7b).

It should be noted that some pixels available in the SS_{90} are not visualized by the t_{90}. There are two cases in which the decaying model works without advantages, leading to the pixel in t_{90} not being visualized. When GOCI pixels (indicated by a red rectangle in Figure 7) have a gap between SS_{max} and $\overline{SS}_{4-7}$, which is close (Figure 3), t_{90} is not visualized due to no decay. When the number of GOCI pixels after t_{max} has fewer than five values, it is also not visualized because of the lack of data (indicated by the black rectangle in Figure 7). Otherwise, the proposed methodology, which uses an exponential temporal decaying model, shows a distinct advantage when GOCI pixels have more than five hourly SS values available after their t_{max}. It is possible to compute the t_{90} with robust results post-typhoon.

4. Discussion

According to previous studies in Taiwan, one possible reason for the SS derived from a typhoon is that it is strongly affected by river discharge. For example, using a case study in Choshui River, Taiwan, and Typhoon Mindulle, the authors of [37] indicated that in a floodplain, more than half of the suspended sediment originating from mountain rivers running into Taiwan coastal waters was generated by river discharge. The authors of [38] observed the Jhoushui River and an adjacent coastal zone in the Taiwan Strait and summarized that the river discharges most of the sediment during the relatively short periods of torrential rain often associated with typhoons. Moreover, the authors of [39,40] indicated that suspended sediment discharge during typhoon events was linked to landslides and rainfall in Taiwan. The authors of [41] considered the impact of typhoons on sediment discharge in Taiwan. River discharge also impacted the change in sediment concentration in the Tamsui River, as discussed in [42]. Even though we have the same opinion, there are no recorded data to support the Typhoon Soudelor case. Therefore, examining the mechanical factors, such as river discharge, related to typhoons is beyond the scope of this study. We only discuss this based on the mean annual discharge data, which were provided by the Taiwan River Restoration Network [28,43].

In terms of the mean annual discharge related to the nine central rivers administered by ATCG, the Tamsui River supplies the largest amount with 7443 m^3/s, followed by the Lanyang and Dajia Rivers with 2773 and 2596 m^3/s, respectively. This is the reason why the Tamsui, Dajia, and Lanyang River mouths act as hotspots with high SS values (SS_{max} above 9 g/m^3) during the two post-typhoon periods ($\overline{SS}_{11-14}$ and $\overline{SS}_{19-22}$). Meanwhile, other rivers show a mean annual discharge lower than 2000 m^3/s, and they influence the coastal regions with SS_{max} values in the range of 6–9 g/m^3. The central rivers are mainly located in the western part of Taiwan, which may be why many GOCI pixels of the west coastal waters are more influenced than those of the east coastal waters, regardless of the effect of Typhoon Soudelor.

Other factors such as tide level, waves, wind speed, and surface currents that impact the spatiotemporal distribution of SS should also be discussed. All of these factors have been considered by many scientists. According to the authors of [44], by using shipboard observations for estimation, transport and tidal currents in the Taiwan Strait were northward (into the East China Sea). This is similar to the ocean surface current from GOCI satellite imagery in summer around the north of Taiwan coastal water being northward [45]. Therefore, we also believe that after Typhoon Soudelor, the ocean surface was a major factor in increasing the outbreak of suspended sediment north of Taiwan. However, due to the limitation of recorded data on Typhoon Soudelor, the transport mechanism responsible for the sediment warrants further investigation.

5. Conclusions

This paper proposes a new approach using the GOCI to monitor the spatial and temporal distribution of suspended sediment in coastal areas affected by typhoons. Remote sensing technology was used instead of other methods such as in situ measurements, numerical models, and station observations to track post-typhoon sediment concentration in Taiwan coastal waters.

The spatial distribution of SS has been highlighted by using the GOCI four-day average image of SS pre- and post-Typhoon Soudelor. As a result, several pixels with an SS above 6 g/m^3 in the west coastal waters were consistently more significant than in the east coastal waters regardless of the typhoon. The Dajia and Tamsui River mouths were hotspots of increased SS and SS_{max} (above 9 g/m^3) during two post-typhoon periods (11–14 and 29–22 August).

According to the GOCI hourly data after the typhoon, SS_{max} was in the range of 6–15 g/m^3, corresponding to t_{max} within four days in the west coastal water, while the east coastal water was 9–12 days. Furthermore, using exponential regression decay to visualize SS_{90} for each GOCI pixel in Taiwan coastal water indicates that SS_{90} was in an asymptotic pre-typhoon state. The corresponding time t_{90} shows that GOCI pixels in both the Tamsui and Heping River mouths generally took the longest time, in a range of 360–480 h.

River discharge could have a significant impact on the post-typhoon sediment characteristics of Taiwan coastal waters. Other factors such as tide level, waves, wind speed, and surface currents could also affect the spatiotemporal distribution of suspended sediment. We suggest that this should be investigated in the future by using a successfully recorded dataset with a new typhoon.

Author Contributions: Conceptualization, C.-K.W.; Data curation, P.M.C.; Formal analysis, P.M.C.; Investigation, P.M.C.; Methodology, P.M.C., C.-K.W. and A.-T.H.; Project administration, C.-K.W.; Software, A.-T.H.; Supervision, C.-K.W.; Validation, P.M.C.; Visualization, P.M.C. and A.-T.H.; Writing—original draft, P.M.C. and A.-T.H.; Writing—review & editing, C.-K.W. All authors have read and agreed to the published version of the manuscript.

Funding: The APC was supported in part by Higher Education Sprout Project, Ministry of Education to the Headquarters of University Advancement at National Cheng Kung University (NCKU).

Institutional Review Board Statement: Not applicable.

Informed Consent Statement: Not applicable.

Data Availability Statement: The data that support the findings of this study are available from the corresponding author upon reasonable request.

Acknowledgments: The authors thank NASA for providing free GOCI satellite images and the SeaDAS application. We also thank the Korea Ocean Satellite Center for providing the GDPS application. The authors would also like to thank the Taiwan River Restoration Network for sharing river system information. Finally, the authors deeply acknowledge anonymous reviewers for their constructive comments and suggestion toward improving this manuscript.

Conflicts of Interest: The authors declare no conflict of interest.

References

1. Snelgrove, P.V.R.; Henry Blackburn, T.; Hutchings, P.A.; Alongi, D.M.; Frederick Grassle, J.; Hummel, H.; King, G.; Koike, I.; Lambshead, P.J.D.; Ramsing, N.B.; et al. The importance of marine sediment biodiversity in ecosystem processes. *Ambio* **1997**, *26*, 578–583.
2. Ouillon, S. Why and how do we study sediment transport? Focus on coastal zones and ongoing methods. *Water* **2018**, *10*, 390. [CrossRef]
3. Casal, G.; Harris, P.; Monteys, X.; Hedley, J.; Cahalane, C.; Casal, G.; Harris, P.; Monteys, X.; Hedley, J.; Cahalane, C.; et al. Understanding satellite-derived bathymetry using Sentinel 2 imagery and spatial prediction models. *GIScience Remote Sens.* **2020**, *57*, 271–286. [CrossRef]
4. Volpe, V.; Silvestri, S.; Marani, M. Remote sensing retrieval of suspended sediment concentration in shallow waters. *Remote Sens. Environ.* **2011**, *115*, 44–54. [CrossRef]
5. Ody, A.; Doxaran, D.; Vanhellemont, Q.; Nechad, B.; Novoa, S.; Many, G.; Bourrin, F.; Verney, R.; Pairaud, I.; Gentili, B. Potential of high spatial and temporal ocean color satellite data to study the dynamics of suspended particles in a micro-tidal river plume. *Remote Sens.* **2016**, *8*, 245. [CrossRef]
6. Malenovský, Z.; Rott, H.; Cihlar, J.; Schaepman, M.E.; García-Santos, G.; Fernandes, R.; Berger, M. Sentinels for science: Potential of Sentinel-1, -2, and -3 missions for scientific observations of ocean, cryosphere, and land. *Remote Sens. Environ.* **2012**, *120*, 91–101. [CrossRef]

7. Li, Y.; Li, D.; Fang, J.; Yin, X.; Li, H.; Hu, W.; Chen, J. Impact of Typhoon Morakot on suspended matter size distributions on the East China Sea inner shelf. *Cont. Shelf Res.* **2015**, *101*, 47–58. [CrossRef]

8. He, X.; Bai, Y.; Chen, C.-T.A.; Hsin, Y.-C.; Wu, C.-R.; Zhai, W.; Liu, Z.; Gong, F. Satellite views of the episodic terrestrial material transport to the southern Okinawa Trough driven by typhoon. *J. Geophys. Res. Ocean.* **2014**, *119*, 1706–1722. [CrossRef]

9. Liu, J.; Cai, S.; Wang, S. Observations of strong near-bottom current after the passage of Typhoon Pabuk in the South China Sea. *J. Mar. Syst.* **2011**, *87*, 102–108. [CrossRef]

10. Miles, T.; Seroka, G.; Kohut, J.; Schofield, O.; Glenn, S. Glider observations and modeling of sediment transport in Hurricane Sandy. *J. Geophys. Res. Ocean.* **2015**, *120*, 1771–1791. [CrossRef]

11. Li, Y.; Wang, A.; Qiao, L.; Fang, J.; Chen, J. The impact of typhoon Morakot on the modern sedimentary environment of the mud deposition center off the Zhejiang-Fujian coast, China. *Cont. Shelf Res.* **2012**, *37*, 92–100. [CrossRef]

12. Li, Y.; Li, H.; Qiao, L.; Xu, Y.; Yin, X.; He, J. Storm deposition layer on the Fujian coast generated by Typhoon Saola (2012). *Sci. Rep.* **2015**, *5*, 1–7. [CrossRef]

13. Chen, S.; Huang, W.; Wang, H.; Li, D. Remote sensing assessment of sediment re-suspension during Hurricane Frances in Apalachicola Bay, USA. *Remote Sens. Environ.* **2009**, *113*, 2670–2681. [CrossRef]

14. Li, Y.; Li, X. Remote sensing observations and numerical studies of a super typhoon-induce d suspende d se diment concentration variation in the East China Sea. *Ocean Model.* **2016**, *104*, 187–202. [CrossRef]

15. Doxaran, D.; Lamquin, N.; Park, Y.J.; Mazeran, C.; Ryu, J.H.; Wang, M.; Poteau, A. Retrieval of the seawater reflectance for suspended solids monitoring in the East China Sea using MODIS, MERIS and GOCI satellite data. *Remote Sens. Environ.* **2014**, *146*, 36–48. [CrossRef]

16. Li, Y.; Xu, X.; Yin, X.; Fang, J.; Hu, W.; Chen, J. Remote-sensing observations of Typhoon Soulik (2013) forced upwelling and sediment transport enhancement in the northern Taiwan Strait. *Int. J. Remote Sens.* **2015**, *36*, 2201–2218. [CrossRef]

17. Li, Y.; Xu, X.; Zheng, B. Satellite views of cross-strait sediment transport in the Taiwan Strait driven by Typhoon Morakot (2009). *Cont. Shelf Res.* **2018**, *166*, 54–64. [CrossRef]

18. Choi, J.K.; Park, Y.J.; Ahn, J.H.; Lim, H.S.; Eom, J.; Ryu, J.H. GOCI, the world's first geostationary ocean color observation satellite, for the monitoring of temporal variability in coastal water turbidity. *J. Geophys. Res. Ocean.* **2012**, *117*, 1–10. [CrossRef]

19. Choi, J.-K.; Park, Y.J.; Lee, B.R.; Eom, J.; Moon, J.-E.; Ryu, J.-H. Application of the Geostationary Ocean Color Imager (GOCI) to mapping the temporal dynamics of coastal water turbidity. *Remote Sens. Environ.* **2014**, *146*, 24–35. [CrossRef]

20. Moon, J.-E.; Ahn, Y.-H.; Ryu, J.-H.; Shanmugam, P. Development of Ocean Environmental Algorithms for Geostationary Ocean Color Imager (GOCI). *Korean J. Remote Sens.* **2010**, *26*, 189–207.

21. Ryu, J.H.; Han, H.J.; Cho, S.; Park, Y.J.; Ahn, Y.H. Overview of geostationary ocean color imager (GOCI) and GOCI data processing system (GDPS). *Ocean Sci. J.* **2012**, *47*, 223–233. [CrossRef]

22. Moon, J.E.; Park, Y.J.; Ryu, J.H.; Choi, J.K.; Ahn, J.H.; Min, J.E.; Son, Y.B.; Lee, S.J.; Han, H.J.; Ahn, Y.H. Initial validation of GOCI water products against in situ data collected around Korean peninsula for 2010–2011. *Ocean Sci. J.* **2012**, *47*, 261–277. [CrossRef]

23. Cheng, Z.; Wang, X.H.; Paull, D.; Gao, J. Application of the Geostationary Ocean Color Imager to mapping the diurnal and seasonal variability of surface suspended matter in a macro-tidal estuary. *Remote Sens.* **2016**, *8*, 244. [CrossRef]

24. Huang, C.; Yang, H.; Zhu, A.X.; Zhang, M.; Lü, H.; Huang, T.; Zou, J.; Li, Y. Evaluation of the Geostationary Ocean Color Imager (GOCI) to monitor the dynamic characteristics of suspension sediment in Taihu Lake. *Int. J. Remote Sens.* **2015**, *36*, 3859–3874. [CrossRef]

25. Qiu, Z.; Zheng, L.; Zhou, Y.; Sun, D.; Wang, S.; Wu, W. Innovative GOCI algorithm to derive turbidity in highly turbid waters: A case study in the Zhejiang coastal area. *Opt. Express* **2015**, *23*, A1179. [CrossRef]

26. Choi, J.-K.; Yang, H.; Han, H.-J.; Ryu, J.-H.; Park, Y.-J. Quantitative estimation of suspended sediment movements in coastal region using GOCI. *J. Coast. Res.* **2013**, *165*, 1367–1372. [CrossRef]

27. Yang, H.; Choi, J.-K.; Park, Y.-J.; Han, H.-J.; Ryu, J.-H. Application of the Geostationary Ocean Color Imager (GOCI) to estimates of ocean surface currents. *J. Geophys. Res. Ocean.* **2014**, 1022–1037. [CrossRef]

28. Taiwan River Restotation Network the List of Central Rivers. Available online: http://trrn.wra.gov.tw/web/index-18.html (accessed on 26 April 2020).

29. National Hurricane Centre Saffir-Simpson Hurricane Wind Scale. *Wind Scale* **2012**, 1–2. Available online: http://www.nhc.noaa.gov/aboutsshws.php (accessed on 15 May 2020).

30. Chen, W.B.; Lin, L.Y.; Jang, J.H.; Chang, C.H. Simulation of typhoon-induced storm tides and wind waves for the northeastern coast of Taiwan using a tide-surge-wave coupled model. *Water* **2017**, *9*, 549. [CrossRef]

31. Typhoon Taiwan's Typhoon Data Base. Available online: http://rdc28.cwb.gov.tw/TDB/ntdb/pageControl/typhoon?year=2015&num=201513&name=SOUDELOR&from_warning=tru (accessed on 31 March 2018).

32. Soudelor Typhoon Soudelor Wiki. Available online: https://en.wikipedia.org/wiki/Typhoon_Soudelor (accessed on 24 April 2018).

33. Soudelor, T. Deadly Typhoon Soudelor's Rainfall Analyzed. Available online: https://pmm.nasa.gov/extreme-weather/deadly-typhoon-soudelors-rainfall-analyzed (accessed on 1 July 2020).

34. Brockmann Consult L3/Binning Tool. Available online: http://www.brockmann-consult.de/beam/doc/help-4.8/binning/BinningAlgorithmDescription.html (accessed on 11 November 2018).

35. Campbell, J.W.; Blaisdell, J.M.; Darzi, M. Level-3 SeaWiFS Data Products: Spatial and Temporal Binning Algorithms. *SeaWiFS Tech. Rep. Ser.* **1995**, *32*, 80.
36. Technologies, M. GDPS Ver.2.0 User's Manual. 2018. Available online: http://kosc.kiost.ac.kr/eng/p30/kosc_p34.html (accessed on 10 November 2018).
37. Milliman, J.D.; Lee, T.Y.; Huang, J.C.; Kao, S.J. Temporal and spatial responses of river discharge to tectonic and climatic perturbations: Choshui River, Taiwan, and Typhoon Mindulle (2004). *Proc. Int. Assoc. Hydrol. Sci.* **2015**, *29*, 11–14. [CrossRef]
38. Chien, H.; Chiang, W.S.; Kao, S.J.; Liu, J.T.; Liu, K.K.; Liu, P.L.F. Sediment Dynamics observed in the Jhoushuei River and Adjacent Coastal Zone in Taiwan Strait. *Oceanography* **2011**, *24*, 122–131. [CrossRef]
39. Lin, G.W.; Chen, H.; Chen, Y.H.; Horng, M.J. Influence of typhoons and earthquakes on rainfall-induced landslides and suspended sediments discharge. *Eng. Geol.* **2008**, *97*, 32–41. [CrossRef]
40. Chen, C.W.; Oguchi, T.; Hayakawa, Y.S.; Saito, H.; Chen, H.; Lin, G.W.; Wei, L.W.; Chao, Y.C. Sediment yield during typhoon events in relation to landslides, rainfall, and catchment areas in Taiwan. *Geomorphology* **2018**, *303*, 540–548. [CrossRef]
41. Hung, C.; Lin, G.-W.; Kuo, H.-L.; Zhang, J.-M.; Chen, C.-W.; Chen, H. Impact of an Extreme Typhoon Event on Subsequent Sediment Discharges and Rainfall-Driven Landslides in Affected Mountainous Regions of Taiwan. *Geofluids* **2018**, *2018*, 1–11. [CrossRef]
42. Liu, W.C.; Chen, W.B.; Cheng, R.T.; Hsu, M.H. Modelling the impact of wind stress and river discharge on Danshuei River plume. *Appl. Math. Model.* **2008**, *32*, 1255–1280. [CrossRef]
43. Water Resources Planning Institute, Water Resources Agency, Ministry of Economic Affairs. Available online: https://www.wrap.gov.tw/pro12.aspx?type=0201000000 (accessed on 23 March 2020). (In Chinese)
44. Wang, Y.H.; Jan, S.; Wang, D.P. Transports and tidal current estimates in the Taiwan Strait from shipboard ADCP observations (1999–2001). *Estuar. Coast. Shelf Sci.* **2003**, *57*, 193–199. [CrossRef]
45. Liu, J.; Emery, W.; Wu, X.; Li, M.; Li, C.; Zhang, L. Computing Coastal Ocean Surface Currents from MODIS and VIIRS Satellite Imagery. *Remote Sens.* **2017**, *9*, 1083. [CrossRef]

remote sensing

Article

Comparison of In Situ and Remote-Sensing Methods to Determine Turbidity and Concentration of Suspended Matter in the Estuary Zone of the Mzymta River, Black Sea

Ksenia Nazirova [1], Yana Alferyeva [2], Olga Lavrova [1,*], Yuri Shur [3], Dmitry Soloviev [4], Tatiana Bocharova [1] and Alexey Strochkov [1]

[1] Space Research Institute of Russian Academy of Sciences, 117997 Moscow, Russia; knazirova@iki.rssi.ru (K.N.); tabo@iki.rssi.ru (T.B.); astro@iki.rssi.ru (A.S.)
[2] Faculty of Geology, Lomonosov Moscow State University, 119991 Moscow, Russia; yanaalf@yandex.ru
[3] Department of Civil and Environmental Engineering, University of Alaska Fairbanks, Fairbanks, AK 99775-5900, USA; yshur@alaska.edu
[4] Marine Hydrophysical Institute of Russian Academy of Sciences, 2990011 Sevastopol, Russia; solmit@mhi-ras.ru
* Correspondence: olavrova@iki.rssi.ru

Abstract: The paper presents the results of a comparison of water turbidity and suspended particulate matter concentration (SPM) obtained from quasi-synchronous in situ and satellite remote-sensing data. Field measurements from a small boat were performed in April and May 2019, in the northeastern part of the Black Sea, in the mouth area of the Mzymta River. The measuring instruments and methods included a turbidity sensor mounted on a CTD (Conductivity, Temperature, Depth), probe, a portable turbidimeter, water sampling for further laboratory analysis and collecting meteorological information from boat and ground-based weather stations. Remote-sensing methods included turbidity and SPM estimation using the C2RCC (Case 2 Regional Coast Color) and Atmospheric correction for OLI 'lite' (ACOLITE) ACOLITE processors that were run on Landsat-8 Operational Land Imager (OLI) and Sentinel-2A/2B Multispectral Instrument (MSI) satellite data. The highest correlation between the satellite SPM and the water sampling SPM for the study area in conditions of spring flooding was achieved using C2RCC, but only for measurements undertaken almost synchronously with satellite imaging because of the high mobility of the Mzymta plume. Within the few hours when all the stations were completed, its boundary could shift considerably. The ACOLITE algorithms overestimated by 1.5 times the water sampling SPM in the low value range up to 15 g/m^3. For SPM over 20–25 g/m^3, a high correlation was observed both with the in situ measurements and the C2RCC results. It was demonstrated that quantitative turbidity and SPM values retrieved from Landsat-8 OLI and Sentinel-2A/2B MSI data can adequately reflect the real situation even using standard retrieval algorithms, not regional ones, provided the best suited algorithm is selected for the study region.

Keywords: river plume; turbidity; suspended particulate matter; ocean color data; satellite remote sensing; in situ measurements; C2RCC; ACOLITE; Landsat-8 OLI; Sentinel-2 MSI; Mzymta River; Black Sea

Citation: Nazirova, K.; Alferyeva, Y.; Lavrova, O.; Shur, Y.; Soloviev, D.; Bocharova, T.; Strochkov, A. Comparison of In Situ and Remote-Sensing Methods to Determine Turbidity and Concentration of Suspended Matter in the Estuary Zone of the Mzymta River, Black Sea. *Remote Sens.* **2021**, *13*, 143. https://doi.org/10.3390/rs13010143

Received: 3 November 2020
Accepted: 29 December 2020
Published: 4 January 2021

Publisher's Note: MDPI stays neutral with regard to jurisdictional claims in published maps and institutional affiliations.

1. Introduction

River discharge into sea plays an important role in the physical, chemical and biological processes in the ocean, especially in the shelf areas, being the main source of suspended and dissolved terrigenous and biogenic substances in the sea, as well as anthropogenic pollution. These substances have significant and in many cases negative effects on coastal ecosystem, including phytoplankton productivity, transport of pollutants in the shelf areas, erosion of coasts, artificial beach formation, nutrient dynamics, etc. [1–3]. Therefore, monitoring the estuarine areas and understanding the dynamics of river water distribution over the sea shelves are important scientific and practical tasks. The influence of a huge number of geographic factors, hydrometeorological conditions and hydrophysical processes, with great complexity and cost of field measurements, create a certain fragmentation of information on the processes of river water spreading in the sea. This problem can be solved only using satellite remote-sensing methods, which provide a unique opportunity to observe almost simultaneously the entire region of interest repeatedly, day after day, for many years.

Reaching a sea, river waters form plumes—mesoscale structures adjacent to the river mouth. Plume water can be distinguished from seawater by its low salinity, temperature and usually by high turbidity and high content of suspended matter and dissolved organics [4,5].

For a river plume area, the main difficulty is obtaining quantitative suspended particulate matter concentration (SPM) estimates, while qualitative information is abundant. In satellite true color images (TCI), plumes can be clearly identified by contrasting differences between muddy river water and relatively clean surrounding seawater [6]. Multiple investigations confirm that river plume boundaries and other turbidity inhomogeneities obtained from contact measurements correlate quite well with satellite observations. For example, a joint analysis of in situ and Aqua Moderate Resolution Imaging Spectroradiometer (MODIS) data allowed tracking propagation of a Vistula plume in coastal waters of the Gulf of Gdansk during intense flooding in May 2010, but only on a qualitative level [7].

There are numerous studies on validation of satellite data using concurrent field measurements [8–18]. This by far incomplete list shows that such works are under way in various regions worldwide, which evidences their importance. Although various methods and techniques of contact measurements and different remote-sensing data are employed, the problem of adequate interpreting satellite data and obtaining products suitable for use instead of expensive in situ data is still far from being solved.

Quantitative estimates of turbidity and SPM can be obtained from satellite remote sensing data using various algorithms that, strictly speaking, should take into account numerous factors, including varying chemical composition of ocean water, coastal shelf waters, water of estuaries and fresh water bodies, geometrical parameters of satellite sounding at a given moment, intrinsic properties of orbital equipment, and current climatic conditions in the study area and much more [14,19–27]. A classic example of a study of river runoff influence on coastal hydrological structure is presented in [28]. Using a set of field measuring instruments, the authors performed a detailed investigation of the properties of vertical hydrological structure of seawater affected by intrusion of fresh river water, as well as sedimentation of suspended matter in the shelf zone. However, satellite data were used only for qualitative consideration as an auxiliary tool.

After numerous comparisons and simultaneous measurements, it was found impossible to develop a universal algorithm for evaluating the standard characteristics of seawater color based only on available data from satellite optical sensors because of extremely diverse set of characteristics and ambiguity in their interpretation under certain observation conditions. As noted in [29], there are three main types of algorithms commonly used to derive SPM from water reflectance: (1) empirical, (2) semi-analytical and (3) analytical algorithms. Empirical single-band and band-ratio models have been commonly used in coastal and estuarine areas [9,15,30]. These types of model are dependent on SPM and water reflectance ranges, and require calibration with regional measurements [29]. Semi-analytical or analytical models are based on the inherent optical properties and provide a more global application [12,31,32].

To date, scientists from different countries have developed a number of specialized algorithms to evaluate characteristics of coastal marine and lake waters [30,33–36]. Originally, some of the standard algorithms were developed for the Sea-Viewing Wide Field-of-View Sensor (SeaWiFS) instrument, then for MODIS and MEdium Resolution Imaging Spectrometer (MERIS) on the Envisat satellite [37,38], which operated for 10 years until 2012. An example of a successful application of the coastal algorithm on MERIS data is described in [39].

The Ocean and Land Colour Instrument (OLCI) instrument on Sentinel-3 (launched in 2016) was developed in part to provide continuity with measurements made previously by MERIS. The algorithms developed for MERIS were adapted for OLCI [40]. Some of them were automated and made available in the specialized BEAM-VISAT software used by a great number of researchers.

Examples of such algorithms are: Case 2 Regional (C2R) [37], FUB/WeW [38], Eutrophic Lake (EUL) and Boreal Lake (BL) [37], as well as the Maximum Chlorophyll Index (MCI) and Fluorescence Line Height (FLH) [41,42]. It was expected that some of these algorithms could be compatible with currently used Sentinel and Landsat sensors.

In this paper, for atmospheric correction, turbidity and SPM estimation the following standard algorithms were used: C2RCC (Case 2 Regional Coast Colour, https://www.brockmann-consult.de/portfolio/water-quality-from-space/) and algorithms provided by the ACOLITE (http://odnature.naturalsciences.be/remsem/software-and-data/acolite) software.

The C2RCC processor was originally developed by Doerffer and Schiller [37] and now is implemented in the European Space Agency (ESA) Sentinel Toolbox SNAP software (https://step.esa.int/main/toolboxes/snap/). The latest development of C2RCC neural networks and the algorithm for optically complex waters are described in [43]. The software calculates marine environment characteristics based on multispectral sensor data from satellites of the latest generation (SeaWiFS, MERIS, MODIS, Visible Infrared Imaging Radiometer Suite (VIIRS), OLCI, Operational Land Imager (OLI), and Multispectral Instrument (MSI)). It is also applicable to historical data from sensors that finished their operation long ago. Thus, it allows "recalculating", for certain purposes, previously calculated parameters to meet current requirements.

Another group of algorithms that we used in our study are implemented in the ACOLITE processor and intended for calculating the main optical parameters. ACOLITE, developed at the Royal Belgian Institute of Natural Sciences (RBINS), is based on the work of a team of researchers led by Dr. Bouchra Nechad and described in detail in [31,44,45]. ACOLITE is specifically developed for marine, coastal, and inland waters and supports free processing of Landsat-8 and Sentinel-2 data [46–48].

Recently, ACOLITE has been frequently used for atmospheric correction of OLI and MSI data [49]. Two atmospheric correction methods are implemented [50]: the Short Wave Infrared (SWIR)-based exponential extrapolation method [51–53] (EXP) and a multi-band dark spectrum fitting technique [50,54] (DSF). The DSF was developed for meter scale resolution sensors and subsequently adapted for the decameter resolution sensors on Landsat and Sentinel [50]. The software is successfully applied both for coastal zones and inner water bodies [46–49,55,56].

Monitoring seawater quality in the northeastern part of the Black Sea is of prime importance since this region is Russia's largest marine recreational area. The motivation of this study was to examine how well the different algorithms can assess turbidity and SPM, key water quality parameters, in such a complex environment as the Black Sea Caucasian coastal zone with multiple mountainous rivers flowing into the sea. The plume area of the Mzymta River, the most affluent river in the region, was chosen as the test site. The main objective was to determine the relationships between water turbidity and SPM obtained by contact and remote sensing methods and compare the performances of the above algorithms. Strong spatial and temporal variability of sub-mesoscale hydrodynamics in the study area required careful selection and comparison of different instruments and techniques for in situ measurements. Coupled with a detailed examination of surface and vertical plume structure, this ensured correct and accurate validation of the satellite algorithms.

Contact measurements were conducted from a small boat using a turbidity sensor mounted on a CTD (Conductivity, Temperature, Depth) probe, a portable turbidimeter and water sampling for further laboratory analysis. Quasi-synchronous satellite data were processed using C2RCC and ACOLITE algorithms proposed by Nechad et al. [31,44,45] and Dogliotti et al. [32].

Quite a number of works, for example [57–59], are devoted to the plume of Mzymta, however, comparison of water quality parameters retrieved from concurrent contact and satellite measurements was performed for the first time.

2. Study Area, Data and Methods

2.1. Study Area

The Mzymta River is the largest river of the Russia's part of the Black Sea coast. It originates on the slopes of the Main Caucasus Range and has a mountainous character for most of its length. The total length of the river is 89 km and the catchment area is 885 km^2 [60]. The river recharge is mixed, including precipitation, melting snow and glaciers and groundwater in the lower part. Mzymta has a high discharge in the warm season, frequent autumn floods, and a stable low water in winter.

Mzymta flow rate varies from 0.4 to 2–3 m/s. The yield of suspended sediment is directly dependent on water runoff: the greater the water discharge, the greater is the yield. The average annual amount of suspended sediment is 488,200 tons and that of bottom sediments is 141,000 tons [60]. The average annual discharge of the river is 45.6 m^3/s [61].

Mzymta plume forms near the city of Adler where the river enters the Black Sea. Plume water is fresher and colder than seawater. Having a highly dynamic character, Mzymta plume is subject to a strong influence of wind and coastal system of currents [2,62], the Coriolis force, the local landscape, and stratification of the ambient sea [58]. Due to a narrow shelf zone in the southeastern Black Sea, the main element of the Black Sea circulation, the Rim Current, is often strongly pressed against the coast, at a distance of ~6 km. Therefore, being involved in the cyclonic structure of the Rim Current, Mzymta plume can spread for many kilometers along the coastline from the river mouth [63]. Interacting with sub-mesoscale and mesoscale vortex structures, the river water acts as a tracer, which aids in the studies of water exchange between the coastal zone and deep sea [64].

The infrastructure developed along the shores of Mzymta can bring a potentially significant anthropogenic impact on its waters. In the upper part of the river, there is the famous Krasnaya Polyana ski resort with a vast complex of hotels. At the mouth of the river, popular tourist attractions include extreme rafting, bungee and BASE (building, antenna, span, and earth) jumping. There are several trout farms in river bends, some of them invite tourists. The hydropower plant in Krasnaya Polyana is also located on the river and supplies electricity to the city of Sochi. The most developed is the lowland part of the river in close proximity to the coast. The cities of Sochi and Adler stretch along the seashore with numerous hotel complexes and swimming beaches. By the 2014 Winter Olympics, the eastern part of the floodplain was densely developed to build the Olympic Village, a yacht port and new artificial sandy beaches. Now, Sochi is the largest resort city in Russia, very popular throughout the year (Figure 1). Such development can dramatically contribute to pollution of the river water and, as a result, the coastal zone [65].

Figure 1. Study area in the northeastern part of the Black Sea. 1 Imereti port; 2 Olympic Park; 3 wastewater outfall; 4 beaches.

A possible attempt to regulate Mzymta flow into the sea and to direct it away from the yacht port in order to decrease its impact on the port protective walls can lead to destruction of the beaches just east of the port (Figure 1). Without new terrigenous material, the beaches will erode and their maintenance will be too expensive. Study, monitoring and control of this area are urgently needed to understand the changes in the coastal ecosystem due to active recreational activity and properly maintain such activity.

2.2. Data and Methods

2.2.1. Boat Measurements

Shipboard measurements were conducted from 23 April to 4 May 2019, in the estuary zone of Mzymta from a small boat called Arabella with Imereti port as point of departure. The route of Arabella within Mzymta plume consisted of 4 legs parallel to the coast, from the river mouth to a visible edge of the plume. Each sailing was organized concurrently with a satellite (Sentinel-2A/-2B MSI, Landsat-7 ETM + and Landsat-8 OLI) passage over the study area. In total, seven boat trips were completed on 23–26, 28 April and 1–2 May 2019. The summary grid of stations included more than 150 points (see Figure 2). At each station, CTD probing was performed from the surface to the bottom using a high-precision instrument RBR-Concerto of the Canadian company Richard Bransker Research Ltd. The main characteristics of the instrument are presented in Table 1. The CTD probe was equipped with a turbidity meter (TM) from Seapoint Ltd. with measurement frequency up to 6 kHz. Additionally, turbidity of the upper layer of water was measured at best possible accuracy using a portable turbidimeter (PT) TN400 from Apera Instruments. At CTD stations and in points of turbidity measurement, water sampling was performed for further evaluation of SPM.

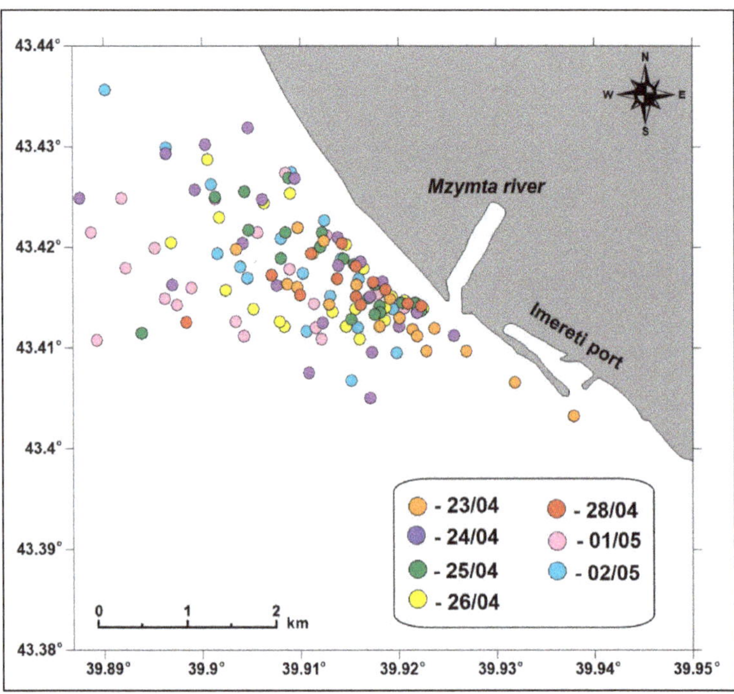

Figure 2. Map of 2019 hydrological stations.

Table 1. The main characteristics of RBR-Concerto CTD instrument.

Parameter	Range	Initial Accuracy	Resolution	Time Constant	Typical Stability/ Per Year	Max. Depth (m)	Sampling Speed (Hz)
Conductivity	0–85 mS */cm	±0.003 mS/cm	0.001 mS/cm	~1 s	0.010 mS/cm	200	2–6
Temperature	−5 °C to 35 °C	±0.002°	0.00005 °C	~1 s	0.002 °C	200	2–6
Depth	0–200 m	±0.05% FS **	0.001% FS	<0.01 s	0.1% FS	200	2–6

*—millisiemens. **—full scale.

TM is an analog sensor that detects scattered light from suspended particles in a specific volume of water placed in front of the optical window of the sensor, at a distance <5 cm. A distinctive feature of the sensor is its ability to detect light scattered from particles smaller in size than the wavelength emitted, which is 880 nm. For suspended particles whose diameters are greater than the wavelength of the light source, light scattering actually occurs through optical processes such as reflection, refraction, and diffraction [66]. A light-scattering pattern after a collision with a particle depends on the relationship between wavelength and particle size. When the particle is larger than the wavelength, light tends to scatter more intensely in the forward direction [67]. TM reports turbidity in nephelometric turbidity units (NTU). TM measuring range is from 0.05 to 15,000 NTU (±2% deviation), operating temperature 0–65 °C (temperature coefficient <0.05%/°C), depth capability 6000 m.

At the CTD stations, TM took measurements from depths of 0.35–0.50 m at best. To improve turbidity data from the upper layer of water (0.10–0.15 m), we also used PT. The instrument is equipped with an infrared light source and uses the nephelometric method that complies with ISO7207 (90° dispersion). PT measurement range is from 0 to 1000 NTU (the instrument is shipped with 4 ready-made calibration standard solutions of high-molecular polymer turbidity: 0.02 NTU, 20.0 NTU, 100 NTU, and 800 NTU), measurement accuracy varies from 0.01 to 1 NTU depending on the selected range. For each sample, two instant measurements were taken; their mean was used as a resulting value. At the same time and at the same stations, samples of water were taken from the upper surface layer for laboratory analysis.

Each optical sensor, in principle, has its own specifics. A detailed discussion can be found, for example, in [68], a work devoted to laboratory experiments on turbidity evaluation by different optical sensors.

During the boat measurements, air temperature, wind speed and wind direction were continuously recorded by the Airmar WeatherStation 150WX weather station along the course of the boat. Table 2 presents its characteristics. The display of weather station parameters was configured and realized by the factory software WeatherCaster ™ Software 3.005. Also, data on air temperature, atmospheric pressure above sea level, wind speed, wind direction and precipitation were obtained from a weather station at the airport of Sochi (https://rp5.ru/). The movement between boat stations was controlled using a chartplotter with a built-in Garmin GPSmap 541s echo sounder.

Table 2. The main characteristics of Airmar WeatherStation 150WX.

Parameter	Range	Accuracy	Resolution
wind speed	0–40 m/s	5%/10 m/s	0.1 m/s
wind direction	0° to 359.9°	±3°/10 m/s	0.1°
air temperature	−40 °C to 80 °C	±1.1 °C/20 °C	0.1 °C
barometric pressure	300 to 1100 hPa	±0.5 hPa	0.1 hPa
pinch and roll	50°	±1° in range of ±30°	0.1°

2.2.2. Laboratory Study

During the field work, 140 water samples were taken from the near-surface layer to evaluate amount and mineral composition of the suspended matter. As mentioned earlier, this was necessary in order to carry out more accurate measurements of Mzymta plume water for subsequent comparison with results derived from remote-sensing data. The volume of each sample was approximately 1.5 L. All samples were weighed in laboratory conditions with an accuracy of 0.01 g. SPM was determined gravimetrically [69,70]. Water was filtered using a vacuum unit Lafil 400-LF30 and fiberglass WHATMAN GF/F filters manufactured from hydrophobic borosilicate glass. These filters are capable to catch fine particles down to 0.7 microns. The filters were preweighed with an accuracy of 0.1 mg, and stored in a desiccator for use within 2 weeks. Water samples were filtered immediately after collection.

To remove sea salt from the suspension, filters were washed with 250 mL of distilled water after filtration. Such an amount of fresh water provided complete dissolution of the salt and its removal from the samples. The samples were stored at −20 °C until further analysis, usually within one month after sampling. Subsequently, all filtered samples were subjected to weight analysis of SPM on high-precision scales in the petrology laboratory of the Moscow State University. Suspension filters were dried for 24 h at 50 °C and reweighed. The accuracy of determining the weight of the suspended particles in the samples was ±0.0001 g. Note that suspended particulate matter includes all organic and mineral material with dimensions over approximately 0.7 mm.

The median and the interquartile range (IQR) were computed for each sample by the protocol detailed in [70]. Observations where the IQR exceeded 45% of the median SPM value were rejected.

To identify the mineral composition of the sediment, 20 samples were chosen for X-ray analysis. The samples were obtained on 26 April, 1 May and 2 May at stations located at different distances from the Mzymta mouth and covering a wide range of turbidities from minimum to maximum NTU.

X-ray scanning and analysis were performed at the Department of Oil and Gas Sedimentology and Marine Geology, College of Geology of Moscow State University. The survey was carried out on a DRON UM 1 powder diffractometer (Co Kα, λ = 1.79021 Å) in the range of angles 2 Θ from 4 to 80° in continuous mode at a speed of 2° per minute. The phases were diagnosed using the MINCRYST crystallographic base for minerals and their structural analogs. The amount of the mineral phase was estimated by comparison of the intensities of the corresponding peaks.

2.2.3. Satellite Observations

All field measurements from the boat were synchronized with satellite data acquisition at best possible accuracy. The source of remote-sensing data was the instruments on board Landsat-8 and Sentinel-2 (A and B), namely, the Operational Land Imager (OLI) and Multispectral Instrument (MSI) multispectral sensors. Also, Sentinel-3 OLCI data were used to compose SPM maps. Because of their 300 m spatial resolution, no comparison with measurements at the boat stations was possible. They were used to highlight the general picture of Mzymta water distribution.

The Landsat-8 satellite, of the National Aeronautics and Space Administration (NASA) and The United States Geological Survey (USGS), is equipped with OLI and Thermal InfraRed Sensor (TIRS) multispectral scanners of medium spatial resolution in the visible and infrared ranges covering a strip about 185 km wide in a continuous mode with a flight frequency once in 7–8 days. The maximum spatial resolution of these sensors is 15, 30 and 60 m, depending on the corresponding spectral range of sensing. The paired satellites Sentinel-2A and Sentinel-2B of ESA are equipped with MSI with a spatial resolution down to 10 m. It continuously covers a strip of the surface about 290 km wide at a frequency of once every 3–10 days for the same region.

Standard algorithms for reconstructing optical parameters based on satellite data, first of all SPM and chlorophyll concentration, were initially designed for open ocean waters with a predominance of phytoplankton and its decay products, so called Case 1 type waters, whereas in our work the study region refers to Case 2 type coastal waters characterized by high turbidity and considerable influence of the coastal zone [71,72].

In the work, we used different software to process satellite data for comparison with the results of in situ measurements. First, we applied the C2RCC (Case 2 Regional Coast Colour) version of a processor originally developed by Doerffer and Schiller [37] and now implemented in the ESA Sentinel Toolbox SNAP software (https://step.esa.int/main/toolboxes/snap/). The latest development of C2RCC neural networks and the algorithm for optically complex waters are described in [43]. Although the current processor version integrates almost all the essential characteristics of the environment and the equipment applied, the algorithm developers leave open a possibility for its users to change certain input parameters and coefficients including those experimentally obtained which need permanent regional correction. They include atmosphere transmittance, reflectance parameters of a specific underlying surface, cloud risk coefficients, air pressure and other.

The recent development of more reliable technologies for the evaluation of key parameters of the marine environment has become possible due to the introduction into widespread use and accessibility of the source code of the algorithm based on the use of neural networks. The most important property of neural networks is the possibility of their training by regular updating the database of correlated input parameters and obtained characteristics of the studied medium by introducing into the model a wide range of results of real contact measurements synchronous with satellite observations in different regions and sequential refinement of the connections between neurons, that is, actually realizing multiple non-linear regression [43].

SNAP processing results that we used in our study can be presented both in a tabular form for easy comparison with in situ measurements and in the form of TCI mapping the retrieved optical characteristic with resolution close to original resolution of the satellite data. TCI do not show numerical values of the optical characteristic, but display its gradients in a way familiar to the human eye, and also carry supplementary information, such as locations of the stations, trajectory of the boat, properties of wind and currents, diurnal displacement of the plume boundary and other. For all processed satellite data, TCI were constructed at the maximum resolution (10 m).

Second, for all Landsat-8 OLI and Sentinel-2 MSI data we performed atmospheric correction with the ACOLITE DSF method. To retrieve turbidity and SPM we used two algorithms developed by a team of researchers led by Dr. Bouchra Nechad and described in detail in [31,44,45]; below they are referred to as Nechad 2009 and Nechad 2015. There were some differences in the Nechad 2009 and Nechad 2015 results, but nothing critical. In contrast, the algorithm proposed by Dogliotti et al. [32] (below referred to as Dogliotti) and intended for highly turbid waters showed rather inconsistent results.

3. Results

3.1. Meteorological Conditions

Knowledge of meteorological conditions is crucial for analyzing the influence of wind on the dynamics of a river plume and the influence of precipitation on the discharge and turbidity of the river. Cloud cover and air temperature affect permeability of the atmosphere, which is important for processing visible remote-sensing data.

Weather information on the days of our field work and adjacent days, from 20 April to 4 May 2019, was available from the weather station at Sochi International Airport (rp5.ru). The prevailing wind directions were: ENE—19 cases, NE—10 cases and E—11 cases in the morning; and W—22 cases in the afternoon. During NE upsurge wind, the area of the river plume reached its maximum. With E/ENE winds, the plume spread strictly westward, being pressed against the coast. In general, wind speeds were moderate and did not exceed 7 m/s. With E/ENE winds, wind speed was in the range of 2–4 m/s. Under NE winds, from 28 April to 1 May, a stronger wind was observed, from 2 to 7 m/s. With predominantly evening W winds (except on 26 April), minimum wind speed was 3 m/s and less. On 2 May and 4 May, wind directions varied. With NE winds, from 28 April to 2 May, the sky was overcast over the study area. Just before our work, on 21–23 April, slight short-term precipitation took place and on 3 May precipitation was up to 30 mm. On 26 April, the weather was cloudless over the area of observation with W winds. Air temperature ranged 8–24 °C. Its gradual increase was observed till 1–2 May. After rain on 3 May, there was a sharp decrease in air temperature to a daily average of 12–13 °C. Air temperature and wind field characteristics are presented in Figure 3.

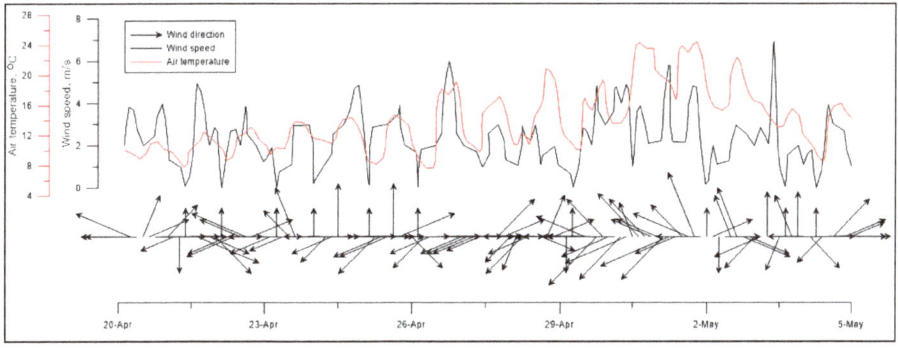

Figure 3. Meteorological conditions during 2019 in situ measurements (rp5.ru).

3.2. Results of In Situ Measurements

During the field studies in April–May 2019, Mzymta plume spread strictly westward, with a sharp eastern boundary, unlike our previous observations in April 2018, when the direction of plume propagation changed depending on coastal currents [73,74].

The spreading of the plume in the western direction was caused by two factors: (1) during our work, the Rim Current jet, which is directed westward at this location, was pressed to the coast and had a high velocity of more than 0.6 m/s [63]; (2) in the eastern part the river mouth, blocks of concrete were laid to limit the spread of water in the eastern direction and to reduce the load on the structures enclosing the port facility.

Due to the relatively small size of the plume, it was possible to cover its entire area with a dense grid of measuring stations (Figure 2). At each station, measurements were taken from the surface to the bottom. For comparison with satellite observations, a special focus was on the near-surface layer.

3.2.1. Variation of Temperature, Salinity and Turbidity in the Near-Surface Layer

Temperature. During the field work, a gradual increase in water temperature in the surface layer occurred, which correlated well with a gradual increase in daily average air temperature in the region. Lowest water temperatures were typically observed close to the river mouth. At a distance of not more than 200–300 m from the mouth, the temperature in the surface layer of water, at a depth of 50–80 cm, varied in the range from 10.8 to 12.7 °C (Table 3). It gradually increased in the direction to the plume boundary with increasing mixing with seawater. The surface temperature of "proper" seawater outside the plume, varied from 12.4 to 17.3 °C. Its highest values were reached by 2 May when air warmed up to 24 °C.

Salinity. Water salinity in the surface layer at the stations closest to the mouth, varied from 6.2 to 11.4 PSU (Table 3). At the plume boundary, salinity was close to seawater, which in the Black Sea is about 18 PSU.

Turbidity. During nine expedition days, turbidity in the surface layer, according to PT measurements, soared more than eight times (Table 3), from 13 NTU in the beginning to 135 NTU in the end. The maximum turbidity was reached on 3 May, when no boat trips were conducted, after a sharp warming in the Mzymta watershed due to melting glaciers and snow in the mountains and precipitation.

Table 3. Maximum and minimum values of temperature, salinity, turbidity and suspended particulate matter concentration (SPM) in the surface water layer on days of boat measurements.

Date	Min Temperature, °C	Max Temperature, °C	Min Salinity, PSU	Max Turbidity, NTU		Max SPM, g/m³
				PT *	TM **	Water Sample
23 April 2019	11.04	12.37	11.36	22	31	18.7
24 April 2019	11.61	13.65	10.95	13	16	18.1
25 April 2019	11.17	13.24	10.38	15	20	16.1
26 April 2019	10.77	13.53	6.23	28	31	22.8
28 April 2019	11.09	14.46	6.55	54	78	46.3
1 May 2019	12.69	15.50	9.24	68	75	64.5
2 May 2019	12.28	17.25	9.80	125	129	106.6

* measured with TN400 portable turbidimeter, Apera Instruments. ** measured with Turbidity Meter, Seapoint Ltd.

3.2.2. Spatial Distribution of Temperature, Salinity and Turbidity in the Plume

As Mzymta water spreads in the sea, changes in the basic parameters of water (temperature, salinity, turbidity) occur unevenly. This depends on speed and discharge of the river flow and on coastal currents, which are highly heterogeneous in this area [58]. As a result, local areas of increasing/decreasing turbidity and salinity are formed in the estuary zone of Mzymta. In addition, in the area of the plume there is a sewage outfall of the city of Adler that evidently influences water parameters. Figure 4 shows maps of the spatial distribution of temperature (Figure 4a), salinity (Figure 4b) and turbidity (Figure 4c) based on boat measurements using CTD and TM in the near-surface layer on 2 May 2019. In Figure 4c, pink dashed line schematically shows the boundary of the plume. Some stations on that day are also shown. They are Stations 119, 129 and 134. Station 129 is located at the outfall of the sewage pipeline. The impact of the sewage outfall is associated with a small region with almost zero turbidity and increased salinity in comparison to surrounding plume waters. This region is easily recognized in the visible satellite imagery.

3.2.3. Depth Distribution of Temperature, Salinity and Turbidity

To solve the problem of satellite data verification with the results of field measurements, it is necessary to know the depth distribution of river water parameters. The distribution of temperature and salinity over depth determines plume water stratification, which impacts the hydrodynamic processes. The thickness of the turbid water layer and turbidity depth distribution determine the depth of the water column contributing to water leaving radiance captured by the satellite sensor. Accordingly, this determines the choice of the techniques and instruments for the field measurements.

In our previous studies [73,74], we found that depth penetration of river water is small and rapidly decreases with distance from the river mouth. This was confirmed again by the observations in April and May 2019. As an example, Figure 5 shows the change in the hydrological characteristics with depth on 2 May 2019. Station 119 was located in close proximity to Mzymta mouth and Station 134 was located at the plume boundary (Figure 4c). By the conventional definition, a plume boundary is a minimum water turbidity location that is the closest to a sharp turbidity gradient. In this example (2 May), water turbidity was about 20 NTU, and outside the plume we observed values close to 0. Changes in turbidity, temperature and salinity with depth at these stations are shown in Figure 5a (Station 119) and 5b (Station 134). At Station 119, the closest to the mouth, the depth of the plume is about 2.5 m, water turbidity in the near-surface layer reaches 125 NTU, and temperature and salinity are much lower than in the underlying layer. This turbidity is the greatest for a given day. At Station 134, which is 250 m more seaward and near the border of the plume, the hydrological section looks different (Figure 5b). The thickness of river water intrusion is not more than 1 m, the turbidity of sea water is about 20 NTU, and the temperature and salinity are almost unchanged with depth.

Thus, it was determined that the depth penetration of river water sharply decreases with distance from Mzymta mouth, from 2.5 to 1 m at the plume boundary; therefore, for comparison with satellite data, all field measurements should be made in the near-surface layer. The thickness of the seasonal thermocline is about 11–12 m.

3.2.4. Results of Portable Turbidimeter (PT) Measurements in the Near-Surface Layer

After comparison of turbidity data obtained with TM and PT in the surface layer of water (see Section 4.1), it was decided to use PT measurements for further comparison with the weight method and remote sensing data, because PT is capable of taking measurements in a thinner surface layer (the first tens of centimeters) than TM.

A change in maximum water turbidity that was found with PT (most pronounced at stations located closest to the river mouth) in a thin surface layer (from 0 to 15 cm) is shown in Figure 6. During the field work, an exponential increase in this parameter was observed. The lowest value was 15.44 NTU and the highest value of 288 NTU was recorded on 4 May: a rise of 18 times in the mouth zone. Most likely, this was due to the cumulative effect of air temperature increase by the end of the field work, triggering active melting of snow and ice in the mountains, and intense (about 30 mm) precipitation on 3 May.

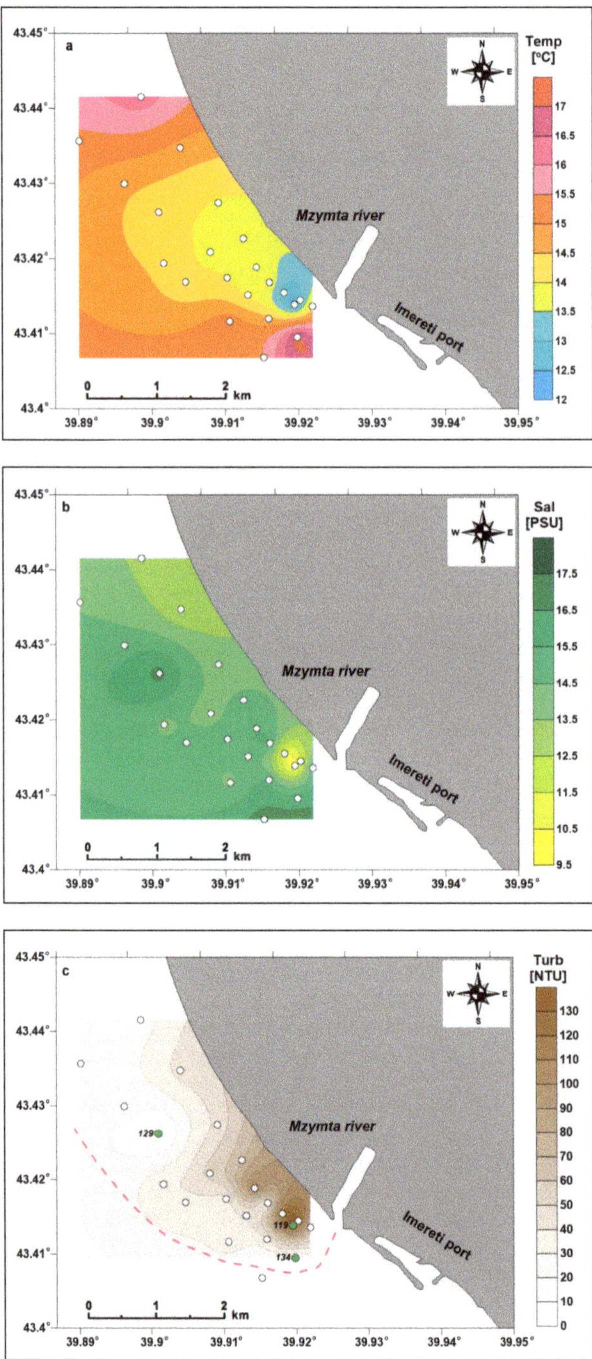

Figure 4. Maps of spatial distribution on 2 May 2019: (**a**) water temperature; (**b**) salinity; (**c**) turbidity from CTD and TM measurements. Pink dashed line indicates the boundary of the river plume.

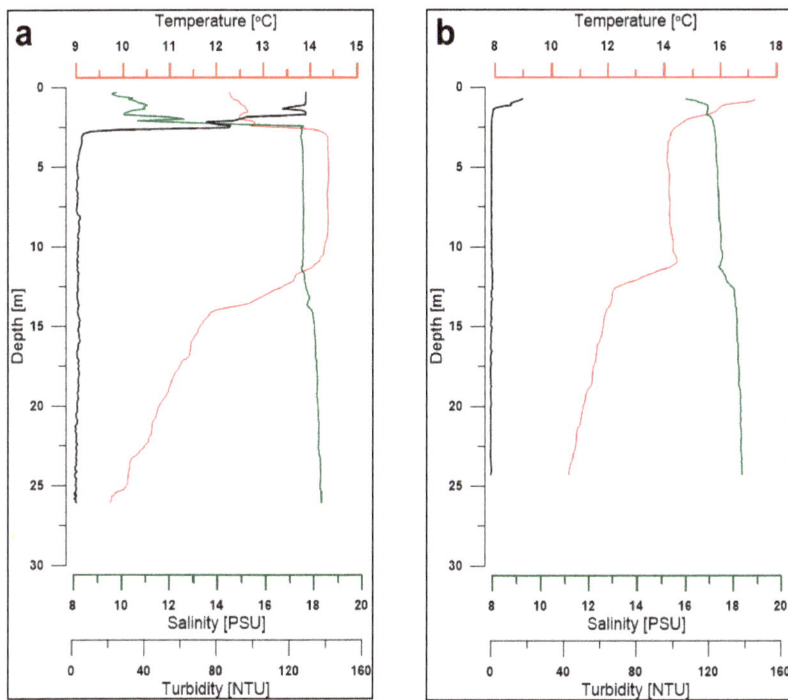

Figure 5. Typical CTD + TM casts: (**a**) Station 119—the closest to the mouth; (**b**) Station 134—near plume boundary (see Figure 4). Red line—temperature, green line—salinity, black line—turbidity.

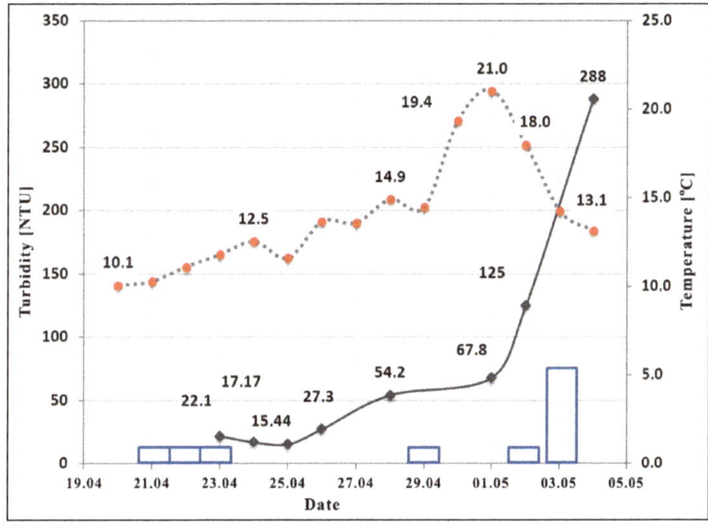

Figure 6. Changes in PT maximum turbidity (black line), daily average air temperature (dashed line) and precipitation (blue columns).

3.2.5. Correlation Analysis of Turbidity and Suspended Particulate Matter Concentration (SPM) from In Situ Measurements

During boat stations, turbidity in the plume was measured with two instruments: Apera Instruments TN400 portable turbidimeter (PT) and Seapoint turbidity meter (TM). At the same stations, water samples were collected at the same depths as PT measurements. After filtering and weighing water samples in accordance with the protocol described in [69], we obtained values of SPM hereinafter referred to as SPM in situ. Note, the water turbidity unit is NTU, while the SPM unit is g/m^3, and no algorithm exists to convert one into another because the two parameters are very different in physical nature and measurement methods. Turbidity strongly depends on particle size and composition of suspended matter. One of the tasks was to define the correlation between turbidity and SPM for the study region in the period of the spring flood. The analysis shows that the SPM in situ is directly proportional to the turbidity determined with PT in the upper surface water layer (Turb in situ, NTU). As shown in Figure 7, the obtained values are well approximated by a straight line: SPM in situ = 0.84 × Turb in situ. The determination coefficient is very high: $R^2 = 0.982$. This is typical of all water samples without exception and does not change with days of measurements or weather conditions. A similar strong relationship was determined between the SPM in situ and turbidity determined with TM, the existing differences will be addressed below in the Discussion (Section 4.1).

With such a high correlation between the PT data and the SPM in situ in this region, it seems reliable to make conversions between the turbidity and SPM units (NTU and g/m^3) using the established empirical equation. The main advantage is the ability to acquire numerous data using only optical turbidity sensors without time-consuming work to determine weight turbidity. Naturally, it is necessary to conduct multiple similar experiments in different seasons and under different meteorological conditions to obtain statistically valid results.

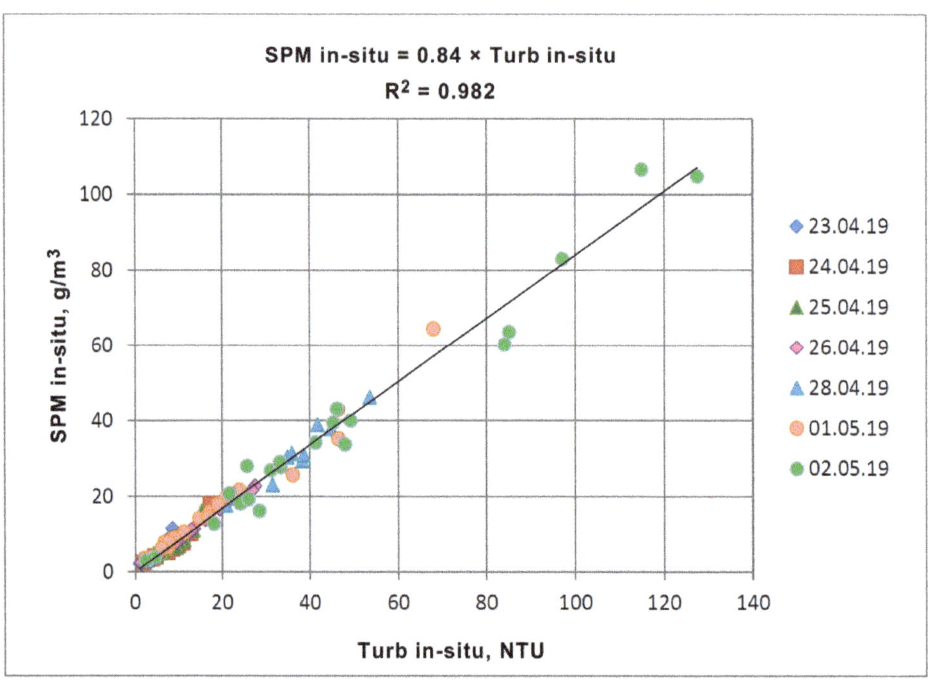

Figure 7. Comparison of SPM in the water samples (SPM in situ) and turbidity determined with PT (Turb in situ) for all measurement days in April–May 2020.

3.2.6. Sampled SPM and Mineral Composition of Suspended Matter

The SPM in water samples collected during the field work ranged from 2 to 106 g/m^3. From 23 April to 2 May, maximum SPM gradually increased from 23 g/m^3 on 23–26 April; to 46 g/m^3 on 28 April; 65 g/m^3 on 1 May; and 106 g/m^3 on 2 May. The minimum SPM values within the plume for the entire period of the study were approximately 2–3 g/m^3.

X-ray phase analysis of mineral composition of the suspended matter showed that in the selected samples: (1) quartz amounted to 16–45% of the suspension mass; (2) feldspars 12–27%; (3) various clay minerals (kaolinite, montmorillonite, chlorite, hydromica, mixed layer minerals) 27–58%; (4) carbonate minerals (calcite, dolomite and aragonite) 0–22% (Table 4).

Table 4. Turbidity in the upper near-surface layer, sampled SPM and mineral composition of suspension.

Date	Station	PT Water Turbidity, NTU	Sampled SPM, g/m^3	Quartz, mas.% *	Feldspars mas.% *	Clay Minerals, mas.% *	Carbonate Minerals, mas.% *	K **
26 April	60_1	28	22.8	28	21	42	9	0.67
26 April	64	16	13.8	30	24	46	0	0.65
26 April	65	7	7.3	19	23	45	10	0.43
26 April	69	20	16.6	23	21	45	10	0.51
26 April	70	15	14.0	22	24	45	9	0.49
26 April	71	11	9.3	16	23	38	22	0.42
26 April	79	7	8.6	17	24	45	9	0.38
01 May	98	68	64.5	45	20	27	8	1.67
01 May	99	47	43.1	38	24	32	6	1.19
01 May	101	21	20.0	31	19	44	6	0.70
01 May	106	17	14.4	27	19	46	6	0.59
01 May	109	7	7.8	22	12	58	8	0.38
02 May	120	97	83.0	35	16	45	4	0.78
02 May	125	48	33.8	30	15	51	4	0.60
02 May	128	26	19.4	24	27	42	6	0.58
02 May	133	41	34.3	30	14	51	4	0.58
02 May	137	33	27.8	28	19	47	6	0.60

* percentage of total weight of the suspended matter. ** K—ratio of the mass of quartz to the mass of clay minerals in suspension.

3.3. Results of Satellite Observations

3.3.1. Satellite Data Processing and Products

Field measurements were carried out concurrently with satellite remote sensing (Table 5). To efficiently compare remote sensing SPM (SPM satellite) with the in situ turbidity and sampled SPM (SPM in situ), it was necessary to use satellite optical data of a sufficiently high spatial resolution. Such data were available from Sentinel-2 MSI, with pixel resolution of 10 m in the visible range, and Landsat-8 OLI, with pixel resolution of 30 m. MSI data were obtained on 23, 26, and 28 April and on 1 May 2019; and OLI on 25 April and 2 May. On 1 and 2 May, there was haze which compromised SPM satellite data, but the plume edge was clearly visible (Figure 8). In total, during the period of field measurements, five images were acquired from Sentinel-2A/-2B MSI; one image from Landsat-7 ETM+; two images from Landsat-8 OLI/TIRS; and four images from Sentinel-3A/-3B OLCI. Based on the satellite data, TCI were composed to highlight the plume boundaries, as well as SPM satellite maps. MODIS data (Aqua/Terra) and NPP VIIRS were used as a source of auxiliary information. All satellite data were swiftly integrated into the

See the Sea (STS) information system [75,76] and analyzed online to supply information for planning the next day of work (define more accurately the coordinates of hydrological stations). The satellite data and products available in the period of the field measurements are listed in Table 5.

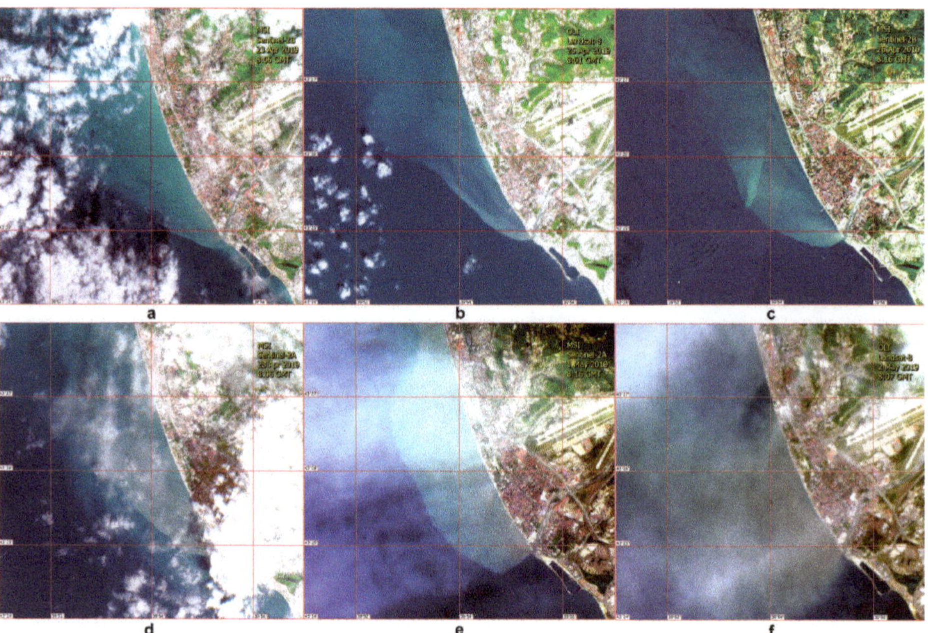

Figure 8. Fragments of satellite images obtained during the measurement period: 23 April 2019, Sentinel-2B Multispectral Instrument (MSI) (**a**); 25 April 2019, Landsat-8 Operational Land Imager (OLI) (**b**); 26 April 2019, Sentinel-2B MSI (**c**); 28 April 2019, Sentinel-2A MSI (**d**); 1 May 2019, Sentinel-2A MSI (**e**); 2 May 2019, Landsat-8 OLI (**f**).

3.3.2. Plume Boundary Detection

One of the main tasks of this work was to compare SPM in situ with SPM satellite Therefore, it was very important to carry out in situ measurements at the same time and for the same points, specifically in the region of maximum turbidity inside the plume and outside it. During boat trips, each station position was clearly defined with respect to the plume: either it was at the plume boundary, inside the plume or outside it. Each measurement cycle took about three–four hours every day to complete all hydrological stations (see Table 5). Because daily boat measurements started approximately at the time of a satellite overflight, they ended 3–4 h after it. When the measurements were completed, on its way back to port the boat followed the plume boundary visible from board, and its path was recorded using a Global Positioning System (GPS) tracker. Subsequently, the plume boundary obtained this way was plotted on the satellite image. The intermediate positions of the eastern boundary of the plume, determined from the corresponding station position records, were also plotted on the satellite image. It was found that the plume boundary changes its position at a rather high velocity, which should be taken into account when comparing the data of contact and remote measurements.

Table 5. Satellite information available during field measurements.

Date	Time UTC	Sensor	Satellite	Pixel Resolution, m	Product	Boat Measurements, Time UTC
23 April	08:17	MSI	Sentinel-2B	10	TCI, SPM	7:42–10:38
24 April	07:59	ETM+	Landsat-7	30	TCI	7:33–11:41
25 April	08:01	OLI/TIRS	Landsat-8	30/60	TCI, SST, SPM, CHL	7:37–11:08
	07:56	OLCI	Sentinel-3B	300	SPM	
26 April	07:30	OLCI	Sentinel-3B	300	SPM	7:27–10:32
	08:27	MSI	Sentinel-2B	10	TCI, SPM	
	10:12	VIIRS	NPP	1000	SST, WLR, CHL	
28 April	08:17	MSI	Sentinel-2A	10	TCI (cloud)	7:37–9:17
30 April	07:26	OLCI	Sentinel-3B	300	SPM	No measurements at stations
01 May	08:27	MSI	Sentinel-2A	10	TCI (cloud)	7:35–10:56
02 May	08:07	OLI/TIRS	Landsat-8	30	TCI (cloud)	7:30–11:07
04 May	08:02	OLCI	Sentinel-3A	300	SPM	No measurements at stations

As an example, Figure 9 shows the positions of the eastern boundary of the plume on 26 April 2019, at 07:30–08:00 UTC (yellow line); 09:11–09:29 UTC (pink line) and 10:30–11:00 UTC when the boat returns to port (green line). Sentinel-2 MSI surveyed the area at 08:27 UTC. From the plume positions determined by the station locations and the plume boundary derived from the satellite image, the velocity of displacement of the plume eastern boundary was estimated to increase from 5 to 13 cm/s.

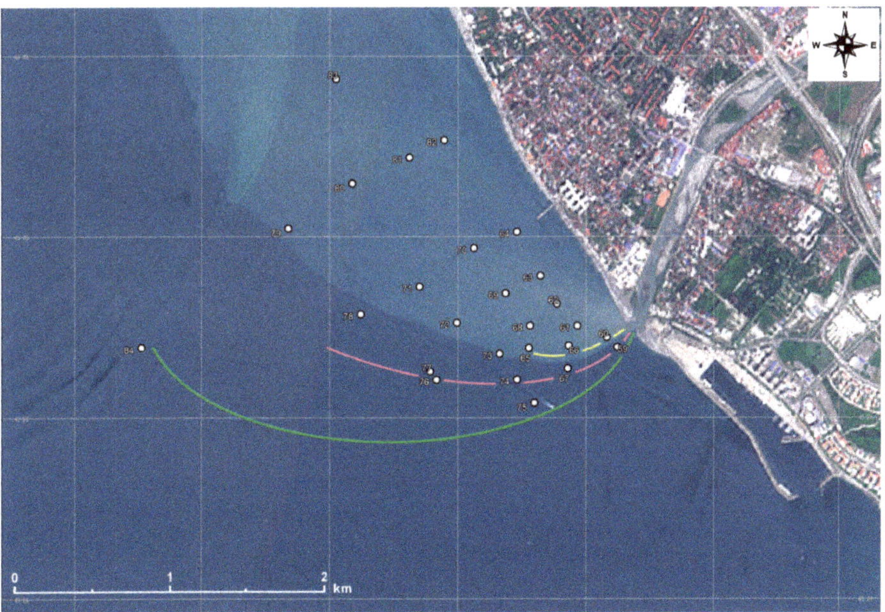

Figure 9. Plume boundary positions on 26 April 2019, at 07:30–08:00 UTC (yellow line); 09:11–09:29 UTC (pink line) and 10:30–11:00 UTC (green line) superimposed on a Sentinel-2B MSI image taken at 08:27 UTC on the same day.

3.3.3. Correlation Analysis of SPM from Contact and Remote-Sensing Data Using Case 2 Regional Coast Color (C2RCC) Algorithm

It is difficult to expect a high correlation between SPM in situ measurements and remote sensing estimations obtained using standard algorithms. As a rule, researchers prefer to develop individual regional algorithms [21]. Nevertheless their performance depends on many factors: season, river discharge, precipitation, etc. Our aim was to compare SPM obtained from in situ measurements and SPM retrieved from satellite data using standard rather than regional algorithms. This section presents C2RCC results.

For a joint analysis of in situ and satellite data, 26 April was selected as the only cloudless day. Figure 10a presents SPM map from the Sentinel-2B MSI data. Here, the C2RCC output is total suspended matter (TSM), a term with the same meaning as SPM and similar wide use in literature. As shown in Section 3.3.2, the plume boundary was rapidly shifting towards open sea (Figure 9). Therefore, some boat stations are inside the plume identified in the satellite image, the others outside it. The MSI data were taken at 08:16 UTC. All stations can be divided into three main groups. The first group includes stations where measurements were taken at a time close to the satellite overflight. These stations were inside the plume, according to visual observations from the boat as well as satellite observations. They are Stations 61, 62, 63, 68, 69, 70, 71, 72 marked blue in Figure 10b depicting the correlation between sampled SPM in situ and SPM satellite. Station 63 failed IQR data control and was excluded from further consideration. The blue marks are approximated by a straight line SPM satellite = $1.353 \times$ SPM in situ with reliability $R^2 = 0.99$. Higher satellite values can be explained by the fact that each of them characterizes a certain volume of water rather than a point in the upper layer.

The second group, marked red in Figure 10b, includes Stations 65 and 66. At the time of in situ measurements they were located at the plume boundary that was quite well visible from the boat, and at the time of satellite imaging inside the plume. Therefore, SPM satellite values are much higher than SPM in situ. Special interest presents Station 60. This station was located at the mouth of the river with an SPM satellite of 51.8 g/m^3. In situ measurements however gave an SPM in situ of only 10.5 g/m^3. This discrepancy is most likely explained by complex conditions in close vicinity of the mouth. The river discharges water at a high speed and it vigorously interacts with seawater causing wave breaking and intense mixing. This can negatively affect both water sampling results and remote-sensing data, for instance, when the sensor captures reflection from the sea bottom in shallow waters. Also, the sample taken at this station could contain less suspended matter due to technical difficulties.

The third group of stations (73, 74, and 77) was inside the plume during water sampling (green marks). However, in situ measurements at these stations were carried out 1–1.5 h after the satellite overflight and, over this time, the plume boundary shifted by almost 250 m relative to its boundary identified in the satellite image. Therefore, in the satellite image, these stations are already outside the plume. Accordingly, at these stations the SPM satellite is lower than the SPM in situ.

Finally, as depicted in Figure 10b, Station 64 (Group 4) was located on the inner border of the plume, which was also displaced, though not as fast as the outer one. Being inside the plume during in situ measurements, Station 64 got practically outside it during satellite imagery. This explains higher values of SPM in situ than SPM satellite attributed to the station.

Thus, even using the standard C2RCC algorithm for determining SPM satellite, we achieve a good agreement with SPM in situ obtained by water sampling, but only for those stations where measurements were taken almost synchronously with satellite imaging.

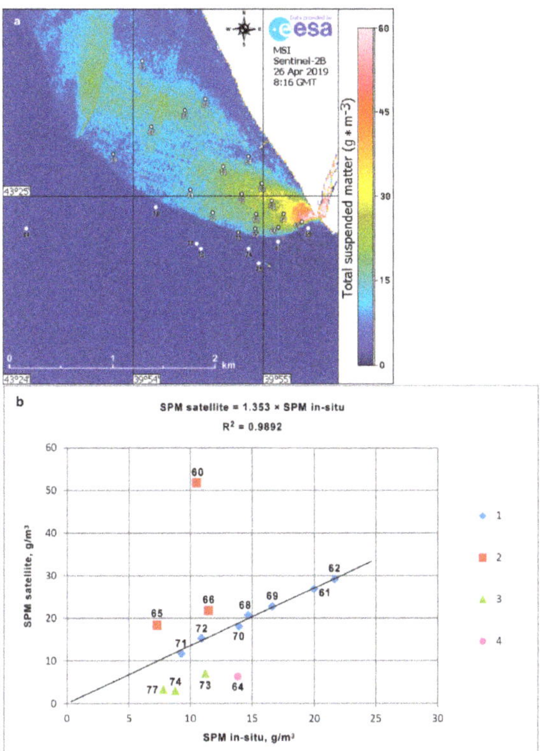

Figure 10. Case 2 Regional Coast Color (C2RCC) performance on Sentinel-2B MSI data of 26 April 2019: (**a**) SPM map with positions of the measurement stations; (**b**) quantitative comparison of SPM satellite and water sampling SPM in situ. In Panel (**b**), Group 1 stations (blue) retained their positions relative to the plume boundary during the time between satellite overflight and water sampling; Group 2 (red) stations were located directly at the plume boundary at the time of water sampling. In the satellite image, they are inside the plume; Group 3 (green) stations were inside the plume at the time of water sampling. In the satellite image, they are outside the plume; Group 4 station (magenta) is located opposite the pier. The station numbers are indicated beside the marks. The trend line is drawn only for Group 1.

3.3.4. Correlation Analysis of SPM and Turbidity from Contact and Remote-Sensing Data Using Different Algorithms

Among the main goals of our work was choosing the best standard algorithm for SPM and turbidity retrieval from remote sensing data, in terms of correlation with in situ measurements. To determine quantitative SPM, in addition to C2RCC we used Nechad 2009 [31] and Nechad 2015 [45]. SPM distribution maps built for 26 April 2019 using the three algorithms are shown in Figure 11a. Qualitatively analyzing these maps, the following conclusions can be drawn. C2RCC results look rather noisy: at low SPM, neighboring values vary significantly. The features of the plume boundary are not pronounced. The results of Nechad 2009 and Nechad 2015 are smoother and all inhomogeneities of the plume boundary can be distinguished. At the same time, the three algorithms give different distributions of the maximum SPM values in the immediate vicinity of Mzymta mouth. On the C2RCC map, the area of maximum SPM values is much larger. For processing by Nechad 2009, Nechad 2015 and comparison with C2RCC, we used only data from those stations that were performed almost synchronously with the imaging (Group 1, Figure 10b) by Sentinel-2B MSI on 26 April 2019. Figure 12a presents comparisons of SPM in situ and

SPM satellite obtained by C2RCC, Nechad 2009 and Nechad 2015. No doubt, the C2RCC results best agree with the in situ data. The straight approximation line for C2RCC goes through the origin of coordinates, the determination coefficient is $R^2 = 0.989$ (Figure 12a) For the two other algorithms, Nechad 2009 and Nechad 2015, the determination coefficients are only 0.943 and 0.941, respectively.

The main advantage of the ACOLITE algorithms is that it is possible to compare their results with in situ turbidity data, for example, measured with PT, without converting the latter to SPM in situ using the obtained dependence (Figure 7) and, more importantly, without water sampling. To quantify turbidity, we used Nechad 2009, Nechad 2015 and Dogliotti algorithms. The turbidity distribution maps compiled using these algorithms for 26 April 2019, are shown in Figure 11b. Since Nechad 2009 and Dogliotti algorithms coincide for low turbidity range, the corresponding patterns of turbidity are the same Near Mzymta mouth, Dogliotti definitely overestimates turbidity. Interestingly, Dogliotti draws a pronounced high turbidity jet westward from the mouth zone. The existence of such a jet is confirmed, for example, by the in situ measurements at Stations 61 and 62 (Figure 10a). Although Station 62 is located somewhat farther from the mouth, but, unlike Station 61, it sits on the jet and, therefore, reports higher SPM and turbidity. A comparison of turbidities obtained by Nechad 2009, Nechad 2015 and Dogliotti (Turb satellite) with turbidity measured in situ (Turb in situ) is presented in Figure 12b. At low turbidity, less than 16 NTU (FNU), Dogliotti yields the same results as Nechad 2009. For turbidities of 20–25 NTU and higher, Dogliotti switches to another method of calculation [32] suitable for extremely turbid waters, but not for our study area (Figure 12b).

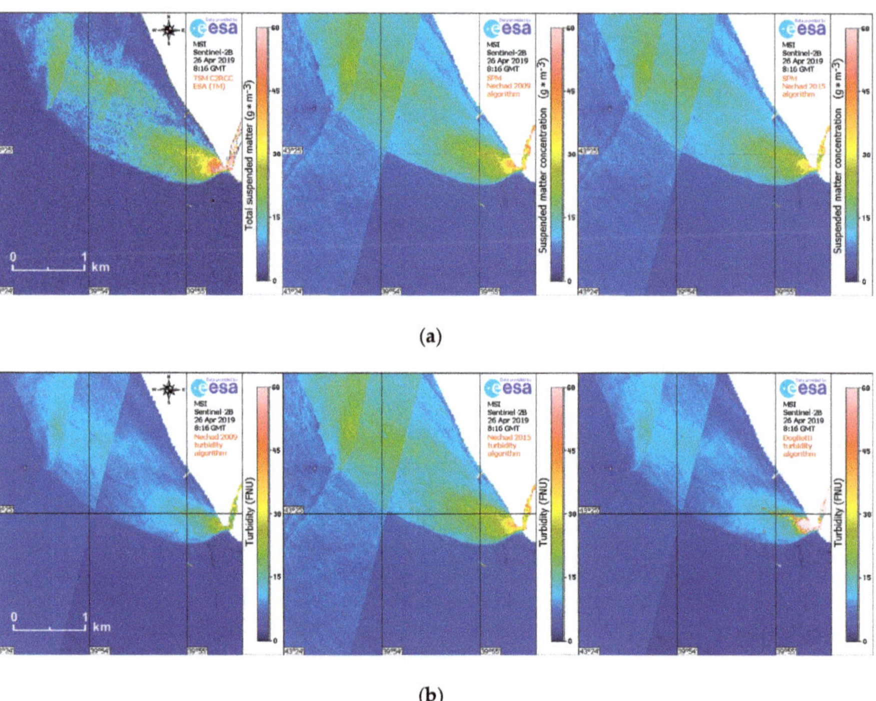

(a)

(b)

Figure 11. Performance of different satellite algorithms to map: (**a**) SPM; (**b**) turbidity, retrieved from Sentinel-2B MSI data of 26 April 2019.

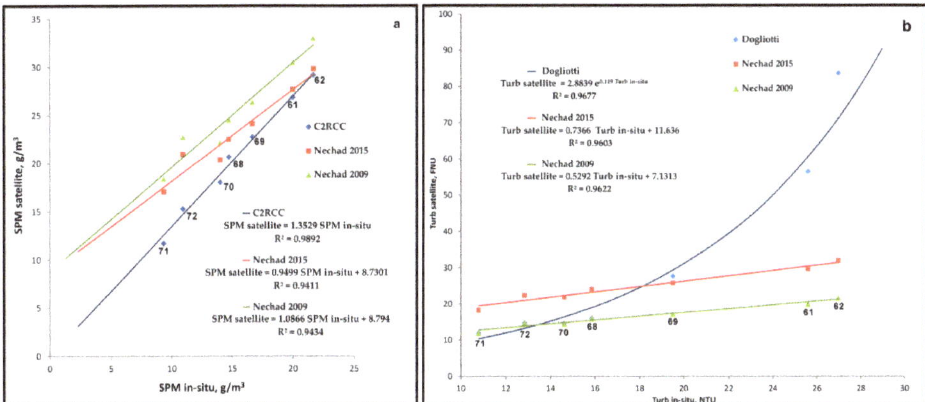

Figure 12. Performance of different satellite retrieval algorithms compared with in situ measurements: (**a**) SPM satellite vs. SPM in situ; (**b**) satellite turbidity (Turb satellite) vs. in situ turbidity obtained with PT (Turb in situ). The station numbers are indicated beside the marks.

The values produced by Nechad 2009 and Nechad 2015 are well approximated by straight lines (Figure 12b). The determination coefficient is $R^2 = 0.96$ in both cases and the lines are almost parallel. It is a puzzling fact that the lines do not go through the origin of coordinates. Obviously, further testing on a much wider array of in situ measurements is required.

4. Discussion

The data obtained by various methods during our work in April–May 2019, in the Mzymta mouth zone can be divided into two groups:

1. Data from two turbidity sensors–an optical turbidity sensor as part of the RBR-concerto CTD probe (TM) and a TN400 portable turbidimeter (PT). Both sensors provide data in NTU units and work roughly on the same principle. A significant difference is that turbidity measurements were taken at different depths, since it is impossible to obtain data in the first centimeters from the surface with the CTD probe.
2. Data on SPM at different points of the plume obtained using different methods. The first method is direct: SPM in situ was measured by weighing water samples. The second method is indirect: SPM satellite was retrieved using the standard algorithms from satellite remote sensing data.

In our work, we aimed to estimate the correlation of the data obtained from different turbidity sensors, find out if there exists a robust dependence between turbidity and SPM measured in situ, and, most importantly, reveal the correlation between satellite and contact measurements.

4.1. Performance of Contact Turbidity Sensors

As expected, turbidities obtained at the same stations with PT and TM in the near-surface horizon of 0.35–0.5 m agreed quite well. The determination coefficient of linear approximation was $R^2 = 0.93$ and, in general, TM turbidities were slightly higher than PM ones, by a factor of 1.042. At high turbidity, a large scatter of the values is observed on both sides of the linear trend. It possibly can be explained by highly unsteady interaction of river and sea waters at the river mouth in shallow water which makes turbidity vary significantly even at close points. Moreover, clapotis and wave breaking often take place. In such conditions, it is not easy to use PT. Several measurements should be done in close points for more reliable results. Sometimes it is technically troublesome, for example, because of the risk to run aground. Meanwhile measuring with TM is much easier in such conditions, besides, each time we get at least 2 turbidity values, when TM is lowered and then raised. This was specifically observed during a sharp increase of the river discharge on May 2 when the velocity of the water flow increased from 2 m/s in the previous days to 6–8 m/s.

For turbidity values from 15 to 40 NTU, the results of measurements with different instruments almost completely coincided, with slightly higher TM values. At turbidities less than 15 NTU, an underestimation by TM can be noted. In fact, as the river waters spread into the sea, both turbidity and plume thickness sharply decrease (Figure 5a). So, the TM may eventually get below the depth of the plume penetration zone.

Considering the relationship of contact turbidity and SPM in situ, naturally, the best correlation was achieved with PT. This is easy to explain. Water samples for determining the SPM were taken at the same depth as measurements with PT, while TM measurements, as mentioned above, took place at lower points. Nevertheless, we also got a linear correlation between TM turbidity and SPM in situ, although with a slightly smaller determination coefficient as compared to PM measurements, 0.924 and 0.982, respectively. The linear relationship obtained makes it possible to convert turbidity values measured in NTU to SPM calculated in g/m^3. Similar linear correlation was determined for the macrotidal estuary of the Gironde [1]. The authors of the work noted that the relationship was specific to the turbidity sensor used, but similar to those established using other instruments in other periods of time. This could be an indication that suspended matter grain size distribution and composition in the estuary did not change significantly in optical terms over the years. Such a hypothesis should be tested for Mzymta plume as well. For this purpose, one need to determine its mineral composition and establish the relationship between, for example, quartz, as the largest suspended matter constituent, and plume turbidity.

Performance of the two turbidity sensors, PT and TM, showed that, in general, it is sufficient to take measurements with only one instrument. A question arises which one suits better, in view of comparison with satellite data. To draw sound conclusions, knowledge of turbidity depth distribution is required. We believe that if river water turbidity is high and its penetration depth is small, it is more reasonable to use PT, since in this case water leaving radiance captured by the satellite sensor is formed in the near-surface horizon. If turbidity is low, water leaving radiance can be formed in a layer down to a few meters deep, so TM appears to be a more suitable instrument. In this case it is obviously necessary to obtain some integral characteristics from TM readings at various depths. This is an interesting and complex problem. As a rule, however, for comparison with satellite data, either data from various PT analogs are used [77], or data from floating spectroradiometers that are widely employed today for validating satellite data obtained during field experiments. They measure absolute spectral irradiance at the sea surface and water leaving radiance immediately under the sea surface [78]. TM data are usually used for estimating river water penetration depth and turbidity profiling.

4.2. Small-Scale River Plume Boundary Dynamics

Our work has demonstrated prime importance of tight synchronization of in situ measurements and satellite survey in the study region of Mzymta River plume. In a thematically close study [23] discussing turbidity characteristics of the Danube plume, it was noted that the maximum time gap considered between in situ SPM and high spatial resolution images acquisition was of 120 h for periods with no substantial river fluctuation and 48 h otherwise. In the case of Mzymta, the maximum time gap should not exceed 30–40 min since, as shown below, the plume boundary can move really fast. Certainly, the plume of the Danube spreads for a much greater distance from the coast, compared to Mzymta, so at considerable distances from its boundary, the plume can be regarded unchanging during a day or two. In studies conducted in the mouth regions of small rivers, such as Mzymta, one has to take into account plume spreading dynamics that is strongly influenced by wind direction [58].

After a series of experiments on 23, 25 and 26 April, small-scale displacement veloci-ties of the plume boundary of Mzymta and their relationship with wind direction were estimated (Figure 13). The essence of the experiment was to follow the plume boundary using a GPS tracker on the way back to port after finishing the in situ measurements. It was found that with weak S/SE winds, the displacement of the plume boundary was very slow. In three hours, it shifted by 250–350 m at a velocity of 0.08–0.12 km/h under the "pressing" winds towards or along the coastal zone (23 and 25 April). The maximum displacement of the plume boundary was noted during W winds on April 26. In three hours, the boundary shifted 900 m seaward. The displacement speed was 0.3 km/h. The results of the experiments confirm the hypothesis of high mobility of the river plume as a whole and its boundary in particular. This rather complicates the comparison of quanti-tative remote sensing and contact data for specific stations, because during several hours of boat measurements, the internal fine structure of the plume can change significantly. Similar results are presented in [58]. Considering the impact of wind on hydrodynamic characteristics of the plume, it is possible to tentatively forecast its spreading velocity and plan more accurately the measurements at those stations whose positions relative to the plume (inside, outside, at the boundary) are not expected to change during the satellite overflight.

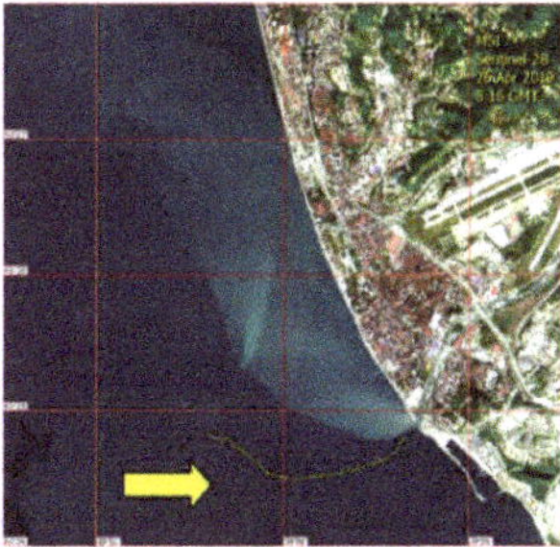

Figure 13. River plume boundary at the time of satellite overflight in the images of 23, 25 and 26 April 2019 (true color images (TCI)). In the images, yellow line is a GPS track along the boundary of the plume 3 h after the satellite overflight. Yellow arrow indicates prevailing wind direction.

4.3. Performance of Satellite SPM and Turbidity Algorithms

Satellite SPM and turbidity were calculated using C2RCC, Nechad 2009, Nechad 2015 and Dogliotti algorithms. Despite its failure for turbidities greater 20 NTU in the Mzymta region, Dogliotti can be nevertheless used to reveal small scale turbidity inhomogeneities that the other two algorithms can hardly detect. The best correlation with the in situ data was achieved with C2RCC. The authors of [77] used C2RCC on Sentinel-3A OLCI L1 data using SNAP and validated the results against dedicated in situ data obtained in the Northwestern Baltic proper. Their validation campaigns took place between 2016 and 2018 in Swedish coastal waters and covered different times of year. On the basis of a large dataset, authors of [77] recommend using C2RCC, but point out the problems of atmospheric correction for pixels close to coast. The problems of adjacency effects or land contamination of satellite sea data are also discussed in [1]. Both works employ low- and medium-resolution ocean color data (MODIS and OLCI), however we do not anticipate any serious complications when using high resolution data, such as MSI and OLI, as we did in this study.

Among the atmospheric correction algorithms employed, ACOLITE DSF appears the most practical and best performing in our study region. Nechad 2009 and Nechad 2015 overestimated SPM by 1.5 times for in situ measurement range up to 15 g/m^3. For the range over 20–25 g/m^3, Nechad 2015 agreed well with in situ data and C2RCC results

(Figure 12a). Data of 1 and 2 May, when Mzymta discharge and turbidity increased significantly, could have been of particular interest, but haze on these days did not allow obtaining meaningful satellite SPM.

4.4. Changes in the Mineral Composition of Suspended Matter Depending on Plume Water

The mineral composition of suspended matter (Table 4) shows significant variations in the content of minerals in different water samples. The quantitative ratio of mineral phases in the suspension composition depends on multiple factors.

Samples with high turbidity values, mainly from the near-mouth zones, contain a large amount of quartz. For example, samples taken on 1 May in the Mzymta estuary zone have high turbidity (up to 68 NTU) and predominance of quartz over clay minerals (see Figure 14). With the distance from the mouth, the amount of suspended matter in the water decreases and its mineral composition changes with a relative increase in clay minerals and a decrease in quartz. Samples with low turbidity values taken at the plume boundary have a significant predominance of clay minerals in the suspension.

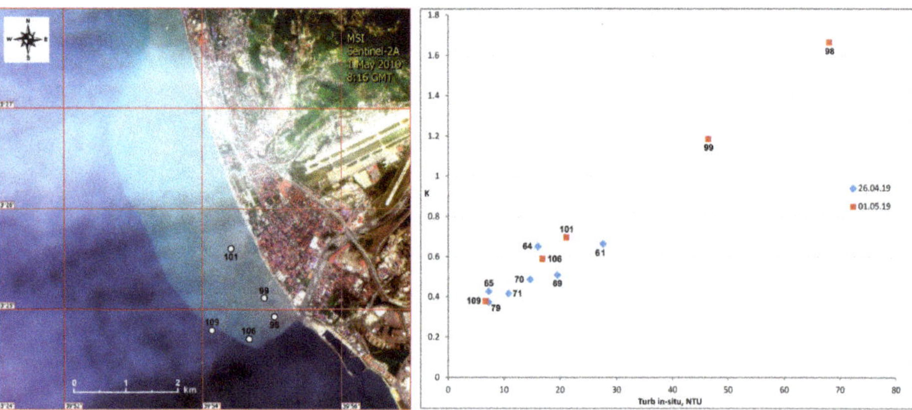

Figure 14. Left: fragment of a Sentinel-2A MSI satellite image of May 1. Marks indicate locations where samples were taken for X-ray phase analysis. Right: graph shows the ratio of the percentage of quartz in the dry matter of the suspension to the percentage of clay minerals (K) depending on turbidity (Turb in situ, NTU). The station numbers are indicated beside the marks. Positions of the stations on 26 April (blue marks) are shown in Figure 10a.

This result is in good agreement with the well-known theory of gravitational differentiation of material, according to which a decrease in the particle size occurs with distance from the coastal zone. Clay has low hardness, is highly susceptible to mechanical weathering, and forms a fine-grained material that can be transported over long distances Quartz is characterized by a higher hardness and has more coarse particles, which are deposited in the immediate vicinity of the river mouth.

As shown in Figure 14 right, the compositional differentiation of the suspension was most pronounced on 1 May. High turbidity values correspond to the substantially quartz composition of the suspension; at low turbidity, the suspension mainly consists of clay minerals. On 26 April, the turbidity of the plume was generally low, not more than 28 NTU The trend of changes in the suspension composition is weakly expressed. At the moment, no unambiguous relationship has been revealed between the value of water turbidity and the mineral composition of the suspension. It is not yet possible to estimate the amount of quartz and other mineral phases in the suspension by remote sensing methods.

5. Conclusions

The paper presents the results of field studies in the northeastern part of the Black Sea in the mouth area of the Mzymta River in April and May 2019. The main objective of the study was to determine the relationships between water turbidity and SPM obtained by contact and remote sensing methods and compare performances of the C2RCC processor and the ACOLITE algorithms Nechad 2009, Nechad 2015 and Dogliotti.

It was shown that the highest correlation between the satellite and the water sampling SPM for the study area in conditions of spring flooding was achieved with C2RCC, but only for those stations where measurements were taken almost synchronously with satellite imaging. For such stations, the comparison of satellite and water sampling SPM showed a linear relationship with a reliability of 0.99. Nechad 2009, Nechad 2015 and Dogliotti overestimated SPM by 1.5 times for in situ measurement range up to 15 g/m^3. For the range over 20–25 g/m^3, Nechad 2015 agreed well with in situ data and C2RCC results, while Dogliotti failed.

In a highly variable environment of the Black Sea northeastern coastal zone, rapidly changing conditions often require a specific choice of both methods and instruments for collecting in situ data suitable for validating the remote-sensing algorithms. Knowledge of depth distribution of the main hydrological parameters is a key prerequisite for the right choice.

When selecting in situ measurements for comparison with satellite data, one should be particularly vigilant with respect to the high mobility of the Mzymta plume: within the 3–4 h when, as a rule, all the stations were completed, its boundary could shift considerably, either being pressed to the coast or driven away from it. The velocity of displacement of the plume boundary was estimated to increase from 5 to 13 cm/s.

A comparison of data on turbidity obtained by a portable turbidity meter and water sampling SPM shows a linear relationship with the reliability of 0.982. This relationship remained stable in time and weather conditions, which makes a portable turbidity meter a valuable tool for fast and multiple measurements. Data obtained by this method can be easily converted to SPM. This new and important result is very promising for in situ SPM evaluation in the sense that expensive and time-consuming water sampling may eventually become redundant. Moreover, without water sampling, the validation of satellite algorithms for SPM retrieval based only on portable turbidity meter data becomes a lot easier and faster: much more shorter stations can be undertaken over the same period of time and no money must be spent on processing of water samples.

Based on X-ray phase analysis of the suspended matter, changes in the total amount of quartz and clay particles were found to be a function of optical turbidity of the water samples. With a decrease in turbidity the mineral composition of the suspension changed with a relative increase in clay minerals and a decrease in quartz.

The authors hope to continue the studies to improve and validate the results presented in this paper.

Author Contributions: Conceptualization, K.N. and O.L.; methodology, K.N., Y.A., O.L. and D.S.; validation, K.N., Y.A., O.L. and Y.S.; investigation, K.N., O.L., and Y.A.; data curation, O.L. and A.S.; writing, K.N., O.L., Y.A., T.B. and Y.S.; visualization, D.S. and A.S. All authors have read and agreed to the published version of the manuscript.

Funding: Field work was funded by the Russian Foundation for Basic Research in the framework of grant No. 17-05-00715. The processing and analysis of satellite and in situ data were carried out within IKI-Monitoring Center for Collective Use using the See the Sea system, that was funded by the Ministry of Science and Higher Education of Russia, Theme "Monitoring", State register No. 01.20.0.2.00164.

Acknowledgments: The authors are grateful to all the participants in the field work: Evgeny Krayushkin and Nikita Knyazev (Space Research Institute), Elena Zhuk (Marine Hydrophysical Institute); and Alexander Dudnikov and Egor Vlasov—the crew of the Arabella boat, from which the measurements were made. The authors are especially grateful to V.L. Kosorukov (Lomonosov

Moscow State University) who carried out X-ray phase analysis of the samples. The authors would like to thank the Academic Editor and four Reviewers for their careful attitude and useful comments and suggestions.

Conflicts of Interest: The authors declare no conflict of interest.

References

1. Doxaran, D.; Froidefond, J.-M.; Castaing, P.; Babin, M. Dynamics of the turbidity maximum zone in a macrotidal estuary (the Gironde, France): Observations from field and MODIS satellite data. *Estuar. Coast. Shelf Sci.* **2009**, *81*, 321–332. [CrossRef]
2. Zavialov, P.O.; Makkaveev, P.N.; Konovalov, B.V.; Osadchiev, A.A.; Khlebopashev, P.V.; Pelevin, V.V.; Grabovskiy, A.B.; Izhitskiy, A.S.; Goncharenko, I.V.; Soloviev, D.M.; et al. Hydrophysical and hydrochemical characteristics of the sea areas adjacent to the estuaries of small rivers of the Russian coast of the Black Sea. *Oceanology* **2014**, *54*, 265–280. [CrossRef]
3. Lavrova, O.Y.; Mityagina, M.I.; Kostianoy, A.G. *Satellite Methods for Detecting and Monitoring Marine Zones of Ecological Risk*; IKI RAN: Moscow, Russia, 2016. (In Russian)
4. Garvine, R.W. A steady state model for buoyant surface plume hydrodynamics in coastal waters. *Tellus* **1982**, *34*, 293–306. [CrossRef]
5. Garvine, R.W.; Monk, J.D. Frontal structure of a river plume. *J. Geophys. Res.* **1974**, *79*, 2251–2259. [CrossRef]
6. Lavrova, O.Y.; Soloviev, D.M.; Strochkov, M.A.; Bocharova, T.Y.; Kashnitsky, A.V. River plumes investigation using Sentinel-2A MSI and Landsat-8 OLI data. In *SPIE Remote Sensing, Proceedings of the Remote Sensing of the Ocean, Sea Ice, Coastal Waters, and Large Water Regions 2016, Edinburgh, UK, 26–29 September 2016*; SPIE—International Society for Optics and Photonics: Bellingham, WA, USA, 2016; Volume 9999, p. 99990G. [CrossRef]
7. Zajączkowski, M.; Darecki, M.; Szczuciński, W. Report on the development of the Vistula river plume in the coastal waters of the Gulf of Gdansk during the May 2010 flood. *Oceanologia* **2010**, *52*, 311–317. [CrossRef]
8. Abascal-Zorrilla, N.; Vantrepotte, V.; Huybrechts, N.; Ngoc, D.D.; Anthony, E.J.; Gardel, A. Dynamics of the Estuarine Turbidity Maximum Zone from Landsat-8 Data: The Case of the Maroni River Estuary, French Guiana. *Remote Sens.* **2020**, *12*, 2173. [CrossRef]
9. Doxaran, D.; Froidefond, J.M.; Castaing, P. A reflectance band ratio used to estimate suspended matter concentrations in sediment-dominated coastal waters. *Int. J. Remote Sens.* **2002**, *23*, 5079–5085. [CrossRef]
10. Babin, M.A.; Morel, V.; Fournier-Sicre, F.F.; Stramski, D. Light scattering properties of marine particles in coastal and open ocean waters as related to the particle mass concentration. *Limnol. Oceanogr.* **2003**, *48*, 843–859. [CrossRef]
11. Constantin, S.; Doxaran, D.; Constantinescu, S. Estimation of water turbidity and analysis of its spatio-temporal variability in the Danube River plume (Black Sea) using MODIS satellite data. *Cont. Shelf Res.* **2016**, *112*, 14–30. [CrossRef]
12. Chen, J.; D'Sa, E.; Cui, T.; Zhang, X. A semi-analytical total suspended sediment retrieval model in turbid coastal waters: A case study in Changjiang River Estuary. *Opt. Express* **2013**, *21*, 13018–13031. [CrossRef]
13. Gernez, P.; Lafon, V.; Lerouxel, A.; Curti, C.; Lubac, B.; Cerisier, S.; Barillé, L. Toward Sentinel-2 High Resolution Remote Sensing of Suspended Particulate Matter in Very Turbid Waters: SPOT4 (Take5) Experiment in the Loire and Gironde Estuaries. *Remote Sens.* **2015**, *7*, 9507–9528. [CrossRef]
14. Ou, S.; Zhang, H.; Wang, D. Dynamics of the buoyant plume off the Pearl River estuary in summer. *Environ. Fluid Mech.* **2009**, *9*, 471–492. [CrossRef]
15. Petus, C.; Chust, G.; Gohin, F.; Doxaran, D.; Froidefond, J.M.; Sagarminaga, Y. Estimating turbidity and total suspended matter in the Adour River plume (South Bay of Biscay) using MODIS 250-m imagery. *Cont. Shelf Res.* **2010**, *30*, 379–392. [CrossRef]
16. Devlin, M.J.; Petus, C.; da Silva, E.; Tracey, D.; Wolff, N.H.; Waterhouse, J.; Brodie, J. Water Quality and River Plume Monitoring in the Great Barrier Reef: An Overview of Methods Based on Ocean Colour Satellite Data. *Remote Sens.* **2015**, *7*, 12909–12941. [CrossRef]
17. Ody, A.; Doxaran, D.; Vanhellemont, Q.; Nechad, B.; Novoa, S.; Many, G.; Bourrin, F.; Verney, R.; Pairaud, I.; Gentili, B. Potential of High Spatial and Temporal Ocean Color Satellite Data to Study the Dynamics of Suspended Particles in a Micro-Tidal River Plume. *Remote Sens.* **2016**, *8*, 245. [CrossRef]
18. Ouillon, S.; Forget, P.; Froidefond, J.-M.; Naudin, J.-J. Estimating suspended matter concentrations from SPOT data and from field measurements in the Rhone River plume. Marine Technology Society. *Mar. Technol. Soc. J.* **1997**, *31*, 15.
19. Berdeal, I.; Hickey, B.; Kawase, M. Influence of wind stress and ambient flow on high discharge river plume. *J. Geophys. Res.* **2002**, *107*, 3130. [CrossRef]
20. Pruszak, Z.; van Ninh, P.; Szmytkiewicz, M.; Ostrowski, R. Hydrology and morphology of two river mouth regions (temperate Vistula Delta and subtropical Red River Delta). *Oceanologia* **2005**, *47*, 365–385.
21. Kopelevich, O.; Sheberstov, S.; Burenkov, V.; Vazyulya, S.; Likhacheva, M. Assessment of underwater irradiance and absorption of solar radiation at water column from satellite data. In *SPIE Remote Sensing, Proceedings of the Current Research on Remote Sensing, Laser Probing, and Imagery in Natural Waters, Moscow, Russian, 1–3 January 2007*; SPIE—International Society for Optics and Photonics: Bellingham, WA, USA, 2007; Volume 6615, p. 661507. [CrossRef]
22. Mulligan, R.P.; Perrie, W.; Solomon, S. Dynamics of the Mackenzie River plume on the inner Beaufort shelf during an open water period in summer. *Estuar. Coast. Shelf Sci.* **2010**, *89*, 214–220. [CrossRef]

23. Güttler, F.N.; Niculescu, S.; Gohin, F. Turbidity retrieval and monitoring of Danube Delta waters using multi-sensor optical remote sensing data: An integrated view from the delta plain lakes to the western–northwestern Black Sea coastal zone. *Remote Sens. Environ.* **2013**, *132*, 86–101. [CrossRef]

24. Lavrova, O.Y.; Soloviev, D.M.; Mityagina, M.I.; Strochkov, A.Y.; Bocharova, T.Y. Revealing the influence of various factors on concentration and spatial distribution of suspended matter based on remote sensing data. In *SPIE Remote Sensing, Proceedings of the Remote Sensing of the Ocean, Sea Ice, Coastal Waters, and Large Water Regions 2015, Toulouse, France, 21–24 September 2015*; SPIE—International Society for Optics and Photonics: Bellingham, WA, USA, 2015; Volume 9638, p. 96380D. [CrossRef]

25. Cai, L.; Tang, D.; Li, X.; Zheng, H.; Shao, W. Remote sensing of spatial-temporal distribution of suspended sediment and analysis of related environmental factors in Hangzhou Bay, China. *Remote Sens. Lett.* **2015**, *6*, 597–603. [CrossRef]

26. Osadchiev, A.A. Estimation of river discharge based on remote sensing of a river plume. In *SPIE Remote Sensing, Proceedings of the Remote Sensing of the Ocean, Sea Ice, Coastal Waters, and Large Water Regions 2015, Toulouse, France, 21–24 September 2015*; SPIE—International Society for Optics and Photonics: Bellingham, WA, USA, 2015; Volume 9638, p. 96380H. [CrossRef]

27. Warrick, J.A.; DiGiacomo, P.M.; Weisberg, S.B.; Nezlin, N.P.; Mengel, M.; Jones, B.H.; Ohlmann, J.C.; Washburn, L.; Terrill, E.J.; Farnsworth, K.L. River plume patterns and dynamics within the Southern California Bight. *Cont. Shelf Res.* **2007**, *27*, 2427–2448. [CrossRef]

28. Many, G.; Bourrin, F.; De Madron, X.D.; Pairaud, I.; Gangloff, A.; Doxaran, D.; Ody, A.; Verney, R.; Menniti, C.; Le Berre, D.; et al. Particle assemblage characterization in the Phone River ROFI. *J. Mar. Syst.* **2016**, *157*, 39–51. [CrossRef]

29. Novoa, S.; Doxaran, D.; Ody, A.; Vanhellemont, Q.; Lafon, V.; Lubac, B. Atmospheric corrections and multi-conditional algorithm for multi-sensor remote sensing of suspended particulate matter in low-to-high turbidity levels Coastal Waters. *Remote Sens.* **2017**, *9*, 61. [CrossRef]

30. Gohin, F. Annual cycles of chlorophyll-*a*, non-algal suspended particulate matter, and turbidity observed from space and in-situ in coastal waters. *Ocean Sci.* **2011**, *7*, 705–732. [CrossRef]

31. Nechad, B.; Ruddick, K.G.; Park, Y. Calibration and validation of a generic multisensor algorithm for mapping of total suspended matter in turbid waters. *Remote Sens. Environ.* **2010**, *114*, 854–866. [CrossRef]

32. Dogliotti, A.I.; Ruddick, K.G.; Nechad, B.; Doxaran, D.; Knaeps, E. A single algorithm to retrieve turbidity from remotely-sensed data in all coastal and estuarine waters. *Remote Sens. Environ.* **2015**, *156*, 157–168. [CrossRef]

33. Vantrepotte, V.; Loisel, H.; Mériaux, X.; Neukermans, G.; Dessailly, D.; Jamet, C.; Gensac, E.; Gardel, A. Seasonal and inter-annual (2002-2010) variability of the suspended particulate matter as retrieved from satellite ocean color sensor over the French Guiana coastal waters. *J. Coast. Res.* **2011**, *64*, 1750–1754.

34. Tavora, J.; Boss, E.; Doxaran, D.; Hill, P. An Algorithm to Estimate Suspended Particulate Matter Concentrations and Associated Uncertainties from Remote Sensing Reflectance in Coastal Environments. *Remote Sens.* **2020**, *12*, 2172. [CrossRef]

35. Knaeps, E.; Ruddick, K.G.; Doxaran, D.; Dogliotti, A.I.; Nechad, B.; Raymaekers, D.; Sterckx, S. A SWIR based algorithm to retrieve total suspended matter in extremely turbid waters. *Remote Sens. Environ.* **2015**, *168*, 66–79. [CrossRef]

36. Han, B.; Loisel, H.; Vantrepotte, V.; Mériaux, X.; Bryère, P.; Ouillon, S.; Dessailly, D.; Xing, Q.; Zhu, J. Development of a Semi-Analytical Algorithm for the Retrieval of Suspended Particulate Matter from Remote Sensing over Clear to Very Turbid Waters. *Remote Sens.* **2016**, *8*, 211. [CrossRef]

37. Doerffer, R.; Schiller, H. The MERIS Case 2 water algorithm. *Int. J. Remote Sens.* **2007**, *28*, 517–535. [CrossRef]

38. Schroeder, T.; Schaale, M.; Fischer, J. Retrieval of atmospheric and oceanic properties from MERIS measurements: A new Case-2 water processor for BEAM. *Int. J. Remote Sens.* **2007**, *28*, 5627–5632. [CrossRef]

39. Gangloff, A.; Verney, R.; Doxaran, D.; Ody, A.; Estournel, C. Investigating Rhône River plume (Gulf of Lions, France) dynamics using metrics analysis from the MERIS 300m Ocean Color archive (2002–2012). *Cont. Shelf Res.* **2007**, *144*, 98–111. [CrossRef]

40. Xue, K.; Ma, R.; Shen, M.; Li, Y.; Duan, H.; Cao, Z.; Wang, D.; Xiong, J. Variations of suspended particulate concentration and composition in Chinese lakes observed from Sentinel-3A OLCI images. *Sci. Total Environ.* **2020**, *721*, 137774. [CrossRef]

41. Gower, J.; Doerffer, R.; Borstad, G.A. Interpretation of the 685 nm peak in water-leaving radiance spectra in terms of fluorescence, absorption and scattering, and its observation by MERIS. *Int. J. Remote Sens.* **1999**, *20*, 1771–1786. [CrossRef]

42. Gower, J.; King, S.; Borstad, G.A.; Brown, L. Detection of intense plankton blooms using the 709 nm band of the MERIS imaging spectrometer. *Int. J. Remote Sens.* **2005**, *26*, 2005–2012. [CrossRef]

43. Brockmann, C.; Doerffer, R.; Peters, M.; Kerstin, S.; Embacher, S.; Ruescas, A. Evolution of the C2RCC Neural Network for Sentinel 2 and 3 for the Retrieval of Ocean Colour Products in Normal and Extreme Optically Complex Waters. In *Living Planet Symposium, Proceedings of the ESA Living Planet Symposium, Prague, Czech Republic, 9–13 May 2016*; ESA-SP: São Paulo, Brazil, 2016; Volume 740, ISBN 978-92-9221-305-3.

44. Nechad, B.; Ruddick, K.G.; Neukermans, G. Calibration and validation of a generic multisensor algorithm for mapping of turbidity in coastal waters. In *SPIE Remote Sensing, Proceedings of the Remote Sensing of the Ocean, Sea Ice, Coastal Waters, and Large Water Regions 2009, Berlin, Germany, 31 August–3 September 2009*; SPIE—International Society for Optics and Photonics: Bellingham, WA, USA, 2009; Volume 7473, p. 74730H. [CrossRef]

45. Nechad, B.; Ruddick, K.; Schroeder, T.; Oubelkheir, K.; Blondeau-Patissier, D.; Cherukuru, N.; Brando, V.; Dekker, A.; Clementson, L.; Banks, A.C.; et al. CoastColour Round Robin data sets: A database to evaluate the performance of algorithms for the retrieval of water quality parameters in coastal waters. *Earth Syst. Sci. Data* **2015**, *7*, 319–348. [CrossRef]

46. Martins, V.S.; Barbosa, C.C.F.; De Carvalho, L.A.S.; Jorge, D.S.F.; Lobo, F.D.L.; Novo, E.M.L.M. Assessment of Atmospheric Correction Methods for Sentinel-2 MSI Images Applied to Amazon Floodplain Lakes. *Remote Sens.* **2017**, *9*, 322. [CrossRef]

47. Warren, M.A.; Simis, S.G.H.; Martinez-Vicente, V.; Poser, K.; Bresciani, M.; Alikas, K.; Spyrakos, E.; Giardino, C.; Ansper, A Assessment of atmospheric correction algorithms for the Sentinel-2A MultiSpectral Imager over coastal and inland waters. *Remote Sens. Environ.* **2019**, *225*, 267–289. [CrossRef]

48. Caballero, I.; Stumpf, R.P. Atmospheric correction for satellite-derived bathymetry in the Caribbean waters: From a single image to multi-temporal approaches using Sentinel-2A/B. *Opt. Express* **2020**, *28*, 11742–11766. [CrossRef] [PubMed]

49. Gernez, P.; Doxaran, D.; Barillé, L. Shellfish Aquaculture from Space: Potential of Sentinel2 to Monitor Tide-Driven Changes in Turbidity, Chlorophyll Concentration and Oyster Physiological Response at the Scale of an Oyster Farm. *Front. Mar. Sci.* **2017**, *4*, 137. [CrossRef]

50. Vanhellemont, Q. Adaptation of the dark spectrum fitting atmospheric correction for aquatic applications of the Landsat and Sentinel-2 archives. *Remote Sens. Environ.* **2019**, *225*, 175–192. [CrossRef]

51. Vanhellemont, Q.; Ruddick, K. Turbid wakes associated with offshore wind turbines observed with Landsat 8. *Remote Sens. Environ.* **2014**, *145*, 105–115. [CrossRef]

52. Vanhellemont, Q.; Ruddick, K. Advantages of high quality SWIR bands for ocean colour processing: Examples from Landsat-8 *Remote Sens. Environ.* **2015**, *161*, 89–106. [CrossRef]

53. Vanhellemont, Q.; Ruddick, K. ACOLITE for Sentinel-2: Aquatic Applications of MSI imagery. In *Living Planet Symposium, Proceedings of the ESA Living Planet Symposium, Prague, Czech Republic, 9–13 May 2016*; ESA-SP: São Paulo, Brazil, 2016; Volume 740.

54. Vanhellemont, Q.; Ruddick, K. Atmospheric correction of metre-scale optical satellite data for inland and coastal water applications. *Remote Sens. Environ.* **2018**, *216*, 586–597. [CrossRef]

55. Ilori, C.O.; Pahlevan, N.; Knudby, A. Analyzing Performances of Different Atmospheric Correction Techniques for Landsat 8: Application for Coastal Remote Sensing. *Remote Sens.* **2019**, *11*, 469. [CrossRef]

56. Bernardo, N.; Watanabe, F.; Rodrigues, T.; Alcântara, E. Atmospheric correction issues for retrieving total suspended matter concentrations in inland waters using OLI/Landsat-8 image. *Adv. Space Res.* **2017**, *59*, 2335–2348. [CrossRef]

57. Korotkina, O.A.; Zavialov, P.O.; Osadchiev, A.A. Submesoscale variability of the current and wind fields in the coastal region of Sochi. *Oceanology* **2011**, *51*, 745–754. [CrossRef]

58. Osadchiev, A.; Sedakov, R. Spreading dynamics of small river plumes off the northeastern coast of the Black Sea observed by Landsat 8 and Sentinel-2. *Remote Sens. Environ.* **2019**, *221*, 522–533. [CrossRef]

59. Lebedev, S.A.; Kostianoy, A.G.; Soloviev, D.M.; Kostianaia, E.A.; Ekba, Y.A. On a relationship between the river runoff and the river plume area in the northeastern Black Sea. *Int. J. Remote Sens.* **2020**, *41*, 5806–5818. [CrossRef]

60. Dzhaoshvili, S. *Rivers of the Black Sea*; Technical Report # 71; European Environment Agency: Copenhagen, Denmark, 2010.

61. Drozhzhina, K.V. Features of the climatic conditions of the Mzymta river basin for recreational activities. *Young Sci.* **2013**, *5*, 196–198.

62. Zavialov, P.O.; Barbanova, E.S.; Pelevin, V.V. Estimating the deposition of river-borne suspended matter from the joint analysis of suspension concentration and salinity. *Oceanology* **2015**, *55*, 832–836. [CrossRef]

63. Lavrova, O.Y.; Soloviev, D.M.; Strochkov, A.Y.; Nazirova, K.R.; Krayushkin, E.V.; Zhuk, E.V. The use of mini-drifters in coastal current measurements conducted concurrently with satellite imaging. *Issl. Zemli Kosmosa* **2019**, *5*, 36–49. [CrossRef]

64. Aibulatov, N.A.; Zavialov, P.O.; Pelevin, V.V. Peculiarities of hydrophysical self-purification of Russian coastal zone of the Black Sea near the river estuaries. *Geoekologiya* **2008**, *4*, 301–310.

65. Nazirova, K.; Lavrova, O.; Mityagina, M.; Krayushkin, E. Influence of vortex structures on the spread of pollution. In Proceedings of the 12th International Conference on the Mediterranean Coastal Environment, Medcoast 2015, Varna, Bulgaria, 6–10 October 2015; pp. 985–996.

66. van de Hulst, H.C. *Light Scattering by Small Particles*; Dover Publications: New York, NY, USA, 1981.

67. Sadar, M. Turbidity Science. Technical Information Series. 1998, Booklet 11. Available online: https://www.hach.com/asset-get download-en.jsa?code=61792 (accessed on 2 November 2020).

68. Merten, G.H.; Capel, P.D.; Minella, J.P.G. Effects of suspended sediment concentration and grain size on three optical turbidity sensors. *J. Soils Sediments* **2014**, *14*, 1235–1241. [CrossRef]

69. Van Der Linde, D.W. Protocol for determination of total suspended matter in oceans and coastal zones. *JRC Tech. Note I* **1998**, *98*, 182.

70. Neukermans, G.; Loisel, H.; Meriaux, X.; Astoreca, R.; McKee, D. In situ variability of mass-specific beam attenuation and backscattering of marine particles with respect to particle size, density, and composition. *Limnol. Oceanogr.* **2012**, *57*, 124–144 [CrossRef]

71. Morel, A.; Prieur, L. Analysis of variations in ocean color. *Limnol. Oceanogr.* **1977**, *22*, 709–722. [CrossRef]

72. Mobley, C.D.; Stramski, D.; Bissett, W.P.; Boss, E. Optical Modeling of Ocean Waters: Is the Case 1–Case 2 Classification Still Useful? *Oceanography* **2004**, *17*, 60–67. [CrossRef]

73. Nazirova, K.; Lavrova, O.; Krayushkin, E. Features of monitoring near the mouth zones by contact and contactless methods. In *SPIE Remote Sensing, Proceedings of the Remote Sensing of the Ocean, Sea Ice, Coastal Waters, and Large Water Regions 2019, Strasbourg, France, 9–12 September 2019*; SPIE—International Society for Optics and Photonics: Bellingham, WA, USA, 2019; Volume 11150, p. 111500H. [CrossRef]
74. Nazirova, K.R.; Lavrova, O.Y.; Krayushkin, E.V.; Soloviev, D.M.; Zhuk, E.V.; Alferyeva, Y.O. Identification features of river plume parameters by in-situ and satellite methods. *Sovr. Probl. DZZ Kosm.* **2019**, *16*, 227–243. [CrossRef]
75. Mityagina, M.I.; Lavrova, O.Y.; Uvarov, I.A. See the Sea: Multi-user information system for investigating processes and phenomena in coastal zones via satellite remotely sensed data, particularly hyperspectral data. In *SPIE Remote Sensing, Proceedings of the Remote Sensing of the Ocean, Sea Ice, Coastal Waters, and Large Water Regions 2014, Amsterdam, The Netherlands, 22–25 September 2014*; SPIE—International Society for Optics and Photonics: Bellingham, WA, USA, 2014; Volume 9240, p. 92401C. [CrossRef]
76. Lavrova, O.Y.; Mityagina, M.I.; Uvarov, I.A.; Loupian, E.A. Current capabilities and experience of using the See the Sea information system for studying and monitoring phenomena and processes on the sea surface. *Sovr. Probl. DZZ Kosm.* **2019**, *16*, 266–287. [CrossRef]
77. Kyryliuk, D.; Kratzer, S. Evaluation of Sentinel-3A OLCI Products Derived Using the Case-2 Regional CoastColour Processor over the Baltic Sea. *Sensors* **2019**, *19*, 3609. [CrossRef]
78. Glukhovets, D.; Kopelevich, O.; Yushmanova, A.; Vazyulya, S.; Sheberstov, S.; Karalli, P.; Sahling, I. Evaluation of the CDOM Absorption Coefficient in the Arctic Seas Based on Sentinel-3 OLCI Data. *Remote Sens.* **2020**, *12*, 3210. [CrossRef]

Article

Improving Stage–Discharge Relation in The Mekong River Estuary by Remotely Sensed Long-Period Ocean Tides

Hongrui Peng [1], Hok Sum Fok [1,2,3,*], Junyi Gong [1] and Lei Wang [4,5]

[1] School of Geodesy and Geomatics, Wuhan University, Wuhan 430079, China; hrpeng@whu.edu.cn (H.P.); jygong@whu.edu.cn (J.G.)

[2] Key Laboratory of Geophysical Geodesy, Ministry of Natural Resources, Wuhan 430079, China

[3] Institute of Marine Science and Technology, Wuhan University, Wuhan 430079, China

[4] Department of Civil, Environmental and Geodetic Engineering, Ohio State University, Columbus, OH 43210, USA; wang.1115@osu.edu

[5] Core Faculty, Translational Data Analytics Institute, Ohio State University, Columbus, OH 43210, USA

* Correspondence: xshhuo@sgg.whu.edu.cn; Tel.: +86-027-6877-8649

Received: 6 October 2020; Accepted: 5 November 2020; Published: 6 November 2020

Abstract: Ocean tidal backwater reshapes the stage–discharge relation in the fluvial-to-marine transition zone at estuaries, rendering the cautious use of these data for hydrological studies. While a qualitative explanation is traditionally provided by examining a scatter plot of water discharge against water level, a quantitative assessment of long-period ocean tidal effect on the stage–discharge relation has been rarely investigated. This study analyzes the relationship among water level, water discharge, and ocean tidal height via their standardized forms in the Mekong Delta. We found that semiannual and annual components of ocean tides contribute significantly to the discrepancy between standardized water level and standardized water discharge time series. This reveals that the long-period ocean tides are the significant factors influencing the stage–discharge relation in the river delta, implying a potential of improving the relation as long as proper long-period ocean tidal components are taken into consideration. By isolating the short-period signals (i.e., less than 15 days) from land surface hydrology and ocean tides, better consistent stage–discharge relations are obtained, in terms of improving the Pearson correlation coefficient (PCC) from ~0.4 to ~0.8 and from ~0.6 to ~0.9 for the stations closest to the estuary and at the Mekong Delta entrance, respectively. By incorporating the long-period ocean tidal height time series generated from a remotely sensed global ocean tide model into the stage–discharge relation, further refined stage–discharge relations are obtained with the PCC higher than 0.9 for all employed stations, suggesting the improvement of daily averaged water level and water discharge while ignoring the short-period intratidal variability. The remotely sensed global ocean tide model, OSU12, which contains annual and semiannual ocean tide components, is capable of generating accurate tidal height time series necessary for the partial recovery of the stage–discharge relation.

Keywords: ocean tidal backwater; stage–discharge relation; ocean tide model; Mekong Delta

---

1. Introduction

Accurate water level (WL) and water discharge (WD) measurements are fundamental to various hydrological applications, including flood forecasting, design and operation of conservancy facilities, as well as water and sediment budget analyses [1,2]. However, due to economy, politics, and topography along a river [3], the spatial distribution of hydrological stations is both sparse and uneven, along with inconsistent and missing datasets [4]. In order to complement the above deficiency of observed datasets, it is a common practice to extend the datasets both in space and time by converting one type of data into another, for instance, estimating WD from WL.

The conversion between WL and WD is referred to as the stage–discharge relation. Under a pure hydrological situation, this relation is represented by a power function, also called rating curve. There are two available methods to obtain the stage–discharge relation. The first method is based on numerical solutions of dynamic models [5–7] that simulates the stage–discharge relation when accurate hydraulic geometry and boundary conditions are available. The second method is based on data-driven models that can be based on the power function fitting, non-linear regression techniques [8–10], or an artificial neural network (ANN) [11–14].

In essence, WD is not only related to WL alone, but also disturbed by water surface slope, channel geometry, bed roughness, flow unsteadiness, lateral flow, and the backwater effect caused by an ocean tidal wave propagating up to estuaries [15,16]. Therefore, the stage–discharge relation becomes more complicated, manifesting as multiple loops [17]. In the river delta, the influence of the ocean tidal wave is a significant factor that distorts the well-established stage–discharge relation [8]. Consequently, the WL and WD data near the estuary mouth at river deltas are used cautiously for research studies, as those data are contaminated by the aforementioned factors. For instance, Sassi et al. [18] quantitatively analyses the contribution of different ocean tidal components (i.e., quarter-diurnal, semidiurnal, diurnal, and fortnightly) to surface water variation. The fluvial-to-marine transition zone of Mekong Delta have been further subdivided into four sections (i.e., fluvial-dominated tide-affected, fluvial-dominated tide-influenced, tide-dominated fluvial-influenced, and tide-dominated fluvial-affected zones), according to salinity, channel morphology, fades/grain size, and the extent of ocean tidal influence by Gugliotta et al. [19]. However, the stage–discharge relation at the river delta corrected by ocean tidal components remains unexplored.

The Mekong Delta (MD) (Figure 1), being home to 19 million people, is an important agricultural and fishing district in Southeast Asia [17,20]. Further anthropogenic stressors are massive river training and construction of a multitude of large hydropower dams and severe sand extraction for concrete production [21–23]. This is characterized by a relatively flat surface with low altitudes and gradients [24,25]. Being a transition zone, WD and WL variability are dynamically affected by both fluvial and marine processes seasonally [26,27]. As a result, reverse flow caused by ocean tidal wave and storm surge can easily propagate along river channels [8]. As a result, salinity intrusion and catastrophic flooding along with rising sea level [28–30] severely threaten the grain production in the MD [31,32]. This also affects hydrological gauge stations within a distance of 200 kilometers away from the estuary mouth. In addition, the Tonle Sap Lake in Cambodia also provides a regulation effect [33–35], before the river runoff delivers to the MD and discharges eventually to the South China Sea through the Bassac River and the Mekong river within the MD [36]. As a consequence, the stage–discharge relation in this region exhibits multiple looping curves along with noisy patterns [33,37].

Despite qualitative explanations, the ocean tidal backwater effect has not been quantified and corrected for. After all, the complex interaction between oceanic and fluvial processes is a cross-disciplinary science among land surface hydrology, estuary, and ocean science. As long as an appropriate method can be introduced to partially recover the stage–discharge relation with good accuracy, the corrected data would be of great usage. For such a purpose, the analysis of the disturbance of the stage–discharge relation by different components of ocean tides, based on a tidal data analysis or a remotely-sensed global ocean tide model, is a prerequisite.

This study aims to demonstrate the potential of incorporating the ocean tidal components into the stage–discharge relation for a partial relation recovery in the MD. The relation among WD, WL, and ocean tidal height data time series are analyzed via their standardized forms. The ocean tidal components generated from remotely sensed OSU12 global ocean tide model are substituted into the resulting model relation generated from the analysis. The fitted model relation is subsequently applied for estimating WD from ocean tidal height and WL. A quantitative evaluation of the estimated WD against the observed hydrological data is also presented.

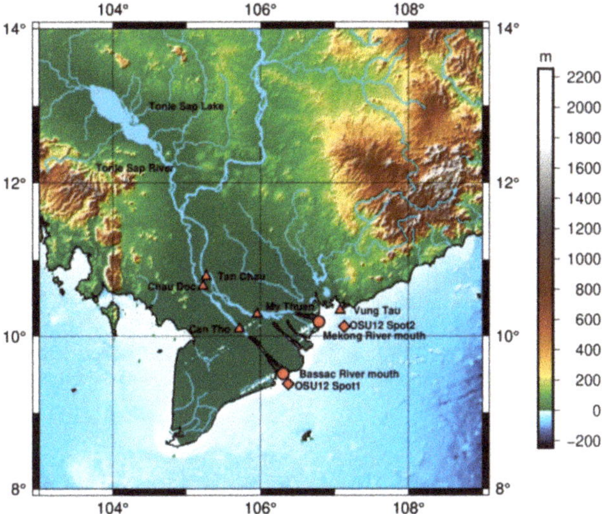

Figure 1. Map of Mekong Delta (MD), with two pairs of hydrological gauge stations (i.e., Can Tho and Chau Doc, and My Thuan and Tan Chau) situated near the estuaries. (The topography dataset, called earth_relief_30s, is a derived product of SRTM15+ [38], which is obtainable from http://mirrors.ustc.edu.cn/gmt/data/).

2. Datasets and Assessment Metrics

In this study, in-situ data from hydrological stations, tidal gauge data, and OSU12 global ocean tide model were analyzed. Table 1 summarizes the essential information about these datasets.

Table 1. The datasets used in this study.

Products	Location	Time Span	Temporal Resolution
In Situ Stations' Water Level Data	Can Tho	2003–2006 2009–2014	Daily average
	My Thuan	2003–2006 2009–2014	
	Chau Doc	2003–2006	
	Tan Chau	2003–2006	
In Situ Stations' Discharge Data	Can Tho	2003–2006 2009–2014	Daily (before 2006) Monthly (after 2009)
	My Thuan	2003–2006 2009–2014	
	Chau Doc	2003–2006	
	Tan Chau	2003–2006	
Tidal Gauge Data	Vung Tau	2003–2014	Hourly
OSU12 Global Ocean Tide Model Data	9.375N, 106.375E 10.125N, 107.125E		Tidal constituents (Sa, Ssa, Mm)

2.1. In-Situ Hydrological Data

Station data time series within the MD were obtained from the Mekong River Commission (MRC) (http://www.mrcmekong.org). Acoustic Doppler Current Profiler (ADCP) was applied to gauge flow velocity for deriving precise discharge, according to MRC [39]. To compare between the two main subdivided branches within the MD, Tan Chau and My Thuan stations along the Mekong River, and Chau Doc and Can Tho stations along the Bassac River were used. Situated at the entrance of the

MD [27], the Tan Chau and Chau Doc stations are, respectively, ~220 and ~240 km away from the estuary mouth. Both stations are in the middle between the Tonle Sap Lake and the estuary mouth, where the regulation effect of the lake and the ocean tidal backwater effect are minimized. Being the closest hydrological stations to the estuary mouth, My Thuan and Can Tho stations are subject to the backwater effect caused by landward ocean tidal propagation, which is clearly shown in the data time series [27]. Hence, the comparison between upper and lower station pairs allows us to further quantify the extent of the ocean tidal backwater effect.

Note that WD data of Tan Chau station were missing in 2001, 2002, and 2007. To be consistent with the time span of other WD data, the station time series spanning from January 2003 to December 2006 were selected for investigation, while those from January 2009 to December 2014 were employed for validation. Given the different temporal resolutions among WL, WD, and in-situ ocean tidal data time series and in order to isolate signals unrelated to hydrology, a Butterworth filter was applied to these time series for suppressing periodic fluctuations shorter than 15 days (e.g., diurnal, semidiurnal, etc.). The mean, maximum, and minimum values of those time series are summarized in Table 2.

Table 2. Maximum, minimum, mean values, and standard deviations of original and processed time series.

Variable	Station	Maximum	Minimum	Mean	Standard Deviation
Original Water Discharge (1×10^4 m^3/s)	My Thuan	1.6500	0.0029	0.7263	0.4036
	Can Tho	1.8400	0.0025	0.7206	0.4416
	Tan Chau	2.2597	0.1190	0.9359	0.6490
	Chau Doc	0.7120	0.0045	0.2625	0.2059
Processed Water Discharge (1×10^4 m^3/s)	My Thuan	1.5345	0.2423	0.7262	0.3109
	Can Tho	1.4666	0.1704	0.7209	0.3236
	Tan Chau	2.1400	0.1600	0.9360	0.6470
	Chau Doc	0.7121	0.0266	0.2626	0.2043
Original Water Level (m)	My Thuan	1.4225	−0.3355	0.4619	0.3522
	Can Tho	1.4591	−0.2707	0.4168	0.3231
	Tan Chau	4.3831	0.0222	1.6820	1.2544
	Chau Doc	4.0036	−0.1486	1.5017	1.1443
Processed Water Level (m)	My Thuan	1.2165	−0.1304	0.4620	0.3267
	Can Tho	1.0358	−0.0685	0.4171	0.2976
	Tan Chau	4.3361	0.2326	1.6825	1.2498
	Chau Doc	3.9558	0.1863	1.5019	1.1396
Original Tide height (m)	Vung Tau	4.3300	−0.4400	2.6433	0.8566
Processed Tide height (m)	Vung Tau	2.9984	2.3413	2.6436	0.1648

Filtered and original time series of the four stations are displayed, showing common characteristics of the WL and WD time series along with their differences (Figure 2a–d). Can Tho and My Thuan station time series show a larger ocean tide backwater effect than those of Chau Doc and Tan Chau stations. By comparing WL with WD time series, WD lags behind WL by approximately a month. This fact is more pronounced for stations closer to the estuary mouth (i.e., Can Tho and My Thuan) than their upper counterparts (i.e., Chau Doc and Tan Chau). Obviously, the annual signal is apparent for all station time series, in which the temporal patterns are highly related to not only seasonal variation of watershed runoff, but also the long-period (e.g., semiannual and annual) ocean tidal components, as shown in Figure 2e. As a consequence, external information obtained from the tide gauge or ocean tide model data near estuaries can be potentially used for removing the effect of long-period ocean tidal components, which is the objective of this study.

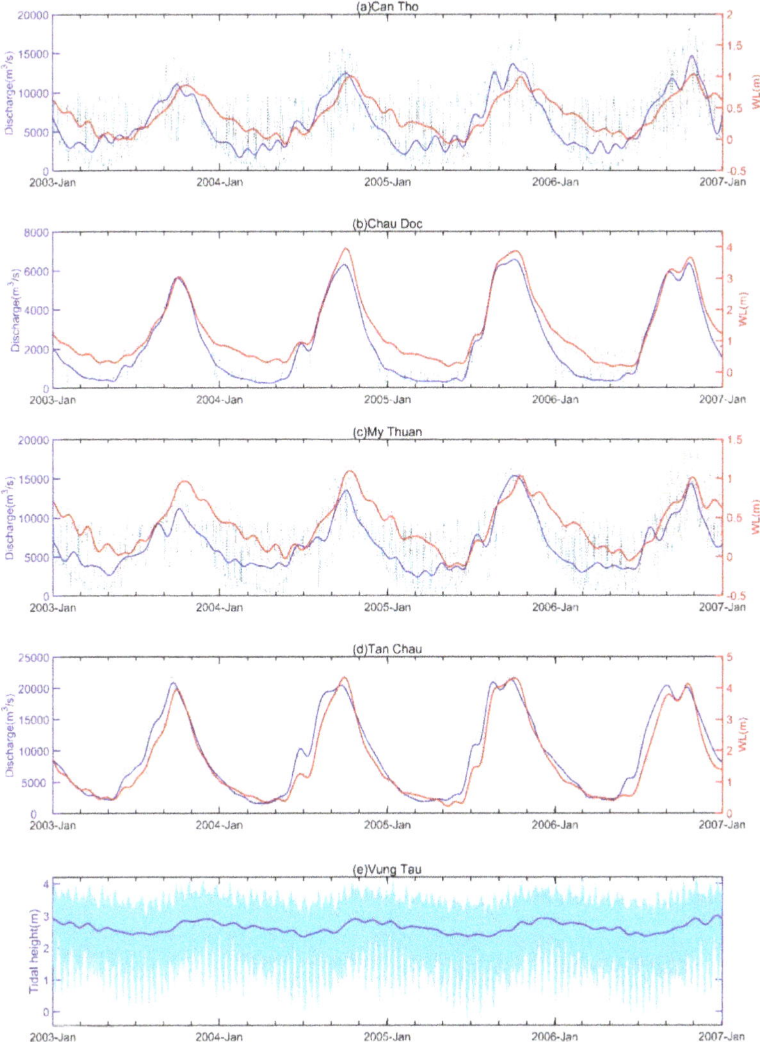

Figure 2. Low-pass filtered (blue) and original (blue dash) time series of water discharge and water level (red) over (**a**) Can Tho, (**b**) My Thuan, (**c**) Chau Doc, and (**d**) Tan Chau stations, respectively, and (**e**) time series of ocean tidal height (sea level) at Vung Tau station spanning from January 2003 to December 2006.

2.2. Sea Level Data from Tide Gauge Station

A tide gauge measures sea level time series at selected locations along the coasts [40]. Vung Tau is the closest tide gauge station to Mekong estuary chosen for relating the long-period ocean tidal variation to WL within the MD. Spanning from 2003 to 2014, the sea level time series at Vung Tau station were recorded on an hourly interval. This dataset is provided by the Hydrological and Environmental station network center in Vietnam and can be obtained from http://www.ioc-sealevelmonitoring.org/station.php?code=vung.

Figure 2e shows filtered and original hourly time series of the tidal gauge data. Fast Fourier transform (FFT) was applied to identify different periodic components of the time series. The highest power spectra are located at both diurnal and semidiurnal ocean tidal components (Figure 3a), which are unrelated to hydrological signals. In order to be consistent with WD and WL time series' daily sampling rate, the hourly tidal height time series are averaged daily after filtering high-frequency components via the Butterworth filter. This process, to a large extent, suppresses or removes the short-period ocean tidal components via the low-pass filtering process, and hence, reducing the effect on long-term ocean tidal components [41–43] (Figure 3b). Compared with the unfiltered time series, only semiannual and annual tide components are apparent in the processed time series.

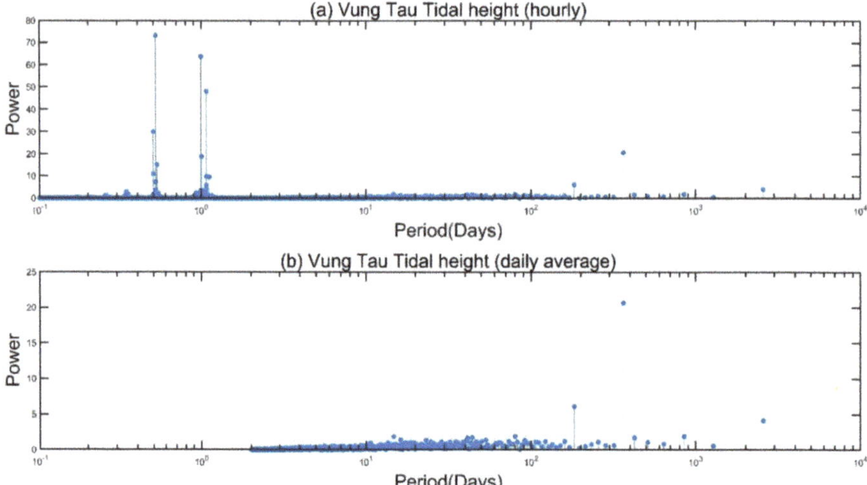

Figure 3. Spectra of the (**a**) hourly and (**b**) daily averaged ocean tidal height time series in Vung Tau tide gauge station.

2.3. Global Ocean Tide Model Data

A global ocean tide model contains gridded in-phase and quadrature amplitudes (or equivalently amplitude and phase) for major tidal constituents, allowing us to generate ocean tidal height in the absence of tide gauge stations along the coasts [44,45]. Although many remotely sensed ocean tide models (e.g., FES2014, GOT4.8, NAO99.b, TPXO8, EOT11a, DTU10, HAMTIDE, OSU12, etc.) are available for the purpose of our study, the OSU12 model, with a 0.25° × 0.25° spatial resolution [46,47], was employed to generate long-period tidal height time series at grid points near Mekong and Bassac river estuaries (Table 3), because it contained long-period tides and was derived purely from remotely sensed satellite altimetry data. Notwithstanding smaller amplitude when compared with semidiurnal and diurnal tides, long-period ocean tidal components are influential to daily and monthly average WL time series. As shown in Figures 2 and 3b, long-period ocean tidal components are likely related to the discrepancies between the pattern of WL and WD time series. It is appropriate to calculate the ocean tidal height time series, $TH(t)$, at time t from the in-phase, H_1, and quadrature amplitudes, H_2, of S_a, S_{sa}, and M_m tides, which can be formulated as:

$$TH(t) = \sum_{i=1}^{3} (H_1)_i \cos\left(\frac{2\pi t}{T_i}\right) + (H_2)_i \sin(\frac{2\pi t}{T_i}), \tag{1}$$

where T_i is the period of each long-period ocean tidal component i. Note that both the in-phase and quadrature amplitudes are with respect to Greenwich Meridian with the starting time, 0:00 AM, 1 January 2002 (UTC +0).

Table 3. Long-period ocean tidal components at two gridded locations close to Mekong and Bassac river estuaries solved at the initial time epoch of 0:00 AM, 1 January 2002 (UTC +0).

Tide Components		Point1 (9.35°N,106.375°E) (in cm)	Point2 (10.125°N, 107.125°E) (in cm)
S_a (365.25 days)	H_1	24.41570	19.06151
	H_2	−1.56798	−3.91959
S_{sa} (182.62 days)	H_1	1.36968	−6.76170
	H_2	3.52620	2.07534
M_m (27.55 days)	H_1	1.30950	0.32715
	H_2	−1.63984	1.72923

2.4. Assessment Metrics

To evaluate the estimated WD against in-situ WD time series in Section 4, three assessment metrics, R-Square, the Pearson correlation coefficient (PCC), and the Nash–Sutcliffe efficiency (NSE) coefficient, are employed.

R-Square, ranging between 0 and 1, describes how much the variation of in-situ WD, WD_g, is explained by the estimated WD, WD_e, generated from the model. The closer the value to 1, the better the model fitting to the WD_g. R-Square is equal to the quotient of sum of squares regression (SSR) divided by sum of squares total (SST), and can be defined as:

$$R - \text{Square} = \frac{\text{SSR}}{\text{SST}} = \frac{\sum_{i=1}^{n}\left(WD_e^i - \overline{WD_g}\right)^2}{\sum_{i=1}^{n}\left(WD_g^i - \overline{WD_g}\right)^2} \tag{2}$$

PCC, ranging between −1 and 1, describe how strong the linear relationship between WD_e and WD_g, which is defined as:

$$PCC = \frac{\sum_{i=1}^{N}\left(WD_e^i - \overline{WD_e}\right)\left(WD_g^i - \overline{WD_g}\right)}{\sqrt{\sum_{i=1}^{N}\left(WD_e^i - \overline{WD_e}\right)^2}\sqrt{\sum_{i=1}^{N}\left(WD_g^i - \overline{WD_g}\right)^2}} \tag{3}$$

where $\overline{WD_e}$ and $\overline{WD_g}$ are the mean of WD_e and WD_g, respectively. Notably, for the power function relating WL to WD, logarithmic transform is applied to obtain the log-linear relation between the two variables in order to assess their correlation. To highlight the difference, PCC was used to represent the linear relationship between WD_e and WD_g, while "correlation coefficient" appeared in each figure of this study referred to the log-linear relation between WD and WL, as shown in Equation (6) below.

NSE coefficient, ranging from −∞ to 1, describes the gain in the performance of WD_e against WD_g. The closer the NSE coefficient to 1, the better the performance of the estimation [48]. It is defined as:

$$NSE = 1 - \frac{\sum_{i=1}^{N}\left(WD_e^i - WD_g^i\right)^2}{\sum_{i=1}^{N}\left(WD_g^i - \overline{WD_g}\right)^2} \tag{4}$$

3. Data Analysis and Methodology

This section explores the relations among ocean tidal height, WL, and WD time series over our study region, so as to illustrate the interaction between fluvial and oceanic factors along with their combined effects on WL and WD data. For an ideal hydrological station location where WL and

WD are purely influenced by the fluvial process, WL and WD are related by a power function [49] expressed as:

$$WD = a*[WL - b]^c \tag{5}$$

where a, b, c are the scaling coefficient, the offset of WL and the exponent of power function, respectively.

However, in reality, the stage–discharge phase diagram between WL and WD appears as random data points with trends (i.e., Can Tho and My Thuan stations) and elliptical curves (i.e., Chau Doc and Tan Chau stations) in the MD (Figure 4).

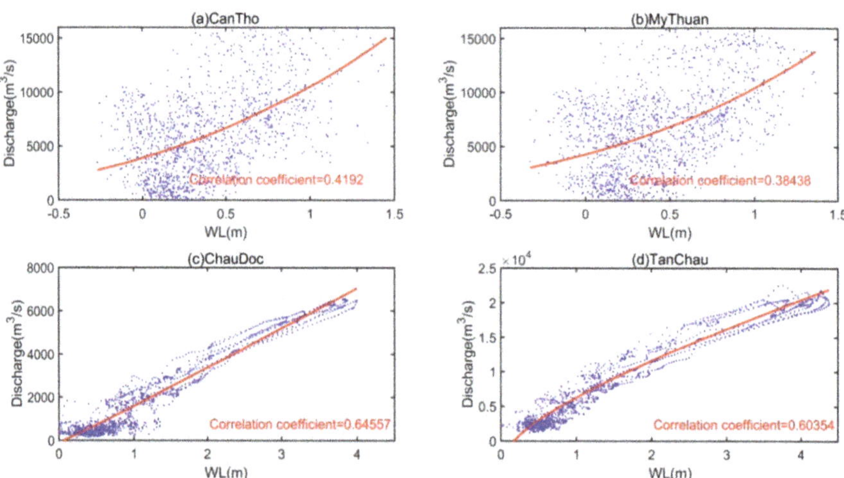

Figure 4. (**a–d**) Relationship between water level (WL) and water discharge (WD) (original daily sampled time series) for the four selected hydrological stations in Mekong Delta.

The logarithmic transform of Equation (5) allows the conversion into log-linear relation, expressed as:

$$ln(WD) = c * ln(WL - b) + ln(a). \tag{6}$$

such that Equation (6) measures a linear relationship between $ln(WD)$ and $ln(WL - b)$. All "correlation coefficients" displayed in all stage–discharge phase diagrams were calculated based on $ln(WD)$ and $ln(WL - b)$, as mentioned in Section 2.4.

Compared to those of the other two stations, the rating curves between WL and WD of Can Tho and My Thuan stations yield lower correlation coefficients because they are more significantly affected by the ocean tidal backwater.

3.1. Data Analysis of Backwater Influence on Water Discharge (WD) and Water Level (WL)

Although the phase diagram between WL and WD in the tide-dominated area appears elliptical, the patterns of the deviation from the rating curves are presumed to be analyzable by different ocean tidal components. Through FFT, the most pronounced periods are 365 days, 182.5 days, and 14.7475 days in both WD and WL time series.

The relative power (to the signal with the largest power) and initial phase of each signal are displayed in Table 4. For an ideal stage–discharge relation (i.e., power function relation), WL and WD are positively correlated. The signals of WD and WL with the same period should have the same initial phase and similar relative power. However, we found that the initial phase of WD and WL time series of the four stations are different from each other.

Table 4. Relative power and initial phases of the three periodic signals in WD and WL time series at the four selected stations with initial phase domain defined between 0° and 360°.

Station		Period: 365 Days		Period: 182.5 Days		Period: 14.7475 Days	
		Relative Power	Initial Phase	Relative Power	Initial Phase	Relative Power	Initial Phase
Can Tho	WD	1	95.2311°	0.3060	173.5607°	0.4727	25.1628°
	WL	1	58.5725°	0.2535	168.3582°	0.2997	244.1391°
My Thuan	WD	1	87.1219°	0.3093	170.8565°	0.5671	25.8326°
	WL	1	55.0654°	0.2664	175.4839°	0.3152	244.2108°
Chau Doc	WD	1	93.8689°	0.3392	196.6511°	0.0324	111.2065°
	WL	1	84.7393°	0.3963	193.8528°	0.0414	274.7407°
Tan Chau	WD	1	97.4019°	0.2385	222.2999°	0.0046	99.6860°
	WL	1	90.3988°	0.3705	202.5387°	0.0336	260.8400°

Firstly, annual signals (i.e., 365-day period) of Can Tho and My Thuan present different initial phases between WD and WL, in particular WL, with its initial phases ~30° lower than that of upper counterparts. This indicates that annual tides can cause around a one-month time lag between the lower and upper stations. A similar situation applies to that of the semiannual signal, but to a lesser extent. Secondly, the initial phase of the half-monthly signal (i.e., 14.7475-day period) of WD and that of WL present the phase difference between 160° and 220°. This shows that the WD is inversely proportional to WL with an additional time lag. In other words, the WD increase (decrease) when the WL decrease (increase), implying that the half-monthly signal of WL and WD interacts with each other seasonally and alternately. This fact further indicates the half-monthly signal is of two origins: land and ocean, which is supported by physical explanations from Guo et al. (2020) [50] and Jay (1991) [51]. Half-monthly signals of the Can Tho and My Thuan stations yields a much larger relative power than their upper counterparts, indicating the damping effect on the amplitude and changing phase when propagating inland via the estuary mouth. Since these half-monthly signals have different changing ratios for inland propagation direction with annual tide components, a band-pass filter (e.g., Butterworth filter) was applied to remove this signal from tidal-influenced time series for consistency.

To further analyze the interaction between oceanic and fluvial effects, the variation of WD, WL, and TH time series are compared via their standardized forms, x_s, expressed as:

$$x_s = \frac{x - \bar{x}}{\sqrt{\frac{\sum (x-\bar{x})^2}{N}}} \tag{7}$$

where $\bar{x}$ is the average value of xx time series, and NN is the number of data in the time series. The standardized WD, WL and TH (i.e., WD_s, WL_s, and TH_s respectively) are compared for the four stations, respectively, in Figure 5.

As shown in Figure 5b,d, it is clear that the standardized WL time series are highly correlated with standardized WD time series, they reach the maximum values in early September and minimum in March and April simultaneously. Influences from ocean tide are minor, and the ocean tidal height series reaches its maximum and minimum values in different months. However, in Figure 5a,c, there exists large deviation between WD and WL time series. In the lower stations, the WL reaches its minimum and maximum value about a month later than WD, consistent with the initial phase difference of around 30° stated above (Table 4). For most cases, WL (red line) is set between WD (blue line) and TH (yellow line), emphasizing the influence of the ocean on WL. Previous studies attribute this phase difference to floods up and down or a time lag caused by tidal propagation [52]. Since this phenomenon is more apparent in stations closer to the estuaries, we speculate it is mainly caused by the mixing of fluvial-dominated and marine-dominated fluctuations at the annual and semiannual scale.

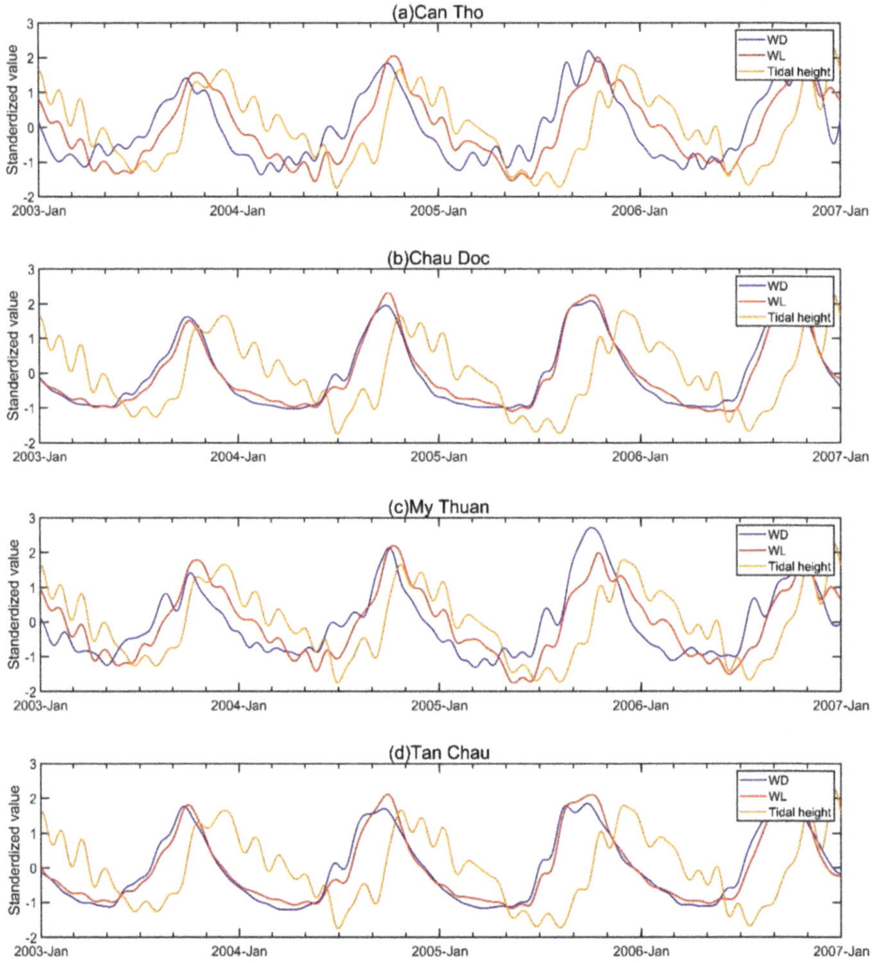

Figure 5. Comparison of standardized WD, WL, and tidal height time series in (**a**) Can Tho, (**b**) Chau Doc, (**c**) My Thuan, and (**d**) Tan Chau station, respectively.

Theoretically, when two signals with the same period (T) are combined, the new signal will have the same period (T) but a different initial phase (ϕ_3), is expressed as:

$$A_1 \cos\left(\frac{2\pi t}{T} + \phi_1\right) + A_2 \cos\left(\frac{2\pi t}{T} + \phi_2\right) = A_3 \cos\left(\frac{2\pi t}{T} + \phi_3\right), \tag{8}$$

where A_1, A_2, and A_3 are three different amplitudes, ϕ_1, ϕ_2, and ϕ_3 are three initial phases, and t refers to time epoch. The proof of Equation (8) is listed in Appendix A.

Since annual and semiannual signals are major components in the WL, WD, TH time series, the fluvial-dominated annual (semiannual) signal and marine-dominated annual (semiannual) signal form a mixed annual (semiannual) WL time series in the MD. For both WL and TH fluctuating in the vertical direction, the WL will be potentially corrected by annual and semiannual ocean tidal components if the TH time series are involved in the power function fitting process. This will be further explored in the next subsection. However, Equation (8) does not work for ocean tidal components shorter than half-monthly one, because of possible non-linear interaction among fluvial factors, bottom topography of an estuarine channel and ocean tidal backwater. This leads to the non-linear change of amplitude and phase during the inland propagation process.

3.2. Incorporating Long-Period Ocean Tidal Components into Rating Curve

Before incorporating long-period ocean tidal components into the rating curve, short-period fluctuations in both WD and WL time series, including short-period diurnal and semidiurnal ocean tides, have to be removed. As mentioned in Section 2.1, this can be achieved by a Butterworth filter that suppresses all high-frequency signals with a period shorter than 15 days. Consistent with the filtered time series (Figure 2), the rating curve with filtered time series of WD and WL at the four stations are plotted in terms of phase diagrams (Figure 6).

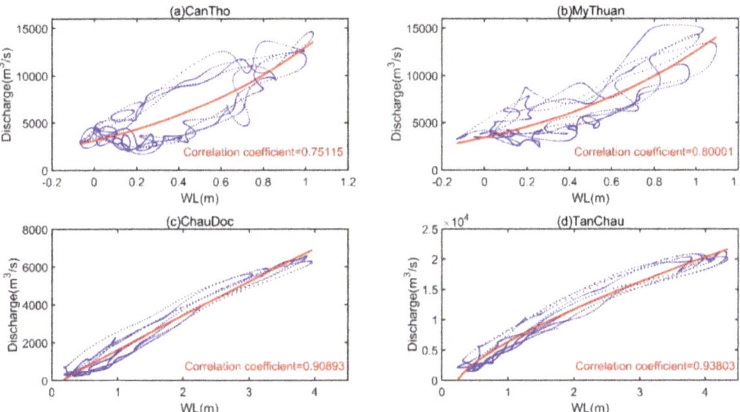

Figure 6. (**a–d**) Relationship between WL and WD (low pass filtered time series) for the selected four hydrological stations in the Mekong Delta.

Compared with those in Figure 4, it is clear that the correlation coefficients have been improved significantly (Figure 6). However, the elliptical loops are still apparent, indicating a time lag between WL and WD time series, as mentioned in Section 3.1.

Given that the relationship between TH_s, WD_s, WL_s has been analyzed in Section 3.1, it is likely that the elliptical looping phenomenon is largely due to semiannual and annual ocean tidal components. For both WL and TH fluctuating in the vertical direction, the TH_s time series were applied to separate the tide-induced fluctuation from the WL time series through a fitting process. The WL free from tide influence, WL_{free}, is defined as:

$$WL_{free} = WL - \alpha \times TH_s. \tag{9}$$

where, α is a coefficient that rescales the TH_s. Consequently, the relationship among WD, WL, TH_s can be represented by:

$$WD = a \times [WL - \alpha \times TH_s - b]^c. \tag{10}$$

where a, b, c, and α are to be determined from the observed WD, WL, and TH time series. Through a non-linear fitting [53,54], a, b, c, and α can be determined, and WL_{free} is obtainable.

4. Results and Discussion

The rating curves of WL_{free} and WD are shown (Figure 7), yielding much higher correlation coefficients when compared to the rating curves of original WD and WL time series. Although rating curves of the Can Tho and My Thuan stations still display lower correlation coefficients than their upper counterparts, significant improvement has been observed. Additionally, the elliptical looping phenomenon related to 'time-lag' between WL and WD is also diminished for all four selected stations. As a countercheck, the time series of WD and WL$_{free}$ for the two stations close to the estuary are shown in Figure 8, revealing no apparent time lag.

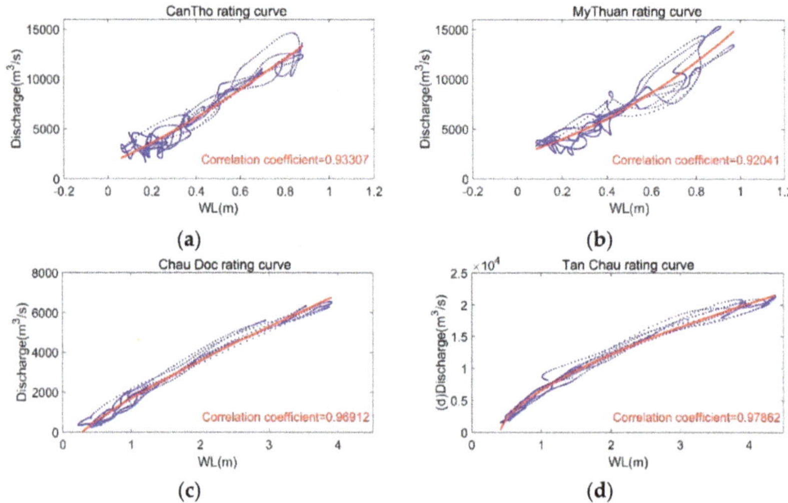

Figure 7. (**a–d**) Relationship between WL$_{free}$ and WD (low pass filtered time series) for the selected four hydrological stations in the Mekong Delta.

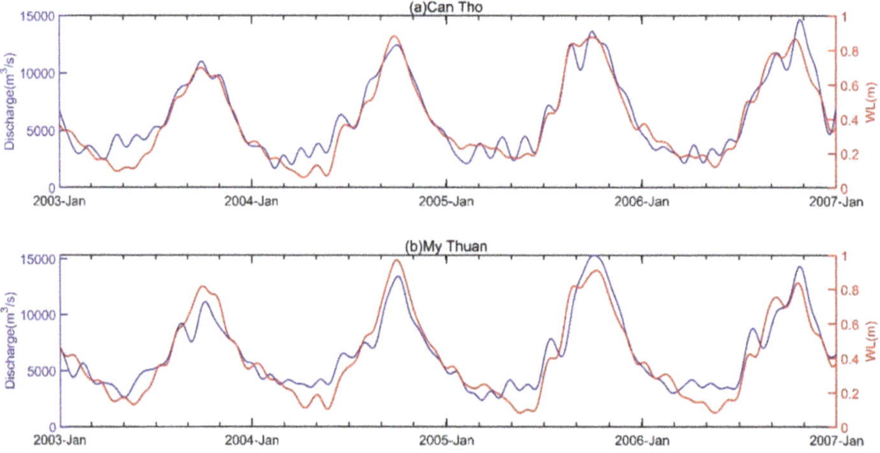

Figure 8. WD and tide-free WL time series from 2003 to 2006 in (**a**) Can Tho and (**b**) My Thuan stations.

In the absence of tide gauge data, the TH time series generated from a global ocean tide model would be a viable alternative, because it can provide ocean tidal height components for the global ocean. The method for obtaining model-derived TH time series has been stated in Section 2.3. The model-derived TH series and in-situ gauged tidal height time series are displayed, manifesting high similarity with each other (Figure 9). Employing the above methodology, the rating curves of the four stations have been recovered using model-derived TH time series (Figure 10).

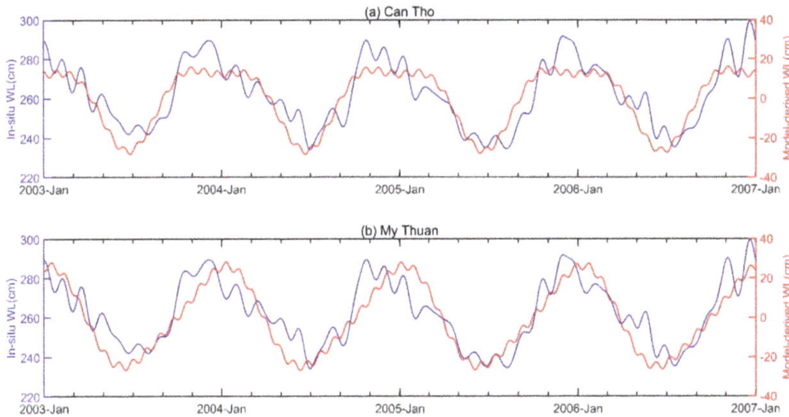

Figure 9. The comparison between OSU12 model-derived WL and in-situ WL at (**a**) Can Tho and (**b**) My Thuan during 2003–2006.

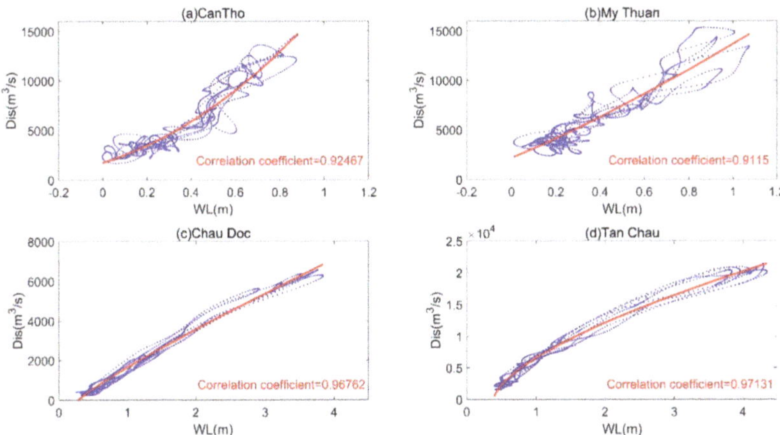

Figure 10. Recovered rating curves at (**a**) Can Tho, (**b**) My Thuan, (**c**) Chau Doc, and (**d**) Tan Chau stations using model-derived ocean tidal height as input.

Although the correlation coefficients of the rating curve fitting generated by model-derived TH time series (Figure 10) are slightly lower than by in-situ tide gauge TH time series (Figure 7), the improvement is considerable when compared to the original unmodified rating curves. This is because most global ocean tide models are derived from satellite altimetry, with the model accuracies lower than that of gauge-derived TH, in particular coastal regions [39]. Given the above results, it is appropriate to use an ocean tidal model to partially recover rating curves over tide-dominated regions.

To evaluate the accuracy of the recovered rating curves, the estimated WD time series via Equation (10) generated from both model-derived and in-situ TH time series are compared with the in-situ WD during 2003–2006. Table 5 lists all determined coefficients of Equation (10) along with the assessment metrics (i.e., R-Square, PCC, NSE) that assess the estimated WD against in-situ WD. For both WD estimated based on in-situ and model-derived TH data, all the assessment metrics yield high-correlation values at all the four stations, suggesting our method can partially recover the tide-free WL for estimating WD. Overall, the recovered stage–discharge relation is capable of predicting a relatively reliable WD. These results also validate the data analysis in Section 3.

Table 5. Assessment of the estimated WD and stage–discharge relation coefficients.

Tidal Height Data	Station	a	b	c	α	R-Square	PCC	NSE
In-situ measured	Can Tho	11683	−0.2230	1.3661	0.1619	0.9291	0.9626	0.9266
	My Thuan	3436.5	−0.8665	2.4063	0.1480	0.8974	0.9468	0.8964
	Chau Doc	2239.8	0.2756	0.8592	0.1577	0.9790	0.9922	0.9843
	Tan Chau	9134.3	0.4052	0.6229	0.1819	0.9809	0.9946	0.9891
OSU12	Can Tho	5948.3	−0.5949	2.3417	0.1722	0.8871	0.9383	0.8804
	My Thuan	9761.2	−0.3079	1.2619	0.1520	0.8631	0.9285	0.8621
	Chau Doc	2201.2	0.2762	0.8911	0.1582	0.9828	0.9934	0.9868
	Tan Chau	8950.3	0.3809	0.6335	0.1314	0.9750	0.9910	0.9820

To highlight the importance of tidal separation by the term $-\alpha \times TH_s$ in Equation (10), the PCC values with different combinations of coefficient b, c, and α were calculated with their best fitted a fixed. Taking Can Tho station as an example (Figure 11), b and c impact the PCC values (i.e., > 0.9) significantly, only if α is ~0.16. The same holds for My Thuan station. Therefore, adding the term $-\alpha \times TH_s$ to the conventional power function (i.e., Equation (5)) is necessary for the improvement of the stage–discharge relation in the MD, which is in the fluvial-to-marine transition zone. In summary, we found that an appropriate α is a prerequisite for the PCC larger than 0.9.

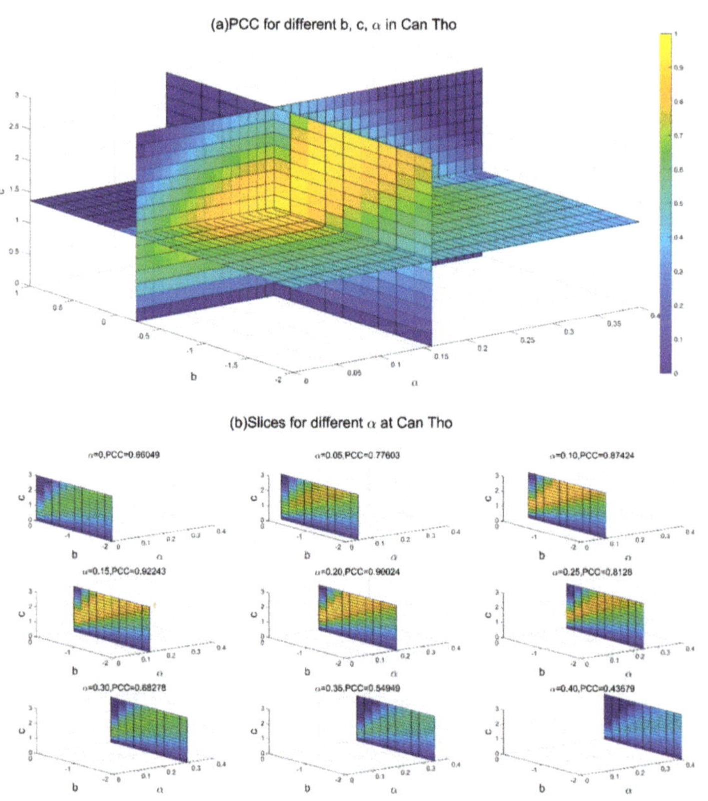

Figure 11. (**a**) Different PCC (presented in color bar) for different b, c and α using time series from Can Tho station, and (**b**) slices of (**a**) for nine chosen α, with maximum PCC for each α shown from the above subplots.

To assess the applicability of the determined coefficients of Equation (10) for Can Tho and My Thuan station time series during 2003–2006, these coefficients were directly employed for the analysis of the WL and TH data time series during 2009–2014. The predicted WD were then compared with the monthly in-situ WD (Figure 12), since only monthly WD are available for Can Tho and My Thuan stations. Hence, WL and WL_f were monthly averaged before the comparison. For both Can Tho and My Thuan stations, tide-free WL, WL_f, leads to higher correlation coefficients and diminishes the looping curve problem to a large degree. This indicates that coefficient α appears to be stable during our study period.

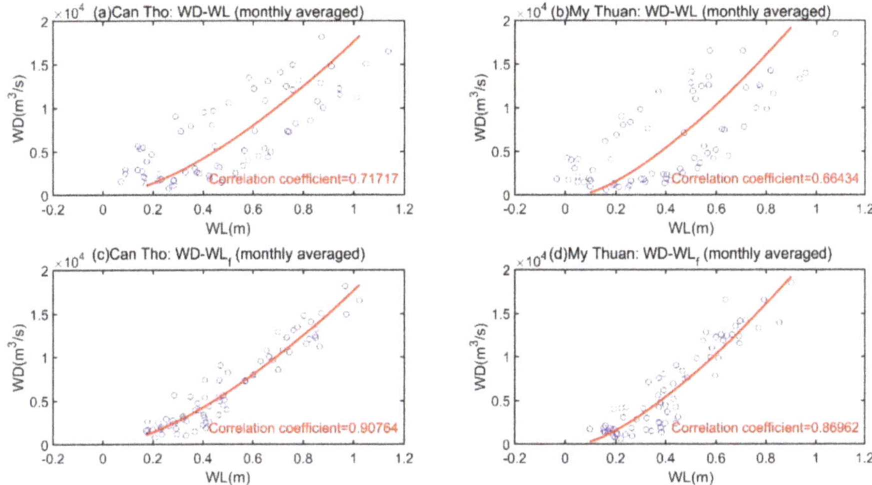

Figure 12. (**a,b**) Stage discharge relation from original WL, and (**c,d**) tide-free WL for Can Tho and My Thuan stations.

Despite a substantial improvement made in this study, small deviations from the ideal power function still exist, particularly for the two stations closest to the estuary mouth. After all, the interaction between fluvial and marine processes are complicated near estuary mouths [55]. Remaining effects cannot be neglected. For instance, WD should pose a non-negligible effect on the tidal propagation along the river channel during the wet season. Overland flows inward or outward from the Tonle Sap Lake would likely be another important factor affecting the stage–discharge relations, because this lake operates as a natural reservoir that regulates Mekong river discharge from the river delta to the coastal ocean [34,56,57]. Erosion and deposition alter hydraulic geometry and increase channel bottom friction and, hence, contribute to the potential instability of the stage–discharge relation. Furthermore, numerous clusters of dams were built along the main stream of Mekong river, which may also alter the stage–discharge relation [21,22]. Sea level rise, which closely connected to salt intrusion and coastal erosion problems may alter the estuarine topography condition, resulting in a secular shift of ocean tidal components [23,26]. Agricultural practices and deforestation also provide additional impact on the evapotranspiration balance of the catchment area. Furthermore, since short-term signals were filtered out or failed to be captured by daily sampling, the short-term variations in WD and WL have not been quantitatively investigated. These considerations represent the current limitations of this study.

5. Conclusions

Instead of seeking a qualitative explanation of the stage–discharge relation influenced by the ocean tidal backwater effect, this study quantitatively analyzes the relations among water discharge

(WD), water level (WL), and ocean tidal components via their standardized forms. We found that annual and semiannual ocean tidal components are significant contributors to the deviation between WL and WD time series. In particular, the annual and semiannual periods of ocean tidal backwater result in the elliptic loop associated with the presence of time lag between WL and WD.

Based on these findings, we adapt the stage–discharge relation to accommodate the effects of annual and semiannual ocean tidal components. It was found that the WD estimated from the de-tided WL yields PCC and NSE values of ~0.9. Although the de-tided WL time series generated based on the TH time series from the OSU12 global ocean tide model are slightly less accurate than that of tide gauge data, the ocean tide model is a viable alternative to partially recover the stage–discharge relation for estuaries in the absence of tide gauge stations.

Further improvement lies in identifying remaining effects contributing to the potential instability of the stage–discharge relation, which include the non-negligible effect of seasonal WD on ocean tidal propagation, the Tonle Sap lake regulation effect on the Mekong river discharge, erosion and deposition effects on the hydraulic geometry, and channel bottom friction. The impact of human activities and artificial structure in the Mekong River area, as well as its interaction with climate change, should also be highlighted. Those factors may introduce a long-term change trend into the WL–WD relationship.

The recent remotely-sensed water balance variables with improved temporal resolutions, such as 8-day MODIS evapotranspiration [58], daily TRMM precipitation [59], and daily GRACE terrestrial water storage data products [60], should enable us to compute tide-free WD, which is independent of in-situ measurements based on the water balance equation [36]. This can serve as a countercheck against the in-situ WD for assessing the first two remaining effects.

Author Contributions: Conceptualization, H.S.F.; methodology, H.P. and J.G.; software, H.P.; validation, H.S.F. and H.P.; formal analysis, H.S.F. and H.P.; investigation, H.P.; resources, H.S.F. and L.W.; data curation, H.S.F. and J.G.; writing—original draft preparation, H.P. and H.S.F.; writing—review and editing, H.S.F., H.P., L.W. and J.G.; visualization, H.P.; supervision, H.S.F.; project administration, H.S.F.; funding acquisition, H.S.F. and L.W. All authors have read and agreed to the published version of the manuscript.

Funding: This research was funded by the National Natural Science Foundation of China (NSFC) (Grant No. 41674007; 41974003; 41374010).

Acknowledgments: We acknowledge the requested water level and discharge data from the Mekong River Commission (MRC) under an agreement, which is purchased using NSFC Grant No.: 41374010.

Conflicts of Interest: The authors declare no conflict of interest.

Appendix A

The mathematic proof of Equation (8) is shown below:

For the convenience of expression, set $\frac{2\pi t}{T} = x$.

$$
\begin{aligned}
A_1 \cos(x + \phi_1) + A_2 \cos(x + \phi_2) &= (A_1 \cos\phi_1) \cos x - (A_1 \sin\phi_1) \sin x \\
&+ (A_2 \cos\phi_2) \cos x - (A_2 \sin\phi_2) \sin x \\
&= (A_1 \cos\phi_1 + A_2 \cos\phi_2) \cos x \\
&- (A_1 \sin\phi_1 + A_2 \sin\phi_2) \sin x
\end{aligned}
\tag{11}
$$

Since A_1, A_2, ϕ_1, ϕ_2 are constant, $(A_1 \cos\phi_1 + A_2 \cos\phi_2)$ and $(A_1 \sin\phi_1 + A_2 \sin\phi_2)$ are also constant. Therefore, we set $C_1 = (A_1 \cos\phi_1 + A_2 \cos\phi_2)$ and $C_2 = (A_1 \sin\phi_1 + A_2 \sin\phi_2)$. Obviously,

$$
C_1 \cos x - C_2 \sin x = \sqrt{C_1^2 + C_2^2} \left(\frac{C_1}{\sqrt{C_1^2 + C_2^2}} \cos x - \frac{C_2}{\sqrt{C_1^2 + C_2^2}} \sin x \right)
\tag{12}
$$

Notice that

$$
\left(\frac{C_1}{\sqrt{C_1^2 + C_2^2}} \right)^2 + \left(\frac{C_2}{\sqrt{C_1^2 + C_2^2}} \right)^2 = 1
\tag{13}
$$

Set $\dfrac{C_1}{\sqrt{C_1^2+C_2^2}} = \cos\phi_3$, and $\dfrac{C_2}{\sqrt{C_1^2+C_2^2}} = \sin\phi_3$. Obviously,

$$C_1 \cos x - C_2 \sin x = \sqrt{C_1^2 + C_2^2} \cos(x + \phi_3) \tag{14}$$

where $\tan\phi_3 = \dfrac{C_2}{C_1}$. If we set $\sqrt{C_1^2 + C_2^2} = A_3$,

$$
\begin{aligned}
A_1 \cos(x + \phi_1) + A_2 \cos(x + \phi_2) = \\
A_3 \cos(x + \phi_3) A_1 \cos(x + \phi_1) + A_2 \cos(x + \phi_2) A_3 \cos(x + \phi_3)
\end{aligned}
\tag{15}
$$

References

1. Zhang, H. The Analysis of the Reasonable Structure of Water Conservancy Investment of Capital Construction in China by AHP Method. *Water Resour. Manag.* **2008**, *23*, 1–18. [CrossRef]
2. Cloke, H.L.; Pappenberger, F. Ensemble flood forecasting: A review. *J. Hydrol.* **2009**, *375*, 613–626. [CrossRef]
3. Jarihani, A.A.; Callow, J.N.; Johansen, K.; Gouweleeuw, B. Evaluation of multiple satellite altimetry data for studying inland water bodies and river floods. *J. Hydrol.* **2013**, *505*, 78–90. [CrossRef]
4. Lv, M.; Lu, H.; Yang, K.; Xu, Z.; Huang, X. Assessment of Runoff Components Simulated by GLDAS against UNH–GRDC Dataset at Global and Hemispheric Scales. *Water* **2018**, *10*, 969. [CrossRef]
5. Fread, D.L. *A Dynamic Model of Stage-Discharge Relations Affected by Changing Discharge*; NOAA Technical Memorandum NWS HYDRO-16; National Weather Service: Silver Spring, MD, USA, 1973.
6. Fread, D.L. Computation of stage-discharge relationships affected by unsteady flow. *Water Resour. Bull.* **1975**, *11*, 213–228. [CrossRef]
7. Schmidt, A.R.; Yen, B.C. Stage-discharge ratings revisited. In *Hydraulic Measurements and Experimental Methods, Proceedings of the EWRI and IAHR Joint Conference, Estes Park, CO, USA, 28 July–1 August 2002*; American Society of Civil Engineers: Reston, VA, USA, 2002.
8. Habib, E.H.; Meselhe, E.A. Stage-Discharge Relation for Low Gradient Tidal Streams Using Data Driven Models. *J. Hydraul. Eng.* **2006**, *132*, 482–493. [CrossRef]
9. Chau, K.W.; Wu, C.L.; Li, Y.S. Comparison of several Flood Forecasting Models in Yangtze River. *J. Hydrol. Eng.* **2005**, *10*, 485–491. [CrossRef]
10. Jain, S.K. Modelling river stage-discharge-sediment rating relation using support vector regression. *Hydrol. Res.* **2012**, *43*, 851–861. [CrossRef]
11. Jain, S.K.; Chalisgaonkar, D. Setting up stage-discharge relations using ANN. *J. Hydrol. Eng.* **2000**, *5*, 428–433. [CrossRef]
12. Sudheer, K.P.; Jain, S.K. Radial basis function neural network for modeling rating curves. *J. Hydrol. Eng.* **2003**, *8*, 161–164. [CrossRef]
13. Dawson, C.W.; Wilby, R.L. Hydrological modelling using artificial neural networks. *Prog. Phys. Geog.* **2001**, *25*, 80–108. [CrossRef]
14. Bhattacharya, B.; Solomatine, D.P. Application of artificial neural network in stage-discharge relationship. In Proceedings of the 4th International Conference on Hydroinformatics, Iowa City, IA, USA, 23–27 August 2000.
15. Simons, D.B.; Stevens, M.A.; Duke, J.H. Predicting stages on sand-bed rivers. *Harb. Coast. Eng. Div.* **1973**, *99*, 231–243.
16. Dawdy, D.R. *Depth-Discharge Relations of Alluvial Stream-Discontinuous Rating Curves. U.S. Geological Survey 1961*; Report No. WSP 1498-C; U.S.G.P.O.: Washington, DC, USA, 1961.
17. Hortle, K.G. Fisheries of the Mekong River Basin. In *The Mekong Biophysical Environment of an International River Basin*; Ian, C.C., Ed.; Academic Press: New York, NY, USA, 2009; pp. 197–239.
18. Sassi, M.G.; Hoitink, A.J.F.; de Brye, B.; Deleersnijder, E. Downstream hydraulic geometry of a tidally influenced river delta. *J. Geophys. Res.* **2012**, *117*, F04022. [CrossRef]
19. Gugliotta, M.; Saito, Y.; Nguyen, V.; Ta, T.K.O.; Nakashima, R.; Tamura, T.; Uehara, K.; Katsuki, K.; Yamamoto, S. Process regime, salinity, morphological, and sedimentary trends along the fluvial to marine transition zone of the mixed-energy Mekong River delta, Vietnam. *Cont. Shelf Res.* **2017**, *147*, 7–26. [CrossRef]

20. Arias, M.E.; Cochrane, T.A.; Norton, D.; Killeen, T.J.; Khon, P. The Flood Pulse as the Underlying Driver of Vegetation in the Largest Wetland and Fishery of the Mekong Basin. *AMBIO* **2013**, *42*, 864–876. [CrossRef]

21. Hecht, J.S.; Lacombe, G.; Arias, M.E.; Dang, T.D.; Piman, T. Hydropower dams of the Mekong River basin: A review of their hydrological impacts. *J. Hydrol.* **2019**, *568*, 285–300. [CrossRef]

22. Li, X.; Liu, J.P.; Saito, Y.; Nguyen, V.L. Recent evolution of the Mekong Delta and the impacts of dams. *Earth Sci. Rev.* **2017**, *175*, 1–17. [CrossRef]

23. Brunier, G.; Anthony, E.J.; Goichot, M.; Provansal, M.; Dussoillez, P. Recent morphological changes in the Mekong and Bassac river channels, Mekong delta: The marked impact of river-bed mining and implications for delta destabilization. *Geomorphology* **2014**, *224*, 177–191. [CrossRef]

24. Xue, Z.; Liu, J.P.; DeMaster, D.; Van, N.L.; Ta, T.K.O. Late Holocene Evolution of the Mekong Subaqueous Delta. *Mar. Geol.* **2010**, *269*, 46–60. [CrossRef]

25. Paul, A.C. The Geology of the Lower Mekong River. In *The Mekong Biophysical Environment of an International River Basin*; Ian, C.C., Ed.; Academic Press: New York, NY, USA, 2009; pp. 13–28.

26. Dang, T.D.; Cochrane, T.A.; Arias, M.E. Future hydrological alterations in the Mekong Delta under the impact of water resources development, land subsidence and sea level rise. *J. Hydrol. Reg. Stud.* **2018**, *15*, 119–133. [CrossRef]

27. Gugliotta, M.; Saito, Y.; Nguyen, V.L.; Ta, T.K.O.; Tamura, T. Sediment distribution and depositional processes along the fluvial to marine transition zone of the Mekong River delta, Vietnam. *Sedimentology* **2019**, *66*, 146–164.

28. Hoa, L.T.V.; Nhan, N.H.; Wolanski, E.; Cong, T.T.; Shigeko, H. The combined impact on the flooding in Vietnam's Mekong River delta of local man-made structures, sea level rise, and dams upstream in the river catchment. *Estuar. Coast. Shelf Sci.* **2007**, *71*, 110–116.

29. Woodruff, J.D.; Irish, J.L.; Camargo, S.J. Costal flooding by tropical cyclones and sea-level rise. *Nature* **2013**, *504*, 44–52. [CrossRef]

30. Werner, A.D.; Bakker, M.; Post, V.E.A.; Vandenbohede, A.; Lu, C.; Ataie-Ashtiani, B.; Barry, D.A. Seawater intrusion processes, investigation and management: Recent advances and future challenges. *Adv. Water Resour.* **2013**, *51*, 3–26. [CrossRef]

31. Minderhoud, P.S.J.; Coumou, L.; Erkens, G.; Middelkoop, H.; Stouthamer, E. Mekong delta much lower than previously assumed in sea-level rise impact assessments. *Nature Comm.* **2019**, *10*, 1–13.

32. Wassmann, R.; Nguyen, X.H.; Chu, T.H.; To, P.T. Sea level rise affecting the Vietnamese Mekong Delta: Water elevation in the flood season and implications for rice production. *Clim. Chang.* **2004**, *66*, 89–107. [CrossRef]

33. Kummu, M.; Tes, S.; Yin, S.; Adamson, P.; Józsa, J.; Koponen, J.; Richey, J.; Sarkkula, J. Water balance analysis for the Tonle Sap Lake–floodplain system. *Hydrol. Process.* **2014**, *28*, 1722–1733. [CrossRef]

34. Tangdamrongsub, N.; Ditmar, P.G.; Steele-Dunne, S.C.; Gunter, B.C.; Sutanudjaja, E.H. Assessing total water storage and identifying flood events over Tonlé Sap basin in Cambodia using GRACE and MODIS satellite observations combined with hydrological models. *Remote Sens. Environ.* **2016**, *181*, 162–173. [CrossRef]

35. Frappart, F.; Biancamaria, S.; Normandin, C.; Blarel, F.; Bourrel, L.; Aumont, M.; Azemar, P.; Vu, P.L.; Le Toan, T.; Lubac, B.; et al. Influence of recent climatic events on the surface water storage of the Tonle Sap Lake. *Sci. Total Environ.* **2018**, *636*, 1520–1533. [CrossRef]

36. Zhou, L.H.; Fok, H.S.; Ma, Z.T.; Chen, Q. Upstream Remotely-Sensed Hydrological Variables and Their Standardization for Surface Runoff Reconstruction and Estimation of the Entire Mekong River Basin. *Remote Sens.* **2019**, *11*, 1064.

37. El-Jabi, N.; Wakin, G.; Sarraf, S. Stage-Discharge Relationship in Tidal Rivers. *Ocean Eng.* **1992**, *118*, 166–174. [CrossRef]

38. Tozer, B.; Sandwell, D.T.; Smith, W.H.F.; Olson, C.; Beale, J.R.; Wessel, P. Global bathymetry and topography at 15 arc seconds: SRTM15+. *Earth Space Sci.* **2019**, 1847–1864. [CrossRef]

39. Koehnken, L. IKMP Discharge and Sediment Monitoring Program Review, Recommendations and Data Analysis. Part 1 and 2. *Tech. Advice Water* **2012**, 2–36.

40. Merrifield, M.; Aarup, T.; Allen, A.; Aman, B.; Caldwel, P.; Fernandes, R.; Hayashibara, H.; Hernandez, F.; Kilonsky, B.; Martin Miguez, B.; et al. The Global Sea Level Observing System (GLOSS). In Proceedings of the Ocean Obs. 09, Ocean information for Society: Sustaining the Benefits, Realizing the Potential, Venice–Lido, Italy, 21–25 September 2009.

41. Godin, G. The daily mean level and short period seiches. *IHR* **1966**, *43*, 75–89.

42. Godin, G. The Propagation of Tides up Rivers with Special Considerations on the Upper Saint Lawrence River. *Estuarine Coas. Shelf Sci.* **1999**, *48*, 307–324. [CrossRef]
43. Godin, G. *The Analysis of Tides*; University of Toronto Press: Toronto, ON, Canada; Liverpool University Press: Buffalo, NY, USA, 1972.
44. Brooks, B.A.; Merrifield, M.A.; Foster, J.; Werner, C.L.; Gomez, F.; Bevis, M.; Gill, S. Space geodetic determination of spatial variability in relative sea level change. *Geophys. Res. Lett.* **2007**, *34*, L01611. [CrossRef]
45. Stammer, D.; Ray, R.D.; Andersen, O.B.; Arbic, B.K.; Bosch, W.; Carrère, L.; Cheng, Y.; Chin, D.S.; Dushaw, B.D.; Egbert, G.D.; et al. Accuracy assessment of global barotropic ocean tide models. *Rev. Geophys.* **2014**, *52*, 243–282. [CrossRef]
46. Fok, H.S.; Iz, H.B.; Shum, C.K.; Yi, Y.; Andersen, O.; Braun, A.; Chao, Y.; Han, G.; Kuo, C.Y.; Matsumoto, K.; et al. Evaluation of Ocean Tide Models Used for Jason-2 Altimetry Corrections. *Mar. Geod.* **2010**, *33* (Suppl. S1), 285–303. [CrossRef]
47. Fok, H.S. Ocean Tides Modeling Using Satellite Altimetry. Ph.D. Thesis, School of Earth Sciences, State University, Columbus, OH, USA, 2012.
48. Gupta, H.V.; Kling, H.; Yilmaz, K.K.; Martinez, G.F. Decomposition of the mean squared error and NSE performance criteria: Implications for improving hydrological modelling. *J. Hydrol.* **2009**, *377*, 80–91. [CrossRef]
49. Mansanarez, V.; Renard, B.; Coz, J.L.; Lang, M.; Darienzo, M. Shift Happens! Adjusting stage-discharge rating curves to morphological changes at known times. *Water Resour. Res.* **2019**, *55*, 2876–2899. [CrossRef]
50. Guo, L.C.; Zhu, C.Y.; Wu, X.; Wan, Y.; Jay, D.A.; Townend, I.; Wang, Z.B.; He, Q. Strong Inland Propagation of Low-Frequency Long Waves in River Estuaries. *Geophys. Res. Lett.* **2020**. [CrossRef]
51. Jay, D.A. Green's Law Revisited: Tidal Long-Wave Propagation in Channels With Strong Topography. *J. Geophys. Res.* **1991**, *96*, 586–598. [CrossRef]
52. Miller, J.L.; Gardner, L.R. Sheet flow in a salt-marsh basin, North Inlet, South Carolina. *Estuaries* **1981**, *4*, 234–237. [CrossRef]
53. Gavin, H.P. The Levenberg-Marquardt Method for Non-Linear Least Squares Curve-Fitting Problems. Department of Civil and Environmental Engineering, Duke University: Durham, NC, USA, 2013; pp. 1–17.
54. Seber, G.A.F.; Wild, C.J. *Non-Linear Regression*; Wiley-Interscience: Hoboken, NJ, USA, 2003.
55. Möller, I.; Kudella, M.; Rupprecht, F.; Spencer, T.; Paul, M.; van Wesenbeeck, B.K.; Wolters, G.; Jensen, K.; Bouma, T.J.; Miranda-Lange, M.; et al. Wave attenuation over coastal salt marshes under storm surge conditions. *Nature Geosci.* **2014**, *7*, 727–731. [CrossRef]
56. Arias, M.E. Impacts of Hydrological Alterations in the Mekong Basin to Tonle Sap Ecosystem. Ph.D. Thesis, Department of Civil and Natural Resources Engineering, University of Canterbury, Christchurch, New Zealand, 2003.
57. Cochrane, T.A.; Arias, M.E.; Piman, T. Historical impact of water infrastructure on water levels of the Mekong River and Tonle Sap system. *Hydrol. Earth Syst. Sc.* **2014**, *18*, 4529–4541. [CrossRef]
58. Running, S.; Mu, Q.; Zhao, M. MYD16A2 MODIS/Aqua Net Evapotranspiration 8-Day L4 Global 500m SIN Grid V006 [Data Set]. NASA EOSDIS Land Processes DAAC. 2017. Available online: https://lpdaac.usgs.gov/products/myd16a2v006/ (accessed on 25 October 2020).
59. Huffman, G.J.; Bolvin, D.T. *Real-Time TRMM Multi-Satellite Precipitation Analysis Data Set Documentation*; NASA Goddard Space Flight Center location: Greenbelt, MD, USA; Science Systems and Applications, Inc.: Hampton, VA, USA, 2015.
60. Gouweleeuw, B.T.; Kvas, A.; Gruber, C.; Gain, A.K.; Mayer-Gürr, T.; Flechtner, F.; Güntner, A. Daily GRACE gravity field solutions track major flood events in the Ganges–Brahmaputra Delta. *Hydrol. Earth Syst. Sc.* **2018**, *22*, 2867–2880. [CrossRef]

Publisher's Note: MDPI stays neutral with regard to jurisdictional claims in published maps and institutional affiliations.

Article

Water Quality Anomalies following the 2017 Hurricanes in Southwestern Puerto Rico: Absorption of Colored Detrital and Dissolved Material

Suhey Ortiz-Rosa [1,*], William J. Hernández [2,3], Stacey M. Williams [4] and Roy A. Armstrong [1]

[1] NOAA CESSRST, Bio-Optical Oceanography Laboratory, Department of Marine Sciences, University of Puerto Rico, Mayagüez 00683, Puerto Rico; roy.armstrong@upr.edu

[2] Environmental Mapping Consultants LLC (EMC LLC), Aguadilla 00603, Puerto Rico; william.hernandez@upr.edu

[3] Department of Electrical and Computer Engineering, University of Puerto Rico, Mayagüez 00603, Puerto Rico

[4] Institute for Socio-Ecological Research, Lajas 00667-3151, Puerto Rico; Iser@isercaribe.org

* Correspondence: suhey.ortiz@upr.edu

Received: 17 September 2020; Accepted: 26 October 2020; Published: 2 November 2020

Abstract: Absorption of colored dissolved organic matter or detrital gelbstoff (aCDOM/ADG) and light attenuation coefficient ($K_d 490$) parameters were studied at La Parguera Natural Reserve in southwestern Puerto Rico, before and following Hurricanes Irma (6–7 September) and María (20–21 September) in 2017. Water quality assessments involving Sentinel 3A ocean color products and field sample data was performed. The estimated mean of ADG in surface waters was calculated at >0.1 m^{-1} with a median of 0.05 m^{-1} and aCDOM443 ranged from 0.0023 to 0.1121 m^{-1} in field samples (n=21) in 2017. Mean ADG443 values increased from July to August at 0.167 to 0.353 m^{-1} in September–October over Turrumote reef (LP6) with a maximum value of 0.683 m^{-1}. Values above 0.13 m^{-1} persisted at offshore waters off Guánica Bay and over coral reef areas at La Parguera for over four months. The ADG443 product presented values above the median and the second standard deviation of 0.0428 m^{-1} from September to October 2017 and from water sample measurement on 19 October 2017. Mean $K_d 490$ values increased from 0.16 m^{-1} before hurricanes to 0.28 right after Hurricane Irma. The value remained high, at 0.34 m^{-1}, until October 2017, a month after Hurricane María. Analysis of the Sentinel (S3) OLCI products showed a significant positive correlation ($r_s = 0.71$, $p = 0.0005$) between $K_d 490\_M07$ and ADG_443, indicating the influence of ADG on light attenuation. These significant short-term changes could have ecological impacts on benthic habitats highly dependent on light penetration, such as coral reefs, in southwestern Puerto Rico.

Keywords: hurricanes; ADG/CDOM colored dissolved organic matter; Sentinel 3; water quality; southwestern Puerto Rico; ocean color; remote sensing; coastal waters

---

1. Introduction

Hurricane María was recorded as the third costliest hurricane in USA history [1]. It is considered the most damaging atmospheric event to have impacted the island in the past 90 years. Hydrological data availability during the study period was limited; nevertheless, estimates suggest that the 24 h-rainfall intensity exceeded 100–250 year values [2]. Severe flooding affected most of the island, and river discharges were at record levels. Hurricane Irma brought maximum inundation levels of 30.48 to 60.96 cm above ground level along Puerto Rico (PR) coastal areas with an estimated storm surge at Magueyes Island, La Parguera (LP), of 0.17 cm and an estimated inundation of 21.33 cm. The total rainfall in the interior mountains was around 254–381 mm between September 5 and 7, 2017 [3].

For this study, we took advantage of the availability of remote sensing imagery such as Sentinel 3 (S3), which provided us with the capacity to monitor water quality parameters remotely and efficiently. Many studies have derived water quality parameters from ocean color radiometry during past decades [4–6]. The most critical water quality parameters that can be derived from satellite ocean color sensors are colored dissolved organic matter (CDOM), chlorophyll-*a* (Chl-*a*), the attenuation coefficient at 490 nm (K_d490), and total suspended matter (TSM). These factors have been historically monitored for water quality assessment and referenced as indicators of coastal and marine ecosystem health [4,7,8]. Chlorophyll (a proxy for phytoplankton abundance) [9] and turbidity (as well as CDOM) contribute to reducing light penetration in the water column [4,5], which has been associated with ecosystem changes, phytoplankton dynamics [9], and growth and distribution of seagrasses [4] and coral reef species [10]. These effects on light penetration and quality can be considered environmental stressors and a water pollutant [11]. Several studies have been conducted on coastal water quality following hurricane events using remote sensing techniques [4,5,12]. These have focused primarily on the continental estuarine and coastal habitats as well lacustrine [5,13–17], with a few studies available for the Caribbean Sea [18–20], where our study area, La Parguera Natural Reserve (LPNR), is located.

LPNR, in southwestern Puerto Rico, was designated to protect fragile tropical marine ecosystems, particularly coral reefs, which are experiencing accelerated degradation and mortality in this and many parts of the world [21]. We have witnessed unprecedented disappearance of coral cover due to coral diseases, bleaching and thermal stress, runoff, anthropogenic uses, and hurricanes [10,22,23]. La Parguera has one of the largest coral reef systems in Puerto Rico, with 10–14 coral species in 100 m^2 located at a diverse bottom type; presenting one of the most diverse benthic habitats on the island, combining coral reefs, seagrasses, mangroves, sandy bottoms, among others [23]. LPNR supports a blue economy around the region with local fisheries, tourism, and recreation. Coral reefs reach their maximum development under oligotrophic conditions but can exist over a wide range of water types under variable coastal influences [10,24]. In coastal waters, light penetration can be subject to sudden changes when specific weather conditions occur. Corals can adapt to light changes compensating energetically and adjusting photosynthetic pigment composition, but this may come at the cost of reduced calcification rates and symbiont tissue habitat [25]. Sporadically, the stress related to water quality can be compounded with coral bleaching [25]. Chronic stress due to changes in water quality can lead to changes in the biodiversity of marine ecosystems [26,27].

CDOM is an optical parameter positively related to light absorption in surface waters [28–31]. The primary source of CDOM is terrestrial runoff, highly influenced by photodegradation [32–34]. CDOM absorption provides a biogeochemical proxy to estimate DOC from optical measurements [6,35,36] and can be used as a tracer of oceanic water masses [31]. CDOM absorption at 412 and 443 nm, while variable, formed a significant component of these wavebands of the total absorption field [37]. It is essential to consider that the absorption of CDOM in the blue-green region increases the uncertainty of Chl-a algorithms leading to over-estimate values [31]. In Puerto Rico, light absorption has been mainly associated with chlorophyll concentrations [10], but CDOM values have not been recorded or related to light availability. K_d490 is another important parameter for water quality since it provides a measure of turbidity related to the total organic and inorganic matter held in solution and suspension in the water column. It can be used to quantify light availability and sediment loading for benthic organisms (i.e., coral reefs and seagrasses) [20,38].

We used remote sensing to monitor water quality trends in the attenuation coefficient at 490 nm (K_d490) and the absorption of colored detrital, and dissolved material (ADG/aCDOM) parameters in waters off southwestern Puerto Rico. We analyzed the changes in Sentinel 3A (S3) products and complemented our records with surface water sample optical analysis to assess water quality in a natural reserve in southwestern Puerto Rico before and following Hurricanes Irma (6–7 September) and María (20–21 September) in 2017.

2. Materials and Methods

2.1. Study Site

The study area includes the region from Guánica Bay (GB) to La Parguera Natural Reserve (LPNR) in southwestern Puerto Rico. LPNR is located about 8 km west of GB and is known for its highly developed coral reefs and extensive seagrass habitats. The average annual water temperature in LPNR is 26.5–30 °C, and the salinity fluctuates from 31 to 36 PS [7].

Coral reefs habitats are shown within the contour lines representing the study area as a region of interest (ROI) (Figure 1). These are delineated using live coral cover classification [39], while the perpendicular lines represent the limits of La Parguera Natural Reserve. The ROI was considered for statistical data analysis. Seven out of the thirteen stations are presented here. Stations GUA4, GUA5, LP12, and LP13 were located offshore and along the insular platform. While sites LP6, LP8, and LP10 were closer to the coast (Figure 1).

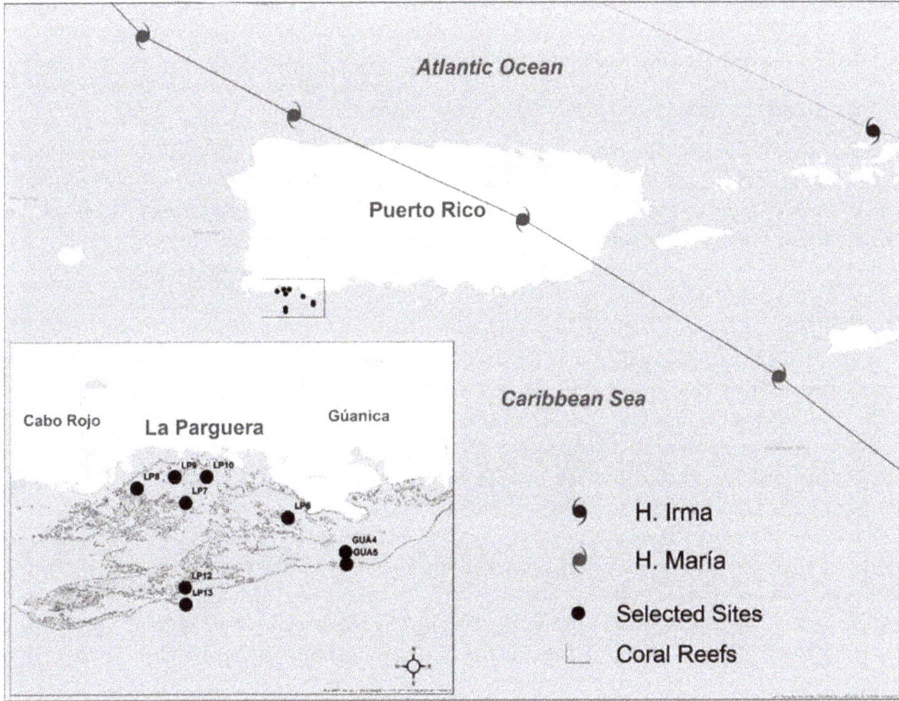

Figure 1. Study area at southwestern Puerto Rico with Hurricanes Irma and María paths. Sampling sites and delimited areas represent submerged coral reef areas that support fisheries and economies around the region. Base map was from ESRI®.

2.2. Satellite Data

The Sentinel 3A (S3A) Ocean and Land Colour Instrument (OLCI) is a push-broom imaging spectrometer with 21 spectral bands in the range of 400–1020 nm [40]. It was launched in February 2016, followed by S3B, launched in March 2017. Their products have a full-spatial resolution of 300 meters and include water-leaving reflectance in 16 bands, algal pigment concentrations [41] and neural network algorithms [42], total suspended matter concentration (TSM), diffuse attenuation coefficient (K_d490_M07) Morel method [41], and absorption of colored detrital and dissolved organic matter

(ADG_443_NN) GSM method [43,44]. The temporal resolution for OLCI is daily with an optimum orbit above the study area every two days.

We obtained the data from the EUMETSAT Copernicus data system. S3A/OLCI data were extracted from the pixel coinciding with our field water monitoring stations and pixels over the ROI (Figure 1). The Sentinel Application Platform (SNAP) tools, developed by the European Space Agency (ESA) for satellite product analysis, were used for obtaining OLCI Level 2 data products. Only Sentinel 3-A data were used in the study.

Water quality products (ADG and K_d490) were extracted from Sentinel 3 OLCI imagery dating from July to December 2017. A subset of three images (out of 20) was evaluated considering the region of interest (ROI). It included the following dates: 22 July, 11 September, and 8 October 2017. This subset was selected to reduce uncertainty due to the following factors: negative reflectance values from bands one to six, sunglint effect, cloud, or land adjacency effect, or products fail/flags. Imagery is visualized with a median (7 × 7) pixel value. The complete set of images was divided into five time-frames, summarizing four images in each period. The time frames included one period previous to the hurricane events, one period immediately after the event and three additional periods after the event to identify the long-term effect on light attenuation. Only the LP6 site data were used in the time frame to avoid negative values and other sensor issues previously mentioned.

2.3. Water Quality Measurements—Field and Laboratory Analysis

Water samples and optical data were collected in 2017. The number of sampling stations varied based on the sea state, environmental parameters, and imagery availability. Sampling was conducted monthly at three to 13 stations in southwestern Puerto Rico (Figure 1). The locations were selected based on depth, bottom type, and habitat in relation to coral reefs. Water samples were obtained from the first meter depth and analyzed in the laboratory for CDOM absorption (aCDOM).

2.3.1. aCDOM

Duplicate samples were collected at each station using gloves, avoiding any contamination with organic matter. They were stored in previously cleaned 250 mL amber glass bottles and transferred to 140 mL bottles after filtration. Sterile membrane filters (0.2 μm pore) were employed (Pall©). The filtration system was rinsed beforehand and between each filtration with a 50 mL portion of sample water and was then discarded [45]. Spectrophotometric analysis was carried out using a Shimadzu 1800-UV diode array instrument. Samples were analyzed in 10 cm path length quartz cells at 0.5 nm intervals over a wavelength range from 250 nm to 800 nm. Milli-Q water absorbance was subtracted from the sample data, and subsequently, the value at 700 nm was subtracted from the entire spectrum [46]. The absorbance values were converted to absorption coefficients, a (λ, m^{-1}), and absorption coefficients at 443 nm (aCDOM$_{443}$ m^{-1}) were reported as quantitative aCDOM. The absorption coefficients an (m^{-1}) were calculated using the following equation:

$$a = 2.303 A_{(\lambda)}/l \qquad (1)$$

where $A_{(\lambda)}$ is the absorbance at a wavelength, and l is the optical path length of the cell in meters.

2.3.2. Satlantic HyperPro

The Satlantic profiling spectroradiometer measures in-water downwelling plane irradiance (Ed) and upwelling radiance (Lu) with 256 spectral bands for a full spectral range of 305–1100 nm [47]. A surface Ed radiometer measures downwelling irradiance above the water surface and is used to normalize the in-water data for fluctuations in the incident light field from passing clouds.

The instrument derives spectral water column attenuation coefficients, including the K_d490 following Aurin and Petzold (1981) in the manufacturer manual [48]:

$$K_{(490)} = 0.0833\,(Lu_{(443)}/Lu_{(550)})^{-1.491} + 0.022 \tag{2}$$

2.4. Statistical Analysis

The SNAP© Sentinel toolbox, pixel extraction, and histogram tools were used to obtain satellite data statistics. S3A data were divided into 5-time frames (July–September, September–October, October–November, November–December, and December 2017) to evaluate the mean and median differences over time. A Spearman correlation was applied to ADG443_NN satellite data, aCDOM field data and K_d490 for field and satellite data to understand the influence of ADG/aCDOM on light attenuation. The analysis was employed using Origin Pro 2016© software.

3. Results

3.1. Satellite Data (ADG443_NN) and In Situ Data (aCDOM443)

Results were based on in situ data for a year (2017) and the last six months (2017) data retrieved from satellite sensor S3A. Before the hurricane events, oligotrophic stations located at the shelf edge showed ADG443_NN values below 0.05 m^{-1}; while values below 0.1 m^{-1} were associated with insular shelf sites. The value of the ADG443_NN was above the median of 0.0435 m^{-1} (prior to the events, over the ROI) for the entire sampling period. On September 11, four days after the first event, amounts above 0.1 m^{-1} were detected at GUA5, LP6, and LP8 for one month (Figure 2). The highest ADG values at offshore waters were detected on Sep 11, after the first hurricane (H. Irma) which was considered less severe because its eye did not make landfall. On the other hand, the effects of H. María on the values were evident on Oct 8th satellite data in most of our study area. It had a similar or lower effect on ADG (GUA5, LP12, and LP13) values at outer shelf waters but the sensor detected higher values at inner shelf waters.

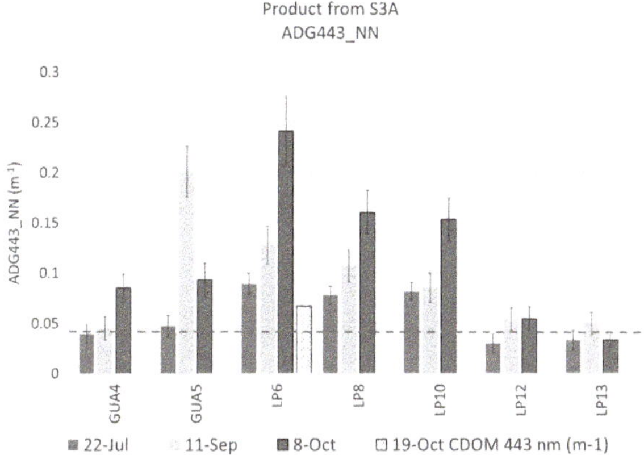

Figure 2. ADG443_NN product from Sentinel 3A for 22 July, 11 September, and 8 October 2017; the figure includes the aCDOM at 443 nm field data on 19 October 2017, at LP6 sampling site. The dashed line represents the second standard deviation (2SD) for aCDOM443 field data.

A view from in situ data showed the aCDOM443 ranging from 0.0023 to 0.1121 m^{-1} in field samples (N = 21) with a 2SD of 0.0428 m^{-1} in 2017 (Figure 3). Before the events, field values were below 0.043 m^{-1}. An unusual value of 0.1121 m^{-1} was observed in the station LP9 on 15 September 2017 (Figure 3A). Another extreme value of 0.068 m^{-1} above the 2SD was present in the field data at the LP6 site (N = 5) during 19 October 2017 (Figure 3B). Station LP6 is located to the southwest of Guánica Bay. The extreme values belong to the sample size and should not be treated as outliers even though a Grubbs' outlier test detect these as such. We can consider them as extreme values as a result of the events.

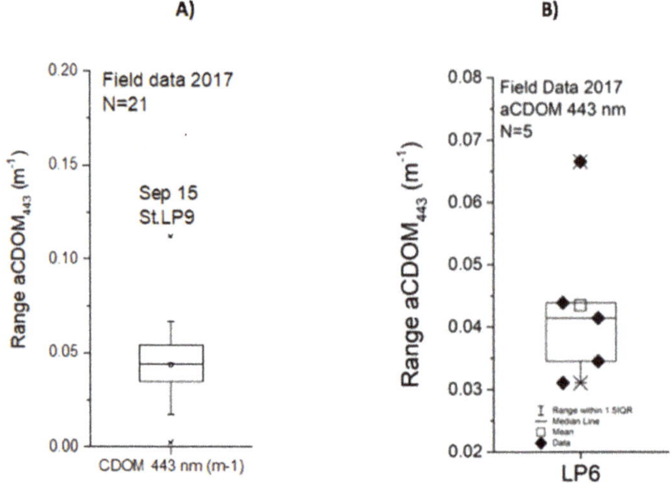

Figure 3. Box plots for field data showing the median, mean and data including the extreme values (**A**) Summary of the absorption coefficient of colored dissolved organic matter (aCDOM) at 443 nm field data for the year 2017. (**B**) The absorption coefficient of colored dissolved organic matter (aCDOM) at 443 nm field data on station LP6, alias Turrumote II, for 2017.

S3A data were divided into five-time frames (July–September, September–October, October–November, November–December, and December 2017) to evaluate the mean and median differences (Table 1). The mean for ADG_443_NN was doubled in the second period from 0.1675 (pre hurricanes) to 0.3536 m^{-1} (September–October). The maximum value of 0.6834 m^{-1} was detected in the same period. The values extracted from S3A started in July with values above the maximum of field data for 2017. Values above 0.13 m^{-1} persisted until December, four months after the events. Moreover, the median showed the same tendency (>0.1 m^{-1}) over four months. River discharges and coastal drainage persist several weeks after the events. No major events took place after September, which may indicate we are seeing the long term effect of the hurricanes in coastal water biogeochemistry.

Table 1. Statistics for five time periods on ADG443_NN product from Sentinel 3A.

S3_ADG443	7/22–9/3	9/11–10/8	10/11–11/27	11/30–12/12	12/16–12/27
Mean	0.1675	0.3536	0.1540	0.1322	0.1716
SD	0.0750	0.2398	0.1108	0.0544	0.0901
SE of mean	0.0375	0.1199	0.0554	0.0272	0.0451
Variance	0.0056	0.0575	0.0123	0.0030	0.0081
Sum	0.6699	1.4143	0.6160	0.5288	0.6863
Minimum	0.0889	0.1277	0.0812	0.0849	0.0973
Median	0.1586	0.3016	0.1093	0.1168	0.1434
Maximum	0.2638	0.6834	0.3162	0.2103	0.3022

Satellite imagery show the absorption of dissolved organic matter over time. Figure 4 shows the S3A ADG443_NN product prior to (July 22), and following (September 11 and Oct. 8) the passage of hurricanes Irma and María over Puerto Rico. Contour lines represent coral reefs as the region of interest (ROI). The ROI was considered for graphs and statistics on Figure 4 and Table 2. The high values of ADG443_NN in Figure 4 correspond to pixels that cover mainly shallow areas and emergent reefs. However, the analysis only considered the extracted values in submerged areas. The histogram for July 22 shows around 46 pixels lower than 0.05 m^{-1} and more than 95 % of pixels with values < 0.1 m^{-1}. The maximum value was 1.0 m^{-1} (Table 2). After the first hurricane event (Irma), an increase in the ADG443_NN values from Guánica Bay to La Parguera was observed (Figure 4) as expected after an event of such magnitude. Approximately, 12% of pixels in the selected area were considered in Table 2. The histogram shows an increment of pixels with values in the range of 0.1 to 0.5 m^{-1} (Figure 4) and shows pixels with up to 4.5 m^{-1}. Table 2 shows the increase in the median ADG443_NN value over time from 0.04 to 0.08 m^{-1}.

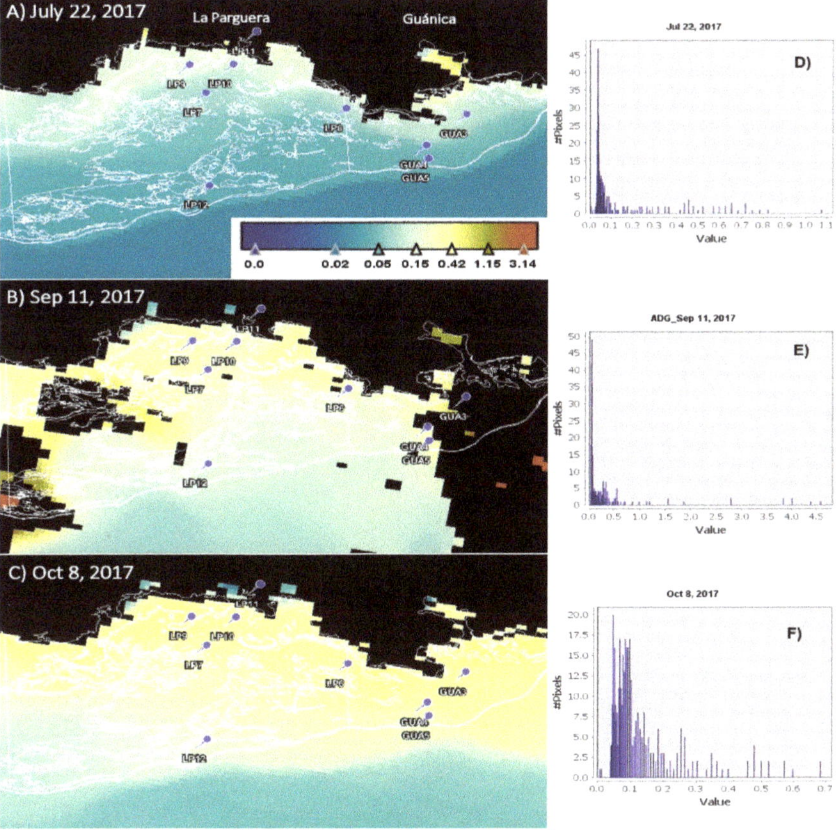

Figure 4. Sentinel 3A ADG_443_NN product (median visualization). Contour lines represent coral reefs as the region of interest (ROI) while the perpendicular lines represent the limits of La Parguera Natural Reserve. Black areas represent land-mask and cloud-mask applied to the imagery (**A**) before the hurricanes on 22 Jul 2017 (**B**) 11 Sep 2017 after Hurricane Irma and (**C**) Oct 8, 2017 after two hurricane events (H. Irma and H. María); (**D**) Histograms for regions of interest representing ADG443_NN pixel values on 22 July 2017, (**E**) 11 September 2017 and (**F**) 8 October 2017.

Table 2. Sentinel 3A Ocean and Land Colour Instrument (OLCI) data statistics for ADG443_NN product over (ROI) coral reef areas around Guánica to La Parguera Natural Reserve.

S3_ADG over Coral Reef Areas	22-Jul	11-Sep	8-Oct
Number of considered pixels	349	339	365
Ratio of considered pixels (%)	12.1138	11.7667	12.3020
Min.	0.0053	0.0343	0.0080
Max.	1.0752	4.5858	0.6834
Mean	0.1097	0.2353	0.1216
SD	0.1650	0.5991	0.1086
CV	1.5048	2.5458	0.8932
Median	0.0435	0.0628	0.0849

To visualize the effect on light attenuation, we chose sampling station (LP6), located between Guánica Bay and LPNR. It is near the coastline but, far enough to be outside the influence of land pixels. Taking a look on satellite data of this site, a spike value was observed on 7 October 2017, for both parameters ADG443_NN and K_d490, with high values on 18 August, 23 October, and 16 December 2017 (Figure 5). These values were concurrent with two heavy rain periods during the last six months of the year 2017. The image from October 8 showed the impact on water quality parameters three weeks after the events. All ADG443_NN values were over 0.04 m^{-1} for the entire sampling period.

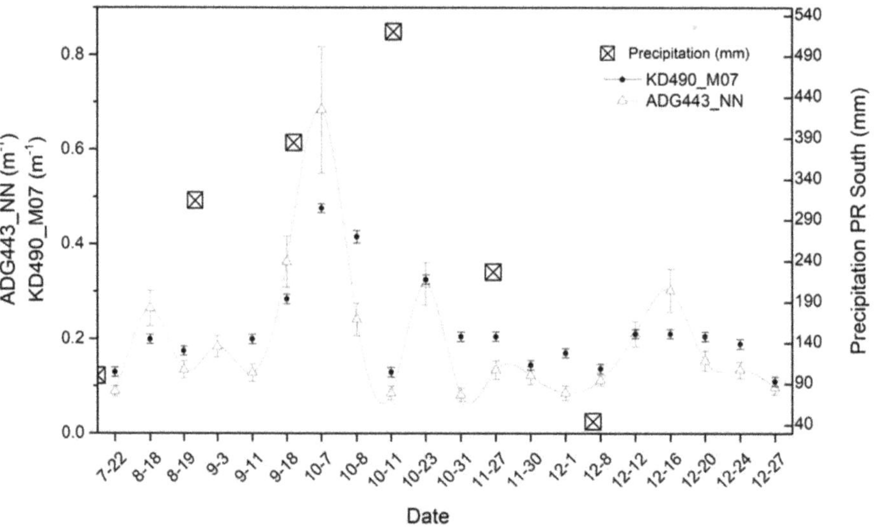

Figure 5. S3A OLCI products from July to December 2017 showing the changes over time for station LP6 known as Turrumote II (located between Guánica Bay (GB) and La Parguera Natural Reserve (LPNR)) for ADG443_NN and K_d490_M07 and monthly precipitation at Guayanilla, Puerto Rico (PR) south station.

The S3A data for the last six months of 2017 shows an ADG mean value of 0.2 m^{-1} (sd = 0.14, n = 20) with a minimum of 0.08 m^{-1} on 31 October 2017, and a maximum of 0.68 m^{-1} on 7 October 2017. This maximum coincides with maximum Chl-a and K_d490 values while the minimum value was not coincident with the minimum of K_d490. From July to December the ADG mean exceeds the 0.0436 m^{-1}

mean and 0.0437 median values found in field data for the year 2017. Considering LP6 field data, it presented 0.0249 m^{-1} as the second standard deviation with a mean of 0.0435 m^{-1} (N = 5) while satellite data showed a mean value of 0.2 m^{-1} (N = 20). The selected images (N = 3) in Figure 2 show 0.13 m^{-1} as a mean value which triplicates the field data mean of aCDOM443 (N = 21).

The highest ADG values among stations were found at station LP6 (mean = 0.1957 m^{-1}) (Figure 6) which may imply influence of the freshwater plume emanating from Guánica Bay reaching the area. The second highest values were observed at LP8, also known as Laurel (mean = 0.1535 m^{-1}) followed by LP10, out of the bioluminescent bay (mean = 0.1023 m^{-1}). These two correspond to shallow inner coral reef areas. Stations GUA4 (mean = 0.0744 m^{-1}) and GUA5 (mean = 0.0753 m^{-1}) were located at the edge of the shelf close to Guánica Bay showing lower values than the inner shelf sites. Sites LP12 and LP13 were located above the border of the insular shelf and closer to La Parguera. These presented the lowest values 0.07342 and 0.0395 m^{-1} (N = 21), respectively.

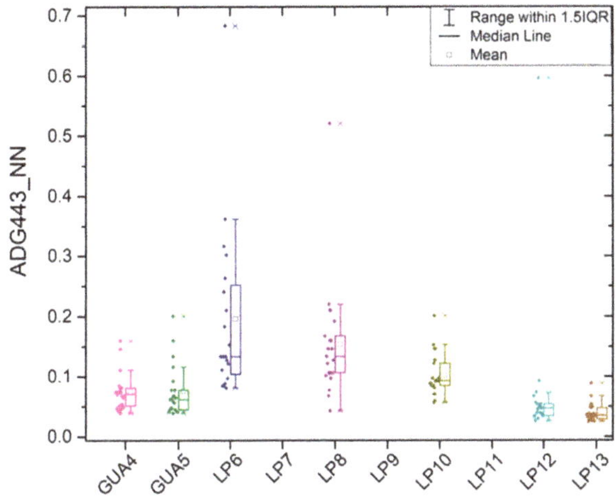

Figure 6. S3A OLCI Product ADG_443_NN from Sentinel 3A (EUMETSAT-Copernicus data) from July to December 2017 at selected sampling sites in southwestern, PR. The graph shows the range per site, median line, mean, data points, and extreme values.

3.2. K_d490 and Correlation with ADG443/aCDOM443

Values for diffuse attenuation coefficient (K_d490) share the ADG443_NN tendencies (Figure 5). Values for K_d490 extracted from S3A for station LP6 show a mean value of 0.22 m^{-1} (sd = 0.1, n = 19) from July to December. The minimum values were on 27 December 2017, and the maximum of 0.48 m^{-1} was detected on 7 October 2017, coincident with other parameters following the events. The stations closer to the shoreline LP6 and LP10 showed values above 0.14 m^{-1} in September and October 2017 (Figure 7). K_d490 median values from S3A varied from 0.15 to 0.34 m^{-1} for six months while the mean values ranged from 0.16 to 0.34 m^{-1} (Table 3). The highest mean value of 0.34 m^{-1} was observed in the period of Sep 11 to Oct 8; during that period, a maximum of 0.48 m^{-1} was detected. The maximum value of K_d490 derived from field data in 2017 was 0.33 m^{-1} on 19 October 2017, four weeks after the last hurricane (Figure 8A,B).

The attenuation coefficient showed a slight variation in outer shelf waters with a greater impact in inner shelf, specifically in LP6, alias Turrumote II (Figure 7). The cumulative effect of biogeochemical processes in production and degradation of organic matter is shown by this increment on October values. K_d490 values reach to the normal between October and November (Table 3).

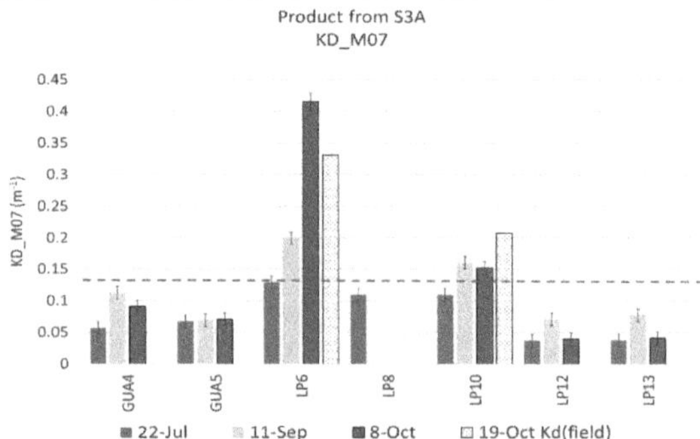

Figure 7. K_d490 product from Sentinel 3A for 22 July, 11 September, and 8 October 2017. The figure includes the K_d490 nm field data on 19 October 2017 at LP6 sampling site. The dashed line represents the second standard deviation (2SD) for K_d490 field data.

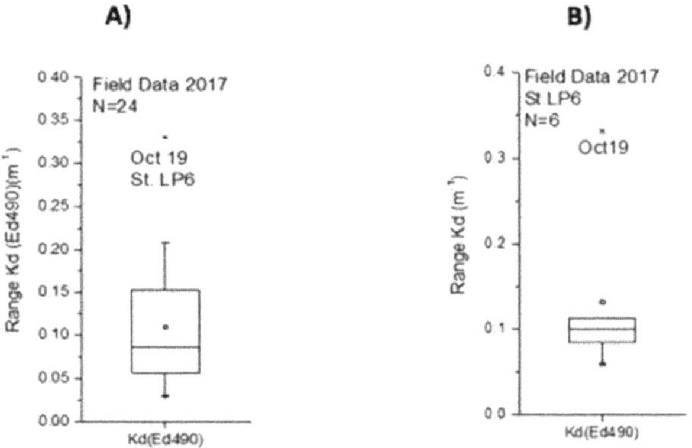

Figure 8. Box plot for light attenuation coefficient (K_d490) at 490 nm field data, showing the value range, median line, mean, and the extreme values (**A**) for the year 2017 from GB to LPNR area and, (**B**) field data on station LP6 known as Turrumote II for 2017.

Table 3. Statistics for five time periods on K_d490 product from Sentinel 3A.

S3_KD490	7/22–9/3	9/11–10/8	10/11–11/27	11/30–12/12	12/16–12/27
Mean	0.1673	0.3433	0.2158	0.1648	0.1783
SD	0.0356	0.1251	0.0811	0.0335	0.0468
SE of mean	0.0205	0.0626	0.0405	0.0167	0.0234
Variance	0.0013	0.0157	0.0066	0.0011	0.0022
Sum	0.5019	1.3734	0.8632	0.6593	0.7131
Minimum	0.1289	0.1992	0.1289	0.1361	0.1095
Median	0.1738	0.3493	0.2047	0.1564	0.1966
Maximum	0.1992	0.4755	0.3249	0.2103	0.2103

The field data for 2017 (N = 21) show a high correlation between K_d490 and aCDOM443 absorption coefficients (r_s = 0.79, p = 0.0003) and a lower but similar correlation between S3 OLCI products, (r_s = 0.71, p = 0.0005). It cannot be interpreted as a sensor validation.

4. Discussion

Caribbean Sea water is mostly oligotrophic with a high light penetration in the water column, although it is seasonally influenced by the Orinoco and Amazon rivers from South America, seasonally [49–51]. Light penetration changes after hurricane events can affect seagrasses [11] and other light-dependent organisms like corals. Previous research shows that coral photo-physiology is altered by light availability [25,52]. García-Sais and collaborators (2017) studied K_d490 and Chl-a trends over individual coral reefs in Puerto Rico using L2 and L3 imagery from SeaWiFS and MODIS Aqua satellite data [10]. A recent publication based on VIIRS data described the tendencies of K_d490 and Chl-a parameters on water quality around PR using a value of 0.1 m^{-1} for K_d490 and 0.45 µg/L for Chl-a as a threshold value for coastal waters [20]. Despite the detrimental effects documented by several authors [26,53–55], an intermittent high turbidity over coral reefs can be photo-protective [10]. García-Sais and collaborators (2017) observed a negative correlation between K_d490 and the percent of coral cover which can be interpreted as a positive light shadow effect during sea surface temperature anomalies [10]. The severity of the damages can be highly influenced by the prevalence of adverse conditions during and after the events.

In 2017, the duration of the abnormal values (above 0.05 m^{-1}) lasted four months as can be seen in Figure 5. Higher values are not necessarily coincident with the events but, rather these were detected two to three weeks later in October. The high number of landslides (>40,000) combined with runoff after hurricanes Irma and María over the Island were unprecedented [56] and washed sediments reached nearshore waters [19]. Miller (2019) documented elevated turbidity values nearshore until February 2018 related to inland hydrological disturbances caused by the hurricanes [19]. Gilbes and collaborators (2001) documented changes in Chl-a due to hurricane Georges up to two and a half weeks after the event [18]. Recently, Hernández and collaborators (2020) documented high K_d490 and Chl-a values from July to December 2017 all around Puerto Rico using VIIRS data; reporting Chl-*a* values above 0.45 µg/L in August and November 2017 [20]. These authors reported anomalous attenuation coefficient values for July 2017 (0.06 m^{-1}) being persistently high until December. Chlorophyll-a is a parameter highly correlated with K_d490 on ocean color data [18,20]. It is important to highlight the oligotrophic water conditions on this study area, being influenced by Guánica Bay dynamics. The values considered in this area can be compared to coral reefs or benthic areas with low influence of rivers.

The aCDOM443 values above 0.05 m^{-1} are not typical for coral reef waters in Puerto Rico. CDOM values with means < 0.043 m^{-1} are the most common values over coral reefs and seagrass beds in the studied area. Otherwise, the values closer to 0.02 m^{-1} are found in offshore waters. An absorption coefficient higher than 0.1 m^{-1} is frequently found on coastal embayments like Guánica Bay or the Bioluminescent Bay surrounded by mangroves [57]. Anomalies like the ones measured in this study lasted for the entire study period.

In terms of attenuation coefficient (K_d490), values above 0.2 m^{-1} corresponded to coastal embayments while values from 0.1 to 0.2 were observed at shallow coral reef or seagrasses areas close to the coast (<1 mile) or closer to the coral cays [20]. The lower values (<0.1 m^{-1}) were found at mid- and outer-shelf coral reef stations. A mean K_d490 value of 0.056 m^{-1} was documented (for 10 y data) in a coral reef site at Guánica by satellite data [10]. Their values are lower than the values reported here from July to December 2017 using OLCI data. A variant of K_d490 parameter, K_dPAR, was measured in situ before and after hurricane events in St. John Island recording the lowest level of light in coral reef in the Caribbean after a hurricane event [58,59]. Certainly, these events had an unprecedent effect on light attenuation over sensitive benthic communities.

These data show the influence of ADG443_NN/aCDOM in light attenuation. However, in estuarine areas, a significant correlation between K_d490 and Chl-a was documented after a hurricane event [13,20]. The high anomalous values of ADG and K_d490 can be related to the unprecedented runoff produced by defoliation and landslides [19,56] followed by biogeochemical oceanographic processes over the coastal waters.

5. Conclusions

As expected from episodic events of this magnitude, significant water quality parameter changes occurred in southwestern Puerto Rico. Sentinel 3A OLCI data was used to extract information on ADG and K_d490 values. These data were compared with in situ data trends and correlated between them. The amount of data acquired during the study period (before N = 5, N = 16 after hurricanes) duplicates the quantity of data obtained from the field (N = 3 after the hurricane) in 2017. Although cloud cover in tropical islands can be high, remote sensing is an accessible and useful tool for short and long-term water quality studies.

Increasing values of satellite-derived water quality parameters were detected with S3 OLCI and field data in southwestern Puerto Rico. The anomalies were observed during the 9/11–10/8 period as expected and extended until December. The ADG values increased throughout all the coral reef zones. The estimated ADG mean in this zone was > 0.1 m^{-1} with a median of 0.05 m^{-1}. The mean values of K_d490 increased from 0.16 m^{-1} before the hurricanes to 0.28 m^{-1} shortly after Hurricane Irma, and 0.34 m^{-1} in October 2017, a month after Hurricane María.

Satellite data are useful for water quality assessment in PR coastal waters with a judicious understanding of their uncertainties and limitations. On the other hand, we cannot conclude the performance of the sensor measurement on ADG443_NN or K_d490 products since we do not have enough in situ data from July to December 2017. Our results represent a pioneering effort in the establishment of tendencies for water quality studies in Puerto Rico. Usually, government agencies' data are a single snapshot influencing the mean values that can be misinterpreted for the establishment of patterns on water quality. These gaps can be addressed with satellite data, as we showed throughout the manuscript. Remote sensing tools can help understand coastal and benthic habitat changes and biogeochemical processes in waters surrounding oceanic islands, especially after extreme weather events. Previous studies have mainly documented the importance of chlorophyll on light attenuation, but this study highlights the importance of detrital and gelbstoff matter on light attenuation coefficient. This is not only done as a historical perspective of the consequences of these events but as an analysis that could be integrated into future efforts aimed at describing the consequences of such events in benthic communities changes on the long run.

Author Contributions: S.O.-R. conducted and performed data processing and analysis; S.O.-R. and W.J.H. performed the experiments; S.M.W. contributed to data analysis; R.A.A. contributed to data collection; S.O.-R. wrote the manuscript, with all authors contributing to manuscript revisions; S.O.-R. and R.A.A. attracted funds for the project. All authors have read and agreed to the published version of the manuscript.

Funding: This study is supported and monitored by The National Oceanic and Atmospheric Administration—Cooperative Science Center for Earth System Sciences and Remote Sensing Technologies (NOAA-CESSRST) under the Cooperative Agreement Grant: NA16SEC4810008. The statements contained within the manuscript/research article are not the opinions of the funding agency or the U.S. government but reflect the author's opinions.

Acknowledgments: The authors would like to thank Vice Admiral Conrad C. Lautenbacher Scholarship, The University of Puerto Rico and NOAA Office of Education, Educational Partnership Program with Minority-Serving Institutions (EPP/MSI) for full fellowship support for Suhey Ortiz-Rosa. S.O.R. thanks to Caribbean Coastal Ocean Observing System (CARICOOS) for lab equipment use; M. Solórzano, J. Corredor and J. Morell for comments and suggestions.

Conflicts of Interest: The authors declare no conflict of interest. The funders had no role in the design of the study; in the collection, analyses, or interpretation of data; in the writing of the manuscript, or in the decision to publish the results.

References

1. Pasch, R.J.; Penny, A.B.; Berg, R. National Hurricane Center (NHC) Tropical Cyclone Report Hurricane María (AL152017) 2019. Available online: https://www.nhc.noaa.gov/data/tcr/AL152017_Maria.pdf (accessed on 2 October 2019).
2. Ramos-Scharrón, C.E.; Arima, E. Hurricane María's precipitation signature in Puerto Rico: A conceivable presage of rains to come. *Sci. Rep.* **2019**, *9*, 15612. [CrossRef] [PubMed]

3. Cangialosi, J.P.; Latto, A.S.; Berg, R. National Hurricane Center (NHC) Tropical Cyclone Report Hurricane Irma (AL112017) 2018. Available online: https://www.nhc.noaa.gov/data/tcr/AL112017_Irma.pdf (accessed on 2 October 2019).

4. Huang, W.; Mukherjee, D.; Chen, S. Assessment of Hurricane Ivan impact on chlorophyll-a in Pensacola Bay by MODIS 250 m remote sensing. *Mar. Pollut. Bull.* **2011**, *62*, 490–498. [CrossRef] [PubMed]

5. Lohrenz, S.E.; Cai, W.J.; Chen, X.; Tuel, M. Satellite assessment of bio-optical properties of northern gulf of Mexico coastal waters following Hurricanes Katrina and Rita. *Sensors (Basel Switzerland)* **2008**, *8*, 4135–4150. [CrossRef] [PubMed]

6. Werdell, P.; McKinna, L.; Boss, E.; Ackleson, S.; Craig, S.; Gregg, W.; Lee, Z.; Maritorena, S.; Roesler, C.; Rousseaux, C.; et al. An overview of approaches and challenges for retrieving marine inherent optical properties from ocean color remote sensing. *Prog. Oceanogr.* **2018**, *160*, 186–212. [CrossRef] [PubMed]

7. Otero, E. Spatial and temporal patterns of water quality indicators in reef systems of southwestern Puerto Rico. *Caribb. J. Sci.* **2009**, *45*, 168–180. [CrossRef]

8. Wild-Allen, K.; Skerratt, J.; Whitehead, J.; Rizwi, F.; Parslow, J. Mechanisms driving estuarine water quality: A 3D biogeochemical model for informed management. *Estuar. Coast. Shelf Sci* **2013**, *135*, 33–45. [CrossRef]

9. Gilbes, F.; López, J.M.; Yoshioka, P.M. Spatial and temporal variations of phytoplankton chlorophyll *a* and suspended particulate matter in Mayagüez Bay, Puerto Rico. *J. Plankton Res.* **1996**, *18*, 29–43. [CrossRef]

10. García-Sais, J.R.; Williams, S.M.; Amirrezvani, A. Mortality, recovery, and community shifts of scleractinian corals in Puerto Rico one decade after the 2005 regional bleaching event. *PeerJ* **2017**, *5*, e3611. [CrossRef]

11. Wetz, M.S.; Yoskowitz, D.W. An 'extreme' future for estuaries? Effects of extreme climatic events on estuarine water quality and ecology. *Mar. Pollut. Bull.* **2013**, *69*, 7–18. [CrossRef]

12. Shi, W.; Wang, M. Observations of a Hurricane Katrina-induced phytoplankton bloom in the Gulf of Mexico. *Geophys. Res. Lett.* **2007**, *34*, 11607. [CrossRef]

13. Peierls, B.L.; Christian, R.R.; Paerl, H.W. Water quality and phytoplankton as indicators of hurricane impacts on a large estuarine ecosystem. *Estuaries* **2003**, *26*, 1329–1343. [CrossRef]

14. D'Sa, E.J.; Joshi, I.; Liu, B. Galveston Bay and coastal ocean optical-geochemical response to Hurricane Harvey from VIIRS ocean color. *Geophys. Res. Lett* **2018**, *45*, 10579–10589. [CrossRef] [PubMed]

15. Liu, B.; D'Sa, E.J.; Joshi, I.D. Floodwater impact on Galveston Bay phytoplankton taxonomy, pigment composition and photo-physiological state following Hurricane Harvey from field and ocean color (Sentinel-3A OLCI) observations. *Biogeosciences* **2019**, *16*, 1975–2001. [CrossRef]

16. Wachnicka, A.; Browder, J.; Jackson, T.; Louda, W.; Kelble, C.; Abdelrahman, O.; Stabenau, E.; Avila, C. Hurricane Irma's impact on water quality and phytoplankton communities in Biscayne Bay (Florida, USA). *Estuaries Coasts* **2020**, *43*, 1217–1234. [CrossRef]

17. Schaeffer, B.A.; Myer, M.H. Resolvable estuaries for satellite derived water quality within the continental United States. *Remote Sens. Lett.* **2020**, *11*, 535–544. [CrossRef]

18. Gilbes, F.; Armstrong, R.A.; Webb, R.M.; Müller-Karger, F.E. SeaWifs helps assess hurricane impact on phytoplankton in Caribbean Sea. *EOS Trans. Am. Geophys. Union* **2001**, *82*, 529–533. [CrossRef]

19. Miller, P.W.; Kumar, A.; Mote, T.L.; Moraes, F.D.S.; Mishra, D.R. Persistent hydrological consequences of Hurricane María in Puerto Rico. *Geophys. Res. Lett.* **2019**, *46*, 1413–1422. [CrossRef]

20. Hernández, W.J.; Ortiz-Rosa, S.; Armstrong, R.A.; Geiger, E.F.; Eakin, C.M.; Warner, R.A. Quantifying the effects of Hurricanes Irma and María on coastal water quality in Puerto Rico using moderate resolution satellite sensors. *Remote Sens.* **2020**, *12*, 964. [CrossRef]

21. Eakin, C.M.; Morgan, J.A.; Heron, S.F.; Smith, T.B.; Liu, G.; Alvarez-Filip, L.; Baca, B.; Bartels, E.; Bastidas, C.; Bouchon, C.; et al. Caribbean corals in crisis: Record thermal stress, bleaching, and mortality in 2005. *PLoS ONE* **2010**, *5*, e13969. [CrossRef]

22. Weil, E.; Croquer, A.; Urreiztieta, I. Temporal variability and impact of coral diseases and bleaching in La Parguera, Puerto Rico from 2003–2007. *Caribb. J. Sci.* **2009**, *45*, 221–246. [CrossRef]

23. Pittman, S.J.; Hile, S.D.; Jeffrey, C.F.G.; Clark, R.; Woody, K.; Herlach, B.D.; Caldow, C.; Monaco, M.E.; Appeldoorn, R. Coral reef ecosystems of Reserva Natural La Parguera (Puerto Rico): Spatial and temporal patterns in fish and benthic communities (2001–2007). In *NOAA Technical Memorandum NOS NCCOS 107*; NOAA: Silver Spring, MD, USA, 2010; p. 202.

24. Hedley, J.D.; Roelfsema, C.M.; Phinn, S.R.; Mumby, P.J. Environmental and sensor limitations in optical remote sensing of coral reefs: Implications for monitoring and sensor design. *Remote Sens.* **2012**, *4*, 271–302. [CrossRef]

25. Cooper, T.; Fabricius, K.E. Coral-based indicators of changes in water quality on nearshore coral reefs of the Great Barrier Reef. In *Unpublished Report to Marine and Tropical Sciences Research Facility*; Reef and Rainforest Research Centre Limited: Cairns, Australia, 2007; p. 31.

26. Rogers, C.S. Responses of coral reefs and reef organisms to sedimentation. *Mar. Ecol. Prog. Ser.* **1990**, *62*, 185–202. [CrossRef]

27. Scheffer, M.; Carpenter, S.; Foley, J.A.; Folke, C.; Walker, B. Catastrophic shifts in ecosystems. *Nature* **2001**, *413*, 591–596. [CrossRef]

28. Daniel, J. Chemical characterization and cycling of dissolved organic matter. In *Biogeochemistry of Marine Dissolved Organic Matter*, 2nd ed.; Dennis, A., Hansell, A., Carlson, C.A., Eds.; Academic Press: London, UK, 2015; Chapter 2; pp. 21–63. ISBN 9780124059405. [CrossRef]

29. Slonecker, E.T.; Jones, D.K.; Pellerin, B.A. The new Landsat 8 potential for remote sensing of colored dissolved organic matter (CDOM). *Mar. Pollut. Bull.* **2016**, *107*, 518–527. [CrossRef]

30. Wei, J.; Lee, Z.; Ondrusek, M.; Mannino, A.; Tzortziou, M.; Armstrong, R. Spectral slopes of the absorption coefficient of colored dissolved and detrital material inverted from UV-visible remote sensing reflectance. *J. Geophys. Res. Oceans* **2016**, *121*, 1953–1969. [CrossRef]

31. Aurin, D.; Mannino, A.; Lary, D.J. Remote sensing of CDOM, CDOM spectral slope, and dissolved organic carbon in the global ocean. *Appl. Sci.* **2018**, *8*, 2687. [CrossRef]

32. IOCCG Protocol Series. Inherent optical property measurements and protocols: Absorption coefficient. In *IOCCG Ocean Optics and Biogeochemistry Protocols for Satellite Ocean Colour Sensor Validation*; Neeley, A.R., Mannino, A., Eds.; IOCCG: Dartmouth, NS, Canada, 2018; Volume 1.0.

33. Vodacek, A.; Blough, N.V.; DeGrandpre, M.D.; Peltzer, E.T.; Nelson, R.K. Seasonal variation of CDOM and DOC in the Middle Atlantic Bight: Terrestrial inputs and photooxidation. *Limnol. Oceanog.* **1997**, *42*, 674–686. [CrossRef]

34. Zhang, Y.; Liu, M.; Qin, B.; Feng, S. Photochemical degradation of chromophoric-dissolved organic matter exposed to simulated UV-B and natural solar radiation. *Hydrobiologia* **2009**, *627*, 159–168. [CrossRef]

35. Castillo, C.D.; Miller, R.L. On the use of ocean color remote sensing to measure the transport of dissolved organic carbon by the Mississippi River plume. *Remote Sens. Environ.* **2008**, *112*, 836–844. [CrossRef]

36. Mannino, A.; Russ, M.E.; Hooker, S.B. Algorithm development and validation for satellite-derived distributions of DOC and CDOM in the U.S. Middle Atlantic Bight. *J. Geophys. Res.* **2008**, *113*, C07051. [CrossRef]

37. D'Sa, E.J.; Miller, R.L. Bio-optical properties in waters influenced by the Mississippi River during low flow conditions. *Remote Sens. Environ.* **2003**, *84*, 538–549. [CrossRef]

38. Kirk, J.T.O. *Light and Photosynthesis in Aquatic Ecosystems*; Cambridge University Press: Cambridge, UK, 2011.

39. Bauer, L.J.; Edwards, K.; Roberson, K.K.W.; Kendall, M.S.; Tormey, S.; Battista, T.A. *Shallow-Water Benthic Habitats of Southwest Puerto Rico*; NOAA Technical Memorandum, NOAA NOS NCCOS 155: Silver Spring, MD, USA, 2012; p. 39.

40. Bonekamp, H.; Montagner, F.; Santacesaria, V.; Nogueira, L.C.; Wannop, S.; Tomazic, I.; O'Carroll, A.; Kwiatkowska, E.; Scharroo, R.; Wilson, H. Core operational Sentinel-3 marine data product services as part of the Copernicus Space Component. *Ocean Sci.* **2016**, *12*, 787–795. [CrossRef]

41. Morel, A.; Huot, Y.; Gentili, B.; Werdell, P.J.; Hooker, S.B.; Franz, B.A. Examining the consistency of products derived from various ocean color sensors in open ocean (Case 1) waters in the perspective of a multi-sensor approach. *Remote Sens. Environ.* **2007**, *111*, 69–88. [CrossRef]

42. Doerffer, R.; Schiller, H. The MERIS Case 2 water algorithm. *Int. J. Remote Sens.* **2007**, *28*, 517–535. [CrossRef]

43. Maritorena, S.; Siegel, D.A. Consistent merging of satellite ocean colour data sets using a bio-optical model. *Remote Sens. Environ.* **2005**, *94*, 429–440. [CrossRef]

44. Maritorena, S.; D'Andon, O.H.F.; Mangin, A.; Siegel, D.A. Merged satellite ocean color data products using a bio-optical model: Characteristics, benefits and issues. *Remote Sens. Environ.* **2010**, *114*, 1791–1804. [CrossRef]

45. Ciotti, A.M.; Bricaud, A. Retrievals of a size parameter for phytoplankton and spectral light absorption by colored detrital matter from water-leaving radiances at SeaWiFS channels in a continental shelf region off Brazil. *Limnol. Oceanogr. Methods* **2006**, *4*, 237–253. [CrossRef]

46. Miller, R.L.; Belz, M.; Del Castillo, C.; Trzska, R. Determining CDOM absorption spectra in diverse coastal environments using a multiple pathlength, liquid core waveguide system. *Cont. Shelf Res.* **2002**, *22*, 1301–1310. [CrossRef]

47. Ondrusek, M.E.; Stengel, E.; AmpoUo, M.; Goode, W.; Ladner, S.; Feinholz, M. Validation of ocean color sensors using a profiling hyperspectral radiometer. In Proceedings of the SPIE—The International Society for Optical Engineering, Baltimore, MD, USA, 23 May 2014; Volume 9111, p. 91110Y. [CrossRef]

48. Austin, R.W.; Petzold, T.J. The determination of the diffuse attenuation coefficient of sea water using the coastal zone color scanner. In *Oceanography from Space. Marine Science*; Gower, J.F.R., Ed.; Springer: Boston, MA, USA, 1981; Volume 13. [CrossRef]

49. Corredor, J.E.; Morell, J.M. Seasonal variation of physical and biogeochemical features in eastern Caribbean Surface water. *J. Geophys. Res.* **2001**, *106*, 4517–4525. [CrossRef]

50. Gilbes, F.; Armstrong, R.A. Phytoplankton dynamics in the eastern Caribbean Sea as detected with space remote sensing. *Int. J. Remote Sens.* **2004**, *25*, 1449–1453. [CrossRef]

51. López, R.; López, J.M.; Morell, J.; Corredor, J.E.; Del Castillo, C.E. Influence of the Orinoco River on the primary production of eastern Caribbean surface waters. *J. Geophys. Res. Oceans* **2013**, *118*, 4617–4632. [CrossRef]

52. Jones, R.; Giofre, N.; Luter, H.M.; Neoh, T.L.; Fisher, R.; Duckworth, A. Responses of corals to chronic turbidity. *Sci. Rep.* **2020**, *10*, 4762. [CrossRef] [PubMed]

53. Ateweberhan, M.; Feary, D.A.; Keshavmurthy, S.; Chen, A.; Schleyer, M.H.; Sheppard, C.R. Climate change impacts on coral reefs: Synergies with local effects, possibilities for acclimation, and management implications. *Mar. Pollut. Bull.* **2013**, *74*, 526–539. [CrossRef]

54. Bessell-Browne, P.; Negri, A.P.; Fisher, R.; Clode, P.L.; Duckworth, A.; Jones, R. Impacts of turbidity on corals: The relative importance of light limitation and suspended sediments. *Mar. Pollut. Bull.* **2017**, *117*, 161–170. [CrossRef] [PubMed]

55. Fourney, F.; Figueiredo, J. Additive negative effects of anthropogenic sedimentation and warming on the survival of coral recruits. *Sci. Rep.* **2017**, *7*, 12380. [CrossRef]

56. Bessette-Kirton, E.K.; Cerovski-Darriau, C.; Schulz, W.H.; Coe, J.A.; Kean, J.W.; Godt, J.W.; Thomas, M.A.; Hughes, K.S. Landslides triggered by Hurricane María: Assessment of an extreme event in Puerto Rico. *GSA Today Geol. Soc. Am.* **2019**, *29*, 4–10. [CrossRef]

57. Ortiz-Rosa, S. Optical Properties and Photochemical Response of Colored Dissolved Organic Matter (CDOM) at Jobos Bay National Estuarine Research Reserve (JOBANERR), Puerto Rico. Master's Thesis, Department of Marine Sciences, University of Puerto Rico, Mayagüez, Puerto Rico, 2010.

58. Edmunds, P.J.; Tsounis, G.; Boulon, R.; Bramanti, L. Acute effects of back-to-back hurricanes on the underwater light regime of a coral reef. Short Article. *Mar. Biol.* **2019**, *166*, 20. [CrossRef]

59. Edmunds, P.J.; Tsounis, G.; Boulon, R.; Bramanti, L. Long-term variation in light intensity on a coral reef. *Coral Reefs* **2018**, *37*, 955–965. [CrossRef]

Publisher's Note: MDPI stays neutral with regard to jurisdictional claims in published maps and institutional affiliations.

Article

The Role of Mean Sea Level Annual Cycle on Extreme Water Levels Along European Coastline

Tomás Fernández-Montblanc [1,2,*], Jesús Gómez-Enri [3] and Paolo Ciavola [2]

[1] Department of Earth Sciences, University of Cádiz, Avda. República Saharaui s/n, 11510 Puerto Real, Cádiz, Spain

[2] Department of Physics and Earth Sciences, Universitá Degli Studi di Ferrara, Via Saragat, 1, 44122 Ferrara, Italy; cvp@unife.it

[3] Department of Applied Physics, University of Cádiz, Avda. República Saharaui s/n, 11510 Puerto Real, Cádiz, Spain; jesus.gomez@uca.es

* Correspondence: tomas.fernandez@uca.es

Received: 10 September 2020; Accepted: 12 October 2020; Published: 18 October 2020

Abstract: The knowledge of extreme total water levels (ETWLs) and the derived impact, coastal flooding and erosion, is crucial to face the present and future challenges exacerbated in European densely populated coastal areas. Based on 24 years (1993–2016) of multimission radar altimetry, this paper investigates the contribution of each water level component: tide, surge and annual cycle of monthly mean sea level (MMSL) to the ETWLs. It focuses on the contribution of the annual variation of MMSL in the coastal flooding extreme events registered in a European database. In microtidal areas (Black, Baltic and Mediterranean Sea), the MMSL contribution is mostly larger than tide, and it can be at the same order of magnitude of the surge. In meso and macrotidal areas, the MMSL contribution is <20% of the total water level, but larger (>30%) in the North Sea. No correlation was observed between the average annual cycle of monthly mean sea level (AMMSL) and coastal flooding extreme events (CFEEs) along the European coastal line. Positive correlations of the component variance of MMSL with the relative frequency of CFEEs extend to the Central Mediterranean (r = 0.59), North Sea (r = 0.60) and Baltic Sea (r = 0.75). In the case of positive MMSL anomalies, the correlation expands to the Bay of Biscay and northern North Atlantic (at >90% of statistical significance). The understanding of the spatial and temporal patterns of a combination of all the components of the ETWLs shall improve the preparedness and coastal adaptation measures to reduce the impact of coastal flooding.

Keywords: storm surge; coastal flooding; marine storms; natural hazards; steric-effect; satellite altimetry

1. Introduction

Coastal areas, prone to be flooded in the case of extreme water levels, are mainly low-elevation territories. In addition, the increase in subsidence rates by anthropogenic actions such as sediment supply reduction by rivers, soil compaction by changes in land use [1], as well as extraction of groundwater [2] or natural gas [3] can exacerbate the vulnerability of coastal areas. Marine flooding threatens coastal areas, causing human casualities and large socio-economic impacts [4]. This is more critical in densely populated zones with a restricted or inadequate adaptive capacity [5].

The European coastline is a densely populated area. In 2011, almost 205 million people (>40% of the European population) lived in coastal regions (<50 km from the sea), and on average, in each country with a coastal border, 36% of the population lived within 5 km from the sea [6]. Moreover, there is a large historical record of marine flooding along the European coast [7–9], either for the North Sea [10]; Bay of Biscay [11]; North Atlantic, Mediterranean Sea or Adriatic sea among other locations [12]. Protection measures have been taken in the last few decades, increasing preparedness for extreme water level impact along the European coast. These actions have reduced the impact and consequences of coastal flooding generated by extreme events (hereafter, CFEEs) along the European coast. Indeed, despite an increase in the exposure in coastal areas, there is a significant decreasing trend in flood fatalities and economic losses for the period 1950–2016 [13]. An outstanding example is the North Sea floods of 1953 and 2013 (storm Xavier), very similar in magnitude but with very different impacts on infrastructures and population. A considerable decline of the damage was observed in the latter [14]. However, the extreme water level is expected to increase in the next century, by the contribution of the mean sea level rise [15,16], and changes in extreme storm surge and wave characteristics [17]. On the other hand, remarkable growth in coastal risk is also expected, associated with socio-economic coastal development [18]. Therefore, there is a continuous necessity for the monitoring and improvement of the forecasting and knowledge of extreme sea-level events and the driven impact on coastal areas to face the present and future challenges.

Traditionally, Extreme Total Water Level (ETWL, henceforth) has been analysed as the sum of tidal level and non-tidal residual. The non-tidal residual includes the so-called surge or meteorological contribution (inverse barometer effect and wind setup), and the non-linear interaction between surges and tides [19,20]. The non-tidal residual can also contain the wave set-up contribution in coastal areas [21], resonance in enclosed basin [22] or contribution of the river runoff in estuaries during extreme discharges [23,24].

ETWLs are dominated by high-frequency signals (tides, surges, waves set-up and run-up), but low-frequency contributions, such as the annual cycle of sea surface height variation and associated anomalies need to be considered too. The annual cycles of the Monthly Mean Sea Level (MMSL, hereinafter) can induce sea-level variations, ranging from few centimetres to up to 0.3 m in some regions (i.e., the Gulf of Carpentaria between Australia and New Guinea) [25]. Several processes and their seasonal variability drive the annual cycle of sea surface height. Thus, the water mass addition/removal from the oceans is a major forcing of the global ocean mean sea level (MSL) variation [26], interannual variability is critical over shelf seas [27], while seasonal variability can dominate in shallower regions [28]. Freshwater runoff contributes also to the annual cycle of MSL [29], and might become dominant near the coast [30]. The effect of river discharge is limited to the areas influenced by the river mouth [31,32]. In the open ocean the annual cycle of MSL is controlled by changes in the density of the water column through the so-called steric component, dominating the sea level variability at annual timescales in the North Atlantic and in the Mediterranean Sea [33]. The steric contribution is mainly driven by the thermal expansion/contraction of the water column (thermos-steric component) associated with changes in temperature of the upper layer of the ocean; the haline expansion/contraction due to salinity changes (halo-steric component) becomes less important [34]. Although the relevance of density changes in sea surface height variability is proportional to the water column depth and can be predominant in the open ocean, it can affect by remote contribution shallower areas [35].

Tide gauges have been used for years for extreme value analysis [36,37]. Some limitations (spatial and temporal coverage) are inherent to in situ measurements [38]. Tide gauge data are also influenced by isostatic adjustment and topographic effect in many locations, hampering the interpretation of sea-level records [39].

Satellite altimetry provides homogeneous and accurate sea level measurements over the open ocean. Apart from data assimilation in the forecast system (e.g., [40]), these measurements have been used for many applications including the contribution of the MMSL seasonal cycle to the extreme water levels [41]. Altimetry observations are available around the world ocean, but the accuracy decreases in coastal zones by land and calm water contamination in the radar footprint and the bad characterisation of some of the range/geophysical corrections [42]. The use of radar altimetry to capture peaks of ETWLs during CFEEs can be limited by its low temporal resolution. Recent studies over Europe [43] show that if two or more satellites (multimission gridded products) are available, more than 90% of the ETWLs events might be captured.

The combination of low-frequency signals (the annual cycle of the sea level variation) with shorter timescale phenomena, such as surges or tides, can contribute to an increase in the ETWLs during CFEEs [44]. Thus, a better characterisation of the high/low-frequency signals can improve our knowledge of the flood risk in coastal areas [45,46]. The main aim of this work was to analyse the contribution of the seasonal cycle of the MMSL (derived from multimission radar altimetry) to the ETWLs during the period 1993–2016. We also analysed the high-frequency signals (surges and tides) contribution to the ETWLs. The combined effect of these signals might result in large flooding with associated impacts on coastal areas. This study focused on the comparison of ETWLs detected with satellite altimetry with a coastal extreme storm impact database at a pan-European scale. The paper is structured as follows; Section 2 describes the methodological approach and dataset used. Section 3 gives the obtained results in terms of ETWLs and its comparison with the storm impact database; in Section 4, the results are discussed and compared with previous studies. Finally, the main conclusions are summarised in Section 5.

2. Datasets and Methodology

2.1. Sea Level Datatasets

The altimeter dataset used was the GLOBAL OCEAN ALONG-TRACK L3 SEA SURFACE HEIGHTS REPROCESSED from CMEMS (Copernicus Marine Environment Monitoring Service) [47]. This product was derived by the DUACS (Data Unification and Altimeter Combination System) multimission altimeter data processing system [48] to provide a consistent, cross-calibrated and homogeneous data for all the altimeter missions: Topex-Poseidon; Topex-Poseidon (interleaved orbit); Jason-1; Jason-1 (interleaved orbit); Jason-1 (geodetic orbit); OSTM/Jason-2; OSTM/Jason-2 (interleaved); Jason-3; Sentinel-3A; ERS-1; ERS-2; Envisat; Envisat (extended phase); Geosat Follow On; CryoSat-2; SARAL; SARAL-DP; HY-2A; HY-2A (geodetic orbit).

The along-track product obtained from CMEMS was the reprocessed Sea Level Anomaly (SLA). Instead of using the fully 1 Hz posting rate (~7 km spatial resolution), we used the filtered and subsampled SLA included in the along-track products of CMEMS. It reduces the residual noise and small scale signals with a posting rate of 0.5 Hz (about 14 km distance between successive measurements), and temporal resolution ranging between 10 and 35 days depending on satellite mission (see [49] for more details). Along-track SLA (referenced to a mean sea surface), includes a set of corrections in order to reduce instrumental noise, range (ionospheric, dry and wet tropospheric effects, and sea state bias correction), and geophysical corrections (tides, inverse barometer and high frequency (<20 days) wind and pressure effects). The inverse barometer and high-frequency wind signal of the atmospheric forcing were removed through the so-called Dynamic Atmospheric Correction (DAC) produced by CLS (Collecte Localisation Satellites) using the Mog2D model [50]. The ocean tide was removed using the FES2014 model (including S1 and S2 components) [51].

2.2. Storm Impact Database

The CFEEs were analysed using the historic and recent coastal flooding extreme events along the European coastline. The database used integrates different systematic coastal flooding and coastal impact databases available at pan-European scale:

- Pan-European HANZE database [8] from 1870 to 2016: 1564 flooding events were recorded including river floods and flash floods. A total of 77 events classified as coastal and compound events (river and coastal contributions to the floodings) were selected.
- Coastal floodings in the United Kingdom [7] from 1915 to 2016: 329 events.
- The RISC-KIT storm impact database for European coastlines [9] from 1806 to 2016: with 298 events.

Information on the impact (if available), location and time of 532 events were recorded in the analysed time period (1993–2016). The database contains records in specific locations and can include more than one event separated by time.

The geographical location of the events was normalised and referred to the European Union statistical regions NUTS3 (Nomenclature of Territorial Units for Statistics Level 3) version 2010.

2.3. Methods

We obtained the total water level (TWL) as the contribution of three components (Equation (1)):

$$TWL = MMSL + SSL + TIDE \tag{1}$$

where MMSL is monthly mean sea level; SSL includes the sea surface variations induced by the meteorological forcing. It includes the contribution of wind and pressure effect on the water level (the so-called surge); and TIDE is the contribution of astronomical tide. The study area focused on the pan-European area covering 32°W–42°E longitude, and 27–74°N latitude (Figure 1). The area of interest was divided in 1° × 1° tiles and time series were obtained by grouping the along-track SLA data inside each tile.

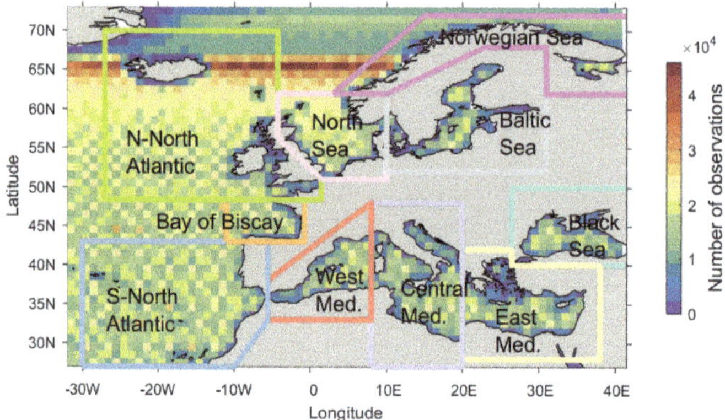

Figure 1. Study area with the 10 oceanographic regions defined for the analysis. The colour scale indicates the number of altimetry observations in each 1° × 1° latitude and longitude cell for the period: 1993–2016.

The events where TWL exceeds the 95th percentile of the TWL have been considered as the extreme total water levels (ETWLs).

2.3.1. MMSL, AMMSL and MSL Anomalies

As mentioned, the study area was divided into regular grids of 1° × 1°. Then, monthly means were constructed with the along-track filtered SLA data inside the grids for the time period analysed; finally, the time series were linearly detrended to obtain the monthly Mean Sea Level (MMSL), which includes the steric and mass components. The MMSL was computed as the monthly mean for each year. The standard deviation (σ) of MMSL for each month representing the interannual variability was also obtained. The average monthly mean sea level (AMMSL), the so-called climatology or average annual cycle, was calculated as the interannual average of the monthly mean sea level for the whole period. Finally, the MSL anomalies (deviation in MSL respect to the mean annual cycle) were estimated subtracting the month value of the AMMSL time average (1993–2016) to the MMSL for the corresponding month following (Equation (2)):

$$MSL\ anomalies = MMSL - AMMSL \tag{2}$$

2.3.2. SSL

The meteorological component was computed following (Equation (3)):

$$SSL = SLA_{DAC} - MMSL \tag{3}$$

where SLA_{DAC} is the monthly mean of the SLA uncorrected by DAC; that is restoring the inverse barometer and high frequency (<20 days) wind and pressure effects removed during the product delivery. The DAC product is available at regular grids ($0.5° \times 0.5°$), and 6 hours of temporal resolution. They were interpolated in time and space to match the altimeter dataset, and subtracted to the along-track SLA. Then, the monthly means were computed and the time series were also detrended (SLA_{DAC}).

2.3.3. TIDE

The ocean tide was calculated using the t_tide package [52] including nodal corrections, and using the amplitudes and phases of 30 tidal components of the FES2014 model (see [51] for further details). The amplitude and phase are provided in a regular grid of $0.0625° \times 0.0625°$ and were interpolated in space to match the altimeter dataset.

The amplitude of the TWL is highly dependent on the phase lag between the surges and the tides. In addition to this, the co-occurrence of surges and spring tides might have a major impact on the floods hitting the coastal area. We analysed the relevance of AMMSL with respect to the neap-spring tidal range calculated from M2 and S2 tidal constituents ([M2-S2 M2+S2]). It is calculated according to Equation (4)

$$Relevance = 100 \cdot \frac{AMMSL}{[M2 - S2 \ M2 + S2]} \tag{4}$$

2.3.4. Correlation of Seasonal MSL with Storm Impact Database

The relationship between the spatio-temporal pattern of MMSL, SSL and TIDE, and the areas affected by coastal flooding registered in the storm impact database was analysed along the European coastline. According to the methodology outlined in the previous section, the MMSL, SSL and TIDE contribution was calculated and assigned for each measurement in the altimetry dataset located in the $1° \times 1°$ tiles closest to the coastline of each region avoiding measurements affected by land contamination.

Afterward, the linear correlation coefficient was calculated between the average monthly fraction of each component variance in ETWLs in the oceanographic region and the relative monthly frequency of the number of CFEEs registered in the storm impacts database at each oceanographic region. The variance of each component is expressed as a fraction of the ETWLs.

Similarly, it was calculated the correlation coefficient between AMMSL and monthly frequency of the storm impact registered on each region.

The relationship of MSL anomalies in the closest $1° \times 1°$ and the storm event registered in the database was assessed through a t-test (alpha = 0.05) to evaluate the hypothesis that positive MSL anomalies are correlated with the storm event recorder. If the CFEEs and MSL anomalies are independent, mean value of MSL anomalies corresponding to the CFEEs should be zero, whereas a positive correlation will produce a mean value of MSL anomalies > 0.

3. Results

3.1. Characterization of the AMMSL and MMSL

The characterisation of AMMSL and MMSL was conducted using the full spatial coverage of the dataset in order to analyse the study area including deep ocean and areas closest to the European coast. The AMMSL is shown in Figure 2. The average annual cycle is not uniform in time and magnitude in the study area. The Mediterranean Sea and the Atlantic areas show minimum in late winter/early spring (−0.12 in the Mediterranean Sea and −0.05 m in the Atlantic) with the maximum of the annual cycle in late summer/early autumn (0.12 m in the Mediterranean and up to 0.07 m along the continental shelf in the N-North Atlantic). Some exceptions are observed in the coastal zones of the United Kingdom. The Bay of Biscay and the North Sea show a similar pattern with the minimum in spring and maximum of AMMSL in late summer/early autumn. In the North Sea, the North-Eastern coast and the German Bight present values > 0.10 m, where positive anomalies extend from September up to January. In the case of the Black Sea, the minimum/maximum (±0.10 m) is given in autumn/spring. Finally, the Baltic Sea shows the minimum/maximum (±0.12 m) in spring/autumn-early winter. Intensifications of positive AMMSL (>0.12 m) are observed in the gulfs of Bothnia and Finland during December and January. The Norwegian Sea is characterised by variations around ±0.06 m with minimum/maximum in late autumn/late spring, and intensification of positive AMMSL (0.08 m) in coastal areas.

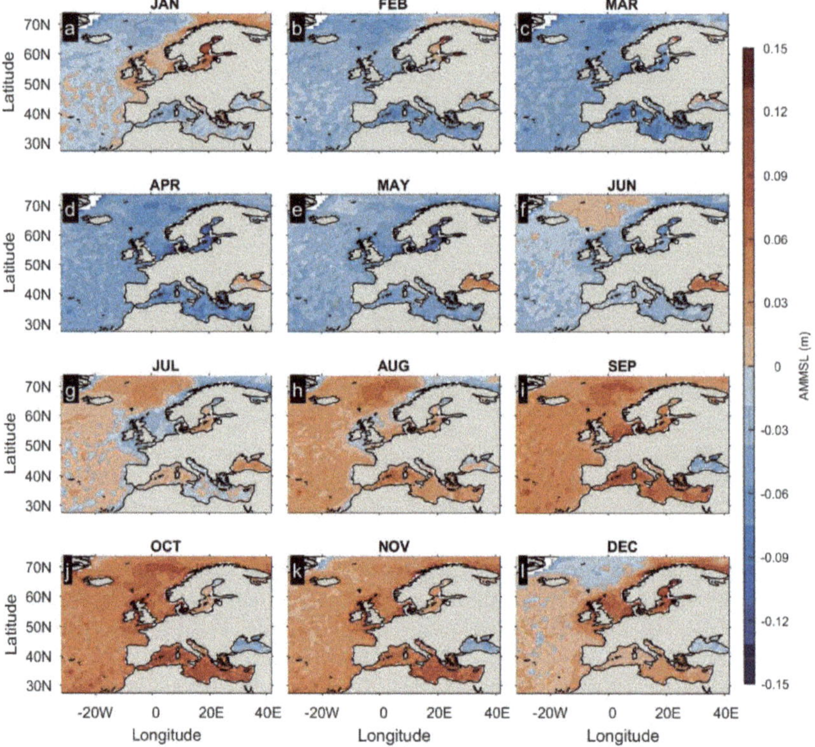

Figure 2. Seasonality of the annual cycle ((**a–l**) January–December) of the average monthly mean sea level (AMMSL) in the study area from 1993 to 2016. Warm/cold colour indicates positive/negative values.

Figure 3 illustrates the standard deviation (σ) of the annual cycle for the analysed period as an indicator of the interannual variability of the MMSL. The values are small (σ < 0.03 m) all around the year and oceanographic regions. Some exceptions with bigger standard deviations are observed in the Baltic, Black and North Seas, especially during autumn/winter seasons. In the North Sea, σ is >0.15 m in the German Bight. The largest σ (up to >0.18 m in February) are in areas located in the head of the Gulf of Bothnia, Finland and the Eastern coast of the Baltic Sea. The deeper ocean of South/North Atlantic shows σ slightly larger (>0.04 m) with respect to the mean value.

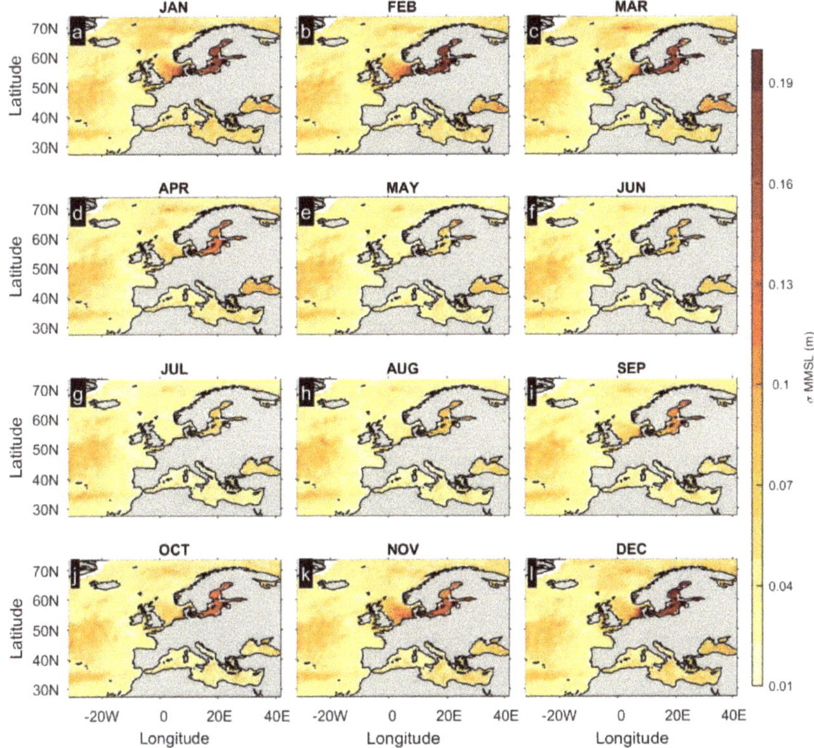

Figure 3. Seasonality of the standard deviation (σ) of monthly mean sea level (MMSL) ((**a–l**) January–December).

The contribution of the AMMSL to the annual cycle obtained with the monthly maximum SLA uncorrected by meteorological forcing (SLA$_{DAC}$) was also analysed. We estimated the percentage of this contribution (Figure 4). The steric and mass components account for almost 45% of the uncorrected seasonal cycle in late autumn/spring in most of the oceanographic regions with the exception of the Black Sea, North Sea, Bay of Biscay and the deeper ocean of South/North Atlantic. Contributions of about 30–35% are observed in the Mediterranean Sea and Black Sea in late summer/autumn.

Regarding the relevance of the AMMSL with respect to the neap-spring tidal range calculated ([M2-S2 M2 + S2]), the results (not shown here) indicate that the AMMSL is more important than the neap-spring range in microtidal areas (Mediterranean Sea (excepting Central Med.), Black Sea and Baltic Sea). In mesotidal (S-North Atlantic and Norwegian Sea) and some macrotidal (Bay of Biscay, N-North Atlantic and Eastern coast of the North Sea) areas the contribution is smaller than 10%.

Figure 5 gives the characterisation by oceanographic region of the monthly AMMSL, its magnitude relative to SLA$_{DAC}$ and to the range of neap-spring tide. The average range of the AMMSL during the annual cycle and its standard deviation are shown in Figure 5a. The largest seasonal range and variability was observed in the Baltic Sea (>±0.10 m), followed by the Mediterranean Sea, and the Norwegian Sea. The weakest variation in the AMMSL was found in the Black Sea, S-North Atlantic, Bay of Biscay and N-North Atlantic. The weight of the AMMSL with respect to the non-tidal residual (SLA$_{DAC}$) (Figure 5b) is, on average, below 50% in all the oceanographic regions ranging from ~40% (Mediterranean Sea) to ~20% in the Bay of Biscay, N-North Atlantic, North Sea and Norwegian Sea. The average ratio of AMMSL and spring-neap range (Figure 5c) points out the major importance of

AMMSL in microtidal areas: Black Sea, Mediterranean Sea and Baltic Sea. This ratio is still high in the North Sea (>60%) and below 10% in the Bay of Biscay.

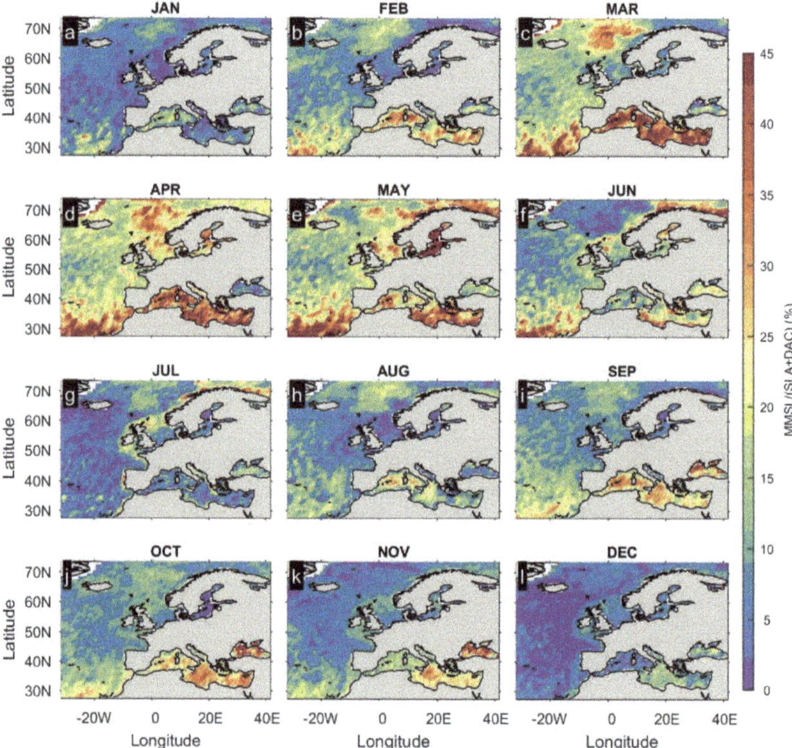

Figure 4. Seasonality of the annual contribution of the seasonal cycle of the MSL to the uncorrected monthly maximum MSL (Sea Level Anomaly (SLA) + Dynamic Atmospheric Correction (DAC)) ((**a–l**) January–December).

The ETWLs are analysed in detail in the coastal area using a subset of the initial data. This subset covers only data contained in the closest $1° \times 1°$ tile to the coast. Figure 6 shows the contribution in terms of the variance of the three components (MMSL, SSL and TIDE) to these extremes. For a more comprehensive visualization, the fraction of components' variance was plotted as ternary plots: SSL is 100% in the bottom left vertex, MMSL is 100% in the upper vertex and TIDE is 100% in the bottom right; the opposite edge of each vertex gives 0% of the corresponding components' fraction. In the Black Sea (Figure 6a), the main component in the ETWLs is the SSL, with the MMSL contribution ranging from 0 to 50%; the TIDE fraction is < 10%. The Mediterranean Sea (Figure 6b–d) shows the larger variability in the contribution of the components, particularly in the West Med. In the East Med. (Figure 6b), ETWLs are characterised by 10–40% of MMSL contribution, 60–90% for SSL, and a smaller relevance of TIDE (<20%). In the Central Med. (Figure 6c), which includes the Adriatic Sea, TIDE is the main component (70–90%), the MMSL contribution is limited to <15%, and SSL is below 30%. In West Med. (Figure 6d) the TIDE contribution ranges from 40 to 60%, MMSL and SSL around 10–30% in the case of the higher ETWLs (red dots), whereas intermediate ETWLs (orange dots) present a wider contribution of MMSL. S-North Atlantic (Figure 6e) is characterised by 70–80% (TIDE), 20–30% (SSL), and about 10% (MMSL). The Bay of Biscay (f) is the oceanographic region where MMSL shows the minor contribution (<5%), so the extreme values are most of the time a combination of SSL (0–20%), and TIDE (80–100%). In N-North Atlantic (Figure 6g) and North Sea (Figure 6h), the MMSL contribution to ETWLs is limited to <10%

and <20% respectively. TIDE and SSL contribute to 20% and 80%, respectively in the N-North Atlantic. The North Sea shows scattered values for TIDE (20–95%) and SSL (0–80%) components. The Baltic Sea is the oceanographic region with the larger contribution of MMSL (20–50%), and minor contribution of SSL 50–80%. ETWLs in the Norwegian Sea are characterised by 80–95% (TIDE), 10–20% (SSL) and <10% (MMSL).

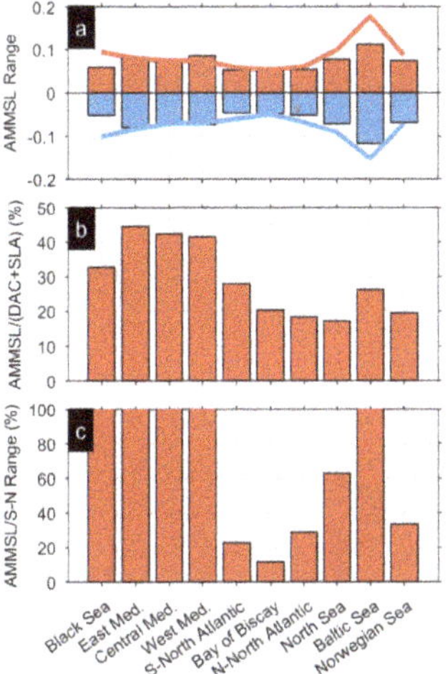

Figure 5. (**a**) Oceanographic region average of range of AMMSL (red/blue lines show the spatially averaged standard deviation of the interannual MMSL positive/negative values, respectively). (**b**) Oceanographic region average of the relative value (%) of AMMSL with respect to the SSL monthly maxima (SLA$_{DAC}$). (**c**) Oceanographic region average of the relevance (%) of mean AMMSL with respect to the neap-spring tidal range following Equation (3). (Exceedance of 100% were represented as 100% in order to facilitate the intercomparison with other regions).

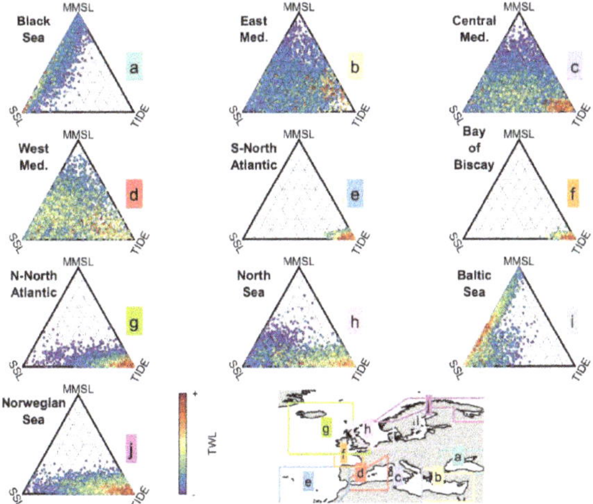

Figure 6. Fraction of components' variance in the extreme total water level (>95th percentile) along the coastline (closest 1° × 1° tile) of each oceanographic region (**a–j**) considering TIDE, SSL and MMSL. Warm/cold colour indicates bigger/smaller extreme total water levels (ETWLs).

3.2. Correlation of AMMSL and MMSL with Storm Impact Database

The relationship between the spatial pattern of AMMSL and the areas affected by coastal floodings registered in the storm impact database is illustrated in Figure 7. The East Med., S-North Atlantic and Norwegian Sea do not have any records in the storm database. The only record in the Black Sea is on the west coast, with the CFEEs mainly registered from December to March (Figure 7a–c,l). The higher frequency of storms is recorded in February (AMMSL < 0.03 m). Central and West Med. are characterised by a peak occurrence of CFEEs during November and December in the Adriatic, Liguria and Catalonia coasts (Figure 7k,l) running into positive AMMSL ([0.03–0.09 m]). The S-North Atlantic is represented only by the southern coast of Portugal, where CFEEs are mainly registered on January-February (Figure 7a,b) during negative AMMSL ([−0.05−−0.03 m]). In the Bay of Biscay, the higher percentage of CFEEs occur during February-March, when the seasonal cycle shows negative AMMSL (−0.05 m). In the N-North Atlantic, most of the CFEEs take place from December to February (Figure 7a,b,l, respectively), corresponding to AMMSL around 0.07 m, 0.05 m and −0.03 m respectively. On the coast of the North Sea, the CFEEs occur mostly in January (Figure 7a) during positive AMMSL (0.07 m), except on the south-eastern English coast. In German Bight, a higher frequency of the CFEEs occur in December, when AMMSL peaks up to 0.11 m. In the Baltic Sea, most CFEEs are observed during January and December in the western coast (Figure 7a,l) running into positive monthly AMMSL (0.06 m). Likewise, the largest frequency in the Gulf of Finland occurs during January concurring with maximum AMMSL (>0.15 m).

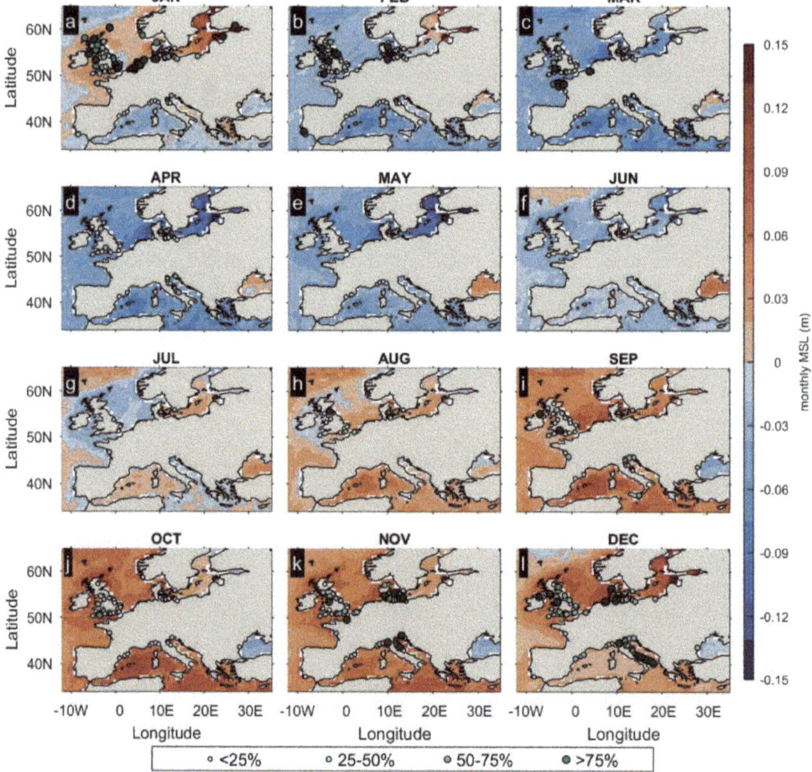

Figure 7. Spatial pattern of the relationship between the CFEEs and the AMMSL ((**a–l**) January–December). The dots indicate the position of the coastal storms and the colour intensity refers to their relative monthly frequency in the affected NUTS3 region.

The monthly average of the fraction of each component (SSL, TIDE and MMSL) variance in the ETWLs and the monthly frequency of the extreme events are illustrated in Figure 8. The correlation coefficient between each component and the relative monthly frequency of CFEEs is shown in Table 1. In the Black Sea and Baltic Sea, the main contributors to the ETWLs are SSL and MMSL. The CFEEs are observed from December to February coinciding with an important contribution of the MMSL (Figure 8a–i). The correlations are not significant in the Black Sea and significant (SSL and MMSL) in the Baltic Sea (Table 1). The contribution pattern of the three components is homogeneous in the Mediterranean (Figure 8b–d). Extreme events are observed from October to January (Central and West Med.). Correlations (Table 1) are significant for TIDE (Central and West Med) and MMSL (Central Med.). The remaining oceanographic areas show a major contribution of TIDE to the extremes. In the N-North Atlantic, North Sea and Norwegian Sea, the contribution of SSL is up to 30%. Significant correlations are observed in N-North Atlantic (SSL and TIDE), and North Sea (SSL, TIDE, and MMSL) (Table 1) where tidal contribution diminishes, increasing MMSL and SSL.

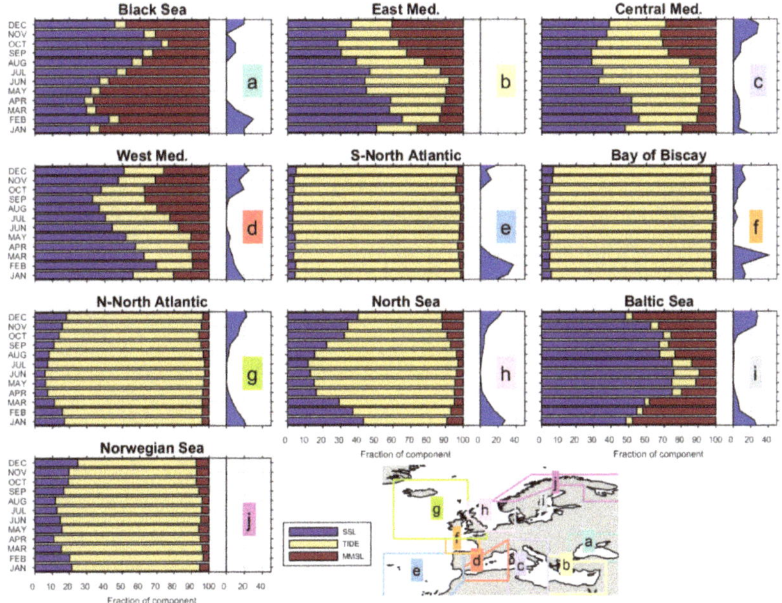

Figure 8. Average monthly fraction of each components' variance in the ETWL along the coastline (closest $1° \times 1°$ tile) of each oceanographic region (**a–j**). The monthly relative frequency of storm impacts at each area (blue polygon) is also shown.

Table 1. Rank correlation coefficient between the average monthly fraction of components' variance of SSL, TIDE and MMSL along the coastline (closest $1° \times 1°$ tile) of each oceanographic region and the relative monthly frequency of the storm events database. *p*-Value is given in brackets.

Region	Black Sea	Central Med.	West Med.	S-North Atlantic	Bay of Biscay	N-North Atlantic	NorthSea	BalticSea
SSL	−0.06 (0.86)	0.03 (0.94)	0.18 (0.57)	0.52 (0.08)	0.37 (0.24)	$0.97\ (3 \times 10^{-7})$	$0.90\ (6 \times 10^{-5})$	$−0.79\ (2 \times 10^{-3})$
TIDE	0.14 (0.66)	$−0.73\ (7 \times 10^{-3})$	$−0.78\ (3 \times 10^{-3})$	−0.16 (0.63)	−0.32 (0.32)	$−0.94\ (4 \times 10^{-6})$	$−0.88\ (2 \times 10^{-4})$	−0.54 (0.07)
MMSL	0.05 (0.87)	$0.73\ (7 \times 10{-3})$	0.29 (0.36)	−0.21 (0.52)	0.12 (0.70)	0.45 (0.14)	0.67 (0.02)	$0.77\ (4 \times 10^{-3})$

Figure 9 provides the temporal variability of the monthly AMMSL averaged along the coastline (closest $1° \times 1°$ tile) over each oceanographic region; the relative monthly frequency of CFEEs registered at each region is also shown. Table 2 gives the correlation coefficient between them. According to the results, no significant correlations were found between the AMMSL in coastal areas and CFEEs.

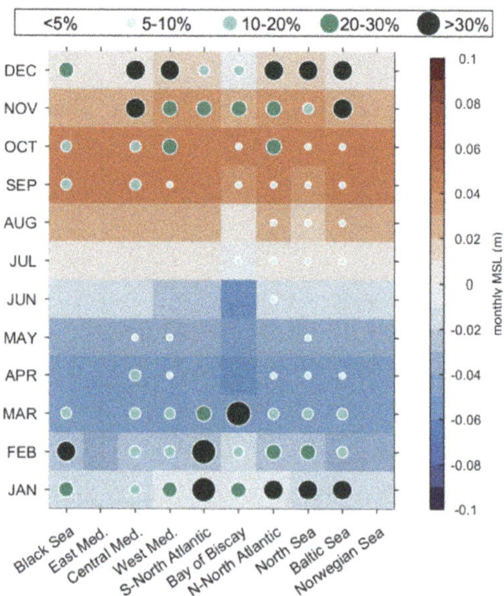

Figure 9. AMMSL along the coastline (closest 1° × 1° tile) averaged by oceanographic regions and monthly frequency of storm impact registered on each region (larger dot size indicate higher frequency).

Table 2. Rank correlation coefficient between the annual cycle ofregionally averaged AMMSL and relative monthly frequency of storm event in the database.

Region	Black Sea	East Med.	Central Med.	West Med.	S-North Atlantic	Bay of Biscay	N-North Atlantic	North Sea	Baltic Sea	Norwegian Sea
Corr. coefficient	−0.1	-	0.28	0.32	−0.22	−0.14	0.30	0.02	0.10	-
p-value	0.75	-	0.38	0.31	0.49	0.67	0.35	0.95	0.76	-

3.3. Correlation of Monthly MSL Anomalies with Storm Impact Database

The potential impact on coastal flooding extreme events derived from changes in the MMSL with respect to the AMMSL, that is the monthly MSL anomalies, was analysed. Figure 10 shows the annual variation of MSL anomalies (ordinate axis) along the European coastline (abscissa axis). Each column represents the data corresponding to the closest 1° × 1° tile to the coast and its location along the coastline is indicated by the ISO country code to identify the coastal region. Each row represents a monthly MSL anomaly for the analysed period. The CFEEs registered in the database are presented as black dots according to the temporal and spatial location of the event. There is a clear correlation between the dates of the CFEEs and positive anomalies of the MSL with the exception of West Med. (Figure 10d) and N-North Atlantic (Figure 10g). This is confirmed by the fact that the frequency curve of MSL anomalies spatially averaged (black curves in Figure 10) and anomalies during CFEEs registered (red curves) are different in the upper tail, indicating a larger number of events during the largest positive anomalies. In the Black Sea (Figure 10a) there were six events with four/two of them during positive/negative MSL anomalies. Similarly, in the Central Med. (Figure 10c) the events that occurred during positive anomalies (51) are almost twice those registered during negative anomalies (32). MSL anomalies are especially large (>0.12 m) in the Adriatic (Central Med.) where there is a most exhaustive record of CFEEs. S-North Atlantic (Figure 10e) and Bay of Biscay (Figure 10e) are under-represented in the storm database, but the storms coincide in time with positive anomalies.

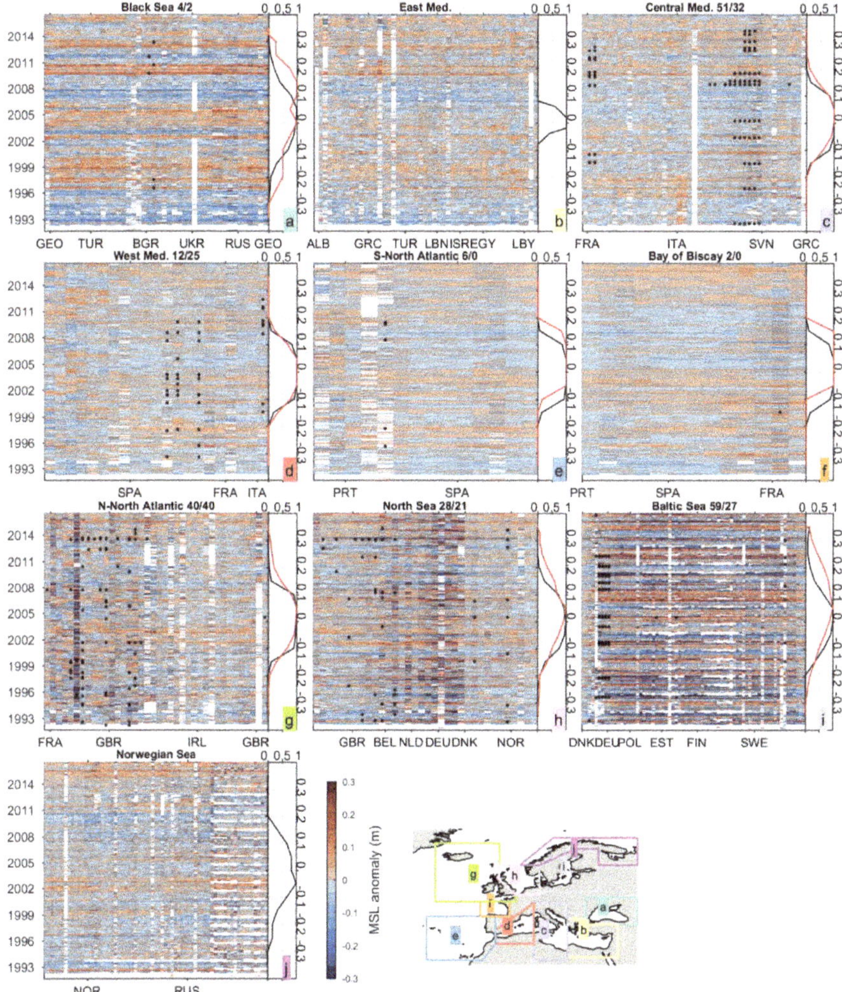

Figure 10. Monthly MSL anomalies along the European coastline (warm/cold colours indicate positive/negative MSL values) for each oceanographic region (**a–j**). The ratio of positive/negative anomalies for the CFEEs is also indicated. The ordinate axis represents the time and the abscissa axis represents the spatial variation along the coast indicated by ISO country code labels. Black dots indicate the date and position of extreme events registered in the database. The red line represents the relative frequency distribution of the monthly anomalies corresponding to the extreme events registered in the database and black line represents the spatially average of monthly anomalies distribution in the region.

In N-North Atlantic (Figure 10g) larger anomalies than ±0.1 m are registered in the English Channel and Irish Sea, but the same number of CFEEs are observed under positive/negative anomalies (40). In the North Sea (Figure 10h) the CFEEs are registered on the west and southwestern coast (GBR and BEL) with lower MSL anomalies than in the German Bight (>0.15 m). The Baltic Sea (Figure 10i) shows strong anomalies (most of them >0.15 m) in the Gulfs of Bothnia and Finland, giving the clearest correlation of positive anomalies and CFEEs recorded in the database.

Table 3 summarises the role of MSL anomalies in each oceanographic region through the results of a *t*-test (alpha = 0.05). The mean value of MSL anomalies is >0 in Central Med., S-North Atlantic, North

Sea and Baltic Sea, indicating a positive correlation (>95% of statistical significance). This positive correlation extends to the Bay of Biscay and N-North Atlantic with >90% of statistical significance.

Table 3. Results of the *t*-test (*p*-value in bracket) to check the hypothesis that the MSL anomaly data comes from a population with a mean greater than zero at the 0.05% significance level. A *t*-test = 1 indicates that the hypothesis is accepted.

Region	Black Sea	Central Med.	West Med.	S-North Atlantic	Bay of Biscay	N-North Atlantic	North Sea	Baltic Sea
t-test (0.05)	0 (0.28)	1 (3×10^{-4})	0 (0. 86)	1 (3×10^{-3})	0 (0.08)	0 (0.07)	1 (4×10^{-2})	1 (3×10^{-7})
Mean MSL anomaly	0.03	0.02	−0.01	0.04	0.04	0.02	0.03	0.07

4. Discussion

Recently, some studies based on tide gauges have analysed the ETWL component during extreme flooding events at the regional–local scale and its relative impact on coastal areas [53,54]. Our study, based on 24 years of satellite altimetry observations, focused on the contribution of the annual variation of MMSL to the ETWL, over the oceanographic regions around Europe. In some of these regions the annual cycle of the MMSL is markedly important as a driving contributor to the ETWLs.

4.1. Time–Space Variations of Seasonal MSL and Interannual Variability

In general, the AMMSL is in agreement with previous studies based on tide gauges and satellite altimeter data. The range of variation of the satellite-derived AMMSL is similar to the range observed by [55] in the Black Sea using tide gauge data, by [56] in the Mediterranean Sea (altimeter observations), and by [35] in the Gulf of Cadiz (tide gauges and altimeter data). We observe slightly larger values with respect to the observations made by [57] in the South and West coast of the Iberian peninsula (derived from tide gauges). The range and spatial pattern of amplification in the continental shelf in the Bay of Biscay are in line with the values reported by [58]. Further north, our results show the spatial pattern of the amplitude of the AMMSL annual cycle, increasing towards the northeastern coast in the German Bight in line with [59] (from altimeter observations).

The amplitude intensification observed from the Danish Straits to the head of the Gulfs of Finland and Bothnia was also noted by [60] using tidal gauge observations. This was also reported by [61] on the Polish coast. Finally, we found a good level of agreement between our results and those obtained by [62] in the Norwegian Sea.

The regional variations originate from different mechanisms. In the Black Sea, the AMMSL seasonal variations are dominated by freshwater balance [63]. In the Mediterranean Sea, the seasonal AMMSL is dominated by steric contribution being not negligible the mass induced by sea-level variation ([64]). The thermosteric effect is also dominant in the S-North Atlantic and South and West coast of the Iberian peninsula [35], Bay of Biscay and the N-North Atlantic. In the North Sea, the seasonal changes are mainly driven by wind, and the contribution of precipitation is not negligible during the autumn season [46]. The local steric contribution is smaller due to the shallow waters; however, long-term AMMSL variability could reflect the steric changes remotely forced [65]. In the Baltic Sea, the seasonal variation of AMMSL is primarily controlled by the direction of the prevailing wind and its role in the water exchange with the North Sea [60,66]. Moreover, seasonal variability is also influenced by river runoff and temperature [67,68].

The interannual variability of MMSL is stronger in the areas with larger amplitudes in the seasonal cycle (German Bight and Baltic Sea) (Figure 3). The variability in the North Sea is larger from December to March (Figure 3) as a result of the stronger atmospheric and meteorological forcing, as noted previously by [46]. The spatial pattern of the intensity of the MSL anomalies in coastal areas of N-North Atlantic and North Sea are in agreement with the results presented in [69]. In the Baltic Sea, the large interannual variation expands to most of the months (excluding July and August). This could be related to the domination of semi-annual variability during some periods [39]. In spite of the meteorological forcing, the sea ice cover (maximum in February and March) could contribute to the increase in interannual variability [66], along with river run-off, with maximum average and deviation values occurring from April to May [39].

4.2. Correlation of Monthly MSL Anomalies with Storm Impact Database

The assessment of the seasonal MMSL with respect to the monthly maximum SLA_{DAC} indicates a relevant contribution of MSL during winter and autumn, especially in semi-enclosed basins (Black Sea, Mediterranean Sea and Baltic Sea). Moreover, in those areas, the seasonal variation of AMMSL exceeds the neap-spring tidal range. Beyond these microtidal areas, the seasonal variation of AMMSL is also significant in the south and eastern coast of the North Sea. Those areas are characterised by the large contribution of the MMSL to the ETWLs in winter and particularly in autumn, during the seasonal peaks of MMSL. Indeed, the large correlation between the annual cycle of MMSL and the CFEEs is observed in Central Med., Baltic region and North Sea. The smaller correlation observed in the West Med. is linked to the earlier drops of MMSL after September. It is also relevant that the largest correlations between the CFEEs registered in the database and seasonally MMSL variations occur in areas with the largest interannual variability (Figures 7 and 10), with the exception of the Central Med. In fact, the correlation of MSL anomalies and CFEEs extend to all regions with the exception of Black Sea and West Med. (Figure 10, Table 3), revealing the potential impact on flood risk derived from changes in the sea-level annual cycle as it was pointed out in previous works (e.g., [45,70]).

Some examples of the contribution of each component during extreme events recorded in the database and captured in the altimetry dataset are plotted in Figure 11. In the analysis, we used the closest satellite track in time and space to the location of the CFEEs. The stronger contribution of the MMSL to the ETWLs (similar in magnitude to the SSL) is observed in the Baltic Sea (Figure 11e), during the storm peak on 07-01-2005. This was also observed during storm Xaver in the North Sea and Halloween storm in the Adriatic Sea (Figure 11a,d respectively) but with a weaker MMSL contribution to the TWL peak. Similarly to the previous areas, the upper tail of the MMSL histogram indicates a higher probability of large contribution, in opposition to the histogram of MMSL in West Med and S-North Atlantic. In fact, the MMSL decreases up to 0.07 m in the West Med (Figure 11c,d) during the storm occurred on 16 December 1997 and drops in the S-North Atlantic event (Figure 11c) where the histogram reveals tidal component as the main contributor to the ETWLs.

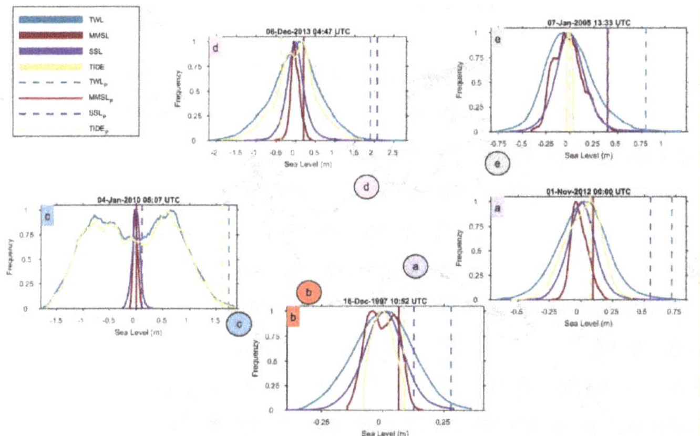

Figure 11. Extreme events registered in the storm impact database and captured by altimetry dataset. Curves represent the histogram of each sea level component (TIDE, SSL and MMSL) in the area and vertical lines mark the magnitude of each component at the peak of the extreme event.

4.3. Limitations and Future Research

According to the results, there is an important contribution of MMSL to the ETWLs. The contribution of MMSL to the ETWLs were calculated and correlated with the storm event impact database in terms of the annual cycle and MSL anomalies. The relative contribution of the MMSL to the extremes is strongly dependent on the concomitance between the storm peak in sea level and the availability of satellite altimeter data. During storms driven by atmospheric perturbations characterised by shorter temporal–spatial scales, the probability of altimeter data availability is reduced. Thus, the extreme events analysed in the Adriatic Sea West Med. (Figure 11a), and Baltic Sea (Figure 11e) underestimated the peak of ETWLs (see Table 4) because the satellite pass was not at the right time. This might be also observed in meso and macrotidal areas (i.e., North Sea) (Table 4) where ETWLs are controlled by tide and surge phase lag. The above-mentioned limitations could bias the final contribution of each component and overestimate the MMSL contribution which is well captured in the altimetry observation.

In addition to this, the availability of accurate altimetry data near the coast might be affected by land/calm water contamination, degraded range and geophysical corrections, producing inaccurate estimations of the sea level in coastal areas [71]. The screening rejects the closest data to the shore, but even though all the components analysed (SSL, TIDE and MMSL), are prone to suffer modifications by several processes (i.e., tide–surge interaction, river discharge, resonance) in the nearshore area changing the final absolute and relative contribution of each component to the TWLs in the coastal area. This limitation could be partially overcome using dedicated coastal altimetry datasets (e.g., X-TRACK, ALES) produced using specific processing techniques to get more accurate estimates of sea level in coastal areas [72,73].

Additionally, as a result of the altimetry limitation in the coastal zone, the wave contribution to the ETWLs is neglected in this study, even though it could be an important component (e.g., [74–76]). This fact could explain part of the differences noted in Table 4, and leads to the overestimation of the MMSL to the ETWLs.

Table 4. Comparison of total water levels (TWLs) peaks captured by altimetry and measured from tide gauges. The data source providing the tide gauge information is indicated in brackets.

	Central Med. [77])	West Med. [78]	S-North Atlantic [78]	North Sea ([53])	Baltic Sea ([79])
TWLp.	0.72 m (1.16 m)	0.28 m (0.46 m)	1.74 m (1.6 m)	1.9 m (~4.67 m)	0.8 m (2.22 m)
SSLp.	0.55 m (0.81 m)	0.13 m	0.12 m	2.08 m (2.67 m)	0.4 m
TIDEp	0.08 m (0.23 m)	0.09 m	1.61 m	−0.38 m (~1.5 m)	0.01 m
MMSLp	0.1 m (~0.12 m)	0.07 m	0.01 m	0.21 m (0.50 m)	0.39
DATEp	01.11.2012 00:00 (31-10-2012 23:30)	16.12.1997 10:52	04.01.2010 05:07	06.12.2013 04:47 (06.12.2013 02:00)	07.01.2019 13:33 (09.01.2009 06:00)

The knowledge of past and present contributions of each individual component to the ETWLs could contribute by reducing the uncertainty of ETWLs forecast, improving the preparedness and reducing damage in the case of coastal flooding. Indeed, very often, large-scale models—especially those with a high resolution on the coast devoted to the ETWL prognosis (e.g., [80,81])—use a 2D barotropic approach neglecting steric effect and mass component sea level variation. The integration of MMSL anomalies from coastal altimetry data assimilation or prognosis through linear regression models (i.e., [69]) could improve the model performance. In light of the achieved results, this could be especially relevant in sensitive areas such as the Baltic Sea, North Sea, and Central Med.

Changes in the magnitude of ETWLs according to climate change scenarios were assessed considering stationary sea level rise (i.e., [82]) omitting the seasonal cycle or monthly MSL anomalies. However, changes in phase and amplitude of the annual cycle or monthly MSL anomalies driven by changes in atmospheric and/or hydrological patterns could modify the extreme water-level projections. Therefore, including seasonality variations of MMSL would contribute by reducing the uncertainty in ETWLs projections improving the rationality of the coastal adaptation measures.

5. Summary and Conclusions

This paper investigates the role of the main sea level components (TIDE, SSL and MMSL) in regard to the ETWL observations along the European seas. Based on 24 years of satellite altimetry, this study evaluates their relative contribution, as well as the correlation of the annual variation of MMSL and MSL anomalies with the extreme events registered in a coastal flooding database along the European coastline.

The largest seasonal range of the AMMSL is observed in the Baltic Sea (±0.11 m), West Med. ([-0.07, 0.09 m]), and the North Sea ([-0.07, 0.08 m]). The smaller MMSL variations are in the Atlantic and Bay of Biscay ([-0.05, 0.06m]). The interannual variability of the MMSL is stronger in the Baltic Sea, Black Sea, North Sea, and Norwegian Sea. The contribution of each component to the ETWLs is subject to important seasonal variations. In microtidal areas (Black Sea, Baltic and Mediterranean Sea) the MMSL contribution is larger than the TIDE most of the time, and its contribution can be at the same order of magnitude of the SSL. In meso and macrotidal areas, the MMSL contribution is <20%, but slightly larger (>30%) in the North Sea.

The comparative analysis of the altimetry data and the storm impact database indicates a non-significant correlation between the AMMSL and the monthly frequency of the CFEEs, since the maximum values of the average annual cycle mostly run on September-October along the European coastline when the low-pressure systems driving SSL are less frequent and intense. However, the average monthly fraction of component variance of MMSL presents significant values of positive correlation with the relative frequency of CFEEs in the Central Med (r = 0.59), North Sea (r = 0.60) and Baltic Sea (r = 0.75). The positive MSL anomalies are correlated with the CFEEs recorded in the database at >90% of the statistical significance in the aforementioned areas, as well as in the Bay of Biscay and N-North Atlantic.

The present contribution demonstrated that there is not a link between the AMMSL and CFEEs along the European coastline. This is caused by the antiphase of the SSL and the AMMSL in most of the oceanographic regions. However, the relationship of MSL anomalies and flooding extreme events indicates a significant and positive correlation between them along the coastline of the Central Med., S-North Atlantic, North Sea and Baltic Sea. In most of these regions, the positive correlation is observed in most of the low-lying areas prone to be flooded. In general, these regions show the largest interannual variability where MSL anomalies are mainly driven by atmospheric and meteorological forcing (North Sea), prevailing wind and the water exchange with another catchment (Baltic Sea, Adriatic Sea). Therefore, the role of MMSL should be considered either for the comprehensive analysis of the past extreme event, or future projection of coastal flooding extreme event.

Satellite altimeter observations provide a valuable and consistent sea-level dataset to analyse the contribution of TIDE, SSL and MMSL to the ETWLs. However, the accuracy of altimeter data close to the coast might be limited. The wave contribution to the ETWLs and the use of accurate sea-level data in the coastal fringe must be taken into consideration in future works. The understanding of every single component of the ETWLs and its spatial and temporal patterns shall improve the preparedness and coastal adaptation measures to reduce the impact of coastal flooding.

Author Contributions: Conceptualization, T.F.-M., J.G.-E., and P.C.; Data curation, T.F.-M.; Formal analysis, T.F.-M.; Funding acquisition, P.C.; Investigation, T.F.-M.; Methodology, T.F.-M. and J.G.-E.; Supervision, J.G.-E. and P.C.; Visualization, T.F.-M., and J.G.-E.; Writing–original draft, T.F.-M.; Writing–review & editing, T.F.-M., J.G.-E., P.C. All authors have read and agreed to the published version of the manuscript.

Funding: This research was funded by EU from the EU-H2020 program under grant agreement number 700099 (ANYWHERE: EnhANcing emergency management and response to extreme WeatHER and climate Events). This work was partially funded by the Spanish Project: Circulation and transport processes in the estuaries of the Gulf of Cadiz: Current situation and projections of future climate change scenarios (TRUCO) (Ref.: RTI2018-100865-B-C22), and by the Coastal Environmental Observatory of the Southwest (OCASO)-Interreg POCTEC EU Project.

Remote Sens. **2020**, *12*, 3419

Acknowledgments: We thank LEGOS (Laboratoire d'Ètudes en Géophysique et Océanographie Spatiales), and Aviso+ by the DAC data distribution with support from CNES (Centre National D'Etudes Spatiales) (https://www.aviso.altimetry.fr/).

Conflicts of Interest: The authors declare no conflict of interest. The funders had no role in the design of the study; in the collection, analyses, or interpretation of data; in the writing of the manuscript, or in the decision to publish the results.

References

1. Tessler, Z.D.; Vörösmarty, C.J.; Grossberg, M.; Gladkova, I.; Aizenman, H.; Syvitski, J.P.M.; Foufoula-Georgiou, E. Profiling risk and sustainability in coastal deltas of the world. *Science* **2015**, *349*, 638–643. [CrossRef] [PubMed]

2. Antonellini, M.; Giambastiani, B.M.S.; Greggio, N.; Bonzi, L.; Calabrese, L.; Luciani, P.; Perini, L.; Severi, P. Processes governing natural land subsidence in the shallow coastal aquifer of the Ravenna coast, Italy. *Catena* **2019**, *172*, 76–86. [CrossRef]

3. Taramelli, A.; Di Matteo, L.; Ciavola, P.; Guadagnano, F.; Tolomei, C. Temporal evolution of patterns and processes related to subsidence of the coastal area surrounding the Bevano River mouth (Northern Adriatic)—Italy. *Ocean Coast. Manag.* **2015**, *108*, 74–88. [CrossRef]

4. Haigh, I.D.; Wadey, M.P.; Wahl, T.; Ozsoy, O.; Nicholls, R.J.; Brown, J.M.; Horsburgh, K.; Gouldby, B. Spatial and temporal analysis of extreme sea level and storm surge events around the coastline of the UK. *Sci. Data* **2016**, *3*, 160107. [CrossRef] [PubMed]

5. Nicholls, R.J.; Cazenave, A. Sea-level rise and its impact on coastal zones. *Science* **2010**, *328*, 1517–1520. [CrossRef]

6. Collet, I.; Engelbert, A. Archive: Coastal Regions—Population Statistics—Statistics Explained. 2013. Available online: https://ec.europa.eu/eurostat/statistics-explained/index.php/Archive:Coastal_regions_-_population_statistics (accessed on 21 February 2019).

7. Haigh, I.D.; Ozsoy, O.; Wadey, M.P.; Nicholls, R.J.; Gallop, S.L.; Wahl, T.; Brown, J.M. An improved database of coastal flooding in the United Kingdom from 1915 to 2016. *Sci. Data* **2017**, *4*. [CrossRef]

8. Paprotny, D.; Morales-Nápoles, O.; Jonkman, S.N. HANZE: A pan-European database of exposure to natural hazards and damaging historical floods since 1870. *Earth Syst. Sci. Data* **2018**, *10*, 565–581. [CrossRef]

9. Ciavola, P.; Harley, M.D.; den Heijer, C. The RISC-KIT storm impact database: A new tool in support of DRR. *Coast. Eng.* **2018**, *134*, 24–32. [CrossRef]

10. Lamb, H.; Frydendahl, K. *Historic Storms of the North. Sea, British Isles and Northwest Europe*; Cambridge University Press: Cambridge, UK, 1991.

11. Bertin, X.; Li, K.; Roland, A.; Zhang, Y.J.; François Breilh, J.; Chaumillon, E. A modeling-based analysis of the flooding associated with Xynthia, central Bay of Biscay. *Coast. Eng.* **2014**, *94*, 80–89. [CrossRef]

12. Garnier, E.; Ciavola, P.; Spencer, T.; Ferreira, O.; Armaroli, C.; McIvor, A. Historical analysis of storm events: Case studies in France, England, Portugal and Italy. *Coast. Eng.* **2018**, *134*, 10–23. [CrossRef]

13. Paprotny, D.; Sebastian, A.; Morales-Nápoles, O.; Jonkman, S.N. Trends in flood losses in Europe over the past 150 years. *Nat. Commun.* **2018**. [CrossRef] [PubMed]

14. Wadey, M.P.; Haigh, I.D.; Nicholls, R.J.; Brown, J.M.; Horsburgh, K.; Carroll, B.; Gallop, S.L.; Mason, T.; Bradshaw, E. A comparison of the 31 January–1 February 1953 and 5–6 December 2013 coastal flood events around the UK. *Front. Mar. Sci.* **2015**, *2*. [CrossRef]

15. Jevrejeva, S.; Jackson, L.P.; Riva, R.E.M.; Grinsted, A.; Moore, J.C. Coastal sea level rise with warming above 2 °C. *Proc. Natl. Acad. Sci. USA* **2016**, *113*, 13342–13347. [CrossRef] [PubMed]

16. Howard, T.; Palmer, M.D.; Bricheno, L.M. Contributions to 21st century projections of extreme sea-level change around the UK. *Environ. Res. Commun.* **2019**, *1*, 095002. [CrossRef]

17. Vousdoukas, M.I.; Mentaschi, L.; Voukouvalas, E.; Verlaan, M.; Feyen, L. Extreme sea levels on the rise along Europe's coasts. *Earth's Future* **2017**, *5*, 304–323. [CrossRef]

18. Vousdoukas, M.I.; Mentaschi, L.; Voukouvalas, E.; Bianchi, A.; Dottori, F.; Feyen, L. Climatic and socioeconomic controls of future coastal flood risk in Europe. *Nat. Clim. Chang.* **2018**, *8*, 776–780. [CrossRef]

19. Wankang, Y.; Baoshu, Y.; Xingru, F.; Dezhou, Y.; Guandong, G.; Haiying, C. The effect of nonlinear factors on tide-surge interaction: A case study of Typhoon Rammasun in Tieshan Bay, China. *Estuar. Coast. Shelf Sci.* **2019**, *219*, 420–428. [CrossRef]

20. Horsburgh, K.J.; Wilson, C. Tide-surge interaction and its role in the distribution of surge residuals in the North Sea C8—C08003. *J. Geophys. Res. Ocean.* **2007**, *112*. [CrossRef]

21. Dietrich, J.C.; Bunya, S.; Westerink, J.J.; Ebersole, B.A.; Smith, J.M.; Atkinson, J.H.; Jensen, R.; Resio, D.T.; Luettich, R.A.; Dawson, C.; et al. A High-Resolution Coupled Riverine Flow, Tide, Wind, Wind Wave, and Storm Surge Model for Southern Louisiana and Mississippi. Part II: Synoptic Description and Analysis of Hurricanes Katrina and Rita. *Mon. Weather Rev.* **2010**, *138*, 378–404. [CrossRef]

22. Bertin, X.; Bruneau, N.; Breilh, J.-F.; Fortunato, A.B.; Karpytchev, M. Importance of wave age and resonance in storm surges: The case Xynthia, Bay of Biscay. *Ocean. Model.* **2012**, *42*, 16–30. [CrossRef]

23. Bevacqua, E.; Maraun, D.; Vousdoukas, M.I.; Voukouvalas, E.; Vrac, M.; Mentaschi, L.; Widmann, M. Higher potential compound flood risk in Northern Europe under anthropogenic climate change. *Sci. Adv.* **2019**, *9*. [CrossRef] [PubMed]

24. Paprotny, D.; Vousdoukas, M.I.; Morales-Nápoles, O.; Jonkman, S.N.; Feyen, L. Pan-European hydrodynamic models and their ability to identify compound floods. *Nat. Hazards.* **2020**, *101*, 933–957. [CrossRef]

25. Vinogradov, S.V.; Ponte, R.M.; Heimbach, P.; Wunsch, C. The mean seasonal cycle in sea level estimated from a data-constrained general circulation model. *J. Geophys. Res.* **2008**, *113*, C03032. [CrossRef]

26. Jordà, G.; Gomis, D. On the interpretation of the steric and mass components of sea level variability: The case of the Mediterranean basin. *J. Geophys. Res. Ocean.* **2013**, *118*, 953–963. [CrossRef]

27. Wu, Q.; Zhang, X.; Church, J.A.; Hu, J. Variability and change of sea level and its components in the Indo-Pacific region during the altimetry era. *J. Geophys. Res. Ocean.* **2017**, *122*, 1862–1881. [CrossRef]

28. Kleinherenbrink, M.; Riva, R.; Frederikse, T.; Merrifield, M.; Wada, Y. Trends and interannual variability of mass and steric sea level in the Tropical Asian Seas. *J. Geophys. Res. Ocean.* **2017**, *122*, 6254–6276. [CrossRef]

29. Tsimplis, M.N.; Woodworth, P.L. The global distribution of the seasonal sea level cycle calculated from coastal tide gauge data. *J. Geophys. Res.* **1994**, *99*, 16031–16039. [CrossRef]

30. Woodworth, P.L.; Melet, A.; Marcos, M.; Ray, R.D.; Wöppelmann, G.; Sasaki, Y.N.; Cirano, M.; Hibbert, A.; Huthnance, J.M.; Monserrat, S.; et al. Forcing Factors Affecting Sea Level Changes at the Coast. *Surv. Geophys.* **2019**, *40*, 1351–1397. [CrossRef]

31. Gómez-Enri, J.; Aboitiz, A.; Tejedor, B.; Villares, P. Seasonal and interannual variability in the Gulf of Cadiz: Validation of gridded altimeter products. *Estuar. Coast. Shelf Sci.* **2012**, *96*, 114–121. [CrossRef]

32. Laiz, I.; Ferrer, L.; Plomaritis, T.A.; Charria, G. Effect of river runoff on sea level from in-situ measurements and numerical models in the Bay of Biscay. *Deep. Res. Part. II Top. Stud. Oceanogr.* **2014**, *106*, 49–67. [CrossRef]

33. Laiz, I.; Gómez-Enri, J.; Tejedor, B.; Aboitiz, A.; Villares, P. Seasonal sea level variations in the Gulf of Cadiz continental shelf from in-situ measurements and satellite altimetry. *Cont. Shelf Res.* **2013**, *53*, 77–88. [CrossRef]

34. Antonov, J.I.; Levitus, S.; Boyer, T.P. Steric sea level variations during 1957-1994: Importance of salinity. *J. Geophys. Res. C Ocean.* **2002**, *107*. [CrossRef]

35. Laiz, I.; Tejedor, B.; Gómez-Enri, J.; Aboitiz, A.; Villares, P. Contributions to the sea level seasonal cycle within the Gulf of Cadiz (Southwestern Iberian Peninsula). *J. Mar. Syst.* **2016**, *159*, 55–66. [CrossRef]

36. Ceres, R.L.; Forest, C.E.; Keller, K. Understanding the detectability of potential changes to the 100-year peak storm surge. *Clim. Chang.* **2017**, *145*, 221–235. [CrossRef]

37. Menéndez, M.; Woodworth, P.L. Changes in extreme high water levels based on a quasi-global tide-gauge data set. *J. Geophys. Res. Ocean.* **2010**, *115*, C10011. [CrossRef]

38. Cid, A.; Castanedo, S.; Abascal, A.J.; Menéndez, M.; Medina, R. A high resolution hindcast of the meteorological sea level component for Southern Europe: The GOS dataset. *Clim. Dyn.* **2014**, *43*, 2167–2184. [CrossRef]

39. Stramska, M.; Kowalewska-Kalkowska, H.; Świrgoń, M. Seasonal variability in the Baltic Sea level. *Oceanologia* **2013**, *55*, 787–807. [CrossRef]

40. De Biasio, F.; Bajo, M.; Vignudelli, S.; Umgiesser, G.; Zecchetto, S. ESA DUE eSurge-Venice project. *Eur. J. Remote Sens.* **2017**, *50*, 428–441. [CrossRef]

41. Woodworth, P.L.; Menéndez, M. Changes in the mesoscale variability and in extreme sea levels over two decades as observed by satellite altimetry. *J. Geophys. Res. Ocean.* **2015**, *120*, 64–77. [CrossRef]

42. Cipollini, P.; Calafat, F.M.; Jevrejeva, S.; Melet, A.; Prandi, P. Monitoring Sea Level in the Coastal Zone with Satellite Altimetry and Tide Gauges. *Surv. Geophys.* **2017**, *38*, 33–57. [CrossRef] [PubMed]

43. Andersen, O.B.; Cheng, Y.; Deng, X.; Steward, M.; Gharineiat, Z. Using satellite altimetry and tide gauges for storm surge warning. In *IAHS-AISH Proceedings and Reports*; Copernicus GmbH: Göttingen, Germany, 2014; Volume 365, pp. 28–34.

44. Passaro, M.; Cipollini, P.; Benveniste, J. Annual sea level variability of the coastal ocean: The Baltic Sea-North Sea transition zone. *J. Geophys. Res. Ocean.* **2015**, *120*, 3061–3078. [CrossRef]

45. Calafat, F.M.; Wahl, T.; Lindsten, F.; Williams, J.; Frajka-Williams, E. Coherent modulation of the sea-level annual cycle in the United States by Atlantic Rossby waves. *Nat. Commun.* **2018**, *9*, 1–13. [CrossRef]

46. Dangendorf, S.; Wahl, T.; Mudersbach, C.; Jensen, J. The Seasonal Mean Sea Level Cycle in the Southeastern North Sea. *J. Coast. Res.* **2013**, *165*, 1915–1920. [CrossRef]

47. Copernicus—Marine Environment Monitoring Service. Available online: https://marine.copernicus.eu/ (accessed on 1 October 2018).

48. Duacs | Altimetry Data for Sea Level Studies & Applications. Available online: https://duacs.cls.fr/ (accessed on 13 January 2018).

49. CMEMS. *QUID forSea Level TAC DUACS Products*; Available online: https://resources.marine.copernicus.eu/ documents/QUID/CMEMS-SL-QUID-008-032-051.pdf (accessed on 21 February 2019).

50. Carrère, L.; Lyard, F. Modeling the barotropic response of the global ocean to atmospheric wind and pressure forcing—Comparisons with observations. *Geophys. Res. Lett.* **2003**, *30*. [CrossRef]

51. Carrere, L.; Lyard, F.; Cancet, M.; Guillot, A.; Dupuy, S.; Carrère, L.F.; Lyard, M.; Cancet, A.; Guillot, N. Picot, 2015: FES2014: A New Tidal Model Onthe Global Ocean with Enhanced Accuracy in Shallow Seas and in the Arctic Region, OSTST2015. 2014. Available online: http://meetings.aviso.altimetry.fr/fileadmin/user_ upload/tx_ausyclssemi (accessed on 1 January 2019).

52. Pawlowicz, R.; Beardsley, B.; Lentz, S. Classical tidal harmonic analysis including error estimates in MATLAB using T_TIDE. *Comput. Geosci.* **2002**, *28*, 929–937. [CrossRef]

53. Dangendorf, S.; Arns, A.; Pinto, J.G.; Ludwig, P.; Jensen, J. The exceptional influence of storm 'Xaver' on design water levels in the German Bight. *Environ. Res. Lett.* **2016**, *11*, 054001. [CrossRef]

54. Serafin, K.A.; Ruggiero, P.; Stockdon, H.F. The relative contribution of waves, tides, and nontidal residuals to extreme total water levels on U.S. West Coast sandy beaches. *Geophys. Res. Lett.* **2017**, *44*, 1839–1847. [CrossRef]

55. Avsar, N.B.; Jin, S.; Kutoglu, H.; Gurbuz, G. Sea level change along the Black Sea coast from satellite altimetry, tide gauge and GPS observations. *Geod. Geodyn.* **2016**, *7*, 50–55. [CrossRef]

56. Criado-Aldeanueva, F.; Del Río Vera, J.; García-Lafuente, J. Steric and mass-induced Mediterranean sea level trends from 14 years of altimetry data. *Glob. Planet. Chang.* **2008**, *60*, 563–575. [CrossRef]

57. Marcos, M.; Tsimplis, M.N. Forcing of coastal sea level rise patterns in the North Atlantic and the Mediterranean Sea. *Geophys. Res. Lett.* **2007**, *34*. [CrossRef]

58. Legeais, J.F.; Ablain, M.; Zawadzki, L.; Zuo, H.; Johannessen, J.A.; Scharffenberg, M.G.; Fenoglio-Marc, L.; Joana Fernandes, M.; Baltazar Andersen, O.; Rudenko, S.; et al. An improved and homogeneous altimeter sea level record from the ESA Climate Change Initiative. *Earth Syst. Sci. Data* **2018**, *10*, 281–301. [CrossRef]

59. Ruiz Etcheverry, L.A.; Saraceno, M.; Piola, A.R.; Valladeau, G.; Möller, O.O. A comparison of the annual cycle of sea level in coastal areas from gridded satellite altimetry and tide gauges. *Cont. Shelf Res.* **2015**, *92*, 87–97. [CrossRef]

60. Medvedev, I.P. Seasonal fluctuations of the Baltic Sea level. *Russ. Meteorol. Hydrol.* **2014**, *39*, 814–822. [CrossRef]

61. Pajak, K.; Kowalczyk, K. A comparison of seasonal variations of sea level in the southern Baltic Sea from altimetry and tide gauge data. *Adv. Sp. Res.* **2019**, *63*, 1768–1780. [CrossRef]

62. Mork, K.A.; Skagseth, Ø. Annual Sea Surface Height Variability in the Nordic Seas. In *The Nordic Seas: An Integrated Perspective: Oceanography, Climatology, Biogeochemistry, and Modeling*; American Geophysical Union: Washintong, DC, USA, 2005; pp. 51–64.

63. Stanev, E.V.; Le Traon, P.-Y.; Peneva, E.L. Sea level variations and their dependency on meteorological and hydrological forcing: Analysis of altimeter and surface data for the Black Sea. *J. Geophys. Res. Ocean.* **2000**, *105*, 17203–17216. [CrossRef]

64. García-García, D.; Chao, B.F.; Boy, J.P. Steric and mass-induced sea level variations in the Mediterranean Sea revisited. *J. Geophys. Res. Ocean.* **2010**, *115*, C12016. [CrossRef]

65. Dangendorf, S.; Calafat, F.M.; Arns, A.; Wahl, T.; Haigh, I.D.; Jensen, J. Mean sea level variability in the North Sea: Processes and implications. *J. Geophys. Res. Ocean.* **2014**, *119*, 6820–6841. [CrossRef]

66. Barbosa, S.M.; Donner, R.V. Long-term changes in the seasonality of Baltic sea level. *Tellus A Dyn. Meteorol. Oceanogr.* **2016**, *68*, 30540. [CrossRef]

67. Hünicke, B.; Zorita, E. Influence of temperature and precipitation on decadal Baltic Sea level variations in the 20th century. *Tellus, Ser. A Dyn. Meteorol. Oceanogr.* **2006**, *58*, 141–153. [CrossRef]

68. Lisitzin, E. *Sea Level Changes*; Elsevier Oceanography Series: Amsterdam, The Netherlands, 1974; ISBN 0444411577.

69. Sterlini, P.; de Vries, H.; Katsman, C. Sea surface height variability in the North East Atlantic from satellite altimetry. *Clim. Dyn.* **2016**, *47*, 1285–1302. [CrossRef]

70. Dangendorf, S.; Wahl, T.; Hein, H.; Jensen, J.; Mai, S.; Mudersbach, C. Mean Sea Level Variability and Influence of the North Atlantic Oscillation on Long-Term Trends in the German Bight. *Water* **2012**, *4*, 170–195. [CrossRef]

71. Benveniste, J.; Cazenave, A.; Vignudelli, S.; Fenoglio-Marc, L.; Shah, R.; Almar, R.; Andersen, O.B.; Birol, F.; Bonnefond, P.; Bouffard, J.; et al. Requirements for a coastal hazards observing system. *Front. Mar. Sci.* **2019**. [CrossRef]

72. Passaro, M.; Rose, S.K.; Andersen, O.B.; Boergens, E.; Calafat, F.M.; Dettmering, D.; Benveniste, J. ALES+: Adapting a homogenous ocean retracker for satellite altimetry to sea ice leads, coastal and inland waters. *Remote Sens. Environ.* **2018**, *211*, 456–471. [CrossRef]

73. Birol, F.; Fuller, N.; Lyard, F.; Cancet, M.; Niño, F.; Delebecque, C.; Fleury, S.; Toublanc, F.; Melet, A.; Saraceno, M.; et al. Coastal applications from nadir altimetry: Example of the X-TRACK regional products. *Adv. Sp. Res.* **2017**, *59*, 936–953. [CrossRef]

74. Marcos, M.; Rohmer, J.; Vousdoukas, M.I.; Mentaschi, L.; Le Cozannet, G.; Amores, A. Increased Extreme Coastal Water Levels Due to the Combined Action of Storm Surges and Wind Waves. *Geophys. Res. Lett.* **2019**, *46*, 4356–4364. [CrossRef]

75. Melet, A.; Meyssignac, B.; Almar, R.; Le Cozannet, G. Under-estimated wave contribution to coastal sea-level rise. *Nat. Clim. Chang.* **2018**, *8*, 234–239. [CrossRef]

76. Serafin, K.A.; Ruggiero, P. Simulating extreme total water levels using a time-dependent, extreme value approach. *J. Geophys. Res. C Ocean.* **2014**, *119*, 6305–6329. [CrossRef]

77. Harley, M.D.; Valentini, A.; Armaroli, C.; Perini, L.; Calabrese, L.; Ciavola, P. Can an early-warning system help minimize the impacts of coastal storms? A case study of the 2012 Halloween storm, northern Italy. *Nat. Hazards Earth Syst. Sci.* **2016**, *16*, 209–222. [CrossRef]

78. RiscKit Storm Database. Available online: http://risckit.cloudapp.net/risckit/#/ (accessed on 16 November 2019).

79. Wolski, T.; Wiśniewski, B.; Giza, A.; Kowalewska-Kalkowska, H.; Boman, H.; Grabbi-Kaiv, S.; Hammarklint, T.; Holfort, J.; Lydeikaite, Z. Extreme sea levels at selected stations on the Baltic Sea coast. *Oceanologia* **2014**, *2*, 259–290. [CrossRef]

80. Fernández-Montblanc, T.; Vousdoukas, M.I.; Ciavola, P.; Voukouvalas, E.; Mentaschi, L.; Breyiannis, G.; Feyen, L.; Salamon, P. Towards robust pan-European storm surge forecasting. *Ocean Model.* **2019**, *133*, 129–144. [CrossRef]

81. Muis, S.; Verlaan, M.; Winsemius, H.C.; Aerts, J.C.J.H.; Ward, P.J. A global reanalysis of storm surges and extreme sea levels. *Nat. Commun.* **2016**, *7*, 11969. [CrossRef] [PubMed]

82. Vousdoukas, M.I.; Mentaschi, L.; Voukouvalas, E.; Verlaan, M.; Jevrejeva, S.; Jackson, L.P.; Feyen, L. Global probabilistic projections of extreme sea levels show intensification of coastal flood hazard. *Nat. Commun.* **2018**, *9*, 2360. [CrossRef] [PubMed]

Publisher's Note: MDPI stays neutral with regard to jurisdictional claims in published maps and institutional affiliations.

Article

Improved Method to Suppress Azimuth Ambiguity for Current Velocity Measurement in Coastal Waters Based on ATI-SAR Systems

Na Yi [1], Yijun He [1,2,*] and Baochang Liu [1]

[1] School of Marine Sciences, Nanjing University of Information Science and Technology, Nanjing 210044, China; nayi@nuist.edu.cn (N.Y.); bcliu@nuist.edu.cn (B.L.)

[2] Laboratory for Regional Oceanography and Numerical Modeling, Qingdao National Laboratory for Marine Science and Technology, Qingdao 266237, China

* Correspondence: yjhe@nuist.edu.cn

Received: 13 August 2020; Accepted: 8 October 2020; Published: 10 October 2020

Abstract: Measurements of ocean surface currents in coastal waters are crucial for improving our understanding of tidal atlases, as well as for ecosystem and water pollution monitoring. This paper proposes an improved method for estimating the baseline-to-platform speed ratio (BPSR) for improving the current line-of-sight (LOS) velocity measurement accuracy in coastal waters with along-track interferometric synthetic aperture radar (ATI-SAR) based on eigenvalue spectrum entropy (EVSE) analysis. The estimation of BPSR utilizes the spaceborne along-track interferometry and considers the effects of a satellite orbit and an inaccurate baseline responsible for azimuth ambiguity in coastal waters. Unlike the existing methods, which often assume idealized rather than actual operating environments, the proposed approach considers the accuracy of BPSR, which is its key advantage applicable to many, even poorly designed, ATI-SAR systems. This is achieved through an alternate algorithm for the suppression of azimuth ambiguity and BPSR estimation based on an improved analysis of the eigenvalue spectrum entropy, which is an important parameter representing the mixability of unambiguous and ambiguous signals. The improvements include the consideration of a measurement of the heterogeneity of the scene, the corrections of coherence-inferred phase fluctuation (CPF), and the interferogram-derived phase variability (IPV); the last two variables are closely related to the determination of the EVSE threshold. Besides, the BPSR estimation also represents an improvement that has not been achieved in previous work of EVSE analysis. When the improved method is used on the simulated ocean-surface current LOS velocity data obtained from a coastal area, the root-mean-square error is less than 0.05 m/s. The other strengths of the proposed algorithm are adaptability, robustness, and a limited user input requirement. Most importantly, the method can be adopted for practical applications.

Keywords: along-track interferometric synthetic aperture radar (ATI-SAR); current line-of-sight (LOS) velocity; coastal waters; azimuth ambiguity; baseline-to-platform speed ratio estimation

---

1. Introduction

Ocean sea surface currents play a key role in air-sea interaction, biological production, and mixing between the upper and lower water layers in coastal areas [1–4]. In addition, their measurement in coastal areas provides important information to fishing and electricity generation industries [5,6].

In coastal waters, tidal currents are one of the most important factors of the sea surface current. Generally, tidal currents are quite deterministic and can also be precisely inferred by in situ measurements. In situ measurement devices, including the acoustic Doppler current profiler (ADCP) and the current meter, however, have limited coverage and are expensive. On the other hand, the along-track interferometric synthetic aperture radar (ATI-SAR) does not have these limitations; meanwhile, ATI performs well for measurements of the sea surface currents, including the tidal currents [7]. Along-track interferometry (ATI) is a powerful tool for the measurement of ocean currents [8–13]. Interferometry was originally proposed in [14], and is based on processing two interferometric SAR images of the same scene obtained with two antennas within a short time [15,16]. Most of the existing studies on the retrieval of surface currents by interferometric SAR [17–19] assume systems with an accurate baseline and constant platform velocity, i.e., a completely accurate baseline-to-platform speed ratio (BPSR). However, in real-life applications, the baseline is often inaccurate; for example, in the commonly used spaceborne SAR data acquisition mode, and the entire antenna is active in pulse transmission but divided into several parts to receive returns. The effective phase center of each receiving channel is assumed to be located in the middle between the physical transmission and the respective receiving phase center, but this method is not accurate [20]. In addition, the accuracy of BPSR is not considered.

While the airborne ATI [21–24] is usually limited by the achievable coverage and complex logistical requirements, the spaceborne ATI [25–27] can illuminate any point of interest during a certain overpass, and obtain wide-swath and high-resolution real-time current observations [28]. Despite the relatively high degree of azimuth ambiguity, spaceborne InSAR systems perform better in ocean current inversion in open sea. Nevertheless, in coastal waters, azimuth ambiguities may have a negative influence on the accuracy of measurements of the velocity of sea surface currents as spaceborne ATI systems with wide bandwidth are particularly prone to azimuth ambiguity, which can produce a "ghost signature" in images. Azimuth ambiguity is mainly caused by under-sampling of a signal, i.e., the signal received by the radar originates not only from the area of interest but also includes ghost signatures from the surrounding areas. In locations, such as coastal waters, the ghost signals of scatterers with strong backscattered powers on land will be shifted in azimuth and superimposed on a relatively weak signal from the water, as shown in Figure 1. In Figure 1, the InSAR signals are modeled within the Doppler baseband—PRF/2 $\leq f_d \leq$ PRF/2 (f_d is the Doppler frequency, PRF is the pulse repetition frequency). In addition, it has a negative impact on the estimation of the baseline-to-platform speed ratio (BPSR), and consequently, on the accuracy of BPSR-based ocean currents measurements [29]. Also, azimuth ambiguities have a strong influence on the accuracy of measured current line-of-sight (LOS) velocities. In coastal waters [30], it is, therefore, necessary to eliminate azimuth ambiguity before estimating BPSR, which necessitates the development of an improved algorithm that not only suppresses azimuth ambiguity but estimates BPSR as well. The two tasks can conveniently be handled using the Doppler interval (for details, see [31]). The Doppler interval specifically refers to the interval without an azimuth-ambiguity Doppler spectrum, which is the Doppler frequency interval with a starting point and ending point in the mathematical sense. Since the Doppler frequency is linearly dependent on the baseline value, according to the definition of the BPSR, the Doppler frequency and BPSR are also linearly dependent, so the former can be used to estimate the BPSR.

Romeiser et al. [7] proposed suppressing azimuth ambiguity through a pixel-value exclusion operation, which eliminates pixels that have an intensity of less than 10 dB at a certain distance. However, this method cannot work reliably in areas with contrast between the land and water. An alternative approach is spectrum filtering and extrapolation [32], but it reduces the azimuthal resolution. The method of analyzing the eigenvalue spectrum entropy (EVSE) proposed by Liu [31] can automatically estimate a usable range of the Doppler domain and needs only limited user inputs, but assumes an accurate baseline and constant velocity of the platform. However, in practical applications, none of these assumptions is true, which motivated us to improve the method.

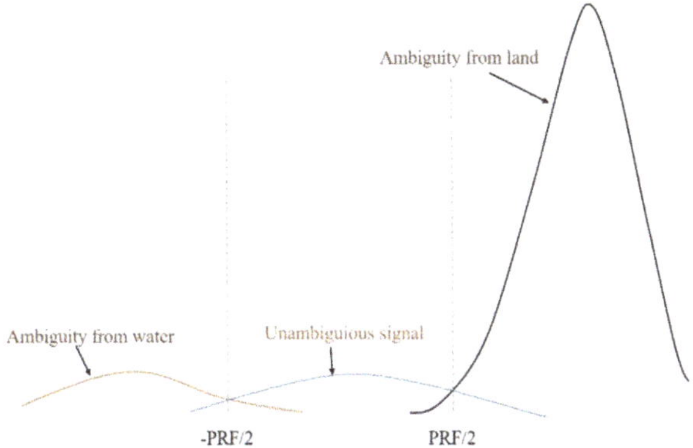

Figure 1. Illustration of the Doppler amplitude patterns of the two azimuth ambiguities and the unambiguous signal part. Modified from Liu [31].

This paper proposes an improved algorithm for both azimuth ambiguity suppression and BPSR estimation, considering both the heterogeneity of the scene and BPSR estimation. Although the azimuth ambiguity of spaceborne SAR is relatively high, it has little influence on the inversion of azimuth ambiguity in open ocean regions with uniform scattering. However, azimuth-ambiguity has a great influence on the performance of spaceborne InSAR current measurement in the non-uniform offshore area. There are two main reasons for this. First, the backscattering coefficient of land radar is usually much larger than that of sea radar. Therefore, the azimuth ambiguity component from the land will be superimposed on the sea surface, resulting in a serious decline in the accuracy of InSAR current measurements. Second, the velocity of land ghosting is different from that of sea-surface ghosting, which will also change the measured value of the sea surface current field. If scene heterogeneity is not taken into account, EVSE analysis will fail when applied to practical situations. Furthermore, the improved method can be adopted for practical applications with only limited user inputs. The remainder of this paper is organized as follows. Section 2 describes the proposed method, including an overview of Liu's method [31], an alternative algorithm, and our innovation. Section 3 presents the results of applying the improved method to simulated and measured data. Finally, a discussion is presented in Section 4, and in Section 5, conclusions are drawn.

2. Methodology

In this section, to improve the accuracy of current LOS velocity estimation, we develop an alternate algorithm for ambiguity suppression and BPSR estimation based on the method of Liu [31]. The surface velocity corresponds more precisely to a mean motion of scattering elements, and the element velocities are weighted by their normalized radar cross section (NRCS) [33]. Considering that the strong NRCS caused by convergence and divergence of the current can lead to large errors [34], we assumed that the ocean surface was smooth so that we could focus more on the suppression of azimuth ambiguity and BPSR estimation.

An overview of the process of ocean current velocity estimation is shown in the flowchart in Figure 2. The process starts with two original SAR images and ends with the estimation of current velocity. As shown in Figure 2, the flowchart mainly includes three parts: SAR image preprocessing (green rectangle in Figure 2), alternating iteration algorithm (blue rectangle in Figure 2), and velocity estimation (orange rectangle in Figure 2). SAR image preprocessing includes SAR image focusing, the interested area extraction of the area of interest, and conversions from the time domain to the frequency domain via the 2D Fourier transform. An alternating iteration algorithm is the focus of our research, and this algorithm is mainly an alternating iterative algorithm that performs azimuth ambiguity suppression and BPSR estimation. Finally, we obtain the surface current velocity, which is the LOS velocity.

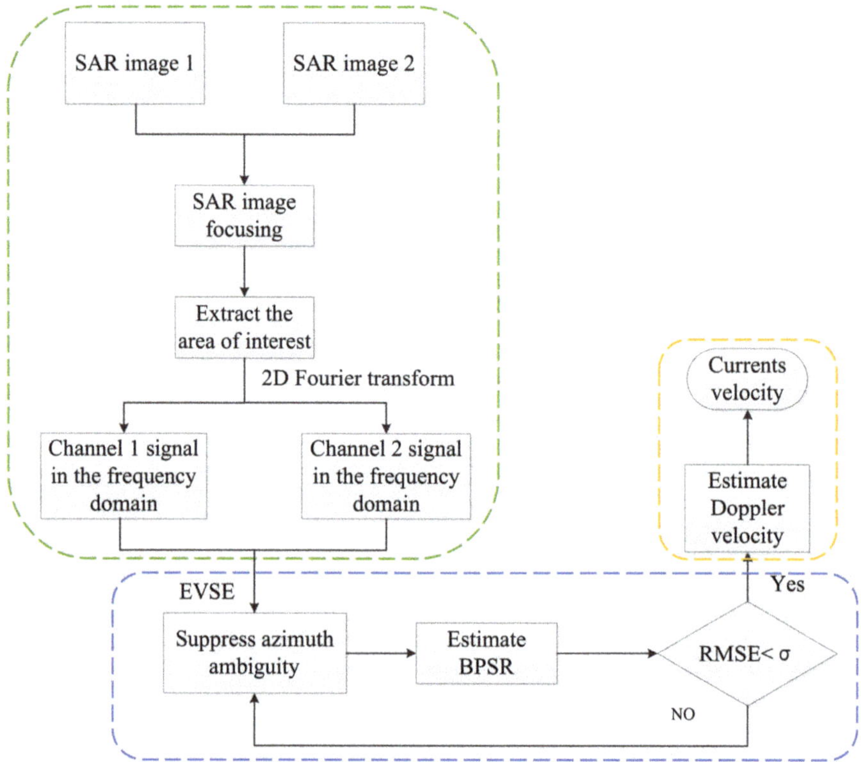

Figure 2. Flowchart of the proposed approach.

Selected key procedures underlying the alternate algorithm for ambiguity suppression and BPSR estimation are introduced in this section. The central part of the process includes alternate iterations of ambiguity suppression and BPSR estimation, which is detailed in Figure 3.

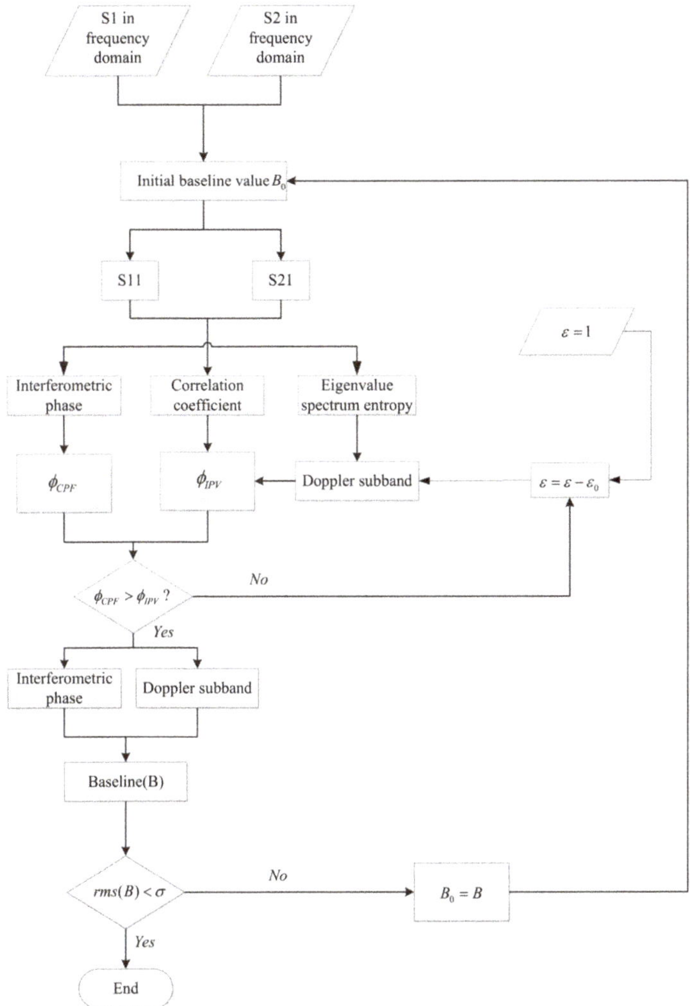

Figure 3. Flowchart of the alternate iterative algorithm.

2.1. Overview of EVSE Analysis

The method proposed in [31] aims to measure ocean surface currents in coastal waters when the problem of azimuth ambiguity is severe. The velocity estimation is conducted for the sea-surface current in coastal waters. The key component of the method is the analysis of EVSE, which is defined as the entropy of the eigenvalue spectrum of the ATI covariance matrix computed in the Doppler domain. It quantifies the degree of mixing among the Doppler components [31], and is a significant parameter for determining the Doppler domain representation of an unambiguous signal.

The first step of the EVES analysis method is to model the SAR signal. Thus, a dual-channel ATI-SAR signal in the 2-D frequency (the range frequency and the Doppler frequency) domain can be modeled as follows [31]:

$$S_1 = S_{una}^0 + S_{amb}^A + S_{amb}^B + S_{N1} \tag{1}$$

$$S_2 = \exp\left\{ j2\pi \cdot \frac{B_e}{V_p} \cdot f_d \right\}$$

$$\times \left[\begin{array}{l} \widetilde{S}^0_{una} \cdot \exp\left\{ j\frac{4\pi}{\lambda} \cdot \frac{B_e}{V_p} \cdot v_r^0 \right\} \\ +\widetilde{S}^A_{amb} \cdot \exp\left\{ j\frac{4\pi}{\lambda} \cdot \frac{B_e}{V_p} \cdot v_r^A \right\} \times \exp\left\{ +j2\pi \cdot \frac{B_e}{V_p} \cdot PRF \right\} \\ +\widetilde{S}^B_{amb} \cdot \exp\left\{ j\frac{4\pi}{\lambda} \cdot \frac{B_e}{V_p} \cdot v_r^B \right\} \times \exp\left\{ -j2\pi \cdot \frac{B_e}{V_p} \cdot PRF \right\} \end{array} \right] + S_{N2} \tag{2}$$

where S_1 and S_2 are the signals received from the two channels of the ATI-SAR system, and S^0_{una} is an unambiguous signal from the fore channel. The parameter PRF represents the pulse repetition frequency. S^A_{amb} and S^B_{amb} are the ambiguous signals from the land and ocean areas, respectively. The tilde-circumflexed signals in $\widetilde{S}^0_{una}$, $\widetilde{S}^A_{amb}$ and $\widetilde{S}^B_{amb}$ are not the same as their uncircumflexed counterparts in Equation (1), because the random motion of the ocean surface affects the received signals. S_{N1} and S_{N2} denote the thermal noise signals of the two channels. B_e is the effective baseline and V_p is the velocity of the radar platform. f_d represents the Doppler frequency and λ denotes the radar wavelength. v_r^0, v_r^A, and v_r^B are the mean line of sight (LOS) surface velocities of the area of interest, the zone I, and the zone IV, respectively, as shown in Figure 1.

From Equations (1) and (2), the covariance matrix R, can be calculated as follows [31]:

$$R = E\left\{ \begin{bmatrix} S_1 \\ S_2 \end{bmatrix} \begin{bmatrix} S_1^* & S_2^* \end{bmatrix} \right\}$$
$$= \left(P^0_{una} + P^A_{amb} + P^B_{amb} \right) \begin{bmatrix} 1 & \rho \\ \rho^* & 1 \end{bmatrix} + P_n \cdot I_{2\times2} \tag{3}$$

where E is the expectation operator, $(\cdot)^*$ denotes the complex conjugate operator, P_n is the noise power, and $I_{2\times2}$ is a two-by-two identity matrix. P^0_{una}, P^A_{amb}, and P^B_{amb} are the powers of the unambiguous signals, the ambiguity of the signal from the land, and the ambiguity of the signal from the ocean, respectively, which can be computed as explained in [31]. Having evaluated the two eigenvalues of the covariance matrix, R, denoted as λ_1 and λ_2, the EVSE, H, of the ATI covariance matrix can be defined as

$$H = -(p_1 \log_2 p_1 + p_2 \log_2 p_2) \tag{4}$$

where p_1 and p_2 are as follows:

$$p_1 = \frac{\lambda_1}{\lambda_1 + \lambda_2}, p_2 = \frac{\lambda_2}{\lambda_1 + \lambda_2} \tag{5}$$

The EVSE quantifies the degree of signal mixing, and is used as the criterion of Doppler domain characterization: the larger the value of EVSE, the higher is the degree of signal component mixing [31].

In Liu's azimuth ambiguity suppression algorithm [31], the EVSE analysis is an important step. As shown in Figure 1, zone II denotes the unambiguous signal. As seen in Figure 4, the fluctuation of the interferometric phase in Doppler frequency is small. Thus, to decide how many Doppler bins should be discarded, a critical EVSE value for the Doppler bins dominated by the unambiguous signal is required. From Liu [31], we can conclude that the determination of zone II depends on three parts: an accurate BPSR, an EVSE curve, and a critical value of EVSE.

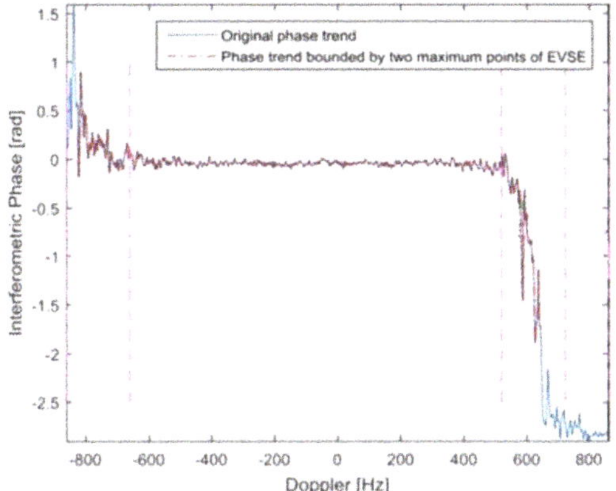

Figure 4. Interferometric phase trend after Doppler bin removal based on the two maximum points of the eigenvalue spectrum entropy (EVSE) curve from [31].

As shown in Figure 4, the Doppler bins falling outside the two maximum points of the EVSE curve can be determined by investigating the EVSE curve to find the two maximum points, and the Doppler bins containing an ambiguous signal are excluded. According to Liu [31], the critical value of EVSE is determined such that this critical value identifies a maximum Doppler sub-band over which the following two parameters are equal.

The first parameter, denoted ϕ_{CPF}, is the coherence-inferred phase fluctuation (CPF), defined as the mean statistical fluctuation of interferometric phases over the Doppler sub-band; and the other parameter, denoted ϕ_{IPV}, is the interferogram-derived phase variability (IPV), defined as the root-mean-square (RMS) variation of interferogram-derived phase over a certain Doppler sub-band. The expressions for the two parameters are as follows:

$$\phi_{CPF} = \frac{1}{\sqrt{2K}} \frac{\sqrt{1 - \hat{\rho}_M^2}}{\hat{\rho}_M} \tag{6}$$

$$\phi_{IPV} = \sqrt{\frac{1}{L}\sum_{l=1}^{L}\left[\widetilde{\phi}(f_d^{(l)}) - \frac{1}{L}\sum_{p=1}^{L}\widetilde{\phi}(f_d^{(p)})\right]^2} \tag{7}$$

where K is the number of averaged range frequency bins, $\hat{\rho}_M$ is the magnitude of the mean coherence in the 2-D frequency domain, $\widetilde{\phi}(f_d^{(l)})$ is the range-frequency-averaged interferogram phase for the l th Doppler bin $(f_d^{(l)})$ of the Doppler sub-band, and L is the size of the Doppler sub- band. Note that K is based on the assumption that the samples are statistically completely independent and uniform in the range frequency images.

However, when BPSR is not accurate, Liu's method [31] is in effective, which limits the practical applications of the algorithm. In addition, the assumption of K is not correct in practice. The improvements aimed at these two problems in this paper, which will be discussed in the next section, are intended to address this shortcoming to make the method better suited for real-life applications.

2.2. Alternate Iteration Algorithm for Azimuth Ambiguity Suppression and BPSR Estimation

The proposed alternate algorithm that can suppress azimuth ambiguity and estimate BPSR, is described below.

The interferometric phase, φ, and the effective baseline, B_e, are related as follows [35]:

$$\varphi = -\frac{2\pi f_d}{V_p} \cdot B_e \tag{8}$$

Based on the above linear relation, the value of the baseline can be obtained from the slope of the phase–frequency curve. Then, BPSR can be shown as

$$BPSR = \frac{B_e}{V_p} \tag{9}$$

As expected, the knowledge of BPSR's accuracy is not sufficient. From Equation (8), we observe that the value of the baseline is related to the Doppler frequency-interference phase. Furthermore, as mentioned in Section 2.1, azimuth ambiguity affects the calculation of the interference phase, and an inaccurate BPSR will result in the failure of the ambiguity suppression algorithm. Therefore, the BPSR and azimuth ambiguity influence each other.

The alternate algorithm for azimuth ambiguity suppression and BPSR estimation are shown in Figure 2, and the detailed flowchart is shown in Figure 3. As the flowchart shows, two adaptive algorithms are executed alternately; one is used to estimate the critical value of EVSE during the process of azimuth ambiguity suppression, and the other is BPSR estimation. There are several key points involved in determining the threshold value of EVSE: first, set ε as a variable ($0 \leq \varepsilon \leq 1$) with an initial value of 1 in order to determine the characteristic spectral entropy that is less than all of its Doppler units and then combine those Doppler units into a Doppler sub-band; second, calculate CPF (ϕ_{CPF}) and IPV (ϕ_{IPV}), when IPV (ϕ_{IPV}) is larger than CPF (ϕ_{CPF}), reduce the value of ε by a certain step size ε_0. Until the condition $\phi_{IPV}(\varepsilon) < \phi_{CPF}$ is established, then the value of ε is determined as the threshold of EVSE. The Doppler sub-band without ambiguous signal is obtained by discarding all the Doppler units whose EVSE is greater than the EVSE threshold. After removing ambiguity by the EVSE analysis, we obtain the Doppler sub-band that contains the unambiguous signal, from which the baseline value can be estimated using the linear relation between the interferometric phase and the baseline. Next, the baseline value can be used to correct the phase of one of the SAR images, after which the BPSR can be estimated. The process is repeated until the BPSR root-mean-square error is reduced below a predefined small number. It can be seen from Figure 3 that this is also an adaptive algorithm.

2.3. Correction of IPV and CPF Based on EVSE Analysis

In the previous sections, an EVSE analysis and an alternate iterative algorithm for the azimuth ambiguity suppression and baseline estimation were discussed. In the current section, we focus on a correction introduced into the method proposed in this paper for non-ideal situations where K in Equation (7) deviates from the original definition in [31].

The correction we added accounts for scene heterogeneity in an SAR image. The heterogeneity of a scene is used to calculate the number of the samples in the range frequency. The so-called effective sample number refers to the number of units of distribution of research objects in an SAR image. In our context, ships and drilling platforms are invalid samples. The sharpness of an SAR image, shp, is used to represent the non-uniformity of the scene, and is defined as follows:

$$shp = \frac{\langle I \rangle^2}{\langle I^2 \rangle} = \frac{\left(\frac{1}{L}\sum\limits_{i=1}^{N} I_i^2\right)^2}{\frac{1}{L}\sum\limits_{i=1}^{N} I_i^4} \tag{10}$$

where I_i is the amplitude of the ith pixel in a range compressed image I, $\langle \cdot \rangle$ denotes the spatial average, N is the number of all samples, and L is the number of effective samples.

Therefore, we take *shp* into account is the interferometric phase induced by across baseline. In the proposed algorithm, we take the effect of sharpness of an SAR image into consideration by modifying the K in the formula of CPF as follows:

$$\phi'_{CPF} = \frac{\sqrt{1 - \rho_0^2}}{\sqrt{2K \cdot shp} \cdot \rho_0} \tag{11}$$

where ρ_0 is the magnitude of the mean coherence, calculated as follows:

$$\rho_0 = \frac{\sum_{i=1}^{L} \langle S_{1i} \cdot S_{2i}^* \rangle}{\sqrt{\sum_{i=1}^{L} \langle S_{1i} \cdot S_{1i}^* \rangle \cdot \sum_{i=1}^{L} \langle S_{2i} \cdot S_{2i}^* \rangle}} \tag{12}$$

where S_{1i} and S_{2i} are the complex values of a corresponding point in S_1 and S_2, respectively, after S_2 has been resampled according to the estimated shift, and L is the number of pixels in the sampling area. Note that the numerator is the interferogram while the denominator is the product of the image amplitudes, not powers. The formula for IPV is altered to

$$\phi'_{IPV} = q \cdot \sqrt{\frac{1}{L} \sum_{i=1}^{L} \left[\widetilde{\phi}(f_d^{(i)}) - \frac{1}{L} \sum_{j=1}^{L} \widetilde{\phi}(f_d^{(j)}) \right]^2} \tag{13}$$

where q is a constant used to relax the condition in the computation of IPV. Similarly, the critical value of EVSE is determined such that this critical value identifies a maximum Doppler sub-band over which the above two parameters are equal. Because the BPSR is not accurate in practice, the harsh condition in [31] is also needed to be revised; after several computations, the BPSR tends to be accurate, and q will be fixed at 1.

The interferometric phase is computed as

$$\varphi_i = \tan^{-1}\left(\sum_{i=1}^{N} \langle S_{1i} \cdot S_{2i}^* \rangle \right) \tag{14}$$

where φ_i is the ith interferometric phase of the corresponding two SAR images. Note that because a difference of 2π may be present between the computed and the true interferometric phase, phase unwrapping may be necessary. If there is a 2π discontinuity in the phase curve, it will cause a large error in the slope of the curve fitted in the Doppler frequency domain and the true interference phase, which will also affect the estimation of BPSR. To alleviate the problem, φ'_i, can be corrected as follows:

$$\varphi'_i = \varphi_i \pm 2\pi \tag{15}$$

After the above series of corrections, or improvements, the algorithm becomes better suited to practical applications.

In the calculation of surface current LOS velocity, the ocean surface is assumed to be composed of scattering objects that constitute a uniform random surface. The ocean surface current LOS velocity can be computed by

$$V_c = \frac{\sum_{i=1}^{N} \langle S_{1i} \cdot S_{2i}^* \rangle}{4\pi \cdot BPSR \cdot \sin(\theta)} \cdot \lambda \tag{16}$$

where θ denotes the incidence angle and N denotes the total number of sample points in the direction of azimuth and range. Because the measured horizontal LOS Doppler velocities are not true current velocities, these measured Doppler velocities for the theoretical contributions of ocean wave motions should be corrected using a numerical model [17].

Liu's method [31] assumes that the value of the effective baseline appearing in Equation (2) is accurate, even though this is often not the case for practical ATI systems. On this basis, we make improvements. Because the algorithm is adaptive, the user only needs to input the SAR data and estimate the platform speed to obtain the BPSR to facilitate the subsequent estimation of the ocean current velocity. In addition, the algorithm is robust and can be applied not only to coastal areas but also land areas, because it estimates the degree of scene heterogeneity. In the following section, the validation data and application results are discussed.

3. Results

To assess the feasibility of the improved method, we applied two different sets of data, simulated coastal area data and measured land data. Because of lack of measured coastal data, we used coastal simulation data, which proved to be reliable in [36]. Note that the measured coastal data exist but were not available for this work. Although we do not have the real data from coastal areas, the real data in land area that we have also validates the alternate algorithm. Besides, the real data is also important for validating the scene. The two sets of data represent airborne and spaceborne data, indicating that the proposed algorithm is applicable to both spaceborne and airborne systems. In addition, it also shows that the algorithm is applicable to different scenes such as coastal and land scenes. Both sets of different data are introduced in this section, and the results of azimuth ambiguity suppression and the BPSR estimation processed by the improved algorithm are also shown.

3.1. Application to Simulated Data

3.1.1. Simulated Data

The simulated raw SAR data of coastal scenes are generated by an inverse omega-k algorithm, whose details can be found in [36] and are not reported here to save space. In the numerical simulation, modulation transfer functions (MTF), including tilt modulation, range modulation, and hydrodynamic modulation, were considered [17]. The simulation parameters were set as in [31], and the key values are listed in Table 1. The range of PRF is about 1000–3000 Hz, and the setting of 1725 Hz is relatively small in this range. However, the selection of PRF is determined by several factors. First, the PRF should satisfy the Nyquist sampling law; second, an excessively large PRF can reduce the unambiguous width and bring range ambiguity; third, PRF selection needs to avoid the echo of sub-satellite point, because this will cause interference in the sampled signal; and lastly, a large PRF comes at the large duty-ratio, which will lead to a large average power and large energy cost. The parameter SNR is the signal-to-noise ratio in ocean surface part and the parameter AASR is the azimuth-ambiguity-to-signal ratio in homogeneous scenes. Note that the effective baseline is 2.4 m and the velocity of the radar platform is 7600 m/s, both of which are closely related to the estimation of BPSR.

Table 1. Key simulation parameters for raw SAR (synthetic aperture radar) data of coastal area.

Parameter	Value
PRF (pulse repetition frequency)	1725 Hz
Polarization	VV
Radar carrier frequency	9.6 GHz
Effective baseline	2.4 m
Radar platform velocity	7600 m/s
SNR (signal-to-noise ratio)	6.5 dB
Mean water-to-land intensity ratio	−12 dB
AASR (azimuth-ambiguity-to-signal ratio)	−20 dB

The simulation processed SAR image of the coastal area is shown in Figure 5. Figure 5a highlights the azimuth ambiguity, and Figure 5b shows the interferogram phase image. As seen in Figure 5a, the bright objects in the land area produce three ghost signatures in the ocean area. The ghost signatures are also observed in Figure 5b, indicated by the yellow spots. The ghost images observed in both figures demonstrate the necessity to suppress azimuth ambiguity before estimating BPSR by the method introduced in Section 2. The results obtained after removing the ghost images and estimating BPSR are presented in the next section. The interferogram amplitude image of the region marked by the rectangle is shown in Figure 6. This sampling area contains more than 200 pixels.

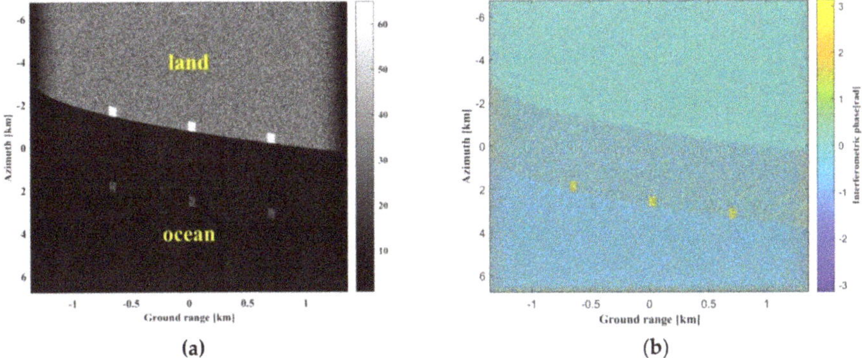

Figure 5. (a) Azimuth ambiguity of the SAR image in the coastal area (note the three bright objects in the land area and their ghost signatures in the ocean area); (b) Interferogram phase image.

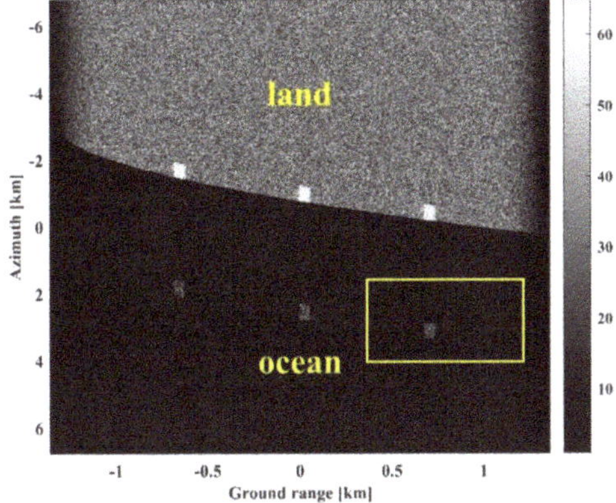

Figure 6. Interferogram amplitude image sampled of the region marked by the rectangle.

3.1.2. Results after Processing of the Simulated Data

After processing the data using the alternate iterative algorithm, the Doppler interval in the Doppler spectrum for estimating BPSR is shown in Figure 7, where the red line indicates the starting point of the Doppler range and the blue line indicates the terminal point. As seen in Figure 7, the starting point line is parallel to the terminal point line after four iterations, meaning that the interval tends to be stable between –580 Hz and 460 Hz. The Doppler interval selected by the EVSE analysis is not only used to suppress ambiguity but can also be adopted for estimating the baseline, improving its accuracy, and consequently, the accuracy of the BPSR estimation. From Equation (16), it can be seen that the value of BPSR is inversely proportional to the LOS velocity of the current. That is, when the

BPSR value decreases by 5×10^{-6} s, the line-of-sight velocity value will increase by about 0.01 m/s. Therefore, it is necessary to consider the effect of the BPSR value on the LOS velocity. The convergence of the BPSR estimate is shown in Figure 8, where the value of BPSR is found to stabilize at 3.15×10^{-4} s after only several iterations. The value of BPSR decreased by 0.17 s compared with the first calculation, so the value of the LOS velocity increased by 0.034 m/s.

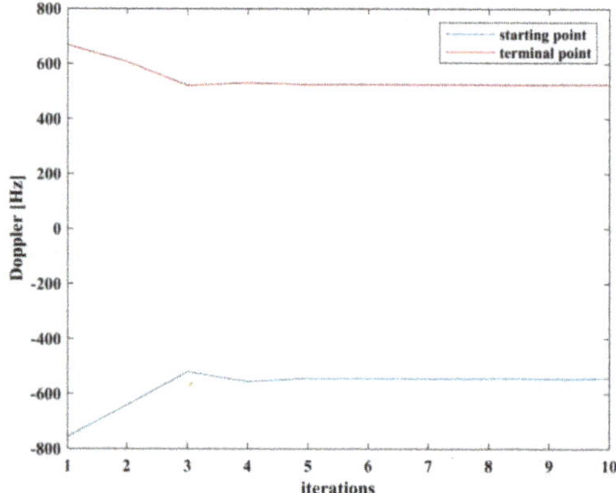

Figure 7. Doppler interval endpoint curves after several iterations using the simulated data (the red line indicates the terminal point of the Doppler range, and the blue line indicates the starting point of the Doppler range).

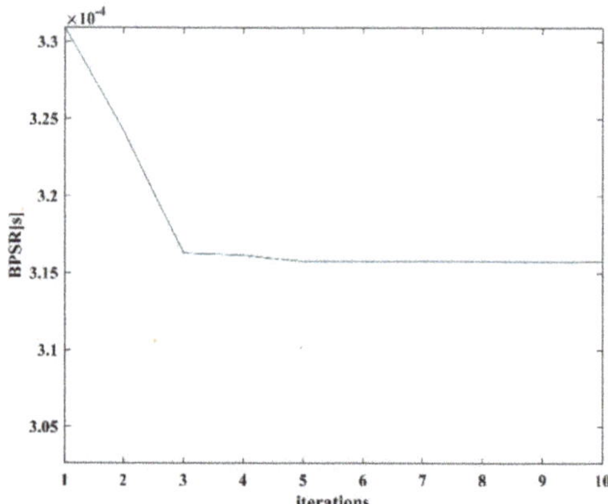

Figure 8. BPSR estimation using simulated data and the proposed algorithm.

To obtain a visual impression of the suppression of ghost signatures, we applied the proposed algorithm to the entire ocean surface. The SAR image and the interferogram phase image after the application of the alternate iterative algorithm for azimuth ambiguity suppression and BPSR estimation are presented in Figures 9a and 9b, respectively. Comparing Figure 5a with Figure 9a, and Figure 5b with Figure 9b, it can clearly be seen that the ghost signatures have been removed.

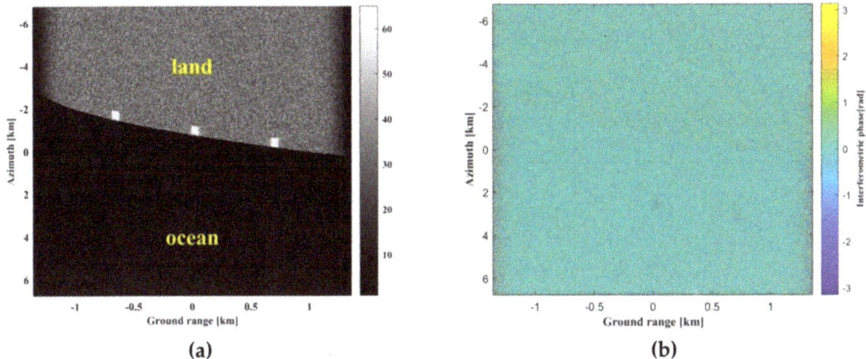

Figure 9. (**a**) SAR image after azimuth ambiguity suppression; (**b**) interferogram phase image after azimuth ambiguity suppression.

The Doppler sub-band after azimuth ambiguity suppression is shown as a two phase-frequency curve in Figure 10. Figure 10a corresponds to the first iteration used for selecting the Doppler sub- band, and Figure 10b shows the final iteration. In both figures, the blue line is the original interferometric phase trend, and the red line is that after the azimuth ambiguity suppression. Comparison of Figures 10a and 10b shows that the length of the Doppler sub-band decreases in the iterative process.

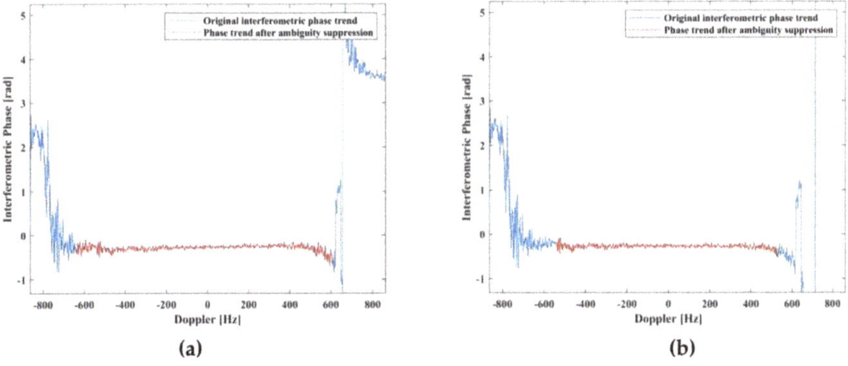

Figure 10. (**a**) Phase-frequency curve comparison in the first iteration of ambiguity suppression using simulated data (the blue line is original interferometric phase trend, and the red line is the phase trend after azimuth ambiguity suppression); (**b**) the phase-frequency curve comparison at the final iteration of ambiguity suppression using simulated data.

An estimation of the current velocity was carried out, and the results are shown in Table 2. Assuming a 20% error in BPSR and the true horizontal LOS current velocity of 3.0 m/s, we obtain an estimated mean LOS current velocity of 3.025 m/s, a mean bias of –0.025 m/s, and a standard deviation (STD) of 0.025 m/s. On the other hand, using Liu's method [31], the estimated mean LOS current velocity is 2.543 m/s and the mean bias is 0.457 m/s. There is a larger error in the sea-surface current velocity estimated by Liu's method [31], as highlighted in Table 2. It can thus be concluded that when the BPSR is not accurate, the proposed improved algorithm demonstrates its robustness for the current velocity estimation. Additionally, the results in Table 2 show the improvement of the proposed method compared with the method of Liu [31].

Table 2. Current velocity estimates.

Method	True Horizontal LOS (Line-of-Sight) Current Velocity	Estimated Mean LOS Current Velocity	Mean Bias	STD
Liu's method [31]	3.0 m/s	2.543 m/s	0.457 m/s	0.457 m/s
Algorithm proposed in this paper	3.0 m/s	3.025 m/s	−0.025 m/s	0.025 m/s

The current LOS Doppler velocity maps before and after the application of the improved method are shown in Figure 11a,b, respectively. Both of them are based on the simulation data of true current velocity of 3 m/s and a 20% margin of error in the baseline to calculate the LOS velocity. Figure 11a shows the result without any algorithm, and Figure 11b shows the result obtained by applying the method proposed in this paper. As shown in Figure 11a, affected by azimuth ambiguity, the LOS velocity of the current between the three "ghost" images and shore is about 4 m/s. In the ambiguous areas, the LOS velocity value of the current is further off to –6 m/s. In the open sea (the lower part of the image), the LOS velocity is 5 m/s. Notes that this is not due to ambiguity but is rather due to a 20% error in the baseline. As explained in the introduction, the azimuth ambiguity affects coastal waters but not the open sea. However, in Figure 11b, to obtain a visual impression of the suppression of ghost signatures, the proposed algorithm was applied to the entire ocean surface, and the baseline error and the azimuth ambiguity were both solved based on the application of the improved method, while the LOS velocity is almost the true value (3 m/s). Thus, Figure 11 shows the efficiency of the improved method.

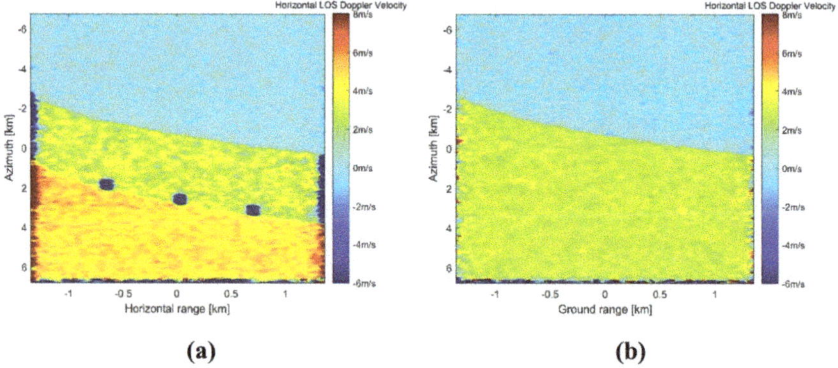

Figure 11. Retrieved horizontal LOS (line-of-sight) current Doppler velocity field based on the simulation data with a true current velocity of 3 m/s. (**a**) is without the algorithm application. (**b**) processed with the improved algorithm.

3.2. Application to Measured Data

3.2.1. Measured Data

The measured data are acquired over a land area but can nevertheless be processed using the proposed approach. The parameters of the data are listed in Table 3. Again, the two parameters to focus on—effective baseline and radar platform speed—have values of 0.2 m and 110 m/s, respectively. The measured data are unfocused in azimuth, as seen in Figure 12. Figure 12 is a range- compressed azimuth-unfocused SAR image, image-formed for a land area, and the vertical axis is the azimuth direction, while the horizontal axis is the ground range direction. Figure 12 is a piece of the land SAR image.

Table 3. Key parameters for the measured data.

Parameter	Value
Wavelength	0.03 m
PRF	830 Hz
Radar carrier frequency	10 GHz
Effective baseline	0.2 m
Radar platform velocity	110 m/s
SNR	18 dB
AASR	−20 dB

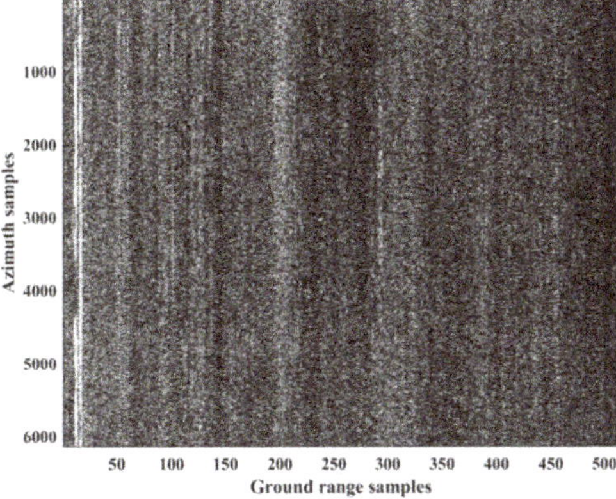

Figure 12. Range-compressed azimuth-unfocused SAR image.

3.2.2. Results after Processing of Measured Data

The Doppler interval endpoint curves are presented in Figure 13, where the Doppler interval converges quickly. From the estimated BPSR curve in Figure 14, the value of BPSR stabilizes at 1.565×10^{-3} s, implying a baseline value of 0.1742 m and a relative error of 1.149×10^{-3}, respectively.

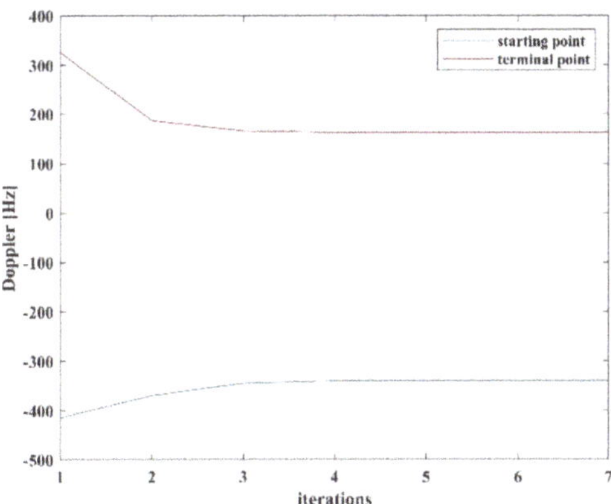

Figure 13. Doppler interval endpoint curve after several iterations using measured data (the red line is the terminal point and the blue line is the starting point of the Doppler range).

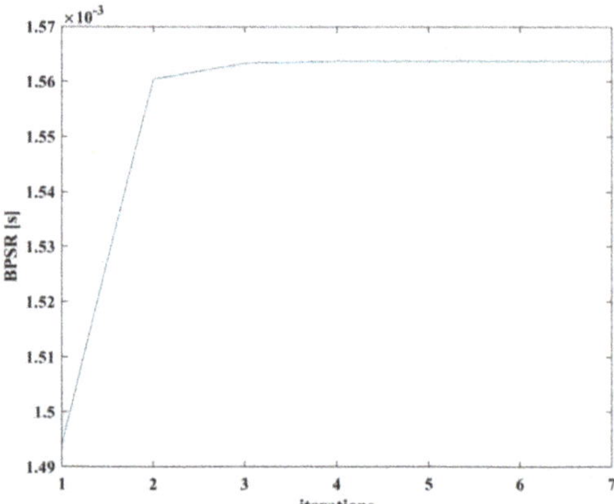

Figure 14. BPSR estimate using measured data and the proposed algorithm.

The above analysis demonstrates that although the measured data are from a land area and there is no azimuth ambiguity, BPSR can be estimated using the proposed approach. When the BPSR estimation is added into the method of Liu [31], the algorithm of Liu did not work due to the lack of a specific baseline value. This also shows the improvement of the proposed method.

The result for the case where scene heterogeneity is not considered is shown in Figure 15; the Doppler interval after ambiguity suppression is so narrow that the alternate iterative algorithm cannot be applied, leading to inaccurate BPSR estimation. Besides, unambiguous signals are discarded. However, when scene heterogeneity is taken into account, a Doppler sub-band can be calculated, as shown in Figure 16a, which shows the frequency chosen at the first iteration of the ambiguity suppression procedure. The oscillating parts at both ends of the curve are suppressed in the middle of the Doppler frequency range, as indicated by the red line. Figure 16b illustrates the Doppler sub-band in the last computation of ambiguity suppression. By comparing Figures 16a and 16b, we can conclude that the algorithm performs well, and that it is self-adaptive.

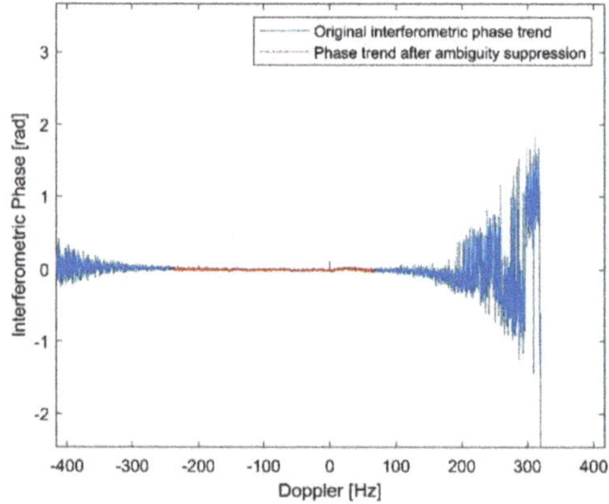

Figure 15. Phase-frequency curve comparison without consideration of scene heterogeneity.

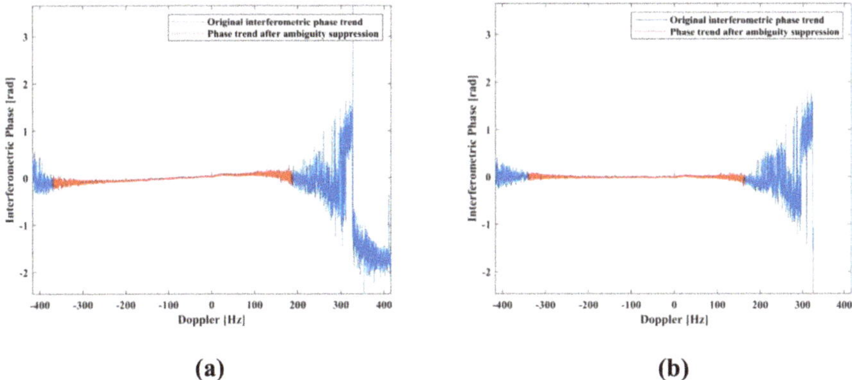

(a)   (b)

Figure 16. Phase-frequency curve comparison in the first iteration of ambiguity suppression using the measured data; (**b**) the phase-frequency curve comparison in the final iteration of ambiguity suppression using the measure data.

4. Discussion

It should be noted that the SAR-ATI phase estimates are almost controlled by ocean surface wave motions, which is called the Wind-wave-induced Artifact Surface Velocity (WASV). The wind speed is 5.5 m/s and the current velocity is 0.7 m/s, the WASV reaches 1.6 m/s, which makes a big contribution to the measured ocean surface motion. Mouche et al. [37] provided the first empirical model of the WASV, and the magnitude of the WASV was quantified by Martin et al. [12]. The removal of the contribution from wind-wave is achieved by simulating the SAR Doppler spectra from wind fields proposed by Elyouncha et al. [38]. However, this article does not aim to provide a detailed discussion of separating the current contribution from the wave-induced contribution to the Doppler velocity. Rather, this article focuses on the SAR system and signals related to measurements of ocean surface motion.

The results of azimuth ambiguity suppression and BPSR estimation using the improved alternate iteration algorithm are shown in Section 3. The method of EVSE analysis proposed by Liu [31] is based on an accurate BPSR, which aims at an ideal situation. For a 20% error in BPSR, the current velocity error calculated by Liu's method [31] is larger than that obtained by the improved algorithm in this paper; thus, for the actual situation, the improved algorithm is more effective. This improved algorithm is of great significance for the calculation of sea surface currents in coastal waters. The algorithm is adaptive, as it can be applied not only to spaceborne platforms but also to airborne platforms. Furthermore, the algorithm is also robust as it can be applied to different scenes with different heterogeneity. As shown in Section 3, the simulated data and real data are in different scenes, namely coastal water and land area, respectively. When the baseline is ambiguous or unknown, the improved method can work. Moreover, the TerraSAR-X [7] satellite based on the divided-antenna InSAR mode has strong azimuth ambiguities, and there is also obvious azimuth ambiguity in the ocean SAR image from the GaoFen-3 [39] satellite with ultra-fine strip mode, which seriously affects the data processing of subsequent marine applications. Therefore, the improved algorithm proposed in this paper can not only provide solutions to these problems but also improve the accuracy of the calculation of coastal current velocity. In the proposed algorithm, we did not consider the interference phase caused by the across baseline, which will be investigated and solved in future studies.

5. Conclusions

This paper proposes improvements in the algorithm for coastal current velocity measurements that consider real-life, non-ideal conditions and increase the precision of the velocity estimates.

The improved algorithm for the alternate azimuth ambiguity suppression and BPSR estimation can be applied to data from the ATI-SAR systems under relaxed conditions. The proposed approach incorporates a measure of scene heterogeneity, and importantly, is applicable to non-ideal situations with an inaccurate BPSR. The algorithm has successfully been tested on simulated and measured data. Because the measured data from a coastal area were not available, we used simulated data instead and measured data from a land area to test the practicability of the method. Note that data processing has no effect on the separation of wave and sea-surface currents in the subsequent estimation of the sea-surface currents. The processing results of the measured data from the land area also show the importance of considering scene heterogeneity. In addition, the algorithm needs only limited user inputs. After the application of an alternate iterative algorithm for ambiguity suppression and BPSR estimation, the current velocity can be estimated with an error of less than 0.05 m/s. This study indicates that the method can also help to increase the measurement accuracy of the current velocity using both airborne and spaceborne systems, even for systems that have limitations.

Author Contributions: Conceptualization: N.Y.; software, N.Y.; validation: N.Y., Y.H. and B.L.; writing-original draft preparation, N.Y.; writing—review and editing: Y.H. and B.L.; supervision: Y.H. All authors have read and agreed to the published version of the manuscript.

Funding: This work was supported in part by the National Key Research and Development Program under Grant 2016YFC1401002, the National Natural Science Foundation under Grant 41620104003, and the National Natural Science Foundation under Grant 41606201.

Conflicts of Interest: The authors declare no conflict of interest.

References

1. Rosen, P.A.; Hensley, S.; Joughin, I.R.; Li, F.K.; Madsen, S.N.; Rodriguez, E.; Goldstein, R.M. Synthetic aperture radar interferometry. *Proc. IEEE* **2000**, *88*, 333–380. [CrossRef]

2. Toporkov, J.; Perkovic, D.; Farquharson, G.; Sletten, M.; Frasier, S. Sea surface velocity vector retrieval using dual-beam interferometry: First demonstration. *IEEE Trans. Geosci. Remote Sens.* **2005**, *43*, 2494–2502. [CrossRef]

3. Prats-Iraola, P.; Scheiber, R.; Reigber, A.; Andres, C.; Horn, R. Estimation of the surface velocity field of the Aletsch glacier using Multibaseline airborne SAR interferometry. *IEEE Trans. Geosci. Remote Sens.* **2008**, *47*, 419–430. [CrossRef]

4. Purkis, S.; Klemas, V. Remote sensing and global environmental change. *Remote Sens. Glob. Environ. Chang.* **2011**, *29*, 216. [CrossRef]

5. Shemer, L.; Marom, M.; Markman, D. Estimates of currents in the nearshore ocean region using interferometric Synthetic Aperture Radar. *J. Geophys. Res. Space Phys.* **1993**, *98*, 7001–7010. [CrossRef]

6. Romeiser, R.; Runge, H.; Suchandt, S.; Sprenger, J.; Weilbeer, H.; Sohrmann, A.; Stammer, D. Current measurements in rivers by Spaceborne along-track InSAR. *IEEE Trans. Geosci. Remote Sens.* **2007**, *45*, 4019–4031. [CrossRef]

7. Romeiser, R.; Suchandt, S.; Runge, H.; Steinbrecher, U.; Grunler, S. First analysis of TerraSAR-X along-track InSAR-derived current fields. *IEEE Trans. Geosci. Remote Sens.* **2009**, *48*, 820–829. [CrossRef]

8. Wollstadt, S.; Lopez-Dekker, P.; De Zan, F.; Younis, M. Design principles and considerations for Spaceborne ATI SAR-based observations of ocean surface velocity vectors. *IEEE Trans. Geosci. Remote Sens.* **2017**, *55*, 4500–4519. [CrossRef]

9. Kim, D.; Moon, W.M. Measurements of ocean surface waves and currents using L- and C-band along-track interferometric SAR. *IEEE Trans. Geosci. Remote Sens.* **2003**, *41*, 2821–2832.

10. Romeiser, R. Current measurements by airborne along-track InSAR: Measuring technique and experimental results. *IEEE J. Ocean. Eng.* **2005**, *30*, 552–569. [CrossRef]

11. Yoshida, T.; Rheem, C.-K. Time-domain simulation of along-track Interferometric SAR for moving ocean surfaces. *Sensors* **2015**, *15*, 13644–13659. [CrossRef] [PubMed]

12. Martin, A.C.H.; Gommenginger, C.; Marquez, J.; Doody, S.; Navarro, V.; Buck, C. Wind-wave-induced velocity in ATI SAR ocean surface currents: First experimental evidence from an airborne campaign. *J. Geophys. Res. Ocean.* **2016**, *121*, 1640–1653. [CrossRef]

13. Martin, A.C.H.; Gommenginger, C.P.; Quilfen, Y. Simultaneous ocean surface current and wind vectors retrieval with squinted SAR interferometry: Geophysical inversion and performance assessment. *Remote Sens. Environ.* **2018**, *216*, 798–808. [CrossRef]

14. Goldstein, R.M.; Zebker, H.A. Interferometric radar measurement of ocean surface currents. *Nat. Cell Biol.* **1987**, *328*, 707–709. [CrossRef]

15. Kersten, P.R.; Toporkov, J.V.; Ainsworth, T.L.; Sletten, M.A.; Jansen, R.W. Estimating surface water speeds with a single-phase center SAR versus an along-track Interferometric SAR. *IEEE Trans. Geosci. Remote Sens.* **2010**, *48*, 3638–3646. [CrossRef]

16. Zeng, Z.; Li, X.; Ren, Y.; Chen, X. Exploratory research on the retrieval of internal wave parameters and sea surface current velocity based on TerraSAR-X satellite data. *Haiyang Xuebao* **2020**, *42*, 90–101. [CrossRef]

17. Romeiser, R.; Thompson, D. Numerical study on the along-track interferometric radar imaging mechanism of oceanic surface currents. *IEEE Trans. Geosci. Remote Sens.* **2000**, *38*, 446–458. [CrossRef]

18. Lombardini, F.; Griffiths, H.D.; Gini, F. Ocean surface velocity estimation in multichannel ATI-SAR systems. *Electron. Lett.* **1998**, *34*, 2429–2431. [CrossRef]

19. Besson, O.; Gini, F.; Griffiths, H.; Lombardini, F. Estimating ocean surface velocity and coherence time using multichannel ATI-SAR systems. *IEEE Proc. Radar Sonar Navig.* **2000**, *147*, 299. [CrossRef]

20. Liu, B.; Wang, T.; Li, Y.; Shen, F.; Bao, Z. Effects of doppler aliasing on baseline estimation in multichannel SAR-GMTI and solutions to address these effects. *IEEE Trans. Geosci. Remote Sens.* **2014**, *52*, 6471–6487. [CrossRef]

21. Biondi, F.; Addabbo, P.; Clemente, C.; Orlando, D. Measurements of surface river doppler velocities with along-track InSAR using a single antenna. *IEEE J. Sel. Top. Appl. Earth Obs. Remote Sens.* **2020**, *13*, 987–997. [CrossRef]

22. Thompson, D.R.; Jensen, J.R. Synthetic aperture radar interferometry applied to ship-generated internal waves in the 1989 Loch Linnhe experiment. *J. Geophys. Res. Space Phys.* **1993**, *98*, 10259. [CrossRef]

23. Graber, H.C.; Thompson, D.R.; Carande, R.E. Ocean surface features and currents measured with synthetic aperture radar interferometry and HF radar. *J. Geophys. Res. Space Phys.* **1996**, *101*, 25813–25832. [CrossRef]

24. Farquharson, G.; Junek, W.N.; Ramanathan, A.; Frasier, S.J.; Tessier, R.; McLaughlin, D.J.; Sletten, M.A.; Toporkov, J.V. A pod-based dual-beam SAR. *IEEE Geosci. Remote Sens. Lett.* **2004**, *1*, 62–65. [CrossRef]

25. Romeiser, R.; Breit, H.; Eineder, M.; Runge, H.; Flament, P.; De Jong, K.; Vogelzang, J. Current measurements by SAR along-track interferometry from a Space Shuttle. *IEEE Trans. Geosci. Remote Sens.* **2005**, *43*, 2315–2324. [CrossRef]

26. Suchandt, S.; Runge, H. Ocean surface observations using the TanDEM-X satellite formation. *IEEE J. Sel. Top. Appl. Earth Obs. Remote Sens.* **2015**, *8*, 5096–5105. [CrossRef]

27. Romeiser, R.; Runge, H.; Suchandt, S.; Kahle, R.; Rossi, C.; Bell, P.S. Quality assessment of surface current fields from TerraSAR-X and TanDEM-X along-track interferometry and doppler centroid analysis. *IEEE Trans. Geosci. Remote Sens.* **2013**, *52*, 2759–2772. [CrossRef]

28. Romeiser, R.; Runge, H. Theoretical evaluation of several possible along-track InSAR modes of TerraSAR-X for ocean current measurements. *IEEE Trans. Geosci. Remote Sens.* **2006**, *45*, 21–35. [CrossRef]

29. Xiao, P.; Wu, P.; Yu, Z.; Li, C. Azimuth ambiguity suppression in SAR images based on compressive sensing recovery algorithm. *J. Radars* **2016**, *5*, 35–41.

30. Klemas, V. Remote sensing of coastal and ocean currents: An overview. *J. Coast. Res.* **2012**, *28*, 576–586. [CrossRef]

31. Liu, B.; He, Y.; Li, Y.; Duan, H.; Song, X. A new azimuth ambiguity suppression algorithm for surface current measurement in coastal waters and rivers with along-track InSAR. *IEEE Trans. Geosci. Remote Sens.* **2018**, *57*, 3148–3165. [CrossRef]

32. Guarnieri, A. Adaptive removal of azimuth ambiguities in SAR images. *IEEE Trans. Geosci. Remote Sens.* **2005**, *43*, 625–633. [CrossRef]

33. Chapron, B.; Collard, F.; Ardhuin, F. Direct measurements of ocean surface velocity from space: Interpretation and validation. *J. Geophys. Res. Space Phys.* **2005**, *110*, 07008. [CrossRef]

34. Kudryavtsev, V.N.; Chapron, B.; Myasoedov, A.G.; Collard, F.; Johannessen, J.A.; Myasoedov, A.G.; Johannessen, J.A. On dual co-polarized SAR measurements of the ocean surface. *IEEE Geosci. Remote Sens. Lett.* **2012**, *10*, 761–765. [CrossRef]

35. Yu, H.; Lee, H.; Cao, N.; Lan, Y. Optimal baseline design for Multibaseline InSAR phase unwrapping. *IEEE Trans. Geosci. Remote Sens.* **2019**, *57*, 5738–5750. [CrossRef]

36. Liu, B.; He, Y. SAR raw data simulation for ocean scenes using inverse omega-K algorithm. *IEEE Trans. Geosci. Remote Sens.* **2016**, *54*, 6151–6169. [CrossRef]

37. Mouche, A.A.; Collard, F.; Chapron, B.; Dagestad, K.-F.; Guitton, G.; Johannessen, J.A.; Kerbaol, V.; Hansen, M.W. On the use of doppler shift for sea surface wind retrieval from SAR. *IEEE Trans. Geosci. Remote Sens.* **2012**, *50*, 2901–2909. [CrossRef]

38. Elyouncha, A.; Eriksson, L.E.; Romeiser, R.; Carvajal, G.K.; Ulander, L.M.H. Wind-wave effect on ATI-SAR measurements of ocean surface currents in the Baltic Sea. In Proceedings of the 2016 IEEE International Geoscience and Remote Sensing Symposium (IGARSS), Beijing, China, 10–15 July 2016; pp. 3982–3985.

39. Long, Y.; Zhao, F.; Zheng, M.; Jin, G.; Zhang, H. An azimuth ambiguity suppression method based on local azimuth ambiguity-to-signal ratio estimation. *IEEE Geosci. Remote Sens. Lett.* **2020**, 1–5. [CrossRef]

Article

Spatial Structure, Short-temporal Variability, and Dynamical Features of Small River Plumes as Observed by Aerial Drones: Case Study of the Kodor and Bzyp River Plumes

Alexander Osadchiev [1,2,*], Alexandra Barymova [3], Roman Sedakov [1], Roman Zhiba [4] and Roman Dbar [4]

[1] Shirshov Institute of Oceanology, Russian Academy of Sciences, Nakhimovskiy prospect 36, 117997 Moscow, Russia; roman.sedakov@gmail.com
[2] Moscow Institute of Physics and Technology, Institusky lane 9, 141701 Dolgoprudny, Russia
[3] Marine Research Center at Lomonosov Moscow State University, Leninskie Gory 1, 119992 Moscow, Russia; alexandra.barymova@marine-rc.ru
[4] Institute of Ecology, Academy of Sciences of Abkhazia, Krasnomayatskaya str. 67, 384900 Sukhum, Abkhazia; romazb@mail.ru (R.Z.); romandbar@mail.ru (R.D.)
* Correspondence: osadchiev@ocean.ru

Received: 20 August 2020; Accepted: 17 September 2020; Published: 20 September 2020

Abstract: Quadcopters can continuously observe ocean surface with high spatial resolution from relatively low altitude, albeit with certain limitations of their usage. Remote sensing from quadcopters provides unprecedented ability to study small river plumes formed in the coastal sea. The main goal of the current work is to describe structure and temporal variability of small river plumes on small spatial and temporal scales, which are limitedly covered by previous studies. We analyze optical imagery and video records acquired by quadcopters and accompanied by synchronous in situ measurements and satellite observations within the Kodor and Bzyp plumes, which are located in the northeastern part of the Black Sea. We describe extremely rapid response of these river plume to energetic rotating coastal eddies. We reveal several types of internal waves within these river plumes, measure their spatial and dynamical characteristics, and identify mechanisms of their generation. We suggest a new mechanism of formation of undulate fronts between small river plumes and ambient sea, which induces energetic lateral mixing across these fronts. The results reported in this study are addressed for the first time as previous related works were mainly limited by low spatial and/or temporal resolution of in situ measurements and satellite imagery.

Keywords: small river plume; aerial drone; coastal processes; frontal zones; internal waves

1. Introduction

Airborne remote sensing of sea surface is constantly expanding during the last ten years due to significant progress in development of aerial drones, especially low-cost quadcopters [1–6]. Many previous works used airborne data to study various marine processes including mapping of coastal topography [7,8] and bathymetry [9–12], surveying of marine flora and fauna [13–22], and monitoring of water quality and anthropogenic pollution [23–28]. Several works used airborne data to study physical properties of sea surface layer including estimation of turbulence [29] and reconstruction of surface currents [30,31]. However, applications of aerial remote sensing are still rare in physical oceanography, especially in comparison with numerous studies based on satellite remote sensing. Studies of river plumes provide a good example of this situation. Hundreds of related works were based on high-resolution [32,33], medium-resolution [34–37], and low-resolution [38–40] optical satellite data, satellite-derived temperature [41–43], salinity [4–46], and roughness [47–49] of sea surface. On the other hand, only several studies used airborne remote sensing to study river plumes [25,26,50–53]. Moreover, we are not aware of any study, which specifically addressed structure, variability, and dynamical features of small plumes using aerial remote sensing data. This point provides the main motivation of the current work.

General aspects of the structure and dynamics of river plumes as well as their regional features were addressed in many previous studies. Nevertheless, these works were mostly focused on large river plumes, while small rivers plumes received relatively little attention. However, small rivers play an important role in global land-ocean fluxes of fluvial water and suspended and dissolved sediments [54–56]. Small rivers form buoyant plumes that have small spatial scales and, therefore, small residence time of freshened water, which is equal to hours and days, due to relatively low volume of river discharge and its intense mixing with ambient sea [57]. Dissipation of freshened water as a result of mixing of a small plume with subjacent saline sea limitedly influences ambient sea and does not result in accumulation of freshwater in adjacent sea area. As a result, small plumes are characterized by sharp salinity and, therefore, density gradients at their boundaries with ambient sea. This feature is not typical for large river plumes and results in significant differences in spreading and mixing between small plumes and large plumes. Sharp vertical density gradient at the bottom boundary of a small plume hinders vertical energy transfer between a small river plume and subjacent sea [57]. This feature strongly affects spreading dynamics of a small plume due to the following reasons. First, the majority of wind energy transferred to sea remains in a small plume, because the vertical momentum flux diminishes at the density gradient between a plume and subjacent sea. Therefore, wind stress is concentrated in a shallow freshened surface layer that causes higher motion velocity and more quick response of dynamics of a small plume to variability of wind forcing, as compared to ambient sea [58,59]. It results in wind-driven dynamics of small plumes, which is characterized by very energetic short-temporal variability of their positions, shapes, and areas [60–64].

Study of structure and variability of small river plumes at small spatial and temporal scales is essential for understanding the fate of freshwater discharge from small rivers to sea and the related transport of suspended and dissolved river-borne constituents. However, high short-temporal variability of small plumes and their small vertical sizes inhibit precise in situ measurements of their thermohaline and dynamical characteristics [57]. Satellite remote sensing also does not provide the necessary spatial resolution and temporal coverage for small river plumes. As a result, many important aspects of structure, variability, and dynamics of small river plumes at small spatial and temporal scales remain unstudied.

Quadcopters are especially efficient in observation of small river plumes because they can continuously observe sea surface with high spatial resolution from relatively low altitude. Quadcopters can be used during overcast sky when optical satellite instruments cannot observe sea surface. The main drawback of their usage is relatively short duration of continuous operation (less than several hours), limited weight of carried instruments, and inability of their operation during strong wind, rain, snow, low temperature, and other inappropriate weather conditions. Despite these limitations, usage of quadcopters provides unprecedented ability to study structure of small river plumes, detect and measure their short-temporal variability, and register various dynamical features of these plumes. Therefore, the main goal of the current work is to describe structure and temporal variability of small river plumes on small spatial (from meters to hundreds of meters) and temporal (from minutes to hours) scales, which are limitedly covered by in situ measurements and satellite imagery and remain almost unaddressed by the previous studies.

In this work we use aerial remote sensing supported by synchronous in situ measurements and satellite observations to study small river plumes formed in the northeastern part of the Black Sea. We show that usage of aerial drones, first, strongly enhances in situ and satellite observations of structure and variability of small plumes, second, provides ability to perform accurate, continuous, and high-resolution measurements of their spatial characteristics and current velocity fields, and, finally, significantly improves operational organization of field measurements. Owing to continuous and high-resolution aerial remote sensing, we report several novel results about spatial structure, short-temporal variability, and dynamical features of small river plumes. These results include strongly inhomogeneous structures of small river plumes manifested by complex and dynamically active internal frontal zones; undulate (lobe-cleft) form of a sharp front between a small river plume and ambient sea; energetic lateral mixing across this front caused by its baroclinic instability; internal waves generated by river discharge near a river estuary and propagating within the inner plume; and internal waves generated by vortex circulation of river plume and propagating within the outer plume. The obtained results reveal significant differences in structure, variability, and dynamics between small plumes and large plumes.

The paper is organized as follows. Section 2 provides the detailed information about the aerial, in situ, and satellite data, as well as the processing methods used in this study. The results derived from aerial observations of small river plumes supported by in situ measurements and satellite observations are described in Section 3. Section 4 focuses on discussion and interpretation of the revealed features of spatial structure, short-temporal variability, and dynamics of small river plumes. The summary and conclusions are given in Section 5.

2. Data and Methods

2.1. Study Area

In this work, we focused on the Kodor and Bzyp river plumes formed in the northeastern part of the Black Sea (Figure 1). These rivers were chosen as the case sites due to the following reasons. First, these rivers have high concentrations of suspended sediments (300-500 g/m^3 in the Kodor River and 100-300 g/m^3 in the Bzyp River) [65], therefore, the turbid Kodor and Bzyp plumes can be effectively detected by optical aerial and satellite imagery. Second, the Kodor and Bzyp rivers are relatively small, their catchment areas are 2000 and 1500 km^2, respectively, and their average annual discharges are approximately 130 and 120 m^3/s, respectively [65]. As a result, the Kodor and Bzyp plumes are small enough to be observed by aerial remote sensing from relatively small altitude (< 200 m). However, both rivers are mountainous with large mean basin altitudes (> 1500 m) and slopes (> 0.02‰), as well as high drainage density (> 0.8 1/km). Therefore, during spring freshet and short-term rain-induced floods the runoffs from the Kodor and Bzyp rivers dramatically increase by 1-2 orders of magnitude. Third, despite their relatively small spatial extents, the Kodor and Bzyp plumes are the largest plumes in the study area. As a result, structure and dynamics of these plumes are not influenced by interaction with other river plumes. Fourth, the Kodor and Bzyp rivers have different mouth morphologies that affect the structure of their plumes. The majority of the Bzyp River runoff inflows to sea from the main river channel, however, a small side-channel is formed during high discharge periods. The Kodor River inflows to sea from three large river channels, which form the Kodor Delta. The mouths of these deltaic branches are located along the 2 km long segment of the coastline. Finally, wind, cloud, and rain conditions in the study area are favorable for aerial and satellite observations of the river plumes during the majority of the year.

The continental shelf at the study area is very steep and narrow. The distance between the coastline and the 500 m isobath is less than 10 km near the Kodor and Bzyp mouths (Figure 1). The main coastline features at the study area are large capes, namely the Iskuria and Pitsunda capes, located to the south from the Kodor and Bzyp deltas, respectively (Figure 1). The local sea circulation from surface to the depth of 200–250 m is governed by alongshore currents due to the current system cyclonically circulating along the continental slope, which is generally referred to as the Black Sea Rim Current [66,67]. Sea surface circulation in the study region is also influenced by nearshore anticyclonic eddies, which are regularly formed between the main flow of the Rim Current and the coast owing to baroclinic instability caused by wind forcing and coastal topography [68–70]. Tidal circulation at the study area is very low and tidal amplitudes are less than 6 cm [71,72]. Salinity in the coastal sea, which is not influenced by river discharge, is 17–18 [67,73].

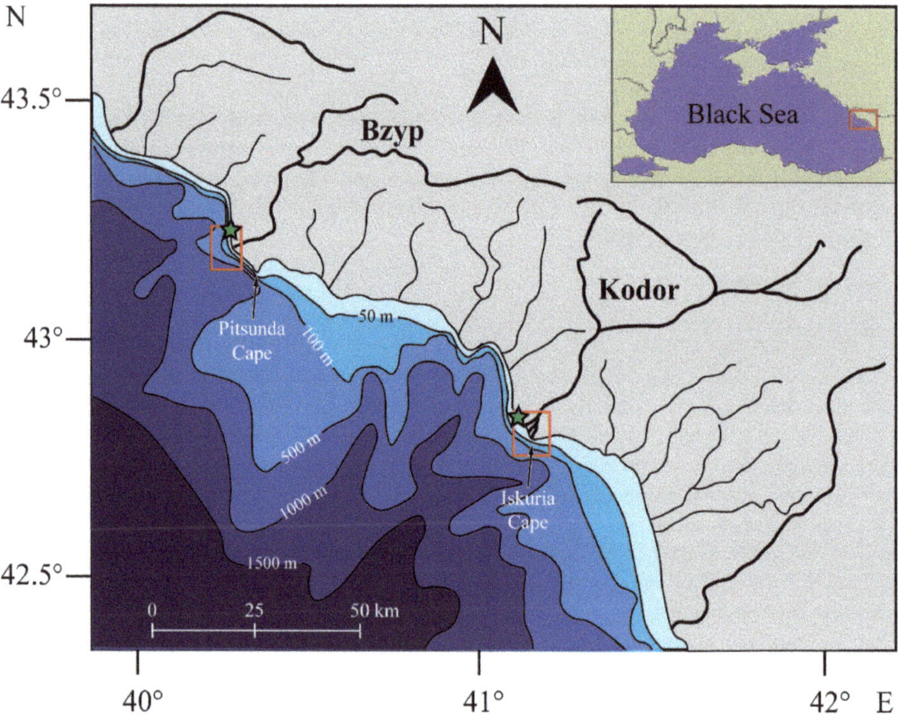

Figure 1. Bathymetry of the study region, locations of the Iskuria and Pitsunda capes, the Bzyp and Kodor rivers, and other smaller rivers of the study region. Location of the study region at the northeastern part of the Black Sea is shown in the inset. Red boxes indicate areas of aerial observations and in situ measurements at the Bzyp and Kodor plumes. Green stars indicate locations of meteorological stations.

2.2. Aerial, In Situ, and Satellite Data

Aerial observations of the Kodor and Bzyp plumes were performed by a quadcopter (*DJI Phantom 4 Pro*) equipped with a 12 MP/4K video camera. Aerial observations of the plumes were supported by ship-borne in situ measurements of salinity, temperature, turbidity, and current velocity within the plumes and the adjacent sea. The size of this quadcopter is small enough to be launched from and landed on a small boat. It provides opportunity for a quadcopter operator to be onboard the research vessel and to effectively coordinate synchronous in situ measurements and water sampling.

Aerial observations and in situ measurements of the Kodor plume were conducted on 1–2 September 2018 and 1–3 April 2019, while aerial observations and in situ measurements of the Bzyp plume were performed on 31 May–1 June 2019. Below we provide the protocols of these aerial surveys according to the scheme suggested by Doukari et al. [74].

The quadcopter was flying over coastal sea areas adjacent to the Kodor and Bzyp river mouths. The take-off and landing spot was located on a vessel/boat that provided opportunity to perform flights at different areas of the plumes without any limitations on their distance to the seashore. The distance between the quadcopter and the research vessel/boat did not exceed 1 km. Quadcopter shooting altitude depended on the spatial scale of the sensing sea surface process and varied from 10–30 m for the small-scale frontal circulation to 150–200 m for detection of plume position and area. Weather conditions during the field surveys were favorable for usage of the quadcopter. Wind forcing during the flights was moderate (< 8 m/s), air temperature varied between 15 and 30° C, and air humidity varied between 60% and 90%. The flights were conducted during no-rain conditions from morning to evening. In case of clear sky conditions, sun glint strongly affected the quality of the aerial data during the daytime. Wave heights were < 0.5 m during the flights.

In situ measurements performed in the study areas were the following. Continuous salinity and temperature measurements in the surface sea layer (0.5–1 m depth) were performed along the ship tracks using a shipboard pump-through system equipped by a conductivity-temperature-depth (CTD) instrument (*Yellow Springs Instrument 6600 V2*) [62,75]. Vertical measurements of salinity, temperature, and turbidity were performed using a CTD-instrument (*Sea-Bird Electronics SBE 911plus*) at 0.2 m spatial resolution. Vertical measurements of current velocity were performed using an acoustic Doppler current profiler (ADCP) (*Teledyne RDI Workhorse Sentinel*) and a CTD-ADCP-instrument (*Aanderaa SeaGuard RCM*). Vertical profiling was performed from sea surface to the depth of 10 m or to seafloor in shallow areas. The positions of individual in situ measurements are given in Section 3. Wind forcing during the field measurements was measured by a compact weather station (*Gill GMX200*) with temporal resolution of 1 minute. The weather station was mounted at the height of 10 m at a pier on a distance of 30 m from the coastline (Figure 1).

The Kodor and Bzyp plumes were also studied using Sentinel-2 Multispectral Instrument (MSI) data collected in 2017–2019. The Sentinel-2 Level-1C products were downloaded from the Copernicus Open Access Hub (https://scihub.copernicus.eu/) (Supplementary Materials). Atmospheric correction was applied to these products using Sen2Cor module version 2.2.1 within the Sentinel-2 Toolbox (S2TBX), Sentinel Application Platform (SNAP) version 5.0.7.

2.3. Processing of Aerial Data

In this study we used an optical flow algorithm to reconstruct velocity fields in the sea surface layer from quadcopter video records [76,77]. The main principle of optical flow algorithms used for calculation of motion from two consecutive pictures is the following. It was assumed that for each point $\vec{x}$ (i.e., pixel) on both frames a certain signal intensity property I (i.e., brightness) was conserved:

$$I\left(\vec{x}, t\right) = I\left(\vec{x} + \Delta\vec{x}, t + \Delta t\right) \tag{1}$$

By linearizing the intensity of the second frame with respect to the intensity of the first frame a gradient constraint equation is obtained in the following way:

$$\nabla I\left(\vec{x}, t\right) \cdot \vec{u} + I_t\left(\vec{x}, t\right) = 0 \tag{2}$$

where $\nabla I = \left(I_x, I_y\right)$ is the spatial partial derivatives of intensity, $\vec{u} = d\vec{x}/dt$ is the velocity, and I_t is the temporal derivative of intensity. The derivatives ∇I and I_t can be directly calculated, while the 2D velocity field $\vec{u}$ is unknown. Therefore, Equation (1) requires an additional constraint and it is assumed that the displacement $\Delta\vec{x}$ is constant in any small neighborhood, i.e., we search for a displacement that minimizes the constraint error:

$$E\left(\vec{x}\right) \;=\; \sum_{\vec{x}} g\left(\vec{x}\right)\left(\nabla I\left(\vec{x},t\right)\cdot\vec{u} + I_t\left(\vec{x},t\right)\right)^2 \tag{3}$$

where $g\left(\vec{x}\right)$ is a weight function. Thus, minimization of $E\left(\vec{x}\right)$ with respect to $\vec{u}$ provides an additional condition for Equation (2). The resulting vector field $\vec{u}$ calculated from Equations (2) and (3) is regarded as an optical flow estimate. In this work, we used the Farneback weight function [78] freely available in the OpenCV computer vision library (https://opencv.org/). This algorithm approximates a neighborhood of a pixel in each pair of frames by a quadratic polynomial function applying the polynomial expansion transform. Therefore, a constraint equation is based on a polynomial approximation of the given signal. On the assumption of small variability of a displacement field, the algorithm minimizes quadratic error of the constraint and calculates the optical flow estimation.

The estimation of surface velocity fields in the study region was performed in two stages. First, we applied the optical flow algorithm with large prescribed sizes of pixel neighborhoods for the reconstruction of motion of distinct plume boundaries and fronts. Second, we reconstructed motion within the river plume using the optical flow algorithm with a reduced neighborhood size. Spatial scale of motion, which is intended to be reconstructed, positively correlates with optimal size of a pixel neighborhood. An algorithm with a small pixel neighborhood more accurately reconstructs small-scale motion, but shows lower quality for large motion patterns, as compared to an algorithm with a large pixel neighborhood. The overall neighborhood size was prescribed according to spatial scales of ocean surface features (e.g., river plume fronts), in which motion is expected to be detected by an optical flow algorithm. Thus, the optimal neighborhood size intended to reconstruct the large-scale motion of river plumes should be equal to the width of distinct plume boundaries and fronts. In this study, the large size of a pixel neighborhood was prescribed equal to 30 m, while the small size of a pixel neighborhood was set equal to 1 m. In case of application of this algorithm to other regions, we suggest prescribing neighborhood sizes equal to relevant spatial scales of the considered ocean surface features.

Due to high resolution of the video camera used and continuous video recording, the optical flow algorithm efficiently detected motion of the distinct frontal zones within the river plumes, as well as motion of foam and floating litter accumulated at these fronts which is indicative of the circulation patterns at the frontal zones. As a result, the reconstructed surface velocity fields showed good accordance with visually inspected shifts of the frontal zones, foam, and floating litter at the video records. Stable positioning of a quadcopter is important for precise motion detection at sea surface. Moderate wind speed during the field surveys did not negatively affect the quality of the obtained aerial data. However, strong wind forcing during camera shooting can hinder accurate reconstruction of surface velocity fields. Sun glint is another important issue that can impede motion detection at aerial video records. Intensity of the sun glint depends on solar elevation angle, camera shooting angle and direction; therefore, it can be reduced by correct selection of quadcopter altitude and position. Usage of polarizing filters for quadcopter camera can reduce glint from water surface, however, its efficiency strongly depends on camera shooting angle.

3. Results

3.1. Spatial Structure and Short-temporal Variability of the Kodor and Bzyp Plumes

The field surveys were performed during spring freshet at the Bzyp River (260 m^3/s) on 31 May–1 June 2019; during drought period at the Kodor River (40 m^3/s) on 1–3 April 2019; and during flash flooding period at the Kodor River (80–150 m^3/s) on 31 August–2 September 2018. Wind forcing was moderate during these field surveys. Average and maximal wind speed registered at weather station in the study regions were 3.1 and 7.6 m/s during 31 August–2 September 2018; 2.4 and 6.2 m/s during 1–3 April 2019; and 2.9 and 5.6 m/s during 31 May–1 June 2019.

Vertical salinity measurements in the study areas revealed that these low-saline plumes are shallow (< 5 m depth) and have distinct vertical salinity gradients with the ambient saline sea. Due to elevated concentrations of terrigenous suspended sediments in the Kodor and Bzyp rivers [65], turbidity within the Kodor and Bzyp plumes was significantly larger than in the ambient sea and showed good correlation with reduced salinity (Figure 2). The Pearson correlation coefficients (r) between salinity and turbidity are equal to −0.87 and −0.71 for the Kodor and Bzyp plumes respectively with p-values equal to 0.0000. These high absolute values of the correlation coefficients at low p-values indicate that the observed relations between salinity and turbidity within the Kodor and Bzyp plumes (low salinity and high turbidity), on the one hand, and the ambient sea water (high salinity and low turbidity), on the other hand, are statistically significant. As a result, surface turbidity structures of the Kodor and Bzyp plumes observed by optical remote sensing are indicative of surface salinity structures of these plumes.

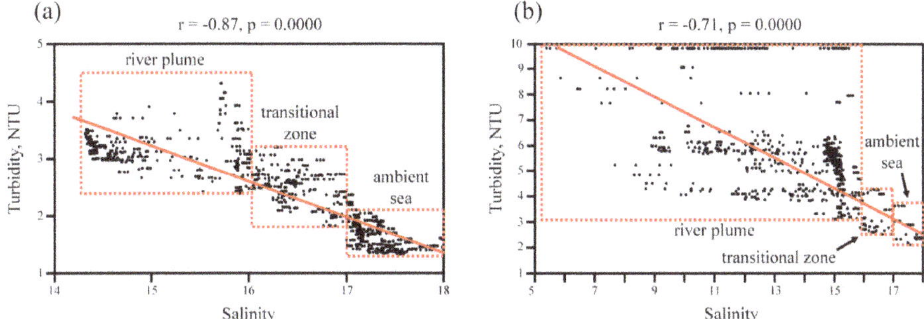

Figure 2. Relations between salinity and turbidity (**a**) within the Kodor plume and the adjacent saline sea on 2–3 April 2019 and (**b**) within the Bzyp plume and the adjacent saline sea on 31 May 2019. Dashed red boxes indicate river plumes, transitional zones, and ambient saline sea. Red lines indicate regression lines. The Pearson correlations coefficients (r) with p-values, which indicate statistical significance of the observed relations, are given above the diagrams.

Aerial remote sensing and satellite imagery showed that the alongshore extents of turbid surface water associated with the considered river plumes during low discharge conditions are 1–5 km. The obtained estimates were consistent with salinity measurements at the study area. However, flooding discharge results in abrupt expanding of these plumes, their extents and areas can exceed 20 km and 50 km^2, respectively. Aerial and satellite images, surface salinity distribution, and vertical salinity profiles obtained on 31 August 2018 in the coastal area adjacent to the Kodor Delta are illustrative of spatial scales, as well as horizontal and vertical structure of the Kodor plume (Figure 3).

Aerial observations and in situ measurements revealed strongly inhomogeneous salinity and turbidity structure of the Kodor plume manifested by complex and dynamically active frontal zones within the plume (Figures 4–6). In particular, surface salinity showed no dependence on the distance to the mouths of the deltaic branches that is regarded typical for river plumes [79–81], especially in numerical modeling studies [82–85]. This inhomogeneous structure is formed due to impact of several different processes including the formation of the Kodor plume by several spatially distributed sources, the large inter-day river discharge variability in response to sporadic rain events, and the bathymetric features that influence spreading of the plume.

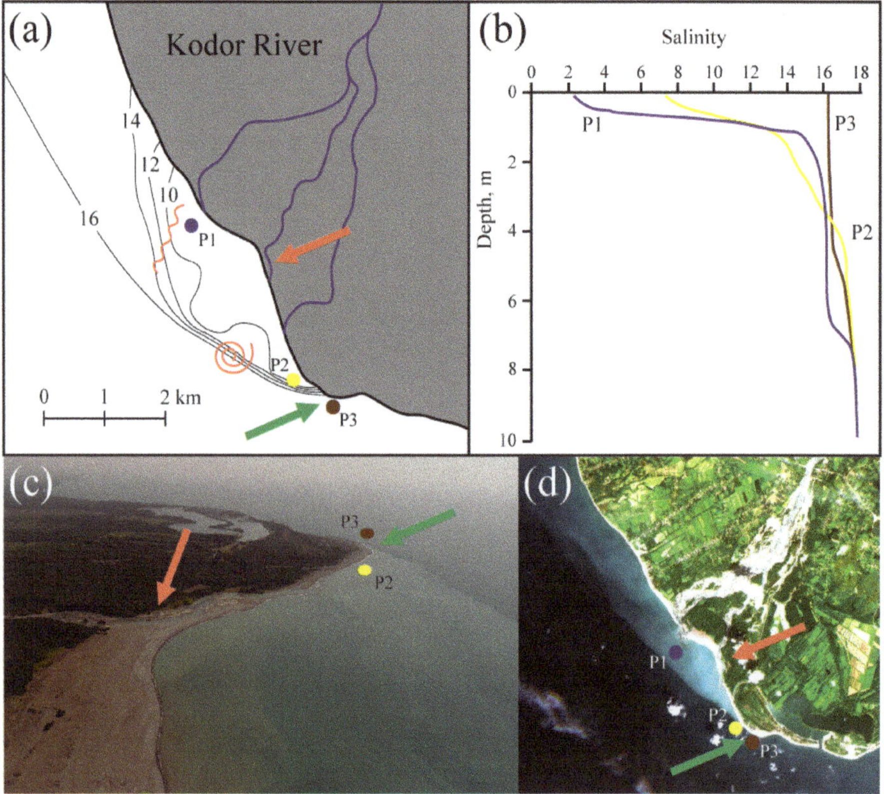

Figure 3. (**a**) surface salinity distribution, (**b**) vertical salinity profiles, (**c**) aerial image (acquisition time 13:29), and (**d**) Sentinel-2 ocean color composite of the Kodor plume from 31 August 2018. Color dots indicate locations of vertical salinity measurements (1, blue—near the river mouth; 2, yellow—near the plume border, and 3, brown—at the ambient saline sea). Red arrows indicate location of the central deltaic branch of the Kodor River, green arrows indicate location of the Iskuria Cape. The red swirl at panel (**a**) indicates location of the eddy detected on 1 September, 2018 (see Figures 7–9). The red wave line at panel (**a**) indicates location of the undulate (lobe-cleft) plume border detected on 1 September 2018 (see Figures 12–15).

The Kodor River inflows to sea from three deltaic branches with different discharge rates. As a result, all three branches form individual river plumes that merge and coalesce into the common Kodor plume. These three river plumes have different structure, spatial characteristics, and dynamics, therefore, they interact as individual water masses and form stable frontal zones observed by aerial imagery (Figure 4a) [86–88]. In situ measurements performed on 2 September, 2018 revealed sharp salinity gradient at the frontal zone between the river plumes formed by the northern and the central deltaic branches of the Kodor River. Surface salinity along the transect that crossed this frontal zone abruptly decreased from 14 to 8–10 on a distance of 5 m (Figure 4b).

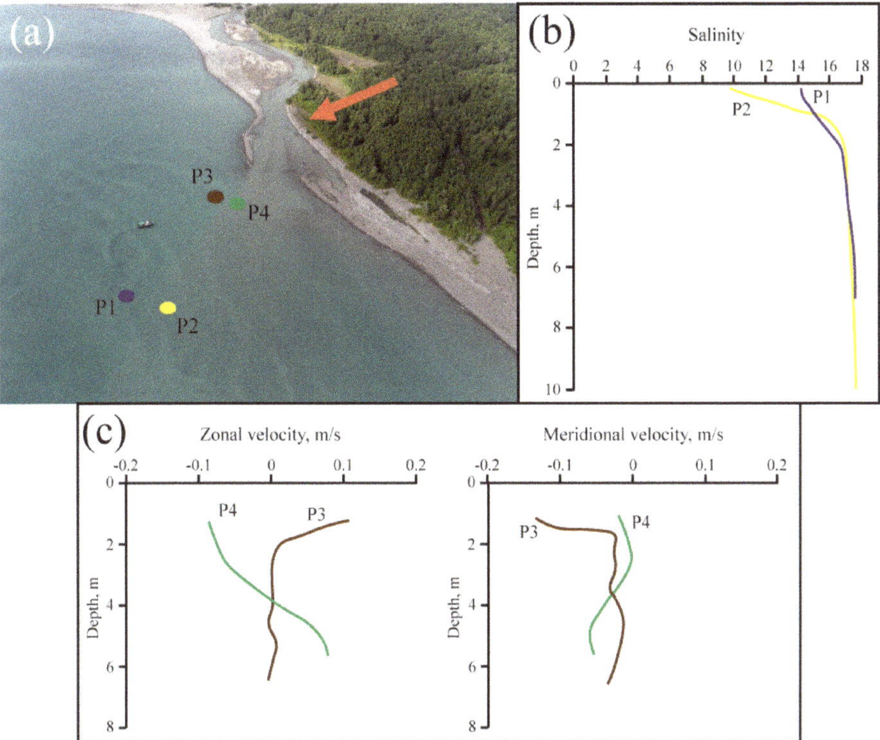

Figure 4. (**a**) aerial image, (**b**) vertical salinity, and (**c**) velocity profiles at the frontal zone between river plumes formed by the northern and central deltaic branches of the Kodor River on 1 September, 2018. Colored dots indicate locations of vertical salinity (P1, blue—the northern plume; P2, yellow—the central plume) and velocity (P3, brown—the northern plume; P4, green—the central plume) measurements. The red arrow in panel (a) indicates location of the central deltaic branch of the Kodor River.

The discharge of the Kodor River shows quick response to precipitation events that is common for small mountainous rivers with small and steep watershed basins. Frequent rains at the mountainous northeastern coast of the Black Sea cause high inter-day variability of the discharge rate of the Kodor River [65,89]. As a result, the area of the Kodor plume can significantly change during less than one day that was observed on 31 August–2 September 2018 during the field survey. Heavy rain that occurred during 6 hours at night on 31 August–1 September (according to the local weather station measurements) caused increase of the river discharge from 80 to 150 m^3/s during several hours. The area of the Kodor plume doubled from 31 August to 1 September in response to the flash flood. Wind direction during 31 August–1 September was stable (southwestern), while wind speed slightly increased from 2–3 m/s to 4–5 m/s. Then river discharge steadily decreased to pre-flooding conditions, which were registered on 2 September, while wind direction changed to eastern and wind velocity decreased to 3–4 m/s. In situ measurements and aerial remote sensing performed on 2 September, i.e., shortly after the flood, observed, first, the large residual plume that was formed on 1 September during the flooding event and did not dissipate yet and, second, the emergent plume that was formed on 2 September after the decrease of river discharge rate (Figure 5). These plumes had different spatial scales, structures, thermohaline, and dynamical characteristics. As a result, similarly to the river plumes formed by different deltaic branches, the residual and the emergent plumes interacted as individual water masses and formed complex frontal zones within the common Kodor plume.

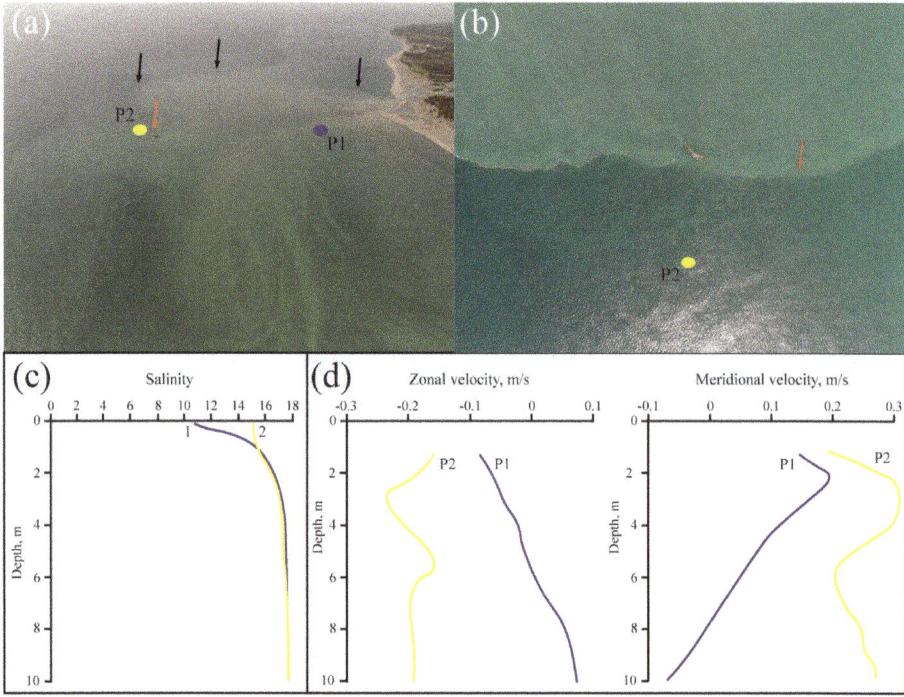

Figure 5. (**a**,**b**) aerial images, (**c**) vertical salinity, and (**d**) velocity profiles at the frontal zone between the emergent and the residual parts of the Kodor plume on 2 September 2018. Colored dots indicate locations of vertical salinity and velocity measurements (P1, blue—the emergent plume; P2, yellow—the residual plume). Arrows in panels (**a**) and (**b**) indicate distinct frontal zones between the emergent and the residual parts of the Kodor plume. Red arrows in panels (**a**) and (**b**) point at the same segment of the frontal zone where in situ measurements were performed.

Interaction between the Kodor plume and the seafloor at the shallow zones is the third process that induces inhomogeneous structure of this plume. Aerial imagery detected the area of reduced turbidity formed behind the shoal, which is located in front of the northern deltaic branch (Figure 6). This low-turbid zone contrasted especially with the surrounding turbid river plume during the flooding discharge on 1 September 2018. In situ measurements showed that surface salinity at this low-turbid zone (15) was significantly greater than at the adjacent turbid part of the plume (12.5–13) (Figure 6c). Surface circulation also differed in these two parts of the plume. The northward flow (10 cm/s) was observed in the low-turbid zone, while the southeastward flow (20 cm/s) dominated in the adjacent turbid part (Figure 6d). The formation of this zone is caused by the interaction of the inflowing river jet with seafloor at the shoal that induces deceleration of the jet and its increased mixing with saline and low-turbid sea water. The stable front bounding this low-turbid and high-saline zone inside the plume was observed on a distance of up to 1 km from the shoal.

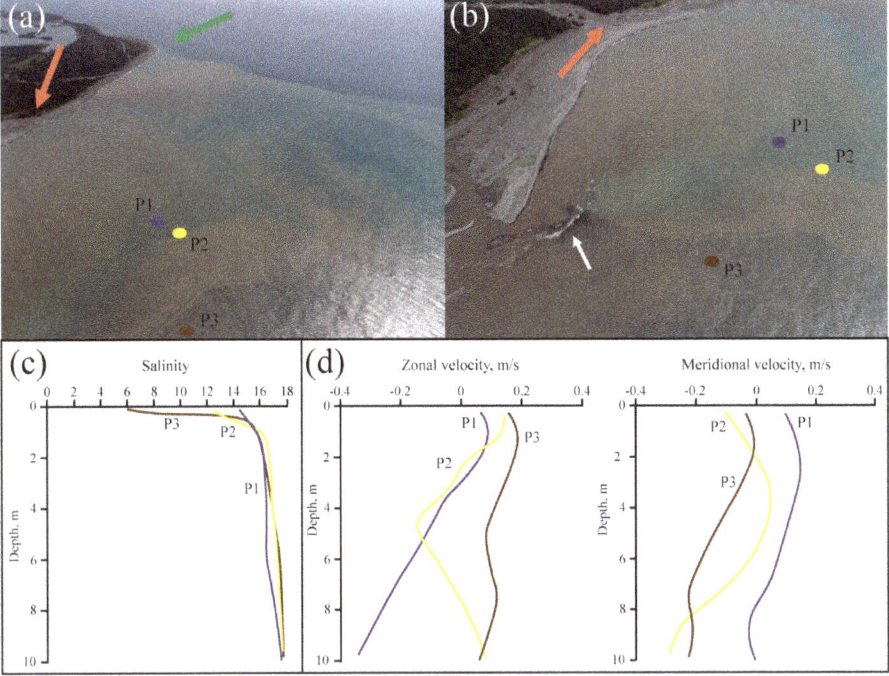

Figure 6. (**a**,**b**) aerial images, (**c**) vertical salinity, and (**d**) velocity profiles at the frontal zone of the Kodor plume formed behind the shoal on 1 September 2018. Colored dots indicate locations of vertical salinity and velocity measurements (P1, blue—the low-turbid zone of the plume; P2, yellow—the frontal zone; and P3, brown—the turbid part of the plume). The white arrow in panel (**b**) indicates location of the shoal, red arrows indicate location of the central deltaic branch of the Kodor River, and the green arrow indicates location of the Iskuria Cape.

3.2. Dynamical Features of the Kodor and Bzyp Plumes

Using aerial remote sensing we detected several dynamical features of the Kodor and Bzyp plumes and measured their spatial characteristics. Based on the surface velocity data reconstructed from the aerial video records, we studied dynamical characteristics of these features and analyzed their physical background. Aerial remote sensing detected a swirling eddy within the Kodor plume on 1 September 2018 (Figure 7). This eddy was formed at the southern part of the emergent plume at its border with the residual plume near the Iskuria Cape. The aerial image of this part of the plume acquired at 12:52 (Figure 7a) showed inhomogeneous structure of the emergent plume without any eddy. The distinct border between the emergent and the residual plumes was stretched from the Iskuria Cape in the northwestern direction. The beginning of formation of the eddy was registered at 14:42 (Figure 7b), then at 15:34 the well-developed eddy was observed (Figure 7c,d). The diameter of the eddy was approximately 500 m, it was rotating in an anticyclonic direction, while its center was moving at an angle of approximately 30° across the border of the emergent plume. Processing of the video record of this eddy provided estimations of velocity of its movement (0.9 m/s) and rotation (0.4 m/s). The aerial observations performed at 16:16 did not show any surface manifestations of the eddy at the study area; therefore, we presume that it shifted off the observation area during less than an hour. Wind conditions were stable during the considered period, wind speed did not exceed 3.5 m/s.

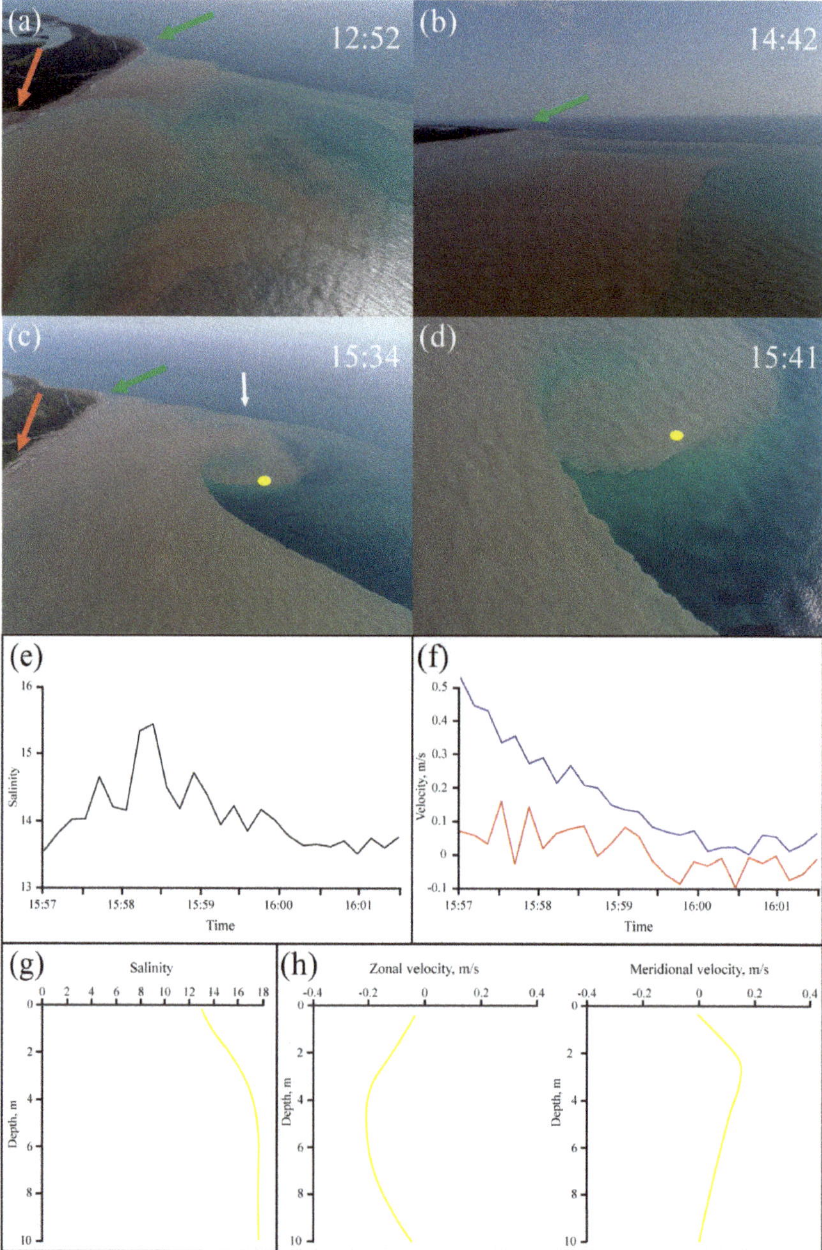

Figure 7. Aerial images of the southern part of the Kodor plume (**a,b**) before and (**c**) during interaction between the plume and the eddy acquired at (**a**) 12:52, (**b**) 14:42, (**c**) 15:34, and (**d**) 15:41 on 1 September, 2018. (**e**) surface salinity, (**f**) zonal (blue) and meridional (red) velocities measured during 15:57—16:01 and (**g**) vertical salinity and (**h**) velocity profiles measured at 16:02 within the eddy. Yellow dots in panels (**c**) and (**d**) indicate location of salinity and velocity measurements. The white arrow in panel (**c**) indicates location of the eddy, red arrows indicate location of the central deltaic branch of the Kodor River, and green arrows indicate location of the Iskuria Cape.

In situ thermohaline and velocity measurements were performed within the eddy at 15:57–16:01 (Figure 7e,f). They included continuous measurements at a depth of 0.7–0.8 m for 4.5 minutes followed by vertical profiling from surface to the depth of 13 m. Note that the measurements were performed at the stable point, while the eddy was moving. As a result, the performed measurements registered salinity and velocity in different parts of the eddy while it was passing the point of measurements. The intense northward flow (55 cm/s) registered in the surface layer at the beginning of the measurements steadily dissipated to <10 cm/s during the first stage of the measurements (Figure 7f). The eastward velocity component was slightly positive during the first two minutes of the measurements (6 cm/s on average with the peak value of 16 cm/s) and then changed to slightly negative (−5 cm/s on average with the peak value of −11 cm/s). It was accompanied by significant variability of salinity that increased from 13.5 to 15.5 during the first 1.5 min of the measurements and then decreased to 13.5 (Figure 7e). The observed variability of velocity and salinity in the surface layer confirms northward propagation and anticyclonic rotation of this eddy observed at aerial video (Supplementary Materials). However, the movement and rotation velocities registered by in situ measurements were twice less than those reconstructed from the aerial video. This difference is caused by the fact that in situ measurements were performed not at the central part of the eddy, but at its periphery. The observed variability of salinity in the surface layer was caused by intrusion of saline water from the ambient sea to the plume induced by the rotation of the eddy (Figure 7d). Vertical profiles of salinity and velocity measured at 16:02, i.e., after the measurements in the surface layer, registered strong northwestward flow in the subjacent saline sea (Figure 7g,h). Its maximal velocity (15–25 cm/s) was observed immediately beneath the plume at depths of 3–5 m, then velocity decreased to 10–15 cm/s at depths of 8–9 m and to <5 cm/s at depths of 10–13 m. This northwestward flow (20–30 cm/s) was also registered along the Iskuria Cape at the previous day that confirms the presence of the northwestward jet behind the Iskuria Cape which is presumed to generate the observed eddy.

Interaction between sub-mesoscale eddies and the Kodor plume was also observed by satellite imagery. The chains of small anticyclonic eddies (300–500 m in diameter) formed behind the Iskuria Cape and interacting with the Kodor plume were registered on 17 July 2018, 21 August 2019, and 26 August 2019 (Figure 8a). Positions, sizes, and shapes of four to five subsequent eddies within these chains indicate that these chains were periodically generated near the Iskuria Cape and propagated in the northwestward direction shortly before the periods of satellite observations. While tracks of the eddies were crossing the Kodor plume, the turbid plume water was twisted into the eddies, which made them visible at satellite imagery. After these eddies propagated off the plume the trapped turbid water remained connected with the plume that illustrated difference in trajectories and velocities of the eddies and the wind-driven far-field part of the plume (Figure 8a).

Satellite images acquired during the periods of field measurements at the Kodor plume did not register interactions between the eddies and the plume due to episodic character of these features, i.e., eddies do not constantly form and propagate at the study area. Therefore, the satellite images presented in Figure 8 are not synchronous with the field surveys. However, sizes and anticyclonic rotation in the northwestward direction were similar for eddies detected at the Kodor plume by aerial and satellite remote sensing. As a result, we presume that we observe the same process and, therefore, can jointly analyze its spatial and temporal characteristics obtained from aerial and satellite measurements. Satellite imagery also observed eddies formed behind the Pitsunda Cape and interacting with the Bzyp plume on 30 July 2017 and 10 October 2019 (Figure 8b). However, in contrast to the eddies registered within the Kodor plume, these eddies were individual, i.e., did not form chains. Moreover, these eddies were much larger (2-4 km in diameter) and were rotating in cyclonic direction. Satellite images acquired during the periods of field measurement at the Bzyp plume also did not register interactions between eddies with the Bzyp plume.

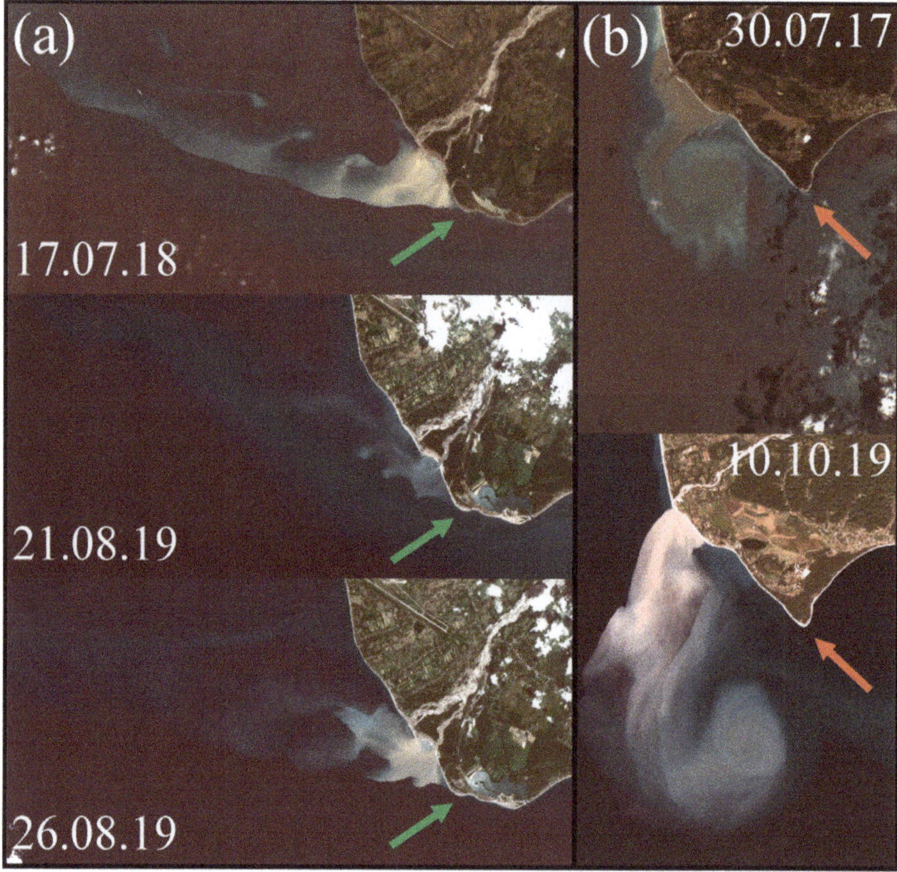

Figure 8. Sentinel-2 ocean color composites (**a**) from 17 July 2018, 21 August 2019, and 26 August 2019 illustrating interactions between eddies and the Kodor plume and (**b**) from 30 July 2017 and 10 October 2019 illustrating interactions between eddies and the Bzyp plume. Green arrows indicate location of the Iskuria Cape and red arrows indicate location of the Pitsunda Cape. Note that images at panels (**a**) and (**b**) are inconsistent, i.e., they show river plumes at different dates.

Satellite image acquired on 10 October 2019 detected packets of internal waves emerging from the rotating eddy and propagating within the Bzyp plume (Figure 9b). Aerial observations on 1 September 2018 also detected a packet of internal waves that emerged from the eddy and was propagating within the outer part of the plume towards the open sea (Figure 10a). Note that the aerial imagery of the Kodor plume (Figure 9a) and the satellite imagery of the Bzyp plume (Figure 9b) are not synchronized and show different river plumes at different dates. Aerial and satellite images acquired during the period of field measurements at the Bzyp plume did not register internal waves within the Bzyp plume. Therefore, in Figure 9 we show airborne images of internal waves at the Kodor plume and satellite images of internal waves at the Bzyp plume.

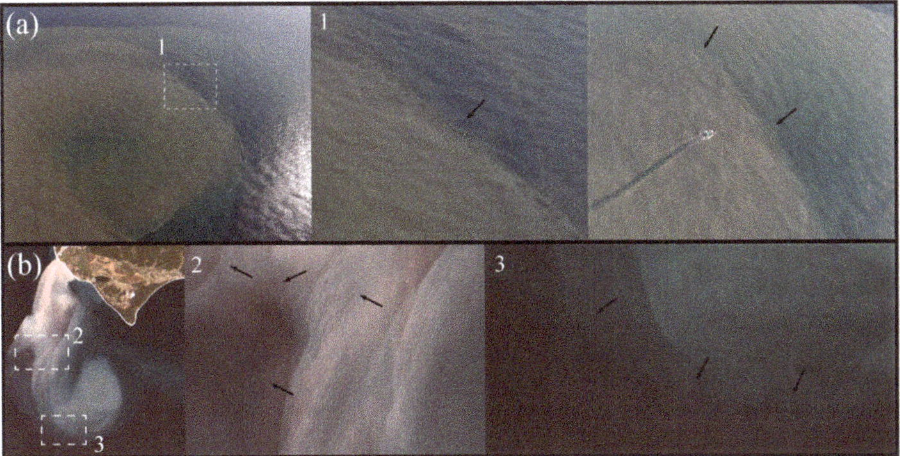

Figure 9. Surface manifestations of high-frequency internal waves generated by the eddies within the (**a**) Kodor and (**b**) Bzyp plumes at (**a**) aerial images acquired on 1 September 2018 and (**b**) satellite images acquired on 10 October, 2019. The central picture at panel (**a**) is the zoomed fragment of the left picture at panel (**a**) indicated by the white dashed rectangle 1. The central and right pictures at panel (**b**) are the zoomed fragments of the left picture at panel (**b**) indicated by white dashed rectangles 2 and 3, respectively. Black arrows indicate surface manifestations of internal waves.

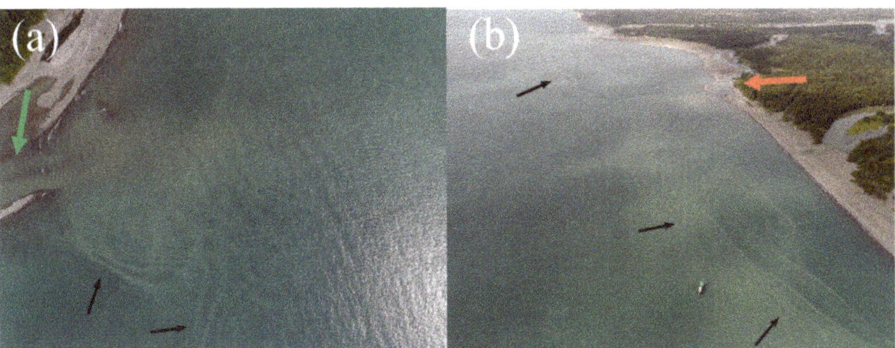

Figure 10. Aerial images of surface manifestations of low-frequency internal waves within the Kodor plume near the mouths of (**a**) the northern and (**b**) the central deltaic branches on 2 September 2018. The green arrow indicates location of the northern deltaic branch of the Kodor River and the red arrow indicates location of the central deltaic branch of the Kodor River. Black arrows indicate surface manifestations of internal waves.

Despite a large difference in coverage and spatial resolution of the aerial and satellite imagery presented in Figure 9, they both distinctly demonstrate propagation of internal waves within the river plumes. Satellite remote sensing has wide spatial coverage and provides information about spatial characteristics of wave packets at different parts of the plumes (Figure 9b). Distances between the wave packets observed at Sentinel-2 satellite images varied from 30 to 150–200 m, while lengths of the wave packets were up to 5-6 km. Satellite images demonstrated that dozens of internal waves were generated within the plume around the rotating eddy. On the other hand, airborne remote sensing provided opportunity to detect individual internal waves with high spatial resolution and to register their velocities (Figure 9a). High-resolution aerial imagery detected that the distances between the individual waves within the wave packet in the Kodor plume were 2–4 m. The length of the wave packet front was approximately 200 m. The number of waves within the wave packet varied from 12 at its northern part to 3 at its southern periphery. Processing of high-resolution video records revealed that velocity of the wave packet was equal to 0.21 m/s.

Aerial remote sensing also detected multiple packets of low-frequency internal waves that propagated within the Kodor plume towards the coast on 2 September 2018 (Figure 10). These packets consisted of 5–15 waves that were stretched along the coast, albeit had complex shapes not related to the shapes of the plume front or the coastline. Distances between individual waves varied from 5 to 70 m in the observed wave packets. Frontal length of these packets varied from ~100 m (Figure 10a) to 2–3 km (Figure 10b), while their speeds were 10–15 cm/s. Wind speed during this period was 2–3 m/s.

Osadchiev [33] described a mechanism of generation of internal waves in small river plumes as a result of rapid deceleration of an inflowing river jet and formation of a hydraulic jump in vicinity of a river mouth. These internal waves propagate offshore and are regularly observed by satellite imagery in many coastal regions in the World [33,90,91]. Using aerial remote sensing we recorded generation and propagation of these internal waves from the mouth of the side-channel of the Bzyp River on 1 July 2019 (Figure 11a). The internal waves were generated at a distance of 40–50 m from the river mouth every 19 seconds on average, i.e., 29 individual waves were generated during a 9-min long video recording of this area. The distances between the waves decreased from 8–10 m near the river mouth to 1–2 m at the distance of 500 m from the river mouth. Wave velocities were equal to 0.27–0.31 m/s. Moderate (2–3 m/s) northern wind was registered during the considered period.

Aerial observations of internal waves in the Bzyp plume described above were supported by in situ salinity and turbidity measurements performed from a flat-bottomed boat with shallow draft to minimize the boat-induced mixing of sea surface layer (Figure 11). Measurements included 15 surface-to-bottom profiles continuously performed from a free-drifting boat starting at the generation area of the internal waves at the distance of 10 m from the river mouth and finishing 90 m far from the starting point (Figure 11a). The obtained data revealed large difference in vertical salinity structure of the Bzyp plume inside and outside this generation area of internal waves. The first half of the hydrological transect was located at the area of formation of the hydraulic jump as a result of abrupt deceleration of the inflowing river jet (Figure 11b). Similarly to the hydraulic jump observed and described by Osadchiev [33] at the inflowing jet of the Mzymta River, we registered anomalously deep penetration of low-saline water at the generation area of the internal waves in the Bzyp plume. Low-saline water (10–14) was observed from surface to the depth of 3–4 m along 0–5 m and 25–35 m of the transect. Vertical salinity structure within this part of the plume was unstable with multiple overturns (reverse salinity difference was up to 1 at vertical distance of 0.1 m) and large salinity gradients. Vertical salinity structure of the Bzyp plume between the areas of the hydraulic jumps, i.e., along the 5–25 m of the transect, showed relatively homogenous salinity (14.5–16) from surface to bottom, albeit it was much higher than within the areas of hydraulic jumps.

Outside the generation area of the internal waves, i.e., along the 35–90 m of the transect, surface salinity was relatively homogenous (14.5–15.5) and vertical salinity structure was stable. Vertical salinity gradient outside the generation area of internal waves was two orders of magnitude less than the largest values registered in the hydraulic jumps. However, salinity measurements did not cover top 0.5 m of the surface layer, where presumably was located the salinity gradient. Vertical turbidity structure, however, did not show large difference inside and outside the generation area of the internal waves (Figure 11c). The turbid layer was observed from surface to the depth of 1–1.5 m along the first part of the transect and then its depth steadily decreased to 0.5 m. This feature shows that salinity and turbidity structure of a river plume can be significantly different in areas of very intense advection and turbulent mixing.

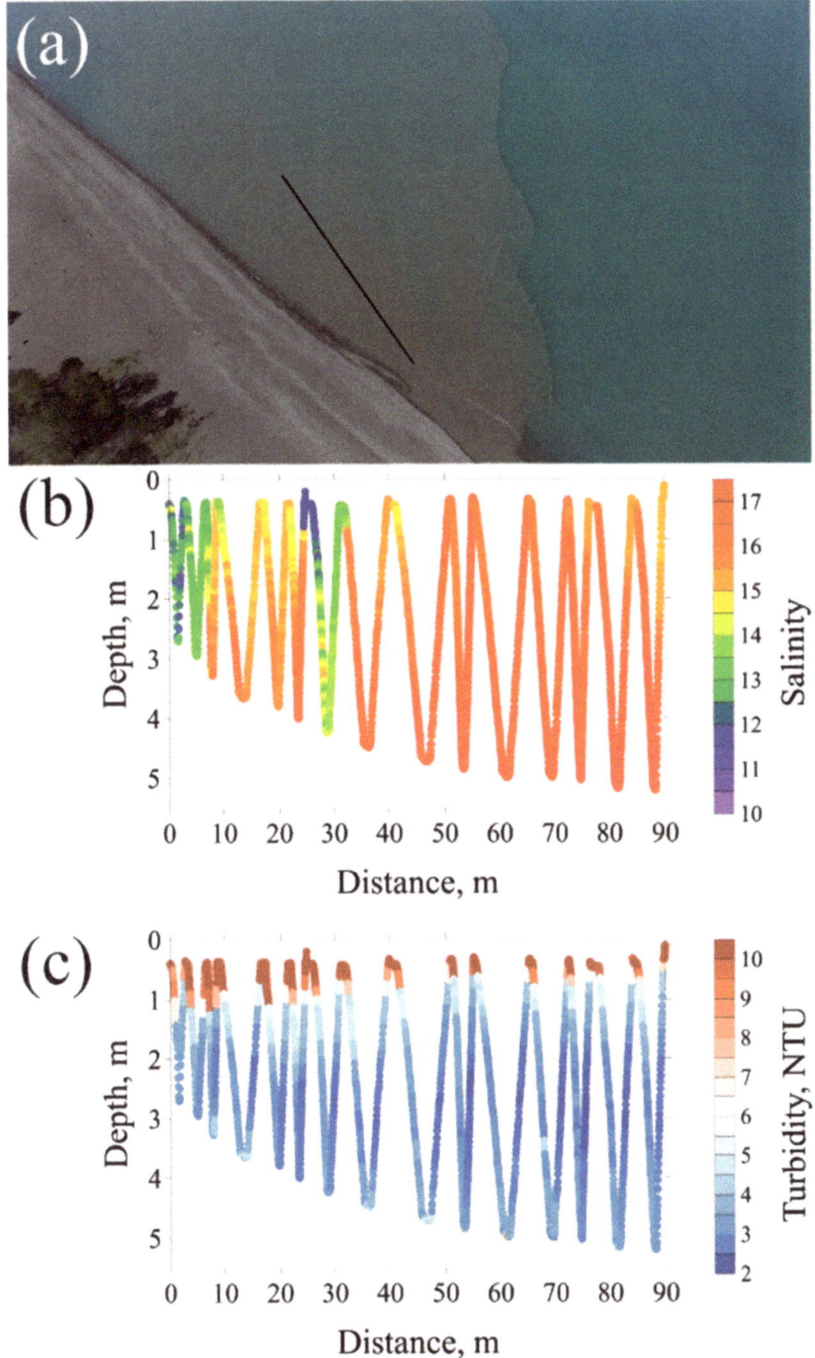

Figure 11. (**a**) aerial image of surface manifestations of internal waves propagating within the Bzyp plume off the river mouth and location of the hydrological transect (black line) on 1 July 2019 (**a**). Vertical (**b**) salinity and (**c**) turbidity profiles along the hydrological transect.

3.3. Undulate Borders of the Kodor and Bzyp plumes

Aerial remote sensing of the Kodor and Bzyp plumes showed undulate structure of long segments of their outer borders manifested by alternation of specific convex and concave segments. These segments are 2–10 m long and up to 2 m wide and hereafter are referred as "lobes" and "clefts" [52,53]. Aerial images of the undulate fronts observed at the Kodor plume border on 1 September, 2018 and at the Bzyp plume border on 1 June 2019 are shown in Figure 12. This lobe-cleft structure was registered only at sharp and narrow frontal zones formed between the emerging plume, on the one hand, and the residual plume or the ambient sea, on the other hand. Lobes and clefts were absent at diffuse fronts, i.e., wide and low-gradient fronts that contour the outer parts of the plumes, which experience intense mixing with the ambient sea. In particular, these undulate fronts commonly extended from the river mouths and bounded the inflowing river jets, i.e., near-field parts of the plumes. These fronts were not observed in the far-field parts of the plumes and in the coastal surf zone during periods of active wave breaking due to intense mixing (Figure 12).

Figure 12. Aerial images (**a**) of undulate fronts at the border of the Kodor plume on 1 September, 2018 and (**b**) at the border of the Bzyp plume on 1 June, 2019. Central and right pictures at panel (**a**) are the zoomed fragments of the left and central pictures at panel (**a**), respectively, indicated by the white dashed rectangles 1 and 2, respectively. Central and right pictures at panel (**b**) are the zoomed fragments of the left and central pictures at panel (**b**), respectively, indicated by the white dashed rectangles 3 and 4, respectively. Black arrows indicate absence of undulate fronts at the surf zone.

We observed significant short-temporal variability of the undulate fronts induced by the following recurrent process (Figure 13). Once a lobe is formed, it starts to increase seaward. Ballooning of neighboring lobes results in their coalescence and the subsequent merging. At the same time the cleft between these lobes is steadily decreasing and transforms into a spot of saline sea (with area of 0.1–0.5 m^2) isolated from the ambient sea, i.e., trapped by the merged lobes within the plume (Figure 13). The merged lobes and the trapped saline sea area finally dissipate, and then the process of formation of new lobes at this part of the plume front restarts. The continuous recurrent process of formation of lobes, their merging, and subsequent dissipationwas observed along the undulate fronts of the Kodor and Bzyp plumes. Residual time of an individual lobe, i.e., from its formation to dissipation, was 1–2 min.

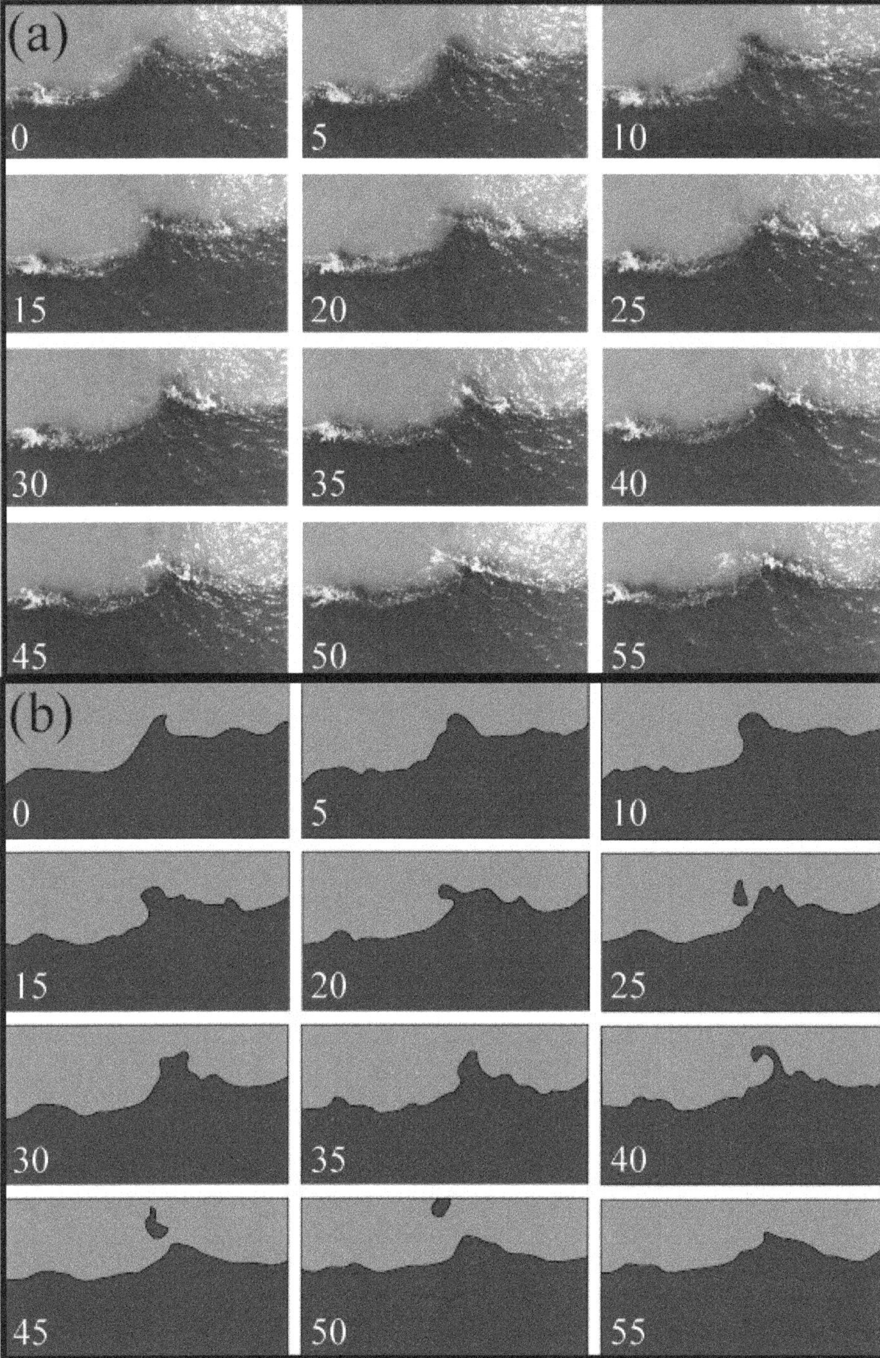

Figure 13. (**a**) aerial images and (**b**) reconstructed shapes of the border of the Kodor plume on 1 September 2018 illustrating merging of lobes and trapping of spots of saline sea. Numbers indicate time intervals in seconds from the beginning of observations.

Due to convergence of surface currents at sharp plume fronts [92], foam and floating litter commonly accumulate at the undulate fronts of the plumes (Figures 13a and 14a). Using optical flow processing of aerial video records, we detected motion of foam and floating litter and reconstructed surface circulation along the undulate fronts of the Kodor and Bzyp plumes (Figure 14). The circulation structure within the lobes consists of pairs of cyclonic and anticyclonic vortices that form, balloon, merge, and dissipate with the lobes (black lines in Figure 14b). The trajectories of foam and floating litter revealed that cyclonic vortices are significantly more prominent and intense, as compared to anticyclonic vortices. Foam and floating litter are mainly accumulated within cyclonic eddies, i.e., in the right parts of the lobes if we look from the sea towards the plume (Figure 14a). Foam and floating litter are rotated by cyclonic eddies within the right parts of the lobes during the majority of time of aerial observations. Once a parcel of foam or floating litter is advected off a cyclonic eddy and enters an anticyclonic eddy in the left part of the lobe, it is transported to the outer part of the lobe and then is trapped by the cyclonic eddy in the neighboring (leftward) lobe (red lines in Figure 14b). As a result, these parcels are skipping leftward between the right parts of lobes. Therefore, foam and floating litter are steadily transported to the left along the plume border. The observed large intensity of cyclonic circulation within the lobes, as compared to anticyclonic circulation, is presumed to have the same background as the dominance of cyclonic spirals at satellite images of sea surface caused by differences between the rotary characteristics of cyclonic and anticyclonic eddies in the sea [93].

Figure 14. (a) aerial image of the undulate border of the Kodor plume on 1 September 2018 and (b) the scheme of the reconstructed circulation within the lobes (black lines) and the transport of foam and floating litter along the plume border (red lines) (b). Black arrows in panel (a) indicate foam accumulated within cyclonic vortexes in the right parts of the lobes.

We presume that the undulate structure of the sharp plume borders is formed due to baroclinic instability between the plumes and the ambient sea. The pressure gradient force across the front is equal to

$$g\frac{\Delta\rho}{\rho_{sea}}\frac{\partial h}{\partial x} \qquad (4)$$

where g is the gravity acceleration, $\Delta\rho$ is the density difference between the plume and the ambient sea, ρ_{sea} is the density of the sea, h is the depth of the plume, and x is the cross-front direction. In situ measurements performed at the undulate fronts showed that surface salinity abruptly increased across these fronts (2–3 m wide) from 10–12 inside the Kodor plume to 17 outside the Kodor plume (Figure 15b) and from 8–10 inside the Bzyp plume to 16–17 outside the Bzyp plume. The depth of the Kodor plume at the narrow frontal zone was 2 m (Figure 15b), the depth of the Bzyp plume was 4 m. As a result, the values of pressure gradient across these frontal zones calculated from Equation (4) are equal to 0.05 and 0.1 m/s^2 for the Kodor and Bzyp plumes, respectively.

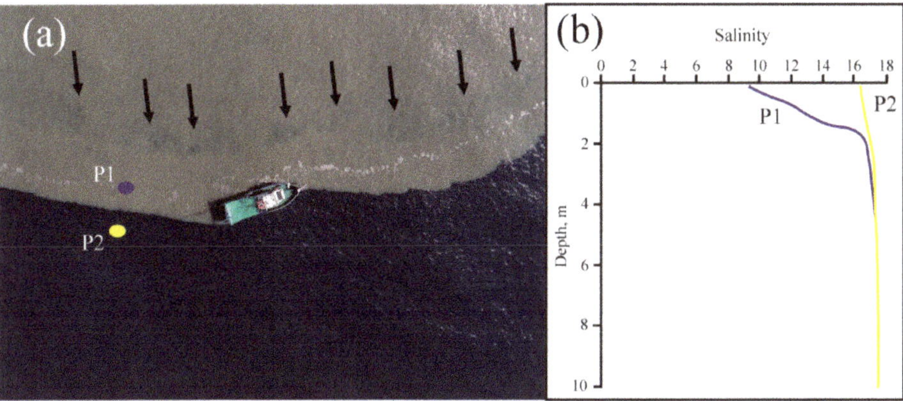

Figure 15. (**a**) aerial image and (**b**) vertical salinity profiles at the undulate border of the Kodor plume on 1 September 2018. Colored dots indicate locations of vertical salinity measurements (P1, blue—the plume; P2, yellow—the ambient saline sea). Black arrows in panel (**a**) indicate a stripe of low-turbid water within the Kodor plume stretched along its border.

This large pressure gradient observed across the plume fronts is the source of potential energy that induces formation of lobes and clefts as follows. Small perturbation of a sharp frontal zone and the subsequent formation of a local convex segment cause increase of local length of the front and, therefore, increase of the cross-front advection induced by the pressure gradient. It results in ballooning of the lobe till it coalesces and merges with the neighboring lobe. Merging of two lobes accompanied by trapping of a spot of saline sea water and its subsequent mixing with the plume water cause a reduction of local salinity anomaly and, therefore, a decrease of local pressure gradient. It hinders formation of a lobe at this segment of the plume, while new lobes are formed at the adjacent segments of the plume front. Therefore, baroclinic instability causes formation, merging, and dissipation of the observed lobe-cleft structures and influences mixing between the river plumes and the ambient sea.

Aerial imagery detected the 3–4 m wide stripe of low-turbid water within the Kodor plume located at the distance of 10–20 m from the undulate border and stretched along this border (Figure 15a). We presume that this low-turbid stripe is formed as a result of continuous trapping of spots of saline sea water by merging lobes. Horner–Devine et al. [53] assumed that the lobe-cleft structure is formed by subsurface vortexes that are propagating from the inner part of the plume towards its border with the ambient sea. However, aerial video records showed stable position and shape of this stripe that evidences absence of any subsurface vortexes described by Horner–Devine et al. [53].

4. Discussion

In this study, we obtained several important results about structure, short-temporal variability, and dynamics of small river plumes. First, we revealed strongly inhomogeneous structure of small plumes manifested by multiple frontal zones between different parts of the plumes. These parts have different structures and dynamical characteristics and interact as individual water masses. Second, we reported fast motion of small plumes caused by interaction with coastal eddies. Third, we observed generation and propagation of different types of internal waves within small plumes. Forth, we described formation of lobe-cleft structures at sharp borders of small plumes and reported intense lateral mixing across these fronts caused by their baroclinic instability. The results listed above are important for understanding spreading and mixing of small plumes, however, they are addressed for the first time as previous related works were mainly limited by low spatial and/or temporal resolution of in situ measurements and satellite imagery. Below we provide physical interpretation of these features observed at the Kodor and Bzyp plumes and discuss importance of their study at other small plumes in the World Ocean.

In general, river plumes are regarded as "smooth" water masses without internal fronts and sharp gradients. This approach is widely used in analytical and numerical modeling studies focused on river plumes, including the fundamental and highly cited papers [82–85,94–96]. Many relevant studies based on in situ and satellite data confirmed that this approach provides realistic results for buoyant plumes formed by large rivers plumes which internal structures indeed are characterized by steady changes of salinity and other characteristics. In this work, we present the results of aerial remote sensing of the Kodor and Bzyp plumes supported by in situ measurements that provide an evidence of strongly inhomogeneous internal structure of small plumes. This structure is manifested by complex internal frontal zones and sharp salinity and turbidity gradients within small plumes. These gradients and frontal zones strongly modify circulation within the plumes, in particular, they hinder cross-frontal advection within the plumes and separate them into semi-isolated, but interacting structures. Therefore, identification and study of the processes that govern formation of frontal zones within small plumes is important for understanding of spreading and mixing of freshwater discharge in the sea and the related transport of river-borne suspended and dissolved material.

The Kodor River inflows to the Black Sea from multiple deltaic branches and forms several river plumes. These plumes are closely located; they interact as individual water masses and coalesce into the common Kodor plume. Interaction, collision, and coalescence of buoyant plumes formed by rivers, which estuaries are located in close proximity, were addressed in several previous studies [86–88,97–99]. Similar processes occur within plumes formed by freshwater discharge from multiple deltaic branches, as was observed for the Kodor plume. Moreover, generally distances between deltaic branches within one deltaic system are smaller than distances between estuaries of neighboring rivers. As a result, interactions between neighboring plumes formed by different rivers generally occur only during high discharge periods [86], while similar interactions between plumes formed by different deltaic branches is a permanent or almost permanent process at many World regions. However, despite a large number of deltaic rivers inflowing to the World Ocean, we are aware of only one related study that was focused on the interaction between the buoyant plumes formed by different deltaic branches of the Pearl River Delta [100].

The Kodor River has very large intra-day and synoptic variability of discharge rate due to morphology and weather conditions at its drainage basin. This variability of discharge rate induces variability of spatial extents of the Kodor plume and residence time of freshened water within the plume. As a result, the Kodor plume formed during high discharge can have different spatial and thermohaline characteristics from those formed during low discharge. In case of abrupt decrease of river discharge rate, the relatively large and mixed residual plume (formed during high discharge period) interacts with the small and freshened emergent plume (formed during the subsequent low discharge period). We report distinct frontal zones and differences in dynamics between the residual and the emergent parts of the Kodor plume. Several previous studies addressed response of river plumes to variable discharge rates [101–107], but limited attention was paid to interaction between parts of an individual river plume formed during different discharge conditions [108]. This feature can strongly affect spreading and mixing of freshwater discharge from small rivers in many World regions and should be considered in the related studies.

Several studies addressed interaction between coastal bathymetry and bottom-advected river plumes, which occupy the whole water column from surface to seafloor and, therefore, experience intense bottom friction [109–111]. In these numerical studies, river plumes were spreading over sea areas with idealized bathymetry, which was steadily sloping in the cross-shore direction and was homogenous in the alongshore direction. Influence of realistic bottom topography on surface-advected river plumes was described by Korotenko et al. [112]. Bottom-generated turbulent mixing induced by coastal circulation penetrates upward and reaches surface layer over shallow zones, therefore, increased local mixing of river plumes occurs at these zones. We presume that a similar mechanism induced intensified mixing of the Kodor plume over the shoal revealed by aerial imagery and in situ measurements. Moreover, we observed that the intense flow of the Kodor plume over this small shoal results in formation of large area within the plume with elevated salinity, which is bounded by the distinct frontal zone. We are not aware of any work describing this effect at river plumes, however, it can be typical for many small plumes with small vertical scales flowing over bathymetric features.

In this study, we address several important dynamical features of small river plumes. Aerial remote sensing revealed a quick motion of the Kodor plume border (~0.5–1 m/s) entrained by the rotating coastal eddy. Such extremely rapid response of a river plume to coastal sea circulation has not been reported before, to the extent of our knowledge. The previous studies showed that general spreading patterns of small plumes are governed by wind forcing, while the impact of ambient circulation was regarded as negligible [113–115]. We demonstrate that energetic features of coastal circulation, e.g., eddies, can induce high velocity motion of plume fronts and, therefore, influence dynamics of a small plume, albeit locally and during short-term periods.

The rotating eddy generated high-frequency internal waves that were propagating within the Kodor plume and dissipated at its border with the ambient sea. Aerial remote sensing also observed multiple long internal waves propagating within the Kodor plume towards the coast, as well as generation of high-frequency internal waves near the mouth of the Bzyp River and their propagation within the Bzyp plume towards the open sea. Internal waves are common features of river plumes in non-tidal seas and their surface manifestations observed by satellite imagery were reported in several previous studies [116,117]. These internal waves can significantly affect mixing of small plumes with subjacent saline sea [33]. In this study, we demonstrate the efficiency of aerial remote sensing in observations of surface manifestations of internal waves, and the ability of aerial remote sensing (in contrast to satellite observations) to measure their spatial and dynamical characteristics and to identify mechanisms of their generation.

Finally, in this study, we address the undulate structure of the sharp borders of the Kodor and Bzyp plumes that were previously observed and reported at other small plumes [52,53,118,119]. Horner–Devine and Chickadel [53] associated formation of the lobe-cleft structures observed at the Merrimack plume with subsurface vortexes that were propagating from the inner part of the plume towards its border with saline sea. Based on processing of aerial video records, we reconstruct surface circulation at the undulate fronts of the Kodor and Bzyp plume and detected similar vortexes within the lobes. However, we observed absence of vortexes outside the frontal zones, i.e., no vortexes were propagating from the inner parts of the plume towards their borders. On the opposite, we observed the recurrent process of formation, merging, and dissipation of lobes, that was not described before. Based on these results, we suggest an alternative mechanism of formation of the undulate fronts caused by baroclinic instability between the plume and the ambient sea and ballooning of local convex segments of the frontal zone in response to its small perturbations. This mechanism is in a good agreement with the reconstructed vortex circulation within the lobes and explains absence of vortexes in the inner parts of the plumes. We reveal intense transport of saline sea water across the undulate plume border as a result of merging of lobes and mechanical trapping of spots of saline sea inside the plume. It can be an important mechanism of mixing between the plume and the saline sea and should be considered together with shear-induced mixing of the plume and the subjacent sea. Satellite imagery reveals that undulate frontal zones are, therefore, the related mixing mechanism are typical for many small plumes in the World. Therefore, study of this mechanism is important in context of transformation and dissipation of freshwater discharge in the sea.

5. Conclusions

In this work, we focused on small buoyant plumes formed by the Kodor and Bzyp rivers located at the northeastern part of the Black Sea. We used quadcopters equipped with video camera to perform aerial remote sensing of these river plumes, which was accompanied by synchronous in situ measurements in the sea. Using an optical flow approach, we reconstructed surface velocity fields within these plumes from the obtained aerial video records. Based on aerial imagery and video records, the reconstructed surface currents, as well as in situ salinity, turbidity, and velocity measurements, we obtained new insights into spatial structure, short-temporal variability, and dynamical features of small river plumes, which are not typical for plumes formed by large rivers.

Based on the obtained aerial and in situ data, we address several different issues, including the methodology and value of the aerial observations of small river plumes, the differences between small and large plumes, the influence of multiple freshwater sources on the structure of a small plume, the influence of bathymetry features on the structure of a small plume, the interaction between small plumes and coastal circulation, the presence of internal waves in river plumes, and the presence of small-scale instabilities along the plume front boundary. The main results obtained in this study are the following. We describe strongly inhomogeneous structure of small plumes, as compared to large plumes. We suggest a new mechanism of mixing of a small plume with ambient sea as a result of baroclinic instability at its outer boundary. We describe internal waves formed within near- and far-filed parts of small plumes, which can strongly influence its mixing with ambient sea. These results are important for understanding the fate of freshwater discharge from small rivers and the related transport of suspended and dissolved river-borne constituents in many coastal sea areas in the World Ocean.

Usage of quadcopters provides ability to perform low-cost aerial remote sensing of coastal sea areas and continuously observe surface manifestations of many coastal processes. In this study, we demonstrate its efficiency in observations of small river plumes characterized by high color contrast with ambient sea, energetic motion, and high short-temporal variability. Aerial imagery can be used for visual detection and tracking of many other processes at small spatial (from meters to kilometers) and temporal (from seconds to hours) scales, which are visible neither from shipboard nor satellite imagery. Spatial scales and motion speeds of the observed processes can be reconstructed from aerial imagery and video records (Supplementary Materials). Therefore, aerial drones can provide quantitative measurements of distances and velocities at sea surface. Finally, aerial remote sensing can be very useful for operational organization of in situ measurements during field surveys, in particular, for selection of places for water sampling and hydrological measurements according to real-time position of the observed sea surface processes. As a result, future studies based on imagery and video records of ocean surface acquired from aerial drones (considering certain important limitations of their usage) and supported by in situ measurements hold promise to significantly improve understanding of various upper ocean features and dynamics.

Supplementary Materials: The aerial images and video records are publicly available at https://doi.org/10.5281/zenodo.3901896. The Sentinel-2 Level-1C products were downloaded from the Copernicus Open Access Hub https://scihub.copernicus.eu/.

Author Contributions: Conceptualization, A.O.; measurements, A.O., A.B., R.S., and R.Z.; formal analysis, A.O.; investigation, A.O., A.B., R.S., R.Z., and R.D.; writing, A.O., A.B., R.S., R.Z., and R.D. All authors have read and agreed to the published version of the manuscript.

Funding: This research was funded by the Ministry of Science and Higher Education of Russia, theme 0149-2019-0003 (collecting and processing of in situ data), the Russian Science Foundation, research project 18-17-00156 (collecting and processing of aerial imagery, study of spreading of river plumes) and the Russian Ministry of Science and Higher Education, research project 14.W03.31.0006 (study of submesoscale dynamics of river plumes).

Acknowledgments: The authors are grateful to the editor Melissa Zhu and five anonymous reviewers for their comments and recommendations that served to improve the article.

Conflicts of Interest: The authors declare no conflict of interest.

References

1. Klemas, V.V. Airborne remote sensing of coastal features and processes: An overview. *J. Coast. Res.* **2013**, *29*, 239–255. [CrossRef]
2. Holman, R.; Haller, M.C. Remote sensing of the nearshore. *Annu. Rev. Mar. Sci.* **2013**, *5*, 95–113. [CrossRef] [PubMed]
3. Colomina, I.; Molina, P. Unmanned aerial systems for photogrammetry and remote sensing: A review. *ISPRS J. Photogramm.* **2014**, *92*, 79–97. [CrossRef]
4. Klemas, V.V. Coastal and environmental remote sensing from unmanned aerial vehicles: An overview. *J. Coast. Res.* **2015**, *31*, 1260–1267. [CrossRef]
5. Floreano, D.; Wood, R.J. Science, technology and the future of small autonomous drones. *Nature* **2015**, *521*, 460–466. [CrossRef]
6. El Mahrad, B.; Newton, A.; Icely, J.D.; Kacimi, I.; Abalansa, S.; Snoussi, M. Contribution of remote sensing technologies to a holistic coastal and marine environmental management framework: A review. *Remote Sens.* **2020**, *12*, 2313. [CrossRef]
7. Casella, E.; Rovere, A.; Pedroncini, A.; Stark, C.P.; Casella, M.; Ferrari, M.; Firpo, M. Drones as tools for monitoring beach topography changes in the Ligurian Sea (NW Mediterranean). *Geo-Mar. Lett.* **2016**, *36*, 151–163. [CrossRef]
8. Topouzelis, K.; Papakonstantinou, A.; Doukari, M. Coastline change detection using unmanned aerial vehicles and image processing techniques. *Fresenius Environ. Bull.* **2017**, *26*, 5564–5571.
9. Holman, R.A.; Holland, K.T.; Lalejini, D.M.; Spansel, S.D. Surf zone characterization from Unmanned Aerial Vehicle imagery. *Ocean Dyn.* **2011**, *61*, 1927–1935. [CrossRef]
10. Turner, I.L.; Harley, M.D.; Drummond, C.D. UAVs for coastal surveying. *Coast. Eng.* **2016**, *114*, 19–24. [CrossRef]
11. Papakonstantinou, A.; Topouzelis, K.; Pavlogeorgatos, G. Coastline zones identification and 3D coastal mapping using UAV spatial data. *ISPRS Int. J. Geo-Inf.* **2016**, *5*, 75. [CrossRef]
12. Holman, R.A.; Brodie, K.L.; Spore, N.J. Surf zone characterization using a small quadcopter: Technical issues and procedures. *IEEE Trans. Geosci. Remote Sens.* **2017**, *9*, 2017–2027. [CrossRef]
13. Ventura, D.; Bruno, M.; Lasinio, G.J.; Belluscio, A.; Ardizzone, G. A low-cost drone based application for identifying and mapping of coastal fish nursery grounds. *Estuar. Coast. Shelf Sci.* **2016**, *171*, 85–98. [CrossRef]
14. Hodgson, A.; Kelly, N.; Peel, D. Unmanned aerial vehicles (UAVs) for surveying Marine Fauna: A dugong case study. *PLoS ONE* **2013**, *8*, e79556. [CrossRef]
15. Burns, J.; Delparte, D.; Gates, R. Takabayashi Integrating structure-from-motion photogrammetry with geospatial software as a novel technique for quantifying 3D ecological characteristics of coral reefs. *PeerJ* **2015**, *3*, e1077. [CrossRef]
16. Casella, E.; Collin, A.; Harris, D.; Ferse, S.; Bejarano, S.; Parravicini, V.; Hench, J.L.; Rovere, A. Mapping coral reefs using consumer-grade drones and structure from motion photogrammetry techniques. *Coral Reefs* **2017**, *36*, 269–275. [CrossRef]
17. Fiori, L.; Doshi, A.; Martinez, E.; Orams, M.B.; Bollard-Breen, B. The use of unmanned aerial systems in marine mammal research. *Remote Sens.* **2017**, *9*, 543. [CrossRef]
18. Murfitt, S.L.; Allan, B.M.; Bellgrove, A.; Rattray, A.; Young, M.A.; Ierodiaconou, D. Applications of unmanned aerial vehicles in intertidal reef monitoring. *Sci. Rep.* **2017**, *7*, 10259. [CrossRef]
19. Torres, L.G.; Nieukirk, S.L.; Lemos, L.; Chandler, T.E. Drone up! Quantifying whale behavior from a new perspective improves observational capacity. *Front. Mar. Sci.* **2018**, *5*, 319. [CrossRef]

20. Papakonstantinou, A.; Stamati, C.; Topouzelis, K. Comparison of true-color and multispectral unmanned aerial systems imagery for marine habitat mapping using object-based image analysis. *Remote Sens.* **2020**, *12*, 554. [CrossRef]

21. Provost, E.J.; Butcher, P.A.; Coleman, M.A.; Kelaher, B.P. Assessing the viability of small aerial drones to quantify recreational fishers. *Fish Manag. Ecol.* **2020**, 1–7. [CrossRef]

22. Fallati, L.; Saponari, L.; Savini, A.; Marchese, F.; Corselli, C.; Galli, P. Multi-Temporal UAV Data and object-based image analysis (OBIA) for estimation of substrate changes in a post-bleaching scenario on a maldivian reef. *Remote Sens.* **2020**, *12*, 2093. [CrossRef]

23. Hakvoort, H.; de Haan, J.; Jordans, R.; Vos, R.; Peters, S.; Rijkeboer, M. Towards airborne remote sensing of water quality in The Netherlands—validation and error analysis. *ISPRS J. Photogramm. Remote Sens.* **2002**, *57*, 171–183. [CrossRef]

24. Klemas, V. Tracking oil slicks and predicting their trajectories using remote sensors and models: Case studies of the Sea Princess and Deepwater Horizon oil spills. *J. Coast. Res.* **2010**, *26*, 789–797. [CrossRef]

25. Svejkovsky, J.; Nezlin, N.P.; Mustain, N.M.; Kum, J.B. Tracking stormwater discharge plumes and water quality of the Tijuana River with multispectral aerial imagery. *Estuar. Coast. Shelf Sci.* **2010**, *87*, 387–398. [CrossRef]

26. Androulidakis, Y.; Kourafalou, V.; Ozgokmen, T.; Garcia-Pineda, O.; Lund, B.; Le Henaff, M.; Hu, C.; Haus, B.K.; Novelli, G.; Guigand, C.; et al. Influence of river-induced fronts on hydrocarbon transport: A multiplatform observational study. *J. Geophys. Res. Oceans* **2018**, *123*, 3259–3285. [CrossRef]

27. Garaba, S.P.; Dierssen, H.M. An airborne remote sensing case study of synthetic hydrocarbon detection using short wave infrared absorption features identified from marine-harvested macro-and microplastics. *Remote Sens. Environ.* **2018**, *205*, 224–235. [CrossRef]

28. Fallati, L.; Polidori, A.; Salvatore, C.; Saponari, L.; Savini, A.; Galli, P. Anthropogenic Marine Debris assessment with Unmanned Aerial Vehicle imagery and deep learning: A case study along the beaches of the Republic of Maldives. *Sci. Total Environ.* **2019**, *693*, 133581. [CrossRef]

29. Savelyev, I.; Miller, W.D.; Sletten, M.; Smith, G.B.; Savidge, D.K.; Frick, G.; Menk, S.; Moore, T.; De Paolo, T.; Terrill, E.J.; et al. Airborne remote sensing of the upper ocean turbulence during CASPER-East. *Remote Sens.* **2018**, *10*, 1224. [CrossRef]

30. Stresser, M.; Carrasco, R.; Horstmann, J. Video-based estimation of surface currents using a low-cost quadcopter. *IEEE Geosci. Remote Sens. Lett.* **2017**, *14*, 2027–2031. [CrossRef]

31. Jung, D.; Lee, J.S.; Baek, J.Y.; Nam, J.; Jo, Y.H.; Song, K.M.; Cheong, Y.I. High temporal and spatial resolutions of sea surface current from low-altitude remote sensing. *J. Coast. Res.* **2019**, *90*, 282–288. [CrossRef]

32. Ouillon, S.; Forget, P.; Froidefond, J.M.; Naudin, J.J. Estimating suspended matter concentrations from SPOT data and from field measurements in the Rhône river plume. *Mar. Technol. Soc. J.* **1997**, *31*, 15.

33. Osadchiev, A.A. Small mountainous rivers generate high-frequency internal waves in coastal ocean. *Sci. Rep.* **2018**, *8*, 16609. [CrossRef] [PubMed]

34. Devlin, M.J.; McKinna, L.W.; Alvarez-Romero, J.G.; Petus, C.; Abott, B.; Harkness, P.; Brodie, J. Mapping the pollutants in surface riverine flood plume waters in the Great Barrier Reef, Australia. *Mar. Poll. Bull.* **2012**, *65*, 224–235. [CrossRef] [PubMed]

35. Brando, V.E.; Braga, F.; Zaggia, L.; Giardino, C.; Bresciani, M.; Matta, E.; Bellafiore, D.; Ferrarin, C.; Maicu, F.; Benetazzo, A.; et al. High-resolution satellite turbidity and sea surface temperature observations of river plume interactions during a significant flood event. *Ocean Sci.* **2015**, *11*, 909. [CrossRef]

36. Nezlin, N.P.; DiGiacomo, P.M. Satellite ocean color observations of stormwater runoff plumes along the San Pedro Shelf (southern California) during 1997 to 2003. *Continent. Shelf Res.* **2005**, *25*, 1692–1711. [CrossRef]

37. Osadchiev, A.A.; Sedakov, R.O. Spreading dynamics of small river plumes off the northeastern coast of the Black Sea observed by Landsat 8 and Sentinel-2. *Remote Sens. Environ.* **2019**, *221*, 522–533. [CrossRef]

38. Nezlin, N.P.; DiGiacomo, P.M.; Stein, E.D.; Ackerman, D. Stormwater runoff plumes observed by SeaWiFS radiometer in the Southern California Bight. *Remote Sens. Environ.* **2005**, *98*, 494–510. [CrossRef]

39. Constantin, S.; Doxaran, D.; Constantinescu, Ș. Estimation of water turbidity and analysis of its spatio-temporal variability in the Danube River plume (Black Sea) using MODIS satellite data. *Cont. Shelf Res.* **2016**, *112*, 14–30. [CrossRef]

40. Gangloff, A.; Verney, R.; Doxaran, D.; Ody, A.; Estournel, C. Investigating Rhône River plume (Gulf of Lions, France) dynamics using metrics analysis from the MERIS 300m Ocean Color archive (2002–2012). *Cont. Shelf Res.* **2017**, *144*, 98–111. [CrossRef]
41. Warrick, J.A.; Mertes, L.A.; Washburn, L.; Siegel, D.A. A conceptual model for river water and sediment dispersal in the Santa Barbara Channel, California. *Cont. Shelf Res.* **2004**, *24*, 2029–2043. [CrossRef]
42. Lihan, T.; Saitoh, S.I.; Iida, T.; Hirawake, T.; Iida, K. Satellite-measured temporal and spatial variability of the Tokachi River plume. *Estuar. Coast. Shelf Sci.* **2008**, *78*, 237–249. [CrossRef]
43. Jiang, L.; Yan, X.H.; Klemas, V. Remote sensing for the identification of coastal plumes: Case studies of Delaware Bay. *Int. J. Remote Sens.* **2009**, *30*, 2033–2048. [CrossRef]
44. Grodsky, S.A.; Reverdin, G.; Carton, J.A.; Coles, V.J. Year-to-year salinity changes in the Amazon plume: Contrasting 2011 and 2012 Aquarius/SACD and SMOS satellite data. *Remote Sens. Environ.* **2014**, *140*, 14–22. [CrossRef]
45. Reul, N.; Quilfen, Y.; Chapron, B.; Fournier, S.; Kudryavtsev, V.; Sabia, R. Multisensor observations of the Amazon-Orinoco river plume interactions with hurricanes. *J. Geophys. Res. Oceans* **2014**, *119*, 8271–8295. [CrossRef]
46. Korosov, A.; Counillon, F.; Johannessen, J.A. Monitoring the spreading of the A mazon freshwater plume by MODIS, SMOS, A quarius, and TOPAZ. *J. Geophys. Res. Oceans* **2015**, *120*, 268–283. [CrossRef]
47. Hessner, K.; Rubino, A.; Brandt, P.; Alpers, W. The Rhine outflow plume studied by the analysis of synthetic aperture radar data and numerical simulations. *J. Phys. Oceanogr.* **2001**, *31*, 3030–3044. [CrossRef]
48. DiGiacomo, P.M.; Washburn, L.; Holt, B.; Jones, B.H. Coastal pollution hazards in southern California observed by SAR imagery: Stormwater plumes, wastewater plumes, and natural hydrocarbon seeps. *Mar. Poll. Bull.* **2004**, *49*, 1013–1024. [CrossRef]
49. Zheng, Q.; Clemente-Colón, P.; Yan, X.H.; Liu, W.T.; Huang, N.E. Satellite synthetic aperture radar detection of Delaware Bay plumes: Jet-like feature analysis. *J. Geophys. Res. Oceans* **2004**, *109*, C03031. [CrossRef]
50. Perez, T.; Wesson, J.; Burrage, D. Airborne remote sensing of the Rio de la Plata plume using STARRS. *Sea Technol.* **2006**, *47*, 31–34.
51. Burrage, D.; Wesson, J.; Martinez, C.; Perez, T.; Moller, O., Jr.; Piola, A. Patos Lagoon outflow within the Río de la Plata plume using an airborne salinity mapper: Observing an embedded plume. *Cont. Shelf Res.* **2008**, *28*, 1625–1638. [CrossRef]
52. Horner-Devine, A.; Chickadel, C.C.; MacDonald, D. Coherent structures and mixing at a river plume front. In *Coherent Flow Structures in Geophysical Flows at the Earth's Surface*; Venditti, J., Best, J.L., Church, M., Hardy, R.J., Eds.; Wiley: Chichester, UK, 2013; pp. 359–369. [CrossRef]
53. Horner-Devine, A.R.; Chickadel, C.C. Lobe-cleft instability in the buoyant gravity current generated by estuarine outflow. *Geophys. Res. Lett.* **2017**, *44*, 5001–5007. [CrossRef]
54. Milliman, J.D.; Syvitski, J.P.M. Geomorphic-tectonic control of sediment discharge to the ocean: The importance of small mountainous rivers. *J. Geol.* **1992**, *100*, 525–544. [CrossRef]
55. Milliman, J.D.; Farnsworth, K.L.; Albertin, C.S. Flux and fate of fluvial sediments leaving large islands in the East Indies. *J. Sea Res.* **1999**, *41*, 97–107. [CrossRef]
56. Milliman, J.D.; Lin, S.W.; Kao, S.J.; Liu, J.P.; Liu, C.S.; Chiu, J.K.; Lin, Y.C. Short-term changes in seafloor character due to flood-derived hyperpycnal discharge: Typhoon Mindulle, Taiwan, July 2004. *Geology* **2007**, *35*, 779–782. [CrossRef]
57. Osadchiev, A.A.; Zavialov, P.O. Structure and dynamics of plumes generated by small rivers. In *Estuaries and Coastal Zones—Dynamics and Response to Environmental Changes*; Pan, J., Ed.; IntechOpen: London, UK, 2019. [CrossRef]
58. Korotkina, O. A.; Zavialov, P.O.; Osadchiev, A. A. Submesoscale variability of the current and wind fields in the coastal region of Sochi. *Oceanology* **2011**, *51*, 745–754. [CrossRef]
59. Korotkina, O. A.; Zavialov, P.O.; Osadchiev, A. A. Synoptic variability of currents in the coastal waters of Sochi. *Oceanology* **2014**, *54*, 545–556. [CrossRef]
60. Xia, M.; Xie, L.; Pietrafesa, L.J. Winds and the orientation of a coastal plane estuary plume. *Geophys. Res. Lett.* **2010**, *37*, L19601. [CrossRef]
61. Xia, M.; Xie, L.; Pietrafesa, L.J.; Whitney, M.M. The ideal response of a Gulf of Mexico estuary plume to wind forcing: Its connection with salt flux and a Lagrangian view. *J. Geophys. Res. Oceans* **2011**, *116*, C8. [CrossRef]

62. Zavialov, P.O.; Makkaveev, P.N.; Konovalov, B.V.; Osadchiev, A.A.; Khlebopashev, P.V.; Pelevin, V.V.; Grabovskiy, A.B.; Izhitskiy, A.S.; Goncharenko, I.V.; Soloviev, D.M.; et al. Hydrophysical and hydrochemical characteristics of the sea areas adjacent to the estuaries of small rivers if the Russian coast of the Black Sea. *Oceanology* **2014**, *54*, 265–280. [CrossRef]

63. Osadchiev, A.A. A method for quantifying freshwater discharge rates from satellite observations and Lagrangian numerical modeling of river plumes. *Environ. Res. Lett.* **2015**, *10*, 085009. [CrossRef]

64. Osadchiev, A.A. Estimation of river discharge based on remote sensing of a river plume. In Proceedings of the SPIE Remote Sensing, Toulouse, France, 14 October 2015. [CrossRef]

65. Jaoshvili, S. *The rivers of the Black Sea*; Chomeriki, I., Gigineishvili, G., Kordzadze, A., Eds.; Technical Report No. 71; European Environmental Agency: Copenhagen, Denmark, 2002.

66. Korotaev, G.; Oguz, T.; Nikiforov, A.; Koblinsky, C. Seasonal, interannual, and mesoscale variability of the Black Sea upper layer circulation derived from altimeter data. *J. Geophys. Res.* **2003**, *108*, 3122. [CrossRef]

67. Ivanov, V.A.; Belokopytov, V.N. *Oceanography of the Black Sea*; ECOSY-Gidrofizika: Sevastopol, Ukraine, 2013.

68. Ginzburg, A.I.; Kostianoy, A.G.; Krivosheya, V.G.; Nezlin, N.P.; Soloviev, D.M.; Stanichny, S.V.; Yakubenko, V.G. Mesoscale eddies and related processes in the northeastern Black Sea. *J. Mar. Syst.* **2002**, *32*, 71–90. [CrossRef]

69. Zatsepin, A.G.; Ginzburg, A.I.; Kostianoy, A.G.; Kremenetskiy, V.V.; Krivosheya, V.G.; Poulain, P.-M.; Stanichny, S.V. Observation of Black Sea mesoscale eddies and associated horizontal mixing. *J. Geophys. Res.* **2003**, *108*, 1–27. [CrossRef]

70. Kubryakov, A.A.; Stanichny, S.V. Seasonal and interannual variability of the Black Sea eddies and its dependence on characteristics of the large-scale circulation. *Deep Sea Res.* **2015**, *97*, 80–91. [CrossRef]

71. Medvedev, I.P.; Rabinovich, A.B.; Kulikov, E.A. Tides in three enclosed basins: The Baltic, Black, and Caspian seas. *Front. Mar. Sci.* **2016**, *3*, 46. [CrossRef]

72. Medvedev, I.P. Tides in the Black Sea: Observations and numerical modelling. *Pure Appl. Geophys.* **2018**, *175*, 1951–1969. [CrossRef]

73. Podymov, O.I.; Zatsepin, A.G. Seasonal anomalies of water salinity in the Gelendzhik region of the Black Sea according to shipborne monitoring data. *Oceanology* **2016**, *56*, 342–354. [CrossRef]

74. Doukari, M.; Batsaris, M.; Papakonstantinou, A.; Topouzelis, K. A protocol for aerial survey in coastal areas using UAS. *Remote Sens.* **2019**, *11*, 1913. [CrossRef]

75. Zavialov, P.O.; Izhitskiy, A.S.; Osadchiev, A.A.; Pelevin, V.V.; Grabovskiy, A.B. The structure of thermohaline and bio-optical fields in the surface layer of the Kara Sea in September 2011. *Oceanology* **2015**, *55*, 461–471. [CrossRef]

76. Baker, S.; Scharstein, D.; Lewis, J.; Roth, S.; Black, M.; Szeliski, R. A database and evaluation methodology for optical flow. *Int. J. Comp. Vis.* **2011**, *92*, 1–31. [CrossRef]

77. Fortun, D.; Bouthemy, P.; Kervrann, C. Optical flow modeling and computation: A survey. *Comput. Vis. Image Underst.* **2015**, *134*, 1–21. [CrossRef]

78. Farneback, G. Two-frame motion estimation based on polynomial expansion. In Proceedings of the 13th Scandinavian Conference on Image Analysis, Halmstad, Sweden, 29 June–2 July 2003; Bigun, J., Gustavsson, T., Eds.; Springer: Berlin/Heidelberg, Germany. [CrossRef]

79. O'Donnell, J.; Ackleson, S.G.; Levine, E.R. On the spatial scales of a river plume. *J. Geophys. Res. Oceans* **2008**, *113*, C4. [CrossRef]

80. Horner-Devine, A.R.; Hetland, R.D.; MacDonald, D.G. Mixing and transport in coastal river plumes. *Ann. Rev. Mar. Sci.* **2015**, *47*, 569–594. [CrossRef]

81. Zavialov, P.O.; Pelevin, V.V.; Belyaev, N.A.; Izhitskiy, A.S.; Konovalov, B.V.; Krementskiy, V.V.; Goncharenko, I.V.; Osadchiev, A.A.; Soloviev, D.M.; Garcia, C.A.E.; et al. High resolution LiDAR measurements reveal fine internal structure and variability of sediment-carrying coastal plume. *Estuar. Coast. Shelf Sci.* **2018**, *205*, 40–45. [CrossRef]

82. Yankovsky, A.E.; Chapman, D.C. A simple theory for the fate of buoyant coastal discharges. *J. Phys. Oceanogr.* **1997**, *27*, 1386–1401. [CrossRef]

83. Fong, D.A.; Geyer, W.R. The alongshore transport of freshwater in a surface-trapped river plume. *J. Phys. Oceanogr.* **2002**, *32*, 957–972. [CrossRef]

84. Whitney, M.M.; Garvine, R.W. Wind influence on a coastal buoyant outflow. *J. Geophys. Res.* **2005**, *110*, C03014. [CrossRef]

85. Choi, B.-J.; Wilkin, J.L. The effect of wind on the dispersal of the Hudson River plume. *J. Phys. Oceanogr.* **2007**, *37*, 1878–1897. [CrossRef]

86. Warrick, J.A.; Farnsworth, K.L. Coastal river plumes: Collisions and coalescence. *Prog. Oceanogr.* **2017**, *151*, 245–260. [CrossRef]

87. Osadchiev, A.A.; Korshenko, E.A. Small river plumes off the north-eastern coast of the Black Sea under average climatic and flooding discharge. *Ocean Sci.* **2017**, *13*, 465–482. [CrossRef]

88. Osadchiev, A.A.; Sedakov, R.O. Reconstruction of ocean surface currents using near simultaneous satellite imagery. In Proceedings of the International Geosciences and Remote Sensing Symposium, Yokohama, Japan, 28 July–2 August 2019; IEEE: New York, NY, USA. [CrossRef]

89. Alexeevsky, N.I.; Magritsky, D.V.; Koltermann, K.P.; Krylenko, I.N.; Toropov, P.A. Causes and systematics of inundations of the Krasnodar territory on the Russian Black Sea coast. *Nat. Hazard. Earth Syst.* **2016**, *16*, 1289–1308. [CrossRef]

90. Marchevsky, I.K.; Osadchiev, A.A.; Popov, A.Y. Numerical modelling of high-frequency internal waves generated by river discharge in coastal ocean. In Proceedings of the 5th International Conference on Geographical Information Systems Theory, Applications and Management, Heraklion, Crete, Greece, 3–5 May 2019; Scitepress: Setubal, Portugal. [CrossRef]

91. McPherson, R.A.; Stevens, C.L.; O'Callaghan, J.M.; Lucas, A.J.; Nash, J.D. The role of turbulence and internal waves in the structure and evolution of a near-field river plume. *Ocean Sci.* **2020**, *16*, 799–815. [CrossRef]

92. O'Donnell, J.; Marmorino, G.O.; Trump, C.L. Convergence and downwelling at a river plume front. *J. Phys. Oceanogr.* **1998**, *28*, 1481–1495. [CrossRef]

93. Zhurbas, V.; Väli, G.; Kuzmina, N. Rotation of floating particles in submesoscale cyclonic and anticyclonic eddies: A model study for the southeastern Baltic Sea. *Ocean Sci.* **2019**, *15*, 1691–1705. [CrossRef]

94. Garvine, R.W. Estuary plumes and fronts in shelf waters: A layer model. *J. Phys. Oceanogr.* **1987**, *17*, 1877–1896. [CrossRef]

95. O'Donnell, J. The formation and fate of a river plume: A numerical model. *J. Phys. Oceanogr.* **1990**, *20*, 551–569. [CrossRef]

96. Hetland, R.D. Relating river plume structure to vertical mixing. *J. of Phys. Oceanogr.* **2005**, *35*, 1667–1688. [CrossRef]

97. Saldias, G.S.; Sobarzo, M.; Largier, J.; Moffat, C.; Letelier, R. Seasonal variability of turbid river plumes off central Chile based on high-resolution MODIS imagery. *Remote Sens. Env.* **2012**, *123*, 220–233. [CrossRef]

98. Saldias, G.S.; Largier, J.L.; Mendes, R.; Perez-Santos, I.; Vargas, C.A.; Sobarzo, M. Satellite-measured interannual variability of turbid river plumes off central-southern Chile: Spatial patterns and the influence of climate variability. *Progr. Oceanogr.* **2016**, *146*, 212–222. [CrossRef]

99. Osadchiev, A.A.; Izhitskiy, A.S.; Zavialov, P.O.; Kremenetskiy, V.V.; Polukhin, A.A.; Pelevin, V.V.; Toktamysova, Z.M. Structure of the buoyant plume formed by Ob and Yenisei river discharge in the southern part of the Kara Sea during summer and autumn. *J. Geophys. Res. Oceans* **2017**, *122*, 5916–5935. [CrossRef]

100. Gong, W.; Chen, L.; Chen, Z.; Zhang, H. Plume-to-plume interactions in the Pearl River Delta in winter. *Ocean Coast. Manag.* **2019**, *175*, 110–126. [CrossRef]

101. Warrick, J.A.; Mertes, L.A.K.; Washburn, L.; Siegel, D.A. Dispersal forcing of southern California river plumes, based on field and remote sensing observations. *Geo-Mar. Lett.* **2004**, *24*, 46–52. [CrossRef]

102. Osadchiev, A.A.; Korotenko, K.A.; Zavialov, P.O.; Chiang, W.-S.; Liu, C.-C. Transport and bottom accumulation of fine river sediments under typhoon conditions and associated submarine landslides: Case study of the Peinan River, Taiwan. *Nat. Haz. Earth Syst. Sci.* **2016**, *16*, 41–54. [CrossRef]

103. Romero, L.; Siegel, D.A.; McWilliams, J.C.; Uchiyama, Y.; Jones, C. Characterizing storm water dispersion and dilution from small coastal streams. *J. Geophys. Res. Oceans* **2016**, *121*, 3926–3943. [CrossRef]

104. Yankovsky, A.E.; Hickey, B.M.; Münchow, A.K. Impact of variable inflow on the dynamics of a coastal buoyant plume. *J. Geophys. Res. Oceans* **2001**, *106*, 19809–19824. [CrossRef]

105. Yuan, Y.; Horner-Devine, A.R.; Avener, M.; Bevan, S. The role of periodically varying discharge on river plume structure and transport. *Cont. Shelf Res.* **2018**, *158*, 15–25. [CrossRef]

106. Yankovsky, A.E.; Voulgaris, G. Response of a Coastal Plume Formed by Tidally Modulated Estuarine Outflow to Light Upwelling-Favorable Wind. *J. Phys. Oceanogr.* **2019**, *49*, 691–703. [CrossRef]

107. Cole, K.L.; MacDonald, D.G.; Kakoulaki, G.; Hetland, R.D. River plume source-front connectivity. *Ocean Model.* **2020**, *150*, 101571. [CrossRef]

108. Horner-Devine, A.R.; Jay, D.A.; Orton, P.M.; Spahna, E.Y. A conceptual model of the strongly tidal Columbia River plume. *J. Mar. Syst.* **2008**, *78*, 460–475. [CrossRef]
109. Avicola, G.; Huq, P. Scaling analysis for the interaction between a buoyant coastal current and the continental shelf: Experiments and observations. *J. Phys. Oceanogr.* **2002**, *32*, 3233–3248. [CrossRef]
110. Lentz, S.J.; Helfrich, K.R. Buoyant gravity currents along a sloping bottom in a rotating fluid. *J. Fluid. Mech.* **2002**, *464*, 251–278. [CrossRef]
111. Pimenta, F.M.; Kirwan, A.D., Jr.; Huq, P. On the transport of buoyant coastal plumes. *J. Phys. Oceanogr.* **2011**, *41*, 620–640. [CrossRef]
112. Korotenko, K.A.; Osadchiev, A.A.; Zavialov, P.O.; Kao, R.-C.; Ding, C.-F. Effects of bottom topography on dynamics of river discharges in tidal regions: Case study of twin plumes in Taiwan Strait. *Ocean Sci.* **2014**, *10*, 865–879. [CrossRef]
113. Ostrander, C.E.; McManus, M.A.; DeCarlo, E.H.; Mackenzie, F.T. Temporal and spatial variability of freshwater plumes in a semi-enclosed estuarine–bay system. *Estuaries Coasts* **2008**, *31*, 192–203. [CrossRef]
114. Osadchiev, A.A.; Zavialov, P.O. Lagrangian model for surface-advected river plume. *Cont. Shelf Res.* **2013**, *58*, 96–106. [CrossRef]
115. Zhao, J.; Gong, W.; Shen, J. The effect of wind on the dispersal of a tropical small river plume. *Front. Earth Sci.* **2018**, *12*, 170–190. [CrossRef]
116. Mityagina, M.I.; Lavrova, O.Y.; Karimova, S.S. Multi-sensor survey of seasonal variability in coastal eddy and internal wave signatures in the north-eastern Black Sea. *Int. J. Remote Sens.* **2010**, *31*, 4779–4790. [CrossRef]
117. Lavrova, O.Y.; Mityagina, M.I. Satellite survey of internal waves in the Black and Caspian seas. *Int. J. Remote Sens.* **2017**, *9*, 892. [CrossRef]
118. Trump, C.L.; Marmorino, G.O. Mapping small-scale along-front structure using ADCP acoustic backscatter range-bin data. *Estuaries* **2003**, *26*, 878–884. [CrossRef]
119. Warrick, J.A.; Stevens, A.W. A buoyant plume adjacent to a headland—Observations of the Elwha River plume. *Continent Shelf Res.* **2011**, *31*, 85–97. [CrossRef]

Article

Investigations into Synoptic Spatiotemporal Characteristics of Coastal Upper Ocean Circulation Using High Frequency Radar Data and Model Output

Lei Ren [1,2,3], Nanyang Chu [4,*], Zhan Hu [4] and Michael Hartnett [5,6]

[1] School of Marine Engineering and Technology, Sun Yat-sen University, Zhuhai 519082, China; renlei7@mail.sysu.edu.cn
[2] Southern Marine Science and Engineering Guangdong Laboratory (Zhuhai), Zhuhai 519082, China
[3] Institute of Estuarine and Coastal Research, Sun Yat-sen University, Zhuhai 519082, China
[4] School of Marine Science, Sun Yat-sen University, Zhuhai 519082, China; huzh9@mail.sysu.edu.cn
[5] School of Engineering and Informatics, National University of Ireland Galway, H91 TH33 Galway, Ireland; michael.hartnett@nuigalway.ie
[6] Ryan Institute, National University of Ireland Galway, H91 TH33 Galway, Ireland
* Correspondence: chuny@mail2.sysu.edu.cn

Received: 20 July 2020; Accepted: 29 August 2020; Published: 1 September 2020

Abstract: Numerical models and remote sensing observation systems such as radars are useful for providing information on surface flows for coastal areas. Evaluation of their performance and extracting synoptic characteristics are challenging and important tasks. This research aims to investigate synoptic characteristics of surface flow fields through undertaking a detailed analysis of model results and high frequency radar (HFR) data using self-organizing map (SOM) and empirical orthogonal function (EOF) analysis. A dataset of surface flow fields over thirteen days from these two sources was used. A SOM topology map of size 4×3 was developed to explore spatial patterns of surface flows. Additionally, comparisons of surface flow patterns between SOM and EOF analysis were carried out. Results illustrate that both SOM and EOF analysis methods are valuable tools for extracting characteristic surface current patterns. Comparisons indicated that the SOM technique displays synoptic characteristics of surface flow fields in a more detailed way than EOF analysis. Extracted synoptic surface current patterns are useful in a variety of applications, such as oil spill treatment and search and rescue. This research provides an approach to using powerful tools to diagnose ocean processes from different aspects. Moreover, it is of great significance to assess SOM as a potential forecasting tool for coastal surface currents.

Keywords: ocean surface circulation; high frequency radar; self-organizing map; empirical orthogonal function; neural networks; synoptic characteristics

1. Introduction

Surface currents primarily driven by winds can flow for thousands of kilometers and can reach depths of hundreds of meters. Their movements carry heat and mass from place-to-place about the Earth system. Understanding of surface current patterns in coastal regions is of great importance for a variety of aspects, such as the development of marine economics and environmental protection [1]. As remote sensing technologies advance, surface currents can be measured not only at a single or few locations by instruments such as an Acoustic Doppler Current Profiler (ADCP), but can be recorded by tools such as radars over large coastal domains with high spatial and temporal resolutions. Understanding, mining, and application of these surface flow field data are a new challenge for researchers [2]. Several researchers have recently undertaken investigations on measured surface flow fields. They have used surface radar data to validate model results, to improve modeling performance through data assimilation, to establish statistical forecasting models, and to characterize the physical process of surface water bodies [3–6].

With the increasing availability of surface current data, investigations into patterns of surface currents have been undertaken using various analysis techniques such as empirical orthogonal function (EOF) (or principal component analysis (PCA)), k-means, and self-organizing map (SOM) to extract patterns of variability in meteorological and oceanographic data. In essence, EOF and PCA are the same, but their focus is different. PCA is the eigenvalue and eigenvector obtained from the covariance matrix, and EOF is computed using the cross matrix of anomaly values, so the calculated eigenvectors are the same, while the latter is n times the former (n is the sample size). PCA, multidimensional scaling (MDS), and SOM are representative unsupervised machine learning techniques. The PCA technique summarizes the dispersion of datasets as a data cloud through converting the original dataset into a set of principal components; the MDS technique takes a set of dissimilarities and returns a set of points such that the distances between points are approximately equal to the dissimilarities [7]. The classical MDS technique is the same as the PCA technique if the input dataset distances are Euclidean [8]. Liu, et al. [9] used EOF and SOM to extract synoptic characteristic patterns of ocean currents at the West Florida Shelf area, and found that flow field patterns extracted by SOM are more accurate and intuitive than those obtained from the leading mode patterns with EOF analysis. This is probably because EOF is a linear extraction tool, whereas SOM is a nonlinear extraction tool; properties of nonlinearity extraction in SOM can better describe nonlinear dynamic oceanic processes. Soto-Navarro, et al. [10] applied EOF decomposition to compare the main flow pattern from model and radar, and found that both systems show satisfactory agreement for the first two EOF modes, while the agreement is less for the third EOF mode. Moreover, because wind data collected in a single station close to the coast were used in the model, correlation of the third principal component (PC) between model and radar was very low (<0.1). They demonstrated that results from that principal components are representative of the entire study domain. Reusch, et al. [11] compared the SOM method with the PCA method for extracting patterns of variability for North Atlantic sea level pressure fields and found that SOM was more robust than PCA. In addition, comparisons between the SOM method and the k-means method were performed by Lobo [12], Lin and Chen [13], and Solidoro, et al. [14]. Their results indicated that SOM generates more accurate patterns than k-means, and SOM in general is more flexible than k-means. Moreover, the SOM method has been broadly applied among disparate range of disciplines such as meteorology (sea level pressure, air temperature, humidity, evaporation, precipitation, cloud, and wind data) and oceanography (satellite ocean color and chlorophyll, biological and geochemical data, sea surface temperature (SST), sea surface height (SSH), and ocean currents) as a data mining and visualization method for complex datasets [15–21]. These studies demonstrate that SOM is a robust, efficient, and concise method to project high dimensional data onto a low dimensional (usually two-dimensional) map for characterizing synoptic patterns. Thus, in this research, SOM was applied to extract synoptic characteristic patterns of coastal ocean flow fields based on hourly HFR measurements and model results at a site off the west coast of Ireland.

Several researchers have previously applied SOM techniques to extract characteristic patterns in physical oceanography from in situ observations and model results. Liu and Weisberg [22] examined patterns of ocean current variability using time series of moored ADCP velocity data based on EOF and SOM. Three characteristic spatial patterns were extracted: (i) spatially coherent southeastward, (ii) northwestward flow patterns with strong currents, and (iii) a transition pattern of weak currents. Based on comparisons between results from EOF and SOM, they found that the SOM had advantages over the EOF in both pattern recognition and description. Subsequently, Liu, et al. [23] applied SOM to extract patterns from a linear progressive sine wave signal, and analyzed the effects of the SOM tunable parameters on the extracted patterns. Effects of varying SOM map size, map lattice structure, and neighborhood function were examined as well. Liu, et al. [23] found that (a) a larger SOM map size led to slightly more accurate mapping, (b) a rectangular lattice was preferable for a small-size SOM and a hexagonal lattice may be useful for larger map sizes, (c) linear initialization provided better SOM results than random initialization, (d) the "ep" (or Epanechikov, see details in Appendix A) type is the best neighborhood function and gave the best results. Liu, et al. [24] also investigated the spatial structure and temporal evolution of distinct physical processes on the West Florida Shelf (WFS) based on patterns of ocean current variability from a joint HFR and ADCP dataset using SOM. Semidiurnal, diurnal, and subtidal frequency bands were separately examined with SOM analysis. Results indicated that SOM is an effective analysis tool for identifying modulated, heterogeneous, anisotropic, three-dimensional coastal ocean current variations observed by HFR and ADCPs [24].

Mihanović, et al. [25] extracted subtidal frequency patterns from HFR surface flow fields in the northern Adriatic using SOM. Since surface current patterns were strongly influenced by local wind forcing, a joint dataset including contemporaneous surface wind data obtained from the operational hydrostatic mesoscale meteorological model ALADIN/HR was used. Their analysis found that the strongest currents observed during energetic bora episodes were represented by several current patterns and another characteristic wind, the sirocco, which was represented by three SOM current patterns. Mihanović, et al. [25] suggested that SOM was a most valuable tool for extracting characteristic patterns of surface flows and forcing functions. Vilibic, et al. [26] applied the SOM method to predict surface currents based on HFR measurements and numerical weather prediction (NWP) data for the northern Adriatic in comparison with operational ROMS-produced surface currents. They found that the SOM-based forecasting system had a slightly better forecasting skill than the ROMS model, especially during strong wind conditions. Liu, et al. [9] applied the SOM method to extract patterns of the loop current system and to identify altimetry sea surface height variability in the eastern Gulf of Mexico. Jin, et al. [27] studied the variability of current patterns near the Karama Gap using outputs from the ocean general circulation model (OGCM) for the earth simulator on the basis of the SOM. Jin, et al. [27] found that the evolution of the four coherent patterns showed a robust cycle characterized as a counterclockwise trajectory in the SOM space. Tsui and Wu [17] applied SOM to study the Kuroshio intrusion into the South China Sea (SCS) through the Luzon Strait using 18 years of archiving, validation, and interpretation of satellite oceanographic (AVISO) mean geostrophic velocity (GSV) data. Results indicate that the Kuroshio intrusion may occur year-round; intrusion is not a major characteristic in the study area and winter intrusion events are more frequent than summer ones, based on seasonal variability.

The SOM technique is in a sense a combination of both PCA and MDS techniques. SOM is a type of cluster analysis, which organizes a dataset of patterns into clusters based on similarity. Grouping a given dataset of unlabeled patterns into meaningful clusters is the main problem that is solved in SOM. Moreover, SOM was used to develop a drought forecast model through a nonlinear mapping of the input domain onto a two-dimensional grid by Barros and Bowden [28]; results indicated that SOM-based data models can be tools of discovery to identify nonlinear diagnostic and prognostic relationships among datasets. Obach, et al. [29] used radial basis function networks combined with a SOM to predict annual abundance of aquatic insects, and found that it is possible to predict the abundance of aquatic insects based on relevant environmental factors. SOM can be employed not only to reduce the size of the dataset by clustering, but also to construct a nonlinear projection of the dataset onto a low dimensional display that are usually of one or two dimensions [30].

Previous studies indicated that SOM is a useful and effective tool for dealing with large datasets. Since there were few records of surface current measurements, in either space or time, available in the Galway Bay area before the deployment of HFR system, previous studies of the bay generally provide little information on spatial patterns of surface currents [31–33]. Herein, both SOM and EOF techniques were applied to investigate synoptic characteristics of surface currents from both numerical models and a HFR system. Additionally, some previous research considered longer term analysis. During this research, because datasets were incomplete, gap filling was used to develop a "synthetic" dataset. The research presented herein considers a small dataset, one of the reasons for this is so that we use only actual data during the analysis and do not introduce extraneous uncertainties; this type of analysis has not previously been carried out. In subsequent research, we will consider longer datasets that have been synthetically enhanced to provide continuous signals and then compare results with those from this current research.

The structure of this paper is as follows. Section 2 presents methodologies, including the research domain, observational data, numerical model, and SOM and EOF methods. Results and discussion are presented in Section 3, followed by research conclusions in Section 4.

2. Methodologies

2.1. Research Domain

Galway Bay is located on the west coast of Ireland; it is a semiclosed bay, as shown in Figure 1. Its length from west to east is approximately 62 km and the mouth of the bay from north to south is approximately 33 km. Regional climate in Galway Bay area is mainly affected by the Atlantic Climate and prevailing southwesterly winds. Tides in this area are semidiurnal, ranging from 5 m during spring tides to 2.5 m during neap tides [34]. The average water depth is approximately 30 m for the area covered by the HFR system.

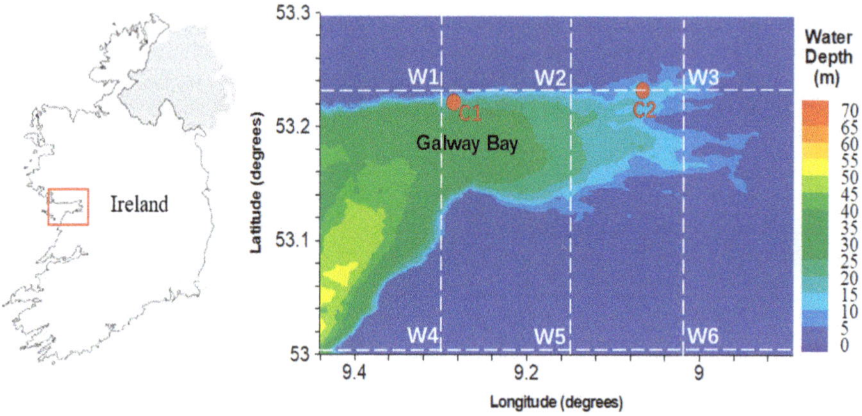

Figure 1. Deployment of the high frequency radar (HFR) system (C1 and C2 indicate deployment location of HFR station).

2.2. Observational Data

Land-based coastal radar systems are capable of monitoring information of surface waters based on the application of high frequency radio wave backscatter [35]. Radars operating in the HF band can measure the Doppler shift of radio waves scattered from ocean surface gravity waves [36–39]. A single radar station determines radial components of surface currents relative to that station, providing current magnitudes and directions toward or away from the station. Surface flow fields are determined by synthesizing radial surface velocity components from two or more radars. The extent of alongshore surface current mapping is limited only by the number of radar stations with overlapping coverage [40]. Spatial coverage of surface currents measured by radars can reach approximately 200 km depending on the radar transmitting frequency. Information obtained from radar has a large number of applications, such as analysis of marine renewable energy resources, oil-spill monitoring [41,42], data assimilation into numerical models [43–46], trajectory forecasts [47], and search and rescue [48].

HFR ocean data quality is affected by several factors such as geometric dilution of precision (GDOP) and signal-to-noise ratio (SNR) [49]. In order to quantitatively assess radar data quality for this case, a commonly used evaluation index, GDOP, describes the quality of a velocity measurement based on the geometrical arrangement between the radar stations and the location being monitored that had been used to assess velocity components. A low value of GDOP indicates ideal geometry and higher values indicate poor geometry where the two velocity components are not highly resolved [50]. O'Donncha, et al. [51] found that the meridional component of surface current flow along the baseline is distorted most by GDOP in Galway Bay, while the zonal component is more accurately resolved, apart from a very small domain along the shoreline due to a slight rotation of the baseline from east–west. Additionally, Ren, et al. [43] compared radial currents between the radar data and ADCP data in the study domain, and found that modest correlation existed between the two datasets.

Two SeaSonde Coastal Ocean Dynamics Applications Radar (CODAR) radars were deployed intermittently at Galway Bay to monitor surface currents and waves since the summer of 2011 [52]. Radar stations are located at Mutton Island (C1 in Figure 1) and Spiddal Pier (C2 in Figure 1); the operating frequency is 25 MHz at both stations. Radial current vector fields from each station are recorded every hour [53]. Data from both radars are routinely transmitted to a combination center that is located in the campus of National University of Ireland, Galway, Ireland. The radar postprocessing software system interpolates surface current data onto a standard orthogonal grid 300×300 m. Measurements of surface currents obtained with the HFR system in Galway Bay have been validated with ADCP data in detail by O'Donncha, et al. [51] and Ren, et al. [54]. The land-based HFR system has provided a powerful method of obtaining synoptic monitoring of surface currents with high temporal and spatial resolutions. Identification of synoptic characteristics of surface flows is a meaningful approach for obtaining good insights into both internal dynamic processes and variations of ocean surface movement.

Coverage of surface currents captured by the HFR system varies in space and time due to variability of the ocean surface roughness. An EOF analysis is consistent and reliable when there are no spatial gaps in the datasets; thus, HFR data with high coverage density in space and time were selected and used in the research. The HFR data at 1117 spatial points between Julian day 220 and Julian day 232, 2013, were used in the following analysis. Because it is the first time that surface currents were obtained at high temporal and spatial resolution using a model and radar system in the study domain, analysis of these short-term (thirteen day) dataset using SOM and EOF can be viewed as a test.

2.3. Numerical Model

The coastal model Environmental Fluid Dynamics Code (EFDC) was applied to simulate the hydrodynamics of Galway Bay. EFDC was developed at the Virginia Institute of Marine Science by the U.S. Environmental Protection Agency (EPA) [55,56]. As a free open source numerical model, EFDC reduces the access threshold for users. EFDC consists of four linked modules: hydrodynamic, water quality and eutrophication, sediment transport, and toxic chemical transport and fate. Only the hydrodynamic module was used to simulate surface flows in this research. This module solves the three-dimensional, vertically hydrostatic, free surface, turbulent averaged equations of motions for a variable density fluid. The hydrodynamic component of EFDC implements a semi-implicit, conservative finite volume solution scheme for the hydrostatic primitive equations with either two- or three-level time stepping [55–57]. The model uses orthogonal curvilinear coordinate or Cartesian rectangular coordinate system and structured grid horizontally, which is suitable for long and straight shoreline with shorter calculation time. The sigma coordinate system is used vertically to avoid the precision difference between deep water area and shallow water area. Additionally, EFDC can simulate one-dimensional, two-dimensional, and three-dimensional hydrodynamic force and water quality of water body. The internal and external mode splitting method is used in the calculation process, and the calculation accuracy of space and time is second order. The model has been successfully applied to a number of modeling studies of rivers, lakes, estuaries, and coastal regions [58–60]. Boas, et al. [61] compared EFDC with WASP (water quality analysis simulation program) and the commercially available software MIKE, and found that because the horizontal scale of most surface water is much larger than the vertical scale, in order to simplify the calculation, the vertical pressure gradient is regarded as the balance with buoyancy, and the vertical acceleration is ignored in EFDC, so EFDC is mostly used in shallow water areas. EFDC has a robust flooding and drying routine which is required in coastal regions. The structured grid also leads to the low adaptation of EFDC to the curved coastline. For these reasons, EFDC was applied to Galway Bay.

A 3D hydrodynamic model of Galway Bay (see Figure 1) was developed using a regular grid coordinate system; a spatial resolution by 150 m in both horizontal directions was employed generating 380×241 grid cells in this research. A bathymetric model of Galway Bay was developed from the recent, high resolution Integrated Mapping for the Sustainable Development of Ireland's Marine Resource (INFOMAR) seabed data program. Variable vertical layer thicknesses were used in the model with a thinner layer at the top and bottom of the water column and thicker layers in the middle, thereby ensuring that wind forcing was not overly damped by tidal forcing. A detailed description on setting up vertical layer structure for the Galway Bay was reported by Ren, et al. [54]. The meteorological forcing parameters including wind, pressure, rainfall, solar radiation, and relative humidity were obtained at a one-minute interval from the Informatics Research Unit for Sustainable Engineering (IRUSE) weather station approximately 5 km from station C2 (see Figure 1). Records of the River Corrib inflows entering Galway Bay close to the north of point C2 were obtained from the Irish Office of Public Works (OPW). Tidal water elevation time series generated from Oregon State University Tidal Inversion Software (OTIS) were used to define the tidal forcing at the western and southern open boundaries in the model [62,63].

2.4. Self-Organizing Map

Kohonen self-organizing maps or self-organizing maps are a type of neural network algorithm proposed by Tuevo Kohonen [64]. The SOM is a kind of unsupervised learning algorithm, which captures patterns in the input data through competitive learning, hence the name "self-organizing". SOM retains a principal "features map" of the input data; this makes SOM very useful. SOM is also considered a "map" projection method. Another intrinsic characteristic of SOM is that vector quantization, which reduces multidimensional data into lower dimensional spaces (usually one or two dimensions), is easier to understand. Additionally, SOM builds relationships that retain information so that any topological relationships developed within the training set are maintained.

Figure 2 presents an example of a 4×3 SOM structure. $X_1, X_2, \ldots, X_n$ are input data, i.e., the two-dimensional surface flow fields in this research, which are projected to each node in the output layer. X_i contains surface velocity components having I data points and J data points in the x and y directions, respectively, over the analysis domain. This indicates that each node of an output layer is linked to each input dataset. Each node of an output layer, as shown in Figure 2, has a specific topological position and contains a vector of weights of the same dimension as the input vectors. If the training data consists of vector X of n dimensions, each node contains a corresponding weight vector of n dimensions. The dotted lines connecting the nodes at the output layer only represent adjacency and do not signify connectivity. There are no lateral connections among nodes on the output layer. The weight vectors adopt an alternative initialization scheme. As the input dataset is processed through the SOM neural network, the summed distance between weights and input dataset are computed at each node. In each successive step the weight vector of the unit having the smallest Euclidian distance is selected as the "winner," the best match unit (BMU) or codebook vectors [30]. The SOM is a neighborhood-preserving vector-quantitative analysis tool working on the winner-take-all rule in a mathematical sense, where the BMU is determined as the most similar node to the input at an instant of time. The key of the SOM algorithm is to update the BMU and its neighborhood nodes concurrently. Input topology of dataset is preserved on the output nodes through performance of such a mapping [65]. Details of the implementation procedure of SOM are described in the Appendix A.

Advantages of the SOM method can be listed as follows: (a) an intuitive approach to building customer segmentation profiles, (b) simple and easy to explain results, and (c) new data points can be mapped to the trained model for predictive purposes. To quantitatively assess the mapping quality, two measures are used. The first is the quantization error (QE, see details in Appendix A), which is used as a metric of the average distance between each data vector and the BMU, whose weight vector has the minimum distance. The second measure is the topographic error (TE, see details in Appendix A), which represents the proportion of all data vectors for which the first and second *BMUs* are not adjacent to each other [66]. Lower values of *QE* and *TE* indicate better reproduction of the patterns of the SOM model. A batch algorithm, a rectangular-lattice structure with a sheet map and the "ep" neighboring function for SOM analysis, as recommended by Liu, et al. [23] and Vilibić, et al. [67], were used in this research.

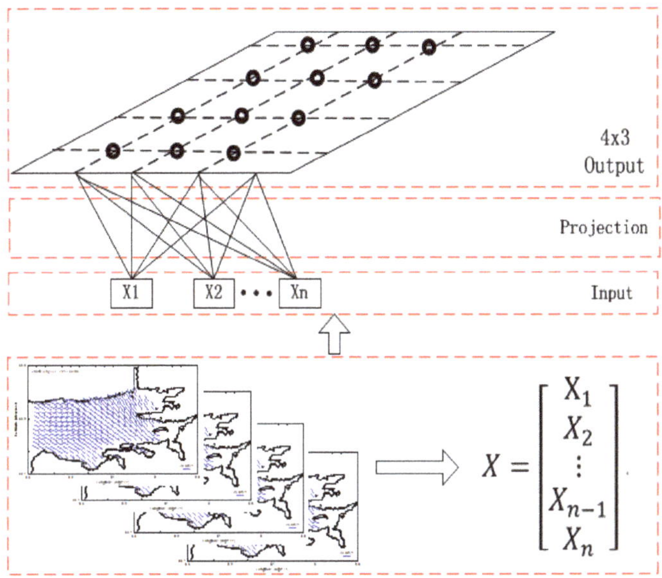

Figure 2. Architecture of a 4×3 self-organizing map (SOM) network.

2.5. Empirical Orthogonal Function

An empirical orthogonal function analysis is a data decomposition tool and it can be used to explain original data in the form of a series of orthogonal base functions with associated coefficients. EOF modes preserve the majority of the variations as much as possible through reducing the dimensionality of an original dataset. One unique attribute of an EOF analysis is that EOF base functions are derived from the original dataset through decomposition. This produces EOF modes that preserve the inherent characteristics of the original dataset and converge rapidly [68]. A detailed description about the EOF analysis method can be found in Hannachi, et al. [69] and Monahan, et al. [70]. In this research, for given vector fields of surface currents in Galway Bay, an investigation of spatial–temporal variations of surface flows was implemented by EOF decomposition. Details of EOF analysis are presented in the Appendix A.

3. Results

Since coverage of HFR surface currents varied in space and time and to ensure reliable analysis in this research, surface currents only at points always covered by the HFR system during the analysis period were selected and used in the following analysis. Here, 312-h surface vector fields with 1117 observation points in total for Galway Bay were used for both SOM and EOF analyses. Surface vector fields extracted from model results at the same points were used. Both observed and simulated surface vector field datasets contained 2234 (1117 for both u and v components) columns and 312 rows.

3.1. SOM Analysis

The goal of the SOM technique is to partition an incoming dataset of arbitrary dimension into a two-dimensional discrete feature map and to display this transformation adaptively in a topologically ordered fashion. Extracted SOM patterns are arranged in a two-dimensional array such that similar patterns are located nearby and dissimilar patterns are distant [71]. To completely represent the characteristic surface flow features and make it small enough for visualization and interpretation, after several test runs, a commonly used SOM size of 4 × 3 was selected and used in this research.

3.1.1. Spatial Variability

The batch of surface flow fields for both model results and HFR data were characterized into 12 typical SOM patterns with corresponding frequencies of occurrence. The 4 × 3 SOM array results of model results and HFR data are shown in Figures 3 and 4, respectively. To quantify the representation of each SOM pattern, the frequency of occurrence of each SOM pattern was computed by summing the number of the *BMU* by the total record lengths (the number of input vectors, 312 here), see details in Appendix B. The relative occurrence frequency for corresponding SOM patterns is shown in the upper left corner in Figures 3 and 4.

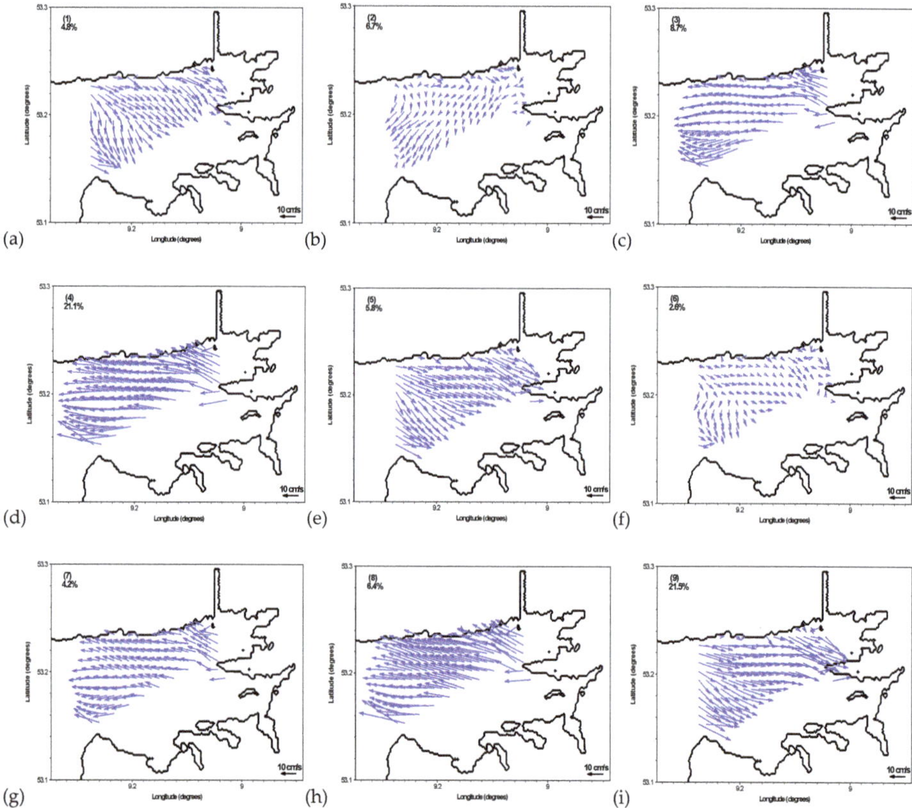

Figure 3. *Cont.*

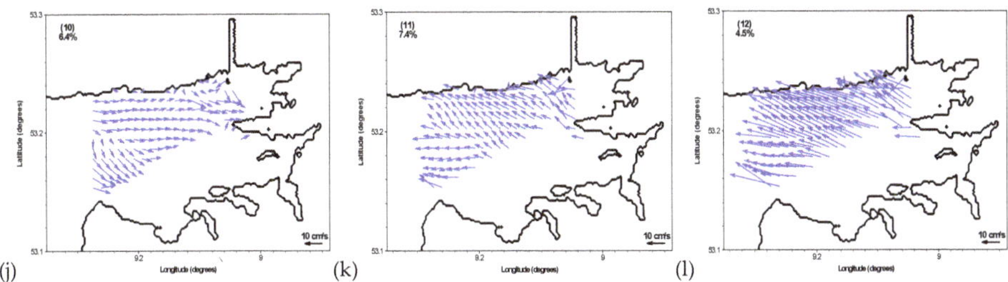

Figure 3. Characteristic spatial patterns of surface currents from model results extracted by a 4 × 3 SOM analysis (subfigures (**a**–**l**) indicates twelve spatial SOM patterns, respectively; the occurrence frequency is given as a percentage number for each pattern at the topleft).

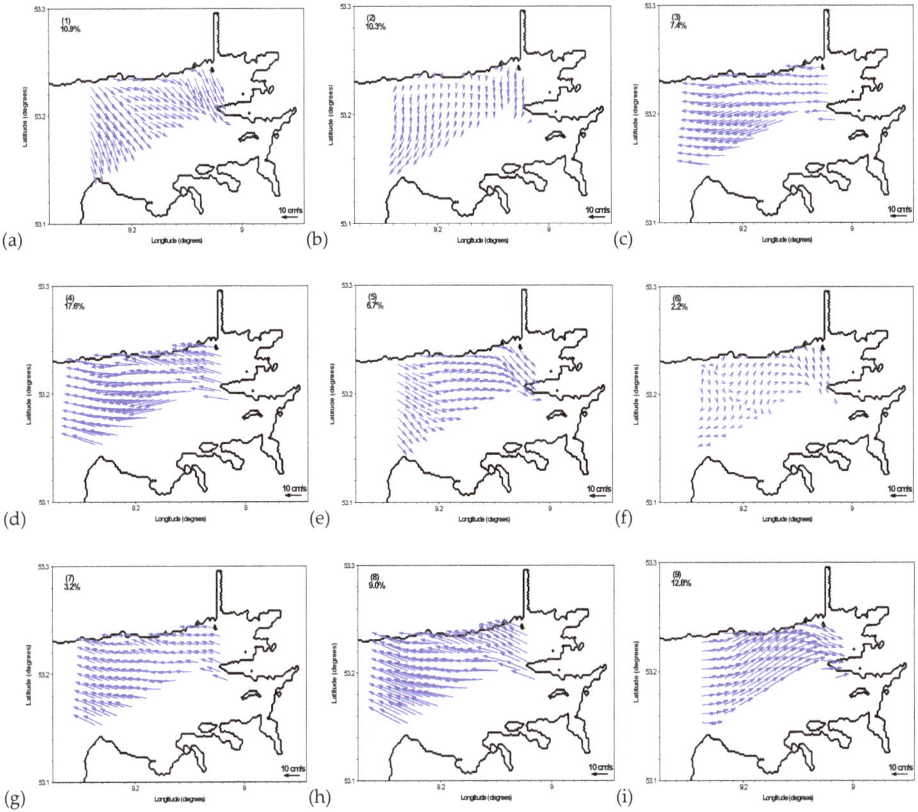

Figure 4. *Cont.*

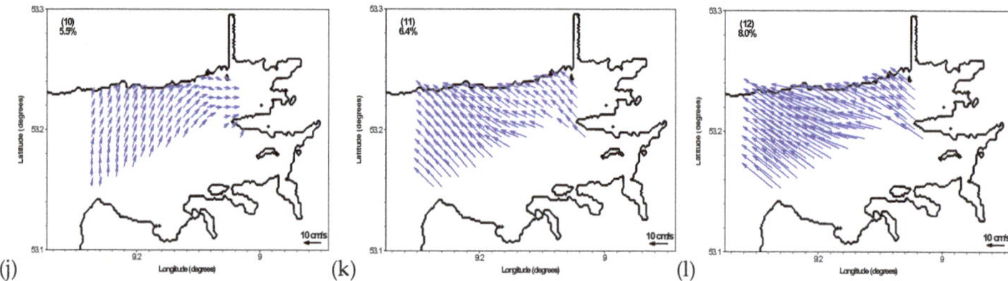

Figure 4. Characteristic spatial patterns of surface currents from radar extracted by a 4 × 3 SOM analysis (subfigures (**a–l**) indicates twelve spatial SOM patterns, respectively; the occurrence frequency is given as a percentage number for each pattern at the topleft).

For SOM surface flow patterns as shown in Figure 3, twelve SOM patterns can be categorized visually into four groups, as presented in Table 1.

Table 1. Categories of SOM patterns for model results.

Group	SOM Pattern	Representative Characteristics	Total Occurrence Frequency (%)
1	1/5/6/9/10	southeastward and eastward flows	41.1
2	3/4/7/8,	western flows	40.4
3	11/12	northwestward flows	11.9
4	2	southwestward and alongshore flows	6.7

Table 1 shows that the occurrence frequency of group 1 consisting of four SOM patterns (1/5/6/9/10) was the highest at 41.1% with southeastward and eastern flows. Surface flow fields categorized as group 2 had the second highest occurrence frequency with westward flows. The total occurrence frequency of groups 1 and 2 was greater than 81%. This indicates that the main patterns of surface flows were southeastern and alongshore flows during the analysis period. Group 3 with an occurrence frequency greater than 11% had a northwestward trend for surface flows. At 6.7%, the occurrence frequency of group 4 was much smaller than groups 1–3. This indicates that SOM pattern 2 occurred with relatively low probability. Additionally, magnitudes of surface flows in SOM pattern 2 were smaller than that of other SOM patterns.

For surface flows of HFR data, twelve SOM patterns can be categorized visually into six groups, as presented in Table 2. Table 2 shows that four SOM modes (3/4/7/8) of surface flows were categorized as group 1 with the highest occurrence frequency at 37.2%. Occurrence frequencies for groups 2, 3, 4, and 6 were greater than 10%, whereas the total occurrence frequency of group 5 was quite low, at 2.2%.

Table 2. Categories of SOM patterns for high frequency radar (HFR) data.

Group	SOM Pattern	Representative Characteristics of Surface Vector Fields	Occurrence Frequency (%)
1	3/4/7/8	western flows	37.2
2	1/5	southeastward and alongshore flows	17.6
3	9/10	northeastward and alongshore flows	18.3
4	2	southern and southwestward flows	10.3
5	6	southwestward and northeastward flows	2.2
6	11/12	northwestward flows	14.4

Coastal currents in Galway Bay are mainly driven by tides and winds. Tides typically propagate in western and eastern directions corresponding to tidal flooding and ebbing, respectively. Figure 3 shows that western currents existed in group 2 by total occurrence frequency 40.4%; eastern currents (patterns 9 and 10) existed in group 1 by total occurrence frequency 27.9%. Surface currents of model results during tidal flooding were stronger than during tidal ebbing. However, SOM patterns having eastern and western current patterns in the radar data were different. Figure 4 shows that eastern currents existed in SOM patterns 5 and 9 by total occurrence frequency 19.5% for radar data; western currents existed in group 1 by total occurrence frequency 37.2%. Surface currents of radar data during tidal ebbing were stronger than during tidal flooding. Since spatially constant winds were used in the modeling, the difference in eastern and western SOM current patterns between model results and radar data may be due to influences of wind variation in space. Thus, wind roses and mean wind vectors based on European Centre for Medium-Range Weather Forecasts (ECMWF) data with 0.15° × 0.15° spatial resolution and 6-h temporal resolution are presented in Figures 5 and 6 to further investigate effects of winds on SOM current patterns.

Figure 5 show that wind speeds and directions varied over the analysis domain. Moreover, occurrence frequencies of winds in the same direction were also different. In general, dominant winds blew from the ocean toward land during the analysis period. Figure 6 shows that the directions of mean wind vectors over higher latitude points had a clockwise trend, while the directions were more uniform (southwest) over lower latitude points. Considering SOM patterns of model results and radar data, group 3, as shown in Figure 3, with total occurrence frequency 11.9% had a northwestern current pattern from the model results; group 6 (SOM patterns 11 and 12) and SOM pattern 10 with total occurrence frequency 19.9% had a landward (northwestern, northern, and northeastern) current pattern in the radar data. The occurrence frequency of landward current patterns extracted by SOM analysis was stronger in the radar data than in the model results. This indicates that surface currents driven by winds were better captured by the radar observation system than the numerical model. As stated before, this is probably due to the fact that spatially constant winds were used in the model, and demonstrates the importance of wind forcing on surface hydrodynamics.

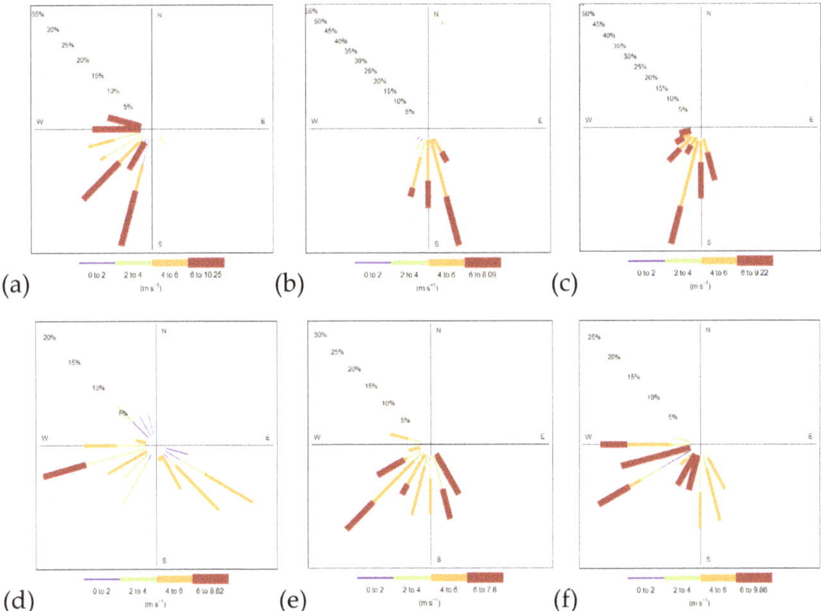

Figure 5. Wind roses during analysis period. (Locations of W1–W6 are shown in subfigures (**a–f**), respectively; direction indicates wind blowing from).

Additionally, infrequent winds blowing from land to ocean (W1, W4, and W5 in Figure 5) occurred during the analysis period. Corresponding vector patterns were also extracted by SOM for both model results and radar data. Southeastern currents, including SOM patterns 1, 5, and 6, had a 13.2% occurrence frequency for model results; offshore SOM current patterns 1 and 5 had a 17.6% occurrence frequency for radar data. In summary, SOM patterns accounting for the effects of tides and winds had occurrence frequencies of 97.8% and 93.3% for radar data and model results, respectively. This indicates that the SOM technique can extract representative synoptic characteristic patterns of surface flows. Additionally, the effects of the main driving forces (tide and wind) on surface flows can be well linked to SOM patterns, especially for the radar data.

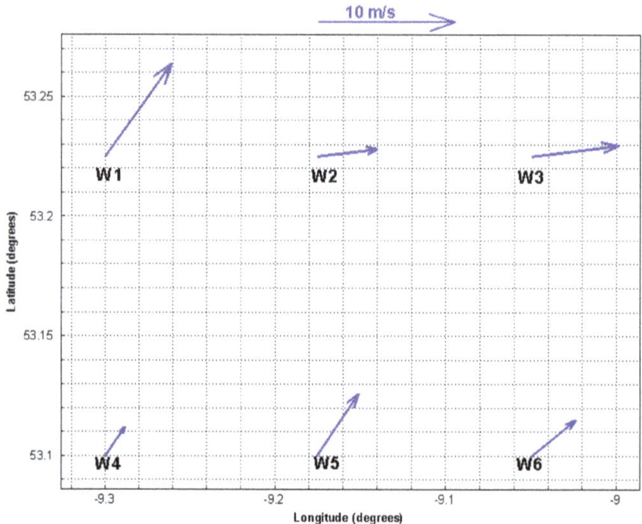

Figure 6. Mean ECMWF (European Centre for Medium-Range Weather Forecasts) wind vectors during analysis period.

3.1.2. Temporal Evolution

To analyze the evolution process of the SOM patterns in time, *BMUs* were computed and are shown in Figure 7. A *BMU* can be found for each input data vector by comparing the 12 SOM patterns with the input data map.

Figure 7a shows that the evolution of *BMU* for model results generally had two patterns: (I) 12→ 4→ 2→ 1→ 9→ 11 and (II) 11→ 8→ 4→ 3→ 2→ 1→ 9→ 10. Evolution processes of both type I and II show that the general variation trend of surface flow fields was anticlockwise, i.e., westward → southwest → southeast → westward, as shown in Figure 3. The corresponding group category pattern is 2→3→1→2, as presented in Table 1.

Figure 7b shows that evolution of *BMU* for HFR data was not as uniform as the model results. The evolution trend was 12→ 8→ 4→ 3→ 2→ 1. The trend of surface flow fields was generally anticlockwise, i.e., northwestward → westward → southward → southeastward, as shown in Figure 4. The corresponding group category pattern was 6→ 1→ 4→ 2, as presented in Table 2.

In general, a similar anticlockwise evolution trend changing from westward to southeastward existed for surface flow fields in both the model results and HFR data. However, the evolution trend of *BMU* was more regular in the model results than in the HFR data. This is again probably due to the fact that a spatially constant wind was applied to force the surface boundary of the models. Thus, tide and wind were the same over simulated domain. However, surface currents monitored by the HFR system appear to have captured more information of surface currents driven from winds.

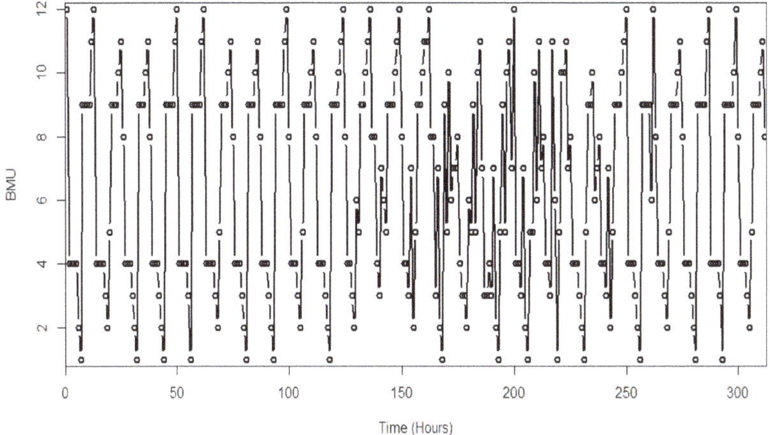

(a)

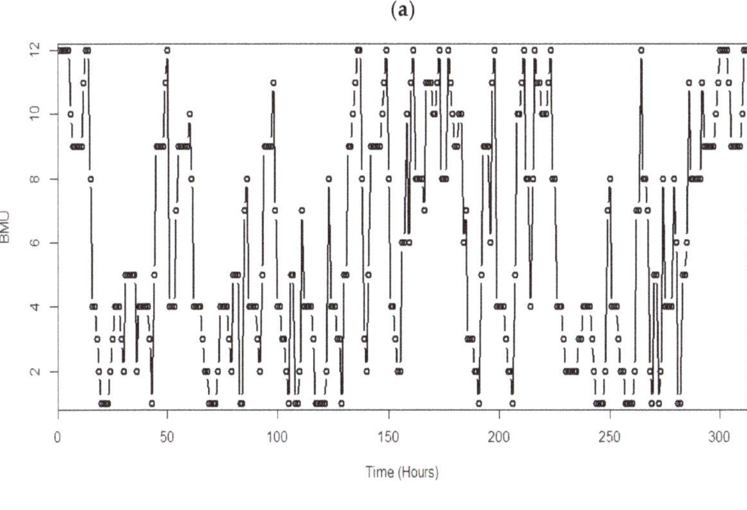

(b)

Figure 7. Time series of best match unit (*BMU*) corresponding to 12 SOM patterns ((**a**) model results and (**b**) radar data).

3.2. Empirical Orthogonal Function

In order to further compare the HFR data and numerical results, an EOF analysis was performed. EOF analysis reduces data dimensionality and represents characteristics of each dataset in a few concise and typical patterns in both space and time. The same dataset as used in the above SOM analysis was used in the following EOF analysis.

3.2.1. Spatial Modes

To investigate synoptic characteristics of coastal flow patterns extracted by EOF, the first six EOF eigenvector modes for both EFDC model results and HFR data, as shown in Figures 8 and 9, respectively, account for more than 95% of the explained variances. Variances explained by the corresponding EOF mode are presented in the left-hand corner in each panel.

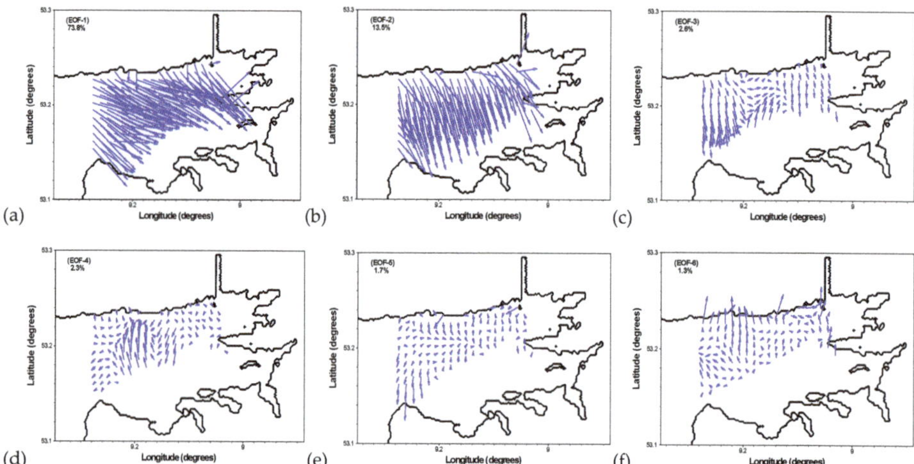

Figure 8. Spatial empirical orthogonal function (EOF) modes of model results (subfigures (**a**–**f**) indicates the first to the sixth EOF eigenvector modes, respectively).

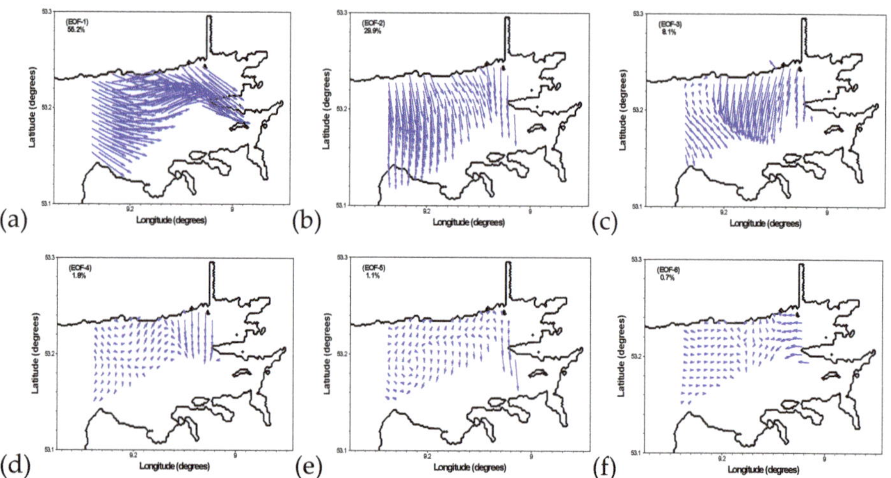

Figure 9. Spatial EOF modes of radar data (subfigures (**a**–**f**) indicates the first to the sixth EOF eigenvector modes, respectively).

The first EOF mode (EOF1) of the model results and the HFR data, as shown in Figures 8 and 9, accounts for 73.8% and 55.2% of the variances separately. This indicates that the possibility of surface flow in EOF1 mode patterns for model results was larger than for HFR data. The general pattern of surface flows in EOF1 mode is southeast; while there are also alongshore surface flows in EOF1 mode of the HFR data. This may result from variation of winds in space, which was not captured by model.

The patterns of surface flows in EOF2 mode were more similar between model results and HFR data. The possibility of flow in EOF2 mode for the HFR data was greater than twice that of the model results. This indicates that patterns of surface currents in EOF2 mode were more likely to occur in the HFR data than in the model results, the trend of surface flows in this mode was in the northern direction. However, surface flow magnitudes of model results were larger than these of HFR data.

The differences of spatial patterns between model results and HFR data in EOF3 mode were more significant than in EOF1 and EOF2 modes. Spatial patterns in HFR data consisted of weak southeastward and strong southwestward flows, which converged around the −45° line. The general spatial pattern of surface currents for model results was southward; a northward trend of surface flows existed in the middle of area covered by the radar. Moreover, magnitudes of surface currents in EOF3 mode for the HFR data were larger than the model results.

Magnitudes of surface currents in EOF4 mode were small and similar between model results and HFR data. Disorder of spatial patterns exists in model results, while a southward trend exists across the right parts of displayed area. Surface flows bifurcate across the left parts with one in the southeastward direction and the other in the eastward direction in HFR data.

There was a clockwise and weak surface flow gyre in EOF5 mode in HFR data while the spatial pattern of EOF5 mode in the model results was generally disordered. Magnitudes of surface currents in this mode were small and comparable between the two datasets.

The spatial patterns of EOF6 mode in HFR data were generally offshore currents and along shore currents, but no such significant type of surface flow trend exists in model results. Additionally, magnitudes of surface currents near coasts were larger than other areas for model results, but magnitudes were generally uniform except for few large surface currents at the right corners for the HFR data.

The first two EOF modes accounting for the majority of the variance (>85%) were relatively similar for model results and HFR data. The differences in the remaining four EOF modes between model results and HFR data were significant, but these modes accounted for a relatively small proportion of the variations.

3.2.2. Variance of Surface Flows Explained by EOF Modes

To further compare EOF analysis of model results and radar data, accumulative explained variances of the first 20 EOF modes and variance values of the first six EOF modes are presented in Figure 10 and Table 3, respectively. The value of variance represents the characteristic strength of the corresponding spatial EOF modes, as shown in Figures 8 and 9; this was computed relative to the entire set of EOF modes. A large value of variance indicates that the corresponding spatial EOF mode was significant; a small value of variance means that the corresponding spatial EOF mode was weak.

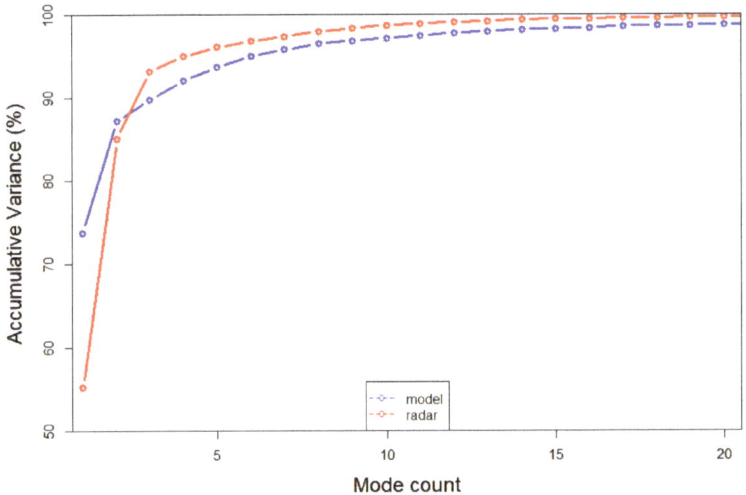

Figure 10. Variance of EOF modes.

Table 3. Summary of variance for the first six EOF modes.

EOF Mode	Variance (%)		Accumulative Variance (%)	
	EFDC	**HFR**	**EFDC**	**HFR**
1	73.8	55.2	73.8	55.2
2	13.5	29.9	87.2	85.1
3	2.6	8.1	89.9	93.2
4	2.3	1.8	92.1	95.0
5	1.7	1.1	93.7	96.1
6	1.3	0.7	95.0	96.8

Note: EFDC and HFR indicate model results and high frequency radar dataset, respectively.

Figure 10 shows that the first two EOF modes explain the majority of variance for both model results and HFR data; accumulative variance of model results at 87.2% was greater than that of HFR data at 85.1%. However, the accumulative variance of HFR data was greater than that of the model results from the third EOF mode. The accumulative variance explained by the first four EOF modes was greater than 90% for both model results and HFR data. This indicates that dominant patterns of surface flow fields can be well represented by only a few EOF modes. Variations of accumulative variances were not significant from the tenth EOF mode onward.

Although the difference of the variances in EOF1 mode between model results and HFR data was significant, the variance of the first two EOF modes (greater than 85%), as presented in Table 3, was comparable. The explained variance from the third EOF mode was much smaller than that of the first and second EOF modes. Additionally, the first six EOF modes accounted for greater than 95% of the total variance for both datasets. This indicated that the first six spatial EOF patterns represented synoptic characteristics of surface currents.

Figure 11 shows the first three EOF PCs over time for both model results and HFR data. A PC represents the time-varying characteristics (i.e., amplitudes) of the corresponding EOF eigenvector spatial distribution modes. The sign of the PC values determines the direction of EOF modes. Positive PC values indicate the same direction as the mode, while negative values indicate the opposite direction. The larger the absolute value of a particular PC, the stronger that PC EOF mode is at that moment. Time series of PC1 and PC2 for both model results and radar data exhibited cyclical trends with periods similar to a tidal period. The correlations between model results and HFR data of the first (PC1) and second PC patterns (PC2) are 0.80 and 0.58, respectively. This indicates that model results and HFR data had high and moderate correlation, respectively, based on categories proposed by Taylor [72]. PC3 represents high frequency wind generated flows. However, correlation between the third principal components (PC3) was quite weak at 0.02. This discrepancy is again due to limitations of the model to generate wind effects at short scales on the surface velocity fields, which would be a consequence of the wind field configuration which forces the model, similar to those obtained by Soto-Navarro, et al. [10].

To provide better insight into the temporal variation characteristics of the EOF PCs between model results and radar data, a spectral analysis was carried out on each EOF PC time series, see Figure 12. The EOF PC1 spectral density peaks were similar for model results and radar data (see Figure 12a), the corresponding frequency is 0.08 cph (cycles per hour). This indicates that the semidiurnal signal (12.5 h) was strongest in both datasets, corresponding to the tidal frequency. However, for EOF PC2 and PC2, the spectral density trends were not as strong between the two datasets. The spectral density peaks of the radar data were weaker in PC2 and PC3 than that in PC1, while the spectral density peaks of the model results were much weaker in PC3 than PC1 and that in PC2. The differences may again result from the radar-derived surface current dataset containing spatially varying wind effects.

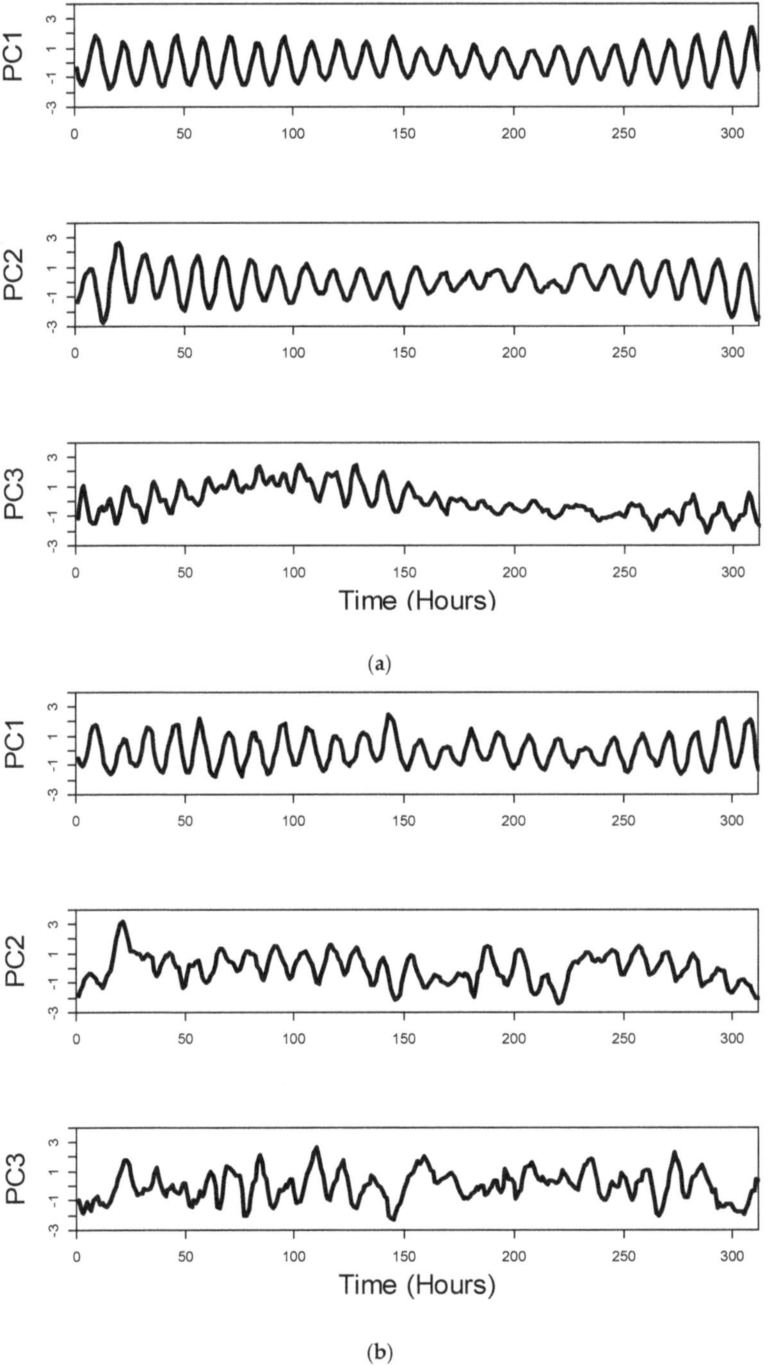

Figure 11. The first three EOF principal components (PCs): (**a**) model results and (**b**) HFR data.

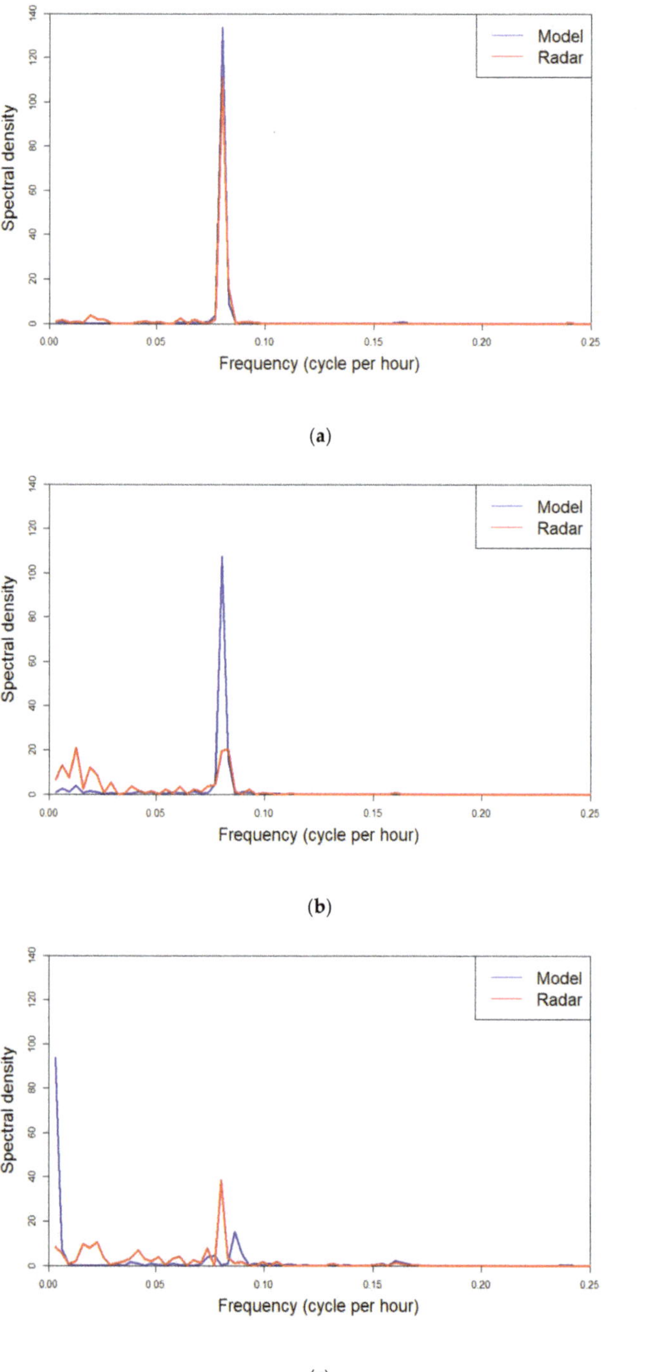

(**a**)

(**b**)

(**c**)

Figure 12. Spectral analysis of EOF PCs between model results and radar data ((**a–c**) indicate spectral analysis for EOF PC1, PC2, and PC3, respectively).

4. Discussion

Characteristic patterns of coastal flows were extracted by SOM (Figures 3 and 4) and EOF (Figures 8 and 9) for both model results and radar data. As stated in the introduction, the SOM technique has more advantages than other conventional data analysis techniques. Based on the SOM analysis, synoptic characteristic patterns of coastal flows were visualized. Additionally, time series of *BMU* can offer evolution trends of characteristic patterns over space; this can better explain spatiotemporal variation of coastal flows. EOF technique can decompose coastal flow fields that change with time into two parts: the spatial modes which are constant in time and the time components (PCs), depending only on time. Although the EOF technique can extract synoptic characteristic patterns of coastal flows for both model results and radar data, it does not offer temporal variation trends corresponding to synoptic characteristic patterns over space. Since EOF is a linear decomposition technique, synoptic characteristics of coastal flow fields in space were less uniform than synoptic patterns extracted by the nonlinear SOM technique. In short, both SOM and EOF techniques offer synoptic characteristic patterns of coastal flow fields for both model results and radar data. However, more detailed information of variation properties for analysis dataset can be provided by an SOM analysis than that of an EOF analysis.

5. Conclusions

This paper presented SOM and EOF analyses of characteristics of surface flow fields in Galway Bay based on data provided by a HFR radar system and output from a numerical model. The main conclusions from this research follow.

Surface flows were categorized into four and six representative synoptic characteristic groups for model results and HFR data using SOM, respectively. The *BMU* time series indicate that the evolution of SOM patterns between model results and HFR data had similar trends varying from west to southeast in the anticlockwise direction.

The total variance explained by the first two EOF modes was comparable, with 87.3% and 85.1% for model results and HFR data, respectively, which underlines the agreement of both datasets in describing the general hydrodynamic characteristics of surface vectors in the region. The difference in the rest of the EOF modes, with relatively low variance, probably results from application of spatially constant wind in model.

Representative synoptic patterns of coastal surface flows were extracted using both EOF and SOM techniques. More detailed spatiotemporal information about coastal flow variation properties can provided by patterns obtained from SOM than from an EOF analysis.

The accuracy of model outputs was also assessed in detail by SOM and EOF type analysis; these analyses illustrate which processes models are good at reproducing and which processes are not well-represented by a model. In this case, the analysis shows that the model does not reproduce wind induced currents well, and so the model must be improved in this regard by forcing the surface with spatially varying wind stresses or through assimilation of the HF radar-derived surface currents [73]. Additionally, flow is dominated by tidal processes and, therefore, 13 days of observations is significantly relevant to those processes. In order to capture multiple (meteorological) synoptic scale events, which would better illuminate the effect of wind forcing, analysis of a longer period using modest spatial filtering could be advantageous and will be included in future research. The use of spectral analysis is very informative when comparing time-varying data that has dominant frequency components in EOF analysis.

This research analyzed a relatively short temporal dataset as this dataset was complete and did not include synthetic data that would introduce further uncertainty into the already difficult intercomparisons. Future research will consider longer-term datasets complemented by gap-filling to give a synthetic continuous dataset; results will be compared with the research presented herein.

In summary, the results above reflect the capability of the EFDC model and HFR system to describe characteristics of surface vector fields of the Galway Bay area. Typical patterns of surface vector fields associating with *BMU* time series can better describe the evolution of the process of surface vector fields. SOM analysis provides more detailed information than that provided by EOF analysis. Such SOM patterns may be useful in a variety of forecasting applications, such as oil spill treatment and search and rescue. Although these results are interesting and useful, and SOM and EOF analysis methods provide a powerful tool to diagnose ocean processes from different aspects, it is also of great significance to further investigate the underlying physical mechanisms such as wind influence in future studies.

SOM has significant potential for surface current forecasting since it is simpler than other techniques, such as data assimilation, and computational costs are much lower. Forecasting of coastal flows based on the SOM technique will be undertaken by the authors in future research.

Author Contributions: Methodology, L.R.; formal analysis, L.R. and N.C.; writing—original draft preparation, L.R. and Z.H.; writing—review and editing, L.R. and M.H.; supervision, M.H., funding acquisition, L.R. and Z.H. All authors have read and agreed to the published version of the manuscript.

Funding: This research was funded by a project from the National Natural Science Foundation of China (NSFC), grant number 51909290; the Science and Technology Program of Guangzhou, China, grant number 201904010430; the Basic Research Program of Sun Yat-sen University, grant number 76170-18841203; the Research and Development Plan in Key Areas in Guangdong Province, grant number 2020B1111020003 and the Joint Research Project: NSFC (51761135022), NWO (ALWSD.2016.026), and EPSRC (EP/R024537/1): Sustainable Deltas.

Acknowledgments: We would like to thank ECMWF for providing wind data and the University of Oregon for providing the OTIS tide prediction software.

Conflicts of Interest: The authors declare no conflict of interest.

Appendix A. Self-Organizing Map

In general, implementation process of SOM algorithm can be summarized as follows:

1. Determine the size and type of the map.
2. Initialize each node's weights W_{ij} at random.
3. Select a vector at random from the training dataset and present to the network. The following Euclidean distance formula is calculated to assess the "best matching unit (BMU)" between each node and all input dataset.

$$DI_j = \sum_{i=0}^{n} \left(X_i - W_{ij}\right)^2 \tag{A1}$$

where

DI_j is the jth distance from all input vector;
X_i is the ith input vector having I data points and J data points in x and y directions, respectively;
W_{ij} is the jth node's weight;
n in the number of weights.

The *BMU* of each node is found based on calculating which nodes' weights are most like the input vector X. The neighborhood function is taken by assuming to maximum when distance is zero. There are four types of neighborhood function available: *"bubble"*, *"gaussian"*, *"cutgauss"* and *"ep"*:

$$\Theta(t)\begin{cases} F(\sigma(t) - d_{ci}) & bubble \\[2mm] \exp\left(\dfrac{-d_{ci}^2}{2\sigma(t)^2}\right) & gaussian \\[2mm] \exp\left(\dfrac{-d_{ci}^2}{2\sigma(t)^2}\right)F(\sigma(t) - d_{ci}) & cutgauss \\[2mm] max\{0, 1, -(\sigma(t) - d_{ci})^2\} & ep \end{cases} \tag{A2}$$

where

$\sigma(t)$ is the neighborhood radius at time t;

d_{ci} is the distance between map units c and i on the map grid;

F is a step function:

$$F(x) = \begin{cases} 0 & (if\ x < 0) \\ 1 & (if\ x \geq 0) \end{cases} \tag{A3}$$

Determine the radius of the neighborhood of the *BMU* (the size of the neighborhood decreases with each iteration).

Tune weights of nodes within the radius of the *BMU* to make them more like the input vector. The closer a node is to the *BMU*, the more its weights are altered.

$$W_i(t+1) = W_i(t) + \Theta(t) \times L(t) \times (I(t) - W_i(t)) \tag{A4}$$

$$L(t) = L_0 e^{\left(\frac{-t}{\lambda}\right)} \tag{A5}$$

where

$L(t)$ is the learning rate.

Repeat steps (A2)–(A5) for N iterations.

Three parameters—number of iterations, learning rate, and neighborhood radius—need to be determined. The quantization error (QE), i.e., average distance between each input data vector X_i and its *BMU* (u_{BMU}) can be expressed as:

$$QE = \sum_{i=1}^{N} \|X_i - u_{BMU}\| \tag{A6}$$

The topographic error (TE) being used to measure the topology preservation can be calculated by the following formula:

$$TE = \frac{1}{N} \sum_{i=1}^{N} u_{BMU}(X_i) \tag{A7}$$

where $u_{BMU}(X_i)$ is 1 if the first and the second *BMUs* are not adjacent to each other, otherwise it is $u_{BMU}(X_i) = 0$.

Appendix B. Empirical Orthogonal Function

In particular, the characteristic patterns of surface flows $H(x, t)$ can be decomposed into a series of linear combinations of temporal and spatial orthonormal function:

$$H(x,t) = \sum_{m=1}^{M} Z_m(x) L_m(t) \tag{A8}$$

where

Z_m (also known as EOFs) are the spatial eigenfunctions representing the dominant spatial patterns of surface flow range variability;

L_m (also known as PCAs) are the temporal eigenfunctions indicating the long-term changes of surface vector fields;

$m = 1$ to M, with M being the number of temporal and spatial samples.

Here, both the EOFs and PCAs should be orthonormal and normalized as follows:

$$\sum_{t=1}^{T} L_i(t)L_j(t) = \begin{cases} 0 & i \neq j \\ 1 & i = j \end{cases} \tag{A9}$$

$$\sum_{x=1}^{X} Z_i(x)Z_j(x) = \begin{cases} 0 & i \neq j \\ \gamma_i & i = j \end{cases} \tag{A10}$$

where

X and T are the maximum values of x and t, respectively;

γ_i is the eigenvalue, which represents the contribution made by the ith EOF mode to the total variance, where the first few largest eigenvalues typically contain the most signals and represent the dominant temporal–spatial patterns of the observed tidal series. The relative contribution of the mth eigenfunction pm can be computed by the following expression:

$$p_m = \left(\lambda_m / \sum_{m=1}^{M} \lambda_m \right) \times 100 \tag{A11}$$

Subsequently, it is possible to reconstruct a matrix $H'(x,t)$ using a subset of the dominant eigenvectors, which explained the maximum variance with the first k EOFs modes:

$$H'(x,t) = \sum_{m=1}^{k} Z_m(x)L_m(t) \tag{A12}$$

References

1. Szuts, Z.B.; Bower, A.S.; Donohue, K.A.; Girton, J.B.; Hummon, J.M.; Katsumata, K.; Lumpkin, R.; Ortner, P.B.; Phillips, H.E.; Rossby, H.T.; et al. The Scientific and Societal Uses of Global Measurements of Subsurface Velocity. *Front. Mar. Sci.* **2019**, *6*, 6. [CrossRef]
2. Farcy, P.; Durand, D.; Charria, G.; Painting, S.J.; Tamminem, T.; Collingridge, K.; Grémare, A.J.; Delauney, L.; Puillat, I. Toward a European Coastal Observing Network to Provide Better Answers to Science and to Societal Challenges; The JERICO Research Infrastructure. *Front. Mar. Sci.* **2019**, *6*. [CrossRef]
3. Vandenbulcke, L.; Beckers, J.-M.; Barth, A. Correction of inertial oscillations by assimilation of HF radar data in a model of the Ligurian Sea. *Ocean Dyn.* **2016**, *67*, 117–135. [CrossRef]
4. Lai, Y.; Zhou, H.; Yang, J.; Zeng, Y.; Wen, B. Submesoscale Eddies in the Taiwan Strait Observed by High-Frequency Radars: Detection Algorithms and Eddy Properties. *J. Atmos. Ocean. Technol.* **2017**, *34*, 939–953. [CrossRef]
5. Hisaki, Y.; Kashima, M.; Kojima, S. Surface current patterns observed by HF radar: Methodology and analysis of currents to the north of the Yaeyama Islands, East China Sea. *Ocean Dyn.* **2016**, *66*, 329–352. [CrossRef]
6. John, M.; Jena, B.K.; Sivakholundu, K.M. Surface current and wave measurement during cyclone phaillin by high frequency radars along the indian coast. *Curr. Sci.* **2015**, *108*, 405–409.
7. Wang, Y.; Ma, X.; Joyce, M.J. Reducing sensor complexity for monitoring wind turbine performance using principal component analysis. *Renew. Energy* **2016**, *97*, 444–456. [CrossRef]
8. Mardia, K.V. Some properties of classical multidimensional scaling. *Commun. Stat. Theory Methods* **1978**, *7*, 1233–1241. [CrossRef]
9. Liu, Y.; Weisberg, R.H.; Vignudelli, S.; Mitchum, G.T. Patterns of the loop current system and regions of sea surface height variability in the eastern Gulf of Mexico revealed by the self-organizing maps. *J. Geophys. Res. Oceans* **2016**, *121*, 2347–2366. [CrossRef]
10. Liu, Y.; Weisberg, R.H.; Shay, L.K. Current Patterns on the West Florida Shelf from Joint Self-Organizing Map Analyses of HF Radar and ADCP Data. *J. Atmos. Ocean. Technol.* **2007**, *24*, 702–712. [CrossRef]

11. Soto-Navarro, J.; Lorente, P.; Fanjul, E.A.; Sanchez-Garrido, J.C.; Garcia-Lafuente, J. Surface circulation at the Strait of Gibraltar: A combined HF radar and high resolution model study. *J. Geophys. Res. Oceans* **2016**, *121*, 2016–2034. [CrossRef]
12. Reusch, D.B.; Hewitson, B.C.; Alley, R.B. Towards ice-core-based synoptic reconstructions of west antarctic climate with artificial neural networks. *Int. J. Clim.* **2005**, *25*, 581–610. [CrossRef]
13. Lobo, V.J.A.S. Application of Self Organizing Maps to the Maritime Environment. In Proceedings of the 4th International Workshop on Information Fusion and Geographical Information Systems, St Petersburg, Russia, 17–20 May 2009; Popovich, V.V., Schrenk, M., Claramunt, C., Korolenko, K.V., Eds.; Springer-Verlag Berlin: St Petersburg, Russia, 2009; pp. 19–36.
14. Lin, G.-F.; Chen, L.-H. Identification of homogeneous regions for regional frequency analysis using the self-organizing map. *J. Hydrol.* **2006**, *324*, 1–9. [CrossRef]
15. Solidoro, C.; Bandelj, V.; Barbieri, P.; Cossarini, G.; Umani, S.F. Understanding dynamic of biogeochemical properties in the northern Adriatic Sea by using self-organizing maps and k-means clustering. *J. Geophys. Res. Space Phys.* **2007**, *112*, 1–13. [CrossRef]
16. Tsai, W.-P.; Huang, S.-P.; Cheng, S.-T.; Shao, K.-T.; Chang, F.-J. A data-mining framework for exploring the multi-relation between fish species and water quality through self-organizing map. *Sci. Total. Environ.* **2017**, *579*, 474–483. [CrossRef]
17. Nkiaka, E.; Nawaz, N.R.; Lovett, J.C. Using self-organizing maps to infill missing data in hydro-meteorological time series from the Logone catchment, Lake Chad basin. *Environ. Monit. Assess.* **2016**, *188*, 1–12. [CrossRef]
18. Tsui, I.-F.; Wu, C.-R. Variability analysis of Kuroshio intrusion through Luzon Strait using growing hierarchical self-organizing map. *Ocean Dyn.* **2012**, *62*, 1187–1194. [CrossRef]
19. Camus, P.; Cofino, A.S.; Mendez, F.J.; Medina, R. Multivariate Wave Climate Using Self-Organizing Maps. *J. Atmos. Ocean. Technol.* **2011**, *28*, 1554–1568. [CrossRef]
20. Kalteh, A.M.; Hjorth, P.; Berndtsson, R. Review of the self-organizing map (SOM) approach in water resources: Analysis, modelling and application. *Environ. Model. Softw.* **2008**, *23*, 835–845. [CrossRef]
21. Liu, Y.; Weisberg, R.H.; He, R. Sea Surface Temperature Patterns on the West Florida Shelf Using Growing Hierarchical Self-Organizing Maps. *J. Atmos. Ocean. Technol.* **2006**, *23*, 325–338. [CrossRef]
22. Reusch, D.B.; Alley, R.B.; Hewitson, B.C. North Atlantic climate variability from a self-organizing map perspective. *J. Geophys. Res. Space Phys.* **2007**, *112*, 1–20. [CrossRef]
23. Liu, Y.; Weisberg, R.H. Patterns of ocean current variability on the West Florida Shelf using the self-organizing map. *J. Geophys. Res. Space Phys.* **2005**, *110*, 1–12. [CrossRef]
24. Liu, Y.; Weisberg, R.H.; Mooers, C.N.K. Performance evaluation of the self-organizing map for feature extraction. *J. Geophys. Res. Space Phys.* **2006**, *111*, 111. [CrossRef]
25. Mihanović, H.; Cosoli, S.; Vilibić, I.; Ivanković, D.; Dadić, V.; Gačić, M. Surface current patterns in the northern Adriatic extracted from high-frequency radar data using self-organizing map analysis. *J. Geophys. Res. Space Phys.* **2011**, *116*, 116. [CrossRef]
26. Vilibić, I.; Šepić, J.; Mihanović, H.; Kalinic, H.; Cosoli, S.; Janeković, I.; Žagar, N.; Jesenko, B.; Tudor, M.; Dadić, V.; et al. Self-Organizing Maps-based ocean currents forecasting system. *Sci. Rep.* **2016**, *6*, 22924. [CrossRef]
27. Jin, B.; Wang, G.; Liu, Y.; Zhang, R. Interaction between the East China Sea Kuroshio and the Ryukyu Current as revealed by the self-organizing map. *J. Geophys. Res. Space Phys.* **2010**, *115*, 1–7. [CrossRef]
28. Barros, A.P.; Bowden, G.J. Toward long-lead operational forecasts of drought: An experimental study in the Murray-Darling River Basin. *J. Hydrol.* **2008**, *357*, 349–367. [CrossRef]
29. Obach, M.; Wagner, R.; Werner, H.; Schmidt, H.-H. Modelling population dynamics of aquatic insects with artificial neural networks. *Ecol. Model.* **2001**, *146*, 207–217. [CrossRef]
30. Malek, S.; Gunalan, R.; Kedija, S.Y.; Lau, C.F.; Mosleh, M.A.A.; Milow, P.; Lee, S.A.; Saw, A. Random forest and Self Organizing Maps application for analysis of pediatric fracture healing time of the lower limb. *Neurocomputing* **2018**, *272*, 55–62. [CrossRef]
31. Booth, D. The Water Structure and Circulation of Killary Harbour and of Galway Bay. Ph.D. Thesis, National University of Ireland, Galway, Ireland, 1975.
32. Fernandes, L. A Study of the Oceanography of Galway Bay, Mid-Western Coastal Waters (Galway Bay to Bralle Bay), Shannon Estuary and the Rive Shannon Plume. Ph.D. Thesis, National University of Ireland, Galway, Ireland, 1988.

33. Wen, L. Three-Dimensional Hydrodynamic Modelling in Galway Bay. Ph.D. Thesis, University College Galway, Galway, Ireland, 1995.

34. Joshi, S.; Duffy, G.P.; Brown, C. Mobility of maerl-siliciclastic mixtures: Impact of waves, currents and storm events. *Estuar. Coast. Shelf Sci.* **2017**, *189*, 173–188. [CrossRef]

35. Paduan, J.D.; Washburn, L. High-Frequency Radar Observations of Ocean Surface Currents. *Annu. Rev. Mar. Sci.* **2013**, *5*, 115–136. [CrossRef] [PubMed]

36. Lipa, B.J.; Barrick, D.E.; Isaacson, J.; Lilieboe, P.M. Codar wave measurements from a north sea semisubmer sible. *IEEE J. Ocean. Eng.* **1990**, *15*, 119–125. [CrossRef]

37. Emery, B.M.; Washburn, L.; Harlan, J.A. Evaluating Radial Current Measurements from CODAR High-Frequency Radars with Moored Current Meters. *J. Atmos. Ocean. Technol.* **2004**, *21*, 1259–1271. [CrossRef]

38. Liu, Y.; Weisberg, R.H.; Merz, C.R.; Lichtenwalner, S.; Kirkpatrick, G.J. HF Radar Performance in a Low-Energy Environment: CODAR SeaSonde Experience on the West Florida Shelf. *J. Atmos. Ocean. Technol.* **2010**, *27*, 1689–1710. [CrossRef]

39. Roarty, H.; Cook, T.; Hazard, L.; George, D.; Harlan, J.; Cosoli, S.; Wyatt, L.; Fanjul, E.A.; Terrill, E.; Otero, M.; et al. The Global High Frequency Radar Network. *Front. Mar. Sci.* **2019**, *6*, 1–26. [CrossRef]

40. Mantovani, C.; Corgnati, L.; Horstmann, J.; Rubio, A.; Reyes, E.; Quentin, C.; Cosoli, S.; Asensio, J.L.; Mader, J.; Griffa, A. Best Practices on High Frequency Radar Deployment and Operation for Ocean Current Measurement. *Front. Mar. Sci.* **2020**, *7*. [CrossRef]

41. Tinis, S.W.; Hodgins, D.O.; Fingas, M. Assimilation of radar measured surface current fields into a numerical model for oil spill modelling. *Spill Sci. Technol. Bull.* **1996**, *3*, 247–251. [CrossRef]

42. Bellomo, L.; Griffa, A.; Cosoli, S.; Falco, P.; Gerin, R.; Iermano, I.; Kalampokis, A.; Kokkini, Z.; Lana, A.; Magaldi, M.G.; et al. Toward an integrated HF radar network in the Mediterranean Sea to improve search and rescue and oil spill response: The TOSCA project experience. *J. Oper. Oceanogr.* **2015**, *8*, 1–13. [CrossRef]

43. Ren, L.; Nash, S.; Hartnett, M. Forecasting of Surface Currents via Correcting Wind Stress with Assimilation of High-Frequency Radar Data in a Three-Dimensional Model. *Adv. Meteorol.* **2016**, *2016*, 1–12. [CrossRef]

44. Marmain, J.; Molcard, A.; Forget, P.; Barth, A.; Ourmières, Y. Assimilation of HF radar surface currents to optimize forcing in the northwestern Mediterranean Sea. *Nonlinear Process. Geophys.* **2014**, *21*, 659–675. [CrossRef]

45. Xu, J.; Huang, J.; Gao, S.; Cao, Y. Assimilation of high frequency radar data into a shelf sea circulation model. *J. Ocean. Univ. China* **2014**, *13*, 572–578. [CrossRef]

46. Ren, L.; Nash, S.; Hartnett, M. Renewable energies offshore. In *Chapter 24 Data Assimilation with High-Frequency (HF) Radar Surface Currents at a Marine Renewable Energy Test Site*; Soares, C.G., Ed.; CRC Press: London, UK, 2015.

47. Solabarrieta, L.; Frolov, S.; Cook, M.; Paduan, J.; Rubio, A.; González, M.; Mader, J.; Charria, G. Skill Assessment of HF Radar–Derived Products for Lagrangian Simulations in the Bay of Biscay. *J. Atmos. Ocean. Technol.* **2016**, *33*, 2585–2597. [CrossRef]

48. Roarty, H.; Glenn, S.; Allen, A. Evaluation of Environmental Data for Search and Rescue. In Proceedings of the OCEANS 2016, Shanghai, China, 10–13 April 2016; IEEE: Piscataway, NJ, USA, 2016; pp. 1–3.

49. Cosoli, S.; Grcic, B.; De Vos, S.; Hetzel, Y. Improving Data Quality for the Australian High Frequency Ocean Radar Network through Real-Time and Delayed-Mode Quality-Control Procedures. *Remote. Sens.* **2018**, *10*, 1476. [CrossRef]

50. Kim, S.Y.; Terrill, E.; Cornuelle, B. Objectively mapping HF radar-derived surface current data using measured and idealized data covariance matrices. *J. Geophys. Res. Space Phys.* **2007**, *112*, 112. [CrossRef]

51. O'Donncha, F.; Hartnett, M.; Nash, S.; Ren, L.; Ragnoli, E. Characterizing observed circulation patterns within a bay using HF radar and numerical model simulations. *J. Mar. Syst.* **2015**, *142*, 96–110. [CrossRef]

52. Rubio, A.; Mader, J.; Corgnati, L.; Mantovani, C.; Griffa, A.; Novellino, A.; Quentin, C.; Wyatt, L.; Schulz-Stellenfleth, J.; Horstmann, J.; et al. HF Radar Activity in European Coastal Seas: Next Steps toward a Pan-European HF Radar Network. *Front. Mar. Sci.* **2017**, *4*. [CrossRef]

53. Ren, L.; Nagle, D.; Hartnett, M.; Nash, S. The Effect of Wind Forcing on Modeling Coastal Circulation at a Marine Renewable Test Site. *Energies* **2017**, *10*, 2114. [CrossRef]

54. Ren, L.; Nash, S.; Hartnett, M. Observation and modeling of tide-and wind-induced surface currents in Galway Bay. *Water Sci. Eng.* **2015**, *8*, 345–352. [CrossRef]

55. Hamrick, J.M. *Efdc Technical Memorandum*; Tetra Tech: Fairfax, VA, USA, 2006.

56. Tetra Tech, Inc. *The Environmental Fluid Dynamics Code Theory and Computation Volume 1: Hydrodynamics and Mass Transport*; Tetra Tech, Inc.: Fairfax, VA, USA, 2007; p. 60.

57. Hamrick, J.M. *A Three-Dimensional Environmental Fluid Dynamics Computer Code: Therotical and Computatonal Aspects*; Virginia Institute of Marine Science, William & Mary: Gloucester Point, VA, USA, 1992.

58. Zou, R.; Carter, S.; Shoemaker, L.; Parker, A.; Henry, T. Integrated Hydrodynamic and Water Quality Modeling System to Support Nutrient Total Maximum Daily Load Development for Wissahickon Creek, Pennsylvania. *J. Environ. Eng.* **2006**, *132*, 555–566. [CrossRef]

59. Jin, K.-R.; Ji, Z.-G. Case Study: Modeling of Sediment Transport and Wind-Wave Impact in Lake Okeechobee. *J. Hydraul. Eng.* **2004**, *130*, 1055–1067. [CrossRef]

60. O'Donncha, F.; Hartnett, M.; Nash, S. Physical and numerical investigation of the hydrodynamic implications of aquaculture farms. *Aquac. Eng.* **2013**, *52*, 14–26. [CrossRef]

61. Bôas, A.B.V.; Ardhuin, F.; Ayet, A.; Bourassa, M.A.; Brandt, P.; Chapron, B.; Cornuelle, B.D.; Farrar, J.T.; Fewings, M.R.; Fox-Kemper, B.; et al. Integrated Observations of Global Surface Winds, Currents, and Waves: Requirements and Challenges for the Next Decade. *Front. Mar. Sci.* **2019**, *6*. [CrossRef]

62. Egbert, G.D.; Erofeeva, S.Y. Effificient inverse modeling of barotropic ocean tides. *J. Atmos. Ocean. Technol.* **2002**, *19*, 183. [CrossRef]

63. Padman, L.; Erofeeva, S. A barotropic inverse tidal model for the Arctic Ocean. *Geophys. Res. Lett.* **2004**, *31*, 1–4. [CrossRef]

64. Kohonen, T. Self-organized formation of topologically correct feature maps. *Boil. Cybern.* **1982**, *43*, 59–69. [CrossRef]

65. Chalasani, R.; Principe, J.C. Self-organizing maps with information theoretic learning. *Neurocomputing* **2015**, *147*, 3–14. [CrossRef]

66. Vilibić, I.; Kalinic, H.; Mihanović, H.; Cosoli, S.; Tudor, M.; Žagar, N.; Jesenko, B. Sensitivity of HF radar-derived surface current self-organizing maps to various processing procedures and mesoscale wind forcing. *Comput. Geosci.* **2015**, *20*, 115–131. [CrossRef]

67. Vilibić, I.; Mihanović, H.; Kušpilić, G.; Ivčević, A.; Milun, V. Mapping of oceanographic properties along a middle Adriatic transect using Self-Organising Maps. *Estuar. Coast. Shelf Sci.* **2015**, *163*, 84–92. [CrossRef]

68. Li, Q.; Chen, P.; Sun, L.; Ma, X. A global weighted mean temperature model based on empirical orthogonal function analysis. *Adv. Space Res.* **2018**, *61*, 1398–1411. [CrossRef]

69. Hannachi, A.; Jolliffe, I.T.; Stephenson, D.B. Empirical orthogonal functions and related techniques in atmospheric science: A review. *Int. J. Clim.* **2007**, *27*, 1119–1152. [CrossRef]

70. Monahan, A.H.; Fyfe, J.C.; Ambaum, M.H.; Stephenson, D.B.; North, G.R. Empirical Orthogonal Functions: The Medium is the Message. *J. Clim.* **2009**, *22*, 6501–6514. [CrossRef]

71. Mau, J.-C.; Wang, D.-P.; Ullman, D.S.; Codiga, D.L. Characterizing Long Island Sound outflows from HF radar using self-organizing maps. *Estuar. Coast. Shelf Sci.* **2007**, *74*, 155–165. [CrossRef]

72. Taylor, R. Interpretation of the Correlation Coefficient: A Basic Review. *J. Diagn. Med. Sonogr.* **1990**, *6*, 35–39. [CrossRef]

73. Paduan, J.D.; Shulman, I. HF radar data assimilation in the Monterey Bay area. *J. Geophys. Res. Space Phys.* **2004**, *109*, 434–446. [CrossRef]

 *remote sensing*

Article

An Approach to Minimize Atmospheric Correction Error and Improve Physics-Based Satellite-Derived Bathymetry in a Coastal Environment

Christopher O. Ilori [1,*] and Anders Knudby [2]

[1] Department of Geography, Simon Fraser University, 8888 University Drive, Burnaby, BC V5A 4S1, Canada
[2] Environment and Geomatics, Department of Geography, University of Ottawa, 60 University, Ottawa, ON K1N 6N5, Canada; aknudby@uottawa.ca
* Correspondence: cilori@sfu.ca; Tel.: +1-77-8929-5350

Received: 22 June 2020; Accepted: 21 August 2020; Published: 25 August 2020

Abstract: Physics-based radiative transfer model (RTM) inversion methods have been developed and implemented for satellite-derived bathymetry (SDB); however, precise atmospheric correction (AC) is required for robust bathymetry retrieval. In a previous study, we revealed that biases from AC may be related to imaging and environmental factors that are not considered sufficiently in all AC algorithms. Thus, the main aim of this study is to demonstrate how AC biases related to environmental factors can be minimized to improve SDB results. To achieve this, we first tested a physics-based inversion method to estimate bathymetry for a nearshore area in the Florida Keys, USA. Using a freely available water-based AC algorithm (ACOLITE), we used Landsat 8 (L8) images to derive per-pixel remote sensing reflectances, from which bathymetry was subsequently estimated. Then, we quantified known biases in the AC using a linear regression that estimated bias as a function of imaging and environmental factors and applied a correction to produce a new set of remote sensing reflectances. This correction improved bathymetry estimates for eight of the nine scenes we tested, with the resulting changes in bathymetry RMSE ranging from +0.09 m (worse) to −0.48 m (better) for a 1 to 25 m depth range, and from +0.07 m (worse) to −0.46 m (better) for an approximately 1 to 16 m depth range. In addition, we showed that an ensemble approach based on multiple images, with acquisitions ranging from optimal to sub-optimal conditions, can be used to estimate bathymetry with a result that is similar to what can be obtained from the best individual scene. This approach can reduce time spent on the pre-screening and filtering of scenes. The correction method implemented in this study is not a complete solution to the challenge of AC for satellite-derived bathymetry, but it can eliminate the effects of biases inherent to individual AC algorithms and thus improve bathymetry retrieval. It may also be beneficial for use with other AC algorithms and for the estimation of seafloor habitat and water quality products, although further validation in different nearshore waters is required.

Keywords: satellite-derived bathymetry; physics-based inversion method; atmospheric correction

---

1. Introduction

Bathymetric information from satellite data is of fundamental importance in optically shallow waters, where the seafloor is visible from space and the water-leaving radiance (L_w) is influenced by reflection off the seafloor. Such information, in the form of maps of water depth, is essential for a wide variety of purposes including offshore activities (e.g., pipeline laying), resource management (e.g., fishery), and defense operations (e.g., navigation). Traditional bathymetric charts are based on soundings obtained during hydrographic surveys. However, as ship-borne surveys are costly and time-consuming, and many shallow-water environments are highly dynamic, it is impossible to survey all areas of interest, and the difficulty in accessing shallow and remote areas means that in practice, up-to-date data are typically only available for limited areas (harbors and main navigation corridors). Airborne light detection and ranging (LiDAR) Bathymetry (ALB) systems, such as CZMIL (Coastal Zone Mapping and Imaging LiDAR) [1], LADS MK 3 (Laser Airborne Depth Sounder MK 3) [2], and EAARL-B (Experimental Advanced Airborne Research LiDAR B) [3] can also be used to map water depth. With these techniques, a vertical accuracy of about ± 15 cm in shallow water is possible [4], although accuracy is affected by turbidity and the LiDAR system. While precise bathymetric mapping of water depth to about 20–70 m depth can be achieved with airborne LiDAR [5,6], costs associated with these systems are relatively high, thus limiting their application over large, or remote, areas.

Passive optical satellite remote sensing can also be used to map bathymetry, typically known as satellite-derived bathymetry (SDB), based on the relationship between the color of a shallow-water area and the depth of the water. SDB can be implemented using empirical or physics-based methods. The empirical methods are based on the simple premise that a statistical relationship can be established between water depth and the remotely sensed radiance of a water body, using regression or similar analysis [2,7–10]. Thus, all empirical approaches require coincident in-situ data on water depth for calibration; ideally, these data should be up to date and have good geographic and depth distribution. Empirical approaches assume that the inherent optical properties (IOPs) of the water, as well as seafloor spectral reflectance, do not vary across the image, and therefore, the results may contain large errors and require manual editing when this is not the case. A key advantage of empirical approaches is the ability to retrieve water depth relatively easily, but their reliance on calibration from coincident field observations means that they cannot be used for systematic regional and global mapping and monitoring. Physics-based methods instead estimate bathymetry on per-pixel basis through the inversion of a radiative transfer model (RTM). As such, they do not assume uniform IOPs and seafloor reflectance, nor do they rely on coincident depth data for calibration. In addition to bathymetry, seafloor reflectance and water IOPs, which can be used to infer substrate and water quality respectively, can be simultaneously retrieved, and per-pixel uncertainties of all these parameters, including water depth, can also be determined. While originally developed for and tested on airborne hyperspectral imagery, physics-based methods for SDB have also been demonstrated for multispectral satellite sensors [11–14]. Physics-based methods can be implemented using either look-up tables (LUTs) [15,16] or semi-analytical optimization methods [17,18]. In the first case, a database of remote sensing reflectance (R_{rs}) spectra is built from an RTM provided with a range of values for water depth, spectral seafloor reflectance, water column optical properties (absorption and backscattering coefficients), and known environmental conditions such as sun angle and wind speed. For the retrieval of parameters (water depth, water IOPs, and seafloor reflectance) in each image pixel, a search is then performed to find the R_{rs} in the LUT that best matches the one observed in the pixel. With semi-analytical optimization methods, the radiative transfer equation is used to estimate water depth by iterative optimization of the same parameters. In both methods, the best match between modeled and observed reflectance is determined using a least squares or similar matching technique.

Despite the advantages of physics-based methods, a substantial challenge is that they rely on precise estimates of absolute radiometry, typically in the form of R_{rs} or L_w. Unlike other optical remote sensing applications, including the empirical approaches to satellite-derived bathymetry, physics-based retrieval algorithms may perform very poorly if R_{rs} is incorrectly estimated, and high-quality R_{rs} data from a robust atmospheric correction (AC) is essential for accurate physics-based water depth estimation. Accordingly, a variety of AC algorithms have been developed for ocean color (OC) products retrieval such as bathymetry, and several studies have validated their performance against in situ data. For example, Pahlevan et al. [19] validated R_{rs} produced from different AC schemes in the Sea-Viewing Wide Field-of-View Sensor (SeaWiFS) Data Analysis System (SeaDAS) with in situ data from the AERONET-OC network. Likewise, Doxani et al. [20] assessed the performance of different AC methods and validated their R_{rs} with match-up datasets over both land and water surfaces in an AC inter-comparison exercise. Warren et al. [21] evaluated the accuracy of a wide range of freely available AC processors by comparing them to reference R_{rs} data from different coastal and inland waters. Similarly, in a more recent AC exercise, Zhang and Hu [22] also analyzed an AC algorithm, comparing its R_{rs} images with those measured over a few sites from the AERONET-OC stations. Collectively, these studies demonstrated that accurate AC remains a challenge for OC remote sensing where precise R_{rs} data are needed. Therefore, it is important to explore ways by which errors in AC outputs, and their effect on the products derived from them, can be minimized. One way to address some of the problems posed by imprecise AC is to assess and quantify the impacts of environmental variables on AC accuracy and then account for this in the atmospherically corrected image. In an earlier study [23], four publicly available AC processors (2 land-based and 2 water-based) for deriving the R_{rs} in coastal waters were compared and validated with 54 R_{rs} match-up datasets from AERONET-OC stations. The study revealed that biases from ACOLITE and SeaDAS, two of the state-of-the-art AC algorithms, are influenced by environmental variables. In this study, we demonstrated the potential of Landsat 8 (L8) data for SDB in US coastal waters and assessed the performance of a commonly used and publicly available water-based AC algorithm (ACOLITE [24]) for physics-based SDB. To minimize the effect of imperfect AC on the bathymetry retrieval, we further used a correction factor to improve the original atmospherically corrected image from ACOLITE. Using a set of 9 images, SDB estimates from these two AC procedures were then compared with LiDAR-derived bathymetry of the area. Lastly, we used an ensemble approach to produce SDB of the study area using all the corrected images.

2. Study Sites and Imagery

2.1. Study Sites

The Florida Keys is a series of islands that extend from the southern end of Florida, USA, to the south–southwest. Their nearshore shallow waters include coral reef tracts, patch reefs, bank reefs, seagrass meadows, and unvegetated hard and soft bottom. This site was chosen because of its relatively clear waters, the good knowledge of seabed features, and availability of LiDAR-derived depth data for validating SDB estimates of water depth. The benthic environment of the section of the Florida Keys used in this study is dominated by extensive seagrass beds, with some patches of reef and unconsolidated sediments. Figure 1 shows this area with the distribution of bathymetric LiDAR data used for validating the SDB estimates in this study.

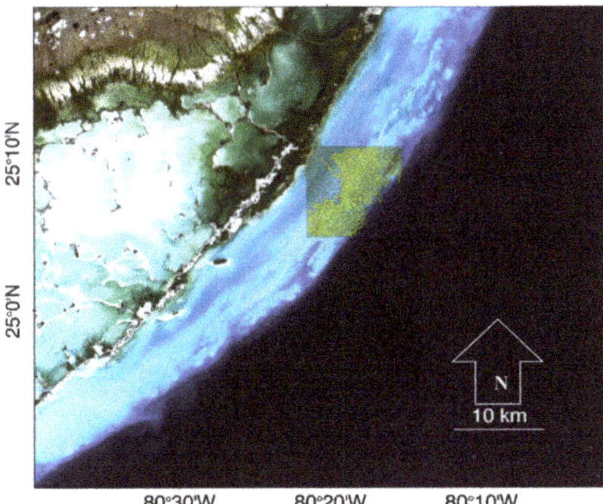

Figure 1. Landsat 8 image showing the upper Florida Keys. Bathymetric LiDAR data used for validation are shown in yellow.

2.2. Satellite Data

Nine L8 images (Figure 2) from the Florida Keys, acquired during both optimal and near-optimal conditions for SDB, were downloaded from the archive of the United States Geological Survey after visually inspecting all available images from May 2013 to May 2019. L8 OLI (Operational Land Imager) collects visible, Near Infrared (NIR) and Short-wave infrared (SWIR) spectral band imagery at 30 m spatial resolution. In addition to the improved positional accuracy of 14 m, compared to 50 m for its predecessors in the Landsat series, L8 includes coastal and aerosol (433–453 nm) and blue (450–515 nm) bands for coastal and bathymetric mapping [25,26].

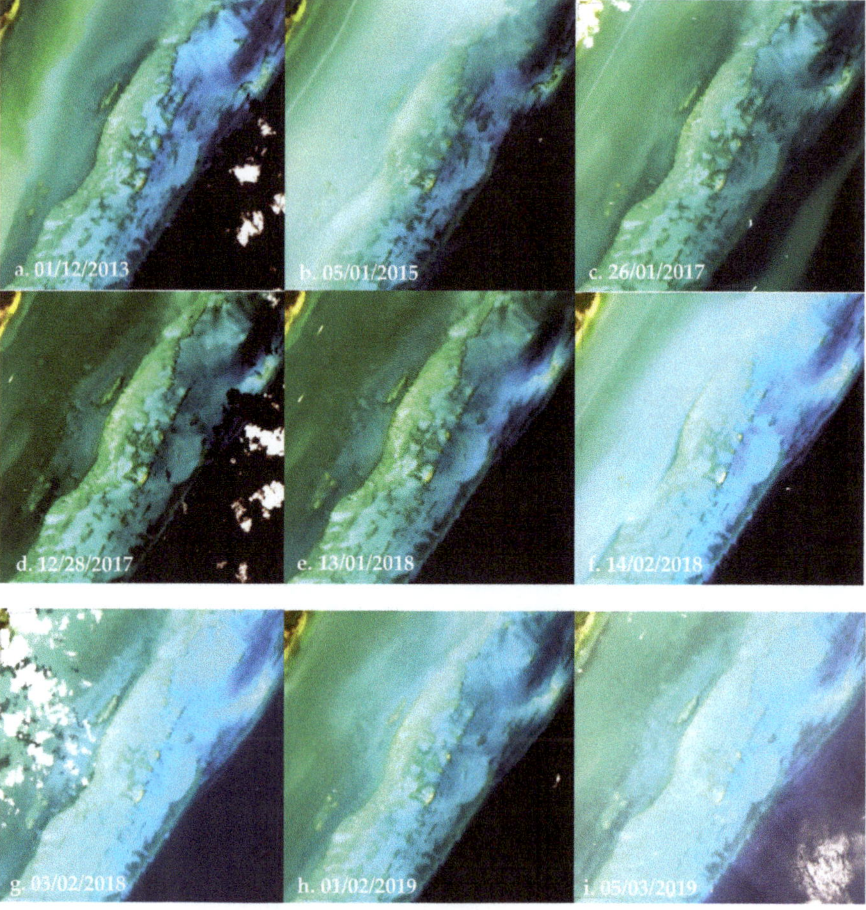

Figure 2. (a–i) A section of Florida Keys image showing the RGB composite of each image used in this study.

2.3. LiDAR Data

To validate the SDB estimates, a bathymetry topographic digital elevation model (DEM) was acquired from the National Oceanic and Atmospheric Administration (NOAA) National Centres for Environment Information (NCEI) coastal LiDAR archive. The LiDAR data collection was conducted in December 2014 over South Florida and the Florida Keys as part of efforts by NOAA to study sea level rise and coastal flooding impacts on US coasts. Several LiDAR sources including topographic and bathymetric LiDAR sensors were used to develop and create a suite of tiled bathymetric–topographic DEMs for South Florida and the Florida Keys [27]. A portion of the DEM tiles covering the study site (Figure 1) was retrieved from the Office of Coastal Managements Data Access Viewer [28] where all DEM data are archived. The DEMs, with a vertical accuracy of approximately 0.5 m, are referenced vertically to the North American Vertical Datum of 1988. Horizontal positions were provided in geographic coordinates and referenced to the North American Datum of 1983 [29,30]. A portion of the collection covering the Florida Keys coastal area was referenced to mean sea level and resampled from 0.3 m to 30 m to match the spatial resolution of L8.

3. Methodology

3.1. Data Preprocessing

3.1.1. Atmospheric Correction

We implemented two types of AC methods for water depth retrieval: (1) we used ACOLITE to process L8 images into R_{rs} values (henceforth Rrs_{raw}) and (2) then applied a correction factor to reduce errors in the original ACOLITE output and create new corrected R_{rs} values (henceforth $Rrs_{corrected}$). ACOLITE [24], specifically designed for AC over water surfaces, is an AC method that estimates L_w by simulating contributions from molecular (Rayleigh) and particulate (aerosol) scattering using a 6SV-based LUT [31]. Based on Ruddick et al. [32], aerosol reflectance is estimated by determining a per-tile aerosol type (or epsilon) from the ratio of reflectances in two bands over water pixels where L_w can be assumed to be zero. Then, the epsilon is used to extrapolate the observed aerosol reflectance to the visible bands to remove atmospheric contributions. ACOLITE was originally designed for processing L8 images, but it has been modified and updated to also process Sentinel-2 data [33]. Furthermore, the most recent version, which can be adapted to commercial sensors such as Pleiades, contains an additional AC scheme (now the default setting) called the dark spectrum fitting (DSF) algorithm, as well as a sun glint correction scheme [34]. In this study, ACOLITE (version 20170113.0) was used to produce all R_{rs} images, which are the direct input into the bathymetry algorithm. The default SWIR option (1609 and 2201 nm band combination) was implemented for all images. This band combination takes advantage of the longest-wavelength SWIR band, where water absorption is the highest. In a previous study [23], in which a range of AC algorithms were compared and validated against in situ L_w from 14 AERONET-OC stations, statistically significant relationships were demonstrated between errors in ACOLITE's R_{rs} estimates for L8's 443 nm and 482 nm bands and three environmental variables: Solar Zenith Angle (SZA), Aerosol Optical Thickness (AOT) at 865 nm (AOT_{865}), and wind speed (u_{10}); probable but statistically non-significant relationships were also demonstrated for the 561 nm and 655 nm bands. Using multiple linear regression, we therefore derived a set of coefficients that were used to estimate the error of ACOLITE's R_{rs} estimates for each of those four bands in each image, as a function of SZA, AOT_{865}, and wind speed. Then, each of the four bands used for depth retrieval in this study was corrected using Equation (1):

$$Rrs_{corrected} = Rrs_{raw} - (a + b*SZA + c*AOT_{865} + d*u10) \tag{1}$$

where $Rrs_{corrected}$ and Rrs_{raw} are the R_{rs} images with and without correction, respectively; and a, b, c, and d are coefficients obtained through fitting a linear model to the data from Ilori et al. [23]. SZA was obtained from the metadata of each L8 scene. AOT_{865} was processed and obtained using the l2gen processor in the SeaDAS software, and an average value used for each image was calculated by randomly sampling multiple pixels over the area of the study site. Wind speed data were obtained from the National Centers for Environmental Prediction Reanalysis project [35], where 6 h global wind speed estimates are archived. Table 1 presents the value of each environmental parameter for each image used in this study.

Table 1. Environmental parameter variables for each image. AOT_{865}: Aerosol Optical Thickness (AOT) at 865 nm, SZA: Solar Zenith Angle.

Scene Date (dd/mm/yyyy)	SZA (Degrees)	AOT_{865}	u_{10} (m/s)
01/12/2013	50.36	0.081	5.29
05/01/2015	52.79	0.088	1.07
26/01/2017	50.13	0.083	2.49
28/12/2017	52.98	0.076	6.45
13/01/2018	52.14	0.142	3.11
14/02/2018	45.63	0.12	4.84
02/03/2018	40.66	0.11	3.21
01/02/2019	49.02	0.122	3.74
05/03/2019	39.74	0.143	4.67

3.1.2. Sun Glint Correction

As sun glint correction is not inherently part of the ACOLITE version used in this study, we implemented the NIR method [36] to remove specular reflection off the sea surface for images where glint was visually obvious. This method assumes that for optically deep areas (where radiation reflected from the seafloor has a negligible influence on L_w), any remaining NIR signal after AC must be due to sea surface reflection. Thus, glint intensity and removal is performed by establishing a linear relationship between the NIR and visible bands over an optically deep area in the image, and that relationship is then used across all water pixels to reduce R_{rs} for the visible bands to its assumed glint-free value.

3.1.3. Estimation of Noise Equivalent Reflectance

Bathymetry model inversion based on least squares optimization techniques is generally sensitive to environmental noise [37,38]; thus, high environmental noise may make images unsuitable for bathymetry extraction. The noise-equivalent difference in reflectance, $NE\Delta R_{rs}$ (sr^{-1}), is a measure of image noise, with contributions from the sensor (e.g., instrument degradation) and the environment (e.g., variability in atmosphere and water surface state) [37,38]. The $NE\Delta R_{rs}$ can be used to assess the suitability of a satellite imagery for aquatic remote sensing applications. For example, it has been used to determine the suitability of the Compact Airborne Spectrographic Imager (CASI) for benthic mapping [11]. Therefore, following AC, we estimated the $NE\Delta R_{rs}$ (sr^{-1}) [39] by calculating the band-wise standard deviation of R_{rs} from a 33×33-pixel window over a homogeneous optically deep area using Equation (2) [40]. This approach assumes that any observed spectral variations in the selected area is due to noise; thus, selected pixels must be as homogenous as possible for a robust standard deviation estimate. Ideally, the $NE\Delta R_{rs}$ should be lower than 0.00025 sr^{-1} in each of the visible bands [41], which was the case for all nine images used in this study. Table 2 shows the per-band value obtained for each of the 9 images used in this study.

$$NE\Delta Rrs = \sigma R_{rs} \tag{2}$$

where σR_{rs} is the standard deviation in each band over an as homogeneous as possible area of optically deep water within the image.

3.1.4. Parameterization of Environmental Properties

To implement the physics-based approach to SDB, values of optical properties and substratum spectral reflectance that are representative of the environment in question are needed. Water inherent optical properties (IOP) (P_{440}, G_{440}, and X_{550}) parameterization for forward modeling for each site was based on assessment from Level 3 OC products from the Visible Infrared Imaging Radiometer Suite Visible Infrared Imaging Radiometer Suite (VIIRS) Generalized Inherent Optical Property (GIOP) algorithms [42]. P_{440} is the phytoplankton absorption coefficient at 440 nm, G_{440} is the absorption of gelbstoff and detrital materials coefficient at 440 nm, and X_{550} is the particulate backscattering of suspended particles coefficient at 550 nm. Using parameter values obtained from these OC products, ranges of values for each parameter were determined by observing the lowest and highest parameter values for all dates from GIOVANNI, which is an online visualization tool for OC products [43]. Then, values slightly lower and higher than the observed lowest and highest values, respectively, were chosen (Table 3). As part of the inversion model, seafloor reflectance spectra are also needed. We used two seafloor spectra (Figure 3), based on the area's benthic description [44]. Depth (Z), which was also needed for forward modeling, was set to 0.1 and 25 m with the understanding that depth penetration greater than 25 m would be difficult.

Table 2. The noise equivalent difference in reflectance (NEΔR$_{rs}$), computed from a kernel of 33 × 33 pixels from an optically deep and homogeneous area, for each image used in this study.

Scene Dates (dd/mm/yyyy)	Band 1	Band 2	Band 3	Band 4
01/12/2013	0.000200	0.000154	0.000096	0.000061
05/01/2015	0.000136	0.000108	0.000084	0.000063
26/01/2017	0.000092	0.000072	0.000057	0.000042
28/12/2017	0.000151	0.000105	0.000081	0.000053
13/01/2018	0.000111	0.000103	0.000069	0.000047
14/02/2018	0.000157	0.000129	0.000108	0.000063
02/03/2018	0.000126	0.000110	0.000069	0.000043
01/02/2019	0.000148	0.000127	0.000100	0.000063
05/03/2019	0.000086	0.000081	0.000059	0.000042

Table 3. Parameter ranges used for forward modeling.

P	G	X	Z
0.006–0.04	0.004–0.04	0.0005–0.006	0.1–25

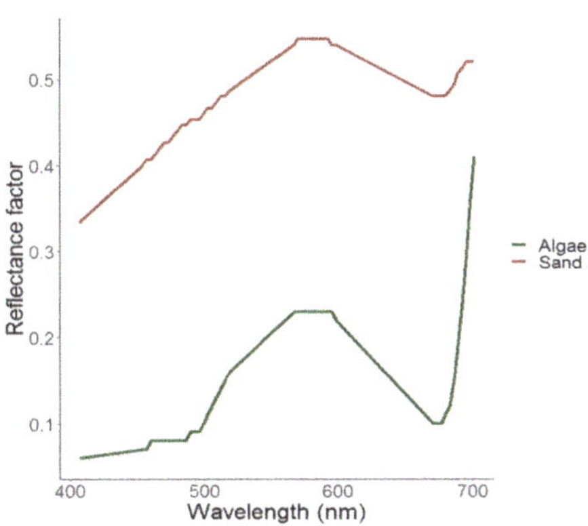

Figure 3. Spectral reflectance of the seafloor used in this study.

3.1.5. Forward Modeling of Remote Sensing Reflectance

To derive water depth, we applied a modified version of the semi-analytical inversion model of Lee et al. [17,18] to the atmospherically corrected images. In this inversion scheme, the sub-surface remote sensing reflectance, r_{rs}, (the ratio of upwelling radiance to downwelling irradiance just below the surface) is related to absorption (a) and backscattering properties (b_b) of the water column, the seafloor reflectance (ρ), and water depth (H). For nadir-viewing satellites, the model can be expressed as:

$$R_{rs} \approx \frac{0.5 r_{rs}}{1 - 1.5 r_{rs}} \tag{3}$$

where r_{rs}, the subsurface remote-sensing reflectance, is expressed as:

$$r_{rs}\left(a, b_b, H, \rho\right) \approx (0.084 + 0.170u)u\left(1 - exp\left\{-\left[\frac{1}{cos(\theta_w)} + \frac{1.03\,\sqrt{1+2.4u}}{cos(\theta_v)}\right]kH\right\}\right)$$
$$+\frac{\rho}{\pi}exp\left\{-\left[\frac{1}{cos(\theta_w)} + \frac{1.04\,\sqrt{1+5.4u}}{cos(\theta_v)}\right]kH\right\} \tag{4}$$

$$u = \frac{b_b}{a + b_b} \tag{5}$$

$$k = a + b_b \tag{6}$$

where θ_w and θ_v are the sub-surface solar zenith and sub-surface sensor viewing angles, respectively. Absorption (a) and backscattering coefficients (b_b) are functions of (1) the absorption coefficient of phytoplankton at 440 nm, P; (2) the absorption coefficient of colored dissolved materials at 440 nm, G; and (3) the backscattering coefficient of suspended particles at 550 nm, X. These are expressed as:

$$a = a_w + Pa^*_{phy} + Ge^{-0.015(\lambda-440)} \tag{7}$$

$$b_b = b_{bw} + X\left[\frac{550}{\lambda}\right]^Y \tag{8}$$

where a_w and b_{bw} are the absorption and backscattering coefficients of pure water, respectively [45], a^*_{phy} is the specific absorption of coefficient of phytoplankton (normalized to a value of 1.0 at 440 nm), λ is the center wavelength, and Y is the spectral shape that depends on the particulate shape and size.

While Lee's inversion model uses the albedo of only one key benthic substrate (sand), our model includes a parameterization to set the seafloor reflectance as a linear mix of the two bottom types (i.e., sand and algae; [46]). To forward model the R_{rs} as a function of water depth, water quality parameters, and the seafloor reflectance, the adaptive look-up table (ALUT) method [11,16] was implemented, which ensures efficient construction and search through the table. In this approach, an LUT consisting of the modeled R_{rs} values of L8 bands 1–4, seafloor reflectance (Figure 3), water optical properties (absorption and scattering characteristics of water), and water depths of the optically shallow zone of the area in question (Table 3) is constructed. With realistic minimum and maximum values of all environmental parameters in the table, the LUT construction is optimized by using a hierarchical structure to efficiently cover the range of expected R_{rs} values while minimizing the under- or over-sampling of spectrally similar regions of environmental space, which is common with discretization by regular intervals in conventional forward models. For example, a small change in depth in shallow water areas will cause a significant/large change in R_{rs}, and will typically result in under-sampling if the depth parameter is discretized by regular intervals. Likewise, oversampling may occur in deep water areas where a small change in R_{rs} is expected [11,16]. To identify the best parameter values to be included in the hierarchy in the ALUT technique, discretization is based on an evenly sampled spectral space and not on an evenly sampled parameter space (e.g., water depth). This approach requires bounded ranges for all the modeled parameters, for which we used the value ranges in Table 3.

3.1.6. Inversion of Remote Sensing Reflectance

Model inversion was subsequently performed using the binary space partitioning (BSP) approach [11,16], as described in Knudby et al. [12]. Briefly, this technique subdivides the LUT created during forward modeling into different nodes. First, the BSP splits the whole LUT into two (left and right child nodes) and subsequently subdivides the nodes into a partitioning tree, which facilitates the optimization of the per-pixel LUT search. After model inversion and depth retrieval, water depths were corrected for tidal height at the time of each image acquisition using tidal height estimates obtained from Oregon State University's tide prediction service [47].

Depth cannot be accurately estimated for optically deep pixels (i.e., where reflection off the seafloor is a negligible component of the radiance measured by the sensor); thus, we estimated a depth threshold value for each scene to distinguish between optically deep and optically shallow areas. These threshold values were obtained by calculating the mean minus 2 standard deviations of depth predictions within a 33 × 33 window of homogeneous optically deep water. Results are reported for each scene both a) for the full depth range, and b) for depths from the surface and down to these scene-specific thresholds.

3.2. Validation of Depth Estimates

Validation of depth estimates from the two AC procedures was performed by comparing the estimated depths to the LiDAR data. The number of depth estimates used for validation (Table 4) varied between the nine images due to differences in the number of pixels for which depth was successfully estimated, as pixels that did not pass the AC's internal quality checks (e.g., due to clouds), pixels with negative depths, and pixels that were visually impacted by boats, wake, or cloud shadows were eliminated prior to validation. Based on the remaining pixels, we used the coefficient of determination (R^2), RMSE (root-mean-squared-error) (Equation (9)) and bias (Equation (10)) to compare the accuracy of the uncorrected and corrected SDB estimates with the LiDAR datasets. The RMSE is used to measure the accuracy of the estimated depth values; and bias is used to indicate overestimation (positive value) or underestimation (negative value):

$$\text{RMSE} = \sqrt{\frac{1}{n}\sum_{i=1}^{n}\left(x^{est} - x^{obs}\right)^2} \tag{9}$$

$$\text{bias} = \frac{\sum_{i=1}^{n}\left(x^{est} - x^{obs}\right)}{n} \tag{10}$$

where n is the number of observations, and x^{est} and x^{obs} are the estimated and measured depths, respectively. Values closer to zero for both error metrics indicate a better result. SDB obtained with Rrs_{raw} and $Rrs_{corrected}$ are hereafter referred to as SDB_{raw} and $SDB_{corrected}$, respectively. These summary statistics (R^2, RMSE, and bias) were calculated both for the full depth range and for depths ranging from the surface down to the per-scene depth threshold (Table 4).

Table 4. Summary validation statistics for satellite-derived bathymetry (SDB) estimates (SDB_{raw} and $SDB_{corrected}$) for the two depth limits in this study. Bold letters in the root-mean-squared-error (RMSE) column indicate where an observed difference between SDB_{raw} and $SDB_{corrected}$ estimates is more than 0.1 m.

Scene Date dd/mm/yyyy	Max Depth (m)	RMSE (m) (SDB_{raw}/$SDB_{corrected}$)	Bias (SDB_{raw}/$SDB_{corrected}$)	R^2 (SDB_{raw}/$SDB_{corrected}$)	Number of Validation Points
01/12/2013	25	**1.96/1.66**	0.83/0.84	0.66/0.79	3102
	13.80	**1.80/1.50**	0.85/0.73	0.58/0.71	2988
05/01/2015	25	2.02/2.03	−0.17/−0.46	0.71/0.71	3317
	11.23	1.84/1.88	−0.46/−0.71	0.53/0.53	3050
26/01/2017	25	1.34/1.25	0.49/0.09	0.85/0.85	3311
	12.03	**1.21/0.95**	0.57/0.23	0.81/0.85	3142
28/12/2017	25	**1.87/1.39**	1.16/0.41	0.74/0.81	3148
	17.54	**1.85/1.39**	1.18/0.42	0.79/0.81	3086
13/01/2018	25	**2.21/2.04**	1.45/1.40	0.66/0.76	2804
	15.21	**2.19/1.89**	1.48/1.35	0.56/0.74	2721
14/02/2018	25	**1.86/1.72**	−1.15/−1.04	0.77/0.79	3316
	14.81	**1.80/1.69**	−1.12/−1.01	0.71/0.74	3247
02/03/2018	25	**2.03/1.65**	0.71/0.68	0.61/0.74	3268
	16.84	**1.98/1.62**	0.69/0.67	0.53/0.71	3226
01/02/2019	25	**1.35/1.05**	0.55/0.62	0.83/0.92	3255
	14.61	**1.33/1.02**	0.54/0.64	0.76/0.90	3187
05/03/2019	25	1.36/1.27	0.76/0.53	0.89/0.89	3312
	16.39	1.32/1.25	0.79/0.55	0.85/0.83	3263

4. Result and Discussion

Scatterplots showing water depth estimates produced from both Rrs_{raw} and $Rrs_{corrected}$ images and the LiDAR depth measurements are shown in Figure 4, and summary statistics (R^2, RMSE, and bias) are listed in Table 4.

The RMSE values for SDB_{raw}/$SDB_{corrected}$ estimates range from 1.35/1.05 m to 2.21/2.04 m for the full depth range, and from 1.21/0.95 m to 2.19/1.89 m when applying the per-scene depth threshold. These values are broadly comparable with other SDB studies (e.g., [10,13,14,41,48]). For example, Dekker et al. [48], who compared one empirical and five physics-based approaches to bathymetry mapping using hyperspectral imagery, reported RMSE values between 0.86 (best) and 4.71 m (worst) for depths less than 13 m for two clear tropical waters in the Bahamas and eastern Australia, suggesting that our results are typical of what should be expected from a physics-based bathymetry method. It is worth mentioning here that impacts from recent hurricanes over parts of the Florida Keys, notably in 2016 and 2017, have resulted in an average increase of approximately +0.3 m in seafloor elevation over different habitat types [49,50]. Such changes were not accounted for in this study and may have had a small effect on the results, although it is worth noting that the best SDB estimate is from 2019 [Figure 4h], after the hurricanes.

Figure 4 shows that accuracy decreases with depth for both SDB_{raw} and $SDB_{corrected}$, particularly beyond approximately 15 m where the proportion of the measured signal originating from reflection at the seafloor becomes very small. In general, for depths shallower than 15 m, $SDB_{corrected}$ points cluster more tightly around the 1:1 line that do the SDB_{raw} points.

4.1. Effects of Image Conditions on Depth Accuracy

4.1.1. Turbidity

Out of the nine Rrs images we applied the correction factor to, eight corrections resulted in negative RMSE changes when considering the two depth limits used in this study, with reductions ranging from 0.09 to 0.48 m (for the full depth range) and 0.07 to 0.46 m (for the per-scene depth threshold). For both depth limits, only one correction resulted in increased RMSE (i.e., the image from 01/05/2015, see Table 4) (RMSE values for SDB_{raw} and $SDB_{corrected}$ will hereafter be referred to as $RMSE_{raw}$ and $RMSE_{corrected}$, respectively). For this image, accurate depth estimates were not possible beyond approximately 14 m (Figure 4b), regardless of correction, and $RMSE_{corrected}$ increased marginally by 0.01 m for the full depth range and by 0.04 m for the per-scene depth threshold (i.e., 0–11.23 m). A visual inspection of this image shows sediment plumes in the study area (Figure 5b), which suggests that turbidity contributed to an underestimation of water depth for both SDB_{raw} and $SDB_{corrected}$ [14], and the image is of marginal use for SDB, regardless of correction. Similarly, one of the two corrected images with the lowest RMSE reduction for the full depth range also has what looks to be a silt plume emerging from nearshore channels in the southwestern portion of the area for which depth was calculated (Figure 5a). For this image, RMSE value marginally decreased by 0.09 m for the full depth range and by 0.26 m for the per-scene depth threshold (i.e., 1 to 12.03 m).

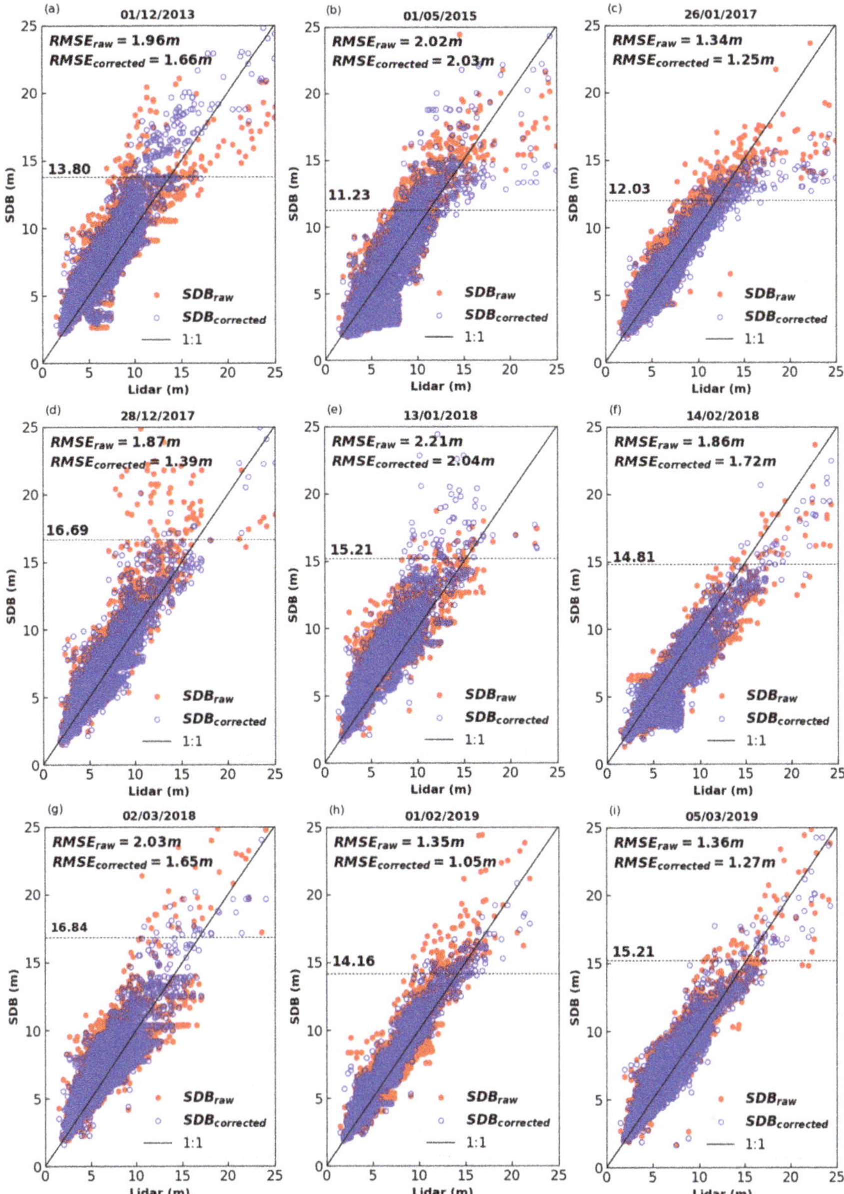

Figure 4. (a–i) Scatterplots of satellite-derived bathymetry estimates versus LiDAR measurements. Red points show water depth estimates obtained from original ACOLITE outputs; blue points show estimates obtained after applying the correction factor. The 1:1 line is shown in black. The dotted horizontal lines and values above them denote threshold values beyond which the water column is deemed optically deep. Threshold value was calculated using Mean − 2 Standard deviation of a 33 × 33 window pixel selected over an optically deep homogeneous area in an image.

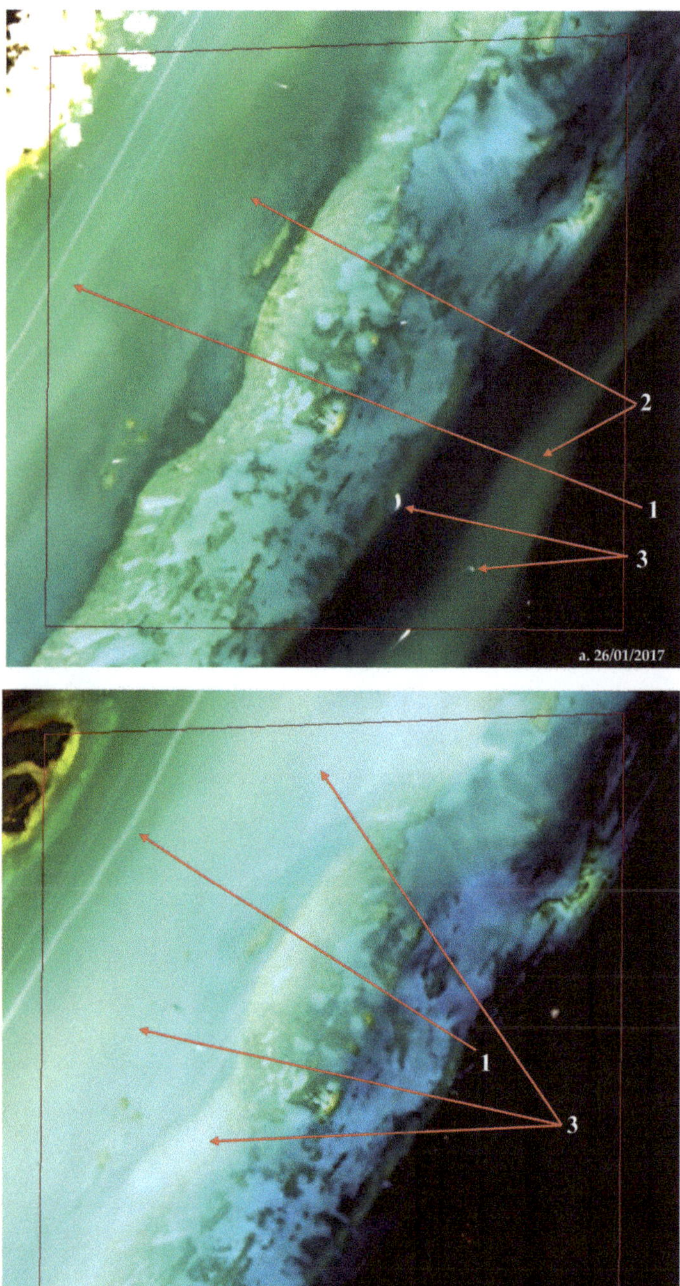

Figure 5. *Cont.*

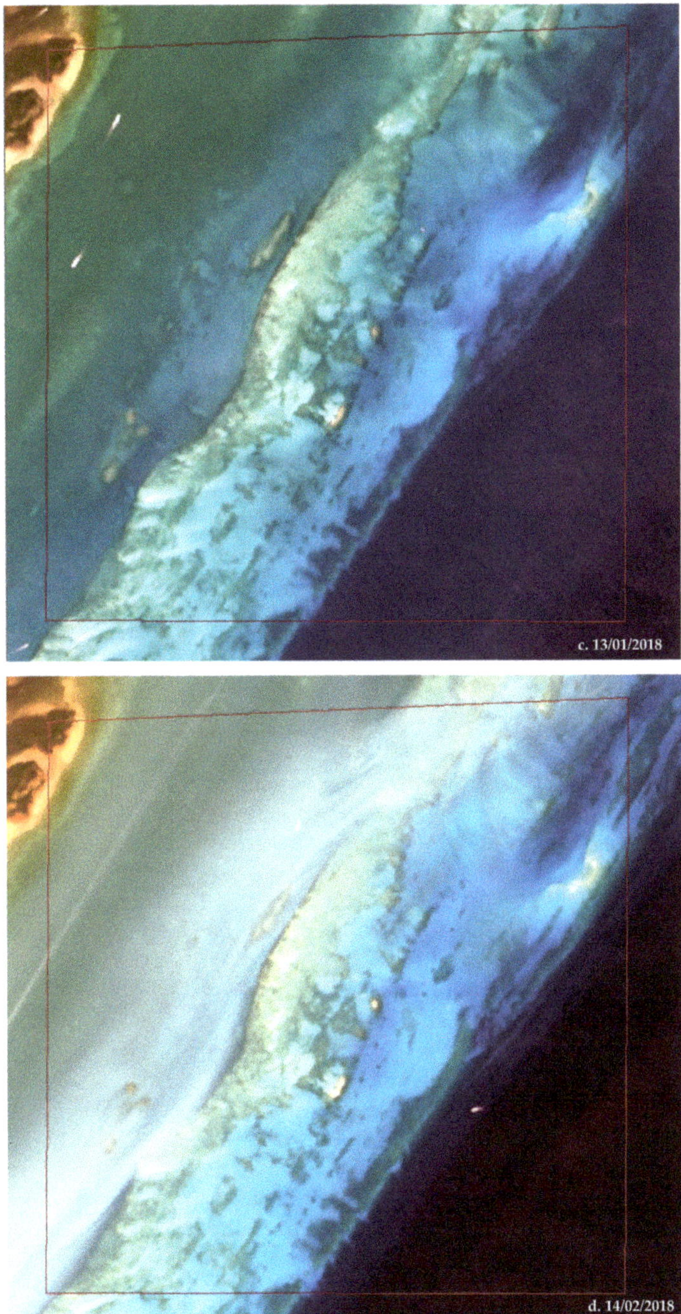

Figure 5. (**a–d**) Maps showing different confounding factors that might have affected SDB estimates from some images. 1: Boat-generated wake. 2: Plume emerging from a near river discharge. 3: Moving boats. Sun glint can be observed in Figure 5c,d as visible texture around the southeastern part of the images.

4.1.2. Glint

With $RMSE_{raw}$ values of 2.21 m and 2.19 m for the full depth range and the per-scene depth threshold, respectively, the image from 13/01/2018 produces the poorest SDB_{raw} and $SDB_{corrected}$ estimates out of the nine images. As shown in Table 4, this image has the highest RMSE and positive bias values. A visual inspection of this image (Figure 5c) indicates the presence of a moderate glint. Glint correction was not performed, as the image did not show any noticeable improvement after the initial testing. Likewise, for the image from 14/02/2018 (Figure 5d), the high negative bias values for both depth ranges, regardless of correction (Table 4), may be attributed to residual sun glint in addition to light turbidity. While an attempt was made to de-glint this image as described in Section 3.1.2, given that the NIR-based de-glinting method [36] implemented (1) relies on manual selection of deep-water pixels to estimate glint contribution, (2) assumes that there are glint-free pixels among those selected [51], and (3) assumes a homogenously low $R_{rs}(NIR)$ across all water pixels, failure to meet these conditions may have resulted in the observed residual glint. For example, $R_{rs}(NIR)$ may be non-negligible in glint-free but very shallow or turbid waters, or where reflective vegetation such as seagrass is close to the upper water column [52]. For these two images, the correction produces slightly reduced RMSE values (Figure 4e,f), substantially so for their respective depth thresholds (Table 4).

4.2. Effect of Wind Speed and SZA on SDB Performance

Out of the eight scenes whose SDB performance improved with the correction for the R_{rs} images, greater corrections were done for five scenes (i.e., Figure 4a,c–f) acquired with SZA > 49° (Table 1), and one scene (i.e., Figure 4g) acquired during high wind speed (3.21 m/s) (Table 1), when considering the per-scene depth thresholds. Likewise, when considering the full depth range, greater corrections were also done for the same number of scenes under similar environmental conditions, with the exception of the image from 26 January 2017, whose change in RMSE is −0.09 m (Table 4). It should be noted that the most noticeable RMSE reduction for $SDB_{corrected}$ images for both depth limits considered in this study (i.e., Figure 4d) was observed for the scene with the highest SZA and highest wind speed. This is supporting evidence for the existence of a relationship between ACOLITE's overestimation of R_{rs} in the first two bands of L8 and these two environmental variables [23], and it gives an idea of the magnitude of its impact on SDB performance. A recent study [53] also found a dependency between AC retrieval accuracy and wind speed in coastal waters. While there is strong evidence to conclude that the correction factor used in this study lowers RMSE values for images with high wind speed, it should be noted that wind speed data used in this study come from 6-h estimates in reanalysis model, and thus, they have their own uncertainty. For this reason, more testing may be needed for a firmer conclusion about the relationship between ACOLITE's error and wind speed.

4.3. Bathymetry Estimates at Different Depth Ranges

Figure 6 shows the performance of SDB_{raw} and $SDB_{corrected}$ for each image, binned to 5 m depth increments for the 1–25 m depth range. The accuracy of SDB estimates decreases with increasing depth for both SDB_{raw} and $SDB_{corrected}$. While higher RMSE values should be expected at deeper depths due to the diminishing signal from seafloor reflectance, it should be noted that the number of LiDAR points for validation is also smaller at deeper depths, leading to increased uncertainty around the RMSE values reported at these depths.

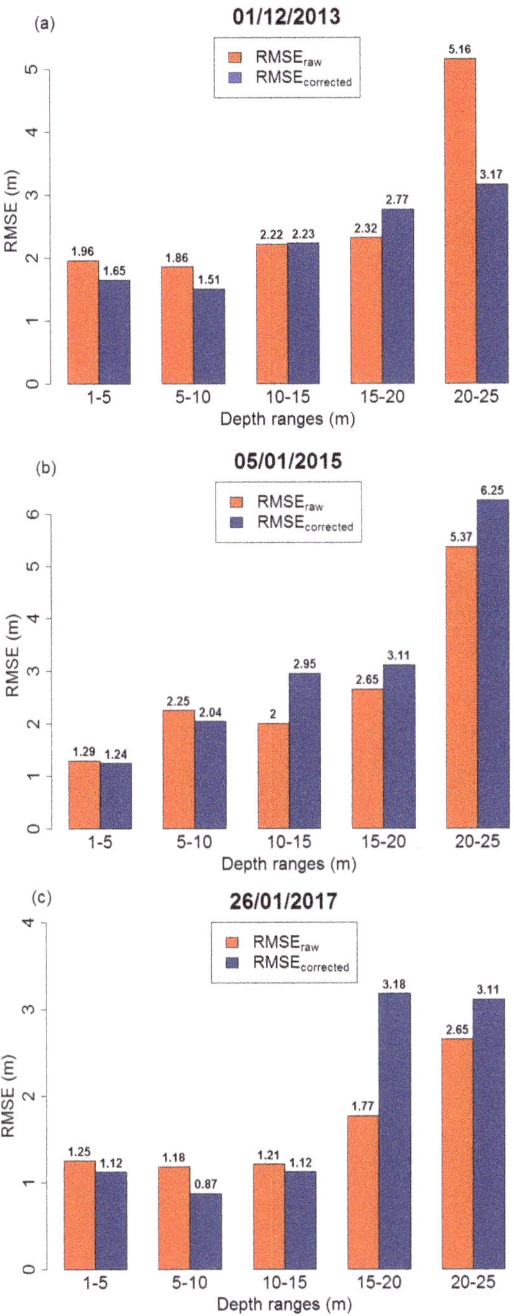

Figure 6. *Cont.*

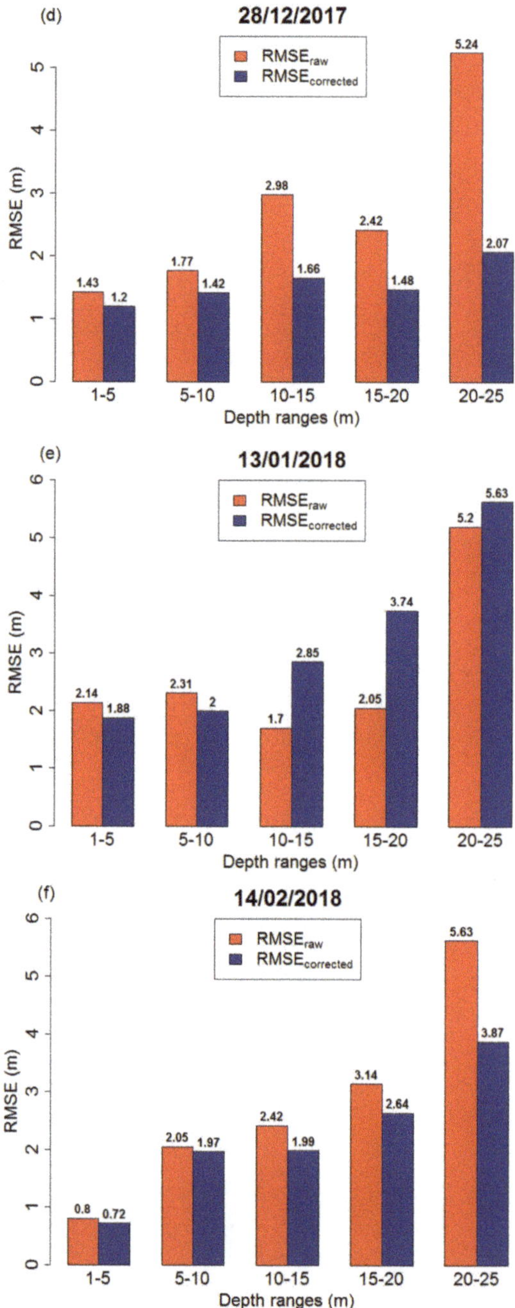

Figure 6. *Cont.*

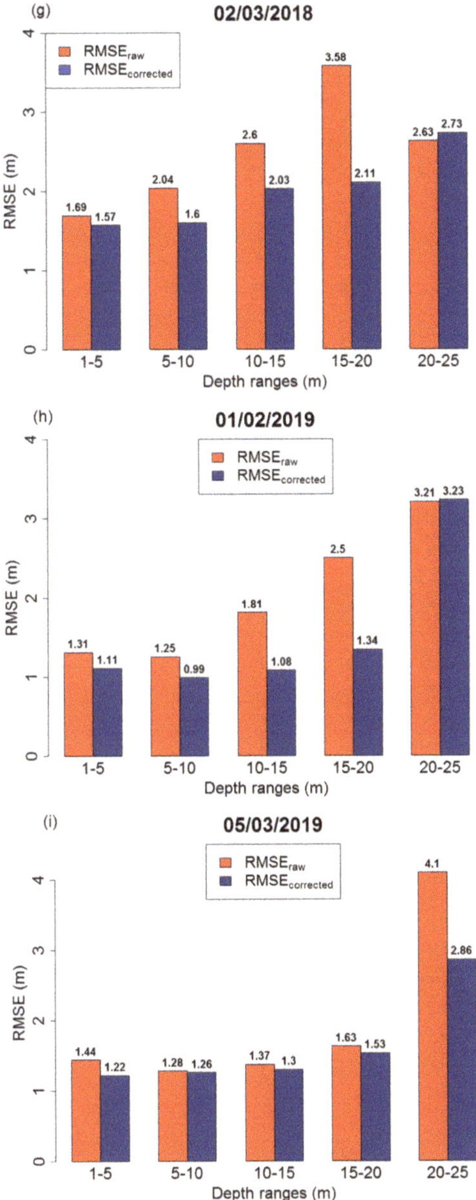

Figure 6. (a–i) RMSE values obtained for SDB_{raw} and $SDB_{corrected}$ estimates at different water depths. Results at higher depth (>15 m) should be interpreted with caution, as the number of depth observations for those depth ranges was comparably lower than those available for shallower depth ranges. Depth observation for each depth range is as follows: 1–5 m: approximately 900 points, 5–10 m: approximately 1800 points, 10–15 m: approximately 300 points, 15–20 m: approximately 40 points, and 20–25 m: approximately 20 points. Note that the y-axes have different ranges for each date to facilitate comparison between $RMSE_{raw}$ and $RMSE_{corrected}$ for each single scene.

4.4. SDB Estimates Using an Ensemble Approach

Most SDB studies are based on a single image for each study area, with researchers typically selecting the best available image using visual inspection [14]. Our results indicate that this may not be a robust approach. To illustrate the problem, we invite readers to visually inspect the nine images used in this study (Figure 2) and identify the one that looks most suitable for SDB. Then, proceed to Table 4 to see if it was indeed the one that produced the best results, as measured by RMSE, bias, or R^2. An informal test among our colleagues, all of whom work on OC remote sensing, suggests that it is not easy to identify the best scene. However, a unique advantage of optical remote sensing is the repetitive acquisition of images over the same area. This is especially important for SDB, where the suitability of a given image is determined by transient environmental factors, such as cloud and aerosols, sea surface state, and turbidity [54]. We explored one way of taking advantage of the multiple images available for the study area by testing an ensemble approach in which we calculated the per-pixel median depth value of all nine corrected images (i.e., $SDB_{corrected}$) used in this study. Then, we compared the resulting depth estimates with those obtained using the best individual image from the analysis in Section 4.1 (i.e., the image from 01/02/2019, see Table 4). Figure 7 shows that the results produced by the ensemble are very similar to those obtained with the best individual image. SDB estimates up to approximately 15 m are similar to the $SDB_{corrected}$ estimates from the 01/02/2019 image, as are the RMSE values for the 1–5, 5–10, and 10–15 m depth ranges (Figure 8). Outliers are noticeably reduced in the ensemble result when compared to any of the sub-optimal images that were also included in its calculation, suggesting that the use of median depth is effective in eliminating noise in the ensemble. Thus, the ensemble approach eliminates the need for the selection of a single best image, while producing SDB results of similar accuracy. In this context, it is noteworthy that one of the two best images (i.e., image from 05/03/2019) is also the one with the lowest $NE\Delta R_{rs}$ in bands 1, 2, and 4, as well as the second-lowest $NE\Delta R_{rs}$ in band 3 (Table 2). This suggests that one effective way to pre-screen images, either for a single best image approach or to determine which images should be included in an ensemble, could be to estimate $NE\Delta R_{rs}$ and select those images with the lowest values across the visible bands.

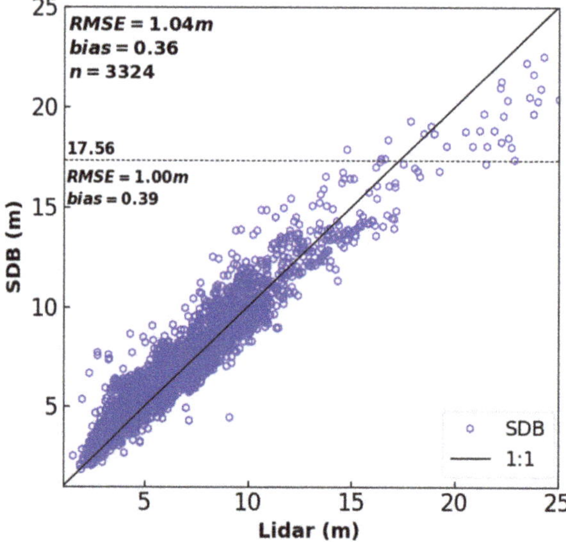

Figure 7. Scatterplots of ensemble-based satellite-derived bathymetry estimates vs. LiDAR measurements. Blue dots show estimates obtained after applying a correction to the R_{rs} images. The solid line represents the 1:1 relationship. The dotted line shows the per-scene depth threshold value and its statistical metrics.

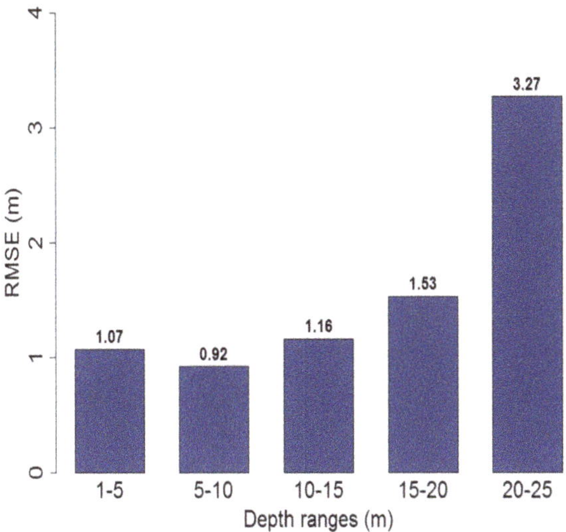

Figure 8. RMSE values obtained for the ensemble-based SDB estimates at different water depths.

5. Conclusions

In this study, we demonstrated the use of Landsat 8 data for physics-based SDB in US coastal waters. A state-of-the-art AC method (ACOLITE) was used to convert per-pixel radiometric units to R_{rs}, and an RTM was inverted to estimate water depth, which was compared to airborne LiDAR validation data. The results showed that ACOLITE can be used to produce SDB from imagery that is free of conditions such as clouds, glint, sediments plumes, boats and wakes, with an accuracy (RMSE 1.34 m for 1–25 m depth range and 1.21 m for approximately 1–15 m depth range) comparable to that reported from empirical and physics-based SDB elsewhere. To account for ACOLITE's known overestimation of R_{rs} for Landsat 8's coastal and blue bands, we applied a correction factor, which was calculated as a function of solar zenith angle, aerosol optical thickness, and wind speed, to obtain a corrected set of R_{rs} images. This correction further improved bathymetry estimates for eight of the nine scenes we tested, with the resulting changes in bathymetry RMSE ranging from +0.01 m (worse) to −0.48 m (better) for a 1 to 25 m depth range, and from + 0.04 m (worse) to −0.46 m (better) for an approximately 1–16 m depth range. Using a total of nine Landsat 8 images, we showed that the correction factor improved SDB results, both on average (ΔRMSE = −0.22 m) and for the best single image (ΔRMSE = −0.30 m) for the 1–25 m depth range. SDB improvements from application of the correction factor were the greatest for images acquired at a high solar zenith angle and at high wind speeds, where ACOLITE is known to have the greatest bias. The correction method demonstrated in this study can be implemented with any appropriate AC algorithm. Finally, we demonstrated that an ensemble approach based on multiple images, with acquisitions ranging from optimal to sub-optimal conditions, can be used to derive bathymetry with a result that is similar to what can be obtained from the best individual image. This is important because it is rarely visually obvious which of several images is best for SDB, and the ensemble approach can be automated to reduce time spent on pre-screening and filtering scenes, and it can potentially also reduce the amount of missing pixels caused by clouds and cloud shadows encountered in any single image. Automating SDB will ultimately facilitate the efficient and operational use of the globally available L8 (and other multispectral) datasets.

Author Contributions: Conceptualization, C.O.I. and A.K.; Methodology, C.O.I. and A.K.; Software, C.O.I. and A.K.; Validation, C.O.I. and A.K.; Formal Analysis, C.O.I.; Resources, C.O.I. and A.K.; Writing—Original Draft Preparation, C.O.I.; Writing—Review & Editing, C.O.I. and A.K.; Supervision, A.K. All authors have read and agreed to the published version of the manuscript.

Funding: This research received no external funding.

Acknowledgments: The authors wish to thank the United States Geological Survey (USGS) for the distribution of Landsat 8 Level-1 data products. We are deeply thankful to Quinten Vanhellemont and Kevin Ruddick for the development of ACOLITE and the NASA Ocean Biology Processing Group for the development of the SeaDAS software. We also thank Chris Roelfsema and his team for making the spectra data used in this study publicly available online.

Conflicts of Interest: The authors declare no conflict of interest.

References

1. Ramnath, V.; Feygels, V.; Kalluri, H.; Smith, B. CZMIL (Coastal Zone Mapping and Imaging Lidar) Bathymetric Performance in Diverse Littoral Zones. In Proceedings of the OCEANS 2015 OCEANS MTS/IEEE, Washington, DC, USA, 19–22 October 2015.

2. Parker, H.; Sinclair, M. The Successful Application of Airborne LiDAR Bathymetry Surveys Using Latest Technology. In Proceedings of the 2012 Oceans—Yeosu, Yeosu, Korea, 21–24 May 2012.

3. Tonina, D.; McKean, J.A.; Benjankar, R.M.; Wright, C.W.; Goode, J.R.; Chen, Q.; Reeder, W.J.; Carmichael, R.A.; Edmondson, M.R. Mapping river bathymetries: Evaluating topobathymetric LiDAR survey. *Earth Surf. Process. Landf.* **2019**, *44*, 507–520. [CrossRef]

4. Wozencraft, J.M.; Lillycrop, W.J. SHOALS Airborne Coastal Mapping: Past, Present, and Future. *J. Coast. Res.* **2003**, *38*, 207–215.

5. Gao, J. Bathymetric mapping by means of remote sensing: Methods, accuracy and limitations. *Prog. Phys. Geogr.* **2009**. [CrossRef]

6. Su, H.; Liu, H.; Heyman, W. Automated derivation of bathymetric information from multi-spectral satellite imagery using a non-linear inversion model. *Mar. Geodesy* **2008**, *31*, 281–298. [CrossRef]

7. Lyzenga, D.R. Passive remote sensing techniques for mapping water depth and bottom features. *Appl. Opt.* **1978**, *17*, 379–383. [CrossRef] [PubMed]

8. Lyzenga, D.R. Remote sensing of bottom reflectance and water attenuation parameters in shallow water using aircraft and landsat data. *Int. J. Remote Sens.* **1981**, *2*, 71–82. [CrossRef]

9. Lyzenga, D.R. Shallow-water bathymetry using combined Lidar and passive multispectral scanner data. *Int. J. Remote Sens.* **1985**, *6*, 115–125. [CrossRef]

10. Chénier, R.; Faucher, M.A.; Ahola, R. Satellite-derived bathymetry for improving Canadian Hydrographic Service charts. *ISPRS Int. J. Geo Inf.* **2018**, *7*, 306. [CrossRef]

11. Hedley, J.; Roelfsema, C.; Koetz, B.; Phinn, S. Capability of the Sentinel 2 mission for tropical coral reef mapping and coral bleaching detection. *Remote Sens. Environ.* **2012**, *120*, 145–155. [CrossRef]

12. Knudby, A.; Ahmad, S.K.; Ilori, C. The Potential for Landsat-Based Bathymetry in Canada. *Can. J. Remote Sens.* **2016**, *42*, 367–378. [CrossRef]

13. Olayinka, I.; Knudby, A. Satellite-Derived Bathymetry Using a Radiative Transfer-Based Method: A Comparison of Different Atmospheric Correction Methods. In Proceedings of the OCEANS 2019 MTS/IEEE SEATTLE, Seattle, DC, USA, 27–31 October 2019. [CrossRef]

14. Casal, G.; Hedley, J.D.; Monteys, X.; Harris, P.; Cahalane, C.; McCarthy, T. Satellite-derived bathymetry in optically complex waters using a model inversion approach and Sentinel-2 data. *Estuar. Coast. Shelf Sci.* **2020**, *241*, 106814. [CrossRef]

15. Mobley, C.D.; Sundman, L.K.; Davis, C.O.; Bowles, J.H.; Downes, T.V.; Leathers, R.A.; Montes, M.J.; Bissett, W.P.; Kohler, D.D.R.; Reid, R.P.; et al. Interpretation of hyperspectral remote-sensing imagery by spectrum matching and look-up tables. *Appl. Opt.* **2005**, *44*, 3576. [CrossRef] [PubMed]

16. Hedley, J.; Roelfsema, C.; Phinn, S.R. Efficient radiative transfer model inversion for remote sensing applications. *Remote Sens. Environ.* **2009**, *113*, 2527–2532. [CrossRef]

17. Lee, Z.; Carder, K.L.; Mobley, C.D.; Steward, R.G.; Patch, J.S. Hyperspectral remote sensing for shallow waters I. A semianalytical model. *Appl. Opt.* **1998**, *37*, 6329–6338. [CrossRef] [PubMed]

18. Lee, Z.; Carder, K.L.; Mobley, C.D.; Steward, R.G.; Patch, J.S. Hyperspectral remote sensing for shallow waters: 2. Deriving bottom depths and water properties by optimization. *Appl. Opt.* **1999**, *38*, 3831–3843. [CrossRef] [PubMed]

19. Pahlevan, N.; Schott, J.R.; Franz, B.A.; Zibordi, G.; Markham, B.; Bailey, S.; Schaaf, C.B.; Ondrusek, M.; Greb, S.; Strait, C.M. Landsat 8 remote sensing reflectance (Rrs) products: Evaluations, intercomparisons, and enhancements. *Remote Sens. Environ.* **2017**, *190*, 289–301. [CrossRef]

20. Doxani, G.; Vermote, E.; Roger, J.C.; Gascon, F.; Adriaensen, S.; Frantz, D.; Hagolle, O.; Hollstein, A.; Kirches, G.; Li, F.; et al. Atmospheric correction inter-comparison exercise. *Remote Sens.* **2018**, *10*, 352. [CrossRef]

21. Warren, M.A.; Simis, S.G.H.; Martinez-Vicente, V.; Poser, K.; Bresciani, M.; Alikas, K.; Spyrakos, E.; Giardino, C.; Ansper, A. Assessment of atmospheric correction algorithms for the Sentinel-2A MultiSpectral Imager over coastal and inland waters. *Remote Sens. Environ.* **2019**, *225*, 267–289. [CrossRef]

22. Zhang, M.; Hu, C. Evaluation of Remote Sensing Reflectance Derived from the Sentinel-2 Multispectral Instrument Observations Using POLYMER Atmospheric Correction. *IEEE Trans. Geosci. Remote Sens.* **2020**, *58*, 1–8. [CrossRef]

23. Ilori, C.O.; Pahlevan, N.; Knudby, A. Analyzing performances of different atmospheric correction techniques for Landsat 8: Application for coastal remote sensing. *Remote Sens.* **2019**, *11*, 469. [CrossRef]

24. Vanhellemont, Q.; Ruddick, K. Advantages of high quality SWIR bands for ocean colour processing: Examples from Landsat-8. *Remote Sens. Environ.* **2015**, *161*, 89–106. [CrossRef]

25. Storey, J.; Choate, M.; Lee, K. Landsat 8 operational land imager on-orbit geometric calibration and performance. *Remote Sens.* **2014**, *6*, 11127–11152. [CrossRef]

26. Czapla-Myers, J.; McCorkel, J.; Anderson, N.; Thome, K.; Biggar, S.; Helder, D.; Aaron, D.; Leigh, L.; Mishra, N. The ground-based absolute radiometric calibration of Landsat 8 OLI. *Remote Sens.* **2015**, *6*, 11127–11152. [CrossRef]

27. Sutherland, M.G.; Amante, C.J.; Carignan, K.S.; Lancaster, M.N.; Love, M.R. NOAA National Centers for Environmental Information Topo-Bathymetric Digital Elevation Modeling: Florida Keys and South Florida. 2016. Available online: https://www.ngdc.noaa.gov/mgg/dat/dems/tiled_tr/florida_keys_tiled_navd88_2016.pdf (accessed on 25 July 2020).

28. NOAA. Data Access Viewer. Available online: https://coast.noaa.gov/dataviewer/#/lidar/search/ (accessed on 13 May 2020).

29. Cooperative Institute for Research in Environmental Sciences. Continuously Updated Digital Elevation Model (CUDEM)—1/9 Arc-Second Resolution Bathymetric-Topographic Tiles. NOAA National Centers for Environmental Information, 2015. Available online: https://doi.org/10.25921/ds9v-ky35 (accessed on 13 March 2020).

30. National Centers of Environmental Information; NESDIS; NOAA; U.S. Department of Commerce. U.S. Coastal Lidar Elevation Data—Including the Great Lakes and Territories, 1996—Present. 2015. Available online: https://catalog.data.gov/harvest/object/abccfcac-9d89-475f-b0f3-58db5d317a86/html (accessed on 25 July 2020).

31. Kotchenova, S.Y.; Vermote, E.F.; Levy, R.; Lyapustin, A. Radiative transfer codes for atmospheric correction and aerosol retrieval: Intercomparison study. *Appl. Opt.* **2008**, *47*, 2215–2226. [CrossRef] [PubMed]

32. Ruddick, K.G.; Ovidio, F.; Rijkeboer, M. Atmospheric correction of SeaWiFS imagery for turbid coastal and inland waters. *Appl. Opt.* **2000**, *39*, 897–912. [CrossRef] [PubMed]

33. Vanhellemont, Q.; Ruddick, K. Acolite for Sentinel-2: Aquatic Applications of MSI Imagery. In Proceedings of the 2016 ESA Living Planet Symposium, Prague, Czech Republic, 9–13 May 2016.

34. Vanhellemont, Q.; Ruddick, K. Atmospheric correction of metre-scale optical satellite data for inland and coastal water applications. *Remote Sens. Environ.* **2018**, *216*, 586–597. [CrossRef]

35. Kalnay, E.; Kanamitsu, M.; Kistler, R.; Collins, W.; Deaven, D.; Gandin, L.; Iredell, M.; Saha, S.; White, G.; Woollen, J.; et al. The NCEP/NCAR 40-Year Reanalysis Project. *Bull. Am. Meteorol. Soc.* **1996**, *77*, 437–472. [CrossRef]

36. Hedley, J.D.; Harborne, A.R.; Mumby, P.J. Simple and robust removal of sun glint for mapping shallow-water benthos. *Int. J. Remote Sens.* **2005**, *26*, 2107–2112. [CrossRef]

37. Botha, E.J.; Brando, V.E.; Dekker, A.G. Effects of per-pixel variability on uncertainties in bathymetric retrievals from high-resolution satellite images. *Remote Sens.* **2016**, *131*, 459. [CrossRef]

38. Jay, S.; Guillaume, M.; Minghelli, A.; Deville, Y.; Chami, M.; Lafrance, B.; Serfaty, V. Hyperspectral remote sensing of shallow waters: Considering environmental noise and bottom intra-class variability for modeling and inversion of water reflectance. *Remote Sens. Environ.* **2017**, *200*, 352–367. [CrossRef]

39. Brando, V.E.; Dekker, A.G. Satellite hyperspectral remote sensing for estimating estuarine and coastal water quality. *IEEE Trans. Geosci. Remote Sens.* **2003**, *41*, 1378–1387. [CrossRef]

40. Wettle, M.; Brando, V.E.; Dekker, A.G. A methodology for retrieval of environmental noise equivalent spectra applied to four Hyperion scenes of the same tropical coral reef. *Remote Sens. Environ.* **2004**, *93*, 188–197. [CrossRef]

41. Brando, V.E.; Anstee, J.M.; Wettle, M.; Dekker, A.G.; Phinn, S.R.; Roelfsema, C. A physics based retrieval and quality assessment of bathymetry from suboptimal hyperspectral data. *Remote Sens. Environ.* **2009**, *113*, 755–770. [CrossRef]

42. Maritorena, S.; Siegel, D.A.; Peterson, A.R. Optimization of a semianalytical ocean color model for global-scale applications. *Appl. Opt.* **2002**, *41*, 2705–2714. [CrossRef] [PubMed]

43. Acker, J.G.; Leptoukh, G. Online analysis enhances use of NASA Earth Science Data. *Trans. Am. Geophys.* **2007**, *88*, 14–24. [CrossRef]

44. NCCOS. Benthic Habitat Mapping of Florida Coral Reef Ecosystems to Support Reef Conservation and Management. 2014. Available online: https://coastalscience.noaa.gov/project/benthic-habitat-mapping-florida-coral-reef-ecosystems/ (accessed on 22 March 2019).

45. Pope, R.M.; Fry, E.S. Absorption spectrum (380–700 nm) of pure water II Integrating cavity measurements. *Appl. Opt.* **1997**, *36*, 8710–8723. [CrossRef]

46. Roelfsema, C.; Phinn, S. Spectral Reflectance Library of Selected Biotic and Abiotic Coral Reef Features in Heron Reef, Bremerhaven, PANGAEA. 2006. Available online: https://epic.awi.de/id/eprint/31865/ (accessed on 1 August 2020).

47. Egbert, G.D.; Erofeeva, S.Y. Efficient inverse modeling of barotropic ocean tides. *J. Atmos. Oceanic Technol.* **2002**, *19*, 183–204. [CrossRef]

48. Dekker, A.G.; Phinn, S.R.; Anstee, J.; Bissett, P.; Brando, V.E.; Casey, B.; Fearns, P.; Hedley, J.; Klonowski, W.; Lee, Z.P.; et al. Intercomparison of shallow water bathymetry, hydro-optics, and benthos mapping techniques in Australian and Caribbean coastal environments. *Limnol. Oceanogr. Methods* **2011**, *9*, 396–425. [CrossRef]

49. Yates, K.Y.; Zawada, D.G.; Arsenault, S.R. Seafloor Elevation Change from 2016 to 2017 at Looe Key, Florida Keys—Impacts from Hurricane Irma: U.S. Geological Survey Data Release. 2019. Available online: https://catalog.data.gov/dataset/seafloor-elevation-change-from-2016-to-2017-at-looe-key-florida-keys-impacts-from-hurricane-irm (accessed on 22 July 2020).

50. Yates, K.Y.; Zawada, D.G.; Arsenault, S.R. Seafloor Elevation Change from 2016 to 2017 at Crocker Reef, Florida Keys—Impacts from Hurricane Irma. 2019. Available online: https://coastal.er.usgs.gov/data-release/doi-P9JI465S/ (accessed on 27 July 2020).

51. Harmel, T.; Chami, M.; Tormos, T.; Reynaud, N.; Danis, P.A. Sunglint correction of the Multi-Spectral Instrument (MSI)-SENTINEL-2 imagery over inland and sea waters from SWIR bands. *Remote Sens. Environ.* **2018**, *204*, 308–321. [CrossRef]

52. Overstreet, B.T.; Legleiter, C.J. Removing sun glint from optical remote sensing images of shallow rivers. *Earth Surf. Process. Landf.* **2017**, *42*, 318–333. [CrossRef]

53. Estrella, E.H.; Grotsch, P.; Gilerson, A.; Malinowski, M.; Ahmed, S. Blue band reflectance uncertainties in coastal waters and their impact on retrieval algorithms. In Proceedings of the SPIE 11420, Ocean Sensing and Monitoring XII, Vancouver, BC, Canada, 27 April–1 May 2020. [CrossRef]

54. Jégat, V.; Pe, S.; Freire, R.; Klemm, A.; Nyberg, J. Satellite-Derived Bathymetry: Performance and Production. In Proceedings of the Canadian Hydrographic Conference, Halifax, NS, Canada, 16–19 May 2016.

Article

Spatio-Temporal Assessment of Land Deformation as a Factor Contributing to Relative Sea Level Rise in Coastal Urban and Natural Protected Areas Using Multi-Source Earth Observation Data

Panagiotis Elias [1], George Benekos [2], Theodora Perrou [2,*] and Issaak Parcharidis [2]

[1] Institute for Astronomy, Astrophysics, Space Applications and Remote Sensing (IAASARS), National Observatory of Athens, GR-15236 Penteli, Greece; pelias@noa.gr
[2] Department of Geography, Harokopio University of Athens, GR-17676 Kallithea, Greece; benekos@hua.gr (G.B.); parchar@hua.gr (I.P.)
* Correspondence: dperrou@hua.gr

Received: 6 June 2020; Accepted: 13 July 2020; Published: 17 July 2020

Abstract: The rise in sea level is expected to considerably aggravate the impact of coastal hazards in the coming years. Low-lying coastal urban centers, populated deltas, and coastal protected areas are key societal hotspots of coastal vulnerability in terms of relative sea level change. Land deformation on a local scale can significantly affect estimations, so it is necessary to understand the rhythm and spatial distribution of potential land subsidence/uplift in coastal areas. The present study deals with the determination of the relative vertical rates of the land deformation and the sea-surface height by using multi-source Earth observation—synthetic aperture radar (SAR), global navigation satellite system (GNSS), tide gauge, and altimetry data. To this end, the multi-temporal SAR interferometry (MT-InSAR) technique was used in order to exploit the most recent Copernicus Sentinel-1 data. The products were set to a reference frame by using GNSS measurements and were combined with a re-analysis model assimilating satellite altimetry data, obtained by the Copernicus Marine Service. Additional GNSS and tide gauge observations have been used for validation purposes. The proposed methodological approach has been implemented in three pilot cases: the city of Alexandroupolis in the Evros Delta region, the coastal zone of Thermaic Gulf, and the coastal area of Killini, Araxos (Patras Gulf) in the northwestern Peloponnese, which are Greek coastal areas with special characteristics. The present research provides localized relative sea-level estimations for the three case studies. Their variation is high, ranging from values close to zero, i.e., from 5–10 cm and 30 cm in 50 years for urban areas to values of 50–60 cm in 50 years for rural areas, close to the coast. The results of this research work can contribute to the effective management of coastal areas in the framework of adaptation and mitigation strategies attributed to climate change. Scaling up the proposed methodology to a continental level is required in order to overcome the existing lack of proper assessment of the relevant hazard in Europe.

Keywords: land subsidence; multi-temporal SAR interferometry; sea-surface height; relative sea level change; satellite altimetry data; GNSS; coastal urban centers; natural protected areas; climate change impact

---

1. Introduction

Sea-level rise (SLR) is one of the most significant effects of climate change that has been the focus of international attention due to its high future projected rates that would impact coastal areas around the world [1]. Particularly, coastal low-lying regions will be affected, and their land will be decreased due to coastal erosion and inundation [1,2]. In the early 1990s, 33.5% of the global population lived within 100 vertical meters of sea level [3], and after a few years, it was estimated that about 23% of the world's population lived within 100 km of coast and up to 100 m above sea-level rise [4]. In a recent study, Vousdoukas et al. [5] indicated that extreme sea levels (ESLs) in the European region could probably rise up to 1 m or more by the end of this century, aggravating the impact of coastal hazards. Potential impacts such as the increase of the frequency of floods, acceleration of coastal erosion, salinization of surface and ground waters, and degradation of coastal habitats such as wetlands are expected to be significant in the future [6]. The impact depends not only on the intensity and extent of coastal changes but also on human response and adaptive capacity to various hazards. Recent research works confirm that the human factor has played a significant role, over the last century, in increasing vulnerability and exposure all around the world, indicating that, in the absence of adaptations, this trend will continue [7].

Coastal areas with very low altitudes, located in inhabited places or in places of highly social and economic activity, are particularly vulnerable to any relative sea level change [8]. In particular, for coastal urban centers, the risk is increased due to high exposure. Coastal protected areas and deltas are also of great interest due to the high conservation and protection of the special characteristics of the natural environment. As stated in the Intergovernmental panel on climate change (IPCC) Special Report [7] SLR will also affect natural protected areas and agriculture mainly through land submergence, soil and fresh groundwater resources, salinization, and land loss due to coastal erosion, with important consequences on production and food security.

Sea-level changes can be driven by either fluctuation in the masses or volume of the oceans or by changes in the land relative to the sea surface [9]. Any sea-level change that is observed with respect to a land reference frame is defined as a relative sea-level (RSL) change [10]. Thus, it is necessary to understand the rhythm and spatial distribution of potential surface deformation. Changes in sea level occur over a wide range of temporal and spatial scales that are not uniform [11], with its contributing factors strongly linked to climate change [12,13]. In many coastal low-lying and delta areas, land subsidence exceeds absolute sea-level rise up to a factor of 10. Without adopting any mitigation measures, parts of many coastal regions will sink below sea level [14]. Land subsidence that can be triggered by groundwater extraction due to population growth and urbanization [15,16] causes a total annual loss of billions of dollars.

Satellite altimetry and tide gauges provide different kinds of sea-level information [17]. As stated in Poitevin et al. [18] altimetry measures sea-surface height (SSH) attached to a well-defined geocentric reference frame, whereas tide gauges record sea-level heights with respect to the land upon which they are grounded [19]. Over the last two centuries, sea-level monitoring was performed mainly by tide gauges [19]. Since the beginning of the 1990s, relative observations are acquired by high-precision satellite altimetry [20]. More specifically, high resolution satellite altimetry was initiated with the launch of the TOPEX/Poseidon and Jason series of spacecraft in the early nineties [7,21]. Satellite altimetry has revolutionized sea-level measurements with its global coverage and a typical repeat cycle of a few days [17]. During the last 30 years, 11 satellite altimeters have been launched providing nearly global sea level measurements (up to ± 82° latitude) [7]. Both satellite radar altimetry and tide gauge techniques require supplemental information on vertical land motion [18] that can be achieved using vertical rate estimates from a permanent global GNSS receiver [22,23].

During recent times, SAR satellite constellations have been used for several coastal zone studies, demonstrating it as a powerful monitoring tool, in terms of spatial and temporal capabilities [24–28]. Apart from mapping the coastlines, SAR data can be exploited to map the land deformation rates, using the methodology of SAR interferometry (InSAR) that has been used in the present study. This methodology provides high spatial and temporal resolution ground displacement measurements due to geophysical processes or even man-made activities. Advanced techniques such MT-InSAR, enable the simultaneous processing of multiple SAR acquisitions in time, increasing the accuracy and the spatial coverage of reliable scatterers on ground. Currently, there are two broad categories of algorithms for processing multiple acquisitions in time, the persistent scatterers interferometry (PSI) and small baseline subsets (SBAS) methods, which are optimized for different models of scattering [29–31].

In this framework, MT-InSAR approaches were extensively and successfully exploited to investigate coastal ground deformation [32–34]. These techniques, such as permanent scatterers (PSs) [35], SBAS [36], and interferometric point target analysis (IPTA) [37] use a medium-to-large dataset of SAR images acquired at different times over the observed area to follow the temporal evolution of a deformation phenomenon, retrieving the mean deformation rate and the time series for each point target. Moreover, these techniques can benefit from the availability of free and open access SAR data archives provided by European and international space agencies.

Although several studies have been conducted for studying land deformation using MT-InSAR [38–40], PSI TERRAFIRMA products [41], and GNSS [42–45] in Greece, only a few have been targeting the coastal hazard [46–48]. Moreover, none of them have used both the aforementioned space geodesy methods in combination with altimetry measurements to define the relative sea-level rise. In this respect, the aim of this study is to improve the understanding of and quantify the coastal relative sea level rise in three coastal regions of Greece by using a combination of multi-source Earth observation—SAR, GNSS, tide gauge, and altimetry data. To that end, MT-InSAR technique will be used by taking advantage of the most recent Copernicus Sentinel-1 data, calibrated by GNSS data and combined with a reanalysis model assimilating satellite altimetry data. Additional GNSS and tide gauge observations have been used for validation purposes. The proposed methodological approach has been implemented in three pilot areas with special characteristics, contributing to a better understanding of the coastal hazards, constraining ground deformation phenomena occurring along urban and natural protected coastal areas.

2. Study Areas

This paper deals with the spatio-temporal assessment of land deformation, as a factor contributing to relative sea level rise, in three pilot cases with different combinations of characteristics (Figure 1). The first pilot case is the city of Alexandroupolis extending up to the Evros Delta region (A), which is a combination of coastal urban area (in the West side) and the unique characteristics of the trans-boundary Evros Delta (on the east side) in terms of the importance of the wetlands. The second pilot case is the coastal zone of the Thermaic Gulf (B) which is a combination of a big urban center (Thessaloniki city on the east side) with its port and airport infrastructures and the unique formation of Delta Axios (in the west side) and the Axios-Loudias-Aliakmonas National Park. The third one is the coastal area of Killini (west side)—Araxos (northeast side) (C) in which there are two important infrastructures (port and airport) and the special Lagoon of Kalogria area (northeast side).

2.1. Coastal Area of Alexandroupolis City—Evros Delta Region

The coastal area from the Alexandroupolis city extending up to the Evros Delta (Gulf of Alexandroupolis) covers an area of about 350 km^2 [49]. Alexandroupolis, the capital of the prefecture of Evros, is the largest city of Thrace and the region of Eastern Macedonia in terms of size and population. This very fast-growing city has an important port and it is a summer resort centre in SE Balkans due to the great location at the center of sea and land routes connecting Greece with Turkey [50]. The coastal zone area under investigation extends to more than 50 km and can be distinguished in the western unity and the eastern unity in terms of geomorphology. The western unity from the Mesimvria coast up to Alexandroupolis is characterized as hilly to mountainous, whereas the eastern one covering the coastal zone from Alexandroupolis to the Evros river mouth is a plain area relief [50].

The Evros Delta region, shared by Greece and Turkey, has been characterized as one of the most important wetlands on a national and European level [51]. A major part of the delta on the Greek side has been characterized as a wetland in special protection area (SPA), as a site of community importance (SCI) in the Natura 2000 Network, and as internationally important under the Ramsar Convention (1971), due to its significant and rare species of plants, fauna and birds [52]. The Turkish part of the delta is also included in the list of wetlands of international importance, while Lake Gala, in close proximity to the delta, has been declared a National Park area [51]. The delta is formed by the Transboundary Evros River alluvial deposits, filled rapidly over millennia and affected by the interaction with the sea [53].

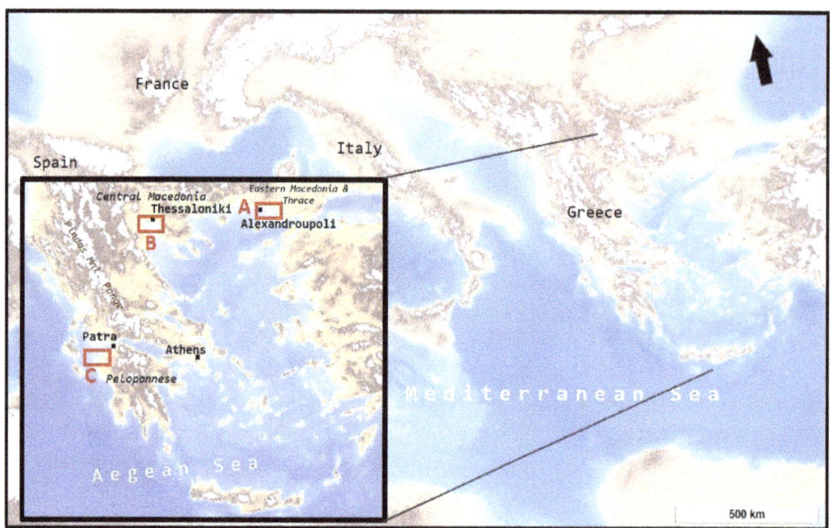

Figure 1. Location map of the three study areas. (**A**) Coastal urban area of the Alexandroupolis city—unique wetland area of the Evros Delta. (**B**) Coastal Zone of Thermaic Gulf; urban center of Thessaloniki city—Axios Delta. (**C**) Coastal Area of Killini—Araxos (Patras Gulf), northwestern Peloponnese.

2.2. Coastal Zone of Thermaic Gulf, City of Thessaloniki—Axios Delta

The City of Thessaloniki, located in Central Macedonia and situated in the inner part of the Thermaic Gulf, is established as the second most important urban center of Greece. It has an extended industrial zone in its suburbs and an international port that constitutes the major center of merchant shipping for the surrounding Balkan countries [54]. The catchment area of the Thermaic Gulf, located in the southern Balkan Peninsula, is approximately 40,000 km^2, and the main rivers are Axios, Aliakmon, Loudias, and Gallikos [55,56]. The Plain of Thessaloniki is formed by the sub-aerial deltaic plains of these rivers, together with River Loudias and the artificially drained lake of Giannitsa [55]. The wetlands of the Axios Delta at the Thessaloniki plain provide a typical example of wetland destruction in Greece. In 1917, 36% of the plain was wetland, but this area now amounts only to 5.5% [56]. This is also confirmed by Psimoulis et al. [57], who stated that the areas of Kalochori and Sindos used to be a delta some thousand years ago. The main environmental pressures such as water discharge decrease, drainage works, urbanization, and pollution negatively affected the ecological character of the deltaic area, leading to the destruction of 70% of the original wetlands during the 20th century [56]. Moreover, the decrease in rainfall in combination with the overuse of water for irrigation, has resulted in severe salinization of the delta area, which has impacted on the flora and fauna of the wetlands [58]. Currently, some of these activities have ceased, and their impacts have already been mitigated [56].

2.3. Coastal Area of Killini—Araxos (Patras Gulf), NW Pelloponese

The coastal area from the port of Killini to the Araxos airport is part of the Gulf of Patras and is located in the northwestern Peloponnese. The Peninsula of Killini, located at the westernmost end of Peloponnese, contains an isolated hilly area of about 130 km^2. It is formed by a morphological rise (Kastron at a highest point of 244 m) connected to mainland Peloponnese in the east by the plain of the Pineios River (Elis plain) [59]. Araxos is a village in the northwestern part of Achaea that is located in the coastal plains near Cape Araxos, which separates the Gulf of Patras from the Ionian Sea. From Araxos to Killini, there are three lagoons—Prokopos, Kalogria, and Kotychi. Part of the area was recognized as a Wetland of International Importance in 1975, when it was included in the 10 wetlands of Greece protected under the Ramsar Convention. A few decades later, parts of the area were recognized as SPA for bird species, in the framework of Directive 2009/147/EE, as well as SCI in line with Directive 92/43/EEC, which led to the formation of the European NATURA 2000 Network of protected areas [60]. Most of the coastal area lies on sand dune formations, while small areas near Kounoupeli, in the eastern area of the Kalogria lagoon and the Mavra Vouna hill, are all composed of hard limestone. The eastern part of the area behind the dunes is covered with clay deposits with depths of a few centimeters to more than 2 m. The seashore consists of unconnected single-grained medium and fine-sized sand with a very small amount of silt [61].

3. Materials and Methods

In this section, the data and methods that were used for this study are presented. More specifically, data specifications are given for all the satellite SAR, altimetry, GNSS, and tide gauge measurements. The mature technology of the MT-InSAR was used for this research work to provide land deformation rates which were set to a reference frame with the use of a GNSS data. The deformation rates were combined with the altimetric SSH rate to calculate the relative sea level change. Additional GNSS and tide gauge observations were used for validation purposes. The workflow of the proposed methodological approach is illustrated in the flowchart of Figure 2.

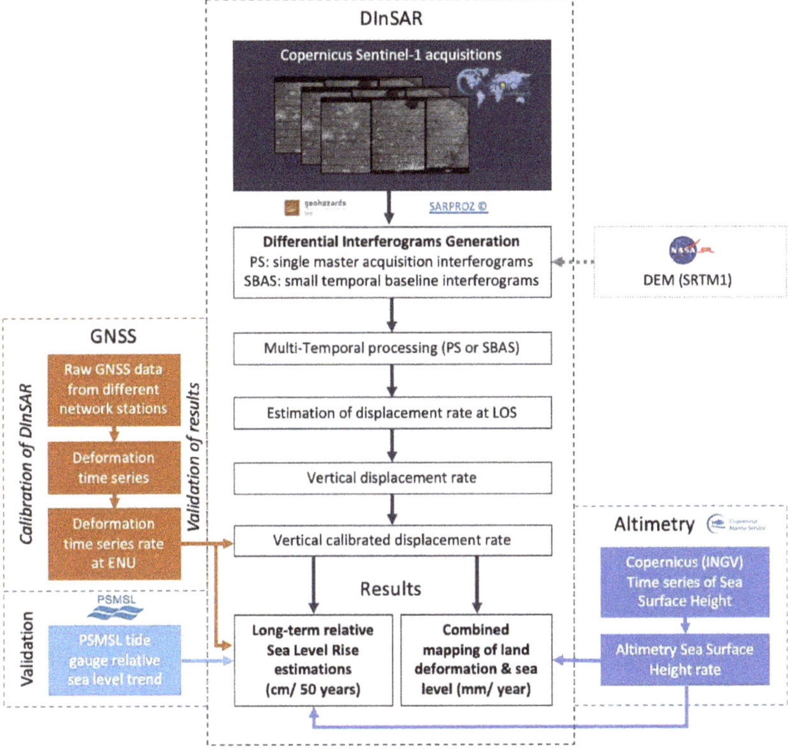

Figure 2. Workflow of the methodological approach. The boxes in the middle represent the intermediate and final products while the peripheral boxes indicate the data used. GNSS stands for global navigation satellite system; DInSAR for differential synthetic aperture radar interferometry; ENU for east-north-up; PSMSL for permanent service for mean sea level; PS for permanent scatterers; SBAS for small baselines subsets; LOS for line of sight; DEM for digital elevation model; INGV for Istituto Nazionale di Geofisica e Vulcanologia.

3.1. Data

This study, apart from coastal urban centers, focuses on natural protected areas. For this reason, the appropriate selection of sensors is important in order to achieve high spatial coverage of surface deformation information. In this case, high temporal resolution is a crucial parameter. Copernicus satellites Sentinel-1 A and B show high spatial and temporal resolution operating at C band. Today, the latest and most advanced SAR mission is the Sentinel-1 constellation. The Sentinel-1A, the first of the twin satellites, was launched in April 2014 by ESA. The Interferometric Wide (IW) swath acquisition mode, has a swath of 250 km and spatial resolution of 5 m in range (dimension perpendicular to the satellite track) and 20 m in azimuth (dimension along the satellite track), with no multilook and a repeat cycle of 12 days. In April 2016, the Sentinel-1B was also successfully placed into orbit, decreasing the temporal resolution for the constellation to six days [62].

In the TOPSAR operation mode (the most common of Sentinel-1), in addition to steering the beam in range as in ScanSAR (the most common mode of ESA's ASAR/ENVISAT and ERS missions), the beam is also electronically steered from backward to forward in the azimuth direction for each burst, resulting in an homogeneous image quality throughout the swath [62]. With this mode, the same coverage and resolution as in ScanSAR is achieved, but with an improved signal-to-noise ratio (SNR). Sentinel acquisitions are available systematically and free of charge through the Copernicus Open Access Hub (https://scihub.copernicus.eu/dhus/#/home).

For the case of Alexandroupolis, Evros Delta, a set of 91 ascending Sentinel-1B, IW, Level-1, single look complex (SLC) acquisitions between 27 September 2016 and 30 October 2019 (Table 1) was used. For the case of the wide area of the Thermaic Gulf, Thessaloniki, Axios Delta, a set of 130 ascending Sentinel-1 A and B, IW, Level-1 SLC acquisitions was collected. The set of Sentinel-1 images was acquired along ascending passes covering the period 12 October 2014 until 24 June 2019 with 70 Sentinel-1A images and 60 Sentinel-1B images (Table 1). Finally, for the case of the coastal area of Killini, Araxos in the northwestern Peloponnese, a set of 98 ascending Sentinel-1A and 60 Sentinel-1 B, Level-1, SLC, IW acquisitions was used. The set of 158 images spans the period from 11 November 2015 to 23 July 2019 (Table 1).

Table 1. Characteristics of the Sentinel-1 synthetic aperture radar acquisitions used in the present study.

Study Area	No Sentinel Images	S1A	S1B	Relative Orbit	First Image	Last Image
Alexandroupolis, Evros Delta	91		91	131	27 September 2016	30 October 2019
Thermaic Gulf; Thessaloniki, Axios Delta	130	70	60	102	12 October 2014	24 June 2019
Kyllini-Araxos, northwestern Pelloponese	158	98	60	175	11 November 2015	23 July 2019

For this research work, the monthly mean "sea-surface height above sea level" of "MEDSEA_REANALYSIS_PHYS_006_004" product at 0.042 degree spatial resolution from Copernicus Marine Environment Monitoring Services (http://marine.copernicus.eu/) was used. This product is the result of a physical re-analysis component from the Nucleus for European Modelling of the Ocean (NEMO) assimilating satellite data. The assimilated dataset includes mono altimeter satellite along-track SSH computed with respect to a seven-year mean [63].

Additionally, 30-sec GNSS data was used to anchor the MT-InSAR vertical rate with the GNSS station one. Three GNSS stations (one for each case), i.e., ALEX (URANUS network) for Alexandroupolis, Evros Delta; AUT1 (AUTH) for Thermaic Gulf; and RLSO (National Observatory of Athens—NOA network, http://www.gein.noa.gr/services/GPS/NOA_GPS/noa_gps_files/rlso.html) for Killini, Araxos were used. The data for the last two GNSS stations was downloaded from the NOA GNSS Network website (http://www.gein.noa.gr/gps.html). Additionally, for validation purposes, three more GNSS solutions were used for the GNSS station THS1 (http://geodesy.unr.edu/NGLStationPages/stations/THS1.sta), located at the Aristotle University of Thessaloniki; AUT1 (http://geodesy.unr.edu/NGLStationPages/stations/AUT1.sta) located south-east of Thessaloniki; and ALE3 (http://geodesy.unr.edu/NGLStationPages/stations/ALE3.sta) located near Alexandroupolis. The network of these three stations was calculated by the Nevada Geodetic Laboratory [64].

Finally, two tide gauges from the Permanent Service for Mean Sea Level (PSMSL) [65], with measurements since 1969, were used also for validation purposes. One is located in Alexandroupolis (https://tidesandcurrents.noaa.gov/sltrends/sltrends_station.shtml?id=290-065), and the other is located in Thessaloniki (https://tidesandcurrents.noaa.gov/sltrends/sltrends_station.shtml?id=290-051). The tide gauge of NOA network named NOA-12 that is also located in Thessaloniki (https://webcritech.jrc.ec.europa.eu/TAD_server/Device/175) was not used in this research as its time series is limited to two years (since September 2018).

3.2. Methodology

The aim of the persistent scatterers technique is to overcome interferometric coherence degradation over time using a set of phase-stable pixels (the persistent scatterers). This technique is performing better in urban areas, where temporal decorrelation is minimized due to the high density of stable structures acting as scatterers (buildings, bridges, etc.). The flowchart of the methodology is illustrated in Figure 2.

MT-InSAR deformation rates from Sentinel-1 data alone suffer from lack of anchor or reference point. A reference PS could be selected in an area located on limestone bedrock with high coherence, being away from deltaic and fluvial deposits or recent seabed outcropped deposits and generally from brittle material. All three GNSS solutions were calculated from NOA GNSS processing station (NGProS) (http://aips.space.noa.gr) using version 6.3 of GIPSY/OASIS II software (http://gipsy-oasis.jpl.nasa.gov), developed by the Jet Propulsion Laboratory of the National Aeronautics and Space Administration (NASA), Pasadena, California, CA, USA [66]. The orbits of the GNSS satellites used were the precise ones in the reference frame system of ITRF2014. The ocean load model that was used is the Goddard/Grenoble Ocean Tide (GOT 4.3) empirical model maintained by NASA-Goddard Space Flight Center. The ocean load values for the three GNSS was calculated from the "free ocean tide loading provider" (http://holt.oso.chalmers.se/loading/). The vertical rate for each GNSS station was calculated using the least squares method (Table 2). The daily time series of the three GNSS stations were calculated and the vertical linear rate values were estimated as shown in Table 2 and Figure S1.

Table 2. Global navigation satellite system stations used, the date ranges, and the estimated vertical linear rates.

Name	Network	φ (WGS84°)	λ (WGS84°)	Vertical Linear Rate (mm year^{-1})	Time Period	Number of Epochs
RLSO	NOA	38.05583	21.46474	−2.1	March 2009–February 2017	2819
AUT1	AUTH (EUREF)	40.56681	23.00371	−1.1	November 2007–July 2018	2245
ALEX	URANUS	40.84916	25.85344	0.8	October 2017–April 2020	783

In the case of both Alexandroupolis—Evros Delta and Kyllini–Araxos, the Parallel SBAS (P-SBAS) [67,68] service under the Geohazards Exploitation Platform (GEP) of ESA was exploited. GEP is a cloud platform that provides a rich set of ready to use EO Data processing services for geohazards analysis and monitoring (https://geohazards-tep.eu/). A threshold of temporal coherence of 0.8 was used. As reference points, the GNSS stations ALEX and RLSO were used accordingly. The P-SBAS service is a parallel version of SBAS algorithm. It was selected for these two cases due to its higher effectiveness in rural areas in comparison with PSI which is more powerful in urban areas such as the Thermaic Gulf surrounded by urban centers.

In the case of the Thermaic Gulf, SAR PROcessing tool by periZ software (SARPROZ, Razer Limited, Hongkong, China) [69] was used as the MT-InSAR processor, extending the standard linear PS technique to estimate non-linear motion having no prior information. Shuttle Radar Topography Mission (SRTM) 1 arc-sec DEM [70] was used for the PSI processing. A master date was chosen for this period based on the minimization of the perpendicular baseline, the temporal baseline and the weather data. After selecting a master image (acquired on 24 March 2017), the proposed algorithm was applied only on pixels showing an inverse amplitude stability (i.e., the ratio between the mean calibrated image amplitudes and its standard deviation) higher than 0.7. The GNSS station AUT1 was chosen as a reference point (Table 2). Once the data was unwrapped, the low pass component was estimated using a three-sample temporal base-line weighted moving average and assumed this residual phase term is an estimation of the non-linear motion contribution. These steps were performed considering only neighboring pixels. The same processing steps were then applied, considering only differences between the reference point and all persistent scatterers including the estimated atmospheric phase screens (APS). Finally, for each PS the linear rate was calculated by using least squares approximation.

All three MT-InSAR datasets are originally in slant range geometry, i.e., along the LOS direction. We assumed that the deformation is totally vertical, a hypothesis that is not valid if an earthquake or slow aseismic slip of tectonic origin or landslide occurs. For the study period and the three study areas according to global Centroid Moment Catalog (https://www.globalcmt.org/) there were no earthquakes greater than Mw = 5.0 and shallower than 15 km within a distance of 40 km. Nevertheless, the earthquake is a sudden event and this kind of deformation has little impact or is filtered out by the MT-InSAR process. The aseismic processes take place in large fault zones (e.g., North Anatolian fault as well as its continuation in the Aegean Sea) or smaller ones (e.g., Gulf of Corinth) and the vast majority of the deformation is concentrated within a narrow zone along the feature. The landslides affect the deformation field only very locally. For the Alexandroupolis city—Evros Delta region the east branch of the north Anatolian fault (a right lateral structure) sits almost 15 km south of the southern coast of Evros Delta. Potentially, spatially horizontal (east–west oriented) deformation due to the continuous deformation of this fault zone is considered negligible due to the long distance involved and the limited spatial coverage of each case. Currently, the distribution of the available GNSS stations, with adequate temporal baselines in the three areas, is not dense enough to differentiate for horizontal deformation velocities. Accordingly, considering the vertical component of the Line of Sight vector for each PS, the MT-InSAR rates were transformed to the vertical ones. In order to anchor them with the GNSS ones the following procedure was used. An average value was calculated, at each of the three GNSS locations, for the PSs (of the corresponding MT-InSAR dataset) closer than 300 m from each location (except for AUT1 that is located in a semi-urban area, where the distance is 600 m). The offset of this value to the GNSS one was subtracted from each one of the three MT-InSAR datasets. Thus finally, all the MT-InSAR products were transformed to vertical direction and anchored to a single reference frame. These products are labelled as "calibrated" in Figure 2. From the SSH, using least squares, we calculated the linear SSH rates as in, e.g., Figure 4. Further, the relative sea-level rise rate was calculated simply by subtracting the vertical deformation rate from the sea-surface height rate. Finally, the additional units of cm per 50 years were used to facilitate the discussion on a basis of longer time periods. For validation purposes, we used the vertical component of GNSS solutions from a Nevada institute as well as tide gauges from PSMSL where they were available. Results of localized deformation rates and thus relative sea-level rise are following in the next section. The GNSS uncertainties are used along with the standard deviation values of the spatially distributed MT-InSAR deformation rates in order to provide a level of uncertainty of each detected localized deformation.

4. Results

In this section, we present the localized deformations detected by MT-InSAR processing and sea-surface height, following the methodological approach detailed in the previous section (Figure 2) in the three Greek coastal areas under investigation effectively covered by Sentinel-1 SAR acquisitions.

The absolute sea level trend across Greece for the period 1993–2019 ranges between 1 mm/year and 3 mm/year and, in some specific areas such as in the southern part of the Peloponnese and in the northeastern part of Crete, to more than 3 mm/year.

4.1. Alexandroupolis, Evros Delta

In the case of the Alexandroupolis, Evros Delta region, the velocity of deformation has been derived using 91 Sentinel-1 acquisitions (Table 1) and the persistent scatterers technique over an area of 212 km^2. The urban areas are adequately covered with natural scatterers, while the lack of them in the vegetated and wetland areas present low coherence (Figure 3). For the period 2016–2019, the characteristic vertical deformation ranges between −1.5 mm/year and 5 mm/year. The city of Alexandroupolis shows relatively stable patterns with a mean rate of 0.4 mm/year, while the Turkish part of the Evros Delta seems to uplift with high rates. In order to investigate this further, the MT-InSAR time series of 199 PSs vertical deformation rate lying in the coast of the Delta (bounded by red dashed polygon in Figure 3), along with its mean values and the linear trend, are plotted in Figure 4. We noted that the seasonal trend is highly visible and that the loading periods of the acquirer (before summer) have a higher gradient than the unloading. Moreover, during the maximum acquirer unloading, the vertical deformation is not reverted to its previous situation but is higher. The mean deformation rate of the 199 PS points is 4.5 mm/year. It is obvious that, for the study period, the rate of the loading/drainage is positive; thus, the superficial land over the acquirer is uplifting. In order to have more concrete results, a wider time range covering drought periods is necessary.

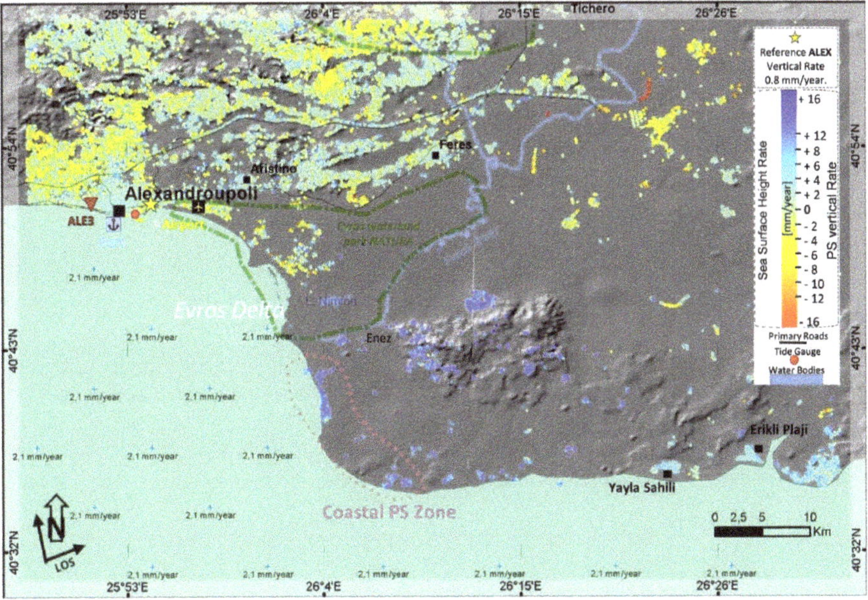

Figure 3. Vertical land deformation rate and sea-surface height rate map of Alexandroupolis, Evros Delta, both represented with the same color scale. Coastal permanent scatterers (PSs) are bounded with a red dashed line. The Global Navigation Satellite System station ALE3 and tide gauge, respectively, used in validation are depicted with reversed red triangles and red circles.

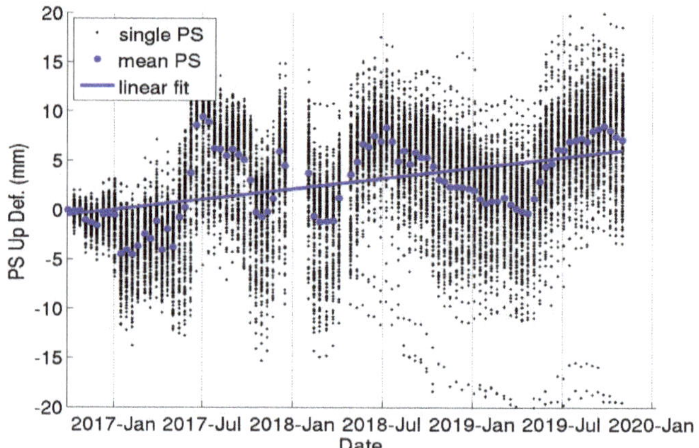

Figure 4. Permanent scatterers' time series of the coastal area (bounded by a red dashed polygon in Figure 3), mean vertical deformation values and the linear trend.

The SSH rate for the coastal area of Alexandroupolis, Evros Delta is ~2.1 mm/year (Figure 3). For this case, we considered six sub-areas—two are bounded by polygons, and four are labelled by their place names. For each one, in Table 3, the statistics of the PSs (i.e., number, mean, minimum, maximum, standard deviation, with confidence level of 85%) were calculated. Additionally, the ALEX GNSS station (set as reference) uncertainty of 1.0 mm/year was added to the standard deviation values. The highest sea level rise rates are located at the airport with a value of 18.5 ± 18 cm in 50 years and at WaterLand Park (green dashed polygon in Figure 3) with 16.5 ± 21 cm in 50 years. Alexandroupolis presents a uniform behavior, with 8.5 ± 11 cm in 50 years. Finally, the areas south of the Delta (i.e., Yayla Sahili and Erickli Plaji) present low values of sea level rise (1.5 ± 10 cm in 50 years).

According to the Nevada Geodetic Laboratory, the vertical deformation rate of the GNSS station, located at the west end of the city center ALE3 (Figure 3), near the ring road, has a value of 1.02 ± 8.74 mm/year. Uncertainty is very high due to the missing data of almost four years, between 2016 and 2020. The mean vertical deformation value of the PSs with a distance less than 200 m from the ALE3 location is 0.4 mm/year, a value close to the GNSS one.

The PSMSL tide gauge (Table 3) located in the harbor (red circle in Figure 3) measures a relative sea level trend of 1.84 ± 0.67 mm/year (with 95% confidence) based on the monthly mean sea level data between 1969 and 2017. The mean vertical deformation rate of selected PSs located in the same deck as the tide gauge is −0.1 mm/year, which corresponds to a relative sea level rate of 2.2 mm/year (or 11 cm in 50 years), a value close to the tide gauge measurement (9 cm in 50 years).

Table 3. Statistics of the sub-areas of Alexandroupolis, Evros Delta. Amongst others, the characteristic (mean) values of vertical deformation rate, the relative sea-level rise, and the uncertainty are shown. The sub-area Coastal Delta zone presents negative sea-level rise, thus sea-level falling. PS stands for permanent scatterers; GNSS for Global Navigation Satellite System; RSL for relative sea level.

Location	Nr. of PSs	Mean (characteristic at 85% Confidence) (mm/year)	Min (mm/year) (85% Confidence)	Max (mm/year) (85% Confidence)	Std (mm/year) (85% Confidence)	Std Counting Also GNSS (±1.0) (mm/year)	RSL Rise (cm in 50 Years)
Alexandroupolis Tide gauge	198	0.4	−5.0	10.0	1.2	2.2	8.5 ± 11 9.2 ± 3
Airport	183	−1.6	−9.0	5.0	2.6	3.6	18.5 ± 18
Waterland Park	630	−1.2	−9.0	9.5	3.1	4.1	16.5 ± 21
Coastal Delta zone	199	5.3	−5.5	8.5	1.8	2.8	−16 ± 14
Yayla Sahili	177	1.8	−1.5	4.5	0.8	1.8	1.5 ± 9
Erikli Plaji	184	1.9	−0.5	4.5	0.9	1.9	1 ± 10

4.2. Thermaic Gulf; Thessaloniki City, Axios Delta

In the case of the Thermaic Gulf, the velocity of deformation was derived using 130 Sentinel-1 images (Table 1) and MT-InSAR technique over an area of 142 km². The land deformation map (Figure 5) shows a very high-density concentration of PSs in the area of Thessaloniki and in the surrounding urban centers. Outside, as in the agricultural fields and the mountains, there is a low density of PSs. A factor indicating the limits of this technique [71]. According to the Nevada Geodetic Laboratory, the vertical deformation rate of the GNSS THS1 station, located in the Aristotle University of Thessaloniki (Figure 5), has a value of 0.2 ± 0.58 mm/year. The mean vertical deformation value of the PSs with a distance less than 200 m from the THS1 location is −0.4 mm/year, a value close to the GNSS value.

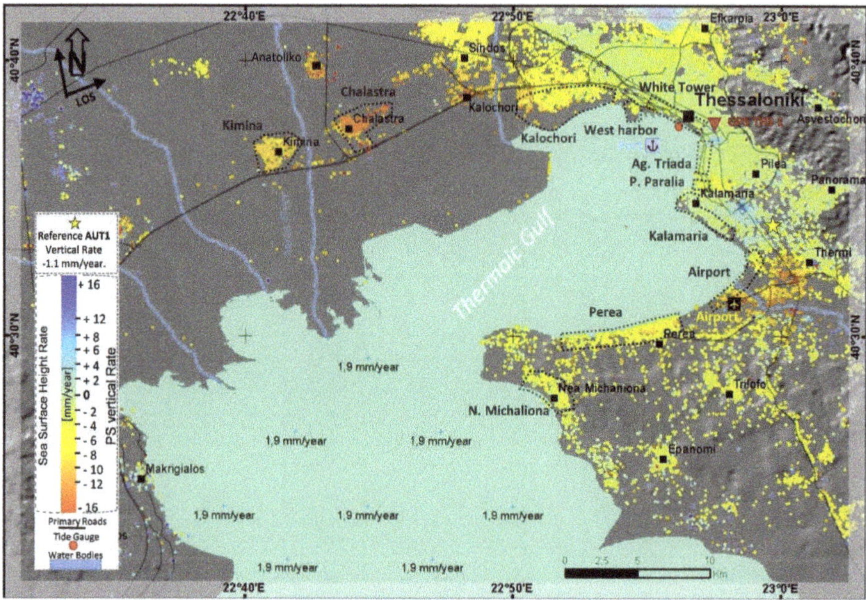

Figure 5. Vertical land deformation rate and sea-surface height rate map of Thermaic Gulf coastal zone; City of Thessaloniki, Axios Delta. The studied sub-areas are bounded with black dashed lines. The Global Navigation Satellite System THS1 station and tide gauge, respectively, used in validation are depicted with a reversed red triangle and a red circle. PS stands for permanent scatterers.

The PSMSL tide gauge located in the harbor (red circle in Figure 5) measures a relative sea level trend of 3.83 ± 0.66 mm/year (with a 95% confidence) based on a monthly mean sea level data from 1969 to 2017. The mean vertical deformation rate of selected PSs located in the same deck as the tide gauge is −2.4 mm/year which corresponds to a relative sea level rate of 4.3 mm/year (or 21 cm in 50 years), a value very close to the tide gauge measurement.

Raucoules et al. [54] using permanent scatterers and stacking interferometry and exploiting 47 ERS acquisitions for the period 1992–2000 produced a deformation map in the slant range (LOS) direction, including a large part of the map of this study. There is an overall consistency with this study except that the areas of maximum subsidence of 1992–2000, i.e., the airport and Kalochori areas. currently present smaller subsidence. For the former area (in the period of 1992–2000), the subsidence was 10–20mm/year in LOS, and currently, it is 4 mm/year in the vertical direction. The latter area (between 1992 and 2000) was characterized by a subsidence of 50 mm/year, and currently, the maximum values reach 10 mm/year in the vertical direction.

Constantini et al. [72], using combined PSI and SBAS approaches and exploiting a number of 20 ASAR/ENVISAT acquisitions for the period of 2004–2010, produced a deformation map in the slant range (LOS) direction, including a large part of the map of this study. There is also an overall consistency with this study. The area of Kalochori (for the period of 2004–2010) presented a subsidence rate of 15 mm/year in the LOS that is also larger than the current rate, estimated at 4 mm/year in vertical direction.

The main reason for the subsidence in the area of Kalochori to Kimina is the over-exploitation of the underground water. It is obvious that the subsiding of these areas has been progressively slowed down since 1992. The most likely explanation is that regulations for water drilling were successfully applied. For the detailed study, the whole area was divided into 11 sub-areas, as shown in Figure 5. These sub-areas were selected as separate locations with a similar deformation rate near the coast or in low altitudes (in case of PSs absence in the coastal zone). For each sub-area, in Table 4, the statistics of the PSs (i.e., number, mean, minimum, maximum, and standard deviation) were calculated. Additionally, the GNSS uncertainty of 0.5 mm/year was added to the standard deviation values. Finally, considering the sea-surface height rate, the relative sea level rise was calculated, together with their uncertainty values. We observe that the maximum values are located at the villages of Chalastra and Kimina (both at an altitude of ~5 m) with a relative sea level rate of 67 ± 23 and 39 ± 14 cm in 50 years. Kalochori. Airport and Perea sub-areas present a rate of 31 ± 17 cm in 50 years. All the others have a rate less than 30 ± 9 cm in 50 years. The lowest values are located in Kalamaria. being 17 ± 7 cm in 50 years.

4.3. Killini, Araxos

For the case of Killini, Araxos, in the northwestern Peloponnese, the velocity of deformation (Figure 6) was derived using 158 Sentinel-1 images (Table 1) and the SBAS technique over an area of 129 km^2. For the time period of 2015–2019, the mean vertical deformation rate for the coastal areas ranges between −3.6 mm/year and 3.8 mm/year. The areas between P. Kalamakiou and Brinia are subsiding, and the rest are uplifting.

The area was divided into five sub-areas. For each one, in Table 5, the statistics of the PSs (i.e., number, mean, minimum, maximum, standard deviation, with a confidence level of 85%) were calculated. Additionally, the GNSS uncertainty of 1.1 mm/year was added to the standard deviation values. Finally, considering the sea-surface height rate, the relative sea level rise was calculated, with their uncertainty values. Specifically, in sub-areas from K. Achaea to P. Kalamakiou and Ioniko, there is a small relative sea level rise. The sub-areas presenting the highest sea level rise rates are P. Kalamakiou to L. Kalogria and Strofilia Forest to Briana, with values of 20 ± 14 cm and 28 ± 21 cm in 50 years. Finally, at Killin, the relative sea level rise is 6 ± 14 cm in 50 years.

Table 4. Statistics of the sub-areas of the Thermaic Gulf; Thessaloniki, Axios Delta. Amongst others, the characteristic (mean) values of vertical deformation rate, the relative sea level rise, and the uncertainty are shown. PS stands for permanent scatterers; GNSS for Global Navigation Satellite System; RSL for relative sea level.

Location	Number of PSs	Mean (Characteristic at 85% Confidence) (mm/year)	Min (mm/year) (85% Confidence)	Max (mm/year) (85% Confidence)	Std (mm/year) (85% Confidence)	Std Counting Also GNSS ($\pm$1.0) (mm/year)	RSL Rise (cm in 50 Years)
White Tower	803	−2.1	−12.7	1.1	0.8	1.3	20 ± 7
Tige gauge							19 ± 3
Kalamaria	422	−0.9	−15.4	1.6	1.0	1.5	14 ± 7
Airport	104	−4.3	−15.1	−0.7	3.0	3.5	31 ± 17
Perea	627	−4.3	−11.4	6.2	1.5	2.0	31 ± 10
N. Michaliona	270	−2.9	−11.6	1.1	1.0	1.5	24 ± 8
Ag. Triada	289	−1.6	−10.5	0.9	0.8	1.3	18 ± 6
Kalochori	1081	−4.2	−19.8	14.4	2.2	2.7	30 ± 14
Chalastra	388	−11.4	−23.8	19.0	4.2	4.7	67 ± 23
Kimina	372	−6.0	−17.7	6.2	2.3	2.8	39 ± 14
P. Paralia	80	−2.6	−4.6	−1.3	0.6	1.1	22 ± 5
West harbor	45	−3.4	−7.8	2.6	1.4	1.9	16 ± 9

Table 5. Statistics of the sub-areas of the Killini, Araxos. Amongst others the characteristic (mean) values of vertical deformation rate, the relative sea level rise and the uncertainty are shown. PS stands for permanent scatterers; GNSS for Global Navigation Satellite System; RSL for relative sea level.

Location	Number of PSs	Mean (Characteristic at 85% Confidence) (mm/year)	Mmin (mm/year) (85% Confidence)	Max (mm/year) (85% Confidence)	Std (mm/year) (85% Confidence)	Std Counting Also GNSS ($\pm$1.0) (mm/year)	RSL Rise (cm in 50 Years)
K. Achaea - P Kalamakiou	508	2.7	−4.0	5.0	2.1	3.1	−3.5 ± 16
P. Kalamakiou, L. Kalogria	310	−1.9	−6.0	4.0	1.8	2.8	19.5 ± 14
Strofilia Forest, Briania	129	−3.6	−8.5	3.0	3.2	4.2	28 ± 21
Killini	152	0.9	−6.5	7.0	1.8	2.8	6 ± 14
Ioniko	105	3.8	−6.0	8.5	2.2	3.7	−8.5 ± 19

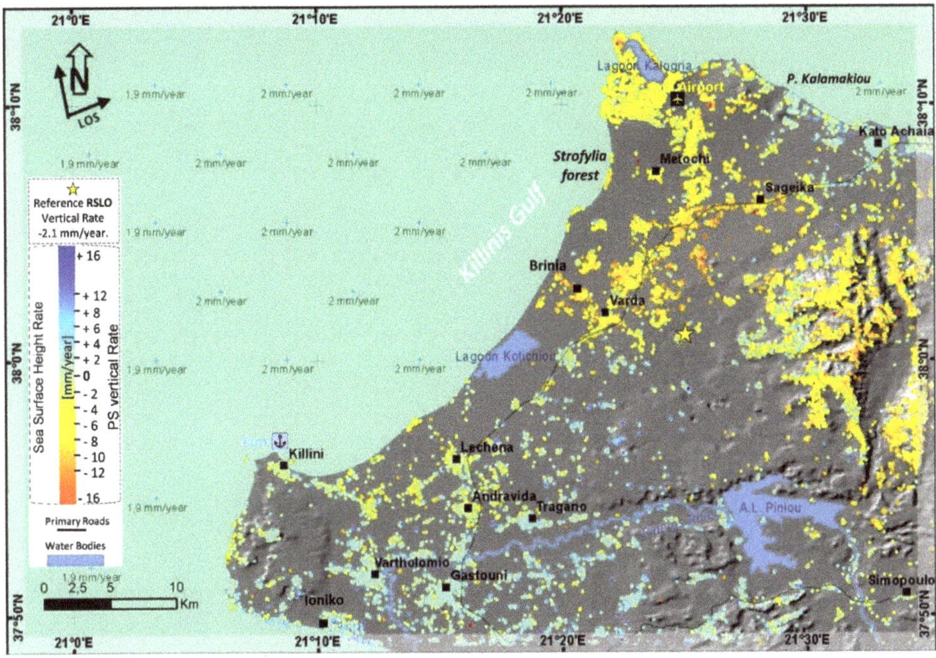

Figure 6. Vertical land deformation rate and sea-surface height rate map of the Killini, Araxos coastal area in the northwestern Peloponnese. PS stands for permanent scatterers.

5. Discussion

In the present work, the relative sea level rise rate was estimated for several coastal areas within the three selected case studies. The MT-InSAR vertical deformation rate data (exploiting Copernicus Sentinel-1 acquisitions) is referenced to the vertical component of GNSS measurements (calculated in ITRF2014 reference frame system) in order for their rate to be "calibrated". Additionally, GNSS measurements not calculated by this study were used for validation purposes. A reanalysis model (Copernicus Marine Environment Monitoring Services) assimilating satellite altimetry dataset that provided sea-surface height time series was used for extracting the respective rate. The "calibrated" MT-InSAR vertical ground displacement rates were combined with the SSH rates to estimate the relative rates. Combined maps of land deformation and SSH rates show the spatial distribution of these measurements. The coastal areas are segmented into sub-areas of similar behavior and their statistics, characteristic velocities, and finally the relative sea level rise rate (expressed in cm over a period of 50 years) are estimated, along with uncertainty values. Tide gauge measurements, which were available from PSMSL network, are in accordance with our findings.

In the three case studies the sea-surface height rate is varied between 1.9 mm/year off-shore the Thermaic Gulf and 2.1 mm/year off-shore the Evros Delta. The characteristic relative sea-level rise rates vary between 1.5 ± 9 and 67 ± 23 cm in 50 years (Figure 7). The case study presenting the smallest characteristic rates is Alexandroupolis and Evros Delta with its values varying between 1 ± 10 and 18.5 ± 18 cm in 50 years (Figure 7). The coastal area of Evros Delta in the southwest seems to be currently in a phase of positive aquifer loading due to high precipitation, presenting high uplifting rates and a relative sea level fall of 16 ± 14 cm in 50 years. Longer temporal study periods are needed to ensure more accurate results in the Delta (on-shore). The highest sea level rise rates are located at the airport and at the Waterland Park northwest of the Delta. In Alexandroupolis, the rate is lower at 9 ± 11 cm in 50 years. In the southeast, the two areas of Evros Delta reach the lowest values. In Killini,

Araxos, the relative sea rise rate varies between 6 ± 14 and 28 ± 21 cm in 50 years (Figure 7) in the sub-areas of Killini and Strofilia Forest-Brinia, respectively. Sea level fall is observed in two sub-areas. Finally, in the Thermaic Gulf, the highest rates observed reach 67 ± 23 and 39 ± 14 cm in 50 years in the villages of Chalastra and Kimina, respectively. The Kalochori. Airport and Perea sub-areas have a rate of ~30 ± 14 cm in 50 years. The city center presents a rate between 14 ± 7 (Kalamaria) and 22 ± 5 cm in 50 years (P. Paralia)

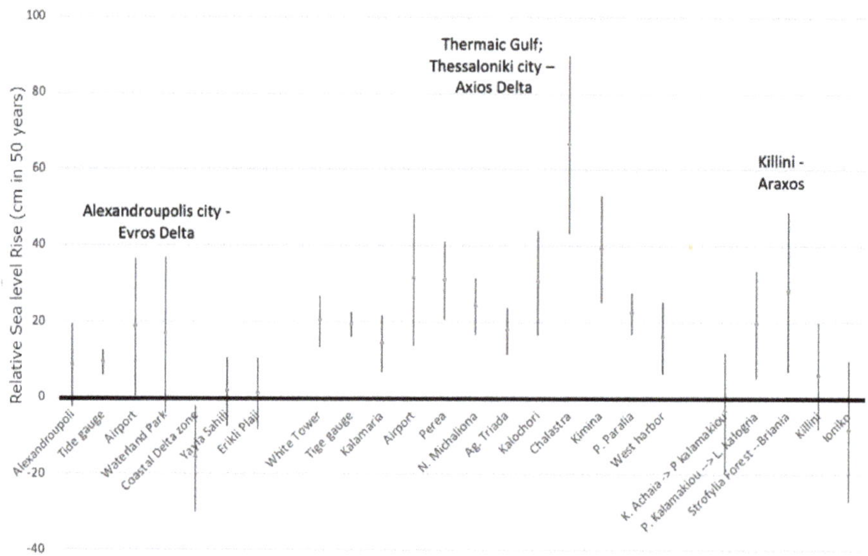

Figure 7. Indicative diagram of relative sea level rise (with uncertainties) of the sub-areas within the three case studies mapped in Figures 3, 5 and 6.

Some of the land deformations in this study are due to the changes in aquifer, i.e., loading and unloading procedures as we have seen in the areas of Evros Delta (uplift) and Kalochori to Kimina (subsidence), respectively, resulting from the equilibrium between rainfall and underground water exploitation. Additionally, since the deltaic areas are prone to subsidence due to their alluvial sediments' compaction, we expect a component similar to what has been observed in many deltaic areas [46,48], which increases toward the shore with a rate of ~2 mm/year per 1 km [73]. Shallow faulting may be another deformation source especially when the delta is located in the hanging-wall of the fault. In this case, the subsidence is the accumulated effect of both deformation sources [46]. Macroscopically, this type of faulting has not been observed, but a more focused analysis, not being within the scope of the current study, may reveal such concurrent effects.

Future efforts will focus on coupling the land deformation rates with a series of rainfall timelines; more accurate relative sea level rise data; a very high-resolution DEM; land use/land cover maps; and wind, storm surges, and tidal levels through modeling in order to identify the most vulnerable areas and thus to contribute to response and mitigation actions. Moreover, instead of the hypothesis of only vertical deformation that we have assumed in this study, a combination of ascending and descending pass acquisitions can be exploited in order to estimate the horizontal deformation rates [46] and accordingly extract the true vertical component. Therefore, a scenario of future relative sea-level and spatio-temporal assessment of vulnerability of coastal areas could be performed. In a very recent study, Vousdoukas et al. [5] presented economically efficient protection scenarios from rising extreme sea levels in coastal areas of the European continent. This very interesting study in *Nature Communications* highlighted the fact that there is a lack of proper assessment of this hazard on a continental scale, which reinforces the need for such research work to be implemented on a larger scale.

6. Conclusions

The combination of measurements. Sensors, platforms, and processing techniques is fundamental for improving the understanding of the ongoing processes along coastal urban and natural protected areas, such as land subsidence and sea-level rise. Satellite data availability and processing methodologies are mature enough to be exploited for relative sea level rise studies. The spatio-temporal assessment of such phenomena could be measured accurately with the available free and openly accessible Copernicus satellite data and the available tools implemented in InSAR processing platforms such as GEP and SARPROZ. The combination of interferometric data with relative sea-level-rise data can be exploited to identify the potential coastal areas at risk and contribute to the decision-making process.

The present research has shown that the variation in the relative sea level in the three case studies is high. ranging from close to zero values i.e. from 5–10 and 30 cm in 50 years for urban areas to characteristic values of 30 and even very locally to 60 cm in 50 years for rural areas close to the coast (Figure 7). This demonstrates the corresponding high variability in the eastern Mediterranean region, implying the need of inclusion of its coastal mapping amongst other hazard properties (like coastal erosion) as an additional feature in risk analysis processes. The new knowledge generated can contribute to the effective management of coastal areas in the framework of adaptation and mitigation strategies attributed to climate change.

The potential upscaling through the extensive exploitation of MT-InSAR (such as Copernicus European Ground Motion Service, https://land.copernicus.eu/user-corner/technical-library/european-ground-motion-service) and GNSS data will assist the study of more coastal areas in Greece, in the Mediterranean Sea, and also in lower latitude areas where the sea level rise is expected to be higher, as in the case of west Africa. The proposed methodological approach is scalable and can be implemented at local, regional, and international scales.

Supplementary Materials: The following are available online at http://www.mdpi.com/2072-4292/12/14/2296/s1, Figure S1. Deformation GNSS time series of ALEX (URANUS), AUT1 (AUTH), RLSO (NOA), processed in this study and their vertical deformation linear rate estimations.

Author Contributions: Conceptualization: I.P. and P.E.; Methodology: I.P., P.E., and G.B.; Software: P.E.; Validation: P.E. and G.B.; Formal analysis: P.E., G.B., and T.P.; Investigation: P.E. and T.P.; Resources: P.E.; Data curation: P.E., G.B., and T.P.; Writing—original draft preparation: T.P., P.E., and G.B.; Writing—review and editing: T.P., P.E., and I.P.; Visualization: P.E., G.B., and T.P.; Supervision: I.P. and P.E.; Project administration: T.P. and I.P.; Funding acquisition: I.P. All authors have read and agreed to the published version of the manuscript.

Funding: This research was funded by Greece and the European Union (European Social Fund- ESF) through the Operational Programme "Human Resources Development, Education and Lifelong Learning 2014-2020" in the context of the project "Spatio-temporal assessment of land deformation as a factor contributing to relative sea level rise in coastal urban and natural protected areas by using space-based Earth observation data" (MIS 5006912). The APC was funded by Greece and the European Union (European Social Fund- ESF) through the Operational Programme "Human Resources Development, Education and Lifelong Learning 2014-2020" in the context of the project "Spatio-temporal assessment of land deformation as a factor contributing to relative sea level rise in coastal urban and natural protected areas by using space-based Earth observation data" (MIS 5006912).

Acknowledgments: The authors would like to thank the European Space Agency and Copernicus program for providing Copernicus Sentinel-1 data and Copernicus Marine Environment Monitoring Services Information. Additionally, the authors would like to thank ESA for providing access to the Geohazard Exploitation Platform (GEP) and Emmanuel Mathot for assisting in the SAR data processing. In addition, we are grateful to Daniele Perissin for providing the SARPROZ software that facilitates our processing work. Finally, we acknowledge the

Aristotle University of Thessaloniki for providing GNSS station AUT1 data, the National Observatory of Athens for station RLSO, and the Tree Company Greece for station ALEX, used as reference stations for the MTInSAR data, as well as the Nevada Geodetic Laboratory for providing solutions for validation purposes of stations AUT1 and ALE3.

Conflicts of Interest: The authors declare no conflict of interest. The funders of the study had no role in the design of the study; in the collection. analysis. or interpretation of data; in the writing of the manuscript. or in the decision to publish the results.

References

1. Mimura, N. Sea-level rise caused by climate change and its implications for society. *Proc. Jpn. Acad. Ser. B* **2013**, *89*, 281–301. [CrossRef] [PubMed]

2. Wong, P.P.; Losada, I.J.; Gattuso, J.-P.; Hinkel, J.; Khattabi, A.; McInnes, K.; Saito, Y.; Sallenger, A. Coastal Systems and Low-Lying Areas. In *Climate Change 2014: Impacts. Adaptation. and Vulnerability. Part A: Global and Sectoral Aspects. Contribution of Working Group II to the Fifth Assessment Report of the Intergovernmental Panel on Climate Change*; Field, C.B., Barros, V.R., Dokken, D.J., Mach, K.J., Mastrandrea, M.D., Bilir, T.E., Chatterjee, M., Ebi, K.L., Estrada, Y.O., Genova, R.C., et al., Eds.; Cambridge University Press: Cambridge, UK; New York, NY, USA, 2014; pp. 361–409. ISBN 978-1107641655.

3. Cohen, J.E.; Small, C. Hypsographic demography: The distribution of human population by altitude. *Proc. Natl. Acad. Sci. USA* **1998**, *95*, 14009–14014. [CrossRef]

4. Small, C.; Nicholls, R.J. A Global Analysis of Human Settlement in Coastal Zones. *J. Coast. Res.* **2003**, *19*, 584–599.

5. Vousdoukas, M.I.; Mentaschi, L.; Hinkel, J.; Ward, P.J.; Mongelli, I.; Ciscar, J.-C.; Feyen, L. Economic motivation for raising coastal flood defenses in Europe. *Nat. Commun.* **2020**, *11*, 1–11.

6. Nicholls, R.J. Planning for the Impacts of Sea Level Rise. *Oceanography* **2011**, *24*, 144–157. [CrossRef]

7. Oppenheimer, M.; Glavovic, B.; Hinkel, J.; van de Wal, R.; Magnan, A.K.; Abd-Elgawad, A.; Cai, R.; Cifuentes-Jara, M.; DeConto, R.M.; Ghosh, T.; et al. Sea Level Rise and Implications for Low Lying Islands. Coasts and Communities. In *IPCC Special Report on the Ocean and Cryosphere in a Changing Climate*; Pörtner, H.-O., Roberts, D.C., Masson-Delmotte, V., Zhai, P., Tignor, M., Poloczanska, E., Mintenbeck, K., Alegría, A., Nicolai, M., Okem, A., et al., Eds.; Cambridge University Press: Cambridge, UK, 2019; pp. 321–445, ISSN 1095-9203 (Electronic) 0036-8075 (Linking).

8. Rodriguez, J.; Saco, P.M.; Sandi, S.G.; Saintilan, N.; Riccardi, G. Potential increase in coastal wetland vulnerability to sea-level rise suggested by considering hydrodynamic attenuation effects. *Nat. Commun.* **2017**, *8*, 16094. [CrossRef]

9. Rovere, A.; Stocchi, P.; Vacchi, M. Eustatic and Relative Sea Level Changes. *Curr. Clim. Chang. Rep.* **2016**, *2*, 221–231. [CrossRef]

10. Kemp, A.C.; Dutton, A.; Raymo, M.E. Paleo Constraints on Future Sea-Level Rise. *Curr. Clim. Chang. Rep.* **2015**, *1*, 205–215. [CrossRef]

11. Meyssignac, B.; Piecuch, C.G.; Merchant, C.J.; Racault, M.F.; Palanisamy, H.; MacIntosh, C.; Sathyendranath, S.; Brewin, R. Causes of the Regional Variability in Observed Sea Level. Sea Surface Temperature and Ocean Colour Over the Period 1993–2011. *Surv. Geophys.* **2017**, *38*, 187–215. [CrossRef]

12. Milne, G.A.; Gehrels, W.R.; Hughes, C.W.; Tamisiea, M.E. Identifying the causes of sea-level change. *Nat. Geosci.* **2009**, *2*, 471–478. [CrossRef]

13. Church, J.; Woodworth, P.L.; Aarup, T.; Wilson, W.S. *Understanding Sea-Level Rise and Variability*; Wiley-Blackwell Publishing: London, UK, 2010; ISBN 9781444334524.

14. Sinking Cities–Global Cities Sinking into the Ocean–LSU Law Center: Climate Change Law and Policy Project. Available online: https://sites.law.lsu.edu/coast/2015/07/sinking-cities-global-cities-sinking-into-the-ocean/ (accessed on 16 November 2019).

15. Erkens, G.; Bucx, T.; Dam, R.; De Lange, G.; Lambert, J. Sinking coastal cities. *Proc. Int. Assoc. Hydrol. Sci.* **2015**, *372*, 189–198. [CrossRef]

16. Bucx, T.H.M.; Van Ruiten, C.J.M.; Erkens, G.; De Lange, G. An integrated assessment framework for land subsidence in delta cities. *Proc. Int. Assoc. Hydrol. Sci.* **2015**, *372*, 485–491. [CrossRef]

17. Gomis, D.; Tsimplis, M.; Marcos, M.; Fenoglio-Marc, L.; Pérez, B.; Raicich, F.; Vilibić, I.; Wöppelmann, G.; Monserrat, S.; Monserrat, S. Mediterranean Sea-Level Variability and Trends. In *The Climate of the Mediterranean Region: From the Past to the Future*; Lionello, P., Ed.; Elsevier Inc.: Hoboken, NJ, USA, 2012; pp. 257–299. ISBN 9780124160422.

18. Poitevin, C.; Wöppelmann, G.; Raucoules, D.; Le Cozannet, G.; Marcos, M.; Testut, L. Vertical land motion and relative sea level changes along the coastline of Brest (France) from combined space-borne geodetic methods. *Remote. Sens. Environ.* **2019**, *222*, 275–285. [CrossRef]

19. Pugh, D.; Woodworth, P. *Sea-Level Science: Understanding Tides. Surges. Tsunamis and Mean Sea-Level Changes*; Cambridge University Press: Cambridge, UK, 2014; ISBN 1107028191.

20. Ablain, M.; Legeais, J.F.; Prandi, P.; Marcos, M.; Fenoglio-Marc, L.; Dieng, H.B.; Benveniste, J.; Cazenave, A. Satellite Altimetry-Based Sea Level at Global and Regional Scales. *Surv. Geophys.* **2017**, *38*, 7–31. [CrossRef]

21. Zheng, Q.; Klemas, V.; Yan, X.-H. 8.01-Volume 8 Overview: Progress in Ocean Remote Sensing. In *Comprehensive Remote Sensing*; Liang, S.B.T.-C.R.S., Ed.; Elsevier: Oxford, UK, 2018; pp. 1–42. ISBN 978-0-12-803221-3.

22. Blewitt, G. Basics of the GPS Technique: Observation Equations. *Geod. Appl. GPS* **1997**, 10–54.

23. Blewitt, G.; Altamimi, Z.; Davis, J.; Gross, R.; Kuo, C.Y.; Lemoine, F.G.; Moore, A.W.; Neilan, R.E.; Plag, H.P.; Rothacher, M.; et al. Geodetic observations and global reference frame contributions to understanding sea-level rise and variability. In *Understanding Sea-Level Rise and Variability*; Church, J.A., Woodworth, P.L., Aarup, P., Wilson, W.S., Eds.; Wiley-Blackwell: Oxford, UK, 2010; pp. 256–284. ISBN 978-1-444-33451-7.

24. Cigna, F. Observing Geohazards from Space. *Geosciences* **2018**, *8*, 59. [CrossRef]

25. Di Paola, G.; Alberico, I.; Aucelli, P.P.; Matano, F.; Rizzo, A.; Vilardo, G. Coastal subsidence detected by Synthetic Aperture Radar interferometry and its effects coupled with future sea-level rise: The case of the Sele Plain (Southern Italy). *J. Flood Risk Manag.* **2018**, *11*, 191–206. [CrossRef]

26. Liu, A.K. Coastal Monitoring by Satellite-Based SAR. In *Remote Sensing of the Changing Oceans*; Springer: Berlin/Heidelberg, Germany, 2011; pp. 195–215.

27. Polcari, M.; Albano, M.; Montuori, A.; Bignami, C.; Tolomei, C.; Pezzo, G.; Falcone, S.; La Piana, C.; Doumaz, F.; Salvi, S.; et al. InSAR Monitoring of Italian Coastline Revealing Natural and Anthropogenic Ground Deformation Phenomena and Future Perspectives. *Sustainability* **2018**, *10*, 3152. [CrossRef]

28. Teatini, P.; Ferronato, M.; Gambolati, G.; Bertoni, W.; Gonella, M. A century of land subsidence in Ravenna, Italy. *Environ. Geol.* **2005**, *47*, 831–846. [CrossRef]

29. Hooper, A.J. A multi-temporal InSAR method incorporating both persistent scatterer and small baseline approaches. *Geophys. Res. Lett.* **2008**, *35*, 1–5. [CrossRef]

30. Crosetto, M.; Garriga, O.T.; Cuevas-González, M.; Devanthéry, N.; Crippa, B. Persistent Scatterer Interferometry: A review. *ISPRS J. Photogramm. Remote Sens.* **2016**, *115*, 78–89. [CrossRef]

31. Lanari, R.; Casu, F.; Manzo, M.; Zeni, G.; Berardino, P.; Manunta, M.; Pepe, A. An Overview of the Small BAseline Subset Algorithm: A DInSAR Technique for Surface Deformation Analysis. *Pure Appl. Geophys.* **2007**, *164*, 637–661. [CrossRef]

32. Montuori, A.; Anderlini, L.; Palano, M.; Albano, M.; Pezzo, G.; Antoncecchi, I.; Chiarabba, C.; Serpelloni, E.; Stramondo, S. Application and analysis of geodetic protocols for monitoring subsidence phenomena along on-shore hydrocarbon reservoirs. *Int. J. Appl. Earth Obs. Geoinf.* **2018**, *69*, 13–26. [CrossRef]

33. Bruno, M.F.; Molfetta, M.G.; Mossa, M.; Morea, A.; Chiaradia, M.T.; Nutricato, R.; Nitti, D.O.; Guerriero, L.; Coletta, A. Integration of multitemporal SAR/InSAR techniques and NWM for coastal structures monitoring: Outline of the software system and of an operational service with COSMO-SkyMed data. In Proceedings of the 2016 IEEE Workshop on Environmental, Energy, and Structural Monitoring Systems (EESMS), Bari, Italy, 13–14 June 2016; pp. 1–16, ISBN 978-1-5090-2371-4.

34. Hu, B.; Chen, J.; Zhang, X. Monitoring the Land Subsidence Area in a Coastal Urban Area with InSAR and GNSS. *Sensors* **2019**, *19*, 3181. [CrossRef]

35. Ferretti, A.; Prati, C.; Rocca, F. Permanent scatterers in SAR interferometry. *IEEE Trans. Geosci. Remote Sens.* **2001**, *39*, 8–20. [CrossRef]

36. Berardino, P.; Fornaro, G.; Lanari, R.; Sansosti, E. A new algorithm for monitoring localized deformation phenomena based on small baseline differential SAR interferograms. *IEEE Trans. Geosci. Remote Sens.* **2002**, *40*, 2375–2383. [CrossRef]

37. Werner, C.; Wegmuller, U.; Strozzi, T.; Wiesmann, A. Interferometric point target analysis for deformation mapping. *Int. Geosci. Remote Sens. Symp.* **2003**, *7*, 4362–4364.

38. Svigkas, N.; Papoutsis, I.; Constantinos, L.; Tsangaratos, P.; Kiratzi, A.; Kontoes, C.H. Land subsidence rebound detected via multi-temporal InSAR and ground truth data in Kalochori and Sindos regions. Northern Greece. *Eng. Geol.* **2016**, *209*, 175–186.

39. Foumelis, M.; Trasatti, E.; Papageorgiou, E.; Stramondo, S.; Parcharidis, I. Monitoring Santorini volcano (Greece) breathing from space. *Geophys. J. Int.* **2013**, *193*, 161–170. [CrossRef]

40. Papoutsis, I.; Papanikolaou, X.; Floyd, M.; Kontoes, C.; Paradissis, D.; Zacharis, V.; Ji, K.H. Mapping inflation at Santorini volcano, Greece, using GPS and InSAR. *Geophys. Res. Lett.* **2013**, *40*, 267–272. [CrossRef]

41. Adam, N.; Gonzalez, F.R.; Parizzi, A.; Liebhart, W. Wide area persistent scatterer interferometry. In Proceedings of the Fringe 2011 Workshop, Frascati, Italy, 19–23 September 2011.

42. Floyd, M.A.; Billiris, H.; Paradissis, D.; Veis, G.; Avallone, A.; Briole, P.; McClusky, S.; Nocquet, J.M.; Palamartchouk, K.; Parsons, B.; et al. A new velocity field for Greece: Implications for the kinematics and dynamics of the Aegean. *J. Geophys. Res. Solid Earth* **2010**, *115*, 1–25. [CrossRef]

43. Briole, P.; Rigo, A.; Lyon-Caen, H.; Ruegg, J.C.; Papazissi, K.; Mitsakaki, C.; Balodimou, A.; Veis, G.; Hatzfeld, D.; Deschamps, A. Active deformation of the Corinth rift, Greece: Results from repeated Global Positioning System surveys between 1990 and 1995. *J. Geophys. Res.* **2000**, *105*, 605–625. [CrossRef]

44. Chousianitis, K.; Ganas, A.; Evangelidis, C. Strain and rotation rate patterns of mainland Greece from continuous GPS data and comparison between seismic and geodetic moment release. *J. Geophys. Res. Solid Earth* **2015**, *120*, 3909–3931. [CrossRef]

45. Avallone, A.; Briole, P.; Agatza-Balodimou, A.M.; Billiris, H.; Charade, O.; Mitsakaki, C.; Nercessian, A.; Papazissi, K.; Paradissis, D.; Veis, G. Analysis of eleven years of deformation measured by GPS in the Corinth Rift Laboratory area. *C. R. Geosci.* **2004**, *336*, 301–311. [CrossRef]

46. Elias, P.; Briole, P. Ground Deformations in the Corinth Rift, Greece, Investigated Through the Means of SAR Multitemporal Interferometry. *Geochem. Geophys. Geosystems* **2018**, *19*, 4836–4857. [CrossRef]

47. Raspini, F.; Bianchini, S.; Moretti, S.; Loupasakis, C.; Rozos, D.; Duro, J.; Garcia, M. Advanced interpretation of interferometric SAR data to detect, monitor and model ground subsidence: Outcomes from the ESA-GMES Terrafirma project. *Nat. Hazards* **2016**, *83*, 155–181. [CrossRef]

48. Parcharidis, I.; Kourkouli, P.; Karymbalis, E.; Foumelis, M.; Karathanassi, V. Time Series Synthetic Aperture Radar Interferometry for Ground Deformation Monitoring over a Small Scale Tectonically Active Deltaic Environment (Mornos, Central Greece). *J. Coast. Res.* **2011**, *29*, 325–338.

49. Pehlivanoglou, K.; Tsirambides, A.; Trontsios, G. Origin and Distribution of Clay Minerals in the Alexandroupolis Gulf, Aegean Sea, Greece. *Estuar. Coast. Shelf Sci.* **2000**, *51*, 61–73. [CrossRef]

50. Xeidakis, G.S.; Delimani, P.; Skias, S. Erosion problems in Alexandroupolis coastline, North-Eastern Greece. *Environ. Geol.* **2007**, *53*, 835–848. [CrossRef]

51. Mentzafou, A.; Markogianni, V.; Dimitriou, E. The Use of Geospatial Technologies in Flood Hazard Mapping and Assessment: Case Study from River Evros. *Pure Appl. Geophys.* **2017**, *174*, 679–700. [CrossRef]

52. Dimitriou, E.; Moussoulis, E.; Mentzafou, A.; Tzortziou, M.; Zeri, C.; Colombari, E.; Markogianni, V. *Environmental Status Assessment for Evros River Basin*; HCMR: Athens, Greece, 2010.

53. Alpar, B. Plio-Quaternary history of the Turkish coastal zone of the Enez-Evros Delta: NE Aegean Sea. *Mediterr. Mar. Sci.* **2001**, *2*, 95–118. [CrossRef]

54. Raucoules, D.; Parcharidis, I.; Feurer, D.; Novalli, F.; Ferretti, A.; Carnec, C.; Lagios, E.; Sakkas, V.; Le Mouélic, S.; Cooksley, G.; et al. Ground deformation detection of the greater area of Thessaloniki (Northern Greece) using radar interferometry techniques. *Nat. Hazards Earth Syst. Sci.* **2008**, *8*, 779–788. [CrossRef]

55. Kapsimalis, V.; Poulos, S.E.; Karageorgis, A.P.; Pavlakis, P.; Collins, M. Recent evolution of a Mediterranean deltaic coastal zone: Human impacts on the Inner Thermaikos Gulf, NW Aegean Sea. *J. Geol. Soc.* **2005**, *162*, 897–908. [CrossRef]

56. Karageorgis, A.P.; Skourtos, M.S.; Kapsimalis, V.; Kontogianni, A.D.; Skoulikidis, N.T.; Pagou, K.; Nikolaidis, N.; Drakopoulou, P.; Zanou, B.; Karamanos, H.; et al. *An Integrated Approach to Watershed Management within the DPSIR Framework: Axios River Catchment and Thermaikos Gulf*; Springer-Verlag: Zurich, Switzerland, 2005; Volume 5, ISBN 1011300400.

57. Psimoulis, P.; Ghilardi, M.; Fouache, E.; Stiros, S. Subsidence and evolution of the Thessaloniki plain, Greece, based on historical leveling and GPS data. *Eng. Geol.* **2007**, *90*, 55–70. [CrossRef]

58. Zalidis, G. Management of river water for irrigation to mitigate soil salinization on a coastal wetland. *J. Environ. Manag.* **1998**, *54*, 161–167. [CrossRef]

59. Maroukian, H.; Gaki-Papanastassiou, K.; Papanastassiou, D.; Palyvos, N. Geomorphological observations in the coastal zone of Kyllini Peninsula. NW Peloponnese-GreecePeloponnesus-Greece. and their relation to the seismotectonic regime of the area. *J. Coast. Res.* **2000**, *16*, 853–863.

60. Management Body of Kotychi and Strofylia Wetlands General Information of the Region–Strofylia National Park. Available online: https://strofylianationalpark.gr/national-park-of-kotychi-strofylia-wetlands/general-information-of-the-region/ (accessed on 4 June 2020).

61. NATURA 2000 Standard Data Form for Special Protection Areas (SPA) for sites eligible for identification as Sites of Community Importance (SCI) and for Special Areas of Conservation (SAC); Site GR2320001, Limnothalassa Kalogrias, Dasos Strofylias kai Elos Lamias, Araxos, 2000. Available online: https://natura2000.eea.europa.eu/Natura2000/SDF.aspx?site=GR2320001 (accessed on 16 July 2020).

62. De Zan, F.; Guarnieri, A.M. TOPSAR: Terrain Observation by Progressive Scans. *IEEE Trans. Geosci. Remote Sens.* **2006**, *44*, 2352–2360. [CrossRef]

63. Simoncelli, S.; Fratianni, C.; Pinardi, N.; Grandi, A.; Drudi, M.; Oddo, P.; Dobricic, S. Mediterranean Sea Physical Reanalysis (CMEMS MED-Physics) [Data set] Copernicus Monitoring Environment Marine Service (CMEMS). 2019. Available online: https://doi.org/10.25423/MEDSEA_REANALYSIS_PHYS_006_004 (accessed on 16 July 2020).

64. Blewitt, G.; Hammond, W.C.; Kreemer, C. Harnessing the GPS Data Explosion for Interdisciplinary Science-Eos. Available online: https://eos.org/science-updates/harnessing-the-gps-data-explosion-for-interdisciplinary-science (accessed on 22 May 2020).

65. Holgate, S.J.; Matthews, A.; Woodworth, P.L.; Rickards, L.J.; Tamisiea, M.E.; Bradshaw, E.; Foden, P.R.; Gordon, K.M.; Jevrejeva, S.; Pugh, J. New Data Systems and Products at the Permanent Service for Mean Sea Level. *J. Coast. Res.* **2012**, *29*, 493–504.

66. Bertiger, W.; Desai, S.D.; Haines, B.; Harvey, N.; Moore, A.W.; Owen, S.; Weiss, J.P. Single receiver phase ambiguity resolution with GPS data. *J. Geod.* **2010**, *84*, 327–337. [CrossRef]

67. De Luca, C.; Cuccu, R.; Elefante, S.; Zinno, I.; Manunta, M.; Casola, V.; Rivolta, G.; Lanari, R.; Casu, F. An On-Demand Web Tool for the Unsupervised Retrieval of Earth's Surface Deformation from SAR Data: The P-SBAS Service within the ESA G-POD Environment. *Remote. Sens.* **2015**, *7*, 15630–15650. [CrossRef]

68. Casu, F.; Elefante, S.; Imperatore, P.; Zinno, I.; Manunta, M.; De Luca, C.; Lanari, R. SBAS-DInSAR Parallel Processing for Deformation Time-Series Computation. *IEEE J. Sel. Top. Appl. Earth Obs. Remote. Sens.* **2014**, *7*, 3285–3296. [CrossRef]

69. Perissin, D. Software Manual–SARPROZ©. Available online: https://www.sarproz.com/software-manual/ (accessed on 22 May 2020).

70. USGS EROS Archive-Digital Elevation-Shuttle Radar Topography Mission (SRTM) 1 Arc-Second Global. Available online: https://www.usgs.gov/centers/eros/science/usgs-eros-archive-digital-elevation-shuttle-radar-topography-mission-srtm-1-arc?qt-science_center_objects=0#qt-science_center_objects (accessed on 19 February 2020).

71. Goel, K.; Adam, N. An advanced algorithm for deformation estimation in non-urban areas. *ISPRS J. Photogramm. Remote Sens.* **2012**, *73*, 100–110. [CrossRef]

72. Costantini, F.; Mouratidis, A.; Schiavon, G.; Sarti, F. Advanced InSAR techniques for deformation studies and for simulating the PS-assisted calibration procedure of Sentinel-1 data: Case study from Thessaloniki (Greece), based on the Envisat/ASAR archive. *Int. J. Remote Sens.* **2016**, *37*, 729–744. [CrossRef]

73. Ford, M.; Williams, E.A.; Malartre, F.; Popescu, S.M. Stratigraphic Architecture, Sedimentology and Structure of the Vouraikos Gilbert-Type Fan Delta, Gulf of Corinth, Greece. In *Sedimentary Processes Environments and Basins. A Tribute to Peter Friend*; Nichols, G., Williams, E., Paola, C., Eds.; Int. Assoc. Sedimentologists, John Wiley & Sons, Inc.: Hoboken, NJ, USA, 2007; pp. 49–90.

 *remote sensing*

Article

Seasonal Variability of Diffuse Attenuation Coefficient in the Pearl River Estuary from Long-Term Remote Sensing Imagery

Chaoyu Yang [1,2], Haibin Ye [3,4,*] and Shilin Tang [3,4]

[1]   South China Sea Marine Prediction Center, State Oceanic Administration, Guangzhou 510300, China; ycy@scsio.ac.cn

[2]   Key Laboratory of Marine Environmental Survey Technology and Application, Ministry of Natural Resources, MNR, Guangzhou 510300, China

[3]   State Key Laboratory of Tropical Oceanography, South China Sea Institute of Oceanology, Chinese Academy of Sciences, Guangzhou 510301, China; sltang@scsio.ac.cn

[4]   Southern Marine Science and Engineering Guangdong Laboratory (Guangzhou), Guangzhou 510301, China

*   Correspondence: hbye@scsio.ac.cn

Received: 3 June 2020; Accepted: 14 July 2020; Published: 15 July 2020

Abstract: We evaluated six empirical and semianalytical models of the diffuse attenuation coefficient at 490 nm ($K_d(490)$) using an in situ dataset collected in the Pearl River estuary (PRE). A combined model with the most accurate performance (correlation coefficient, $R^2 = 0.92$) was selected and applied for long-term estimation from 2003 to 2017. Physical and biological processes in the PRE over the 14-year period were investigated by applying satellite observations (MODIS/Aqua data) and season-reliant empirical orthogonal function analysis (S-EOF). In winter, the average $K_d(490)$ was significantly higher than in the other three seasons. A slight increasing trend was observed in spring and summer, whereas a decreasing trend was observed in winter. In summer, a tongue with a relatively high $K_d(490)$ was found in southeastern Lingdingyang Bay. In Eastern Guangdong province (GDP), the relatively higher $K_d(490)$ value was found in autumn and winter. Based on the second mode of S-EOF, we found that the higher values in the eastern GDP extended westward and formed a distinguishable tongue in winter. The grey relational analysis revealed that chlorophyll-a concentration (C_{chla}) and total suspended sediment concentration (C_{tsm}) were two dominant contributors determining the magnitude of $K_d(490)$ values. The C_{tsm}-dominated waters were generally located in coastal and estuarine turbid waters; the C_{chla}-dominated waters were observed in open clear ocean. The distribution of constituents-dominated area was different in the four seasons, which was affected by physical forces, including wind field, river runoff, and sea surface temperature.

Keywords: Pearl River estuary; diffuse attenuation coefficient; MODIS; S-EOF

1. Introduction

The light diffuse attenuation coefficient ($K_d(\lambda)$) in aquatic systems is defined by the exponential decrease in the irradiance with depth [1,2]. $K_d(\lambda)$ is an ecologically important water property that provides an estimate of the availability of light to underwater communities, which influences ecological processes and biogeochemical cycles in natural waters [3,4]. The estimation of $K_d(\lambda)$ is also critical for understanding physical processes such as sediment resuspension and heat transfer in the upper layer of the ocean [5–7].

The in situ $K_d(\lambda)$ is traditionally measured by the ocean color scientific community at 490 nm, $K_d(490)$, following the primary studies in the 1970s [8]. Traditional field measurement of $K_d(\lambda)$ is costly and time consuming, but recent advances in satellite sensors have provided synoptic and frequent measurements of various bio-optical products on large scales, considerably improving spatial and temporal resolution compared to in situ data [9]. Today, several empirical and semianalytical models of $K_d(490)$ are commonly used to derive the $K_d(490)$ maps from satellite sensors such as the Sea-Viewing Wide Field-of-View Sensor (SeaWiFS) [10,11], Moderate Resolution Imaging Spectroradiometer (MODIS) [4,12], and the Medium Resolution Imaging Spectrometer (MERIS) [6,13].

However, no K_d model can be applied globally. For example, no model developed for Case 1 open ocean waters can be used in turbid coastal environment [2,11,13].

The Pearl River is well known for its complex river networks, low lying terrain, and intense rainfall events. The water composition varies widely both spatially and temporally in the Pearl River estuary (PRE). Given the need to understand the light environment in the PRE waters, we aimed to evaluate the accuracy of six total empirical or semianalytical models for $K_d(490)$ retrieval. Brief descriptions of the models are given in Section 2.2.4. The model that performed best was selected to construct $K_d(490)$ maps in PRE based on long-term MODIS/Aqua imagery. Seasonal variability and spatial distribution of $K_d(490)$ were analyzed by applying season-reliant empirical orthogonal function (S-EOF) analysis. The dominant water constituents in different regions were determined using grey relational analysis (GRA). The influences of physical factors on the $K_d(490)$ were also discussed.

2. Materials and Methods

2.1. Study Area

The PRE is located in the northern South China Sea (NSCS), known as a subtropical and high biological productivity estuary. The PRE is characterized by a complicated hydrodynamic system regulated by many physical factors, including bottom topography, river discharge, wind field, and a coastal current [14]. The PRE is influenced by the East Asia monsoon system, characterized by prevailing northeasterly and southwesterly winds in winter and summer, respectively [15,16]. In this study, season refers to those for the northern hemisphere, for example, summer refers to June, July, and August. As China's third largest river, the Pearl River flows into the PRE through eight main outlets [17], carrying a large amount of organic and inorganic suspended matter, with an annual average discharge of 10^5 m^3·s^{-1} [18]. With increasing human activity, the PRE is contaminated by industrial pollution, agricultural runoff, and domestic sewage [19,20].

2.2. Data Sources and Processing

2.2.1. In Situ Measurements

A cruise was conducted on 5 June 2012 to collect water samples and the water spectrum. Positions for all sampling stations are plotted in Figure 1. The field spectral measurements were composed of two parts: the above-water remote sensing reflectance (R_{rs}) and the downwelling irradiance within the water column. To obtain the background water column conditions, water samples from the 15 sampling stations were used for measurement of chlorophyll-a (C_{chla}), total suspended sediment (C_{tsm}), absorption coefficient for phytoplankton ($a_p(\lambda)$), and colored dissolved organic matter (CDOM, $a_g(\lambda)$) (Table 1).

The above-water R_{rs} was measured using a spectroradiometer (USB4000, Ocean Optics, Inc., Dunedin, FL, USA) following the National Aeronautics and Space Administration (NASA) ocean-optics standard protocol [21]. The upward radiance (L_u), downward sky radiance (L_{sky}), and radiance from standard spectra on a reference plaque (L_{pla}) were measured, and R_{rs} was calculated using the following equation:

$$R_{rs}(\lambda) = \rho_{pla}(\lambda)\left[L_u(\lambda) - \rho_f(\lambda)L_{sky}(\lambda)\right] / \left[\pi L_{pla}(\lambda)\right] \tag{1}$$

where λ is the wavelength, ρ_{pla} is the reflectance of the plaque provided by the manufacturer (Ocean Optics, Inc., Dunedin, FL, USA), ρ_f is the water surface Fresnel reflectance, where a value of 0.028 was taken for wind speeds of less than 5 m·s^{-1}.

To evaluate the MODIS-based $K_d(490)$ retrieval models, in situ R_{rs} was aggregated to simulate MODIS/Aqua R_{rs} according to the following equation [22–24]:

$$R_{rs}(Bi) = \frac{\int_{\lambda_m}^{\lambda_n} RSR(\lambda) * R_{rs\_meas}(\lambda)d\lambda}{\int_{\lambda_m}^{\lambda_n} RSR(\lambda)d\lambda} \tag{2}$$

where $R_{rs}(Bi)$ denotes the simulated R_{rs} for the ith band of MODIS/Aqua, with integration from λ_m to λ_n; $R_{rs\_meas}(\lambda)$ denotes the field-measured $R_{rs}(\lambda)$; and $RSR(\lambda)$ denotes the MODIS/Aqua spectral response function.

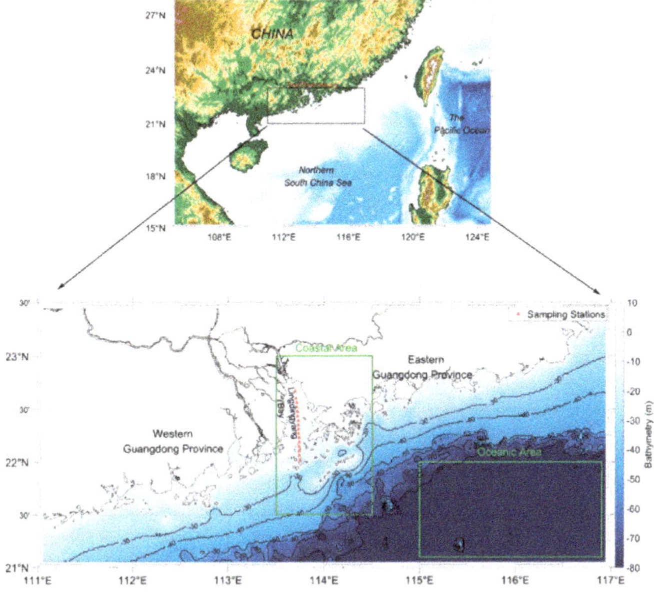

Figure 1. Study area and the location of sampling stations during the survey on 5 June 2012.

Table 1. Background Pearl River estuary (PRE) water column conditions from field measurements. The level of absorption coefficients in the PRE are represented by $a_p(443)$ and $a_g(443)$.

Period	n	$a_p(443)$ (m^{-1})	$a_g(443)$ (m^{-1})	C_{tsm} (g·m^{-3})	C_{chla} (mg·m^{-3})
5 June 2012	15	0.31–1.61	0.12–0.58	4.16–25.70	1.52–9.67

Downwelling irradiance within the water column was measured with a TriOS-RAMES hyperspectral spectroradiometer (TriOS GmbH, Oldenburg, Germany). The spectroradiometer recorded irradiance signal in the range of 320 to 950 nm with a wavelength resolution of 3.3 nm. The TriOS-RAMES instrument was slowly hand-lowered at a stable speed from the surface to a water depth of about 5 m and set to a sampling rate of one sample every five seconds. Meanwhile, a pressure sensor recorded the corresponding depth of water. By releasing the TriOS-RAMES instrument (TriOS GmbH, Oldenburg, Germany) into water twice, two profiles of the downwelling irradiance were collected. The two profiles were averaged to minimize the effect of near-surface wave focusing. The natural logarithm of the measured irradiance was plotted against depth, and an estimate of $K_d(\lambda)$ was acquired from the resulting slope [25]:

$$K_d(\lambda, z) = \ln[E_d(\lambda, z)/E_d(\lambda, z + \Delta z)]/\Delta z \tag{3}$$

where λ is the wavelength, $E_d(z)$ is the downwelling irradiance at depth z, and Δz is the infinitesimal thickness at depth z.

2.2.2. MODIS/Aqua Imagery

The Level-1B MODIS/Aqua ocean color dataset and the geolocation dataset from 2003 to 2017 were obtained from the Level-1 and Atmosphere Archive and Distribution System (LAADS) Distributed Active Archive Center (DAAC). Imagery was preprocessed using the SeaWiFS data analysis system (SeaDAS, version 7.5.1). The Management Unit of the North Seas Mathematical Models (MUMM)-based atmospheric correction [26] and an iterative f/Q Bidirectional Reflectance Distribution Function (BRDF) correction [27–30] were used to acquire accurate R_{rs} values. Flags were used to mask contamination from land, clouds, sun glint, and other potential disturbances to the radiance signal.

2.2.3. Ancillary Data

The wind field dataset was obtained from the National Centers for Environmental Prediction (NCEP) Climate Forecast System Version 2 (CFSv2). The model is fully coupled, representing the Earth's atmosphere, oceans, land, and sea ice [31]. The mixed layer depth (MLD), defined as the depth where the density is equal to the sea surface density plus an increase in density equivalent to 0.8 °C, was acquired from the global ocean Argo gridded dataset (BOA_Argo, provided by the China Argo Real-time Data Center, ftp://data.argo.org.cn/pub/ARGO/BOA_Argo/) [32]. The monthly river runoff was acquired from the Chinese River Sediment Bulletin. The Level-3 MODIS/Aqua sea surface temperature (SST) dataset was obtained from the Ocean Color Website (https://oceancolor.gsfc.nasa.gov/l3/), a website that provides the derived geophysical variables that have been aggregated/projected onto a well-defined spatial grid during a well-defined time period.

2.2.4. Models for K_d(490) Retrieval

At present, the standard methods for $K_d(490)$ estimation are roughly classified into three types: (1) empirical relationship between $K_d(490)$ and apparent optical properties (AOP), including water-leaving radiance or reflectance [11,33,34]; (2) empirical relationship between $K_d(490)$ and chlorophyll-a based on regression analyses [35]; and (3) semianalytical approaches based on radiative transfer models [1,36]. These three types of models, six models in total (Table 2), were evaluated in the PRE waters using the in situ dataset.

Table 2. Description of different algorithms for $K_d(490)$ retrieval, where nL_w denotes normalized water-leaving radiance, θ_a denotes above surface solar zenith angle, a denotes absorption coefficient, b_b denotes backscattering coefficient, $K_d^{clear}(490)$ denotes the model for open clear water, and $K_d^{turbid}(490)$ denotes the model for coastal turbid water (AOP refers to apparent optical properties).

Type	Form of Algorithm	Reference
Empirical model with AOP	$K_d(490) = 0.016 + 0.15645[nL_w(490)/nL_w(555)]^{-1.5401}$	Mueller, 2000
Empirical model with C_{chla}	$K_d(490) = 0.01666 + 0.0773C_{chla}^{0.6715}$	Morel et al., 2001
Semianalytical model	$K_d(490) = (1 + 0.005\theta_a)a(490) + 4.18\left(1 - 0.52e^{-10.8a(490)}\right)b_b(490)$	Lee et al., 2005
Empirical model with AOP	$IF R_{rs}(490)/R_{rs}(555) \geq 0.85 K_d(490) = 10^{(-0.843-1.459X-0.101X^2-0.811X^3)}$ $with \quad X = \log_{10}[R_{rs}(490)/R_{rs}(555)]$ $ELSEIF \quad R_{rs}(490)/R_{rs}(555) < 0.85 \quad K_d(490) = 10^{(0.094-1.302X+0.247X^2-0.021X^3)}$ $with \quad X = \log_{10}[R_{rs}(490)/R_{rs}(665)]$	Zhang and Fell, 2007
Semianalytical model	$K_d(490) = (1-W)K_d^{Clear}(490) + WK_d^{Turbid}(490)$ $with \quad W = -1.175 + 4.512R_{rs}(670)/R_{rs}(490)$	Wang et al., 2009
Empirical model with AOP	$K_d(490) = 0.011405 + 0.92[R_{rs}(670)/R_{rs}(490)]$	Tiwari et al., 2014

2.2.5. S-EOF and Grey Relational Analyses

The S-EOF analysis, proposed by Wang and An (2005) [37], was applied here to detect the spatial patterns and temporal variability of $K_d(490)$ in different seasons. The processing steps of S-EOF analysis are as follows: Firstly, the time series of seasonal $K_d(490)$ anomaly was calculated. Secondly, the EOF was analyzed based on the matrix composed of the four seasons. Finally, each S-EOF mode containing four spatial modes, which represented the spatial patterns of $K_d(490)$ in the four seasons, and a corresponding principal component time series were obtained.

GRA is an important part of grey system theory, which is used to determine the relational degree among factors according to the similarities in their geometry [38]. The GRA was applied here to identify the dominant water constituents (total suspended matter, phytoplankton, and dissolved matter) affecting the spatial distribution and temporal variation of $K_d(490)$. In GRA, the reference series of $K_d(490)$ and comparison sequences (water constituents, including C_{tsm}, C_{chla}, and $a_{dg}(443)$) were constructed in advance to calculate the grey relational grade (GRG), which is a measure of similarity between the reference sequence and comparison sequences. Details about the calculation of GRG were described by Liu and Lin (2005) [39] and Wan et al. (2019) [40].

2.2.6. Performance Assessment

To compare the performance of different $K_d(490)$ retrieval models, several statistical parameters were used: the determination coefficient (R^2), root mean square error ($RMSE$), mean absolute difference (MAD), and mean absolute percentage difference ($MAPD$), which are calculated as:

$$R^2 = 1 - \frac{\sum_{t=1}^{N}\left(x_{mt} - x_{pt}\right)^2}{\sum_{t=1}^{N}\left(x_{mt} - \overline{x_m}\right)^2}, \tag{4}$$

$$RMSE = \sqrt{\frac{1}{N}\sum_{t=1}^{N}\left(x_{mt} - x_{pt}\right)^2}, \tag{5}$$

$$MAD = \frac{\sum_{t=1}^{N}\left|x_{mt} - x_{pt}\right|}{N}, \tag{6}$$

$$MAPD(\%) = \frac{100}{N}\sum_{t=1}^{N}\left|\frac{x_{mt} - x_{pt}}{x_{mt}}\right|, \tag{7}$$

where x_m and x_p denote the measured and predicted samples, respectively; $\overline{x_m}$ denotes the mean value of the measured samples; and N is the number of samples.

3. Results

3.1. Model Performance

We evaluated the six different models with MODIS/Aqua spectral bands or C_{chla}. The evaluation was based on the comparison of the model-derived $K_d(490)$ with in situ measured $K_d(490)$ collected from the PRE on 5 June 2012. Figure 2 shows scatterplots between the in situ measured and different models' $K_d(490)$ retrievals, and Table 3 lists the statistical parameters. The results provided by both Mueller's and Morel's models constantly underestimated the $K_d(490)$ compared with the in situ dataset for the PRE, with $RMSE$s higher than 1.1 m^{-1}, MADs close to 1.0 m^{-1}, and $MAPD$s up to 70%. The Morel (empirical model with C_{chla}) and Mueller (empirical model with water-leaving radiance) models not only underestimated the in situ values of the PRE, they had little to no sensitivity along a broad gradient of in situ values.

By comparison, the other four models appeared to be more effective when applied in the PRE waters. These four models performed well with R^2 values higher than 0.9, $RMSE$s ranging from 0.31 to 0.70 m^{-1}, MADs ranging from 0.27 to 0.54 m^{-1}, and $MAPD$s ranging between 25.51% and 37.10%. We found that Wang's model, combining Lee's algorithm for turbid waters and Mueller's algorithm for clear waters, was a better choice for $K_d(490)$ retrieval in the PRE waters. Comparison of Wang's model to the other models showed that Wang's model had considerably lower *RMSE* and *MAD* values and outperformed the other models, especially at relatively higher $K_d(490)$ levels.

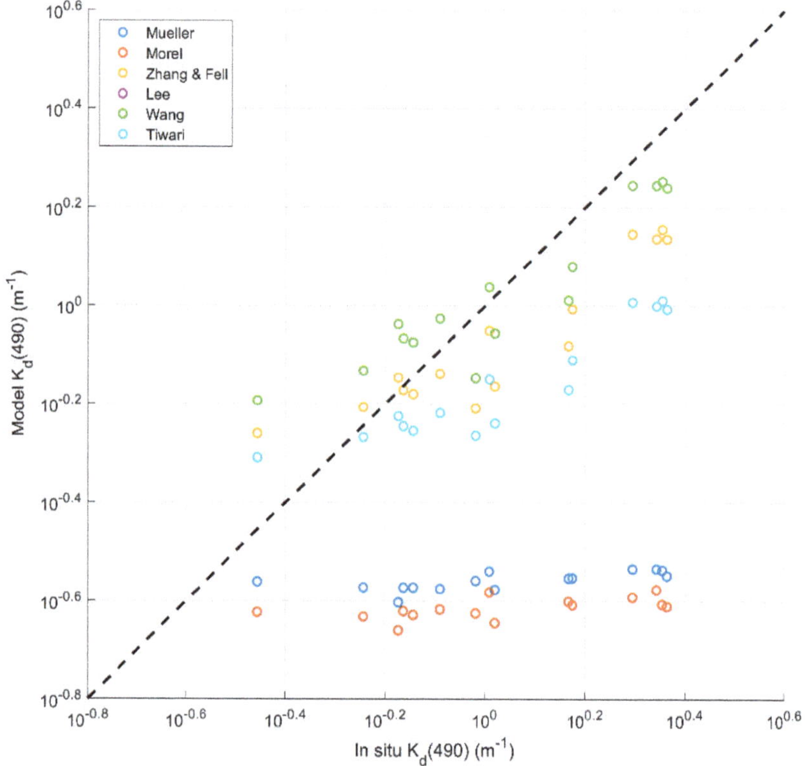

Figure 2. Scatterplots of in situ and model retrieval $K_d(490)$ values.

Table 3. Statistical parameters between the in situ measured and different model-retrieved $K_d(490)$; models with best performing values are in bold.

Algorithm	Slope	Intercept	R^2	*RMSE*	*MAD*	*MAPD* (%)
Mueller	0.01	0.26	0.56	1.15	0.96	70.32
Morel	0.01	0.23	0.38	1.18	0.99	73.90
Zhang	0.47	0.33	0.92	0.49	0.37	26.52
Lee	**0.60**	**0.39**	**0.91**	**0.31**	**0.27**	**25.51**
Wang	**0.60**	**0.39**	**0.91**	**0.31**	**0.27**	**25.51**
Tiwari	0.28	0.36	0.92	0.70	0.54	37.10

3.2. Spatial Distribution and Temporal Variation

Given its superior performance of the six considered models, the long-term MODIS $K_d(490)$ products were derived based on Wang's model. Significant seasonal variation was identified over the entire study area from 2003 to 2017 (Figure 3). The mean values for the entire study area were 0.13 m^{-1} in spring, 0.12 m^{-1} in summer, 0.14 m^{-1} in autumn, and 0.21 m^{-1} in winter. In the coastal area, the relatively high $K_d(490)$ was observed in Lingdingyang Bay (LB) and western Guangdong Province (GDP), where the highest value exceeded 4.0 m^{-1}. In summer, the river plume extends from LB southeastward into the coastal region, resulting in a wider distribution of high-value $K_d(490)$. The plume waters formed a tongue along the eastern GDP (located 113–115°E, 22–22.5°N) in some specific years, though this feature was not so remarkable in the seasonal climatological imagery due to long-term average smoothing. The influence of the terrestrial input of nutrients from the Pearl River is highest in summer. In addition, a southeast wind prevails and rainfall mainly occurs in summer. The winds blow from PRE to the middle shelf. The winds control the spatial pattern of $K_d(490)$ distribution in the PRE. In the eastern GDP, relatively higher $K_d(490)$ values were found in autumn and winter, whereas lower values were observed in spring and summer. This phenomenon is closely correlated with the coastal upwelling along the eastern GDP coast [41,42]. Chen et al. (1982) [43] reported that a radiating current could generate an upwelling in winter near the Jieshi Bay in the eastern GDP. In open ocean areas, the average $K_d(490)$ values in winter and spring were higher than that in summer and autumn. The prevailing northeasterly monsoon is stronger in the northern South China Sea in winter, so the MLD was deeper. The mixing effects are relatively stronger in winter. Figure 3 shows that the distribution of $K_d(490)$ reveals the significant seasonal variation over the entire study area during 2003 to 2017.

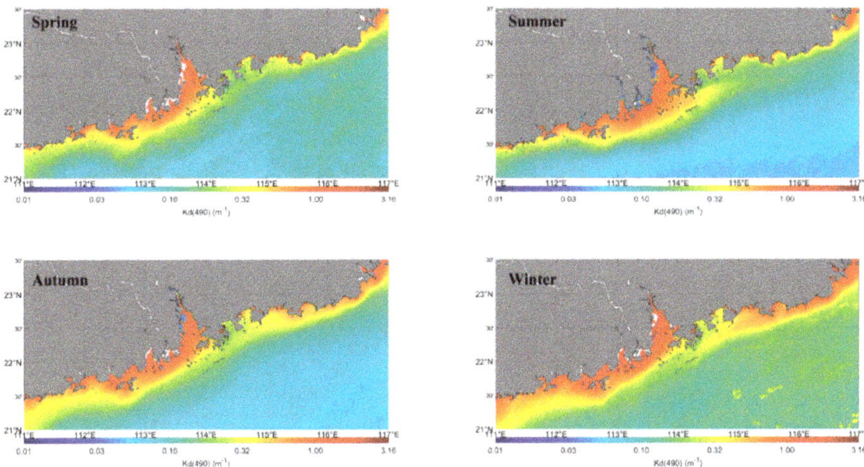

Figure 3. Seasonal distribution of $K_d(490)$ (m^{-1}) from 2003 to 2017.

Due to the different physical factors affecting the variability of $K_d(490)$ in nearshore and offshore regions, we separated the two regions and analyzed the separate regions' trends rather than averaging the entire region for time series analysis (Figure 4). Two subregions, representing the turbid coastal waters and the clear open ocean waters, were chosen (marked in Figure 1 by green boxes). During the period from 2003 to 2017, the trend lines of the annual average in coastal and oceanic areas were around 0.3 and 0.1 m^{-1}, respectively. No significant increasing or decreasing trend was observed. However, in terms of seasonal variability, the average nearshore and offshore $K_d(490)$ showed some differences. The average $K_d(490)$ in the coastal area was constantly high in the four seasons, with values ranging between 0.2 and 0.4 m^{-1}. An increasing trend was observed in spring, with a slope of approximately 0.006 m^{-1} per year. Compared to the coastal region, significant seasonal variability was observed in the open ocean region. Average $K_d(490)$ values during spring and winter were found to be higher than during summer and autumn. Trend lines of spring and winter ranged from 0.1 to 0.2 m^{-1}, whereas those in summer and autumn ranged from 0.04 to 0.08 m^{-1}. In winter, the average showed a significant decreasing trend, with a slope of approximately −0.005 m^{-1} per year.

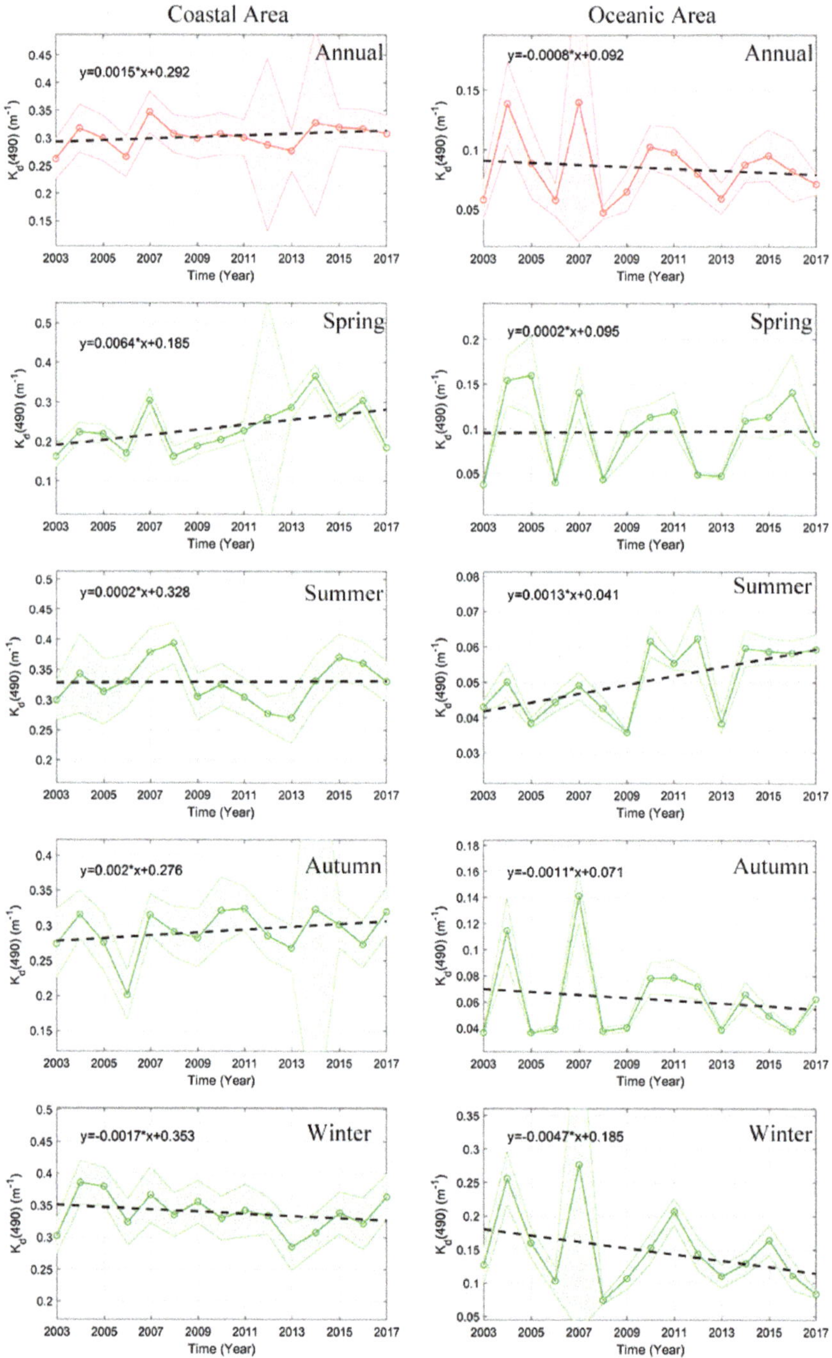

Figure 4. Annual and seasonal average $K_d(490)$ of the PRE waters between 2003 and 2017.

3.3. GRG of Water Constituents

The optical properties were determined using the absorption or backscattering of different water constituents. Here, three types of water constituents were considered; the monthly average C_{tsm}, C_{chla}, and $a_{dg}(443)$ were obtained based on the same atmospheric-corrected MODIS/Aqua R_{rs} dataset that was used for $K_d(490)$ retrieval. A band ratio algorithm was adopted for C_{tsm} retrieval [17]. The OC3M algorithm was used for C_{chla} retrieval [44]. The generalized inherent optical property (GIOP) model was applied for $a_{dg}(443)$ retrieval [45,46]. The GRGs, which can be used to measure the relationships between the $K_d(490)$ and the three water constituents, were calculated pixel by pixel for the four seasons (Figure 5).

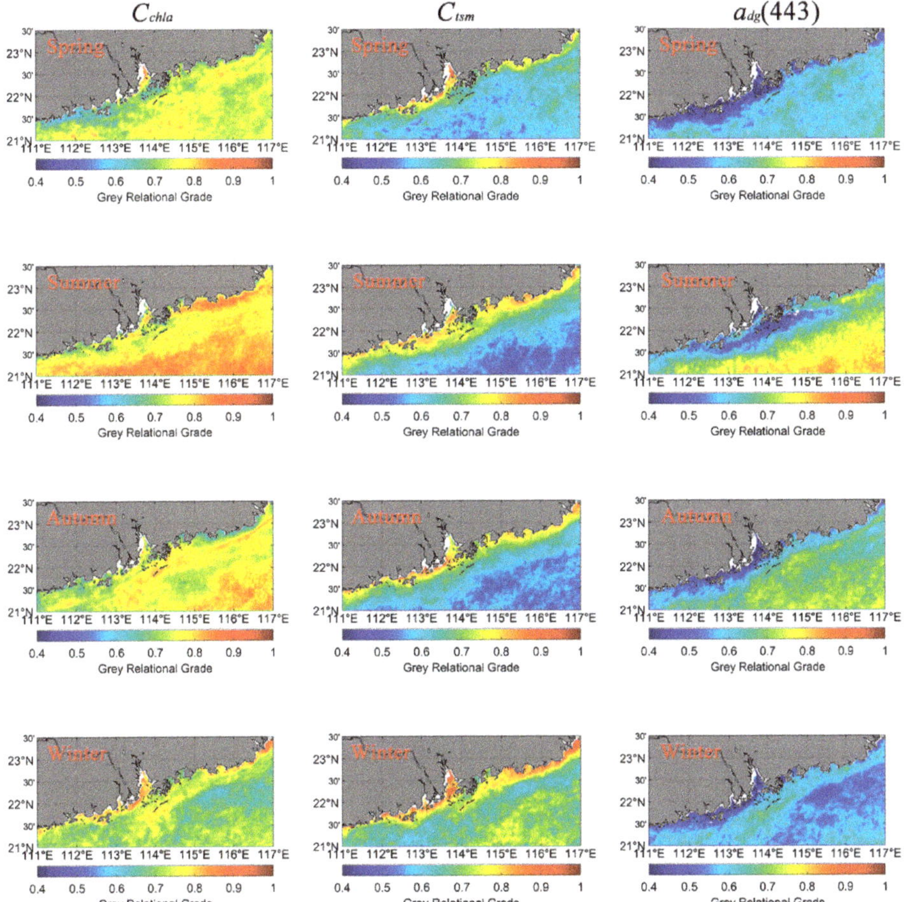

Figure 5. Grey relational grades (GRGs) between $K_d(490)$ and water constituents in four seasons.

Spatially, GRGs between $K_d(490)$ and C_{chla} or $a_{dg}(443)$ were higher in the clear open ocean region than in the coastal region, but the values contrasted between $K_d(490)$ and C_{tsm}. The GRGs gradually decreased from nearshore to offshore, similar to the distribution of C_{tsm}. Seasonally, the GRG was higher in summer and autumn than in spring and winter between $K_d(490)$ and C_{chla}, with most of the pixels' values being above 0.8. Similar phenomena were observed in the GRGs between $K_d(490)$ and $a_{dg}(443)$, although the average value was lower than between $K_d(490)$ and C_{chla}.

3.4. S-EOF Analysis

An S-EOF analysis was performed after subtracting the long-term monthly climatological average $K_d(490)$. The first two modes and the corresponding principal components (PC) were separated, which accounted for approximately 81.16% of the total variance (Table 4).

Table 4. Variance of the first three season-reliant empirical orthogonal function (S-EOF).

S-EOF Mode	Single Contribution Rate	Cumulative Contribution Rate
1	56.67	56.67
2	24.49	81.16

Figure 6 shows the PC time series of the first two S-EOF modes. All the values of PC1 were positive, indicating that the seasonal fluctuation was stable. The strength of fluctuation was related to the magnitude of the positive values. We observed a significant increasing trend during 2003 to 2006, whereas a slight decline was observed during 2007 to 2009. From the beginning of 2010 to 2014, PC1 reached its highest value. After that, the values began to decline again. PC2 was characterized by negative values during 2010 to 2014 and positive values in other years.

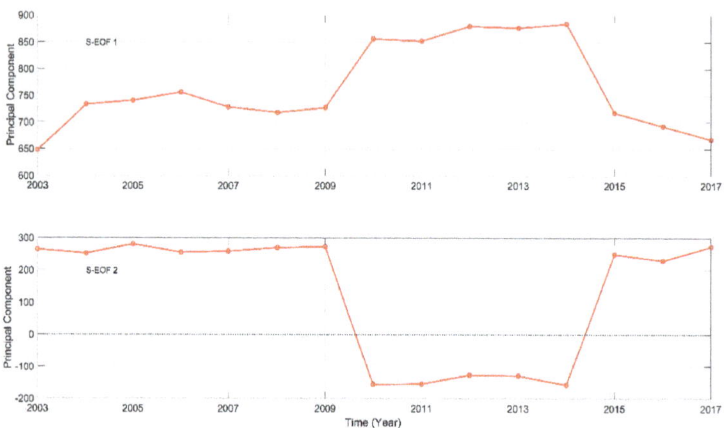

Figure 6. Principal components (PCs) of the first three S-EOF modes of $K_d(490)$ in the PRE waters.

Figure 7 shows the spatial distribution of the first mode of S-EOF, which explained approximately 56.7% of the total variance. Relatively high values were observed in LB and the western coast of GDP in the four seasons, and the average values for spring were significantly lower than in the other seasons. In summer, the high value area tended to expand to the southeastern LB. From autumn to winter, we observed a high value area along the east coast of the GDP, whereas this high value area disappeared in spring and summer.

The second mode of S-EOF explained 24.5% of the total variance. A relatively higher value area was observed in LB during the four seasons (Figure 8). In summer, the high value area extended eastward and formed a distinguishable tongue. Compared with spring and summer, higher values were observed along the whole coastal zone of the PRE. Based on the second mode of S-EOF, we found that the higher values in the eastern GDP extended westward and formed a distinguishable tongue in winter.

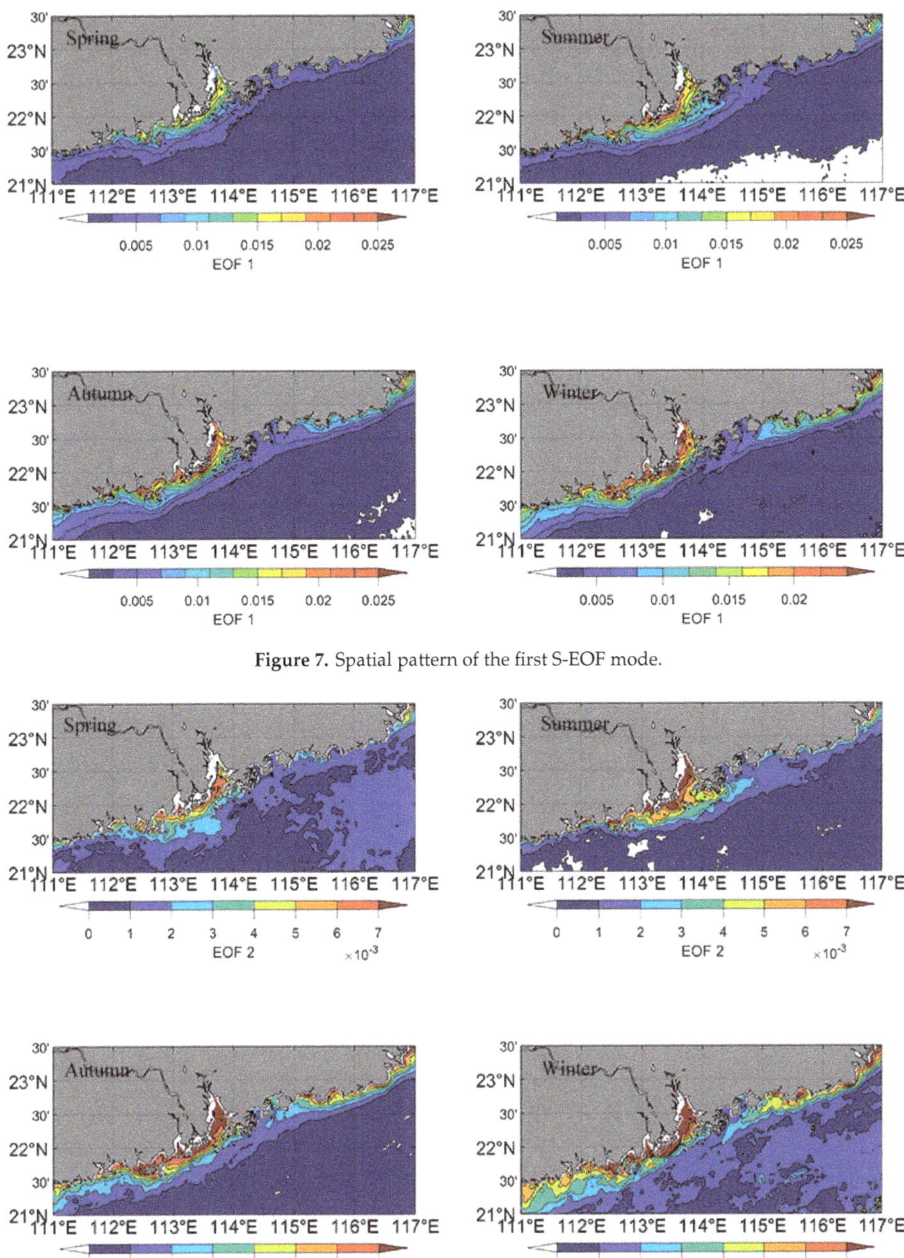

Figure 7. Spatial pattern of the first S-EOF mode.

Figure 8. Spatial pattern of the second S-EOF mode.

4. Discussion

4.1. Evaluation of $K_d(490)$ Models

Both Mueller and Morel's models were unsuitable for the PRE waters because these two models were established for clear waters and only use the spectral information from the blue and green bands. For clear waters where the downwelling attenuation is mainly determined by phytoplankton, the blue–green band ratio is sensitive to the variability of the C_{chla}, resulting in a high accuracy for $K_d(490)$ retrieval. However, for turbid waters where the optical properties are more complex, the blue–green band ratio demonstrates a lower sensitivity to the variability in $K_d(490)$. The strong absorption of phytoplankton and CDOM could lead to relatively smaller R_{rs} in the blue and green bands [47,48]. In the PRE, the water constituents are from river inputs and coastal erosion. The C_{chla}, CDOM, and C_{tsm} are very high, which may result in both Mueller and Morel's models being inapplicable in the PRE.

Tiwari's model uses the reflectance ratio at 490 and 670 nm, $R_{rs}(490)/R_{rs}(670)$, to derive $K_d(490)$. Zhang's model is composed of two independent algorithms: one based on the ratio $R_{rs}(490)/R_{rs}(555)$ for clear waters and another based on the ratio $R_{rs}(490)/R_{rs}(665)$ for turbid waters. When tested with the independent PRE dataset, the predictions of these two models were statistically better compared to both Mueller's and Morel's models. However, the two models also showed a pronounced underestimation for higher $K_d(490)$ values (>1.0 m^{-1}), which might be due to the strong backscattering of suspended sediments in the more turbid waters in the PRE. Lee's model produced a suitable estimation, which uses a relationship relating the backscattering coefficient at 490 nm to the irradiance reflectance just beneath the surface within the red band. The performance of Wang's model was the same as Lee's, which was attributed to the same approach used in both models for highly turbid waters. In Wang's model, the retrieval method switched to Mueller's in clear waters, and the bridging of the two types of models is based on a certain weighting function (W). The spectral information within the red band cannot be ignored when retrieving $K_d(490)$ for turbid waters. However the values of $R_{rs}(670)/R_{rs}(490)$ tended to be very low and, therefore, values of $K_d(490)$ for clear waters were inaccurate. Therefore, Wang's model, which uses a combination of different algorithms for clear and turbid waters, is a better choice for $K_d(490)$ retrieval in the PRE waters.

4.2. Dominant Contributor to $K_d(490)$ of Water Constituents

Attenuation of light in water depends on concentrations of particulate matter and dissolved matter, which can be expressed by C_{tsm}, C_{chla}, and the absorption coefficient of CDOM [7,49]. The contribution of these constituents varies for different types of water and within the same water body in different seasons [50–52]. Since the calculated GRGs between $K_d(490)$ and $a_{dg}(443)$ were significantly lower than the other two water constituents, only the GRGs of C_{tsm} and C_{chla} were considered. Figure 9 depicts the subtraction of both GRGs. Positive values indicate the GRGs of C_{tsm} were higher than those of C_{chla}, which means that C_{tsm} played a dominant role in $K_d(490)$ variability. In contrast, negative values indicate that the C_{chla} had a greater influence. The C_{tsm}-dominated were waters generally located in coastal and estuarine turbid areas, whereas the C_{chla}-dominated waters were observed in open clear ocean. Notably, waters dominated by $a_{dg}(443)$ were rare. The strong absorption of CDOM in the blue bands influenced the variability of $K_d(490)$, particularly in waters with high CDOM concentrations. The major sources of CDOM in the PRE were the river water and the human and industrial sewage [53,54]. However, in coastal or estuarine areas with highly turbid waters, C_{tsm} can reach over 100 g·m^{-3}. During the survey conducted on 5 June 2012, the range of measured $a_g(443)$ was 0.12 to 0.58 and the range of measured $a_p(443)$ was 0.31 to 1.61. The latter was approximately three orders higher than the former, indicating that the influence of total suspended sediments on $K_d(490)$ was far greater than that of CDOM.

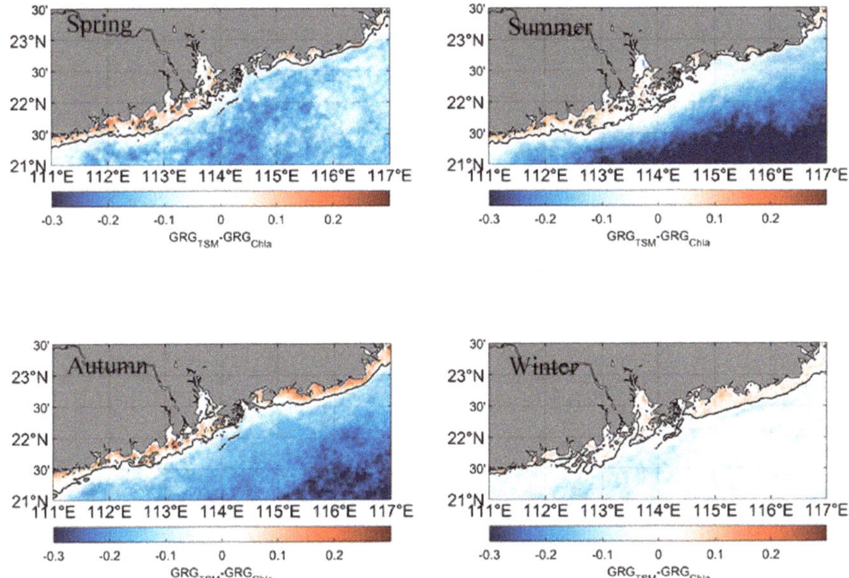

Figure 9. Distribution of dominant water constituents in the four seasons.

The distribution of dominant constituents showed some seasonality. In spring and summer, the C_{tsm}-dominated waters were mainly distributed in LB and the western GDP. The C_{tsm}-dominated area was confined close to the nearshore areas in the eastern GDP, indicating the impact of C_{chla} can extend from offshore to nearshore regions. Compared with other seasons, the most significant feature in summer was the southward extension of C_{tsm} from LB to the open ocean, which can probably be attributed to the increase in river runoff. In autumn and winter, the C_{tsm}-dominated area was wider in the eastern GDP than in spring and summer. The underlying reason for the change in area still requires future research. Currently, the change in area in the eastern GDP during autumn and winter might be indirectly caused by the decrease in C_{chla} rather than the variability of C_{tsm}. In autumn and winter, the entire eastern GDP is influenced by monsoons. The northeasterly wind-induced downwelling appears to decrease the amount of resuspension, resulting in the slight decrease in surface C_{tsm}, which seems to contradict the expansion of the C_{tsm}-dominated area. However, the downwelling also inhibits the growth of phytoplankton. The decrease in surface C_{chla} may prevent it from becoming the primary factor affecting the variability of $K_d(490)$.

4.3. Influence of Physical Factors on $K_d(490)$ Variability

Figure 3 shows that the $K_d(490)$ values in the PRE waters were markedly different in different regions and in different seasons, and Figure 9 shows that the spatial variations can be attributed to the changes in C_{chla} and C_{tsm}.

To understand the mechanism through which the seasonal $K_d(490)$ varies, correlation analysis was performed between several types of physical factors, including wind field, river runoff, MLD, SST, and seasonal average $K_d(490)$. The results showed that the average $K_d(490)$ was highly correlated with the wind speed (u-component) in summer, with an R^2 of about 0.69. During winter, we found a significant negative correlation between $K_d(490)$ and SST, with an R^2 of –0.66 (Figure 10). Seasonal anomalies were also obtained by subtracting the seasonal climatological average. In 2007 and 2015, when the wind speed anomaly (u-component) reached its peak, a distinguishing tongue of $K_d(490)$ anomaly was observed near the southeastern LB (Figure 11). Inside this tongue region, the $K_d(490)$

anomaly in the west was higher than in the east, indicating that the variability can be attributed to the high turbid river plume waters in the surface layer, which are driven by the intense eastward wind.

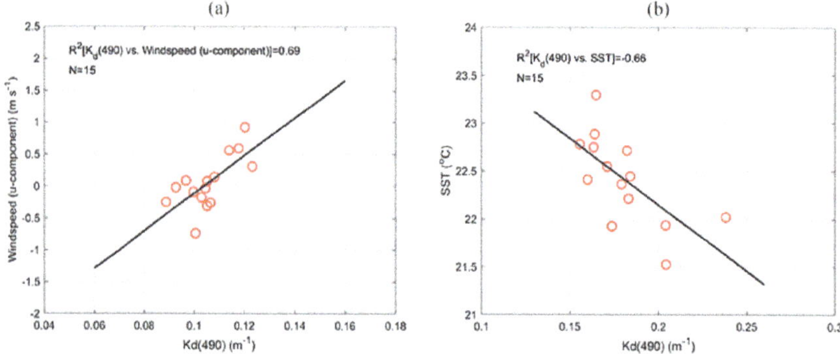

Figure 10. (**a**) Scatterplots of average $K_d(490)$ and wind speed (u-component) in summer, (**b**) scatterplots of average $K_d(490)$ and sea surface temperature (SST) in winter.

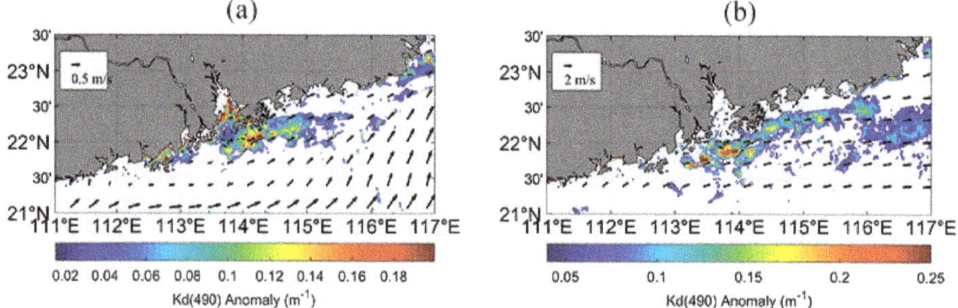

Figure 11. $K_d(490)$ and wind field anomalies during summer in (**a**) 2007 and (**b**) 2015.

The winter SST cooling in 2004 was the most significant during the whole study period, and was located in the southeastern PRE, which was about 0.4 °C cooler than the winter climatological average. Within these cooling regions, $K_d(490)$ values higher than the average values were observed, with anomalies ranging approximately from 0.1 to 0.35 m^{-1}. The observed winter variations in $K_d(490)$ in the southeastern PRE were strongly consistent with the changes in SST anomalies, and higher values coincided with lower SST (Figure 12).

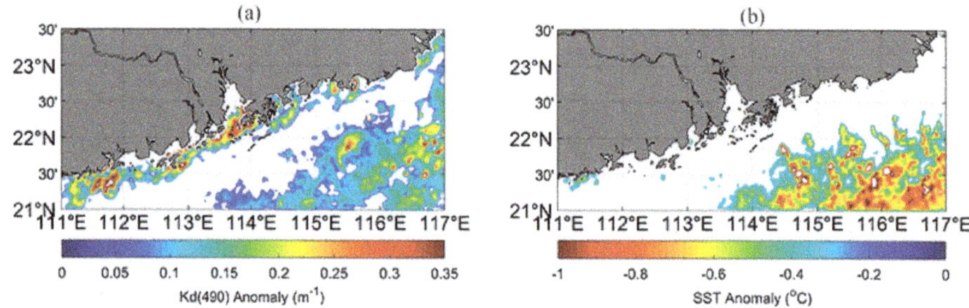

Figure 12. (**a**) $K_d(490)$ anomaly during winter 2004, (**b**) SST anomaly during summer 2004.

The variability of $K_d(490)$ in the southeastern PRE was mainly determined by C_{chla}. The average values of $K_d(490)$ were higher in winter than in summer. This seasonal variability might be attributed to the deepening of MLD in winter. In marine systems, MLD is generally deeper in winter than in summer [55]. Nutrients are brought from the bottom of the ocean to the surface or subsurface, which may enhance phytoplankton growth. The strong mixing in winter was demonstrated by the deepening of MLD, and a significant relationship between SST and MLD provided evidence that nutrients were supplied from the bottom waters (Figure 13).

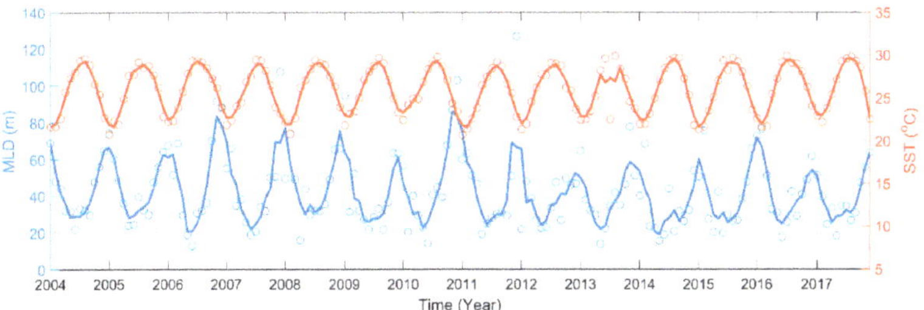

Figure 13. Time-series of average SST and MLD in the PRE from 2004 to 2017.

5. Conclusions

Accurate estimation of $K_d(490)$ using ocean color remote sensing imagery is challenging in turbid coastal waters due to the optical complexity of the water. Several approaches, including empirical and semianalytical models, were applied to retrieve the $K_d(490)$ in PRE water. The results showed that Wang's model was more accurate and is most suitable for PRE water, which uses a combination of different algorithms for clear and turbid waters. Hence, Wang's model was selected for deriving $K_d(490)$ products from long-term MODIS/Aqua imagery.

Derived from long-term MODIS/Aqua imagery, the temporal variability and spatial distribution of $K_d(490)$ were tracked using S-EOF analysis. The results of GRA showed that both phytoplankton and suspended sediments were the two dominant contributors to the variability in $K_d(490)$. The C_{tsm}-dominated waters were generally located in coastal and estuarine turbid area, whereas the C_{chla}-dominated waters were observed in clear open ocean. The influence of wind field on the variability of $K_d(490)$ was significant near the coastal and estuarine regions in summer. With the strengthening of the eastward wind, a water tongue of relatively higher $K_d(490)$ values formed in the southeastern PRE. In winter, the location of the negative SST anomaly and positive $K_d(490)$ anomaly was strongly consistent, indicating that the sea surface cooling was related to the positive $K_d(490)$ anomaly. The winter variability might be attributed to the strong mixing, which brought nutrients from the bottom layer to the surface to enhance phytoplankton growth.

Estuarine and coastal regions are complex ecosystems. To better examine the biogeochemical responses to physical events, a combination of remote sensing and coupled hydrodynamic–biological models should be applied in future research.

Author Contributions: Conceptualization, S.T. and H.Y.; implementation, H.Y. and C.Y.; validation, H.Y. and C.Y.; writing—original draft preparation, H.Y. and C.Y.; writing—review and editing, S.T., H.Y., and C.Y.; funding acquisition: S.T. All authors have read and agreed to the published version of the manuscript.

Funding: This research was funded by the Strategic Priority Research Program of the Chinese Academy of Sciences (No. XDA13010404 and XDA19060501), the Key Special Project for Introduced Talents Team of Southern Marine Science and Engineering Guangdong Laboratory (Guangzhou) (No. GML2019ZD0302, 2019BT2H594), the National Key Research and Development Program of China (No. 2018YFD0900901), the Joint Funds of the National Natural Science Foundation of China (No. U1901215), and Key Research and Development Project of GuangXi (GUI-KE: AB16380339).

Acknowledgments: The authors would like to thank the NASA Goddard Space Center for providing MODIS data and the NASA OBPG group for providing the SeaDAS software package. Our colleagues in the Ocean Color group of the South China Sea Institute of Oceanology, Chinese Academy of Sciences, are thanked for their effort in collecting and processing the samples.

Conflicts of Interest: The authors declare no conflict of interest.

References

1. Wang, M.; Son, S.H.; Harding, L.W., Jr. Retrieval of diffuse attenuation coefficient in the Chesapeake Bay and turbid ocean regions for satellite ocean color applications. *J. Geophys. Res.* **2009**, *114*. [CrossRef]
2. Zhang, Y.; Liu, X.; Yin, Y.; Wang, M.; Qin, B. A simple optical model to estimate diffuse attenuation coefficient of photosynthetically active radiation in an extremely turbid lake from surface reflectance. *Opt. Express* **2012**, *20*, 20482–20493. [CrossRef] [PubMed]
3. Robert, P.B.; Alexander, S.; Kirill, Y.K. *Optical Properties and Remote Sensing of Inland and Coastal Waters*, 1st ed.; CRC Press: New York, NY, USA, 1995.
4. Zhao, J.; Barnes, B.; Melo, N.; English, D.; Lapointe, B.; Muller-Karger, F.; Schaeffer, B.; Hu, C.M. Assessment of satellite-drived diffuse attenuation coefficients and euphotic depths in south Florida coastal waters. *Remote Sens. Environ.* **2013**, *131*, 38–50. [CrossRef]
5. Wu, Y.; Tang, C.C.; Sathyendranath, S.; Platt, T. The impact of bio-optical heating on the properties of the upper ocean: A sensitivity study using a 3-D circulation model for the Labrador Sea. *Deep. Sea Res. Part II Top. Stud. Oceanogr.* **2007**, *54*, 2630–2642. [CrossRef]
6. Saulquin, B.; Hamdi, A.; Gohin, F.; Populus, J.; Mangin, A.; D'Andon, O.F. Estimation of the diffuse attenuation coefficient KdPAR using MERIS and application to seabed habitat mapping. *Remote Sens. Environ.* **2013**, *128*, 224–233. [CrossRef]
7. Shi, K.; Zhang, Y.; Liu, X.; Wang, M.; Qin, B. Remote sensing of diffuse attenuation coefficient of photosynthetically active radiation in Lake Taihu using MERIS data. *Remote Sens. Environ.* **2014**, *140*, 365–377. [CrossRef]
8. Jerlov, N.G. *Optical Oceanography*; Elsevier: New York, NY, USA, 1976.
9. Clavano, W.; Boss, E.; Karp-Boss, L. Inherent optical properties of non-spherical marine-like particles-From theory to observation. *Oceanogr. Mar. Biol.* **2007**, *45*, 1–38. [CrossRef]
10. McClain, C.R.; Feldman, G.C.; Hooker, S.B. An overview of the SeaWiFS project and strategies for producing a climate research quality global ocean bio-optical time series. *Deep. Sea Res. Part II Top. Stud. Oceanogr.* **2004**, *51*, 5–42. [CrossRef]
11. Mueller, J.L. *SeaWiFS Algorithm for the Diffuse Attenuation Coefficient K(490) Using Water-Leaving Radiance at 490 and 555 nm*; SeaWiFS Postlaunch Calibration and Validation Analyses, Part 3; Center for Hydro-Optics and Remote Sensing/SDSU: San Diego, CA, USA, 2000; Chapter 3.
12. Shi, W.; Wang, M. Characterization of global ocean turbidity from Moderate Resolution Imaging Spectroradiometer ocean color observations. *J. Geophys. Res. Oceans* **2010**, *115*. [CrossRef]
13. Kratzer, S.; Brockmann, C.; Moore, G. Using MERIS full resolution data to monitor coastal waters—A case study from Himmerfjärden, a fjord-like bay in the northwestern Baltic Sea. *Remote Sens. Environ.* **2008**, *112*, 2284–2300. [CrossRef]
14. Wong, L.A.; Heinke, G.; Chen, J.C.; Xue, H.; Dong, L.X.; Su, J.L. A model study of the circulation in the Pearl River Estuary (PRE) and its adjacent coastal waters: 1. Simulations and comparison with observations. *J. Geophys. Res. Oceans* **2003**, *108*, 3156. [CrossRef]
15. Wyrtki, K. *Physical Oceanography of the Southeast Asian Waters: Scientific Results of Marine Investigations of the South China Sea and the Gulf of Thailand*; Naga Report 2; Scripps Institution of Oceanography: San Diego, CA, USA, 1961; 195p.
16. Gan, J.; Li, H.; Curchitser, E.N.; Haidvogel, D.B. Modeling South China Sea circulation: Response to seasonal forcing regimes. *J. Geophys. Res.* **2006**, *111*, C06034. [CrossRef]
17. Ye, H.; Chen, C.; Tang, S.; Tian, L.; Sun, Z.; Yang, C.; Liu, F. Remote sensing assessment of sediment variation in the Pearl River Estuary induced by Typhoon Vicente. *Aquat. Ecosyst. Health Manag.* **2014**, *17*, 271–279. [CrossRef]
18. Cai, W.J.; Dai, M.H.; Wang, Y.C.; Zhai, W.D.; Huang, T.; Chen, S.T.; Zhang, F.; Chen, Z.Z.; Wang, Z.H. The biogeochemistry of inorganic carbon and nutrients in the Pearl River estuary and the adjacent Northern South China Sea. *Cont. Shelf Res.* **2004**, *24*, 1301–1319. [CrossRef]
19. Chen, C.; Tang, S.; Pan, Z.; Zhan, H.; Larson, M.; Jonsson, L. Remotely sensed assessment of water quality levels in the Pearl River Estuary, China. *Mar. Pollut. Bull.* **2007**, *54*, 1267–1272. [CrossRef]
20. Zhao, J.; Cao, W.; Wang, G.; Yang, D.; Yang, Y.; Sun, Z.; Zhou, W.; Liang, S. The variations in optical properties of CDOM throughout an algal bloom event. *Estuar. Coast. Shelf Sci.* **2009**, *82*, 225–232. [CrossRef]

21. Mueller, J.L.; Fargion, G.S. *Ocean Optics Protocols for Satellite Ocean Color Sensor Validation;* SeaWiFS Technical Report Series; NASA Center for AeroSpace Information: Linthicum Heights, MD, USA, 2002.

22. Sun, D.; Hu, C.; Qiu, Z.; Shi, K. Estimating phycocyanin pigment concentration in productive inland waters using Landsat measurements: A case study in Lake Dianchi. *Opt. Express* **2015**, *23*, 3055–3074. [CrossRef]

23. Zheng, Z.; Li, Y.; Guo, Y.; Xu, Y.; Liu, G.; Du, C. Landsat-Based Long-Term Monitoring of Total Suspended Matter Concentration Pattern Change in the Wet Season for Dongting Lake, China. *Remote Sens.* **2015**, *7*, 13975–13999. [CrossRef]

24. Li, Y.; Zhang, Y.; Shi, K.; Zhu, G.; Zhou, Y.; Zhang, Y.; Guo, Y. Monitoring spatiotemporal variations in nutrients in a large drinking water reservoir and their relationships with hydrological and meteorological conditions based on Landsat 8 imagery. *Sci. Total Environ.* **2017**, *599*, 1705–1717. [CrossRef]

25. Pierson, D.; Kratzer, S.; Strömbeck, N.; Håkansson, B. Relationship between the attenuation of downwelling irradiance at 490 nm with the attenuation of PAR (400 nm–700 nm) in the Baltic Sea. *Remote Sens. Environ.* **2008**, *112*, 668–680. [CrossRef]

26. Ruddick, K.G.; Ovidio, F.; Rijkeboer, M. Atmospheric correction of SeaWiFS imagery for turbid coastal and inland waters. *Appl. Opt.* **2000**, *39*, 897–912. [CrossRef] [PubMed]

27. Morel, A.; Gentili, B. Diffuse reflectance of oceanic waters: Its dependence on Sun angle as influenced by the molecular scattering contribution. *Appl. Opt.* **1991**, *30*, 4427–4438. [CrossRef] [PubMed]

28. Morel, A.; Gentili, B. Diffuse reflectance of oceanic waters. II. Bidirectional aspects. *Appl. Opt.* **1993**, *32*, 6864–6879. [CrossRef] [PubMed]

29. Morel, A.; Gentili, B. Diffuse reflectance of oceanic waters III Implication of bidirectionality for the remote-sensing problem. *Appl. Opt.* **1996**, *35*, 4850–4862. [CrossRef]

30. Gordon, H. Normalized water-leaving radiance: Revisiting the influence of surface roughness. *Appl. Opt.* **2005**, *44*, 241–248. [CrossRef]

31. Saha, S.; Moorthi, S.; Wu, X.; Wang, J.; Nadiga, S.; Tripp, P.; Behringer, D.; Hou, Y.-T.; Chuang, H.-Y.; Iredell, M.; et al. The NCEP Climate Forecast System Version 2. *J. Clim.* **2014**, *27*, 2185–2208. [CrossRef]

32. Lu, S.L.; Liu, Z.H.; Li, H.; Li, Z.Q.; Wu, X.F.; Sun, C.H.; Xu, J.P. *User Manual of Global Ocean Argo Gridded Dataset (BOA_Argo);* Second Institute of Oceanography, MNR: Hangzhou, China, 2020; 28p.

33. Zhang, T.; Fell, F. An empirical algorithm for determining the diffuse attenuation coefficient Kd in clear and turbid waters from spectral remote sensing reflectance. *Limnol. Oceanogr. Methods* **2007**, *5*, 457–462. [CrossRef]

34. Tiwari, S.P.; Shanmugam, P. A Robust Algorithm to Determine Diffuse Attenuation Coefficient of Downwelling Irradiance from Satellite Data in Coastal Oceanic Waters. *IEEE J. Sel. Top. Appl. Earth Obs. Remote Sens.* **2013**, *7*, 1616–1622. [CrossRef]

35. Morel, A.; Maritorena, S. Bio-optical properties of oceanic waters: A reappraisal. *J. Geophys. Res. Ocean* **2001**, *106*, 7163–7180. [CrossRef]

36. Lee, Z.; Du, K.; Arnone, R. A model for the diffuse attenuation coefficient of downwelling irradiance. *J. Geophys. Res. Ocean* **2005**, *110*. [CrossRef]

37. Wang, B.; An, S.-I. A method for detecting season-dependent modes of climate variability: S-EOF analysis. *Geophys. Res. Lett.* **2005**, *32*, 15710. [CrossRef]

38. Deng, J.L. Introduction of grey system theory. *J. Grey Syst.* **1989**, *1*, 1–24.

39. Liu, S.; Lin, Y. *Grey Information: Theory and Practical Applications;* Springer: London, UK, 2005.

40. Wan, S.; Chang, S.-H. Crop classification with WorldView-2 imagery using Support Vector Machine comparing texture analysis approaches and grey relational analysis in Jianan Plain, Taiwan. *Int. J. Remote Sens.* **2018**, *40*, 8076–8092. [CrossRef]

41. Xu, J.D.; Cai, S.Z.; Xiong, L.L.; Qiu, Y.; Zhu, D.Y. Study on coastal upwelling in eastern Hainan Island and western Guangdong in summer, 2006. *Acta Oceanol. Sin.* **2013**, *35*, 11–18, (In Chinese with English Abstract).

42. Xu, J.D.; Cai, S.Z.; Xiong, L.L.; Qiu, Y.; Zhou, X.W.; Zhu, D.Y. Observational study on summertime upwelling in coastal seas between eastern Guangdong and southern Fujian. *J. Trop. Oceanogr.* **2014**, *33*, 1–9.

43. Chen, J.Q.; Fu, Z.L.; Li, F.X. A study of upwelling over Minnan-Taiwan shoal fishing gournd. *J. Oceanogr. Taiwan Strait* **1982**, *2*, 5–13.

44. Werdell, P.J.; Bailey, S.W. An improved in-situ bio-optical data set for ocean color algorithm development and satellite data product validation. *Remote Sens. Environ.* **2005**, *98*, 122–140. [CrossRef]

45. Werdell, J. Global Bio-optical Algorithms for Ocean Color Satellite Applications: Inherent Optical Properties Algorithm Workshop at Ocean Optics XIX; Barga, Italy, 3–4 October 2008. *Eos* **2009**, *90*, 4. [CrossRef]

46. Werdell, P.J.; Franz, B.; Bailey, S.W.; Feldman, G.C.; Boss, E.; Brando, V.E.; Dowell, M.; Hirata, T.; Lavender, S.; Lee, Z.; et al. Generalized ocean color inversion model for retrieving marine inherent optical properties. *Appl. Opt.* **2013**, *52*, 2019–2037. [CrossRef] [PubMed]

47. Li, S.J.; Wu, Q.; Wang, X.J.; Piao, X.Y.; Dai, Y.N. Correlation between reflectance spectra and contents of Chlorophyll-a in Chaohu Lake. *J. Lake Sci.* **2002**, *14*, 228–234.

48. Chen, J.; Cui, T.; Tang, J.; Song, Q. Remote sensing of diffuse attenuation coefficient using MODIS imagery of turbid coastal waters: A case study in Bohai Sea. *Remote Sens. Environ.* **2014**, *140*, 78–93. [CrossRef]

49. Kirk, J.T.O. *Light and Photosynthesis in Aquatic Ecosystems*, 3rd ed.; Cambridge University Press: Cambridge, UK, 2011.

50. Christian, D.; Sheng, Y. Relative influence of various water quality parameters on light attenuation in Indian River Lagoon. *Estuar. Coast. Shelf Sci.* **2003**, *57*, 961–971. [CrossRef]

51. Lund-Hansen, L.C. Diffuse attenuation coefficients Kd(PAR) at the estuarine North Sea–Baltic Sea transition: Time-series, partitioning, absorption, and scattering. *Estuar. Coast. Shelf Sci.* **2004**, *61*, 251–259. [CrossRef]

52. Zhang, Y.; Zhang, B.; Ma, R.; Feng, S.; Le, C. Optically active substances and their contributions to the underwater light climate in Lake Taihu, a large shallow lake in China. *Fundam. Appl. Limnol.* **2007**, *170*, 11–19. [CrossRef]

53. Huang, M.F.; Wang, D.F.; Xing, X.F.; Wei, J.A.; Liu, D.; Zhao, Z.L.; Wang, W.X.; Liu, Y. The research on remote sensing mode of retrieving ag(440) in Zhujiang River Estuary and its application. *Haiyang Xuebao* **2015**, *37*, 67–77.

54. Zhao, J.; Cao, W.; Xu, Z.; Ai, B.; Yang, Y.; Jin, G.; Wang, G.; Zhou, W.; Chen, Y.; Chen, H.; et al. Estimating CDOM Concentration in Highly Turbid Estuarine Coastal Waters. *J. Geophys. Res. Oceans* **2018**, *123*, 5856–5873. [CrossRef]

55. Mann, K.H.; Lazier, J.R.N. *Dynamics of Marine Ecosystem: Biological-Physical Interactions in the Oceans*; Blackwell Science, Inc.: Cambridge, MA, USA, 1996.

 *remote sensing*

Article

Morphological Band Registration of Multispectral Cameras for Water Quality Analysis with Unmanned Aerial Vehicle

Wonkook Kim [1], Sunghun Jung [2,*], Yongseon Moon [3] and Stephen C. Mangum [4]

[1] Department of Civil and Environmental Engineering, Pusan National University, Busan 43241, Korea; wonkook@pusan.ac.kr

[2] Department of Smart Mobile Convergence System, Chosun University, Gwangju 61452, Korea

[3] Department of Electric Engineering, Sunchon National University, Suncheon 57922, Korea; moon@sunchon.ac.kr

[4] MicaSense, Inc., Seattle, WA 98103-7901, USA; stephen.mangum@micasense.com

* Correspondence: jungx148@chosun.ac.kr

Received: 30 May 2020; Accepted: 22 June 2020; Published: 24 June 2020

Abstract: Multispectral imagery contains abundant spectral information on terrestrial and oceanic targets, and retrieval of the geophysical variables of the targets is possible when the radiometric integrity of the data is secured. Multispectral cameras typically require the registration of individual band images because their lens locations for individual bands are often displaced from each other, thereby generating images of different viewing angles. Although this type of displacement can be corrected through a geometric transformation of the image coordinates, a mismatch or misregistration between the bands still remains, owing to the image acquisition timing that differs by bands. Even a short time difference is critical for the image quality of fast-moving targets, such as water surfaces, and this type of deformation cannot be compensated for with a geometric transformation between the bands. This study proposes a novel morphological band registration technique, based on the quantile matching method, for which the correspondence between the pixels of different bands is not sought by their geometric relationship, but by the radiometric distribution constructed in the vicinity of the pixel. In this study, a Micasense Rededge-M camera was operated on an unmanned aerial vehicle and multispectral images of coastal areas were acquired at various altitudes to examine the performance of the proposed method for different spatial scales. To assess the impact of the correction on a geophysical variable, the performance of the proposed method was evaluated for the chlorophyll-a concentration estimation. The results showed that the proposed method successfully removed the noisy spatial pattern caused by misregistration while maintaining the original spatial resolution for both homogeneous scenes and an episodic scene with a red tide outbreak.

Keywords: band registration; morphological registration; multispectral camera; water quality; Micasense Rededge-M

---

1. Introduction

In multispectral images, precise registration of multispectral bands is critical for subsequent quantitative data analysis, which relies on the "spectrum" of the signal (e.g., radiance or reflectance) from the target. If the bands are not perfectly aligned with each other, due to the reasons such as lens distortion, displacement in lens location, and inaccurate geometry transformation between the locations, spectral radiometric values in a fixed pixel location may originate from different targets. Algorithms that depend on the band ratios, or the band difference, are particularly sensitive to the quality of the band registration and may produce significant errors for an inhomogeneous target area if misregistration exists.

There are various sources for misregistration: a difference in the lens locations for each band, a difference in the image acquisition timing accompanied by fast-moving targets, etc. Commercial multispectral cameras typically include individual lenses for the multi-bands and have different exposure times to maximize the effective radiometric range for various targets. The differences in the lens locations for the multi-bands can be modeled via a projective transformation if the image is assumed to be free of nonlinear image distortions, such as radial distortion; the differences in the viewing geometry can be corrected to a reference band using band-to-band projective transformations that specifies eight parameters for rotation (1), translation (2), isotropic scaling (1), anisotropic scaling (1), skew (1), and perspective shortening (2) (figures in the parenthesis denotes the number of free variables for the quantity) [1,2]. The methods based on Fourier transform does not require the time-consuming process of finding matching points between two images, and effectively register multiple bands solely based on its spatial frequency pattern [3].

However, such transformation approaches that rely on geometric characteristics of the scene cannot effectively address the cases having non-rigid body target deformation, where the forms of targets may vary between the bands. This issue is prominent when analyzing the color of water where the targets (i.e., water surface) move or deform quickly during a short time interval (<1 s) between the acquisition of different bands. As shown in Figure 1, multispectral band images for ocean surface in the normal coastal area in Korea reveal that the differences in water reflectance, and its spatial pattern, clearly do not correspond to rigid-body transformation; thus, the differences cannot be resolved by a projective transformation. Note that the band images shown in Figure 1 have already undergone band-to-band registration through a projective transformation.

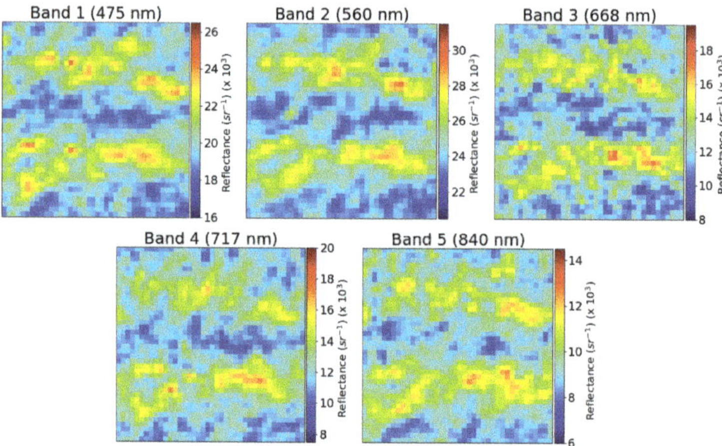

Figure 1. A subset of a multispectral image acquired in a coastal area of Korea targeted on the ocean surface with normal states (low chlorophyll and suspended particle concentration), showing the differences in the spatial pattern for the five spectral bands.

This type of image registration, which involves a non-rigid body transformation, has been investigated for morphological image registration [4–7]. It is applicable when the images of two different targets, or of one target object that experienced a non-rigid body deformation, are expected to have a similar internal structure and most of the image contents have common features. The basic mathematical tools for morphological image transformation include calculus of variations, optimization with regularization or constraints, and the derivation of invariant features. The methods are typically used in medical image processing, such as computed tomography and magnetic resonance imaging, for diagnostic purposes; however, applications to remote sensing are rare, particularly for the band registration, because typical observation targets in optical remote sensing are stationary objects, such as land surfaces and man-made structures. However, because water surfaces move fast, owing to ocean currents and wave motion, multispectral images from ships and low-altitude platforms, such as unmanned aerial vehicles (UAVs), clearly exhibit locational errors.

In this study, we develop a novel morphological band registration technique, designed for high-resolution water quality analysis, which preserves the true spectrum of fast-moving targets. The proposed method exploits the quantile plots between the bands to accurately determine the radiometric correspondence of the pixels in different bands. For multispectral images, a Micasense Rededge-M camera was operated onboard a UAV, DJI Inspire-2, and ocean surface images were acquired from coastal areas of Korea, of various altitudes and biological conditions. The specification and radiometric properties of the multispectral camera and the image data are described in Section 2 (Materials). The radiometric/geometric preprocessing, derivation of the "remote sensing reflectance" for water quality analysis are presented in Section 3 (Methodology and Analysis), along with the analysis on the adverse impact of the residual misregistration on the water quality variable estimation. In Section 4 (Algorithm Development and Assessment), the proposed morphological registration scheme is described in detail and the development of the entire correction procedure is presented. The algorithm results are demonstrated for multiple test images, taken at various altitudes. In Section 5 (Discussion and Conclusion), the correction results and remaining tasks are discussed.

2. Materials

2.1. Micasense Rededge-M and Data

The Rededge-M camera has five spectral bands, the center wavelengths of which are located at 475, 550, 668, 717, and 840 nm (Figure 2). The red (668 nm) and near-infrared (NIR) (717 nm) bands are designed to capture the red edge feature, which is salient in vegetation, and to quantify the photosynthetic pigments via indices, such as the normalized differenced vegetation index [8] and soil-adjusted vegetation index [9]. The blue band (475 nm) is useful when quantifying pigment absorption in water when it is referenced with respect to the green band (550 nm) [10]. In water color analysis, the last NIR band (840 nm) is particularly useful for quantifying atmospheric scattering between the target and the sensor because clear water theoretically has zero water-leaving radiance in the 840 nm band [11]. For turbid waters, the radiance at 840 nm is often large and can thus be used to detect the existence of suspended sediments in water [12]; however, as a result, the estimation of atmospheric effects becomes more complicated [13,14]. The radiometric sensitivity of the Rededge-M was tested for water color analysis in Kim et al. (2019) [15] in a brief experiment in which the radiometric data from a Rededge-M camera was compared with that from a hyperspectral radiometer, TriOS RAMSES. The study showed that the Rededge-M was able to retrieve a comparable spectral shape for a water body (which typically has a low radiance level compared to terrestrial targets) when calculated using remote sensing reflectance (R_{rs}).

In this study, four Rededge-M image sets from multiple field campaigns were used for the development of a morphological registration algorithm. Table 1 shows the dates, locations, and altitudes of the camera images that were used for the analysis, and the study site is presented in Figure 3a. All images were captured by a drone, DJI Inspire-2. The camera body and the downwelling irradiance sensor were installed on the drone using a simple bracket and a damper (Figure 3b). The Rededge-M camera was installed with a fixed viewing zenith angle of 40° and the camera attitude was controlled to head north to constrain the relative azimuth angle within 90–135° with respect to the sun direction, to minimize the surface reflectance [16,17]. The Zenmuse-X5s camera was installed in front of the Rededge-M to capture a wider-angle overview of the target scene with higher spatial resolution. RGB images of water reflectance are presented in Figure 4 for all four scenes.

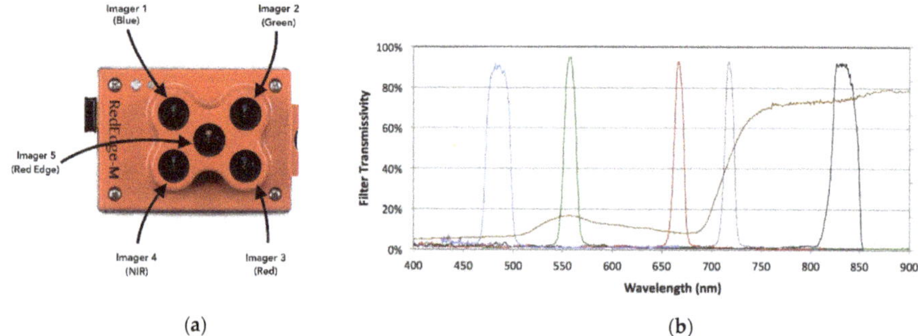

(a)   (b)

Figure 2. (**a**) Lens configuration of Rededge-M camera and (**b**) the spectral response function of the bands overlaid with a typical vegetation spectrum (source: Rededge-M User Manual).

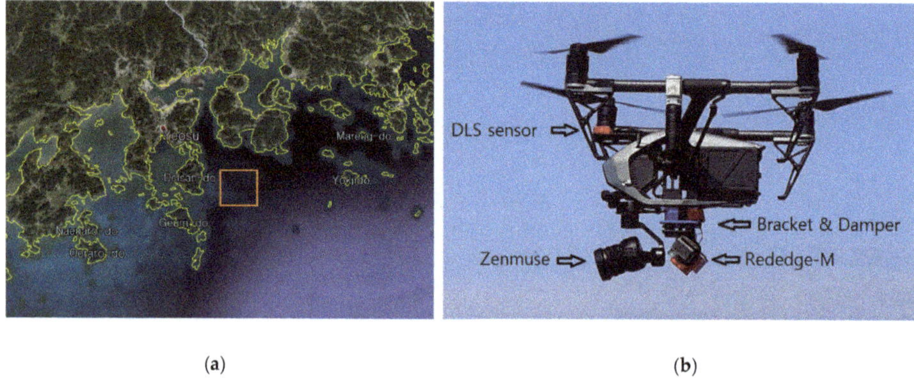

(a)   (b)

Figure 3. (**a**) The map of study site near Yeosu, a southern coast of Korea, and the bounding box (orange) for the area that the unmanned aerial vehicle (UAV) was operated for, and (**b**) photographs of the multispectral-UAV system configured with the downwelling light sensor (DLS), the RGB camera, and the multispectral camera.

Table 1. Data list used in this study.

Scene ID	Time	Location	Altitude (m)	Scene Description
A	26–07–2019 15:23	Sumoon	85.4	Coastal Area
B	31–08–2019 12:54	Yeosu	8.1	From Ship
C	31–08–2019 13:29	Yeosu	196	Coastal Area
D	31–08–2019 13:17	Yeosu	390	Red Tide

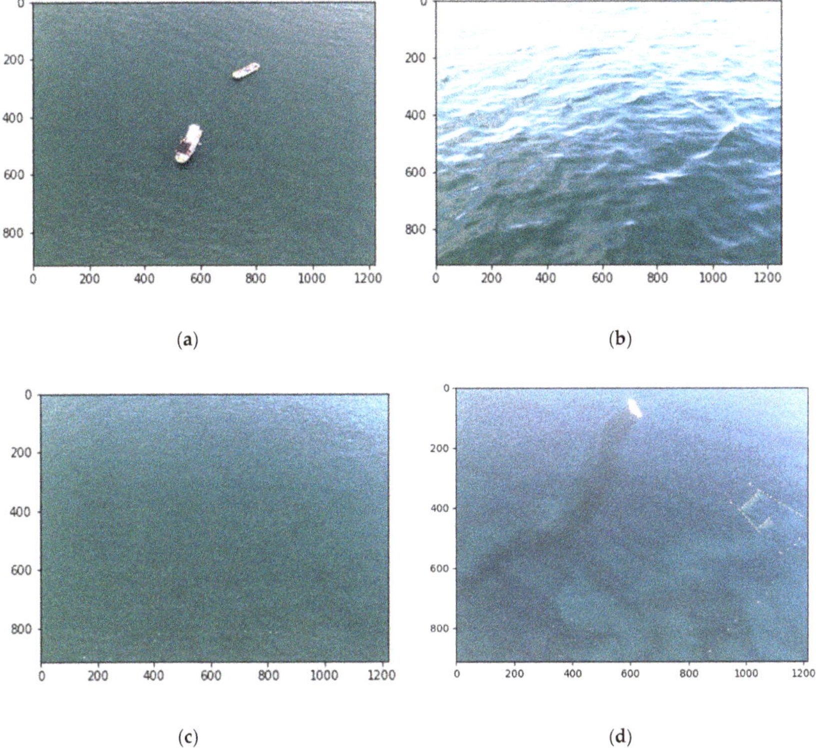

Figure 4. Water reflectance image (RGB: Band 3, Band 2, Band 1) acquired by the Rededge-M for (**a**) Scene-A (85.4 m), (**b**) Scene-B (8.1 m), (**c**) Scene-C (196.2 m), and (**d**) Scene-D (390.5 m, red tide), with the UAV altitudes in the parenthesis.

2.2. Acquisition Time Difference in Rededge-M Band Images

The Rededge-M employs a proprietary Auto Gain Control (AGC) algorithm that works to minimize the number of overexposed pixels, but the number of overexposed pixels will never be zero because the AGC also wants there to be a maximal number of properly exposed pixels. The AGC optimizes the gain (ISO) and exposure of each capture for each of the five imagers such that the resulting picture is properly exposed. The Rededge-M has five imagers that trigger the top of each frame together, and one "capture" is created during each triggering, which is represented by five different frames, one for each wavelength band. Each imager has a different filter as to only capture data from the wavelength band of interest. Because of the AGC, the ISO Speed and exposure time (which can be inspected in the metadata of each frame) may vary by band on an individual capture, and different captures taken during the same flight will also vary. The small differences between the exposure times among frames of a capture usually don't make a difference. However, motion blur may occur when there is a large difference (i.e., 1 ms vs. 5 ms) in exposure time between multiple frames in a single capture. If the camera is mounted on an aircraft and is in motion, the frames with longer exposures will have motion blur, and will have a slightly different geometric offset compared to the shorter exposure time. The geometric offset between frames with different exposure times on a single capture will be a function of the angular rate of the camera and the respective exposure times of the frames. This will manifest as motion blur, and will result in a difference in average pointing angle of $\theta_2 - \theta_1$, where θ_2 is the pointing angle from a frame with a longer exposure time and θ_1 is the pointing angle from a frame with a shorter exposure time. With all this information in mind, while each frame will have had the top of the frame triggered at the same instant, the exposure time metadata may be different, resulting in slightly different exposure times among a single capture. For example, the exposure time of the 5 band images of Scene-A are 1/741, 1/585, 1/780, 1/367, and 1/356 s, respectively, for 475, 550, 668, 717, and 840 nm, causing the targets to be captured in different status.

3. Methodology and Analysis

3.1. Radiometric and Geometric Calibration

Basic radiometric preprocessing was performed using the processing modules provided on the Micasense Github page [18]. A Vignette correction was first performed for each band image and these Vignette-corrected band images were input to the radiometric calibration process to produce the radiance data. The radial and tangential distortion were corrected according to following formula,

$$u_{radial-corrected} = u\left(1 + k_1 r^2 + k_2 r^4 + k_3 r^6\right) \tag{1}$$

$$v_{radial-corrected} = v\left(1 + k_1 r^2 + k_2 r^4 + k_3 r^6\right) \tag{2}$$

$$u_{tangential-corrected} = u + \left(2p_1 uv + p_2(r^2 + u^2)\right) \tag{3}$$

$$v_{tangential-corrected} = v + \left(2p_2 uv + p_1(r^2 + 2u^2)\right) \tag{4}$$

where u and v are image coordinates, $r = \sqrt{u^2 + v^2}$, k_1, k_2, k_3 are coefficients for radial distortion, and p_1, p_2 are for tangential distortion. As shown in Figure 1, the Rededge-M acquires radiance at five wavelengths, through five individual lenses, inevitably leading to misalignment in the band images. The misalignment can be corrected using a projective transformation, constructed by matching numerous matching points between two band images. The projective transformation between the bands is

$$x' = Mx, \tag{5}$$

where x and x' are 3×1 homogeneous vectors of image coordinates in two band images and M is a 3×3 non-singular matrix for the projective transformation.

3.2. Water Color Analysis

To conduct water color analysis for the estimation of in-water constituents (e.g., chlorophyll-a concentration), remote sensing reflectance must first be derived from the radiance measurements. Remote sensing reflectance (R_{rs}) can be calculated as

$$R_{rs} = \frac{L_{wT} - \rho L_{sky}}{E_d}, \tag{6}$$

where L_{wT} is the total radiance from water, L_{sky} is the downward radiance from the sky, E_d is the downward irradiance, and ρ is the Fresnel reflectance factor [19,20]. Setting aside ρ, to derive R_{rs} in the field, three radiometric measurements are required for each scene: L_{wT}, L_{sky}, and E_d. The first two radiance variables—L_{wT} and L_{sky}—are acquired by capturing the water surface and sky using the Rededge-M, following the measurement protocol suggested in the ocean color analysis [16]. For both observations, the recommended azimuth angle is 135°, with respect to the sun azimuth, and the recommended zenith angles are 45° and −45° for the water and sky, respectively.

The measurement protocol is intended to minimize the variation in ρ, which varies from 0.02 to 0.07, depending on wind speed and sun–sensor–target geometry [17], where the factor is confined to an approximate range of 0.02 to 0.025 when the aforementioned measurement protocol is observed. However, high-altitude drone images with a wide viewing angle lead to a wide range of viewing zenith and azimuth angles, in which case, ρ varies significantly outside the 0.02 to 0.025 range, requiring a pixel-based adaptive ρ estimation. A simple method to determine ρ adaptively for different locations is by exploiting the fact that the total water radiance at 840 nm is solely attributed to the surface-reflected radiance (not to the water-leaving radiance from the water body). This assumption holds when the water is clear; thus, the water-leaving radiance at 840 nm is nearly zero,

$$L_{wT} = L_w + L_{src} \tag{7}$$

$$L_{sfc} = \rho L_{sky}, \tag{8}$$

where L_w and L_{sfc} denote the water-leaving radiance and the surface-reflectance radiance, respectively. If $L_w(840) = 0$, then $L_{wT}(840) = L_{sfc}$, leading to $\rho = L_{wT}(840)/L_{sky}(840)$.

There are two options to determine the downwelling irradiance: (1) using the DLS and (2) the reference panel. The irradiance from the panel reflectance can be calculated as

$$E_d = \frac{L_{ref}}{\pi \, r_{panel}}, \tag{9}$$

where L_{ref} is the radiance from the reference panel and r_{panel} is the reflectance of the panel. If the two instruments are located and perfectly calibrated, the results of the two calculations should theoretically match. In this experiment, the DLS was attached to the UAV and the reference panel measurements were made on the ship, causing a difference in the altitude. In this experiment, the irradiance difference caused by the atmospheric conditions (water vapor, aerosol, etc.) was approximately 15–20%. Because our focus is on the water surface, we opted to use the irradiance measured at the ship level, via the reference panel.

After R_{rs} is obtained, it can be used to derive bio-geochemical variables, such as chlorophyll-a(Chla). To examine the effect of misregistration between the bands and the performance of the proposed morphological registration method, retrieval results are computed for chlorophyll concentrations, which is one of the most central biological quantities in the water quality analysis. For Chla concentrations in a non-turbid ocean condition, the OC2 algorithm, which utilizes one blue and one green band, was used [10,21].

$$\log_{10} \text{Chla} = a_0 + \sum_{i=1}^{4} a_i \left(\log_{10} \frac{R_{rs}(\lambda_{blue})}{R_{rs}(\lambda_{green})} \right) \tag{10}$$

where $R_{rs}(\lambda)$ is the remote sensing reflectance for the wavelength λ and a_i's are the algorithm coefficients.

The OC2 coefficients for the Landsat-8 operational land imager were used (482 nm for blue and 561 nm for green) for the Rededge-M, whose blue band is centered at 475 nm and the green at 550 nm [21]. For the tested scenes with a red tide outbreak, the red-to-blue ratio (RBR) algorithm [22], developed for the geostationary ocean color imager (490 nm for blue and 680 nm for green), was used to retrieve the chlorophyll contents in the bloom. It is important to note that the algorithm coefficients for OC2 and RBR were not specifically tuned to the Rededge-M in this study because the focus of the study is not on the precise retrieval of Chla concentrations but the analysis of the impact of misregistration (particularly spatial pattern). The band centers in the original OC2 and RBR algorithms do not significantly differ from those of the Rededge-M, from which we can reasonably assume that the spatial anomaly pattern would be similar, even after the fine calibration of the algorithms to Rededge-M.

3.3. The Impact of Pixel Misregistration on Water Quality Analysis

Figure 5a,b shows the RGB images of R_{rs} for Scene-A, for the fixed Fresnel factor ($\rho = 0.025$) and adaptive Fresnel factor cases, respectively. R_{rs} with a fixed ρ demonstrates that slant viewing angles cause a high surface reflectance in the upper right corner of the image. It can be observed that wave facets of different surface normals also led to varying viewing geometry, resulting in a variation of the

R_{rs} estimation, which exhibits residual sky reflectance on the surface. On the contrary, the adaptive approach demonstrates that the variation caused by the viewing geometry is significantly reduced with less across-image R_{rs} variation and smaller residual surface reflectances by the wave facets.

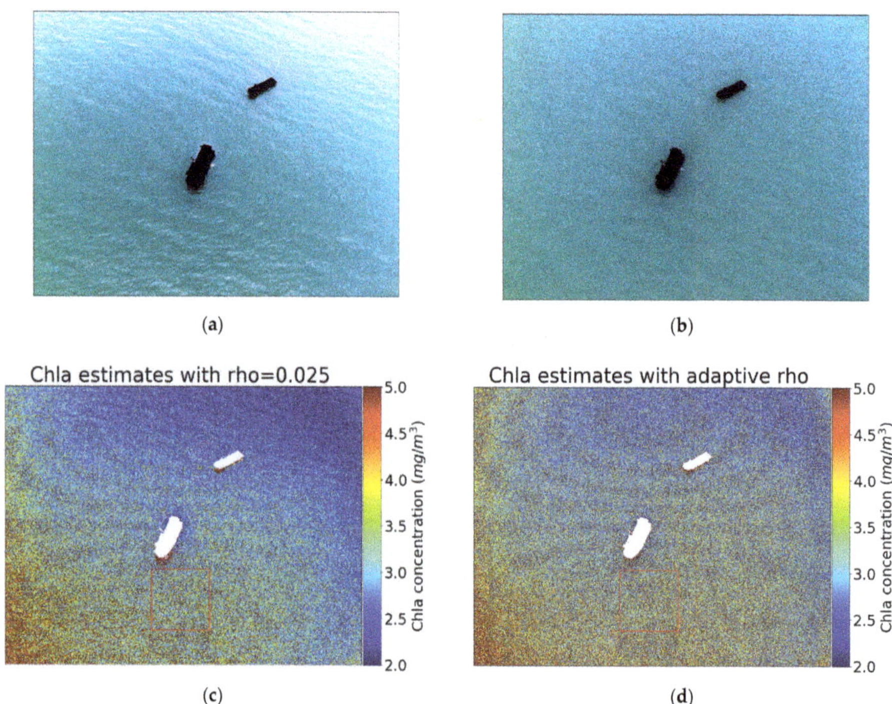

Figure 5. Images of remote sensing reflectance (**a**,**b**) and the Chla estimates from the respective R_{rs} images (**c**,**d**) for Scene-A. A fixed Fresnel reflectance factor was applied to panel (**a**) and the adaptive approach was used for panel (**b**).

Figure 5c,d are the OC2 Chla estimates, derived from the R_{rs} data, with a fixed ρ and adaptive ρ, respectively. The large residual surface reflectance caused by the slant viewing angles in the upper right corner led to significant underestimates of Chla and inflation in the bottom left corner. The anomalies are less significant in the adaptive ρ case; however, they have not been completely removed, even with the adaptive scheme. This implies that the surface-reflectance mechanism is more complex than what is described by the adaptive scheme model (e.g., the existence of a nonlinear band-by-band behavior). The phenomenon to focus on here is the large and noise-like Chla variation in a small-scale window. For a more detailed analysis, the subset areas marked by the red rectangles in Figure 5c,d were displayed in Figure 6. The Chla subset images show that the pixel-to-pixel variation is significantly large for both the fixed and adaptive approaches and such a high-frequency pattern is not caused by the real Chla spatial variation in the field. The adaptive approach exhibited a similar degree of variation to the fixed approach, which reveals that the noise-like pattern is not from the variation of wave facets. The images of Band 4 and Band 5 support this interpretation because the spatial pattern of the reflectance in the two bands are consistent with each other and it reflects the wave facet distribution (note that the two NIR bands have R_{rs} values of almost zero; therefore, the residual surface reflectance mostly contributes to the reflectance of the bands). The spatial variation of Chla and NIR R_{rs} appears to have no clear spatial correlation, implying that the high Chla variation in the images is not caused by the residual surface reflectance (equivalently, the viewing geometry) but from a factor related to the image quality or pixel registration.

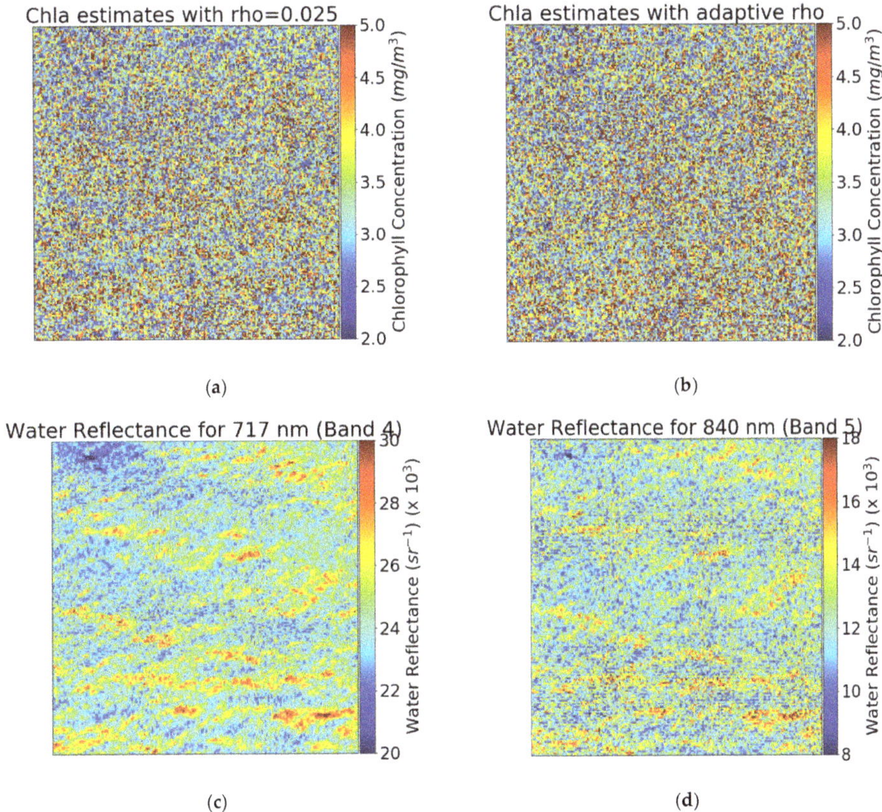

(**a**)   (**b**)

(**c**)   (**d**)

Figure 6. Chla images for the subset area marked in Figure 5, with (**a**) the fixed ρ and (**b**) the adaptive ρ, and water reflectance images for Band 4 (**c**) and Band 5 (**d**).

A regression analysis was performed on a further subset area (40×40 pixels) of the scene. Figure 7 shows the scatter plots of the pixel-to-pixel water reflectance of four bands (Bands 1 and 3–5), with respect to the reference band, Band 2. In all bands, a general linear relationship was identified; however, it showed a low correlation ($R^2 < 0.75$). To assess the spatial pattern of the misregistration, the Band 2 image was regressed to Band 1, using the slope and offset estimated in the regression analysis. The difference between the original Band 1 reflectance and the regressed Band 1 image (Figure 7b) clearly shows a pixel-wise mismatch and the spatial patterns differ from that of the wave facets. It can be observed that the Chla estimation from the mismatch data (Figure 7b) shows a similar spatial variation to that of the difference image.

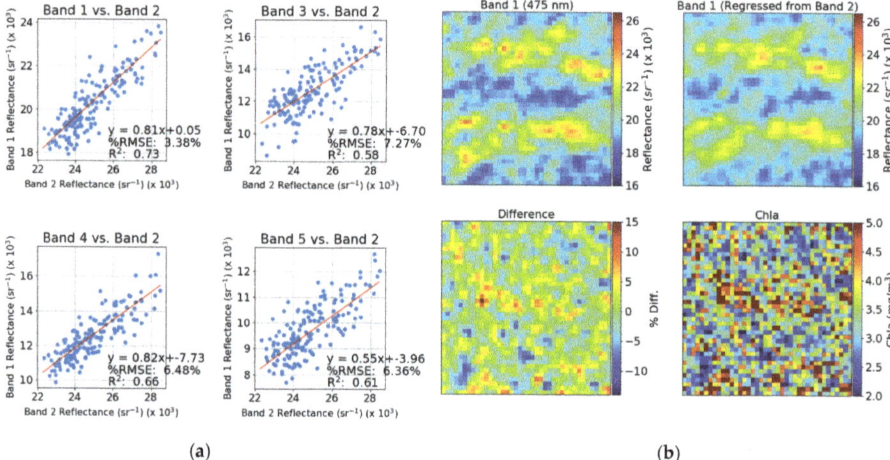

Figure 7. (**a**) Regression results between the water reflectance for multispectral bands, with respect to Band 2, and (**b**) water reflectance for a 40×40 pixel subset area of Band 1, Band 1 estimates regressed from Band 2, the difference, and Chla estimates from the two bands.

3.4. Proposed Approach for the Morphological Registration

Because the surface of the water is not a rigid body, no geometric transformation can find its appropriate pixel-to-pixel correspondence. For a solution, the overall reflectance distribution must be conserved in a sufficiently small area (referred to as "window" hereafter), even if we do not know which pixel in a window corresponds to a pixel in the other band. The images of all five bands contain five instances of the scene at slightly different timings. Consequently, it can be assumed that the distribution of physical quantity, such as the radiance, does not significantly vary in the short period. By setting the image acquisition time of one band as a reference time frame, the radiometric values of the other bands can be matched to the reference band, according to the radiometric distribution, not the pixel location. Figure 8 shows the quantile-to-quantile plot (QQ plot) of the four bands, with respect to Band 2, exhibiting that the radiometric relationship can be established with a nearly perfect correlation when the reflectance of the two bands are compared based on the quantile in reflectance, not on the pixel location. A comparison of the QQ plot with the previous pixel-to-pixel scattered plots (Figure 7) reveals how the pixel-to-pixel misregistration, based on the location, degraded the correlation between the bands. In all four cases of the QQ plots, R^2 is nearly one, producing band-dependent slopes and offsets for the linear relationship. Using this new linear model, the Band 2 image was regressed to Band 1 and compared with the previous results from the regression analysis. Figure 9 shows that Band 1, regressed from the QQ plot, exhibits reflectance levels that are more similar to the original Band 1 image than the location-based regression case. The application of the linear models, derived by the QQ plots, to all four bands, serves as the morphological registration between bands and the subsequent Chla estimation produces a significant improvement in the image quality (Figure 10). Figure 10 shows that the noise pattern for Chla in the original data was significantly reduced and the corrected Chla image contains only the spatial variation caused by wave facets, without being affected by the pixel-to-pixel misregistration. The mean and median values are comparable between the two results; however, the standard deviation and the coefficient of variation reduce from 1.03 to 0.28 and from 29% to 8%, respectively.

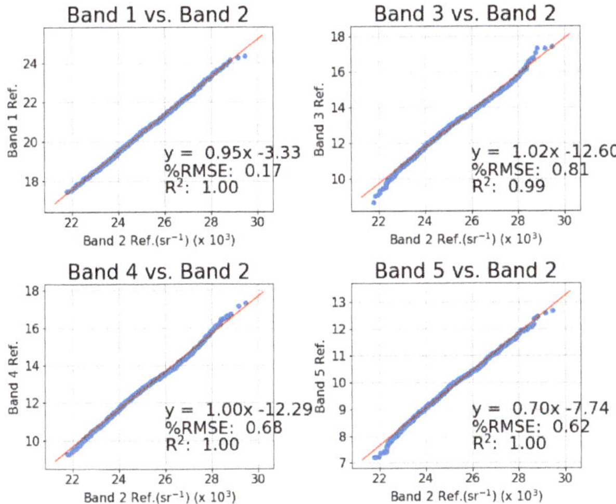

Figure 8. The QQ plots of water reflectance for the four spectral bands, with respect to Band 2 and the corresponding regression results.

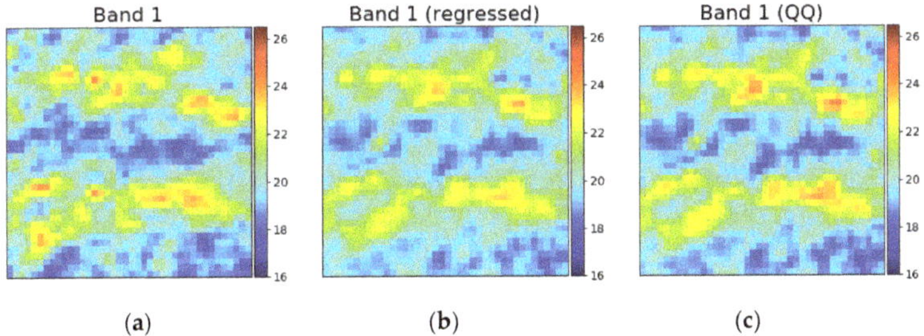

(**a**)   (**b**)   (**c**)

Figure 9. Water reflectance of Band 1 for the subset area in Scene-A: (**a**) original data, (**b**) scaled from Band 2 through the location-based regression, and (**c**) scaled from Band 2 with the QQ-based regression.

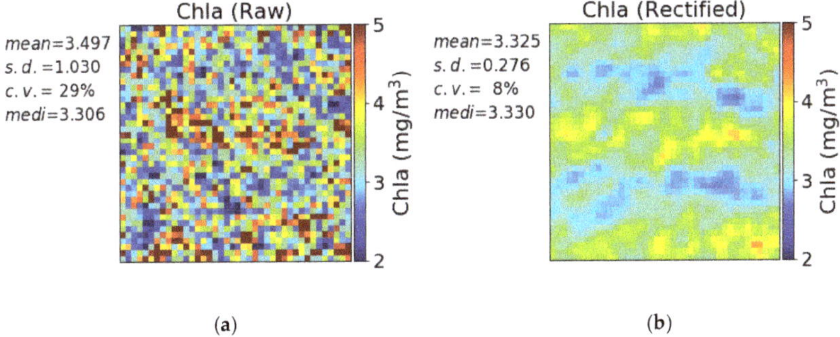

(**a**)   (**b**)

Figure 10. Chla estimation from R_{rs} for the subset area in Scene-A: (**a**) original R_{rs} and (**b**) R_{rs} composited through the QQ-based regression.

Because the determination of the window size can be critical for this approach, the sensitivity of the window size has been investigated. Six different window sizes—15, 25, 51, 101, 251, and 501 pixels—were tested for the QQ plots between Bands 1 and 2 (Figure 11). The window areas for the various sizes were displayed in the corrected Band 1 reflectance image. The plots showed that a high correlation between

Bands 1 and 2 was maintained throughout all window sizes; however, the derived linear relationships were different for different window sizes. As the window size increased, the slopes increased and the y-intercepts decreased. This reveals that the reflectance distribution may change depending on the areas used for the QQ calculation and the local characteristics (in a small window) may be lost when the QQ is derived for a large area. To achieve the goal of the proposed morphological registration, the window size must be kept as small as the local distribution because fetching the reflectance value from a distant location may not guarantee that the two values are from the continuum of targets.

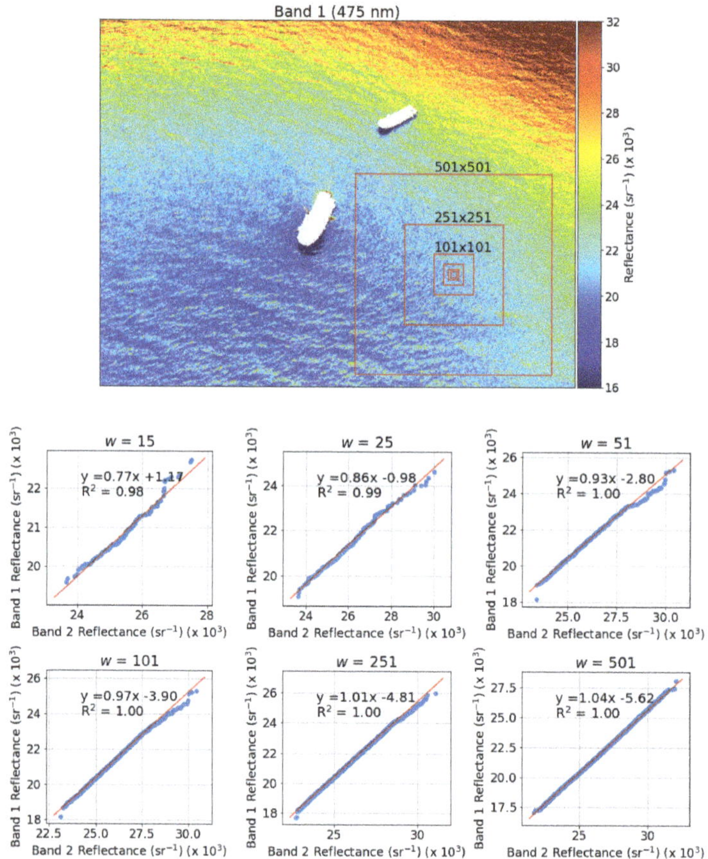

Figure 11. Regression results for varying sizes of the QQ plot calculation window. The extent of each window is marked in the corrected Band 1 image.

4. Algorithm Application and Results

4.1. Algorithm Development for the Entire Image

The QQ plot approach is iterated over the image dimension to process the entire image. However, the calculation of quantiles, which is essentially an order statistics, for all pixels requires exhaustive computation with the complexity of $O(n{\cdot}log\,n)$. A Rededge-M image consists of 1280×960 pixels, which totals to ~1.2×10^6 pixels. Thus, we employ an alternate fast approach, where the QQ calculation is performed for subsampled pixels (e.g. every n-th pixel), and the resultant linear model coefficients (i.e., slope and y-intercept) are propagated to the vicinity of the subsampled pixels with distanced weights assigned by a 2-dimensional Gaussian filter. Figure 12a,b displays the slope and the y-intercept images that were calculated at every pixel, which were then compared with the images obtained with the subsampling scheme, involving the Gaussian filers (Figure 12c,d) (the window size of the Gaussian filter (w_{gauss}) was 25, and the step size (d_{step}) was 12). While the spatial patterns do not significantly deviate from each other, the computation time scales down from 12 min to 1 min per band, saving more than 90% of the computation time (computation done with Intel®Core™ i-5-8265U CPU@1.60GHz, and 8GB RAM). The overall flow chart of the algorithm is presented in Figure 13.

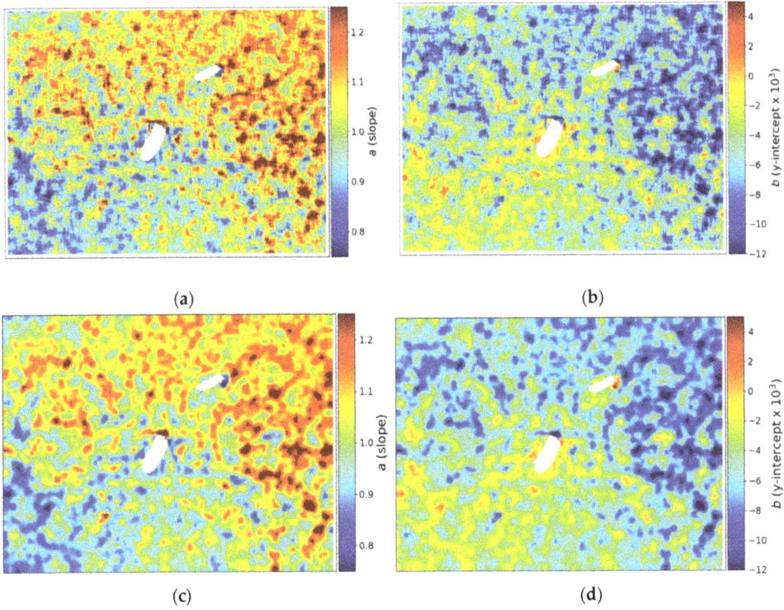

Figure 12. Images of regression slopes (**a**) and y-intercepts (**b**) that are calculated at every pixel, and the slope images (**c**) and the y-intercept images (**d**) computed in every 12 pixels and convolved with a Gaussian filter.

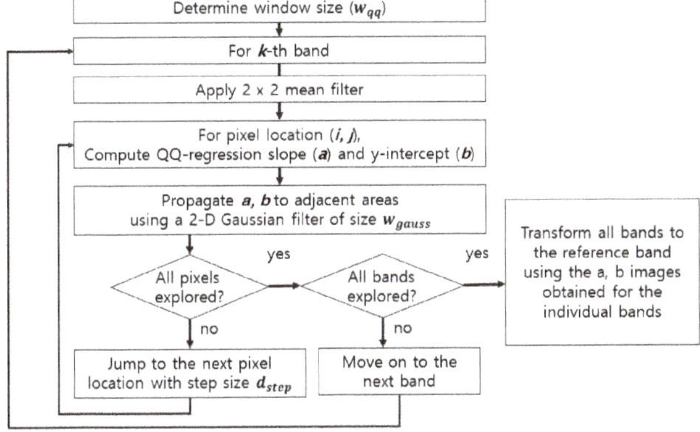

Figure 13. The flow chart of the proposed algorithm.

4.2. Results

The proposed algorithm was applied to the four data sets listed in Table 1. Figure 14 shows the comparison between the Chla estimates, before and after the application of the algorithm for

Scene-A, taken at an altitude of 85.4 m. The high-frequency Chla variation before the correction was significantly and consistently reduced after the correction, over all areas of the image, enhancing the sharpness of the wave features on the surface. Note that the extremely high Chla values under the ship is caused by the ship shadows that made the blue-to-green ratio significantly altered compared to the sun shed areas, thus the anomalously high Chla values are not artifacts of the proposed morphological registration method.

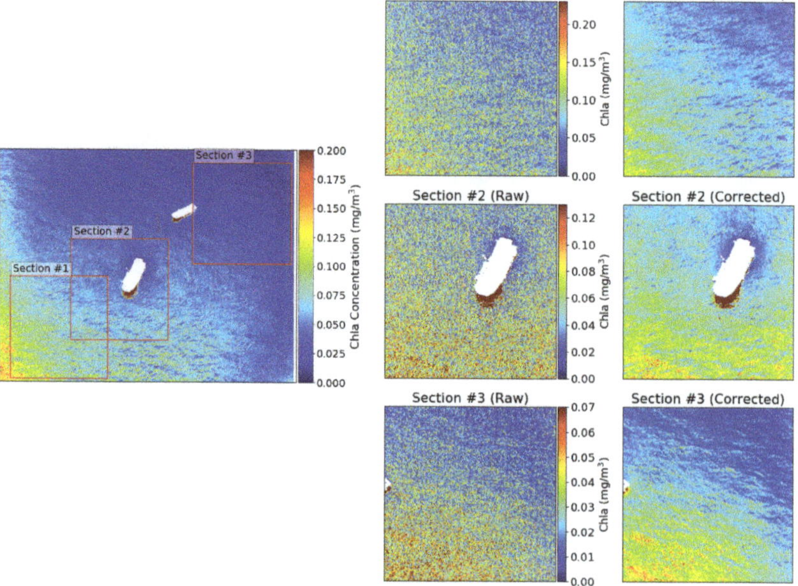

Figure 14. Comparison of resultant Chla estimates for Scene-A (altitude 85.4 m), before and after the application of the proposed morphological band registration method. Three subsections were magnified for improved visual evaluation.

The evaluation of images captured at various altitudes is important because the effect of misregistration may vary with the spatial frequency of surface reflectance features. Data sets from 8.1 m (Scene-B) and 196 m (Scene-C) were tested and the results are displayed in Figures 15 and 16, respectively. The figures display four types of Chla estimates, each of which is derived from (1) R_{rs} data before the correction, (2) R_{rs} data after low pass filtering with a 2 × 2 average window, (3) R_{rs} data after low pass filtering with a 32 × 32 average window, and (4) R_{rs} data after correction using the proposed morphological registration. Figure 15a shows the results for Scene-B (altitude 8.1 m), where the Chla estimates before the correction exhibits a very noisy spatial pattern, which remains even after the 2 × 2 mean filter. The noisy pattern was not minimized until the size of the mean filter increased to 32 × 32, as shown in the figure. The Chla estimates, with the morphological registration, exhibited a noise-free retrieval while maintaining the sharpness of the image. Note that the 32 × 32 mean filter removed the noise at the expense of losing spatial details, or sharpness. For a more detailed evaluation, a boxed area, marked in the figure, is displayed in Figure 15b. The figure demonstrates that the misregistered pixels caused many spikes with Chla estimates exceeding 3.0 mg/m³ in the original resolution. In the 2 × 2 mean filter case, the Chla values of the corrected estimates are in the level of approximately 2.0 mg/m³, with the lowest value at approximately 1.0 mg/m³. It is neither realistic that Chla varies with a factor of three in such a small area (< 1 m²) nor true that the viewing or reflecting geometry changes with such a high frequency. The scatter statics such as standard deviation and coefficient of variation decreased significantly after the morphological registration compared to the 2×2 mean filter (s.d.: 601 to 0.2, c.v.: 14224% to 15%), while the median value stays similar (1.745 to 1.768), implying that the outliers caused by misregistration significantly deteriorated the image quality. It is shown that the mean values were also significantly affected by the noise (4.228 to 1.745).

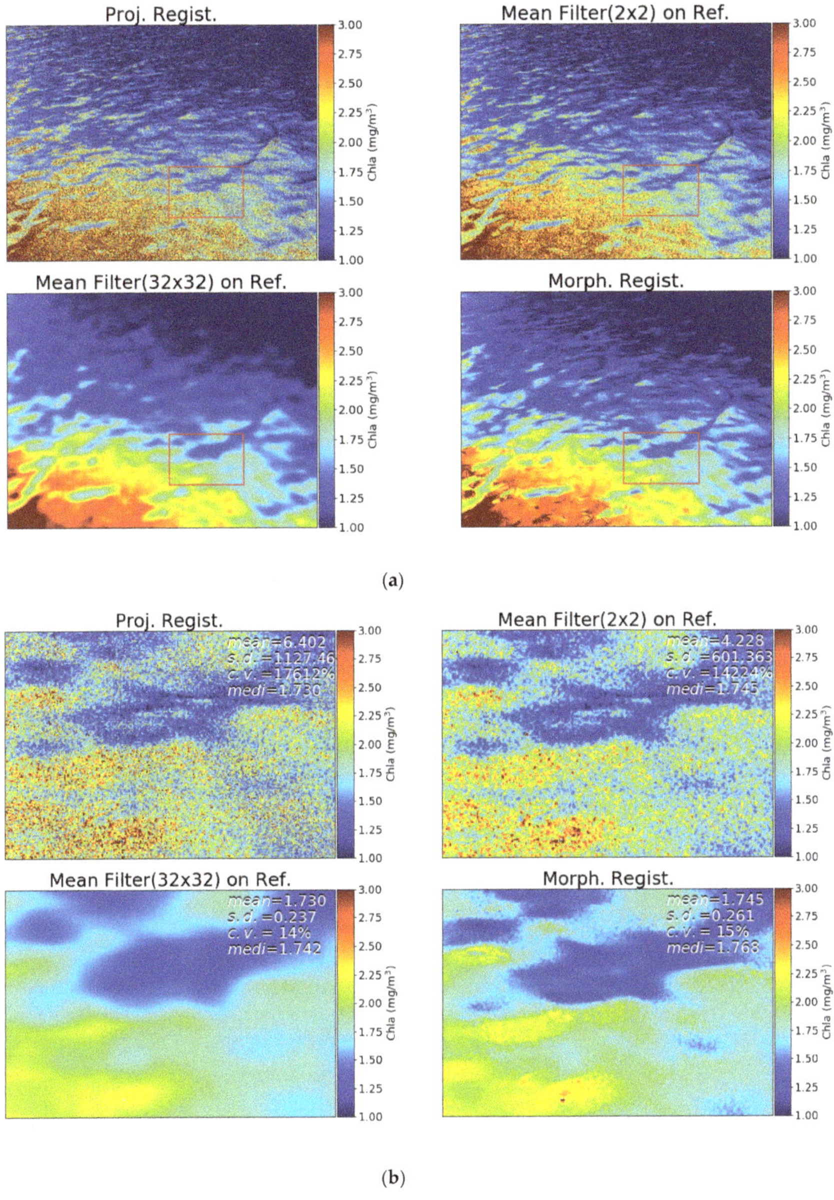

Figure 15. Scene-B, acquired on 2019.08.31 at an altitude of 8.1 m. (**a**) Chla estimates from four different processings of R_{rs} data; (**upper left**) before morphological registration, (**upper right**) after 2×2 mean filter on water reflectance, (**bottom left**) after 32×32 mean filter on water reflectance, and (**bottom right**) after morphological registration. (**b**) Magnified figures for the subset area marked in a red box in panel (**a**).

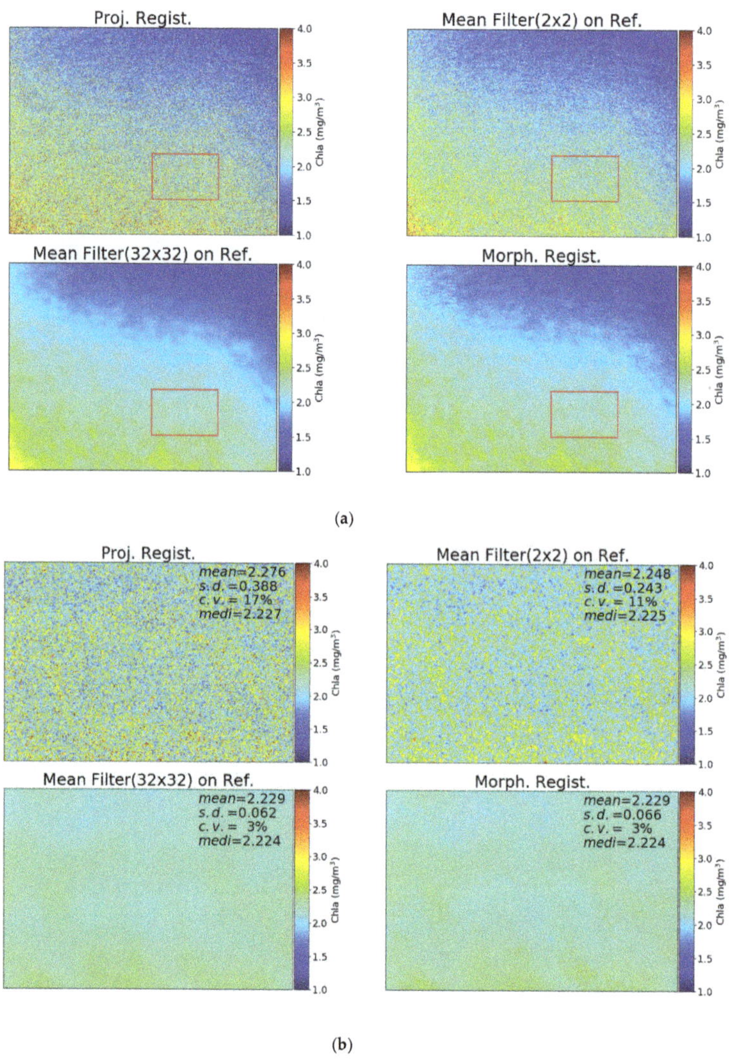

Figure 16. Scene-C acquired on 2019.08.31 at an altitude of 196 m. (**a**) Chla estimates from four different processings of R_{rs} data; (**upper left**) before morphological registration, (**upper right**) after 2×2 mean filter on water reflectance, (**bottom left**) after 32×32 mean filter on water reflectance, and (**bottom right**) after morphological registration. (**b**) Magnified figures for the subset area marked in a red box in panel (**a**).

The results for the scene of much higher altitude (Scene-C (196 m)) are presented in Figure 16. It shows that the scene is in general more homogeneous than the previous low-altitude image (Scene-B (8.1 m)), and it exhibits a gradual radiance change across the scene with the boxed area occupied with fairly homogeneous radiance values. However, even if the surface features have a significantly smaller scale than the low-altitude images, the noisy pattern is still strong in all areas of the image as well as in the boxed area, which was effectively removed by the morphological band registration. In the boxed area, while the mean and the median values stay similar (mean: 2.248 to 2.229, media: 2.225 to 2.224), the scatter statics greatly improved (s.d.: 0.243 to 0.0066, c.v.: 11% to 3%) (Figure 16b). The proposed algorithm was also tested on a scene with an event in a part of the image (Figure 17). Scene-D was acquired for a strong red tide outbreak of *Cochlodinium polykrikoides* (the species was confirmed from microscopic analysis during the field campaign) and the Chla contents were estimated with the RBR algorithm. The morphological registration reduced the number of pixels that had extremely high Chla estimates (> 50 mg/m^3), which are considered to be generated misregistration pixels. While the overall extent and concentration level in the morphological registration is comparable to that of the 32×32 filter results, a spatial pattern of a higher frequency is still observable in the improved result.

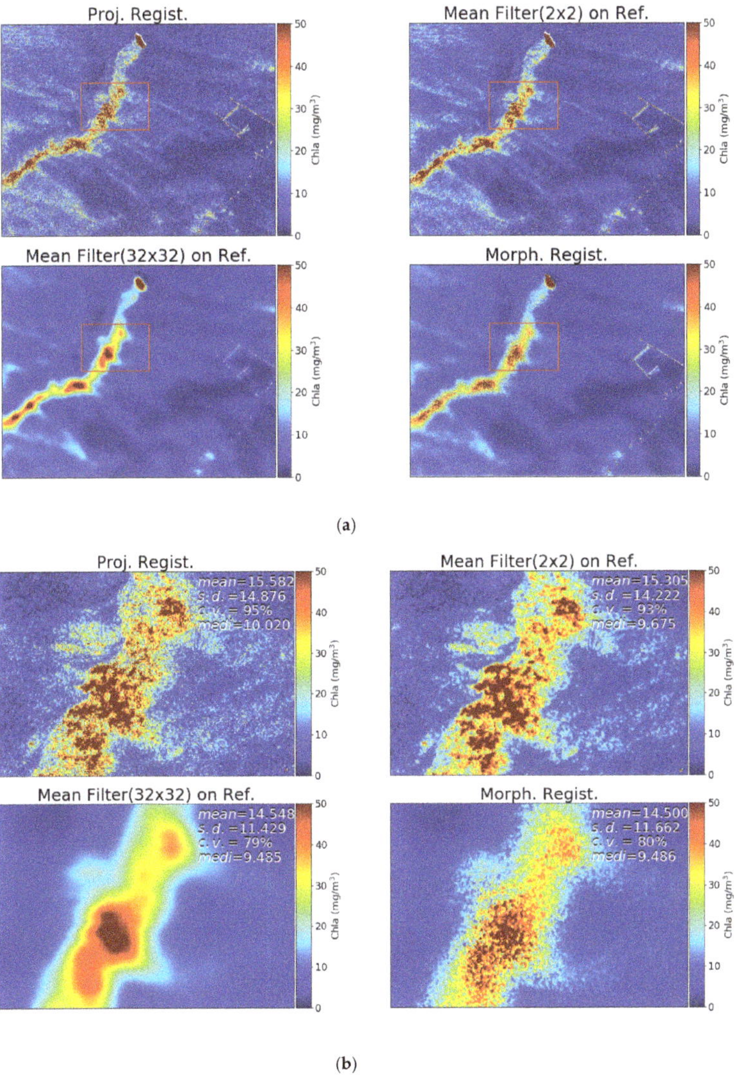

Figure 17. Scene-D acquired on 2019.08.31 at an altitude of 390 m. (**a**) Chla estimates from four different processings of R_{rs} data; (**upper left**) before morphological registration, (**upper right**) after 2×2 mean filter on water reflectance, (**bottom left**) after 32×32 mean filter on water reflectance, and (**bottom right**) after morphological registration. (**b**) Close-up figures for the subset area marked in a red box in panel (**a**).

5. Discussion and Conclusions

This study analyzed the impact of residual misregistration between the bands of a multispectral camera, which is particularly problematic for fast-moving targets such as the water surface. The analysis suggested that the residual misregistration is difficult to correct using geometric coordinate transformation (e.g., projective transformation). It produces abnormal band ratios and band differences, resulting in significant anomalies in the water quality variables, such as the Chla concentration. The proposed registration algorithm succeeded in effectively removing noisy spatial patterns caused by the misregistration while maintaining the original spatial resolution of the image, unlike the smoothing approach which significantly degrades the sharpness of the images. Contrary to the intuition that high-altitude images will be less affected by pixel-level misregistration (because a scene appears more homogeneous when observed from far), the test results for various altitudes showed that the residual misregistration exist at all tested altitudes (8–390 m). This suggests that there exist various frequencies of the surface reflectance feature on the water surface. The proposed algorithm was robust to local events that occurred in a partial section of the image, which may have distinct spectral characteristics of the remaining image area (usually normal water surface), as shown in the red tide image.

The proposed method is expected to improve estimation of other water quality variables such as colored dissolved organic matter (CDOM) and total suspended sediments (TSM) as many of the CDOM and TSM retrieval algorithms rely on the band ratio of reflectance.

Future work includes analysis on the effects on other water quality variables, and also the further correction of residual sky reflectance that is often caused by the spatially varying normals of the wave facets. The residual sky reflectance still existed even after the morphological band registration. The research requires a comprehensive understanding of the reflecting mechanism and more detailed modeling of the surface reflectance, associated with the analysis on the downward sky radiance incident on the water surface.

Author Contributions: Data Curation, W.K., S.J. and Y.M.; Formal Analysis, W.K.; Validation, W.K.; Writing—Original draft, W.K.; Funding Acquisition, S.J.; Resources, S.J. and S.C.M.; Project Administration, Y.M.; Supervision, Y.M.; Writing—Review & editing, S.C.M. All authors have read and agreed to the published version of the manuscript.

Funding: Funding: This work was performed with the support of Jeollannam-do through the research project titled "Development of a monitoring system for damage reduction in aquacultures using marine environment observation drones" (B0080802000131) operated by the Jeonnam Technopark (JNTP), and was also supported by the National Research Foundation (NRF) project "Developing an atmospheric correction module for retrieving water quality from hyperspectral images" (NRF-2019R1F1A1062585).

Acknowledgments: The authors would like to thank Hyung-Jun Kim and Yoo-deok Boo for participating in the field campaign.

Conflicts of Interest: The authors declare no conflicts of interest

References

1. Holtkamp, D.J.; Goshtasby, A.A. Precision registration and mosaicking of multicamera images. *IEEE Trans. Geosci. Remote Sens.* **2009**, *47*, 3446–3455. [CrossRef]

2. Jhan, J.-P.; Rau, J.-Y.; Huang, C.-Y. Band-to-band registration and ortho-rectification of multilens/multispectral imagery: A case study of MiniMCA-12 acquired by a fixed-wing UAS. *ISPRS J. Photogramm. Remote Sens.* **2016**, *114*, 66–77. [CrossRef]

3. Anuta, P.E. Spatial registration of multispectral and multitemporal digital imagery using fast Fourier transform techniques. *IEEE Trans. Geosci. Electron.* **1970**, *8*, 353–368. [CrossRef]

4. D'Agostino, E.; Maes, F.; Vandermeulen, D.; Suetens, P. A Viscous Fluid Model for Multimodal Non-rigid Image Registration Using Mutual Information. *Med. Image Anal.* **2003**, *7*, 565–575. [CrossRef]

5. Droske, M.; Rumpf, M. A Variational Approach to Nonrigid Morphological Image Registration. *SIAM J. Appl. Math.* **2004**, *64*, 668–687. [CrossRef]

6. Matsopoulos, G.K.; Mouravliansky, N.A.; Delibasis, K.K.; Nikita, K.S. Automatic retinal image registration scheme using global optimization techniques. *IEEE Trans. Inf. Technol. Biomed.* **1999**, *3*, 47–60. [CrossRef] [PubMed]

7. Soille, P. Morphological image compositing. *IEEE Trans. Pattern Anal. Mach. Intell.* **2006**, *28*, 673–683. [CrossRef]

8. Tucker, C.J. Red and Photographic Infrared Linear Combinations for Monitoring Vegetation. *Remote Sens. Environ.* **1978**, *8*, 127–150.

9. Huete, A.; Huete, A.R. A soil-adjusted vegetation index (SAVI). *Remote Sens. Environ.* **1988**, *25*, 295–309. [CrossRef]

10. O'Reilly, J.E. Ocean color chlorophyll algorithms for Sea WiFS. *J. Geophys. Res. Oceans* **1998**, *103*, 24937–24953. [CrossRef]

11. Siegel, D.A.; Wang, M.; Maritorena, S.; Robinson, W. Atmospheric correction of satellite ocean color imagery: The black pixel assumption. *Appl. Opt.* **2000**, *39*, 3582–3591. [CrossRef] [PubMed]

12. Doxaran, D.; Froidefond, J.-M.; Castaing, P. A reflectance band ratio used to estimate suspended matter concentrations in sediment-dominated coastal waters. *Int. J. Remote Sens.* **2002**, *23*, 5079–5085. [CrossRef]

13. Hu, C.; Carder, K.L.; Muller-Karger, F.E. Atmospheric correction of SeaWiFS imagery over turbid coastal waters: A practical method. *Remote Sens. Environ.* **2000**, *74*, 195–206. [CrossRef]

14. Ruddick, K.G.; Ovidio, F.; Rijkeboer, M. Atmospheric correction of SeaWiFS imagery for turbid coastal and inland waters. *Appl. Opt.* **2000**, *39*, 897–912. [CrossRef] [PubMed]

15. Kim, W.; Roh, S.-H.; Moon, Y.; Jung, S. Evaluation of Rededge-M Camera for Water Color Observation after Image Preprocessing. *J. Korean Soc. Surv. Geod. Photogramm. Cartogr.* **2019**, *37*, 167–175.

16. Mueller, J.L.; Davis, C.; Arnone, R.; Frouin, R.; Carder, K.; Lee, Z.P.; Steward, R.G.; Hooker, S.; Mobley, C.D.; McLean, S.; et al. Above-water radiance and remote sensing reflectance measurements and analysis protocols. *Ocean Opt. Protoc. Satell. Ocean Color Sens. Valid. Revis.* **2000**, *2*, 98–107.

17. Mobley, C.D. Estimation of the remote-sensing reflectance from above-surface measurements. *Appl. Opt.* **1999**, *38*, 7442–7455. [CrossRef] [PubMed]

18. MicaSense GitHub. Available online: https://github.com/micasense/imageprocessing (accessed on 30 May 2020).

19. Lee, Z.; Carder, K.L.; Mobley, C.D.; Steward, R.G.; Patch, J.S. Hyperspectral remote sensing for shallow waters I A semianalytical model. *Appl. Opt.* **1998**, *37*, 6329. [CrossRef] [PubMed]

20. Lee, Z.; Ahn, Y.-H.; Mobley, C.; Arnone, R. Removal of surface-reflected light for the measurement of remote-sensing reflectance from an above-surface platform. *Opt. Express* **2010**, *18*, 26313–26324. [CrossRef] [PubMed]

21. NASA. Ocean Biology Processing Group, Algorithm Theoretical Basis Documents for Chlorophyll-a Concentration Product. Available online: https://oceancolor.gsfc.nasa.gov/atbd/chlor_a/ (accessed on 30 May 2020).

22. Noh, J.H.; Kim, W.; Son, S.H.; Ahn, J.-H.; Park, Y.-J. Remote quantification of *Cochlodinium polykrikoides* blooms occurring in the East Sea using geostationary ocean color imager (GOCI). *Harmful Algae* **2018**, *73*, 129–137. [CrossRef] [PubMed]

Article

Effects of Spring–Neap Tidal Cycle on Spatial and Temporal Variability of Satellite Chlorophyll-A in a Macrotidal Embayment, Ariake Sea, Japan

Mengmeng Yang [1], Joaquim I. Goes [2], Hongzhen Tian [3], Elígio de R. Maúre [4] and Joji Ishizaka [1,*]

[1] Institute for Space-Earth Environmental Research (ISEE), Nagoya University, Furo-cho, Chikusa-ku, Nagoya, Aichi 464-8601, Japan; mengmeng.yang14@gmail.com

[2] Lamont-Doherty Earth Observatory, Marine Biology and Paleo Environment, Columbia University, 61 Route 9W—P.O. Box 1000, Palisades, NY 10964-8000, USA; jig@ldeo.columbia.edu

[3] School of Management and Economics, Tiangong University, Tianjin 300387, China; tianhongzhen@vip.163.com

[4] Department of Research and Study, Northwest Pacific Region Environmental Cooperation Center, Toyama 930-0856, Japan; eligiomaure@gmail.com

* Correspondence: joji.ishizaka@gmail.com

Received: 4 May 2020; Accepted: 5 June 2020; Published: 8 June 2020

Abstract: We investigated the spatio-temporal variability of chlorophyll-a (Chl-a) and total suspended matter (TSM) associated with spring–neap tidal cycles in the Ariake Sea, Japan. Our study relied on significantly improved, regionally-tuned datasets derived from the ocean color sensor Moderate Resolution Imaging Spectroradiometer (MODIS) Aqua over a 16-year period (2002–2017). The results revealed that spring–neap tidal variations in Chl-a and TSM within this macrotidal embayment (the Ariake Sea) are clearly different regionally and seasonally. Generally, the spring–neap tidal variability of Chl-a in the inner part of the Ariake Sea was controlled by TSM for seasons other than summer, whereas it was controlled by river discharge for summer. On the other hand, the contribution of TSM to the variability of Chl-a was not large for two areas in the middle of Ariake Sea where TSM was not abundant. This study demonstrates that ocean color satellite observations of Chl-a and TSM in the macrotidal embayment offer strong advantages for understanding the variations during the spring–neap tidal cycle.

Keywords: chlorophyll-a variability; spring–neap tides; Ariake Sea; MODIS-Aqua; total suspended sediment; river discharge

---

1. Introduction

The spring–neap tidal cycle is an important factor for the variability of chlorophyll-a (Chl-a) in macrotidal ecosystems [1–4]. During a spring–neap tidal cycle, which is about 15 days long, sea level increases (decreases) and tidal mixing is enhanced (weakened) during spring (neap) tide. In some macrotidal embayments, it has been suggested that the concentration of total suspended sediment (TSM) increases (decreases) in spring (neap) tide, which consequently influences the variability of Chl-a during a spring–neap tidal cycle. This may be explained by the phenomenon whereby strong tidal mixing during spring tide induces the resuspension of sediments in shallow water, causing high turbidity, which reduces light availability for the growth of phytoplankton. On the other hand, the stratification that occurs during neap tide reduces the resuspension of sediments, which increases light availability and thus promotes phytoplankton growth [3,5,6]. However, most previous studies were primarily based on in situ data from only a few observation stations and on short-term time scales, which makes it difficult to understand the mechanisms of spring–neap tidal variability of Chl-a for a whole embayment and for longer time scales. Satellite ocean color products are now routinely used to investigate variations in phytoplankton biomass and productivity, both in coastal and open ocean systems. One significant advantage of satellite remote sensing over traditional shipboard measurements is their broad synoptic coverage and frequency of observations. Thus, satellite ocean color products have been used extensively for the detection and monitoring of phytoplankton biomass indicated by Chl-a as well as water turbidity and TSM concentrations in coastal waters [7–10]. However, at present, only a few studies have focused on the variation in satellite ocean color during the spring–neap tidal cycle. One example is the study by Shi et al. [11], which investigated the spring–neap tidal effects on Moderate Resolution Imaging Spectroradiometer (MODIS) Aqua-derived normalized water leaving radiance spectra (nLw(λ)), water diffuse attenuation coefficient at 490 nm (Kd(490)), and TSM in Bohai Sea, Yellow Sea, and East China Sea between 2002 and 2009. Another example is the study by Su et al. [12], which investigated the relationship between variation in net phytoplankton growth and tidal resuspended events using the daily Medium Resolution Imaging Spectrometer (MERIS) data from 2003 to 2004 in the German Bight. The authors proposed that spring–neap tidal resuspension supplied nutrients and thus enhanced phytoplankton growth, which was different from the findings in the above-mentioned studies [3,5,6], indicating that the effect of spring–neap tidal cycle on the variability of Chl-a varies in regions with different characteristics.

The Ariake Sea is a macrotidal embayment (~20 km wide and 10 km long) located in the Kyushu Island of Japan (Figure 1). It is a shallow bay with an average depth of ~ 15 m and a depth of ~5 m in the onshore area. The range of spring–neap tides in the Ariake Sea is the largest among the Japanese coastal waters, and it can reach to ~6 m during the spring tide in the inner part of the bay [13]. The large tidal range produces strong tidal currents and large tidal flat areas. As a result, strong tidal currents lead to high turbidity zones around the tidal flat areas. Some rivers also discharge into Ariake Sea and supply large amounts of nutrients and suspended sediments to the Sea [14]. The largest river (143 km long and 2860 km^2 in area) that empties into the Ariake Sea is the Chikugo River which connects to the northern part of the bay (Figure 1). The spring–neap tidal range and tidal mixing varies in different regions, and they are largest off Saga, located in the northern part of Ariake Sea. Isahaya Bay, which is the small bay at the western side of the Ariake Sea, used to be known as one of the largest tidal flats in Japan, but its topography was changed following the Isahaya Reclamation Project (Figure 1) [15,16]. For the off-Kumamoto area, which is in the middle and eastern part of Ariake Sea, is surrouded by the two main rivers discharge (Figure 1), and the tidal flat is mostly sandy [13].

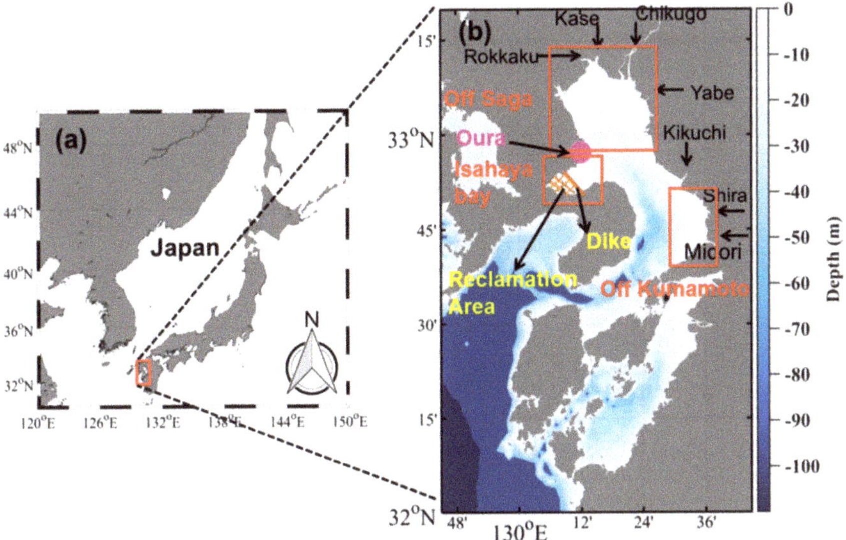

Figure 1. Location of Ariake Sea, Japan (**a**). The water depth of the bay is shown in light to dark blue (**b**). The seven main rivers, Rokkaku, Kase, Chikugo, Yabe, Kikuchi, Shira, Midori, and Kuma, are indicated by the arrows. The three regional areas—i.e., off Saga, Isahaya Bay, and off Kumamoto—are highlighted by the red boxes. The dike and the reclamation area within Isahaya Bay are represented by the brown line and meshed lines, respectively. The observation station for tidal level data of Ariake Sea, named Oura, is represented by the magenta filled circle.

There have been previous attempts to study the influence of the spring–neap tidal cycle on Chl-a variability in Ariake Sea. For instance, in the study by Tanaka et al. [17], the variation forced by the spring–neap tidal cycle in phytoplankton biomass was measured by Chl-a fluorescence and turbidity data at four stations in the northern part of inner Ariake Sea. The authors reported that phytoplankton biomass increased during the neap tide and decreased during the spring tide. They also suggested that the increase and decrease was the result of changes in available light by tidally resuspended TSM. However, it is not known whether the impact of spring–neap tides is significant over the larger area of Ariake Sea and over the whole year.

Additionally, river discharge was reported to be an important factor for the seasonal variability of Chl-a for the whole Ariake Sea based on the standard Sea-Viewing Wide Field-of-View Sensor (SeaWiFS) Chl-a data from May 1998 to December 2001 [10]. However, the standard in-water algorithms of satellite ocean color sensors, such as SeaWiFS and MODIS-Aqua, for this region are prone to errors, and the remote sensing blue band reflectance (Rrs) values are inaccurate. In a previous study [18], we showed that the accuracy of MODIS-Aqua Rrs and Chl-a for the Ariake Sea could be significantly improved by a Rrs recalculation method and a local Chl-a switching algorithm. The Rrs recalculation method first estimated the value of MODIS Rrs(412), from which the standard MODIS Rrs(412) was subtracted to obtain the error in MODIS Rrs(412). Then, the errors in MODIS Rrs(λ) (λ = 443, 488 nm) were calculated based on the assumption that they were linear to the error in MODIS Rrs(412) between 412 and λ nm. Finally, the error in MODIS Rrs(λ) was added to the standard MODIS Rrs(λ) to obtain the recalculated MODIS Rrs(λ). This Rrs recalculation method is simple and effective to reduce the errors in the standard MODIS Rrs(λ) (λ = 412, 443, 488) and therefore Chl-a. In addition, a local Chl-a switching algorithm was developed, which was based on the in situ Chl-a and the maximum blue-to-green band ratio with Rrs(443), Rrs(488), and Rrs(547), for the turbid (Rrs(667) > 0.005 sr^{-1}) and non-turbid (Rrs(667) <= 0.005 sr^{-1}) waters of the Ariake Sea. The local Chl-a switching algorithm significantly improved the Chl-a estimates over that possible by the standard MODIS-Aqua in-water algorithm

(OC3M). Moreover, it was superior to the near-infrared to red band ratio [19] and the red-to-green band ratio [20] algorithms in terms of the accuracy of the estimated Chl-a.

In this study, we hypothesized that there were regional and seasonal differences in the spring–neap tidal variability of Chl-a associated with TSM and river discharge for Ariake Sea. Therefore, we investigated the spring–neap tidal variability of MODIS-Aqua Chl-a for the three regional areas—i.e., off Saga, Isahaya Bay, and off Kumamoto—as well as the whole Ariake Sea, from 2002 to 2017. The impact of TSM and river discharge on the spring–neap tidal variability of Chl-a was quantitatively evaluated using locally tuned MODIS data.

2. Materials and Methods

2.1. Satellite Data and Preprocessing

For our study, reprocessed (2018.1) MODIS-Aqua level 2 products (July 2002–December 2017) were downloaded from the NASA Ocean Biology Processing Group data portal at https://oceancolor. gsfc.nasa.gov/. The spatial and temporal resolution was 1 km and daily, respectively. Before data processing, data quality control was carried out to exclude some of the questionable data. The data flagged by LAND, HIGLINT, HILT, HISATZEN, CLDICE, HISOLZEN, LOWLW, MAXAERITER, and NAVFAIL, (https://oceancolor.gsfc.nasa.gov/atbd/ocl2flags/), were discarded. We also eliminated the data at the edge of satellite view because it is known that they are influenced by a long atmospheric path and that they form a larger pixel size. Besides, we did not use the coverage of less than 20% of the study area because of the possible noise from the cloud edge. With this data quality control and filtering approach, all available daily MODIS-Aqua images (1582) were reduced to 899 images, which were then processed by the Rrs recalculated method and the local Chl-a switching algorithm [18]. Following the reprocessing, pixel values of Chl-a which were spotty and more than three times higher than the adjacent pixel values were defined as outliers and were masked. Then, pixel values of Chl-a more than 100 mg m^{-3} were set to be 100 mg m^{-3} because the maximum in situ Chl-a for the algorithm development was around this value.

Finally, the data frequency and number of observations in each pixel based on all the daily data from 2002 to 2017 were calculated to evaluate the spatial distribution of all the data. The data number was lower inshore and increased to the middle of the bay (Figure S1). This different distribution of data numbers may cause bias for the later data analysis, and thus the areas where the data number was less than 450 were masked.

Regarding the estimation of TSM, an empirical TSM algorithm was developed based on the relationship between in situ TSM and Rrs(667)/Rrs(547) [21] (Figure 2a). The in situ data were the same as those used in [18]. MODIS-derived TSM was obtained by applying the TSM algorithm to the recalculated MODIS-Aqua Rrs, and then validated by comparing it with matching in situ TSM data which were different from the data set used for the development of the TSM algorithm. The matches were derived with the same matching criteria as that used in [18]. The formulas of RMSE and bias were the same as those for Chl-a in [18]. Besides, we also calculated the mean absolute percentage error (MAPE) for the estimated TSM, and the formula was expressed as follows:

$$\text{MAPE} = \frac{1}{N}\sum_{t=1}^{N}\left|\frac{A_t - F_t}{A_t}\right| \tag{1}$$

where N is the data number and A_t and F_t represent the in situ and estimated TSM concentrations (in log-scale), respectively.

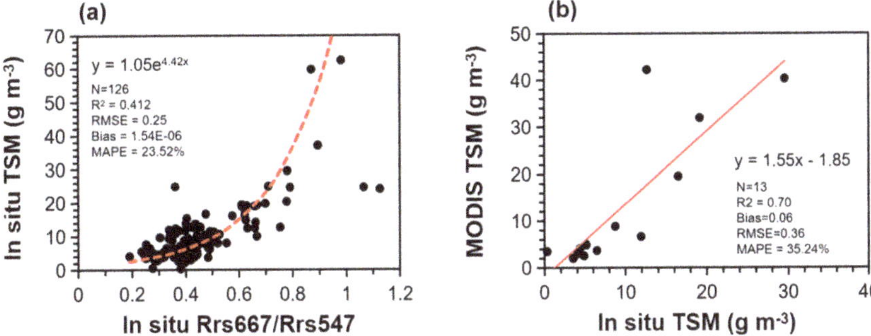

Figure 2. (**a**) Regression of the total suspended matter (TSM) algorithm based on the in situ TSM and Rrs667:Rrs547, and (**b**) the comparison between the in situ and Moderate Resolution Imaging Spectroradiometer (MODIS)-derived TSM data. The red lines are the regressions of the data in each plot, and the red dashed line represents the regression of the TSM algorithm. Equations of each regression and error statistics are also shown.

2.2. Tidal Level Data

The hourly tidal level data (cm) over the same time period as the whole satellite data set was downloaded from the Japan Oceanographic Data Center (2002–2010; https://www.jodc.go.jp/jodcweb/JDOSS/index_j.html) and the Japan Meteorological Agency (2011–2017; https://www.data.jma.go.jp/gmd/kaiyou/db/tide/suisan/index.php). We used the data from observation station Oura (Figure 1). Based on the tidal level data, the time periods of each tidal cycle from 2002 to 2017 were identified, and each spring–neap tidal cycle was divided into four tidal stages—namely spring to neap (SN), neap (N), neap to spring (NS), and spring (S) tide—by the tidal range. The tidal level decreased during SN tide, decreased further during N tide, then increased during NS tide, and further increased during S tide. Therefore, N and S tides were the trough and peak of the tidal range, respectively, and SN and NS tides were the transitional tides during each spring–neap tidal cycle.

2.3. Satellite Composite Data

The Chl-a and TSM composites were initially made for the four tidal stages (SN, N, NS, and S tides) for all the individual events of spring–neap tidal cycles (2002-2017). Subsequently, composites of the four tidal stages were made for the annual and seasonal climatology data. The procedure of producing the satellite composite data is described in a schematic flow (Figure 3) following steps 1 to 4 below.

(1) From the daily data, composites were made for each tidal stage of each individual spring–neap tidal cycle to derive all the individual spring–neap tidal cycle data (four tidal stages (per tidal cycle) × two tidal cycles (per month) × 12 months × 16 years).

(2) The individual spring–neap tidal cycle data was averaged for each month of each year, and then the data in the same month were averaged for all the years to obtain the monthly climatology data of each tidal stage (four tidal stages × 12 months).

(3) Meanwhile, the individual spring–neap tidal cycle data were averaged for each year first, and then the data were averaged for all the years to derive the annual climatology data of each tidal stage (four tidal stages).

(4) An average of the annual climatology of each tidal stage's data was made to obtain the annual climatology data (one data point).

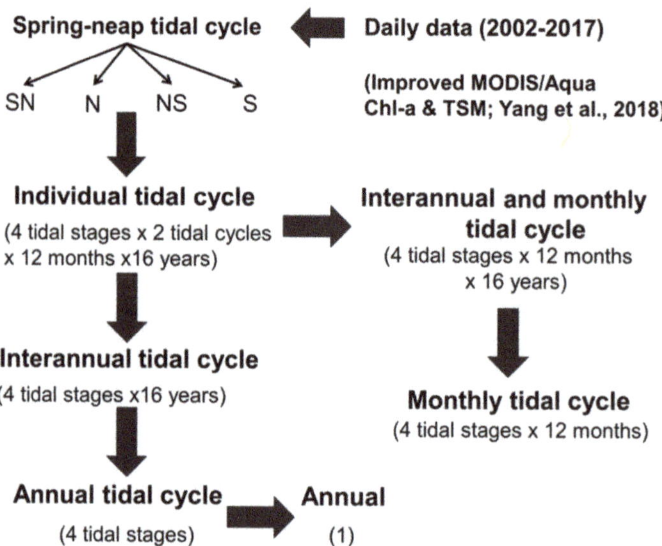

Figure 3. Schematic flow of satellite composite data processing. The individual spring–neap tidal cycle, monthly and annual climatology of chlorophyll-a (Chl-a) and TSM were obtained from the MODIS-Aqua data set (2002–2017). Data of interannual and monthly, and interannual tidal cycle were the intermediate data. SN: spring to neap; N: neap; NS: neap to spring; S: spring.

Furthermore, the spatially-averaged Chl-a and TSM were calculated and compared over the spring–neap tidal cycle for the whole sea (Figure 4), and the three regional areas; i.e., off Saga, Isahaya Bay, and off Kumamoto (Figure 1). The accuracies of the MODIS-Aqua Chl-a were discussed in [18]. To understand the resuspension mechanism of Chl-a and TSM over the spring–neap tidal cycle for the Ariake Sea, the ratio of Chl-a to TSM (Chl-a:TSM) was analyzed [8,22]. In [22], the authors proposed that TSM was phytoplankton-dominated when the ratio was between 1:600 and 1:275 and was suspended sediment-dominated when the ratio was lower than 1:600 for Tokyo Bay, which was once applied to the East China Sea (ECS) to study the seasonal and interannual dynamics of Chl-a and TSM [8].

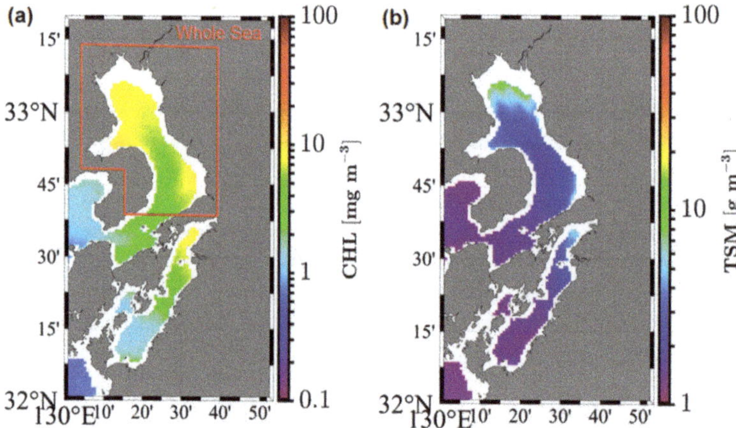

Figure 4. Satellite imagery of annual climatology of chlorophyll-a (Chl-a) (**a**) and total suspended sediment (TSM) (**b**). The whole sea, where spatially-averaged Chl-a and TSM were calculated, is marked in red.

2.4. River Discharge Data

Daily river discharge data was downloaded from the website of Water Information System of the Japanese Ministry of Land, Infrastructure, Transport and Tourism (http://www1.river.go.jp/). Twelve rivers were selected for the seven main river systems, namely Chikugo, Yabe, Rokkaku, Kase, Kikuchi, Shirakawa, Midori and Kuma Rivers (Figure 1). In addition, total daily and monthly river discharge data were calculated. The total daily river discharge data were derived as the sum of the daily data from all the rivers, and then they were averaged for each month of all years from 2002 to 2017 to obtain the total monthly river discharge data. Missing daily data for a certain river were estimated based on the correlation between the daily river discharge from that river and that from the other rivers ($0.891 < R^2 < 0.998$). However, for several years (2005, 2007, 2013, 2014 and 2016), large amounts of missing data diminished the usefulness of this data, and these years were therefore excluded from further analysis. Interannual variations of the monthly river discharge were quantified by the variation bars (Figure S2), and they were small in terms of the standard deviation of the interannual monthly river discharge except for June and July.

3. Results

3.1. Annual Climatology of Chl-a and TSM

The annual climatology data showed higher Chl-a in the three regional areas—i.e., off Saga, Isahaya Bay, and off Kumamoto—than that in the middle part of the Ariake Sea (Figure 4a). However, the difference in the magnitude of the spatially-averaged Chl-a for the areas off Saga (7.66 mg m^{-3}), Isahaya Bay (7.55 mg m^{-3}), and off Kumamoto (7.59 mg m^{-3}) was small. In contrast, the averaged TSM was much higher (4.93 g m^{-3}) in the area off Saga than that in Isahaya Bay (2.61 g m^{-3}) and off Kumamoto (3.06 g m^{-3}) (Figure 4b). In addition, spatially-averaged Chl-a (6.77 mg m^{-3}) and TSM (3.20 mg m^{-3}) values were calculated for the whole sea by excluding the southern part of the sea where the validation of the satellite Chl-a and TSM was missing.

Differences in Chl-a and TSM over the spring–neap tidal cycle were observed (Figures 5 and 6a). For Chl-a, the variability was larger in the three regional areas—i.e., off Saga, Isahaya Bay, and off Kumamoto—than that in the middle-western areas, and they were slightly higher during NS and S tide than in SN and N tide. The TSM was also higher during NS and S tide than that during SN and N tide, and the variability of TSM was larger especially off Saga than in other areas. Note that the standard deviations of the spatial averages of all the interannual data were calculated to assess the significance of the difference of the spatial average of the annual data during each tidal stage (Figure 6a). All the standard deviations were smaller than the difference of the spatial averages of the annual data, indicating that the differences of the spatial averages were significant.

Furthermore, the relation between annual Chl-a and TSM over the spring–neap tidal cycle was investigated (Figure 6). For the whole sea, and especially for off Saga (Figure 6a), TSM was low in SN and N, and then dramatically increased in NS and S tide. In contrast, for Isahaya Bay and off Kumamoto (Figure 6a), the variability of TSM was not consistent with the spring–neap tidal cycle. Besides, the ratios of Chl-a:TSM were lower compared over the spring–neap tidal cycle (Figure 6b). For the whole sea and off Saga, the ratio increased from SN to N, then decreased to NS, and then further to S tide. Moreover, the ratios were lower than 1:600 except for N tide for off Saga due to the high concentration of TSM (>4 g m^{-3}) during the spring tide in this area. For the whole sea, the ratios were all between 1:600 and 1:275. For Isahaya and off Kumamoto, the variability of the ratios was between 1:600 and 1:275 and much smaller than off Saga, and TSM almost linearly increased with Chl-a.

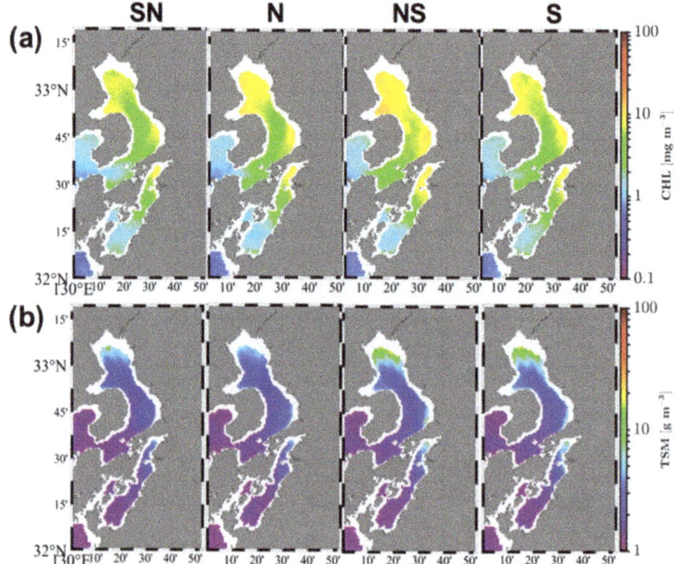

Figure 5. Satellite images of annual climatology of Chl-a (**a**) and TSM (**b**) over the spring–neap tidal cycle.

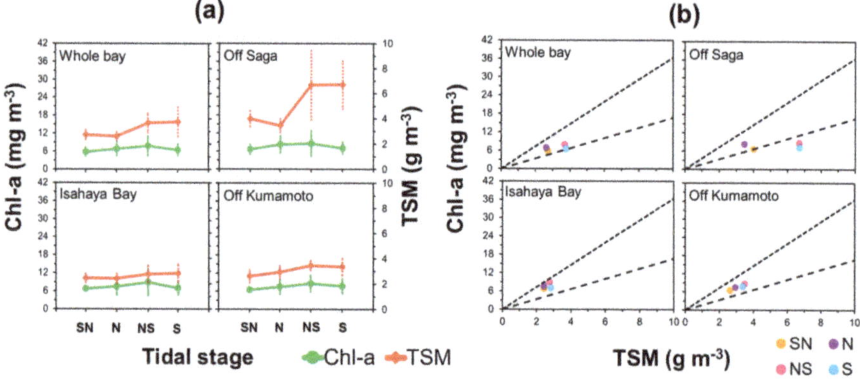

Figure 6. Change of spatially-averaged annual climatology of Chl-a and TSM over the spring–neap tidal cycle (**a**) and the scattering plot (**b**) for the whole sea, off Saga, Isahaya Bay and off Kumamoto. The vertical lines in (**a**) represent the standard deviations of the spatial averages, and the dashed and dotted lines in (**b**) represent the Chl-a:TSM ratios of 1:600 and 1:275, respectively.

3.2. Monthly Climatology of Chl-a and TSM

The magnitude of Chl-a showed seasonal variations: it was generally lowest in winter (December, January, February), increased in spring (March, April, May), reached its height in summer (June, July, August), and then decreased in autumn (September, October, November) (Figure S3). For each season, the spring–neap tidal variability of monthly Chl-a showed similar patterns. Therefore, the middle months of each season were chosen as representative months (Figure 7). In winter and spring, Chl-a was much higher (>7 mg m^{-3}) for the areas off Saga, Isahaya Bay and off Kumamoto, whereas in summer, high Chl-a expanded to the whole sea, and then began to be restricted to the areas off Saga, Isahaya Bay and off Kumamoto in October and November. The seasonal variability of TSM was much less than that of Chl-a. In addition, TSM was generally much higher (>6 g m^{-3}) off Saga than that in other areas of the sea (<4 g m^{-3}).

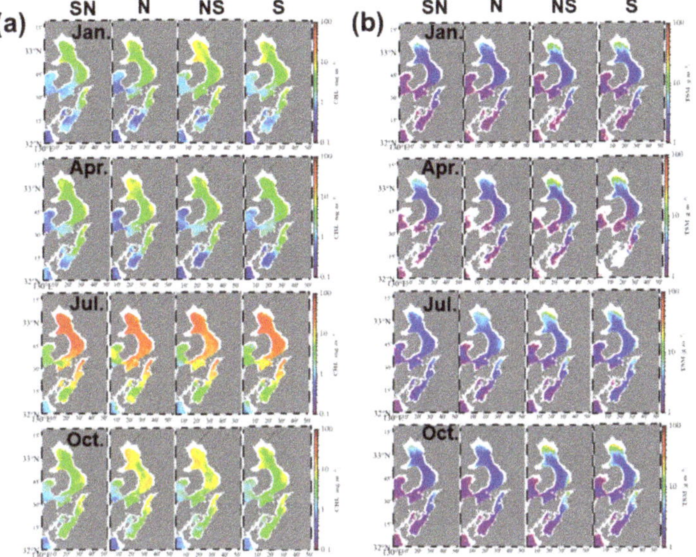

Figure 7. Satellite images of monthly climatology of Chl-a (**a**) and TSM (**b**) over the spring–neap tidal cycle. The four months—i.e., Jan., Apr., Jul., and Oct.—represent winter, spring, summer, and autumn, respectively.

The relationship between the monthly Chl-a averaged over the spring–neap tidal cycle and the monthly river discharge were also examined (Figure 8) because river discharge was suggested to be one of the important factors for seasonal variation [10]. The monthly Chl-a was strongly and positively correlated with the monthly river discharge from all the rivers for the whole sea, off Saga, Isahaya and off Kumamoto (R^2 = 0.88, 0.89, 0.78 and 0.87; $p < 0.05$). The magnitude of Chl-a was much higher in summer and highest in July when the river discharge was highest. This suggests that river discharge could be one of the important factors for the large seasonal variability of Chl-a in all of the regions.

The relationship between monthly Chl-a and TSM over the spring–neap tidal cycle was separately investigated for the four areas (Figure 9; Figure 10; Table 1). All the standard deviations were smaller than the difference of the spatial averages of the monthly data, suggesting that the differences of the spatial averages were significant. The variability of monthly Chl-a showed clear seasonal differences for all the four areas. The magnitude of Chl-a over the spring–neap tidal cycle was much higher (Figure 9) in July, which represented summer, than in other months, which corresponded to a higher river discharge in summer, especially in June (Figure 8). The Chl-a peaks within each tidal cycle occurred at SN or NS tides for summer, whereas the Chl-a peaks generally occurred at N or NS tides for other seasons, for all areas (Table 1). In contrast, the Chl-a peaks all occurred at N or NS tides for the annual climatology data. For the variability of monthly TSM over the tidal cycle, seasonal differences were relatively small compared with the regional difference as well as the monthly variation in Chl-a (Figure 9), and the TSM peaks all occurred at NS or S tide (Table 1).

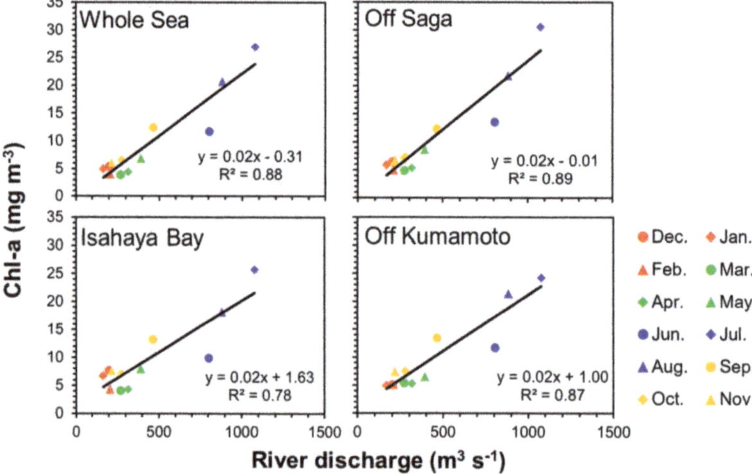

Figure 8. Monthly climatology of Chl-a averaged over the spring–neap tidal cycle against the monthly climatology of river discharge. The data in winter, spring, summer, and autumn are represented by red, green, blue, and yellow markers, respectively. The lines are regression lines.

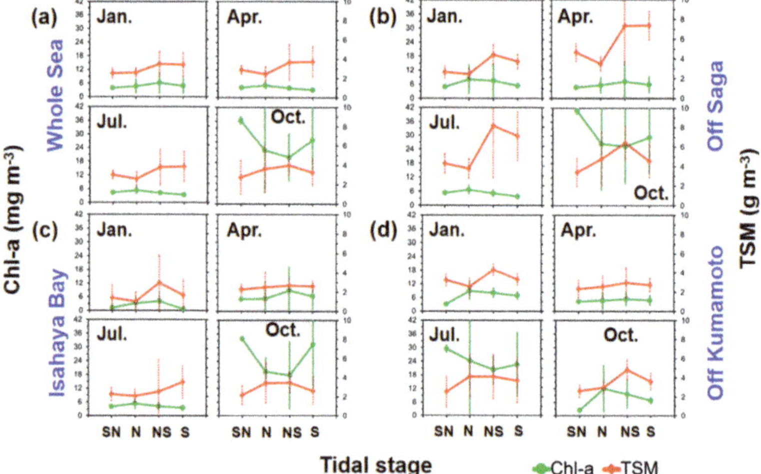

Figure 9. The time-series of monthly climatology of spatially-averaged Chl-a over the spring–neap tidal cycle for (a) the whole Araike Sea, (b) off Saga, (c) Isahaya Bay, and (d) off Kumamoto. The vertical lines in each plot are standard deviations of spatial averages of all the interannual data.

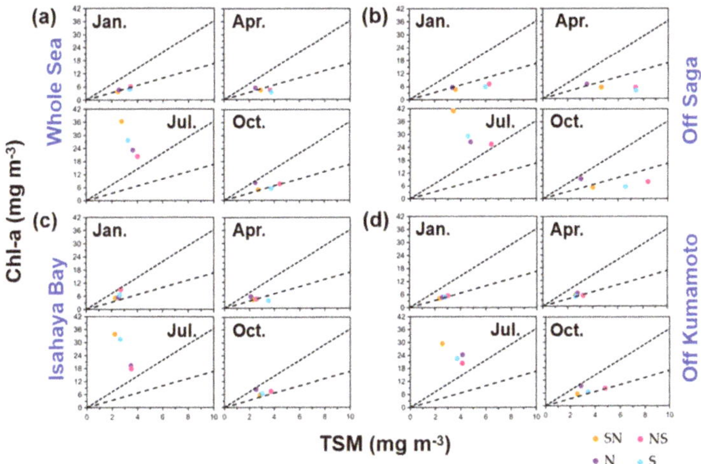

Figure 10. Scatter plots of monthly climatology of Chl-a and TSM over the spring–neap tidal cycle for the whole Sea (**a**), off Saga (**b**), Isahaya Bay (**c**) and off Kumamoto (**d**). The data for the four tidal stages—SN, N, NS and S tides—are represented by yellow, purple, magenta and light blue markers, respectively. The dashed and dotted line represents the Chl-a:TSM ratios of 1:600 and 1:275, respectively.

Table 1. Statistics of the peaks within the spring–neap tidal cycles made for the monthly climatology of Chl-a and TSM for the whole sea, off Saga, Isahaya Bay, and off Kumamoto. A value of "1" indicates the occurrence of peaks.

| Chl-a Peaks | Whole Bay | | | | Off Saga | | | | Isahaya Bay | | | | Off Kumamoto | | | |
Month	SN	N	NS	S	SN	N	NS	S	SN	N	NS	S	SN	N	NS	S
Dec.			1			1					1			1		
Jan.			1				1				1			1		
Feb.				1		1				1						1
Mar.				1				1			1			1		
Apr.		1				1				1				1		
May			1			1					1			1		
Jun.			1				1				1			1		
Jul.	1				1					1			1			
Aug.	1						1			1			1			
Sep.			1				1				1				1	
Oct.		1				1				1				1		
Nov.			1				1					1			1	

| TSM Peaks | Whole Bay | | | | Off Saga | | | | Isahaya Bay | | | | Off Kumamoto | | | |
Month	SN	N	NS	S	SN	N	NS	S	SN	N	NS	S	SN	N	NS	S
Dec.			1				1				1				1	
Jan.			1				1				1				1	
Feb.				1				1				1				1
Mar.				1				1				1				1
Apr.				1				1				1			1	
May				1				1			1					1
Jun.				1				1				1				1
Jul.			1				1				1			1		
Aug.			1				1					1			1	
Sep.			1				1					1			1	
Oct.			1				1				1				1	
Nov.				1				1			1				1	

The variability of the Chl-a:TSM ratios also showed seasonal and regional differences (Figure 10). For summer, the ratios were much higher and were almost above 1:600 for all areas, which was due to the high Chl-a (7.74~41.08 mg m^{-3}). For other seasons, the ratios were generally highest at N tide and lowest at S tide and were almost below 1:600 for the whole sea and off Saga (Figure 10a,b), which was due to the high TSM (2.02~9.51 g m^{-3}); in contrast for Isahaya and off Kumamoto (Figure 10c,d), the variability of the ratios over the tidal cycle was small, and the ratios were mostly between 1:600 and 1:275 due to the lower TSM.

3.3. Individual Events of Spring–Neap Tidal Cycle Variability of Chl-a

The annual and monthly composite analysis indicated that there were strong spring–neap tidal cycles in Chl-a variation, which further varied regionally and seasonally. Therefore, the individual events of spring–neap tidal cycles from 2002 to 2017 were investigated. Because much of the Chl-a data were missing over the spring–neap tidal cycles, only 10 individual events had data available for all spring–neap tidal stages (Table 2). The data were missing in some of the three target regional areas (Figure 11); thus, spatially-averaged Chl-a and TSM values were only calculated for the whole sea (Table 2).

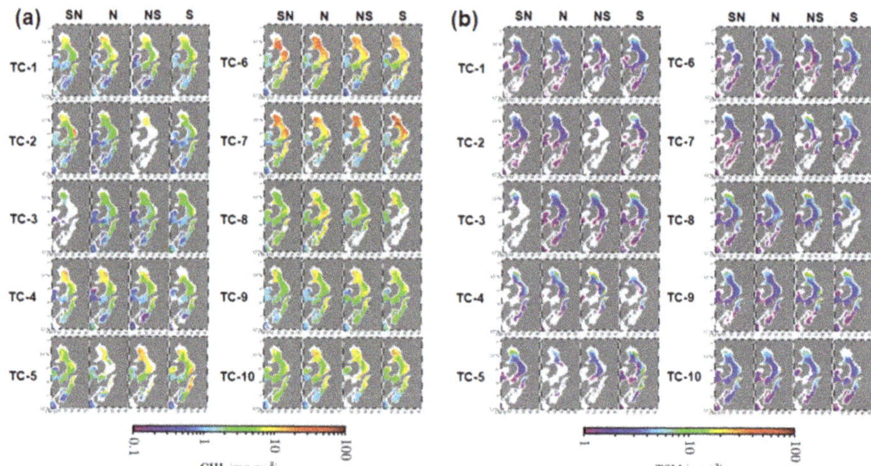

Figure 11. Composite Chl-a (**a**) and TSM (**b**) images of the spring–neap tidal stages for the ten selected individual tidal cycles, i.e., TC-1 to TC-10. TC-1 to TC-3 were from winter, TC-4 to TC-5 were from spring, TC-6 to TC-7 were from summer, and TC-8 to TC-10 were from autumn.

Table 2. Statistics of the spatially-averaged Chl-a and TSM values from the individual events of spring–neap tidal cycles, and river discharge during all ten tidal cycles. SD stands for standard deviation.

Tidal Cycle ID	Time Period	Month	Chl-a (mg m^{-3})						TSM (g m^{-3})						Chl-a:TSM				River Discharge (m^3 s^{-1})		
			SN	N	NS	S	Mean	SD	SN	N	NS	S	Mean	SD	SN	N	NS	S	Mean	Peak	Peak/Mean
TC-1	11/27/2003-12/11/2003	Dec.	6.06	7.72	7.85	6.53	7.04	0.88	3.07	2.64	5.27	4.58	3.89	1.41	1:472	1:365	1:511	1:495	198.43	311.33	1.57
TC-2	1/26/2004-2/10/2004	Feb.	8.06	3.73	8.66	4.04	6.12	2.60	2.45	2.29	2.27	2.78	2.45	0.10	1:304	1:613	1:262	1:689	138.46	161.54	1.17
TC-3	2/11/2004-2/23/2004	Feb.	2.73	3.36	3.23	3.49	3.20	0.33	3.59	2.52	3.34	4.79	3.56	0.56	1:1316	1:750	1:1034	1:1373	130.77	173.24	1.32
TC-4	3/26/2012-4/10/2012	Apr.	8.45	7.38	5.51	4.10	6.36	1.94	3.87	3.88	5.84	3.19	4.19	1.13	1:457	1:526	1:1059	1:777	429.34	1359.58	3.17
TC-5	4/28/2005-5/10/2005	May	3.98	7.39	11.29	8.19	7.71	3.00	4.15	3.17	3.15	5.83	4.07	0.57	1:1042	1:429	1:279	1:711	426.58	1533.01	3.59
TC-6	8/14/2010-8/28/2010	Aug.	30.81	20.96	12.54	12.20	19.13	8.78	2.64	3.29	2.57	3.97	3.12	0.40	1:86	1:157	1:205	1:325	255.59	428.56	1.68
TC-7	7/24/2008-8/5/2008	Jul.	17.58	12.78	17.89	32.97	20.31	8.76	2.26	2.68	4.98	3.47	3.35	1.46	1:129	1:210	1:278	1:105	289.92	639.89	2.21
TC-8	9/29/2003-10/12/2003	Oct.	4.94	8.17	6.07	6.67	6.46	1.35	5.11	3.55	5.58	5.49	4.93	1.06	1:1036	1:435	1:918	1:824	186.76	200.17	1.07
TC-9	10/13/2003-10/27/2003	Oct.	5.37	6.22	7.45	5.82	6.21	0.89	3.41	2.62	6.47	4.65	4.29	2.03	1:634	1:421	1:869	1:800	174.64	226.01	1.29
TC-10	11/16/2004-11/28/2004	Nov.	5.42	8.20	7.37	8.51	7.37	1.39	3.56	2.93	3.70	3.08	3.32	0.41	1:656	1:357	1:502	1:362	257.59	428.62	1.66

The individual events of the spring–neap tidal cycles also suggested the seasonal variability of Chl-a over the spring–neap tidal cycle (Figure 11). In other words, the magnitude of Chl-a was generally low in winter (TC-1 to TC-3), spring (TC-4 to TC-5), and autumn (TC-8 to TC-10) and high in summer (TC-6 to TC-7); higher Chl-a (>7 mg m^{-3}) generally occurred in the area off Saga, Isahaya Bay and off Kumamoto in winter and spring, expanded to the whole sea in summer, and was reduced to the area off Saga, Isahaya and off Kumamoto in autumn. This seasonal variability of Chl-a over the spring–neap tidal cycle was similar to that of the monthly climatology data.

The spring–neap tidal cycle variability of Chl-a and TSM was observed for the whole area (Figure 12; Table 2). In general, the spring–neap tidal variation was much smaller in winter, spring, and autumn than that in summer for Chl-a in terms of standard deviation. Daily river discharge data from two weeks before the first tidal stage (SN tide) was also used to investigate its influence on Chl-a (Figure 12). To quantify the variation of the river discharge, the ratio of the maximum:average of river discharge was calculated for each tidal cycle (Table 2). For TC-1 to TC-3, and TC-8 to TC-10, the river discharge peaks were low (ratio < 1.67), and Chl-a generally increased from SN to N or NS tide, and decreased during S tide; TSM was generally low during SN and N tide, and increased during NS or S tide. This variability of Chl-a and TSM was similar to that of the spatially-averaged monthly climatology data for the region off Saga and for the whole sea except for summer.

For the other tidal cycles, i.e., TC-4 and TC-7, the river discharge peaks were high (ratio > 1.67), and the river discharge peaks occurred before SN tide for TC-4, TC-6, and TC-7, while a peak occurred between NS and S tides for TC-5. For TC-5, the temporal variability of Chl-a and TSM was similar to that for TC-1 to TC-3, and TC-8 to TC-10, and the high river discharge had no influence on the Chl-a variation during the tidal stages before S tide. On the other hand, for TC-4, TC-6, and TC-7, after the high river discharge peak, Chl-a decreased continuously from SN to S tide except for TC-7. This may be due to the increase of Chl-a associated with high river discharges before the SN tide. TSM increased slightly from SN to N tide, then peaked at NS or S tide, which was similar to that exhibited for other tidal cycles. Furthermore, for TC-6 and TC-7, the Chl-a:TSM ratios (Table 2) were much higher than for other tidal cycles, indicating the dominance of phytoplankton in TSM for these two tidal cycles. These results indicate that high river discharge (ratio > 1.67) influenced the variation of Chl-a and TSM after during summer (TC-6 to TC-7), and that the occurrence of high river discharges can also influence the individual tidal cycles of Chl-a for other seasons (TC-4).

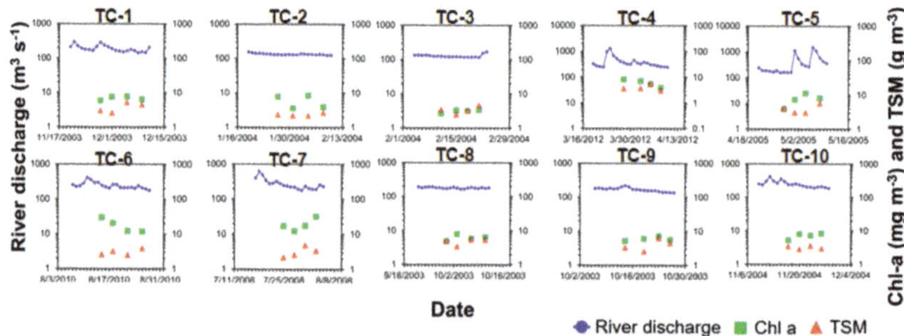

Figure 12. Time series of whole area-averaged Chl-a and TSM, and the daily river discharge for the ten representative individual events of tidal cycles: TC-1 to TC-3 (winter), TC-4 to TC-5 (spring), TC-6 to TC-7 (summer), and TC-8 to TC-10 (autumn). Chl-a and TSM in the four tidal cycles were represented by red and green markers, respectively. The river discharge, from two weeks before the tidal cycle to the end of the tidal cycle, is represented by the blue marked line. The average data was the average of all the daily data in each tidal stage. For each tidal cycle, the four Chl-a and TSM data from left to right were for SN, N, NS and S tides, respectively.

4. Discussion

4.1. Use of Satellite Data to Investigate the Spring–Neap Tidal Variations in Chl-a and TSM

Previous studies of the variability of Chl-a and TSM over spring–neap tidal cycles have primarily relied on in situ observations [3,4,17,23], and there are few studies based on satellite data [11,12,24]. In addition, previous studies of the Ariake Sea were only focused on the area off Saga, where the tidal range is larger than in other regions of the sea. Because of the observation method, those studies were limited to a few stations and a few spring–neap tidal cycles. Our study is different in that it relies on ocean color satellite data (MODIS-Aqua) to investigate the variability of Chl-a over the spring–neap tidal cycle, and we examine the tidal impacts on Chl-a for the whole Ariake Sea and specifically focus on three regional areas; i.e., off Saga, Isahaya Bay, and off Kumamoto. We used ocean color data from 2002 to 2017 to understand the annual and seasonal climatology as well as some events corresponding to spring–neap tidal cycles. This approach reveals that there are significant regional and seasonal differences in the Chl-a variability influenced by spring–neap tidal cycles. One of the known difficulties in using the standard MODIS Chl-a product for shallow coastal waters, such as the Ariake Sea, is the inadequacy of atmospheric correction schemes for obtaining accurate satellite-based Rrs that can be used to calculate Chl-a. It is also known that the present standard in-water algorithms for Chl-a estimates are biased in turbid waters. Here, we applied the Rrs recalculation method used in [18,25] and a local switching in-water algorithm for MODIS data to improve Chl-a retrievals in the Ariake Sea [18]. We also developed a TSM algorithm suitable for this area in this study. We have shown previously [18] that this recalculation method for Rrs and the improved empirical in-water algorithm significantly enhance the accuracy of the Chl-a retrievals from MODIS-Aqua. Our results showed the independent behavior of Chl-a and TSM over the spring–neap tidal cycle, which was consistent with previous studies for off Saga [13,14,17], suggesting that the influence of TSM on the satellite estimation of Chl-a was minimal, and the accuracy of our algorithms was adequate for our objective although imperfect.

To understand the influence of the spring–neap tidal cycle on Chl-a and TSM variability, we divided the spring–neap tidal cycle into four tidal stages (SN, N, NS and S tides). For each tidal stage, we produced annual and monthly climatology data and individual events of spring–neap tidal cycles of MODIS-Aqua Chl-a and TSM (2002–2017). This analysis made it possible for us to understand the seasonal and regional variations of tidal cycles of Chl-a and TSM, although there were much missing data in the spatial and temporal scales for many individual events of spring–neap tidal cycles over the 16 years.

4.2. Spatial and Seasonal Variability of the Spring–Neap Tidal Cycle

Tidal currents have been reported to be an important factor for the resuspension and transport of Chl-a and TSM into macrotidal environments, such as embayments, estuaries and tidal flats [4,23,26–28]. Using in situ data collected at several stations in the inshore area off Saga (October 2002–April 2003), in [17], the authors reported that Chl-a increased from N to NS and decreased during S tide, whereas TSM increased during S and decreased during N tide in the northern part of the Ariake Sea. The variability of TSM was explained by the resuspension of the sediment caused by the strongest tidal current during the S tide and re-sedimentation due to the weakest tidal current during N tide. The increase of Chl-a was explained by the increased light availability due to the reduction of TSM for the phytoplankton growth during N tide and the reduced light availability due to the increase of TSM during S tide [4,17].

Our results showed that the SN tidal resuspension in the Ariake Sea varied in space and time. Chl-a increased from SN to NS and decreased in S when TSM was high off Saga during fall, winter and spring when river discharge was low (Figure 7; Figure 8). This tidal cycle of Chl-a and TSM is consistent with a previous study [17]. The high TSM in this area reduces light availability and therefore limits phytoplankton growth, specifically during the spring tide when the resuspension of the sediment increased. This relation between the light availability and TSM is supported in a study by

Ooshima et al. [29], in which it was reported that the attenuation coefficient in the surface water of the Ariake Sea was strongly and positively correlated with suspended sediment in winter, suggesting that light availability declines when the suspended sediment is higher. Moreover, the low Chl-a:TSM ratio (Figure 10b) during the spring tide indicates that the non-phytoplankton particles in particular suspended sediments are the dominant TSM during the S tide. In contrast, in [12], the authors reported that phytoplankton growth was enhanced due to the increased nutrients supplied by the spring tidal resuspension in German Bight. This suggests that the spring–neap tidal variability of Chl-a varied by regions. In Isahaya Bay and off Kumamoto also, Chl-a increased from SN to NS and decreased in S during fall, winter and spring, and the variation was still similar to that in the area off Saga (Figure 9c,d). However, the variations in TSM in terms of magnitude over the spring–neap tidal cycle were smaller than off Saga, suggesting that there may not be a large-scale resuspension of the sediment. This is consistent with the spatial variation of TSM (Figure 4; Figure 5; Figure 7), which shows a high TSM off Saga, with relatively low values in Isahaya Bay and off Kumamoto. The lower TSM in Isahaya Bay and off Kumamoto may be related to the reduction of the tidal flat of this area by the construction of a dike and sandy tidal flat caused by the lower tidal current, respectively [17]. The Chl-a:TSM ratio was higher than that off Saga, and Chl-a and TSM were highly correlated during the spring–neap tidal cycles (Figure 10). This indicates that the small variations in TSM were mostly composed of phytoplankton and that resuspended TSM was not the controlling factor for phytoplankton variation. Therefore, there might be different mechanisms, such as the advection of phytoplankton, which explain the spring–neap tidal variation in Chl-a and TSM for those regional areas. For example, in [27], it was reported that diatom blooms during winter were advected from the estuaries connected to off Saga to the middle parts of the Ariake Sea. In order to understand the mechanism for the spring–neap tidal variations in Chl-a for those two areas, further investigations may be required.

For the whole sea, the spring–neap tidal variability of Chl-a and TSM was similar to that off Saga (Figures 9 and 10). This is probably due to the fact that both Chl-a and TSM values were higher off Saga than that in Isahaya Bay and off Kumamoto as well as in the middle-western areas. The similar spring–neap tidal variability of Chl-a and TSM between the whole sea and off Saga indicates that the influence of the tidal cycle-induced TSM was mostly important off Saga, whereas tidal cycle-driven variations in Chl-a were important over the whole sea. Even in the enclosed bay, it is clear that the tidal influence of the variability of Chl-a was different in each region.

4.3. Seasonal Influence of River Discharge to the Spring–Neap Tidal Variations in Chl-a

In estuaries and coastal systems, river discharge containing nutrients and suspended sediments can either positively or negatively influence Chl-a. The positive and negative relationship between river discharge and Chl-a is largely dependent on the dominant influence of either nutrients or irradiance on phytoplankton growth [30–35].

Our results showed that the monthly climatology data of river discharge were strongly and positively correlated with the spatially-averaged monthly climatology data of Chl-a over the spring–neap tidal cycle for the whole sea and regional areas (Figure 7). River discharge was also an important factor for the variability of Chl-a over the individual events of spring–neap tidal cycles, as our results showed that river discharge was probably the major driver of the variability of Chl-a in TC-4 and TC-6 (Figure 11). This suggests that river discharge promotes high phytoplankton growth in the Ariake Sea, which is consistent with previous studies [10,13].

The effects of spring–neap tidal cycle on the variability of Chl-a and TSM was not very clear during summer when the river discharge was high. The variation caused by the high river discharge during summer can mask the tidal cycle variation of Chl-a and TSM. The high correlation between Chl-a and river discharge indicates that the increase in Chl-a was caused by the possible nutrient input from the events of river discharge. We also observed that the Chl-a:TSM ratio was extremely high after the river discharge, and the values were often higher than 1:275–1:600, which is the range of phytoplankton-dominated water in Tokyo Bay [22]. The high ratio also reflects possible differences in the physiological conditions of phytoplankton after the river discharge.

5. Conclusions

As the spring–neap tidal variability of satellite Chl-a associated with TSM has not been investigated for a broad area and with a long-term data set, we investigated the spring–neap tidal variability of Chl-a on the basis of annual and monthly climatology data and individual events of spring–neap tidal cycles using an improved MODIS-Aqua data set (2002-2017). Spatially-averaged Chl-a and TSM and daily and monthly river discharge values were calculated to quantify the influence of TSM and river discharge on Chl-a for the whole sea and three regional areas (off Saga, Isahaya Bay and off Kumamoto).

The errors in MODIS-Aqua-derived Rrs and Chl-a for the Ariake Sea were effectively reduced by applying the methods in [11]. Therefore, we recalculated the 16-year MODIS-Aqua Rrs and Chl-a data with the same methods developed for this area used in [11]. Moreover, a local TSM algorithm was developed in this study, and then it was applied to the improved MODIS-derived Rrs to obtain the MODIS-derived TSM. The variability of Chl-a and TSM over the spring–neap tidal cycle off Saga was reasonable and was consistent with the field-based observations in previous studies, suggesting that the recalculated Chl-a and MODIS-derived TSM were separable.

The results of this study suggested seasonal and regional differences in the factors controlling the variability of Chl-a over the spring–neap tidal cycle. In general, the variability of Chl-a over the tidal cycle was controlled by river discharge during summer. In other seasons, it was controlled by the tidally resuspended TSM for off Saga and possibly by direct tidal transportation and tidal mixing in Isahaya Bay and off Kumamoto, respectively. In summary, this study suggests that satellite ocean color data offers an effective means for understanding the mechanisms of seasonal and regional Chl-a variability in coastal ecosystems that come under the influence of tides and river discharge.

This study also reveals that satellite ocean color data can discern the effects of spring–neap tidal cycles on Chl-a and TSM. However, we found different correlations between Chl-a and TSM in the Ariake Sea, confirming that the feature of spring–neap tidal cycles can vary in different areas. In addition, this is the first study to investigate the spring–neap tidal variability of Chl-a and TSM using satellite ocean color data in the Ariake Sea. Given the broad coverage and frequent sampling by satellites, our results reveal that satellite ocean color data can contribute significantly to our knowledge and understanding of the environmental dynamics caused by spring–neap tidal cycles in the Ariake Sea. The availability of these datasets offers the potential for the better management of the water quality of enclosed embayments such as the Ariake Sea.

Supplementary Materials: The following are available online at http://www.mdpi.com/2072-4292/12/11/1859/s1.

Author Contributions: Conceptualization, M.Y. and J.I.; methodology, M.Y. and J.I.; software, M.Y. and E.d.R.M..; validation, M.Y. and J.I.; formal analysis, M.Y. and J.I.; investigation, M.Y.; resources, M.Y. and J.I.; data curation, M.Y.; writing—original draft preparation, M.Y.; writing—review and editing, M.Y., J.I., J.I.G., H.T. and E.d.R.M.; visualization, M.Y.; supervision, J.I.; project administration, J.I.; funding acquisition, J.I. All authors have read and agreed to the published version of the manuscript.

Funding: This research was supported by Fisheries Agency of Japan and by the GCOM-C project of the Japan Aerospace Exploration Agency.

Acknowledgments: We are thankful to the Ocean Ecology Laboratory and Ocean Biology Processing Group of NASA Goddard Space Flight Center for providing long-term (July 2002 to present) Moderate Resolution Imaging Spectroradiometer (MODIS) Aqua L2 data. We also thank Dr. Masataka Hayashi in Science and Technology, Inc. for helping with satellite data processing.

Conflicts of Interest: The authors declare no conflict of interest.

References

1. Aoki, K.; Onitsuka, G.; Shimizu, M.; Matsuo, H.; Kitadai, Y.; Ochiai, H.; Yamamoto, T.; Furukawa, S. Interregional difference in spring neap variations in stratification and chlorophyll fluorescence during summer in a tidal sea (Yatsushiro Sea, Japan). *Estuar. Coast. Shelf Sci.* **2016**, *180*, 212–220. [CrossRef]
2. Azhikodan, G.; Yokoyama, K. Spatio-temporal variability of phytoplankton (Chlorophyll-a) in relation to salinity, suspended sediment concentration, and light intensity in a macrotidal estuary. *Cont. Shelf Res.* **2016**, *126*, 15–26. [CrossRef]
3. Cloern, J.E.; Powell, T.M.; Huzzey, L.M. Spatial and temporal variability in south San Francisco Bay (USA), II, Temporal changes in salinity, suspended sediments, and phytoplankton biomass and productivity over tidal time scales. *Estuar. Coast. Shelf Sci.* **1989**, *28*, 599–613. [CrossRef]
4. Koh, C.H.; Khim, J.S.; Araki, H.; Yamanishi, H.; Mogi, H.; Koga, K. Tidal resuspension of microphytobenthic Chlorophyll-a in a Nanaura mudflat, Saga, Ariake Sea, Japan: Flood−ebb and spring−neap variations. *Mar. Ecol. Prog. Ser.* **2006**, *312*, 85–100. [CrossRef]
5. Monbet, Y. Control of phytoplankton biomass in estuaries: A comparative analysis of microtidal and macrotidal estuaries. *Estuaries* **1992**, *15*, 563–571. [CrossRef]
6. Wofsy, S.C. A simple model to predict extinction coefficients and phytoplankton biomass in eutrophic waters. *Limnol. Oceanogr.* **1983**, *28*, 1144–1155. [CrossRef]
7. Feng, L.; Hu, C.; Chen, X.; Song, Q. Influence of the Three Gorges Dam on total suspended matters in the Yangtze Estuary and its adjacent coastal waters: Observations from MODIS. *Remote Sens. Environ.* **2014**, *140*, 779–788. [CrossRef]
8. Yamaguchi, H.; Ishizaka, J.; Siswanto, E.; Baek Son, Y.; Yoo, S.; Kiyomoto, Y. Seasonal and spring interannual variations in satellite-observed chlorophyll-a in the Yellow and East China Seas: New datasets with reduced interference from high concentration of resuspended sediment. *Cont. Shelf Res.* **2013**, *59*, 1–9. [CrossRef]
9. Gons, H.J.; Auer, M.T.; Effler, S.W. MERIS satellite chlorophyll mapping of oligotrophic and eutrophic waters in the Laurentian Gt Lakes. *Rem. Sens. Environ.* **2008**, *112*, 4098–4106. [CrossRef]
10. Ishizaka, J.; Kitaura, Y.; Touke, Y.; Sasaki, H.; Tanaka, A.; Murakami, H.; Suzuki, T.; Matsuoka, K.; Nakata, H. Satellite detection of red tide in Ariake Sound, 1998–2001. *J. Oceanogr.* **2006**, *62*, 37–45. [CrossRef]
11. Shi, W.; Wang, M.; Jiang, L. Spring-neap tidal effects on satellite ocean color observations in the Bohai Sea, Yellow Sea, and East China Sea. *J. Geophysic. Res.* **2011**, *116*. [CrossRef]
12. Su, J.; Tian, T.; Krasemann, H.; Schartau, M.; Wirtz, K. Response patterns of phytoplankton growth to variations in resuspension in the German Bight revealed by daily MERIS data in 2003 and 2004. *Oceanologia* **2015**, *57*, 328–341. [CrossRef]
13. Tsutsumi, H. Critical events in the Ariake Sea ecosystem: Clam population collapse, red tides, and hypoxic bottom water. *Plankton Benthos Res.* **2006**, *1*, 3–25. [CrossRef]
14. Hayami, Y.; Maeda, K.; Hamada, T. Long term variation in transparency in the inner area of Ariake Sea. *Estuar. Coast. Shelf Sci.* **2015**, *163*, 290–296. [CrossRef]
15. Unoki, S. Why did the tide and the tidal current decrease in Ariake Sea? *Oceanogr. Japan* **2002**, *12*, 85–96. (In Japanese) [CrossRef]
16. Unoki, S. The results of re-examining the recent decay of tide in Ariake Sea, based on smoothed data of observations. *Oceanogr. Japan* **2003**, *12*, 307–313. (In Japanese) [CrossRef]
17. Tanaka, K.; Kodama, M. Effects of resuspended sediments on the environmental changes in the inner part of Ariake Sea, Japan. *Bull. Fish Res. Agency* **2007**, *19*, 9–15.
18. Yang, M.M.; Ishizaka, J.; Goes, J.I.; Gomes, H.R.; Maúre, E.R.; Hayashi, M.; Katano, T.; Fujii, N.; Saitoh, K.; Mine, T.; et al. Improved MODIS-Aqua chlorophyll-a retrievals in the turbid semi-enclosed Ariake Sea, Japan. *Remote Sens.* **2018**, *10*, 1335. [CrossRef]
19. Gitelson, A.A.; Schalles, J.F.; Hladik, C.M. Remote chlorophyll-a retrieval in turbid, productive estuaries: Chesapeake Bay case study. *Remote Sens. Environ.* **2007**, *109*, 464–472. [CrossRef]
20. Le, C.; Hu, C.; Cannizzaro, J.; Duan, H. Long-term distribution patterns of remotely sensed water quality parameters in Chesapeake Bay. *Estuar. Coast. Shelf Sci.* **2013**, *128*, 93–103. [CrossRef]
21. Binding, C.E.; Bowers, D.G.; Mitchelson-Jacob, E.G. Estimating suspended sediment concentrations from ocean color measurements in moderately turbid waters; The impact of variable particle scattering properties. *Remote Sens. Environ.* **2005**, *94*, 373–383. [CrossRef]

22. Kishino, M.; Tanaka, A.; Ishizaka, J. Retrieval of Chlorophyll-a, suspended solids and colored dissolved organic matter in Tokyo Bay using ASTER data. *Remote Sens. Environ.* **2005**, *99*, 66–74. [CrossRef]

23. Ito, Y.; Katano, T.; Fujii, N.; Koriyama, M.; Yoshino, K.; Hayami, Y. Decreases in turbidity during neap tides initiate late winter large diatom blooms in a macrotidal embayment. *J. Oceanogr.* **2013**, *69*, 467–479. [CrossRef]

24. Valente, A.S.; da Silva, J.C.B. On the observability of the fortnightly cycle of the Tagus estuary turbid plume using MODIS ocean colour images. *J. Mar. Sci.* **2009**, *75*, 131–137. [CrossRef]

25. Hayashi, M.; Ishizaka, J.; Kobayashi, H.; Toratani, M.; Nakamura, T.; Nakashima, Y.; Yamada, S. Evaluation and improvement of MODIS and SeaWiFS-derived chlorophyll-a concentration in Ise-Mikawa Bay. *J. RSSJ* **2015**, *35*, 245–259, (In Japanese with English Abstract).

26. Demers, S.; Therriault, J.C.; Bourget, E.; Bah, A. Resuspension in the shallow sublittoral zone of a macrotidal estuarine environment: Wind influence. *Limnol. Oceanogr.* **1987**, *32*, 327–339. [CrossRef]

27. Black, K.S. Suspended sediment dynamics and bed erosion in the high shore mudflat region of the Humber Estuary, UK. *Mar. Pollut. Bull.* **1998**, *37*, 122–133. [CrossRef]

28. DeJonge, V.N.; Vanbeusekom, J.E.E. Wind-and-tide-induced resuspension of sediment and microphytobenthos from tidal flats in the Ems estuary. *Limnol. Oceanogr.* **1995**, *40*, 766–778.

29. Ooshima, I.; Abe, K. Estimation method for the attenuation coefficient in the surface layer of the Ariake Sea. *Oceanogr. Japan* **2005**, *14*, 593–600, (In Japanese with English Abstract). [CrossRef]

30. Lohrenz, S.E.; Dagg, M.J.; Whitledge, T.E. Enhanced primary production at the plume/oceanic interface of the Mississippi River. *Cont. Shelf Res.* **1990**, *10*, 639–664. [CrossRef]

31. Smith, W.O.; DeMaster, D.J. Phytoplankton biomass and productivity in the Amazon River plume: Correlation with monthly river discharge. *Cont. Shelf Res.* **1996**, *16*, 291–319. [CrossRef]

32. Dortch, Q.; Whitledge, T.E. Does nitrogen or silicon limit phytoplankton production in the Mississippi River plume and nearby regions? *Cont. Shelf Res.* **1992**, *12*, 1293–1309. [CrossRef]

33. Cloern, J.E.; Cole, B.E.; Wong, R.L.; Alpine, A.E. Temporal dynamics of estuarine phytoplankton: A case study of San Francisco Bay. *Hydrobiologia* **1985**, *129*, 153–176. [CrossRef]

34. DeMaster, D.J.; Knapp, G.B.; Nittrouer, C.A. Effect of suspended sediments on geochemical processes near the mouth of the Amazon River: Examination of biogenic silica uptake and the fate of particle-reactive elements. *Cont. Shelf Res.* **1986**, *6*, 107–125. [CrossRef]

35. Cole, J.C.; Caraco, N.F.; Peierls, B.L. Can phytoplankton maintain a positive carbon balance in a turbid, freshwater, tidal estuary? *Limnol. Oceanogr.* **1992**, *37*, 1608–1617. [CrossRef]

 *remote sensing*

Article

A Novel Method Based on Backscattering for Discriminating Summer Blooms of the Raphidophyte (*Chattonella* spp.) and the Diatom (*Skeletonema* spp.) Using MODIS Images in Ariake Sea, Japan

Chi Feng [1], Joji Ishizaka [2,*], Katsuya Saitoh [3], Takayuki Mine [4] and Hirokazu Yamashita [5]

[1] Graduate School of Environmental Studies, Nagoya University, Furo-cho, Chikusa-ku, Nagoya, Aichi 464-8601, Japan; feng.chi@a.mbox.nagoya-u.ac.jp
[2] Institute for Space-Earth Environmental Research (ISEE), Nagoya University, Furo-cho, Chikusa-ku, Nagoya, Aichi 464-8601, Japan
[3] Japan Fisheries Information Service Center, 4-5, Toyomicho, Toyomishinko Bldg. 6F., Chuo-ku 104-0055, Japan; ksaitoh@jafic.or.jp
[4] Saga Ariake Fisheries Promotion Center, 1-1-59, Jonai, Saga 840-8570, Japan; mine-takayuki@pref.saga.lg.jp
[5] Kumamoto Prefectural Fisheries Research Center, 1-11-1, Higashisakura, Higashi Ward, Nagoya, Aichi 461-0005, Japan; yamashita-h-dw@pref.kumamoto.lg.jp
* Correspondence: jishizaka@nagoya-u.jp; Tel.: +86-052-789-3487

Received: 3 April 2020; Accepted: 4 May 2020; Published: 8 May 2020

Abstract: The raphidophyte *Chattonella* spp. and diatom *Skeletonema* spp. are the dominant harmful algal species of summer blooms in Ariake Sea, Japan. A new bio-optical algorithm based on backscattering features has been developed to differentiate harmful raphidophyte blooms from diatom blooms using MODIS imagery. Bloom waters were first discriminated from other water types based on the distinct spectral shape of the remote sensing reflectance $R_{rs}(\lambda)$ data derived from MODIS. Specifically, bloom waters were discriminated by the positive value of Spectral Shape, SS (645), which arises from the $R_{rs}(\lambda)$ shoulder at 645 nm in bloom waters. Next, the higher cellular-specific backscattering coefficient, estimated from MODIS data and quasi-analytical algorithm (QAA) of raphidophyte, *Chattonella* spp., was utilized to discriminate it from blooms of the diatom, *Skeletonema* spp. A new index $b_{bp-index}$ (555) was calculated based on a semi-analytical bio-optical model to discriminate the two algal groups. This index, combined with a supplemental Red Band Ratio (RBR) index, effectively differentiated the two bloom types. Validation of the method was undertaken using MODIS satellite data coincident with confirmed bloom observations from local field surveys, which showed that the newly developed method, based on backscattering features, could successfully discriminate the raphidophyte *Chattonella* spp. from the diatom *Skeletonema* spp. and thus provide reliable information on the spatial distribution of harmful blooms in Ariake Sea.

Keywords: harmful algal blooms; *Chattonella* spp.; *Skeletonema* spp.; backscattering; MODIS; Ariake Sea

---

1. Introduction

Harmful Algal Blooms (HABs) are becoming more frequent in the coastal environment causing significant harm to fisheries, the environment and economies. Some HABs produce toxins, some of them consume nutrients used in seaweed aquaculture and, they often discolor the water.

Remote sensing is an effective method for bloom detection, because algal groups can show a distinct remote sensing reflectance $R_{rs}(\lambda)$ signature which can be then related to large algal accumulation at the surface [1]. Algal blooms are associated with anomalously high chlorophyll-a concentrations, which influence the signal of $R_{rs}(\lambda)$ in green and red bands, prominently. Methods like Fluorescence Line Height (FLH), Maximum Chlorophyll Index (MCI) and Floating Algae Index (FAI) have successfully mapped bloom distribution in global open oceans by taking advantage of the distinct characteristic in $R_{rs}(\lambda)$ [2–4]. Other studies have been conducted to detect specific bloom species in coastal waters. *Trichodesmium* spp. blooms have been detected based on multispectral patterns of $R_{rs}(\lambda)$ in the North Atlantic [5] while *Karenia brevis* blooms have been captured using the *Karenia brevis* bloom index (KBBI) in the Gulf of Mexico [6]. *Cochlodinium polykrikoides* blooms have been quantified by a novel red tide quantification algorithm in the coastal waters of the East China Sea [7], while *Microcystis aeruginosa* blooms in the Laurentian Great Lakes have been distinguished from other phytoplankton by an index denoted as Spectral Shape (SS) [8]. The harmful species *Karenia mikimotoi* has been discriminated from other types of blooms in the Seto-Inland Sea of Japan using the spectral slope difference [9].

Other algorithms use Inherent Optical Properties (IOPs) to identify harmful algal blooms. Methods like those of Shang et al. [10] for the East China Sea (ECS) differentiate dinoflagellates from diatom blooms by a combination of total absorption coefficient at 443 nm and a Bloom Index (BI). Tao et al. [11] developed a Green–Red Index (GRI), indicating absorption at 510 nm of bloom waters, to distinguish *Prorocentrum donghaiense* from diatom blooms. Backscattering properties of bloom waters have also been used in red tide detection. Cannizzaro et al. [12] detected the toxic *Karenia brevis* from diatom blooms in the Gulf of Mexico by its featured lower backscattering. A *Coccolithophorid* bloom in the Barents Sea was captured by the unusual sharp increase in backscattering [13]. Lei et al. [14] differentiated dinoflagellate blooms from diatom blooms in the East China Sea by the difference in backscattering coefficient ratios. However, more work still needs to be done, because of the large variations in the IOPs of different harmful algal species, especially in coastal waters.

Recently, there have been frequent reports of HABs outbreaks in the Ariake Sea, an enclosed small bay in the southwest of Japan, which result in great damage to aquaculture farms and fisheries [15–17]. HABs in summer are especially serious in this area due to the strong solar radiation and elevation of the water temperature after the rainy season. The raphidophyte *Chattonella* spp. and the diatom *Skeletonema* spp. are the dominant harmful species of summer blooms.

Raphidophytes like *Chattonella antiqua* can secrete toxic compounds causing a large reduction in shellfish, while the less harmful diatom species like *Skeletonema* spp. may produce arsenite and dimethylarsenic which block important biochemical pathways in other algae [18–20]. It is empirically known that raphidophyte blooms alternate with diatom blooms when the surface water lacks nitrogen and silicate [17,21]. Thus, it becomes important to distinguish harmful algal species from the non-bloom conditions to evaluate the possible damage and to provide related information for protection of the marine ecosystem and the economies that they support.

The objective of this study is to develop techniques to distinguish between raphidophyte and diatom dominated blooms in optically complex, coastal waters of Ariake Sea using MODIS data. If successful, the method can provide effective bloom information for the coastal monitoring by local fisheries institutions.

2. Materials and Methods

2.1. Study Area

Ariake Sea is located along the west coast of Kyushu in western Japan. It is approximately 1700 km^2 in area, 20 m in average depth, and 34 billion m^3 in volume. Inland rivers take about 8×10^9 m^3 freshwater into the sea every year. Among the rivers, the Chikugo River is the largest one discharging 50% of the freshwater inflow. Tidal flats of this semi-enclosed shallow sea are the largest in Japan covering 40% of the total tidal flat area in Japan [22]. Seaweed and shellfish, once plentiful in this area, have dramatically decreased in recent decades, while the number of red tide events has increased to more than 20 times per year since 1985. Blooms events generally occur during summer when conditions are ideal for phytoplankton growth. Field observations have been conducted in every summer to study these blooms and their associated hydrography for fishery management of the Ariake Sea.

2.2. Field Data

Field data during the summer season (June to September) were collected from Saga Fisheries Promotion Center and Kumamoto Fisheries Research Center (from 2011–2014). Figure 1 shows the sampling site where data on chlorophyll-a concentration (Chl a), phytoplankton species and their cellular abundance were collected. As no in situ $R_{rs}(\lambda)$ data was collected, remotely sensed MODIS $R_{rs}(\lambda)$ were used for algorithm development, which is further explained in Section 2.3.

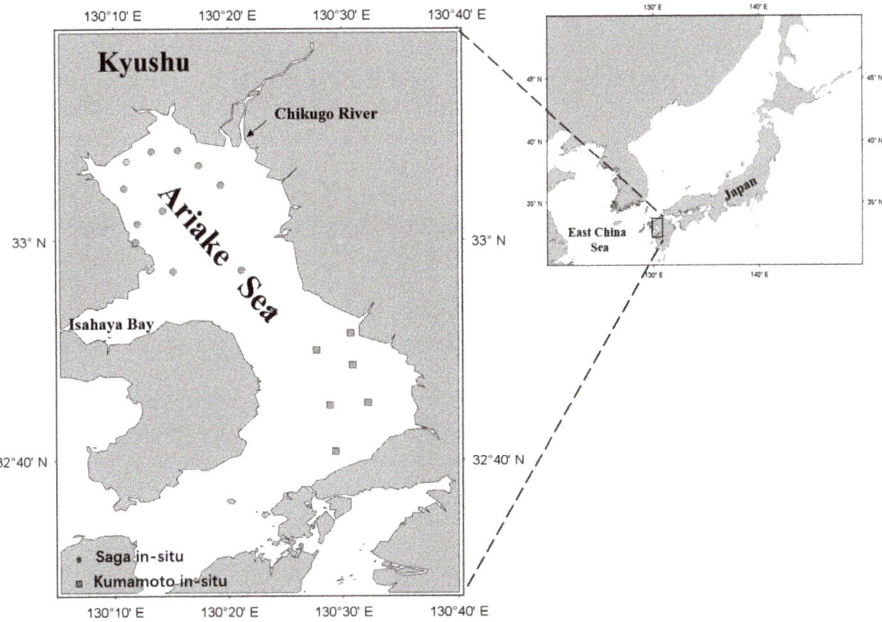

Figure 1. Map of Ariake Sea, Japan. Solid points show the field sampling locations in 2011–2014. The gray circle and square indicate the position of data taken by Saga Fisheries Promotion Center and Kumamoto Fisheries Research Center, respectively.

For data analysis, field data from 2011 to 2014 were divided into three data groups according to the bloom conditions (Table 1). Specifically, a diatom bloom was confirmed when the cell numbers of *Skeletonema* spp. were >10,000 cells mL^{-1} (N = 126 for 41 days), and a raphidophyte bloom was confirmed when the cell numbers of *Chattonella* spp. were > 1000 cells mL^{-1} (N = 12 for 3 days). Non-bloom data was collected (N = 280 for 70 days). The cellular abundance threshold of bloom conditions was defined according to previous red tide reports from local fisheries institutions. Diatom blooms were dominated by *Skeletonema* spp. both at the Saga (N = 113) and Kumamoto sampling sites (N = 13), while raphidophyte blooms of *Chattonella* spp. were only found at Saga sampling sites (N = 12). The environmental conditions for the three groups are shown in Table 1.

Table 1. Environmental conditions of diatom and raphidophyte blooms and non-bloom waters. All parameters were averaged within each group and standard deviations are provided.

Group	Chl a (mg m^{-3})	Salinity	Temperature (°C)	Abundance (cells mL^{-1})
Diatom Bloom	30.03 ± 43.35	20.04 ± 7.06	27.42 ± 2.10	25,964 ± 33,209
Raphidophyte Bloom	210.47 ± 178.48	26.27 ± 1.81	29.68 ± 0.81	2217 ± 1965
Non-bloom	8.95 ± 11.64	25.77 ± 5.19	25.93 ± 2.61	–

In addition to the field data, a number of bloom reports were also obtained from the Japan Red Tide Online site (http://akashiwo.jp/) for the period 2015–2018 for algorithm validation. Sampling site, occurrence times, phytoplankton species and cellular abundance were recorded.

2.3. Satellite Data

As no situ optical measurements were made, MODIS-Aqua level 2, $R_{rs}(\lambda)$ values (downloaded from https://oceancolor.gsfc.nasa.gov/) were extracted from locations where field data was collected in 2011–2014 using Windows Image Manager (WIM) software. The routine wam_match within WIM was used to find matches between in situ measurements and satellite image data. The point sample was within a rectangular window of 3 × 3 pixels, centered at the nearest matching pixel. Mean value of the valid pixels within the 3 × 3 windows were used as the final remote sensing reflectance $R_{rs}(\lambda)$. Flags, HIGLINT, CLDICE, HISOLZEN, CHLFAIL, ATMFAIL (flags information can be found at https://oceancolor.gsfc.nasa.gov/atbd/ocl2flags/), were used to control the quality of the MODIS $R_{rs}(\lambda)$ data. As there were limited cloud-free satellite images coincident with the field sampling, the time difference between satellite data and in-situ data was extended by 1.5 days during bloom events. Finally, there were four matched data points for raphidophyte bloom (from one image), eight for diatom bloom (from four images) and 23 for non-bloom water (from five images) in 2011–2014. Additionally, in 2015–2018, six images during the bloom period (three images from raphidophyte bloom and three images from diatom bloom) were also used to validate the algorithm based on bloom locations obtained from the Japan Red Tide Online. One image in 2018 was also selected when no bloom occurred (Table 2 shows the summary of match up results).

Table 2. Summary of MODIS matches with field data (2011–2014) and with bloom reports from Japan Red Tide Online (2015–2018).

Name of Bloom	Training Data	Validation Data
Diatom bloom	29 August 2011 26 July 2012 2 August 2012 29 August 2013	12 July 2018 13 July 2018 3 September 2018
Raphidophyte bloom	9 August 2013	07 September 2015 18 August 2016 29 July 2018
Non-bloom	11 August 2011 11 June 2012 20 August 2012 17 June 2013 28 August 2013	29 August 2018

Since there was some underestimation in the short bands of the MODIS data, from an error in atmosphere correction that resulted from the difficulties in estimating aerosol type and optical thickness [23–25], $R_{rs}(\lambda)$ showing negative value in the short bands was discarded. We also decided not to use the short bands to develop our algorithm as the complex pigment composition of algal species makes it hard to distinguish phytoplankton groups in that range [26,27], while $R_{rs}(\lambda)$ at green and red bands has shown good agreement with in-situ $R_{rs}(\lambda)$ in Ariake Sea as confirmed by Yang et al. [25].

Again due to the lack of field observations, inherent optical properties (IOPs) like total absorption, $a_t(\lambda)$, and particle backscattering, $b_{bp}(\lambda)$, were derived using MODIS $R_{rs}(\lambda)$ as input into the Quasi-Analytical Algorithm (QAA Version 6) (details are in Lee et al. [28]). Although the short bands of MODIS are not reliable for our work, it has been shown that $a_t(\lambda)$ can be derived by QAA with high accuracy since it is not sensitive to errors in $R_{rs}(\lambda)$ in the short bands [10,29]. Because of particle size variations in the field samples, a 20% additional error could be introduced in the retrieval of $b_{bp}(\lambda)$ [30,31]. In spite of these uncertainties and considering the optically complex Ariake Sea, it is still meaningful to use QAA-derived IOPs to derive the spectral shape for bloom discrimination.

3. Results

3.1. Detection of Bloom Waters

The first step was to correctly identify blooms from other optically dominant water types. MODIS $R_{rs}(\lambda)$ coincident with in-situ data showed considerable variability in both spectral shape and magnitude, indicating different water types in the observations (Figure 2). By comparing the spectral shape of $R_{rs}(\lambda)$ and Chl a concentrations for the three data groups (Table 1), the coastal area of Ariake Sea could be roughly separated into four bio-optical water types: (1) clear waters in the northeastern coast of Ariake Sea with low Chl a ($<$6.32 mg m^{-3}), and relatively high blue reflectance compared to the longer wavelength green band where no peak was observed; (2) turbid waters within estuary area, which exhibited extremely high reflectance at longer wavelength because of the high suspended sediments; (3) bloom waters, typically located in the northwestern part of Ariake Sea with moderate to high Chl a, and a spectral shape of reflectance typical of phytoplankton blooms with minimal values in the blue region and high values near 550 nm and 678 nm; and (4) mixed water defined as water with a middle range of Chl a concentration and a peak in green bands.

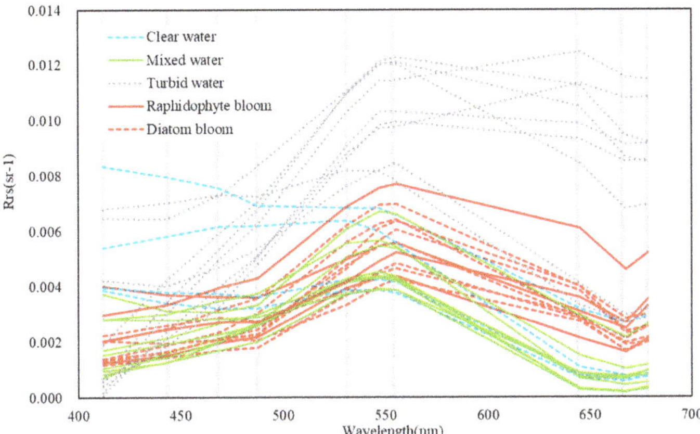

Figure 2. MODIS $R_{rs}(\lambda)$ spectra (N = 35) using the match-up method (See Section 2.3). Blue dashed line (clear water); green solid line (mixed waters); gray dotted line (turbid water), red solid and dashed line indicates raphidophyte and diatom bloom water, respectively. The gray lines indicate location of MODIS bands.

Based on the differences observed, we could separate clear water if the $R_{rs}(\lambda)$ peak was at wavelengths shorter than 555 nm, and turbid water if the $R_{rs}(\lambda)$ peak was $>$ 0.008 sr^{-1}. However, mixed waters with moderate Chl a concentration could not be separated from bloom waters based on this simple method since the former/latter also has a spectral peak at 555 nm ($<$0.008 sr^{-1}). So more detailed characteristics of the spectral shape were considered. Bloom waters showed prominent shoulders near 645 nm compared to mixed water (see Figure 2), which might be caused by strong backscattering of phytoplankton particles and weak absorption at this wavelength. To better characterize this feature, changes in the curvature of $R_{rs}(645)$ were compared to determine whether bloom waters could be distinguished from mixed water using this approach. The spectral shape algorithm (SS) of Wynne et al. [8], equivalent to the 2nd derivative when the bands are evenly distributed, provides an SS index which can describe the spectral variations useful in bloom detection [32,33]. In this study, normalized water leaving radiance (nL_w) is replaced with $R_{rs}(\lambda)$. The defined SS is as:

$$SS(\lambda) = R_{rs}(\lambda) - R_{rs}(\lambda^-) - \left(R_{rs}(\lambda^+) - R_{rs}(\lambda^-)\right) * \frac{(\lambda - \lambda^-)}{(\lambda^+ - \lambda^-)}, \tag{1}$$

where λ is the central band of the shape (645 nm), λ^- is the next lower band (555 nm) and λ^+ is the next higher band (667 nm). SS (645) of bloom waters showed positive values while mixed waters exhibited negative values (Figure 3). Turbid water also showed positive SS (645), and was distinguishable by the threshold at $R_{rs}(555)$ (>0.008 sr^{-1}).

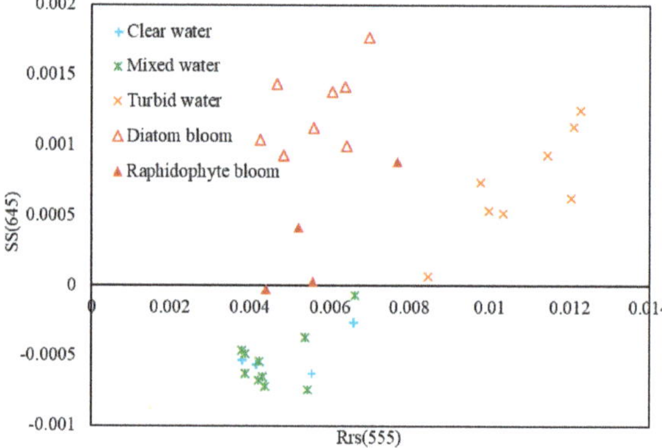

Figure 3. Scatter plot of MODIS $R_{rs}(555)$ and SS (645) generated based on the MODIS match-up pairs.

Together, algal bloom waters were differentiated from clear, turbid, and mixed waters by a combination of $R_{rs}(555)$ and SS (645). In the relationship between $R_{rs}(555)$ and SS (645) (Figure 3), clear and mixed waters were distinguished by negative SS (645) while bloom waters were identified by a positive SS (645) and low $R_{rs}(555)$ (<0.008 sr^{-1}). Even though turbid water also showed a positive SS (645), $R_{rs}(555)$ was higher than in bloom waters. The scatter plot shown in Figure 3 indicates that all observations with positive SS (645) and $R_{rs}(555)$ less than 0.008 sr^{-1} could be characterized as algal bloom waters.

In order to verify the utility of satellite-derived SS (645) and $R_{rs}(555)$ for detecting blooms, three independent MODIS images were selected to coincide with summer bloom reports by the Japan Red Tide Online from 2015–2018: (a raphidophyte bloom on 29 July 2018; a diatom bloom on 13 July 2018; one non-bloom day on 29 August 2018). Scatter plot of satellite derived SS (645) versus $R_{rs}(555)$ is shown in Figure 4. The MODIS Chl a image in late summer on 29 August 2018, when no bloom event was reported in Ariake Sea, showed no sign of high Chl a in most of the area (Figure 4a). High values were seen only near Kumamoto coasts and Isahaya Bay. The corresponding scatter plot (Figure 4g) showed that the surface waters was roughly divided into clear, turbid and mixed waters. Conversely, relatively high Chl a was observed in the MODIS images (Figure 4b,c) in association with blooms of the raphidophyte (*Chattonella* spp.) and the diatom (*Skeletonema* spp.), which had been confirmed by field observations (Figure 4h,i).

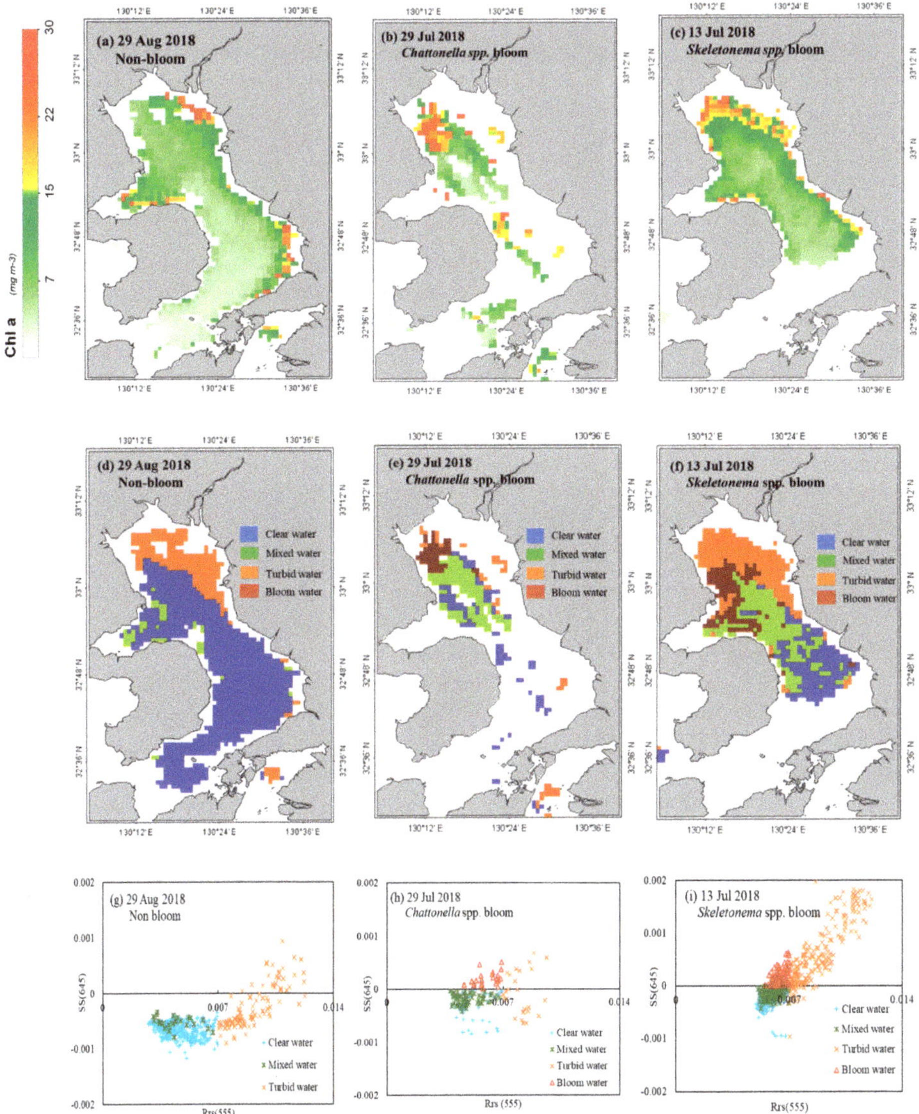

Figure 4. (**a–c**) Standard Chl a images from MODIS level-2 for Ariake Sea. Non-bloom (**a**), raphidophyte bloom (**b**) and diatom bloom (**c**) as confirmed by bloom reports from Japan Red Tide Online. (**d–f**) Water types derived using MODIS $R_{rs}(\lambda)$ and our newly developed method. (**g–i**) Scatter plot of $R_{rs}(555)$ and SS (645) derived from the MODIS $R_{rs}(\lambda)$ extracted from scenes shown in (**a–c**). Only areas with positive $R_{rs}(\lambda)$ are shown.

It is to be noted that negative values of standard MODIS $R_{rs}(\lambda)$ at 412 nm and 645 nm were observed for pixels associated with high Chl a, and such high Chl a retrievals are inaccurate. Thus, for the bloom distinguishing method these pixels were excluded. Other than this limitation, the bloom distinguishing method can be applied to MODIS data and the combination of $R_{rs}(555)$ and SS (645) can serve as the first step to classify blooms from space.

3.2. Discrimination of Harmful Algal Groups

To differentiate dominant algal blooms, inherent optical properties should be first compared. The spectral shape of $R_{rs}(\lambda)$ for bloom water is influenced by absorption and backscattering coefficients $(a_t(\lambda), b_b(\lambda))$ [34,35]. They can be expressed as:

$$a_t(\lambda) = a_w(\lambda) + a_{dg}(\lambda) + a_{ph}(\lambda), \tag{2}$$

$$b_b(\lambda) = b_{bw}(\lambda) + b_{bp}(\lambda), \tag{3}$$

where $a_w(\lambda)$ and $b_{bw}(\lambda)$ are the absorption and backscattering coefficients of pure water which are constants [36,37]. $a_{dg}(\lambda)$ and $a_{ph}(\lambda)$ represent non-algal and algal absorption, respectively. Non-algal absorption includes absorption by non-algal particles (NAP) and dissolved chromophoric dissolved organic matter (CDOM). $b_{bp}(\lambda)$ is the suspended particle backscattering coefficient, and includes backscattering by phytoplankton and inorganic particles.

Ideally, if we could derive $a_{ph}(\lambda)$ accurately by QAA, then we would be able to discriminate harmful algal species accurately. However, it has been shown that there is much uncertainty in partitioning $a_t(\lambda)$ into $a_{ph}(\lambda)$ and $a_{dg}(\lambda)$ for high absorption waters [29,38]. During the bloom, except for absorption by water, both algal particles and non-algal particles influenced variations in $R_{rs}(\lambda)$ because of the optical complexity of coastal waters. So, here we assume that the variations in absorption by bloom waters can be represented as the difference between $a_t(\lambda)$ and $a_w(\lambda)$. To prevent confusion, we use:

$$a_{bloom}(\lambda) = a_t(\lambda) - a_w(\lambda), \tag{4}$$

where $a_{bloom}(\lambda)$ represent the absorption by bloom waters.

In addition, the contribution of $b_{bw}(\lambda)$ was much smaller than suspended particle backscattering by bloom waters, and extremely high $b_{bp}(\lambda)$ in turbid water was confirmed by the threshold of $R_{rs}(555)$ (>0.008 sr^{-1}). Additionally, previous studies have shown that the summer bloom occurs when the water column stratifies with higher nutrients and lower turbidity in the Ariake Sea [39–41]. Thus $b_{bp}(\lambda)$ of bloom water was mainly contributed by organic matters rather than inorganic particles.

Figure 5 shows the Chl a-specific $a_{bloom}(\lambda)$ and cell-specific $b_{bp}(\lambda)$, as normalized to Chl a concentrations and cellular abundances, respectively. Both $a_{bloom}(\lambda)$ and $b_{bp}(\lambda)$ were derived from MODIS $R_{rs}(\lambda)$ by QAA V6. Figure 5a shows that the Chl a-specific absorption $a_{bloom}(\lambda)$ for the raphidophyte bloom was lower than the diatom bloom, which might be caused by the higher intracellular pigment concentration of raphidophyte (*Chattonella* spp.) than that of the diatom (*Skeletonema* spp.). Chl a cell^{-1} was 0.0811 for raphidophytes and 0.001 for diatoms. Additionally, when compared with longer wavelengths, there was large difference in the short bands, possibly due to variations in CDOM and NAP.

In contrast to $a_{bloom}(\lambda)$, $b_{bp}(\lambda)$ showed less spectral dependence (within 8%). The cell-specific $b_{bp}(\lambda)$ of raphidophyte bloom water was about 10 times than that of the diatom bloom water. The difference in cell-specific $b_{bp}(\lambda)$ can be attributed to cell size, cell shape, cell structure and particulate organic carbon content [42]. Specifically, the cell diameter of *Chattonella* spp. (raphidophyte) (30–100 μm) is 5 times that of *Skeletonema* spp. (diatom) (2–12 μm) and raphidophytes carbon content is much higher than that of diatoms [43–45]. Besides, *Skeletonema* spp. is a chain forming diatom while *Chattonella* spp. is present as single cells during a bloom, which may also be responsible for the backscattering feature.

To better understand the significance of the difference in Chl a-specific $a_{bloom}(\lambda)$ and cellular-specific $b_{bp}(\lambda)$ of *Chattonella* spp. and *Skeletonema* spp., we plotted the relationship between in situ Chl a and $a_{bloom}(443)$ because absorption from multiple components overlap at this band. Very little variation was found for the diatom bloom (R^2 = 0.004), whereas a trend (R^2 = 0.68) was observed for raphidophyte bloom waters (Figure 6). The former could have arisen because of lower Chl a concentration per unit cell and the package effect. Additionally, the invariant relationship between in situ Chl a and $a_{bloom}(443)$ in diatoms (Figure 6a) could have been due to absorbance in this band by CDOM and NAP. A similar pattern was also found in the relationship between cellular abundance and $b_{bp}(555)$. The $b_{bp}(555)$ of the raphidophyte bloom increased with cellular abundance ($R^2 = 0.86$) while that of the diatom bloom did not. This indicates lower backscattering per unit cell in the diatom bloom (Figure 6b). Based on the above, $b_{bp}(\lambda)$ appears to be a better indicator to discriminate the two algal groups, especially considering the uncertainties associated with CDOM and NAP absorption in these coastal waters.

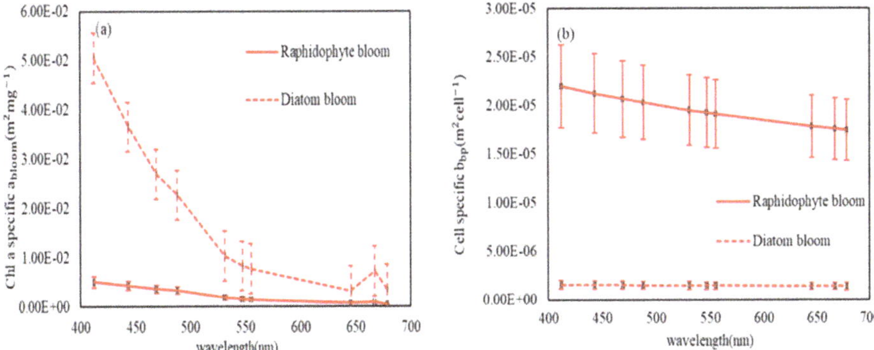

Figure 5. Inherent optical properties of diatom and raphidophyte bloom waters: (**a**) Chl a-specific absorption of bloom water $a_{bloom}(\lambda)$ where $a_{bloom}(\lambda)$ is normalized to in situ Chl a concentration; (**b**) Cell-specific backscattering of suspended particles $b_{bp}(\lambda)$. The spectra are normalized to cellular abundance. The $a_{bloom}(\lambda)$ and $b_{bp}(\lambda)$ were derived by the MODIS $R_{rs}(\lambda)$ match up results using QAA V6. N = 4 for raphidophyte bloom and N = 8 for diatom bloom. The solid and plot line represent raphidophyte and diatom bloom, respectively.

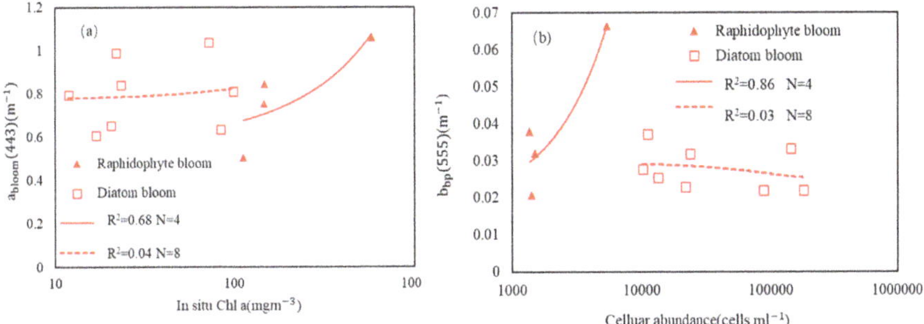

Figure 6. (**a**) Scatterplot of in situ Chl a and $a_{bloom}(443)$ of the raphidophyte and diatom blooms; (**b**) scatterplot of cellular abundance and $b_{bp}(555)$. The triangles and squares depict raphidophyte (N = 4) and diatom bloom (N = 8), respectively. Linear regression line was drawn on log transformed data. The data was from the match-up pairs of MODIS in 2011–2014. The $a_{bloom}(443)$ and $b_{bp}(555)$ were derived from MODIS $R_{rs}(\lambda)$ by QAA V6.

Therefore, based on Tao et al. [11], an index $b_{bp-index}(555)$ was developed using $R_{rs}(\lambda)$ from green to red bands to replace the QAA derived $b_{bp}(555)$, and thus avoiding the uncertainties associated with using the short waveband in QAA. The 555 nm waveband was chosen because both of NAP and water absorption is relatively low at this band [36].

It is known that $R_{rs}(\lambda)$ can be expressed in terms of absorption and backscattering [35,46]:

$$R_{rs}(\lambda) = \frac{f(\lambda)}{Q(\lambda)} \frac{b_b(\lambda)}{a(\lambda) + b_b(\lambda)}, \tag{5}$$

The $f(\lambda)$ refers to the irradiance reflectance within water while $Q(\lambda)$ describes the angular distribution of upwelling radiance, and their ratio is relatively stable [47]. Here we assume the $f(\lambda)/Q(\lambda)$ is spectrally invariant between 500 to 670 nm. From Equations (2) and (4), we can then derive the following relationship:

$$R_{rs}(\lambda) \propto \frac{f(\lambda)}{Q(\lambda)} \frac{b_{bp}(\lambda)}{a_w(\lambda) + a_{bloom}(\lambda) + b_{bp}(\lambda)}, \tag{6}$$

Figure 5a shows that the difference in $a_{bloom}(\lambda)$ between 555 and 667 nm was very small. Thus, the following assumption can be made for each species:

$$a_{bloom}(555) = a_{bloom}(667), \tag{7}$$

Additionally, the spectral dependence of $b_{bp}(\lambda)$ was small, and it showed only a small variation, within 8% (Figure 5b). So $b_{bp}(\lambda)$ was assumed to be equal at bands 555 and 667 nm:

$$b_{bp}(555) = b_{bp}(667), \tag{8}$$

Finally, $b_{bp}(555)$ was derived from Equation (5) using the reciprocal of $R_{rs}(\lambda)$ at 555 and 667 nm, as:

$$\frac{1}{R_{rs}(667)} - \frac{1}{R_{rs}(555)} \propto \frac{a_w(667) - a_w(555)}{b_{bp}(555)}, \tag{9}$$

where $a_w(667) - a_w(555) = 0.37 \text{ m}^{-1}$. The variations in $b_{bp}(555)$ can thus be expressed by the variations in $R_{rs}(\lambda)$ at 555 and 667 nm:

$$b_{bp}(555) \propto 0.37 \times \frac{R_{rs}(555)R_{rs}(667)}{R_{rs}(555) - R_{rs}(667)} = b_{bp-index}(555), \tag{10}$$

To differentiate from $b_{bp}(555)$, henceforth we use $b_{bp-index}(555)$ described in the equation above.

Although bloom vs. non-bloom conditions could be detected by the positive SS (645) (Section 3.1), it was not possible to differentiate between raphidophyte and diatom blooms based on the combination of SS (645) and $b_{bp-index}(555)$. Considering the difference in magnitude of Chl a concentrations during the two blooms, a supplementary index, the Red Band Ratio (RBR) that accounts for Chl a concentrations was used to identify algal types. RBR utilizes the ratio of $R_{rs}(678)$ and $R_{rs}(667)$ to describe the high fluorescence emission around red bands caused by Chl a [48]. The ratio is characterized as:

$$\text{RBR} = \frac{R_{rs}(678)}{R_{rs}(667)}, \tag{11}$$

The two algal species were thus classified by the distribution of $b_{bp-index}(555)$ and the RBR for MODIS data collocated with field data (Figure 7a). For a given RBR value, the raphidophyte blooms showed higher $b_{bp-index}(555)$ than diatom blooms. Figure 7b presents a more distinct relationship between $b_{bp-index}(555)$ and RBR using the independent MODIS $R_{rs}(\lambda)$ data shown in Figure 4b,c for bloom waters. Although some raphidophyte points overlapped with diatoms, the plot shows two distinct relationships for raphidophytes and diatoms. An exponential curve was fit to the data (Figure 7b), which can be expressed as:

$$b_{bp-index}(555) = 0.0019RBR^{-2.261} \tag{12}$$

Accordingly, a bloom can be classified as a raphidophyte (*Chattonella* spp.) bloom if the $b_{bp-index}(555)$ is higher than the value calculated from the RBR value using Equation (12), and conversely as a diatom (*Skeletonema* spp.) bloom if it $b_{bp-index}(555)$ is lower than that calculated from the RBR value.

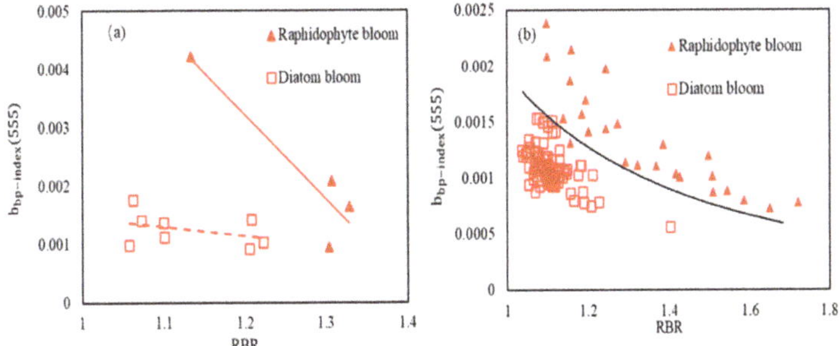

Figure 7. (**a**) Scatter plot of RBR and $b_{bp-index}(555)$ using MODIS $R_{rs}(\lambda)$ coincident with field data for 2011–2014. (**b**) Scatter plot of RBR and $b_{bp-index}(555)$ derived from the points flagged as bloom in Figure 4b,c. The solid line in Figure 7b represents the function expressed by Equation (12) separating raphidophyte blooms from diatom blooms.

MODIS images from 2015 to 2018 were examined to verify the discrimination between raphidophyte and diatom blooms, using the independent data of bloom reports from Japan Red Tide Online. It included the raphidophyte (*Chattonella* spp.) blooms on 7 September 2015 and 18 August 2016 (Figure 8a,b), and the diatom (*Skeletonema* spp.) blooms occurred on 12 July 2018 and 3 September 2018 (Figure 8c,d). The corresponding scatter plot of $b_{bp-index}(555)$ and RBR is shown in Figure 8e. It confirmed that the combination of $b_{bp-index}(555)$ and RBR could successfully distinguish raphidophyte blooms from diatom blooms in MODIS images.

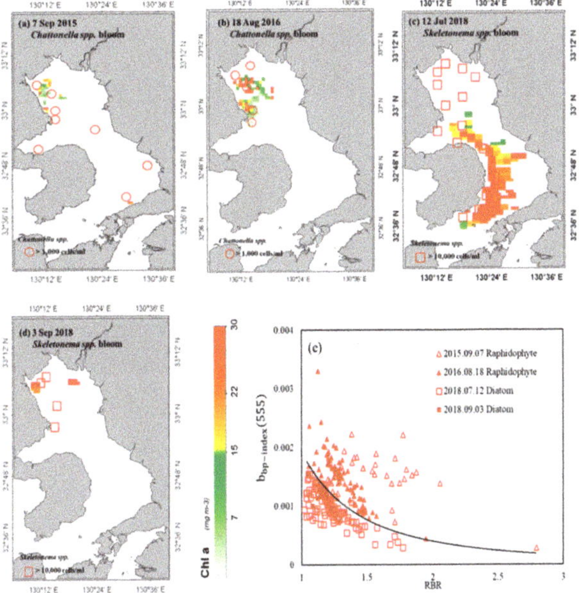

Figure 8. (**a–d**) MODIS Chl a images from standard level_2 products showing bloom distribution confirmed by reports from Japan Red Tide Online. Only pixels positively flagged as bloom waters are shown in color. The red circles and squares indicate the location of raphidophyte and diatom blooms, respectively. (**e**) Scatterplot of $b_{bp-index}(555)$ and RBR derived from the bloom pixels in (**a–d**), indicating distinct algal groups. Red triangle and squares indicate raphidophyte and diatom blooms, respectively.

4. Discussion

The outline of the new method proposed in this study, which can distinguish raphidophyte blooms, diatom blooms, and non-bloom waters in the Ariake Sea, is illustrated in Figure 9. The method is simple, but effective, in discriminating harmful algal blooms, and it does offer several novel findings over previous studies.

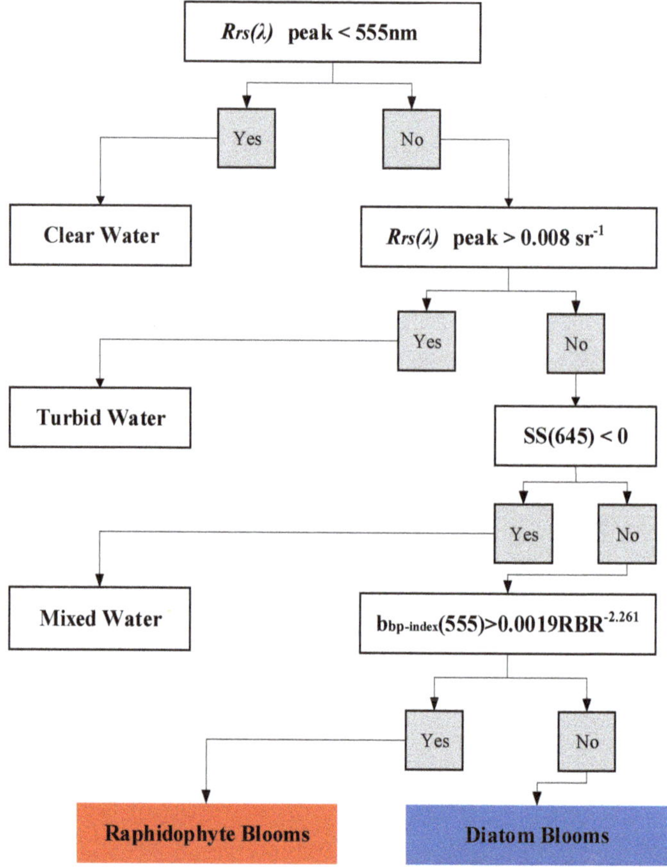

Figure 9. Multispectral method for the identification of raphidophyte and diatom, and non-bloom waters.

It is widely known that phytoplankton blooms are associated with high Chl a concentration and a peak in $R_{rs}(\lambda)$ in the green band [2,16,49], but our method shows that high Chl a using the standard MODIS algorithm is not always related to blooms (Figures 4a–f and 8a–d). Many previous methods detect harmful blooms based on the peak at 555 nm of satellite derived $R_{rs}(\lambda)$ [7,50,51]. However, both sediments and CDOM might influence the accuracy of $R_{rs}(\lambda)$ at the green peak observed by satellite, as well as inaccurate atmospheric correction in coastal waters [23–25] which may make the peak in the green bands and satellite retrieved Chl a concentrations unreliable for bloom detection. A suitable local-based atmosphere correction is needed to overcome the uncertainty in short bands. Our novel method captured the unique $R_{rs}(\lambda)$ shoulder at 645 nm in bloom waters, which successfully distinguished raphidophyte and diatom blooms in the Ariake Sea. Although this method used $R_{rs}(555)$ in the SS (645) calculation, it was used only as a baseline to describe the spectral shape of $R_{rs}(\lambda)$ at 645 nm rather than as the dominant component, thus the uncertainty in SS (645) can be much smaller than the green band based algorithm. Compared with previous studies, our method detects harmful blooms effectively using the spectral features at longer wavelengths and without the uncertainty of short bands.

The 645 nm shoulder of $R_{rs}(\lambda)$ might be influenced by low Chl a and backscattering of algal particles [52], which means pigment absorption is weak around 645 nm, and algal particle backscatter dominated the $R_{rs}(\lambda)$. Mixed water was influenced strongly by water absorption at 645 nm, which results in negative values of SS (645) even though there is a peak in the green band. Although turbid waters also show a spectral shoulder at 645 nm, which is caused by the strong scattering by non-algal particles, we can exclude it by placing a threshold for $R_{rs}(555)$. This newly developed method can detect the bloom precisely and provide accurate information related to areas where blooms are occurring.

Many studies have been developed to use applied inherent optical properties to distinguish harmful algal species [10–12,53]. Even though these studies showed excellent results for their study regions, it has not been applicable to other regions such as the Ariake Sea (results were not show here). One reason might be differences in water conditions. FLH or Chl a concentration is needed as a precondition in methods like those of Cannizzaro et al. [12,53] and Shang et al. [28] to constrain the use of the developed bloom index ($b_{bp}(\lambda)$ ratio and bloom index (BI), respectively), which may vary for different regions. The Green–Red Index of Tao et al. [11] did not work for the Ariake Sea probably because of the different pigment composition of algal species. Although both the East China Sea (ECS) and the Ariake Sea are dominated by phytoplankton groups like diatoms and flagellates [15,54,55], the species can be very different. Consequently, one method which works well in one place may not work in other locations.

Existing methods have not been successful in discriminating raphidophytes, the more common, non-diatom, bloom forming organism in the Ariake Sea, from diatoms, while the newly developed method in the present study showed potential for algal species distinction. Considering the frequent and alternately occurring diatom and raphidophytes HABs in the Ariake Sea, our method is highly advantageous as it is able to discriminate *Chattonella* spp. and *Skeletonema* spp. blooms. As can be seen in Figure 8a–b, the newly developed method captured the blooms on 7 September 2015 and 18 August 2016, and this was in accord with the field observations on the day, which showed cells count of >1000 cells mL^{-1} of *Chattonella* spp. The MODIS pixels indicated as algal blooms were classified as *Chattonella* spp. blooms in the scatter plot of Figure 8e. In Figure 8c–d, the blooms captured in the MODIS image of 12 July 2018 and 3 September 2018 were confirmed to contain high concentrations of *Skeletonema* spp. cells as per local bloom reports. Accordingly, pixels from the bloom areas were classified as diatom blooms as shown in Figure 8e. In spite of the lack of in situ measurements of inherent optical properties, the exciting results in Figure 8e encourage us to pay more attention to backscattering features in harmful algae discrimination in the future. This demonstrates how backscattering can be used in combination with Chl a for bloom detection and harmful algal discrimination.

In summary, our method has several advantages over previous methods. This method can be used directly on MODIS $R_{rs}(\lambda)$ products. Additionally, satellite short waveband data was excluded to avoid possible errors arising from incorrect atmospheric corrections and the influence of non-phytoplankton particles.

5. Conclusions

A novel multispectral approach using MODIS-derived $R_{rs}(\lambda)$ and based on an algal backscattering feature was developed to detect raphidophyte and diatom blooms in the Ariake Sea. As a first step, this method uses the $R_{rs}(\lambda)$ spectral shape signature in the red band to detect HABs. The bloom waters are successfully differentiated by a positive SS (645) and the water can be divided into clear, turbid, mixed and bloom waters. For the next step, indices of $b_{bp-index}(555)$ were developed and used with RBR for discriminating raphidophyte and diatom blooms, based on the distinct optical properties of backscattering between the two algal species. Comparison with the red tide report in 2015–2018 showed that this new method could provide reliable spatial distribution of the raphidophyte and diatom blooms, which may provide a better understanding of harmful algal bloom distributions in the Ariake Sea.

Since the coastal environment is optically complex and varies temporally, more field measurements are needed to better understand the unique backscattering feature that allowed us to distinguish *Chattonella* spp. from *Skeletonema* spp. blooms. Moreover, additional efforts are required to apply this method to other coastal areas with similar algal constituents. Besides, satellites like GOCI and GCOM-C will be utilized in the future to check its applicability for investigation high temporal variability of these blooms over larger areas.

Author Contributions: Conceptualization, C.F. and J.I.; methodology, C.F.; software, C.F.; validation, C.F.; formal analysis, C.F.; resources, K.S., T.M. and H.Y.; writing—original draft preparation, C.F.; writing—review and editing, C.F. and J.I.; visualization, C.F.; supervision, J.I. All authors have read and agreed to the published version of the manuscript.

Funding: This research was partly supported by Fisheries Agency of Japan and by Global Change Observation Mission-Climate (GCOM-C) project of Japan Aerospace Exploration Agency (JAXA).

Acknowledgments: This work was supported by Fisheries Agency of Japan and by GCOM-C project of the Japan Aerospace Exploration Agency. We wish to thank Saga Ariake Fisheries Promotion Center and Kumamoto Prefectural Fisheries Research Center for providing in situ data of Ariake Sea. And we also thank the NASA Goddard Space Flight Center, the Ocean Ecology Laboratory and Ocean Biology Processing Group for providing the Moderate-resolution Imaging Spectroradiometer (MODIS) Aqua L2 data.

Conflicts of Interest: The authors declare no conflict of interest.

References

1. Xi, H.; Hieronymi, M.; Röttgers, R.; Krasemann, H.; Qiu, Z. Hyperspectral differentiation of phytoplankton taxonomic groups: A comparison between using remote sensing reflectance and absorption spectra. *Remote Sens.* **2015**, *7*, 14781–14805. [CrossRef]
2. Gower, J.; King, S.; Goncalves, P. Global monitoring of plankton blooms using MERIS MCI. *Int. J. Remote Sens.* **2008**, *29*, 6209–6216. [CrossRef]
3. Shanmugam, P. A new bio-optical algorithm for the remote sensing of algal blooms in complex ocean waters. *J. Geophys. Res. Oceans* **2011**, *116*. [CrossRef]
4. Hu, C. A novel ocean color index to detect floating algae in the global oceans. *Remote Sens. Environ.* **2009**, *113*, 2118–2129. [CrossRef]
5. Westberry, T.; Siegel, D.; Subramaniam, A. An improved bio-optical model for the remote sensing of *Trichodesmium* spp. blooms. *J. Geophys. Res. Oceans* **2005**, *110*. [CrossRef]
6. Amin, R.; Zhou, J.; Gilerson, A.; Gross, B.; Moshary, F.; Ahmed, S. Novel optical techniques for detecting and classifying toxic dinoflagellate *Karenia brevis* blooms using satellite imagery. *Opt. Express* **2009**, *17*, 9126–9144. [CrossRef]
7. Noh, J.H.; Kim, W.; Son, S.H.; Ahn, J.H.; Park, Y.J. Remote quantification of *Cochlodinium polykrikoides* blooms occurring in the East Sea using geostationary ocean color imager (GOCI). *Harmful Algae* **2018**, *73*, 129–137. [CrossRef]
8. Wynne, T.; Stumpf, R.; Tomlinson, M.; Warner, R.; Tester, P.; Dyble, J.; Fahnenstiel, G. Relating spectral shape to cyanobacterial blooms in the Laurentian Great Lakes. *Int. J. Remoe Sens.* **2008**, *29*, 3665–3672. [CrossRef]
9. Siswanto, E.; Ishizaka, J.; Tripathy, S.C.; Miyamura, K. Detection of harmful algal blooms of *Karenia mikimotoi* using MODIS measurements: A case study of Seto-Inland Sea, Japan. *Remote Sens. Environ.* **2013**, *129*, 185–196. [CrossRef]
10. Shang, S.; Wu, J.; Huang, B.; Lin, G.; Lee, Z.; Liu, J.; Shang, S. A new approach to discriminate dinoflagellate from diatom blooms from space in the East China Sea. *J. Geophys. Res. Oceans* **2014**, *119*, 4653–4668. [CrossRef]
11. Tao, B.; Mao, Z.; Lei, H.; Pan, D.; Bai, Y.; Zhu, Q.; Zhang, Z. A semianalytical MERIS green-red band algorithm for identifying phytoplankton bloom types in the East China Sea. *J. Geophys. Res. Oceans* **2017**, *122*, 1772–1788. [CrossRef]

12. Cannizzaro, J.P.; Carder, K.L.; Chen, F.R.; Heil, C.A.; Vargo, G.A. A novel technique for detection of the toxic dinoflagellate, *Karenia brevis*, in the Gulf of Mexico from remotely sensed ocean color data. *Cont. Shelf Res.* **2008**, *28*, 137–158. [CrossRef]

13. Burenkov, V.; Kopelevich, O.; Rat'kova, T.; Sheberstov, S. Satellite observations of the coccolithophorid bloom in the Barents Sea. *Oceanology* **2011**, *51*, 766. [CrossRef]

14. Lei, H.; Pan, D.; Bai, Y.; Chen, X.; Zhou, Y.; Zhu, Q. HAB detection based on absorption and backscattering properties of phytoplankton. In Proceedings of the Remote Sensing of the Ocean, Sea Ice, Coastal Waters, and Large Water Regions 2011, Prague, Czech Republic, 19–22 September 2011; p. 81751F.

15. Ishizaka, J.; Kitaura, Y.; Touke, Y.; Sasaki, H.; Tanaka, A.; Murakami, H.; Suzuki, T.; Matsuoka, K.; Nakata, H. Satellite detection of red tide in Ariake Sound, 1998–2001. *J. Oceanogr.* **2006**, *62*, 37–45. [CrossRef]

16. Sasaki, H.; Tanaka, A.; Iwataki, M.; Touke, Y.; Siswanto, E.; Tan, C.K.; Ishizaka, J. Optical properties of the red tide in Isahaya Bay, southwestern Japan: Influence of chlorophyll a concentration. *J. Oceanogr.* **2008**, *64*, 511–523. [CrossRef]

17. Aoki, K.; Onitsuka, G.; Shimizu, M.; Yamatogi, T.; Ishida, N.; Kitahara, S.; Hirano, K. *Chattonella* (Raphidophyceae) bloom spatio-temporal variations in Tachibana Bay and the southern area of Ariake Sea, Japan: Interregional displacement patterns with *Skeletonema* (Bacillariophyceae). *Mar. Pollut. Bull.* **2015**, *99*, 54–60. [CrossRef]

18. Khan, S.; Arakawa, O.; Onoue, Y. A toxicological study of the marine phytoflagellate, *Chattonella antiqua* (Raphidophyceae). *Phycologia* **1996**, *35*, 239–244. [CrossRef]

19. Andreae, M.O.; Klumpp, D. Biosynthesis and release of organoarsenic compounds by marine algae. *Environ. Sci. Technol.* **1979**, *13*, 738–741. [CrossRef]

20. Howard, A.; Comber, S.; Kifle, D.; Antai, E.; Purdie, D. Arsenic speciation and seasonal changes in nutrient availability and micro-plankton abundance in Southampton water, UK. *Estuar. Coast. Shelf Sci.* **1995**, *40*, 435–450. [CrossRef]

21. Imai, I. Distribution of diatom resting cells in sediments of Harima-Nada and northern Hiroshima Bay, the Seto Inland Sea, Japan. *Bull. Coast Oceanogr.* **1990**, *28*, 75–84.

22. Azad, M.A.K.; Ohira, S.-I.; Oda, M.; Toda, K. On-site measurements of hydrogen sulfide and sulfur dioxide emissions from tidal flat sediments of Ariake Sea, Japan. *Atmos. Environ.* **2005**, *39*, 6077–6087.

23. Zhang, M.; Carder, K.; Muller-Karger, F.E.; Lee, Z.; Goldgof, D.B. Noise reduction and atmospheric correction for coastal applications of Landsat Thematic Mapper imagery. *Remote Sens. Environ.* **1999**, *70*, 167–180. [CrossRef]

24. Hu, C.; Muller-Karger, F.E.; Andrefouet, S.; Carder, K.L. Atmospheric correction and cross-calibration of LANDSAT-7/ETM+ imagery over aquatic environments: A multiplatform approach using SeaWiFS/MODIS. *Remote Sens. Environ.* **2001**, *78*, 99–107. [CrossRef]

25. Yang, M.M.; Ishizaka, J.; Goes, J.I.; Gomes, H.d.R.; Maúre, E.D.R.; Hayashi, M.; Katano, T.; Fujii, N.; Saitoh, K.; Mine, T. Improved MODIS-Aqua chlorophyll-a retrievals in the turbid semi-enclosed Ariake Bay, Japan. *Remote Sens.* **2018**, *10*, 1335. [CrossRef]

26. Majchrowski, R.; Ostrowska, M.A. Influence of photo-and chromatic acclimation on pigment composition in the sea. *Oceanologia* **2000**, *42*, 157–175.

27. Bricaud, A.; Claustre, H.; Ras, J.; Oubelkheir, K. Natural variability of phytoplanktonic absorption in oceanic waters: Influence of the size structure of algal populations. *J. Geophys. Res. Oceans* **2004**, *109*. [CrossRef]

28. Lee, Z.; Huot, Y. On the non-closure of particle backscattering coefficient in oligotrophic oceans. *Opt. Express* **2014**, *22*, 29223–29233. [CrossRef]

29. Lee, Z.; Carder, K.L.; Arnone, R.A. Deriving inherent optical properties from water color: A multiband quasi-analytical algorithm for optically deep waters. *Appl. Opt.* **2002**, *41*, 5755–5772. [CrossRef]

30. Ronald, J.; Zaneveld, V. Remotely sensed reflectance and its dependence on vertical structure: A theoretical derivation. *Appl. Opt.* **1982**, *21*, 4146–4150. [CrossRef]

31. Jerome, J.; Bukata, R.; Miller, J. Remote sensing reflectance and its relationship to optical properties of natural waters. *Remote Sens.* **1996**, *17*, 3135–3155. [CrossRef]

32. Stumpf, R.P.; Wynne, T.T.; Baker, D.B.; Fahnenstiel, G.L. Interannual variability of cyanobacterial blooms in Lake Erie. *PLoS ONE* **2012**, *7*, e42444. [CrossRef] [PubMed]

Remote Sens. **2020**, *12*, 1504

33. Soto, I.M.; Cannizzaro, J.; Muller-Karger, F.E.; Hu, C.; Wolny, J.; Goldgof, D. Evaluation and optimization of remote sensing techniques for detection of *Karenia brevis* blooms on the West Florida Shelf. *Remote Sens. Environ.* **2015**, *170*, 239–254. [CrossRef]

34. Morel, A.; Prieur, L. Analysis of variations in ocean color 1. *Limnol. Oceanogr.* **1977**, *22*, 709–722. [CrossRef]

35. Gordon, H.R.; Brown, O.B.; Evans, R.H.; Brown, J.W.; Smith, R.C.; Baker, K.S.; Clark, D.K. A semianalytic radiance model of ocean color. *J. Geophys. Res. Atmos.* **1988**, *93*, 10909–10924. [CrossRef]

36. Pope, R.M.; Fry, E.S. Absorption spectrum (380–700 nm) of pure water. II. Integrating cavity measurements. *Appl. Opt.* **1997**, *36*, 8710–8723. [CrossRef]

37. Morel, A. Optical properties of pure water and pure sea water. *Opt. Asp. Oceanogr.* **1974**, *1*, 1–24.

38. Lee, Z.; Arnone, R.; Hu, C.; Werdell, P.J.; Lubac, B. Uncertainties of optical parameters and their propagations in an analytical ocean color inversion algorithm. *Appl. Opt.* **2010**, *49*, 369–381. [CrossRef]

39. Tanaka, K.; Kodama, M. Effects of resuspended sediments on the environmental changes in the inner part of Ariake Bay, Japan. *Bull. Fish. Res. Agency Jpn.* **2007**, *19*, 9.

40. Imai, I.; Yamaguchi, M. Life cycle, physiology, ecology and red tide occurrences of the fish-killing raphidophyte *Chattonella*. *Harmful Algae* **2012**, *14*, 46–70. [CrossRef]

41. Arai, K. Prediction Method for Large Diatom Appearance with Meteorological Data and MODIS Derived Turbidity and Chlorophyll-A in Ariake Bay Area in Japan. *Inter. J. Adv. Comput. Sci. Appl.* **2017**, *8*, 39–44. [CrossRef]

42. Vaillancourt, R.D.; Brown, C.W.; Guillard, R.R.; Balch, W.M. Light backscattering properties of marine phytoplankton: Relationships to cell size, chemical composition and taxonomy. *J. Plankton Res.* **2004**, *26*, 191–212. [CrossRef]

43. Tomas, C.R. *Identifying Marine Phytoplankton*; Elsevier: Amsterdam, The Netherlands, 1997.

44. Strathmann, R.R. Estimating the organic carbon content of phytoplankton from cell volume or plasma volume 1. *Limnol. Oceanogr.* **1967**, *12*, 411–418. [CrossRef]

45. Menden-Deuer, S.; Lessard, E.J. Carbon to volume relationships for dinoflagellates, diatoms, and other protist plankton. *Limnol. Oceanogr.* **2000**, *45*, 569–579. [CrossRef]

46. Carder, K.L.; Chen, F.; Lee, Z.; Hawes, S.; Kamykowski, D. Semianalytic Moderate-Resolution Imaging Spectrometer algorithms for chlorophyll a and absorption with bio-optical domains based on nitrate-depletion temperatures. *Limnol. Oceanogr.* **1999**, *104*, 5403–5421.

47. Morel, A.; Gentili, B. Diffuse reflectance of oceanic waters. II. Bidirectional aspects. *Appl. Opt.* **1993**, *32*, 6864–6879. [CrossRef]

48. Gitelson, A. The peak near 700 nm on radiance spectra of algae and water: Relationships of its magnitude and position with chlorophyll concentration. *Int. J. Remote Sens.* **1992**, *13*, 3367–3373. [CrossRef]

49. Hu, C.; Muller-Karger, F.E.; Taylor, C.J.; Carder, K.L.; Kelble, C.; Johns, E.; Heil, C.A. Red tide detection and tracing using MODIS fluorescence data: A regional example in SW Florida coastal waters. *Remote Sens. Environ.* **2005**, *97*, 311–321. [CrossRef]

50. Ahn, Y.-H.; Shanmugam, P.; Ryu, J.-H.; Jeong, J.-C. Satellite detection of harmful algal bloom occurrences in Korean waters. *Harmful Algae* **2006**, *5*, 213–231. [CrossRef]

51. Lou, X.; Hu, C. Diurnal changes of a harmful algal bloom in the East China Sea: Observations from GOCI. *Remote Sens. Environ.* **2014**, *140*, 562–572. [CrossRef]

52. Hoepffner, N.; Sathyendranath, S. Effect of pigment composition on absorption properties of phytoplankton. *Mar. Ecol. Prog. Ser* **1991**, *73*, 11–23. [CrossRef]

53. Cannizzaro, J.P. Detection and Quantification of *Karenia brevis* Blooms on the West Florida Shelf from Remotely Sensed Ocean Color Imagery. Master's Thesis, University of South Florida, Tampa, FL, USA, 2004.

54. Tsutsumi, H. Critical events in the Ariake Bay ecosystem: Clam population collapse, red tides, and hypoxic bottom water. *Plankton Benthos Res.* **2006**, *1*, 3–25. [CrossRef]

55. Tao, B.; Mao, Z.; Lei, H.; Pan, D.; Shen, Y.; Bai, Y.; Zhu, Q.; Li, Z. A novel method for discriminating *Prorocentrum donghaiense* from diatom blooms in the East China Sea using MODIS measurements. *Remote Sens. Environ.* **2015**, *158*, 267–280. [CrossRef]

 *remote sensing*

Article

Quantifying the Effects of Hurricanes Irma and Maria on Coastal Water Quality in Puerto Rico using Moderate Resolution Satellite Sensors

William J. Hernández [1,*], Suhey Ortiz-Rosa [2], Roy A. Armstrong [2], Erick F. Geiger [3,4], C. Mark Eakin [3] and Robert A. Warner [5]

[1]   NOAA-EPP/Center for Earth System Science and Remote Sensing Technologies (CESSRST) City College, City University of New York, Post-doctoral Fellow; New York, NY 10031, USA

[2]   Bio-Optical Oceanography Laboratory, Department of Marine Sciences, University of Puerto Rico-Mayagüez, NOAA-EPP/CESSRST Fellow, Mayagüez, PR 00683, USA; suhey.ortiz@upr.edu (S.O.-R.); roy.armstrong@upr.edu (R.A.A.)

[3]   Coral Reef Watch NOAA/NESDIS/STAR, College Park, MD 20740, USA; erick.geiger@noaa.gov (E.F.G.); Mark.Eakin@noaa.gov (C.M.E.)

[4]   Earth System Science Interdisciplinary Center/Cooperative Institute for Climate and Satellites-Maryland, University of Maryland, College Park, MD 20740, USA

[5]   NOAA/NOS/NCCOS, Silver Spring, MD 20910, USA; robert.a.warner@noaa.gov

*   Correspondence: william.hernandez@upr.edu

Received: 17 January 2020; Accepted: 12 March 2020; Published: 17 March 2020

Abstract: Coastal, benthic communities, such as coral reefs, are at particular risk due to poor water quality caused by hurricanes. In addition to the physical impacts from wave action and storm surge, hurricanes bring significant rainfall resulting in increased runoff from land. Hurricanes Irma and Maria caused record or near-record floods at many locations across Puerto Rico and resulted in major impacts on coastal and benthic ecosystems from heavy rainfall and river discharge. In this study, we use imagery from the moderate resolution Visible Infrared Imaging Radiometer Suite (VIIRS) satellite to quantify the impacts of hurricanes Irma and Maria, which struck Puerto Rico during September 2017, on the water quality of the coastal waters of Puerto Rico using the chlorophyll-*a* (Chl-*a*) and the diffuse attenuation coefficient at 490 nm (K_d490) products. The objectives include: (1) quantify the water quality and light attenuation after the hurricanes; (2) compare this event to the climatology of these parameters, and 3) evaluate long-term exposure and exceedances of various coastal areas to low levels of turbidity. The Chl-*a* inner shelf values increased in 2017 during the months of June (8% above baseline), July (17%), August (5%), September (8%), October (19%), and November (28%) when compared to 2012–2016 baseline data. The values for Chl-*a* concentration reached and exceeded 0.45 µg/L by August 2017 and persisted above that value until December 2017. The K_d490 inner shelf values for 2017 increased (in percent) for the months of June (4% above baseline), July (9%), August (10%), September (5%), October (12%), and November (7%) when compared to 2012–2016 baseline data. The values of K_d490 in August, September, and December 2017 were the highest seen during 2012–2017. Even with the limitations of spatial resolution and loss of data to cloud cover, the 6-year imagery time-series analysis can provide a useful evaluation of the effects of these two hurricanes on the coastal water quality in Puerto Rico, and quantify the exposure of benthic habitats to higher nutrient and turbidity levels.

Keywords: ocean color; hurricanes; remote sensing; water quality; Puerto Rico

1. Introduction

Hurricanes can produce sudden and massive disturbances in estuaries, coastal aquatic, and terrestrial ecosystems around the world [1,2]. The most noticeable impacts to benthic organisms are physical damage. Seagrass can be scoured and uprooted by strong currents, causing them to be transported offshore [3]. For corals, hurricanes can cause breakage, particularly for branching corals, abrasion of the living surface of the corals through the movement of coarse sand and rolling of rubble, and burial of corals through sediment redistribution.

In addition to the physical impacts from wave action, hurricanes bring significant rainfall that leads to increased runoff from land. As development has increased in Puerto Rico, reduced vegetation increases the likelihood of sediments, nutrients, and hazardous substances that can be eroded into coastal waters [4], especially after extreme rain events like hurricanes. Excess sediment, nutrients, and other pollutants can negatively affect seagrass and reef environments principally by decreasing light availability and thereby reducing the photosynthetic capacity for growth [4,5]. Benthic organisms, especially sessile animals, are at particular risk due to poor water quality caused by hurricanes [6]. Coral cover showed a strong correlation with light attenuation, suggesting that deterioration in water quality due to anthropogenic activity could result in reef degradation [7].

Two powerful hurricanes, Irma and Maria, struck Puerto Rico in 2017. At its closest point, Hurricane Irma tracked about 92.6 km (50 nautical miles) to the north of the northern shore of Puerto Rico, delivering rainfall totals between 25.4 cm and 38.1 cm over high elevations in the central portion of the island between September 5–7, 2017 [8] (Figure 1). Hurricane Maria struck Puerto Rico on September 20, 2017 as a Category 4 hurricane (250 kmh (155 mph)) and crossed the island from the southeast to the northwest [9] (Figure 1). Heavy rainfall included one location with nearly 96.5 cm (38 in) of rain. River discharges caused record or near-record floods at many locations across all regions of the island. In addition, major power, transportation, and communication infrastructure were lost.

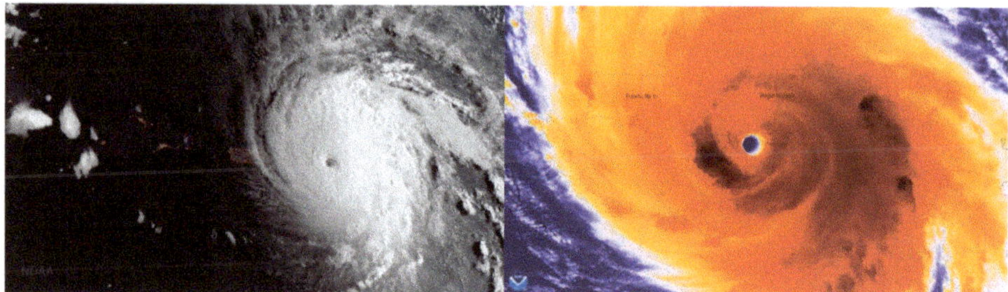

Figure 1. Visible Infrared Imaging Radiometer Suite (VIIRS) satellite images of hurricanes Irma (left) and Maria (right) using I-band 5 (11 um). Images courtesy of NOAA National Environmental Satellite, Data, and Information Service (NESDIS).

Satellite ocean color data can provide critical information on coastal water quality conditions after these episodic events. Remote sensing is a cost-effective tool for monitoring large-scale effects [10–13] of hurricanes in the water quality conditions before and after the events. In the case of hurricanes Irma and Maria in Puerto Rico, satellite ocean color data provided the only source of information on water quality and light availability due to lost or damaged in situ sensors and lack of field observations after the storm. These satellite-derived water quality products include the chlorophyll-*a* concentration (Chl-*a*) [14–16] and the water diffuse attenuation coefficient at the wavelength 490 nm (K_d490) [17–19]. The Chl-*a* concentration provides a measurement of phytoplankton biomass, which is related to nutrient status (i.e., productivity), and can be used as an index of water quality. Chl-*a* can also be described as organic material in the water column contributing to light attenuation. K_d490 is an important parameter for water quality since it provides a measure of turbidity (related to the total organic and inorganic matter held in solution and suspension) in the water column and can be used to quantify light availability and sediment loading for benthic organisms (i.e., coral reefs and seagrasses) [20]. The Chl-*a* algorithm used is based on the ocean color index (OCI) [14] which provides data retrievals for both coastal and oceanic waters. The K_d490 algorithm used (Wang et al., 2009) is particularly useful for turbid coastal and inland waters, when compared with in situ measurements.

In this study, we use moderate-resolution Visible Infrared Imaging Radiometer Suite (VIIRS) satellite images to quantify the impacts of hurricanes Irma and Maria on the quality of coastal waters of Puerto Rico from K_d490 and chlorophyll-*a* products. The objectives include: (1) quantify the water quality and light attenuation after the hurricanes; (2) compare this event to the climatology of these parameters, and (3) evaluate long-term exposure and exceedances of various coastal areas to low levels of turbidity.

2. Data and Methods

2.1. Satellite Data Analysis

This study was focused on the waters surrounding Puerto Rico and used the satellite-derived ocean color products Chl-*a* concentration and K_d490 from VIIRS. This sensor provides daily images at a spatial resolution of 750 m (Figure 2). The study area was divided into four cardinal coastal geographical regions (e.g., North, South, East, and West) to quantify the effects of the hurricanes on water quality in these regions (Figure 3). This segmentation of the study area allowed a refined characterization of the major watersheds, precipitation rates, and important coastal habitats located in those regions. Time-series analysis provided a baseline of these water quality parameters for the regions from 2012–2016 and 2017, to compare directly with the effects of hurricanes.

Figure 2. VIIRS satellite map of monthly mean chlorophyll-*a* (Chl-*a*) concentration during September 2017.

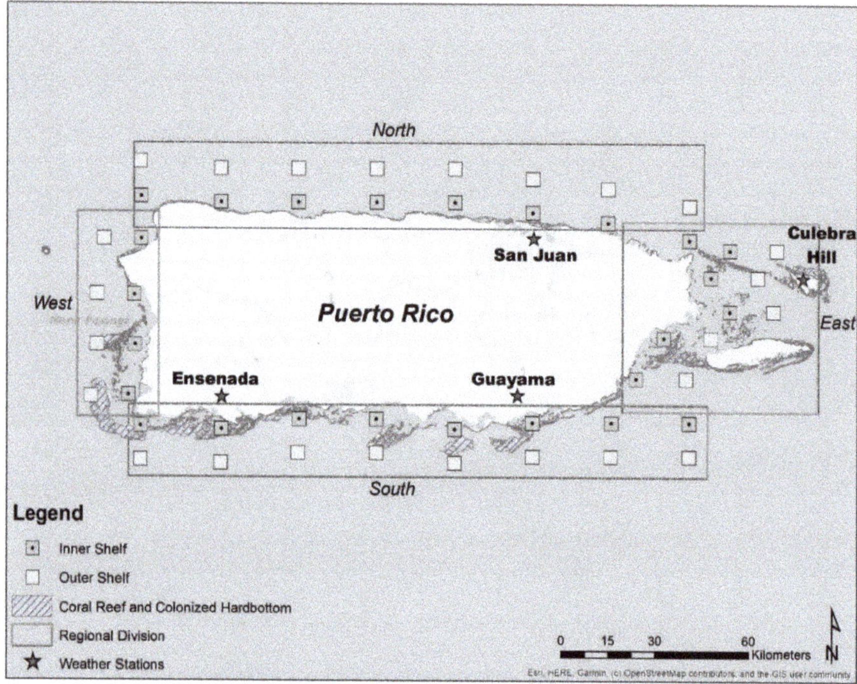

Figure 3. Map of Puerto Rico showing the inner and outer shelf locations, coral reef, and colonized hardbottom [21], regional area divisions, and weather station locations.

Point (pixel) locations were established to characterize the inner shelf and outer shelf contribution to the regional and overall values of Chl-a and K_d490, and to quantify these values over coral reef and hard-bottom areas. These points were expanded using a 5 × 5 pixel box to maximize coverage of the areas within the regions and obtain values from the inner and outer shelf (Figure 3).

The VIIRS images were obtained through the NOAA Coast Watch website (https://coastwatch.noaa.gov/) at Level 2 Science Quality accessed in March 2018. These images include a land-mask and a cloud-mask and were gridded and cropped to include only the Puerto Rico regions and further co-registered to ensure pixel overlaps for the time-series. A total of 1825 daily images from January 2012 to December 2017 were analyzed and the images were organized into 72 monthly composites. Chl-a concentration and K_d490 values were extracted from the images using a gridded point selection within the regions (Figure 3). The images were processed and stored in NetCDF (.nc) format and exported in GeoTiff (.tif) format for use in other GIS mapping platforms.

The 2012–2016 monthly means for the Chl-a concentrations and K_d490 were used as the baseline values for these parameters and then compared with the monthly averages from 2017 to evaluate potential anomalous water quality areas around Puerto Rico.

In addition to the changes to water quality produced by extreme events, the values were analyzed based on the coastal water quality standards that have been adopted by both national and international jurisdictions. No coastal water quality standards have been adopted by Puerto Rico for the chlorophyll concentration or light attenuation so the State of Hawai'i [22] and the Great Barrier Reef Marine Park (GBRMP) [23] standards were used for reference. The threshold values of chlorophyll concentration for open coastal waters were established at 0.15 to 0.30 µg/L for Hawai'i coastal waters and 0.45 µg/L for the GBRMP. For the K_d490 values, only State of Hawai'i provides values for light attenuation (K_d) at 0.1 m^{-1} for open coastal waters. The values of Chl-a 0.45 µg/L and K_d490 0.1 m^{-1} were used as threshold values for this study and values above these thresholds are recognized globally as adverse for coral reefs [23].

2.2. Precipitation Analysis

Precipitation data were obtained from the NOAA National Center for Environmental Information (NCEI) Global Summary of the Month product that provided a global summary of the precipitation and temperature data. Four stations were chosen to represent the regions selected. For the North (San Juan, Station ID: RQW00011641), for the East (Culebra Hill, Station ID: RQC00666343), for the South (Guayama, Station ID: RQC00664193), and for the West (Ensenada, Station ID: RQC00665693). The selection was based on the locations of the hurricane-impacted habitats including corals, seagrass beds, mangroves, and other benthic ecosystems, and data availability for the stations from 2012–2017 (Figure 3). Some precipitation values for 2013 were absent for Culebra. For any absent data, the monthly average for 2012 to 2016 was calculated excluding the missing data points. Precipitation data were then correlated to Chl-*a* and K_d490 concentration values across Puerto Rico.

3. Results

3.1. Precipitation Vvalues

During September and October 2017 there was significantly higher precipitation than the monthly average in all four regions over 2012–2016 (Figure 4). This higher rainfall was mainly due to the impact of hurricanes Irma and Maria during September 5, 2017 and September 20, 2017, respectively. Increased rainfall throughout the month of October also contributed to the peaks observed, particularly for the South region.

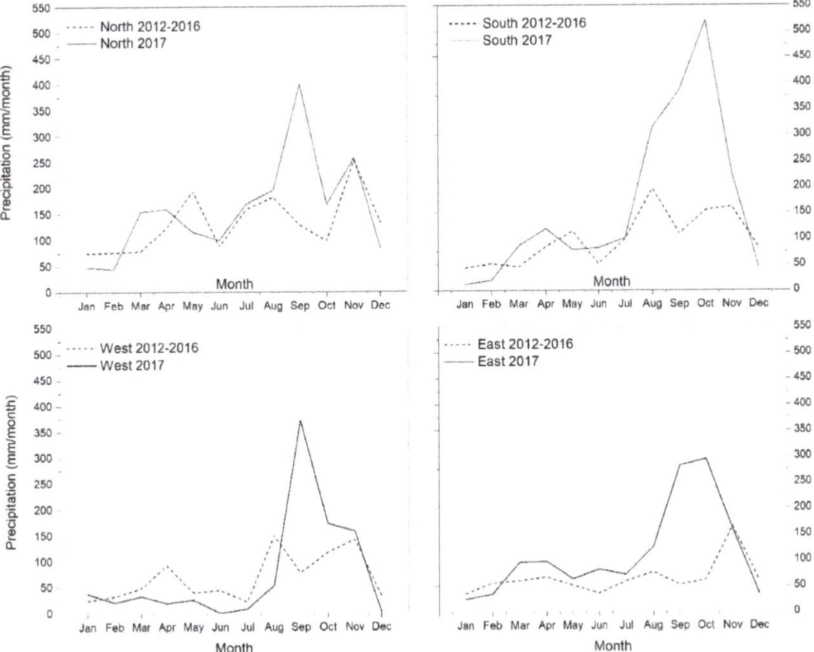

Figure 4. Monthly averaged precipitation for the North, South, West, and East during 2012–16 baseline period and for 2017.

The South region showed the highest amount of precipitation, followed by North, East, and West. The North and West regions experienced a significant increase in precipitation starting in August that began to decrease in October 2017. The East and South regions also experienced a major increase in rainfall in August but did not decrease until November. The West region showed lower precipitation values from February 2017 to early August 2017 compared to 2012–2016 (Figure 4).

3.2. Regional and Monthly Pre- and Post-Hurricane Water Quality

3.2.1. Chl-a Concentrations

An increase in Chl-*a* concentration was observed from August–November 2017 for all regions (Figure 5). Chl-*a* concentration spikes may not have been observed in the East and North because baseline values were relatively high. When all regions are considered, an increase in Chl-*a* concentration for 2017 positively correlates with an increase in precipitation ($r_s = 0.52$; $p = 0.080$). The East region showed the highest contribution to the overall chlorophyll-*a* value for the island, followed by the North, West, and South regions (Figure 5) for both 2012–16 and 2017. The West region showed lower precipitation values from February 2017 to the beginning of August 2017 as compared to 2012–2016; during this time, Chl-*a* concentrations were lower than the average for 2012–2016 as well. The East region had higher rainfall values in 2017 overall as compared to 2012–2016; this may have contributed to increased Chl-*a* concentration for 2017. Precipitation influences the Chl-*a* concentration for the north region (2017 $r_s = 0.73$; $p = 0.006$, 2012–16 $r_s = 0.60$; $p = 0.038$) and the south region (2017 = 0.40; $p = 0.191$, 2012–16 $r_s = 0.60$; $p = 0.036$).

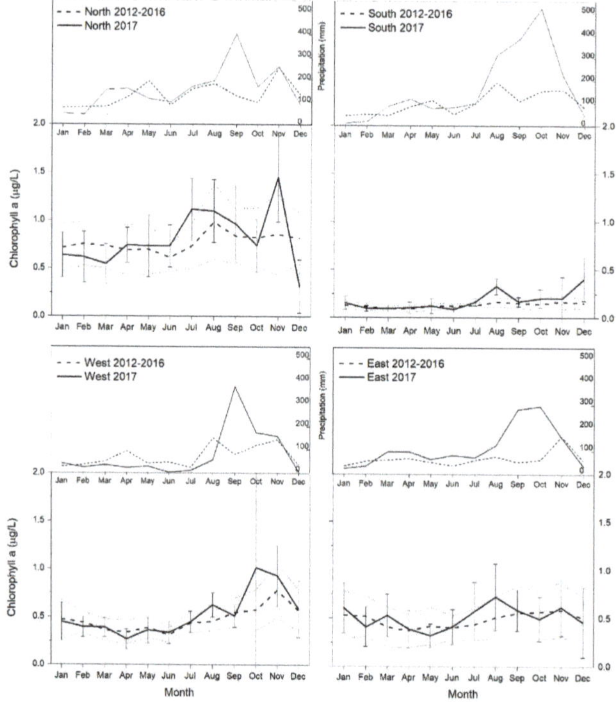

Figure 5. Monthly averaged precipitation and Chl-*a* concentration with standard deviation by regions for 2012–16 baseline period and 2017.

Chl-*a* concentrations were analyzed monthly for each year to identify the differences between years and the seasonal trends (Figure 6). The higher Chl-*a* values were observed from July to December 2017 when considering the average of all regions. Chl-*a* concentration exceeded 0.45 µg/L by August 2017 and persisted until December 2017. Chl-*a* concentration values above 0.45 µg/L were also present in previous years but never exceeded this threshold for more than 5 months. Average Chl-*a* concentration for Puerto Rico showed an increase in 2017 when compared with previous values from 2012–2016 especially in the peak of the rainy season (August–November) (Figure 6). The months that showed an increase over the previous baseline values are July (17% higher than baseline), August (36%), September (20%), October (9%), November (14%), and December (13%).

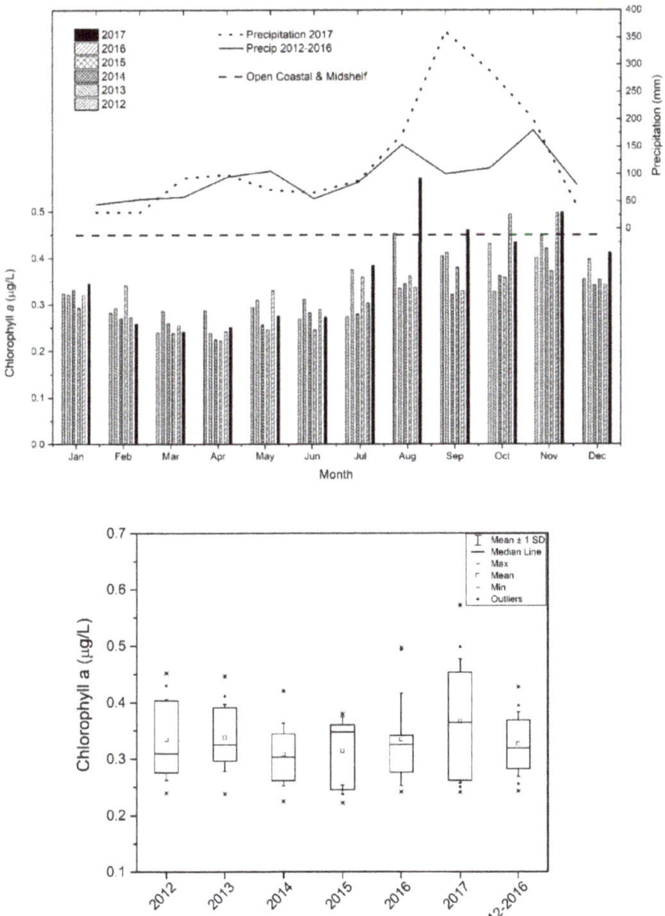

Figure 6. Monthly comparison of the Chl-*a* concentration for all regions (µg/L) showing the yearly distribution from 2012–2017 with associated precipitation data (top). The year 2017 is filled in black. The dashed line for 0.45 µg/L represents the chlorophyll threshold for open coastal and mid-shelf waters [23]. (Bottom) Box plot showing yearly statistics of the Chl-*a* concentration for all regions.

3.2.2. K_d490

The K_d490 values were analyzed by month per year to compare the differences between years and seasonal trends. The East region showed the highest contribution to the overall K_d490 value for the island, followed by the North, West, and South regions (Figure 7). Precipitation contributes to a greater degree to the K_d490 values in the South (2012–16, $r_s = 0.79$; $p = 0.001$) and North (2017 $r_s = 0.68$; $p = 0.014$) regions. The East region had higher rainfall values in 2017 overall as compared to 2012–2016; this correlated with increased K_d490 values for 2017 ($r_s = 0.44$; $p = 0.459$). Overall, an increase in K_d490 correlated with an increase in precipitation (2017, $r_s = 0.46$; $p = 0.124$), 2012–16, $r_s = 0.53$; $p = 0.075$) Additionally, there is a strong correlation between the Chl-*a* and K_d490 ($r_s = 0.99$; $p = 0$) products for the complete time series.

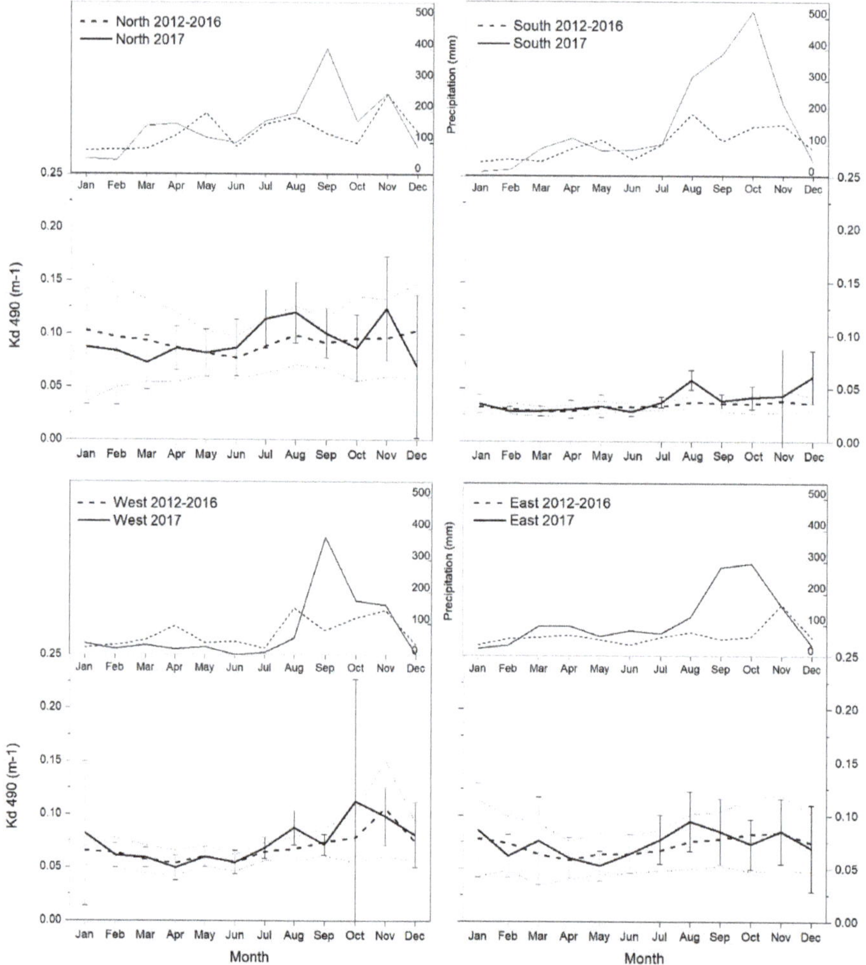

Figure 7. Monthly averaged precipitation and K_d490 with standard deviation by regions for 2012–16 baseline period and 2017.

Higher values were present from July to December 2017 when considering all regions (Figure 7). The values for K_d490 reached and exceeded 0.06 m^{-1} by July 2017 and persisted by December 2017. K_d490 values above the 0.06 m^{-1} were also present in previous years, but the 2017 values for the months of August, September, and December are the highest for all the time series from 2012–2017. The values of K_d490 for Puerto Rico show an increase for 2017 when compared with values from 2012–2016, especially in the peak of the rainy season (August–November) (Figure 8). The months that show an increase from the previous baseline values are July (10% higher than baseline), August (28%), September (15%), October (5%), November (7%), and December (12%). The values for K_d490 concentration from August to December 2017 were all above 0.06 m^{-1}.

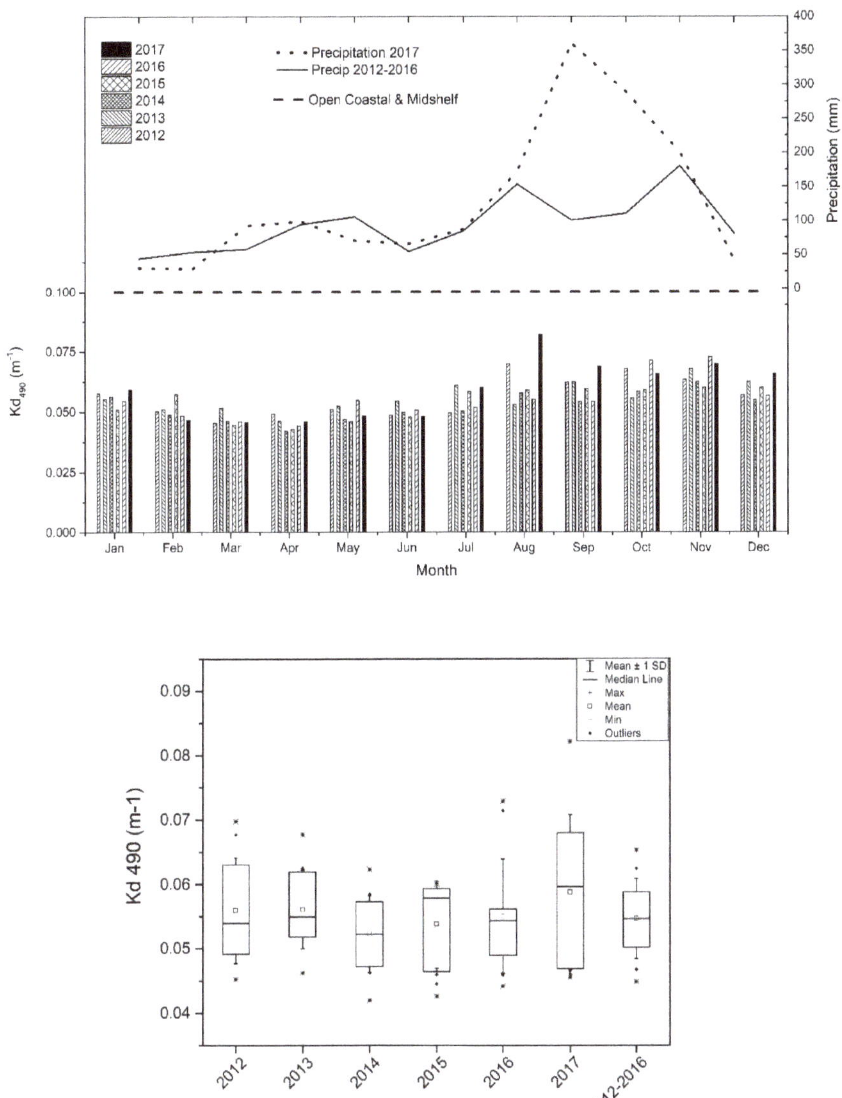

Figure 8. Monthly comparison of K_d490 values showing the yearly distribution from 2012–2017 with associated precipitation data (top). The year 2017 is filled in black. The dashed line for 0.1 m^{-1} represents the K_d490 value threshold for open coastal waters [22]. (Bottom) Box plot showing yearly statistics of K_d490 for all regions.

3.3. Inner Shelf vs Outer Shelf Pixels

Pixel point locations were evaluated to quantify the difference from coastal to oceanic waters around Puerto Rico. The pixel locations are broken out into inner shelf and outer shelf regions (Figures 9 and 10). Approximately 74% of the average value of Chl-*a* for Puerto Rico (0.55 µg/L) was driven by the inner shelf pixel locations, where 26% was attributed to the outer shelf locations. A small variation (± 3%) was found in the contribution of the inner vs. outer shelf to the average value of Chl-*a* for all Puerto Rico, even when considering regional and yearly distributions. These pixel locations were also analyzed by month and compared between 2012–2016 and 2017 data (Figure 9). Chl-*a* values for outer shelf locations remained below 0.45 µg/L except for the months of July and August of 2017. The values for the inner shelf pixels remained below the 0.8 µg/L value from January to June for both 2017 and 2012–2016 data. The inner shelf values for 2017 displayed an increase (in percent) for the months of June (8% from baseline), July (17%), August (5%), September (8%), October (19%), and November (28%) when compared to 2012–2016 data.

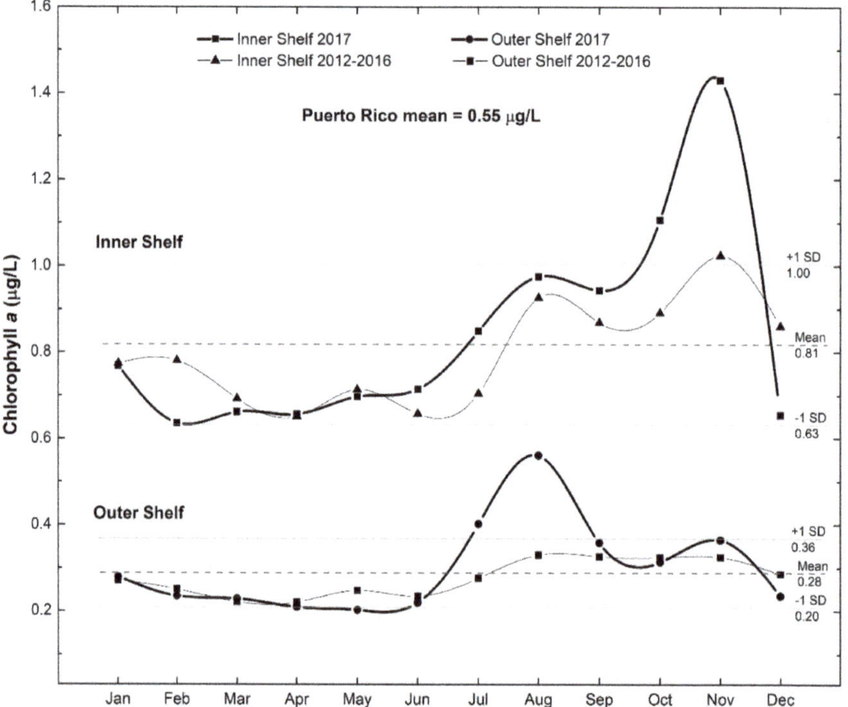

Figure 9. Chlorophyll-*a* concentration for inner shelf and outer shelf of Puerto Rico waters.

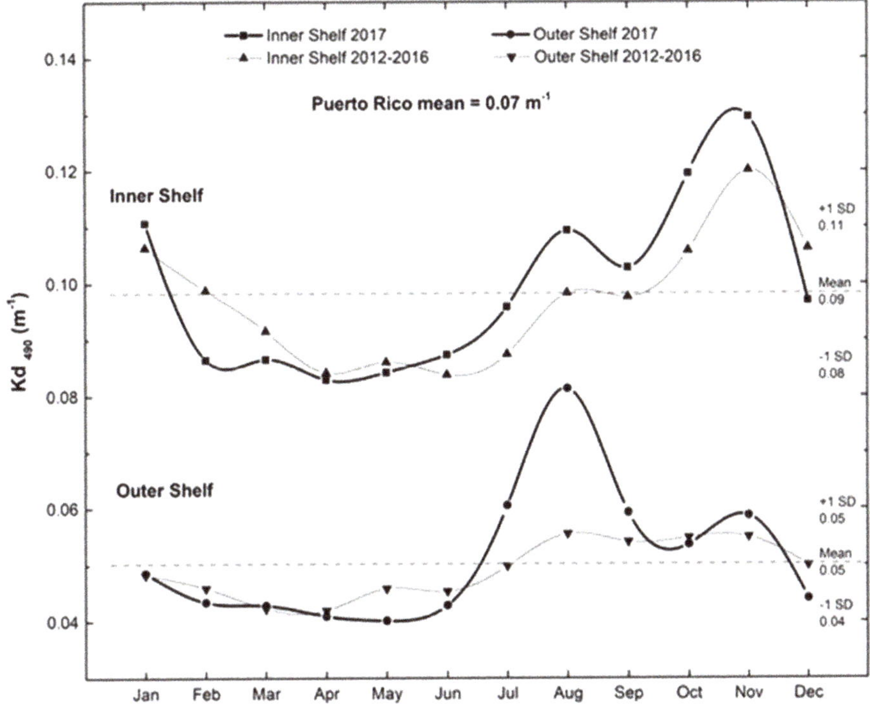

Figure 10. K_d490 for inner shelf and outer shelf of Puerto Rico waters.

For the K_d490 values, approximately 70% of the average value for Puerto Rico (0.07 m^{-1}) were driven by the inner shelf pixel locations, while the remaining 30% was from outer shelf pixel locations. There was little variation (± 3%) in the contribution of the inner vs. outer shelf to the average value of K_d490 for all Puerto Rico, even when considering regional and yearly distributions. These pixel locations were also analyzed by month and compared from 2012–2016 to 2017 data (Figure 10). The values for outer shelf locations remained below 0.06 m^{-1} except for the months of July and August of 2017. The inner shelf values remained below the 0.10 m^{-1} value from February to July for both 2017 and 2012–2016 data. The inner shelf values for 2017 increased (in percent) during the months of June (4% from baseline), July (9%), August (10%), September (5%), October (12%), and November (7%) when compared to 2012–2016 data.

4. Discussion

Hurricanes Irma and Maria Effects on Water Quality

Hurricane Maria's 24-hr rainfall intensity was undoubtedly the highest for any tropical cyclone in Puerto Rico since 1898 [24]. This event represented a 13% increase in the island-wide 24-hour rainfall rates, which is within the range of predicted increases associated to climate change both locally and worldwide [24]. The precipitation data for our selected stations also shows an increase in 2017, especially during the months of August to October. Such extreme precipitation events from tropical cyclones can alter the coastal water quality regimes [25]. Higher mean, median, and maximum values were observed for both Chl-*a* and K_d490 when compared to previous years (Figures 6 and 8).

The highest values of both Chl-*a* and K_d490 were present in August 2017, just before the hurricane impact, which corresponds with an increase in precipitation during that same month when compared with 2012–2016 values. Regional values also show a dramatic increase by the month of August 2017 when compared to 2012–2016. According to National Weather Service (NWS) San Juan weather report for 2017, July and August showed an increase in monthly precipitation of 53.3 mm and 61.5 mm respectively from the normal (1981–2010) [26] which explain the higher values for both Chl-*a* and K_d490.

In addition to increased light attenuation, sediment reaching the coast from runoff can smother corals and has been shown to have a detrimental effect on coral recruitment, decrease calcification, decrease net productivity of corals, and reduce rates of reef accretion [27]. In addition, introducing nutrients and high turbidity to what is generally an oligotrophic system, combined with decreased grazing due to overfishing, can promote the growth of macroalgae which may then outcompete corals for space on reefs [28]. To put these values into context, impairment thresholds for Chl-*a* and water clarity from the GBRMP and State of Hawai'i water quality standards were shown in previous figures (Figures 5 and 7–9). Thresholds of 0.45 µg/L were used for Chl-*a*, and 0.1 m^{-1} for K_d490 coefficients for open coastal waters. For both indicators, inner shelf points showed values above these thresholds before and after the hurricanes and these values exceeded thresholds globally recognized as adverse for coral reefs [23]. One alarming factor is that the mean Chl-*a* value for Puerto Rico for the complete time-series (2012–2017) was 0.55 µg/L, higher than the established threshold even when considering inner/outer shelf pixels, suggesting chronically impaired water quality. These data can provide key information for management to establish water quality thresholds for coastal waters, as well as prioritize restoration efforts of watershed and coral reef areas.

When comparing the inner and outer shelf pixel values of Chl-*a* and K_d490, this follows the distinct neritic to oceanic gradients of water turbidity observed for the oceanic waters surrounding Puerto Rico [5]. This suggests that areas closer to shore experienced higher levels of degraded water. 9 of the 25 inner shelf pixels were over coral reef and colonized hardbottom areas (Figure 3), where higher values of both Chl-*a*, and K_d490 were present and persisted for various months. Miller [10] used total suspended matter (TSM) from the Sentinel 3 Ocean Color Land Instrument (OLCI) to find that island-wide mean TSM increased 2.2 times (5.57 mg/L to 12.39 mg/L) between before and two weeks after Hurricane Maria's strike, rapidly dropping by November (5–6 mg/L), and reaching normal levels by February 2018 (2.14 mg/L). Degraded water quality conditions persisted for months after the hurricanes as shown by our Chl-*a* and K_d490 values. No TSM products are available from VIIRS to compare these results. Additionally, the North and East regions contributed to the high values of both Chl-*a* and K_d490 since those regions contain Puerto Rico's major rivers, while the smallest rivers are mainly found along the south coast [29].

Our results for Chl-*a* and K_d490 did not find strong overall positive correlations with precipitation values. This may be due to a non-linear relationship between precipitation and surface runoff as a result of reduced foliage interception due to hurricane defoliation, which did not exceed 1.0 kg/m^2 in the month preceding Maria, but reached 3 times this value before comparatively drier conditions arrived in December [10]. Warne [29] estimated that 57 % of the mean annual precipitation (911 mm/yr) in Puerto Rico is discharged to the coast as runoff due to steep gradients, relatively shorter river lengths, and low water holding capacity. These surface hydrology characteristics and the fact that both the precipitation and satellite data were summarized into monthly values, reduce the potential lag between precipitation and satellite ocean color measurements. In addition, there were few stations with reliable historical records of precipitation data per region that could be correlated with the 2012–2017 time series. This may have contributed to the low correlation values for some regions, due to an incomplete representation of the total precipitation values for the regions.

In addition, the peak values for both Chl-*a* and K_d490 for September-October 2017 may have been underestimates due to limited numbers of retrievals resulting from high cloud cover. According to Mikelsons and Wang [30], the most significant limiting factors in satellite ocean color data retrievals is cloud cover, and large cyclone systems can prevent ocean color data retrievals over vast areas for several days. In fact, a prolonged period of cloudy conditions followed these storms. This reduction in cloud-free imagery is magnified by the afternoon timing of overpass of satellites flying the VIIRS instrument. During that season, cloud formation is generally high due to local trade winds and orographic effects. "Blended" Chl-*a* and K_d490 products from a combination of moderate resolution sensors (i.e., SeaWiFS, MODIS, JPSS, OLCI) at full resolution (≤ 1 km), with different satellite observation times may improve the amount of cloud-free observations in near-shore environments, as would the presence of ocean color instruments on geostationary satellites.

5. Conclusions

Moderate-resolution satellite imagery, such as VIIRS can provide a reliable method to evaluate potential habitat exposure to degraded water quality without conducting extensive in-situ water quality monitoring. Even with the limitations of spatial resolution and loss of data due to cloud cover, moderate-resolution imagery time-series analysis has provided a useful evaluation of the effects of the hurricanes on the coastal water quality in Puerto Rico, and the potential exposure of benthic habitats to higher turbidity waters. The regions around Puerto Rico experienced extreme and prolonged levels of pollution exceeding established thresholds for coastal and open ocean areas that contain coral and seagrass habitats. This pollution came from multiple sources including sediment from the extensive landslides and untreated sewage from the persistent losses of power across the island and use of combined sewer systems that collect rainwater runoff, domestic sewage, and industrial wastewater into a single system. Depending on the magnitude and duration of the pollution, and the condition of the habitat prior to the hurricane, these exposures likely led to a range of habitat-scale impacts including, but not limited to inhibition of light penetration needed to support photosynthesis, physical smothering of the habitat by sediment, and excessive algal growth, which will outcompete coral reefs and seagrass. Those habitats that were previously impaired due to chronic pollution are particularly susceptible to this threat.

Water quality exceedances and corresponding habitat exposures varied across the inner shelf and outer shelf locations (i.e., coastal, oceanic) and regions (i.e., North, South, East, West). For inner shelf locations, many of the observed baseline and post-hurricane values for Chl-*a* and K_d490 are above thresholds for impairment recognized by coral jurisdictions around the globe. Outer shelf locations generally show lower values. Degraded coastal water quality has the highest potential of impact and these were present close to shore, where coral reef and other critical habitats are located. In addition, some regions suffered more severe hurricane impact than others. In the east region, for instance, turbidity was higher than in other regions prior to the hurricanes, therefore post-hurricane differences in degraded water quality were masked in our change analysis. The results from this project can be used as a guideline to establish local thresholds for water quality of these parameters in the coastal areas of Puerto Rico taking into consideration suggested long-term increases in precipitation from altered extreme weather scenarios.

Author Contributions: W.J.H. conceived and designed the study, W.J.H. and S.O.-R. performed the experiments, conducted and performed data processing and analysis; R.A.A., C.M.E., E.F.G., and R.A.W. contributed to data collection; W.J.H. wrote the manuscript, with all authors contributing to manuscript revisions; W.J.H. and R.A.A. attracted funds for the project. All authors have read and approved the manuscript.

Funding: This study was supported in part by The National Oceanic and Atmospheric Administration –Cooperative Science Center for Earth System Sciences and Remote Sensing Technologies (NOAA-CESSRST) under the Cooperative Agreement Grant # NA16SEC4810008. The authors would like to thank The City College of New York, NOAA-CESSRST program and NOAA Office of Education, Educational Partnership Program for full fellowship support. The statements contained within the manuscript/research article are not the opinions of the funding agency or the U.S. government but reflect the author's opinions. NOAA Coral Reef Watch staff were fully supported, and this study was partially supported, by NOAA grant NA19NES4320002 (Cooperative Institute for Satellite Earth System Studies) at the University of Maryland/ESSIC. The scientific results and conclusions, as well as any views or opinions expressed herein, are those of the author(s) and do not necessarily reflect the views of NOAA or the Department of Commerce.

Acknowledgments: The authors would like to thank Lisa Vandiver, Anne Kitchell and Roberto Viqueira for their contributions and support to this project.

Conflicts of Interest: The authors declare no conflict of interest. The funders had no role in the design of the study; in the collection, analyses, or interpretation of data; in the writing of the manuscript, or in the decision to publish the results.

References

1. Greening, H.; Doering, P.; Corbett, C. Hurricane impacts on coastal ecosystems. Estuaries and coasts, part A: Hurricane impacts on coastal ecosystems. *J. Coast. Estuar. Res. Fed.* **2006**, *29*, 877–879. [CrossRef]
2. Valiela, I.P.; Peckol, C.; D'Avanzo, J.; Kremer, D.; Hersh, K.; Foreman, K.; Lajtha, B.; Seely, W.R.; Geyer, I.; Crawford, R. Ecological effects of major storms on coastal watersheds and coastal waters: Hurricane Bob on Cape Cod. *J. Coast. Res.* **1998**, *14*, 218–238.
3. Hernandez-Cruz, L.R.; Purkis, S.J.; Riegl, B. Documenting decadal spatial changes in seagrass and *Acropora palmatta* cover by aerial photography analysis in Vieques, Puerto Rico: 1937–2000. *Bull. Mar. Sci.* **2006**, *2*, 401–414.
4. Hernández-Delgado, E.A. The emerging threats of climate change on tropical coastal ecosystem services, public health, local economies and livelihood sustainability of small islands: Cumulative impacts and synergies. *Mar. Pollut. Bull.* **2015**, *101*, 5–28. [CrossRef]
5. García-Sais, J.R.; Williams, S.M.; Amirrezvani, A. Mortality, recovery, and community shifts of scleractinian corals in Puerto Rico one decade after the 2005 regional bleaching event. *PeerJ* **2017**, *5*, e3611. [CrossRef]
6. Mallin, M.A.; Corbett, C.A. How Hurricane attributes determine the extent of environmental effects: Multiple hurricanes and different coastal systems. *Estuaries Coasts* **2006**, *29*, 1046–1061. [CrossRef]
7. Bejarano, I.; Apeldoorn, R. Seawater turbidity and fish communities on coral reefs of Puerto Rico. *Mar. Ecol. Prog. Ser.* **2013**, *474*, 217–226. [CrossRef]
8. Cangialosi, J.P.; Latto, A.S.; Berg, R. National Hurricane Center Tropical Cyclone Report Hurricane Irma (AL112017). 2018. Available online: https://www.nhc.noaa.gov/data/tcr/AL112017_Irma.pdf (accessed on 20 August 2018).
9. Pasch, R.J.; Penny, A.B.; Berg, R. National Hurricane Center Tropical Cyclone Report Hurricane María (AL152017). 2018. Available online: https://www.nhc.noaa.gov/data/tcr/AL152017_María.pdf (accessed on 20 August 2018).
10. Miller, P.W.; Kumar, A.; Mote, T.L.; Moraes, F.D.S.; Mishra, D.R. Persistent hydrological consequences of Hurricane María in Puerto Rico. *Geophys. Res. Lett.* **2019**, *46*, 1413–1422. [CrossRef]
11. Hu, T.; Smith, R.B. The Impact of Hurricane María on the Vegetation of Dominica and Puerto Rico Using Multispectral Remote Sensing. *Remote Sens.* **2018**, *10*, 827. [CrossRef]
12. Shi, W.; Wang, M. Observations of a Hurricane Katrina-induced phytoplankton bloom in the Gulf of Mexico. *Geophys. Res. Lett* **2007**, *34*, 11607. [CrossRef]
13. Gilbes, F.; Armstrong, R.A.; Webb, R.M.; Müller-Karger, F.E. SeaWifs helps assess hurricane impact on phytoplankton in caribbean sea. *Eos Trans. Am. Geophys. Union* **2001**, *82*, 529. [CrossRef]
14. Wang, M.; Son, S. VIIRS-derived chlorophyll-a using the ocean color index method. *Remote Sens. Environ.* **2016**, *182*, 141–149. [CrossRef]
15. Hu, C.; Lee, Z.; Franz, B.A. Chlorophyll-a algorithms for oligotrophic oceans: A novel approach based on three-band reflectance difference. *J. Geophys. Res.* **2012**, *117*. [CrossRef]
16. O'Reilly, J.E.; Maritorena, S.; Mitchell, B.G.; Siegel, D.A.; Carder, K.L.; Garver, S.A.; Kahru, M.; McClain, C.R. Ocean color chlorophyll algorithms for SeaWiFS. *J. Geophys. Res.* **1998**, *103*, 24937–24953. [CrossRef]
17. Wang, M.; Son, S.; Harding, L.W., Jr. Retrieval of diffuse attenuation coefficient in the Chesapeake Bay and turbid ocean regions for satellite ocean color applications. *J. Geophys. Res.* **2009**, *114*. [CrossRef]
18. Morel, A.H.Y.; Gentili, B.; Werdell, P.J.; Hooker, S.B.; Franz, B.A. Examining the consistency of products derived from various ocean color sensors in open ocean (Case 1) waters in the perspective of a multi-sensor approach. *Remote Sens. Environ.* **2007**, *111*, 69–88. [CrossRef]
19. Lee, Z.P.; Du, K.; Arnone, R. A model for the diffuse attenuation coefficient of downwelling irradiance. *J. Geophys. Res.* **2005**, *110*. [CrossRef]
20. Kirk, J.T.O. *Light and Photosynthesis in Aquatic Ecosystems*; Cambridge University Press: Cambridge, UK, 2011.

21. Kendall, M.S.C.R.; Kruer, K.R.; Buja, J.D.; Christensen, M.; Finkbeiner, R.A.; Warner Monaco, M.E. NOAA Technical Memorandum NOS NCCOS CCMA 152 (On-line). Methods Used to Map the Benthic Habitats of Puerto Rico and the U.S. Virgin Islands. 2001. Available online: http://biogeo.nos.noaa.gov/projects/mapping/caribbean/startup.htm (accessed on 21 November 2019).

22. Water Quality Standards (WQS). Hawaii Administrative Rules (HAR). Amendment and Compilation of Chapter 11–54. State of Hawai'i Department of Health. Available online: https://www.epa.gov/wqs-tech/water-quality-standards-regulations-hawaii (accessed on 22 November 2019).

23. (GBRMPA) Great Barrier Reef Marine Park Authority. *Water Quality Guidelines for the Great Barrier Reef Marine Park*; Great Barrier Reef Marine Park Authority: Townsville, Australia, 2010; p. 99.

24. Ramos-Scharrón, C.E.; Arima, E. *Hurricane María's Precipitation Signature in Puerto Rico: A Conceivable Presage of Rains to Come*; Scientific Reports; Nature Research: London, UK, 2019.

25. Keellings, D.; Hernández Ayala, J.J. Extreme rainfall associated with Hurricane María over Puerto Rico and its connections to climate variability and change. *Geophys. Res. Lett.* **2019**, *46*, 2964–2973. [CrossRef]

26. NWS San Juan. Climate Review for Puerto Rico and the U.S. Virgin Islands. Climate Reports from the Weather Forecast Office San Juan. 2017. Available online: https://www.weather.gov/media/sju/climo/monthly_reports/2017/2017.pdf (accessed on 8 February 2020).

27. Rogers, C.S. Responses of coral reefs and reef organisms to sedimentation. *Mar. Ecol.* **1990**, *62*, 185–202. [CrossRef]

28. Scheffer, M.S.; Carpenter, J.A.; Foley, C.; Folke Walker, B. Catastrophic shifts in ecosystems. *Nature* **2001**, *413*, 591–596. [CrossRef]

29. Warne, A.G.; Webb, R.M.T.; Larsen, M.C. Water, sediment, and nutrient discharge characteristics of rivers in Puerto Rico, and their potential influence on Coral Reefs. Available online: http://pubs.usgs.gov/sir/2005/5206/ (accessed on 30 January 2007).

30. Mikelsons, K.; Wang, M. Optimal satellite orbit configuration for global ocean color product coverage. *Opt. Express* **2019**, *27*, A445–A457. [CrossRef] [PubMed]

Article

Wind-Driven Coastal Upwelling near Large River Deltas in the Laptev and East-Siberian Seas

Alexander Osadchiev [1,2,*], **Ksenia Silvestrova** [1] and **Stanislav Myslenkov** [1,3,4]

[1]   Shirshov Institute of Oceanology, Russian Academy of Sciences, Nakhimovskiy prospect 36, 117997 Moscow, Russia; silvestrova.kp@ocean.ru (K.S.); myslenkov@nral.org (S.M.)

[2]   Institute of Geology of Ore Deposits, Petrography, Mineralogy and Geochemistry, Russian Academy of Sciences, Staromonetny pereulok 35s2, 119017 Moscow, Russia

[3]   Faculty of Geography, Moscow State University, Leninskie Gory 1, 119992 Moscow, Russia

[4]   Marine Forecast Division, Hydrometeorological Research Centre of the Russian Federation, B. Predtechenskiy pereulok 11-13, 123242 Moscow, Russia

*   Correspondence: osadchiev@ocean.ru

Received: 8 February 2020; Accepted: 4 March 2020; Published: 5 March 2020

Abstract: The Lena, Kolyma, and Indigirka rivers are among the largest rivers that inflow to the Arctic Ocean. Their discharges form a freshened surface water mass over a wide area in the Laptev and East-Siberian seas and govern many local physical, geochemical, and biological processes. In this study we report coastal upwelling events that are regularly manifested on satellite imagery by increased sea surface turbidity and decreased sea surface temperature at certain areas adjacent to the Lena Delta in the Laptev Sea and the Kolyma and Indigirka deltas in the East-Siberian Sea. These events are formed under strong easterly and southeasterly wind forcing and are estimated to occur during up to 10%–30% of ice-free periods at the study region. Coastal upwelling events induce intense mixing of the Lena, Kolyma, and Indigirka plumes with subjacent saline sea. These plumes are significantly transformed and diluted while spreading over the upwelling areas; therefore, their salinity and depths abruptly increase, while stratification abruptly decreases in the vicinity of their sources. This feature strongly affects the structure of the freshened surface layer during ice-free periods and, therefore, influences circulation, ice formation, and many other processes at the Laptev and East-Siberian seas.

Keywords: coastal upwelling; wind forcing; river plume; MODIS; Arctic Ocean

---

1. Introduction

The Arctic Ocean covers an area of about 3% of the World Ocean area and holds only 1% of its volume, but receives approximately 11% of world continental discharge [1,2]. This enormously large freshwater runoff forms large freshened water masses at the Arctic shelf and induces strong vertical stratification that plays a crucial role in the variability of ice cover and regional albedo [3–5]. As a result, the spreading and mixing of freshwater runoff in the Arctic Ocean influences global climate processes. Freshened water masses also significantly affect many local processes in the Arctic Ocean, especially in coastal and shelf areas where the impact of freshwater discharge is the strongest [6–14].

The Lena, Kolyma, and Indigirka rivers are among the largest rivers that inflow to the Arctic Ocean. Annual discharges of the Lena, Kolyma, and Indigirka rivers are estimated as 530, 130, and 60 km^3 and they provide approximately 70% and 75% of the total freshwater discharge to the Laptev and East-Siberian seas, respectively [15,16]. The majority of this discharge inflows to the sea during the ice-free period in June–September and forms the Lena, Kolyma, and Indigirka river plumes [17]. These buoyant plumes occupy hundreds of thousands square kilometers in the Laptev and East-Siberian seas and are among the largest freshwater reservoirs in the Arctic Ocean [17–20]. Spreading and transformation of these river plumes determine vertical stratification and, therefore, strongly affect circulation and ice formation in the Laptev and East-Siberian seas, as well as many other physical, geochemical, and biological processes [21–32].

In this study we focus on upwelling events which regularly occur at coastal areas adjacent to the deltas of the Lena, Kolyma, and Indigirka rivers. Surface manifestations of these upwelling events are visible on ocean color satellite imagery due to elevated turbidity and on sea surface temperature (SST) satellite imagery due to reduced temperature. However, correct identification of the origin of SST and turbidity features observed on satellite imagery is not a straightforward task. SST features in the study region are formed as a result of interaction between water masses with different temperature, namely, warm river plumes and cold saline sea water, and are associated with spreading of river plumes, mixing of surface layer with subjacent sea, and ice melting. Areas of elevated sea surface turbidity in coastal and shelf regions are commonly associated with four different processes: spreading of turbid river plumes, coastal erosion, resuspension of bottom sediments penetrated to sea surface, and algal blooms [33]. The first three processes are common features of the Laptev and East-Siberian seas [25,27,34,35], while algal blooms do not occur in these seas [32,36–38]. Turbid regions associated with river plumes are adjacent to river estuaries and deltas. Spatial and temporal variability of these regions is defined mainly by river discharge rate, turbidity of river water, and local wind forcing [39–47]. Coastal erosion in the Laptev and East-Siberian seas is extremely intense due to active thermal abrasion. It provides large land-ocean fluxes of terrigenous sediments whereby total eroded sediment volume exceeds river sediment discharge [31,33,48]. Turbid regions associated with coastal erosion are adjacent to long segments of sea coast, but it does not cause elevated turbidity in offshore areas. Finally, resuspension of bottom sediments occurs in shallow areas and can be caused by upwelling events, tides, and wind waves [24,49–53]. In the latter case, turbulence induced by breaking surface waves penetrates from the surface layer to the sea bottom, causing resuspension of bottom sediments and their subsequent upward convection to surface layer. Tidal circulation and coastal upwelling, conversely, initially induce turbulence in the bottom layer, which penetrates upward and can reach the surface layer carrying resuspended sediments.

Interaction between river plumes and coastal upwelling were addressed in many previous works. Stratification in the coastal area affects the depth of the mixed layer and alters wind-driven cross-shore circulation [54–57]. Upwelling-favorable winds induce offshore transport of river plumes and their detachment from the sea shore [58–62]. As a result, a sharp salinity gradient is formed between the saline and low-stratified near-shore area and offshore located river plume [36]. Upwelling winds also cause intense mixing of a river plume with subjacent saline sea due to increased velocity shear and an Ekman straining mechanism [58,63]. Therefore, upwelling events along coastal areas influenced by freshwater discharge significantly affect spreading and mixing of river plumes, as well as the local nutrient cycle, biological consumption, food webs, and biological productivity [56,64–68].

Many previous works addressed wind-driven coastal upwelling events in the Arctic Ocean [19,69–74]. However, interaction between river plumes and coastal upwelling in the Arctic Ocean remain mainly unstudied. We are aware of only a few studies focused on coastal upwelling influenced by large freshwater discharge, namely, the Mackenzie River [75–78]. In this study we report coastal upwelling events that occur over wide areas adjacent to deltas of the Lena, Indigirka, and Kolyma rivers. Using satellite imagery and atmospheric reanalysis fields, we reveal that these upwelling events are regularly induced by wind forcing. We show that they strongly affect the thermohaline and turbidity properties of the sea surface layer and influence spreading and mixing of the large Lena, Kolyma, and Indigirka plumes.

The paper is organized as follows. Section 2 provides detailed information about the study area, the satellite and wind reanalysis data, and the methods of detection of upwelling events used in this study. Section 3 describes spatial and temporal characteristics of wind-driven coastal upwelling events that occur near the Lena, Kolyma, and Indigirka deltas. Frequency and duration of these coastal upwelling events are assessed and their influence on the spreading and mixing of the Lena, Kolyma, and Indigirka plumes is analyzed in Section 4, followed by the conclusions in Section 5.

2. Study Area, Data, and Methods

2.1. Study Area

The Laptev and East-Siberian seas are located at the east of the Eurasian part of the Arctic Ocean. These seas are semi-enclosed by the Siberian coast and large archipelagos and islands (Severnaya Zemlya, New Siberian Islands, and Wrangel Island) in the south, east, and west, and only in the north they are open to the central part of the Arctic Ocean (Figure 1). Half of the Laptev Sea and almost the whole area of the East-Siberian Sea rest on the continental shelf. The distance between the sea shore and the continental slope increases from 100 to 200 km at the western part of the Laptev Sea and to 1000 km at the eastern part of the East-Siberian Sea. Average sea depths of the Laptev and East-Siberian seas are 580 and 45 m, respectively.

General circulation in the Laptev and East-Siberian seas is governed by river runoff and zonal water exchange with the Kara Sea [79], the Chukchi Sea [80], and the deep basin of the Arctic Ocean [12]. The Laptev and East-Siberian seas receive a large volume of continental discharge, approximately 800 km^3 to the Laptev Sea and 250 km^3 to the East-Siberian Sea annually, which accounts for approximately a quarter of the total freshwater runoff to the Arctic Ocean [1,15,81]. Spatial and temporal variability of river plumes formed in these seas are mainly governed by river discharge rates and wind forcing conditions [21,23,25,27,80–83]. Tidal circulation in the Laptev and East-Siberian seas is dominated by a lunar semidiurnal tidal wave that propagates from the North Atlantic to the Arctic Ocean. Tidal amplitudes in these seas are generally low, as compared to the World Ocean, and do not exceed 0.5 m in the majority of their area [84–86].

The Laptev and East-Siberian seas are frozen during the majority of a year. The southern parts of the seas adjacent to the Lena, Indigirka, and Kolyma deltas are covered by landfast ice (1.5–2 m thick) from the end of October to June–July. The ice regime in the study areas is significantly influenced by continental runoff [17,25,27] and the Great Siberian Polynya [87]. Summer and autumn ice coverage of these seas shows large inter-annual variability. The northernmost position of the edge of the sea ice was located at a distance of 200–300 km from the Siberian shore during certain years (e.g., 2013, 2014, 2018), so the central and northern parts of these seas were covered by ice during the whole year. On the other hand, these seas can be totally free of ice at the end of August–beginning of October during the years of reduced ice coverage (e.g., 2012, 2017, 2019).

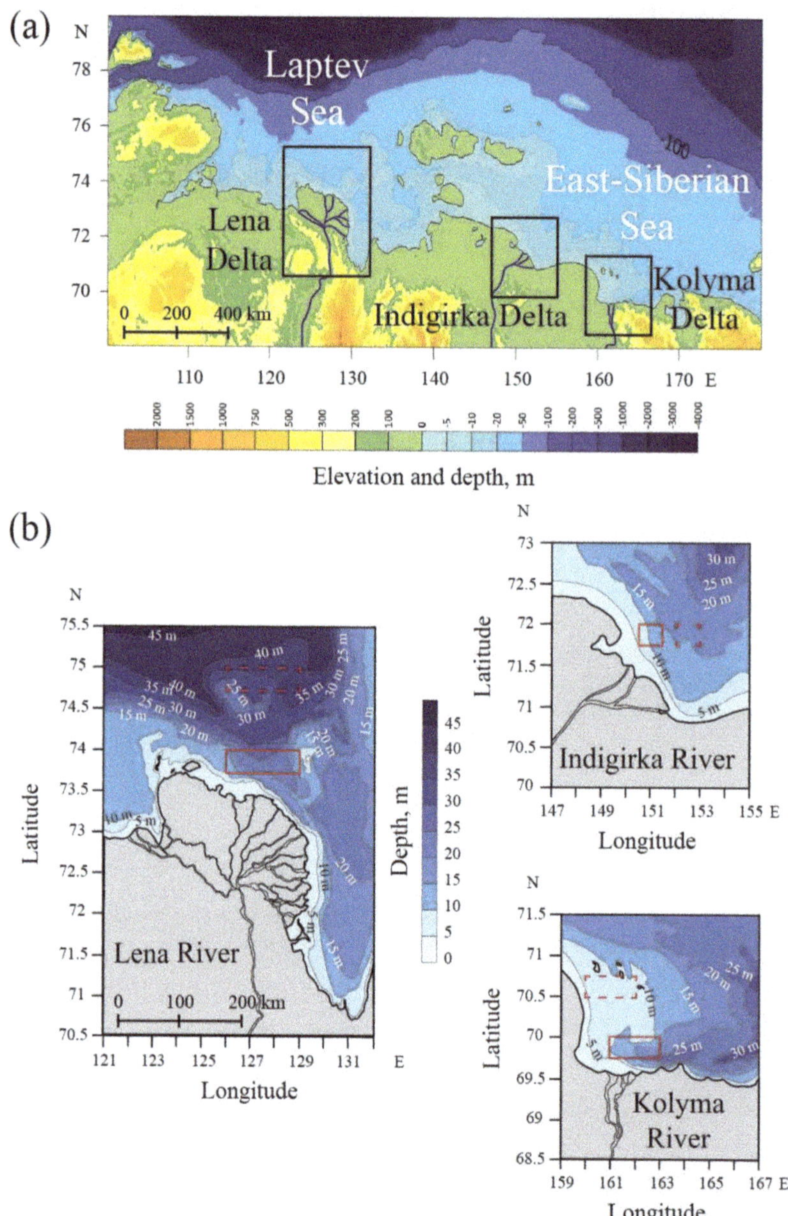

Figure 1. (**a**) Bathymetry and topography of the study region locations of the Lena, Indigirka, and Kolyma deltas in the Laptev and East-Siberian seas; (**b**) bathymetry of the areas adjacent to the Lena, Indigirka, and Kolyma deltas. The graphic scales correspond to the latitude of 72°. Red boxes indicated in panel (**b**) show locations of reference areas in the upwelling regions (dashed contours) and the ambient sea (solid contours) used to identify upwelling events.

2.2. Data and Methods

Satellite data used in this study include Terra/Aqua Moderate Resolution Imaging Spectroradiometer (MODIS) satellite imagery for the period 2000–2019 provided by the National Aeronautics and Space Administration (NASA). MODIS L1b calibrated radiances including MODIS bands 1 (red), 3 (blue), 4 (green), and daytime 31 (thermal) were downloaded from the NASA web repository (https://ladsweb.modaps.eosdis.nasa.gov/). We used ESA BEAM software for retrieving maps of sea surface distributions of corrected reflectance and brightness temperature at the study areas with spatial resolutions of 100 m and 1 km, respectively. Wind forcing conditions were examined using NCEP/CFSR/CFSv2 atmospheric reanalysis with a ~0.3° (1979–2010) and ~0.2° (2011–2019) spatial and hourly temporal resolution [88,89]. The reanalysis data were downloaded from the National Climatic Data Center of the National Oceanic and Atmospheric Administration (NCDC NOAA) web repository (https://www.ncdc.noaa.gov/). Visual inspection of all satellite images of three study areas (Figure 1b) acquired during ice-free seasons (July–October) of 2000–2019 was performed to detect cloud-free and ice-free satellite images. The resulting 252 images were used to detect upwelling events near the Lena, Indigirka, and Kolyma deltas in the following way. For every considered region we identified a pair of reference areas, namely, the upwelling area adjacent to the delta and the ambient sea area not affected by upwelling events. The pairs of these reference areas are shown in Figure 1b by red boxes, while their coordinates are given in Table 1. Then for every cloud-free and ice-free satellite image we calculated differences in average brightness temperature within the pairs of reference areas. If the temperature of an upwelling area was smaller than the temperature of an ambient sea area by >2 °C, we regarded this case as a "cold event" bounded by a "distinct" frontal zone which is a surface manifestation of upwelling.

Table 1. Coordinates of reference areas used to identify upwelling events near the Lena, Indigirka, and Kolyma deltas.

	Lena Delta		Indigirka Delta		Kolyma Delta	
	Upwelling Area	Ambient Sea	Upwelling Area	Ambient Sea	Upwelling Area	Ambient Sea
Longitude, °E	126–129	126–129	150.5–151.5	152–153	161–163	160–162
Latitude, °N	73.75–74	74.75–75	71.75–72	71.75–72	69.75–70	70.5–70.75

Due to the complexity of coastal processes that govern the temperature of the sea surface and the absence of specific regional algorithms for retrieving SST in the study areas with very limited in situ measurements, we did not used the standard SST product of MODIS. Instead, we used a brightness temperature product that does not provide an accurate temperature of the sea surface, but shows relative temperature differences, which can be used to detect upwelling events. The qualitative routine for detection of upwelling events was based on the brightness temperature values, which was followed by assessment of surface turbidity during upwelling and non-upwelling events. Due to the absence of specific regional algorithms for retrieving total suspended matter in the study area influenced by multiple processes (resuspension of bottom sediments, river discharge, coastal erosion), we did not apply quantitative assessment of surface turbidity, but performed qualitative visual inspection that identified elevated turbidity at the upwelling area in all cases during and shortly after upwelling events and mainly normal turbidity during non-upwelling periods. As a result, hereafter in the text, we regard "cold events" as "cold and turbid events".

3. Results

3.1. Coastal Upwelling near the Lena Delta in the Laptev Sea

Optical satellite imagery regularly reveals events of increased sea surface turbidity and reduced sea surface temperature at the area located to the north from the Lena Delta (Figure 2). To study this feature, we analyzed all MODIS Terra and MODIS Aqua satellite images of the study region taken in 2000–2019 during July–October when the southern part of the Laptev Sea was free of ice. Due to common cloudy weather conditions, we detected only 25 periods (1–6 days long) when the area adjacent to the Lena Delta was clearly seen in optical satellite images and the structure of surface turbidity and temperature could be identified. Cold and turbid sea to the north from the Lena Delta was observed in 12 cases of the 25 considered periods. The other 13 cases were characterized by relatively homogenous turbidity and temperature at the study area, without any distinct frontal zones.

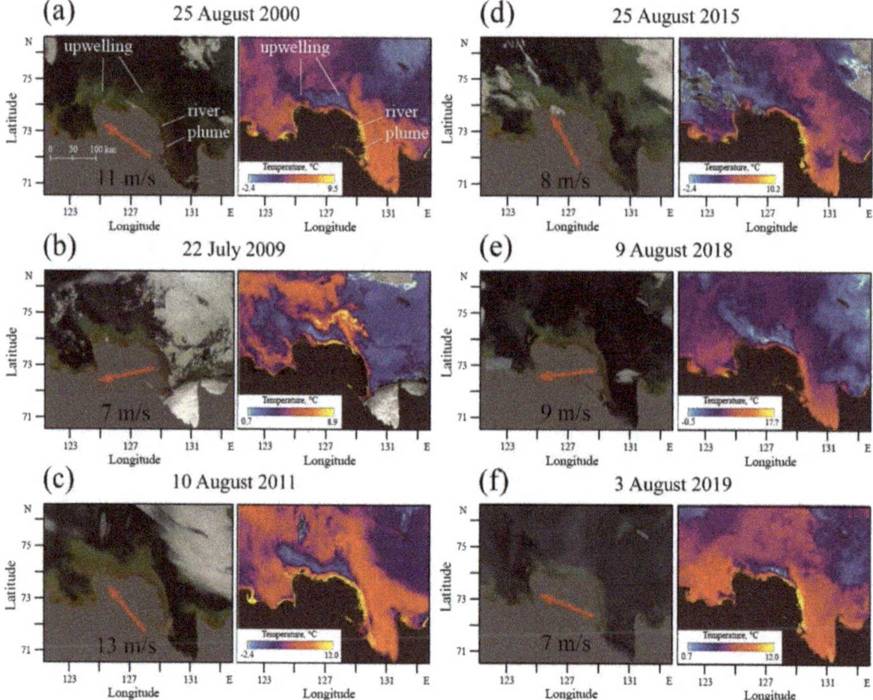

Figure 2. Corrected reflectance (left) and brightness temperature (right) from MODIS (Moderate Resolution Imaging Spectroradiometer) Terra and MODIS Aqua satellite images of the Laptev Sea acquired on (**a**) 25 August 2000, (**b**) 22 July 2009, (**c**) 10 August 2011, (**d**) 25 August 2015, (**e**) 9 August 2018, and (**f**) 3 August 2019 indicating the location of upwelling events to the north of the Lena Delta induced by wind forcing (arrows) and manifested by elevated sea surface turbidity and reduced sea surface temperature. Surface manifestations of upwelling events and river plumes are indicated in panel (**a**).

Typical examples of the cold and turbid events observed during six different days in 2000–2019 are shown in Figure 2. Sharp sea surface temperature gradients are formed between the cold area located to the north from the Lena Delta and the surrounding warm sea. This cold area is bounded by the distinct frontal zone whose location and shape is stable on satellite images taken on different days. The location and shape of this thermal frontal zone show good agreement with local bathymetry (Figure 3, right panels). The southern and eastern parts of the thermal frontal zone are located over the isobaths of 5–10 m stretched along the northern coast of the Lena Delta and the large shoal located to the northeast from the Lena Delta. The northern part of the thermal frontal zone is generally located over the isobath of 30 m, but its position was less stable. The cold area typically occupies a large part of the coastal sea (15,000–17,000 km^2), apart from two days, 11 September 2005 and 2–4 August 2019, when this area was relatively small and the northern part of the thermal frontal zone had shifted in a southeasterly direction (Figure 3a,b, right panels).

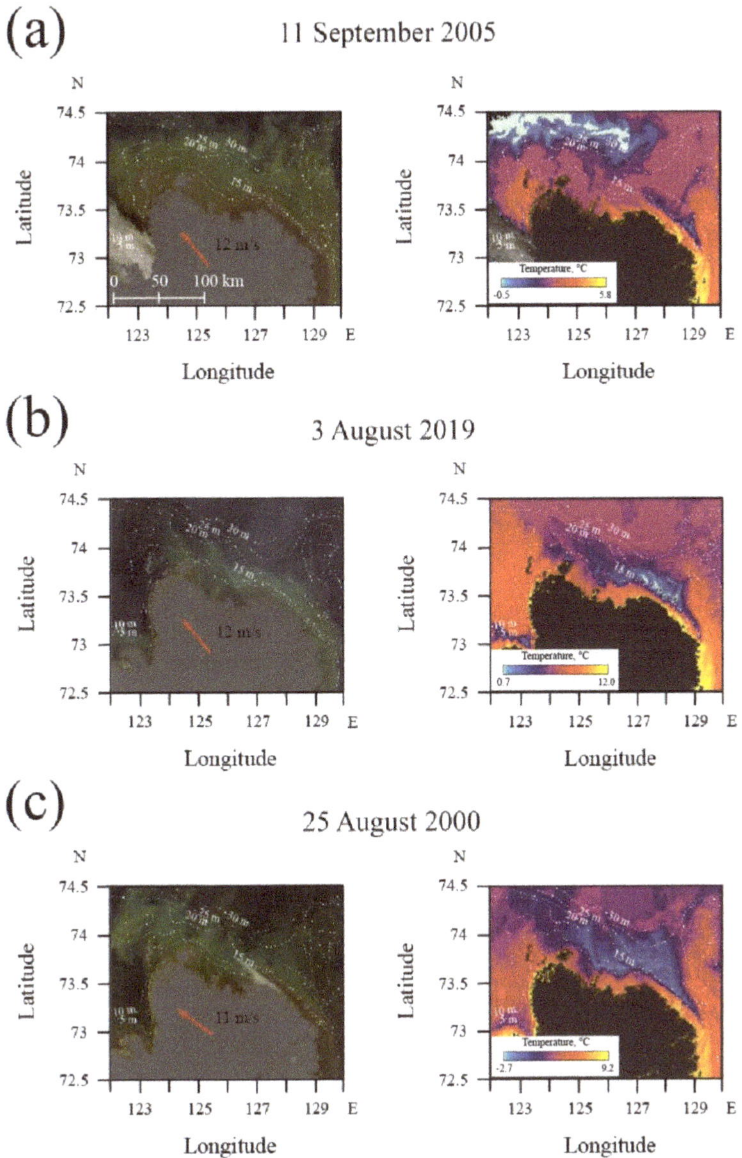

Figure 3. Corrected reflectance (left) and brightness temperature (right) from MODIS Terra and MODIS Aqua satellite images of the area adjacent to the Lena Delta acquired on (**a**) 11 September 2005, (**b**) 3 August 2019, and (**c**) 25 August 2000 illustrating initial (**a**), middle (**b**), and well-developed (**c**) stages of formation of upwelling events in response to wind forcing (arrows).

The surface turbidity structure of the study region during the cold and turbid events was more complex than the surface temperature structure. Surface turbidity was elevated to the north from the Lena Delta at the area occupied by cold surface water (Figure 2). Elevated turbidity was also registered along the eastern part of the Lena Delta. We associate elevated turbidity to the north of the delta with upwelling events and elevated turbidity along the eastern part of the delta with the Lena plume, due to the following reasons. Around 80–90% of freshwater and sediment discharge of the Lena River inflows to the Laptev Sea from the eastern part of the Lena Delta, while its northern part accounts only for 5–8% [90]. As a result, a large turbid and warm river plume is formed only along the eastern part of the Lena Delta. Therefore, areas of elevated turbidity located to the north from the Lena Delta are not likely to be formed by turbid river discharge.

As was described in Section 1, discharge-induced, erosion-induced, and resuspension-induced turbidity events can have similar sea surface manifestations on optical satellite imagery that hinders detection of their origin. However, these processes can be distinguished using other characteristics of sea water. River plumes generally have different salinity, temperature, concentrations of chlorophyll a and dissolved organic matter, as compared to adjacent sea water [91–94]. In particular, river plumes formed in the Laptev and East-Siberian seas during summer and early autumn are significantly warmer than surrounding sea due to the large temperature difference between river and sea water [17,25,27]. The surface temperature in sea areas influenced by coastal erosion is also greater than in surrounding sea due to the absorption of heat from sunlight by suspended particles in the absence of vertical convection. Sea areas influenced by bottom resuspension, on the other hand, are colder than the surrounding sea during the warm season due to mixing of the warmer surface layer with colder bottom water. As a result, the cold and turbid zones observed to the north of the Lena Delta are caused by bottom resuspension, while warm and turbid zones along the coast of Lena Delta are caused by spreading of turbid river plumes.

As was discussed in Section 1, upwelling events, tides, and wind waves are the three possible processes that induce bottom resuspension and form the considered cold and turbid zone. Tidal circulation is very low in the central part of the Laptev Sea and limitedly affects mixing at the study area [62,65]. Coastal upwelling events are commonly manifested by cold and turbid zones in satellite imagery in many world regions [95–98]. Distributions of sea surface temperature observed during coastal upwelling events show significant dependence on local bathymetry. Shapes of cold surface areas formed by upwelling are consistent with isobaths and cores of cold areas are commonly detached from the sea shore [99]. This is the case of the cold and turbid area observed to the north of the Lena Delta which is stably located between isobaths of 5–10 and 30 m. Therefore, this area is not formed as a result of mixing by wind waves, because this process does not depend on bathymetry and can cause surface mixing over both shallow and deep sea areas.

Wind forcing in the study area obtained from the NCEP/CFSR/CFSv2 wind reanalysis confirms that the cold and turbid zone to the north of the Lena Delta is formed by wind-driven upwelling events. Figure 4 shows wind direction (measured in a clockwise direction from north) and wind speed at the area of formation of cold and turbid events. Daily averaged wind forcing conditions are shown in Figure 4 for the days of satellite observations (filled symbols) and for the preceding days (empty symbols). All cold and turbid events detected on satellite imagery (red squares) occurred either during strong southeast winds or shortly after their secession, i.e., average wind speed exceeded 8 m/s and average wind direction was between 120° and 170° on the day of satellite observation or on the preceding day. In particular, if a filled red square is located outside the black dashed rectangle (indicating the upwelling-favorable conditions) in Figure 4, its corresponding empty red square is located inside the dashed rectangle. Therefore, we consider these events as residual upwelling events, i.e., upward penetration of cold and turbid water that did not dissipate shortly after secession of an upwelling wind. On the other hand, there were no cases when an absence of a cold and turbid zone occurred during (filled blue triangles) or shortly after (empty blue triangles) strong upwelling winds were detected, i.e., all triangles in Figure 4 are located outside the black dashed rectangle. Figure 4 shows an asymmetry in wind conditions with almost absent wind forcing between 135° and 225°. This feature is presumably caused by the dependence of cloud coverage of the considered coastal areas on wind direction. Offshore areas of the Laptev Sea are mostly constantly covered by clouds due to intense evaporation, while the land is often cloud-free. As a result, wind that blows from sea to land induces the transport of clouds from the open sea to coastal areas, which hinders optical satellite observations of the sea surface. As a result, there are almost no wind forcing conditions between 135° and 225° among the relatively small sets of cloud-free satellite images of the considered deltaic area.

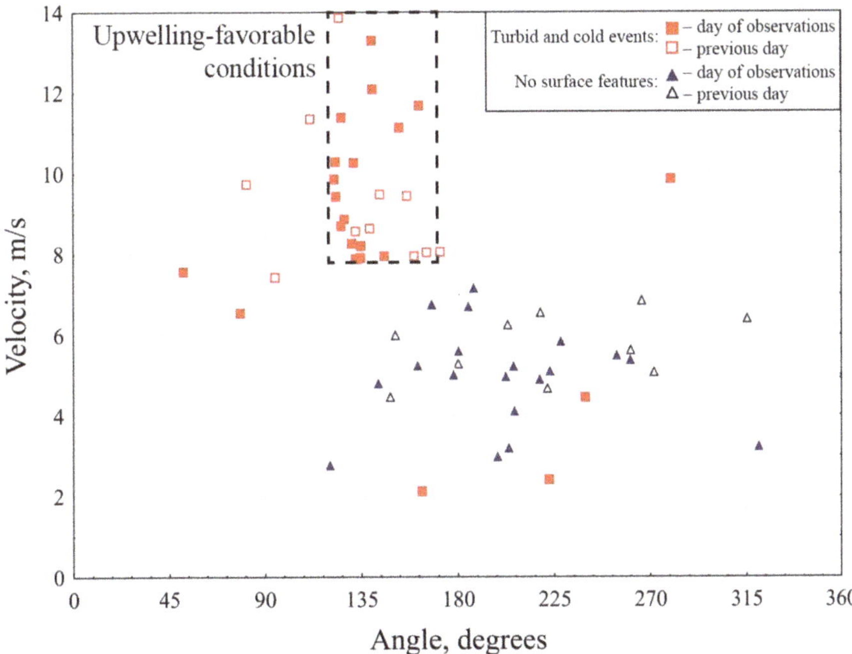

Figure 4. Wind forcing conditions at the Lena Delta region during periods of presence (red squares) and absence (blue triangles) of cold and turbid events detected on satellite imagery. For each satellite image, averaged wind forcing conditions are shown during the day of satellite observation (filled symbols) and during the preceding day (empty symbols). The black dashed rectangle indicates upwelling-favorable wind forcing conditions.

Joint analysis of satellite imagery and wind forcing conditions revealed different stages of formation and dissipation of upwelling events in response to changes of wind forcing regimes (Figures 3 and 5). Early stage of formation of upwelling events characterized by a small area of the cold and turbid zone was detected twice, namely, on 9–11 September 2005 and 2–4 August 2019. South wind forcing was prevailing in the study region on 8–10 September 2005 and changed its direction to southeast (12 m/s) on 11 September 2005. No upwelling manifestations were observed on satellite images acquired on 9 and 10 September 2005. On the next day, 11 September 2011, a relatively small cold area was detected at the isobaths of 5–10 m between the northeastern coast of the Lena Delta and the large shoal, indicating the beginning of formation of the upwelling event (Figure 3a, right panel). Large cold area located northwestward from the Lena Delta on 11 September 2005 was formed by ice melting and does not relate to vertical mixing processes. Satellite imagery acquired on 2–4 August 2019 shows the development of the upwelling event in response to strong southeast wind (7–8 m/s), which started dominating in the study region on 1 August 2019. As on 11 September 2011, the cold and turbid upwelling zone initially was formed at the northwestward coast of the Lena Delta and steadily propagated westward along the isobaths of 5–10 m and then northward towards the isobath of 30 m. The upwelling area steadily increased from 2000 km^2 on 2 August to 4500 km^2 on 4 August (Figure 3b).

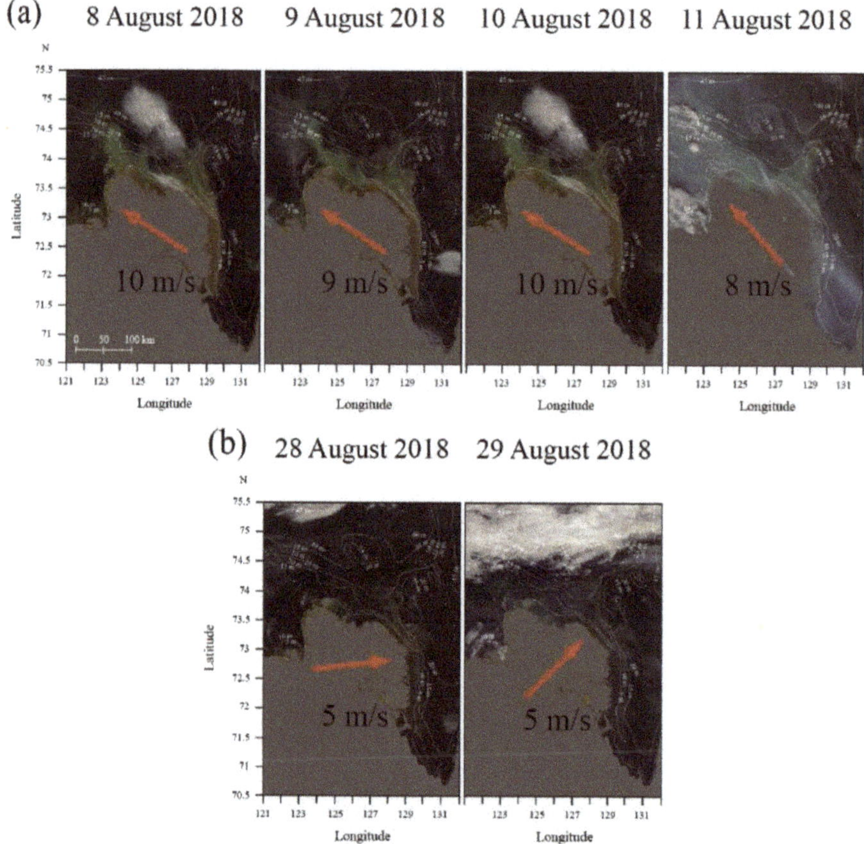

Figure 5. Corrected reflectance from MODIS Terra and MODIS Aqua satellite images of the area adjacent to the Lena Delta acquired on (**a**) 8–11 August 2018, and (**b**) 28–29 August 2018 and wind forcing (arrows) during upwelling (**a**) and non-upwelling (**b**) events.

Well-developed upwelling events that resulted in formation of a cold and turbid zone up to the isobath of 30 m were registered after 4–5 days of upwelling winds. In particular, this case was observed on 25 August 2000 after 4 days of strong southeasterly wind (7–11 m/s) (Figure 3c). After the development of an upwelling event, the cold and turbid zone remained stable and did not spread offshore. Satellite images acquired on 5, 6, 8, 9, 10, and 11 August 2018 during upwelling wind forcing showed that the area of the fully developed upwelling zone did not change (Figure 5a). However, after secession of upwelling wind, this cold and turbid area dissipated and was not observed on satellite imagery acquired on 28–29 August 2018 (Figure 5b). Steady dissipation of the cold and turbid area was also registered at the end of August 2015. A week of strong easterly winds on 15–23 August 2015 caused formation of an upwelling event, whereby surface manifestation bounded by a distinct frontal zone was observed on satellite imagery acquired on 24 August 2015. After secession of upwelling winds on 24 August 2015, sharp temperature and turbidity gradients between the upwelling zone and the adjacent sea steadily dissipated. A satellite image of the study area acquired on 28 August

2015 after 4 days of non-upwelling winds revealed that surface turbidity at the upwelling area had significantly decreased, however, remained relatively high, as compared to the adjacent sea.

3.2. Coastal Upwelling near the Indigirka and Kolyma Deltas in the East-Siberian Sea

Cold and turbid events similar to those observed to the north from the Lena Delta in the Laptev Sea were regularly registered near the large Indigirka and Kolyma deltas in the East-Siberian Sea (Figure 6). We analyzed all MODIS Terra and MODIS Aqua satellite images of the study region taken in 2000–2019 during July–October when the southern part of the East-Siberian Sea was free of ice. We detected 40 and 62 periods when the areas adjacent to the Indigirka and Kolyma deltas, respectively, were free of clouds and the structure of surface turbidity and temperature could be identified.

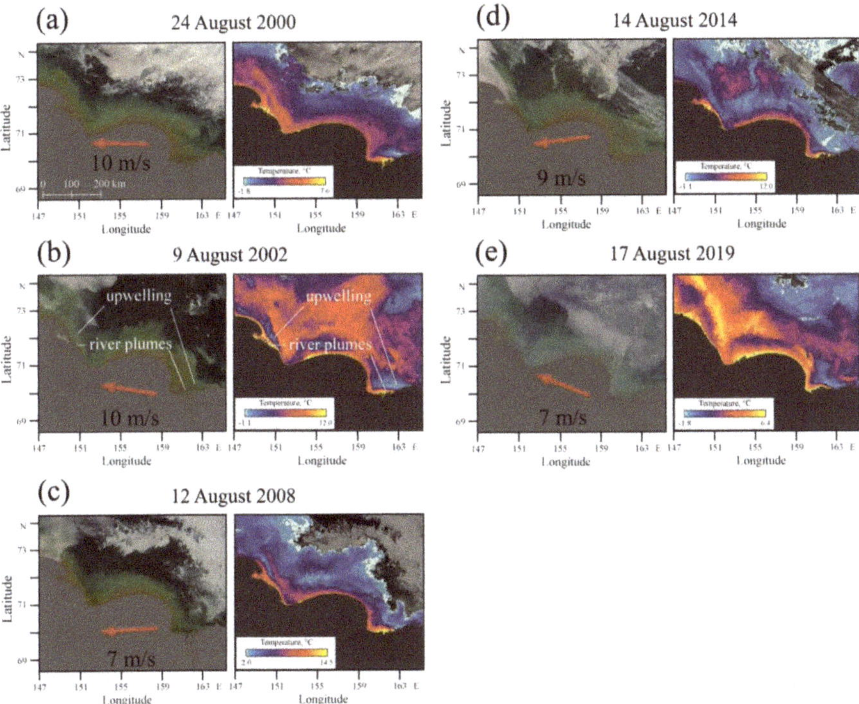

Figure 6. Corrected reflectance (left) and brightness temperature (right) from MODIS Terra and MODIS Aqua satellite images of the East-Siberian Sea acquired on (**a**) 24 August 2000, (**b**) 9 August 2002, (**c**) 12 August 2008, (**d**) 14 August 2014 and (**e**) 17 August 2019, indicating location of upwelling events to the north of the Indigirka and Kolyma deltas induced by wind forcing (arrows) and manifested by elevated sea surface turbidity and reduced sea surface temperature. Surface manifestations of upwelling events and river plumes are indicated at panel (**b**).

Similarly to upwelling events near the Lena Delta, the periods of formation of the cold and turbid area near the Indigirka and Kolyma deltas show very good agreement with the periods of upwelling-favorable wind forcing (Figures 7 and 8). The reanalysis wind data reveals that a strong (>6 m/s) easterly and southeasterly wind (100°–160°) for the Indigirka Delta (Figure 7) and strong (>7 m/s) easterly wind (60°–120°) for the Kolyma Delta (Figure 8) were dominating local atmospheric circulation several days before and/or during all periods when the cold and turbid zones were observed on satellite imagery. Black dashed rectangles in Figures 7 and 8 indicate the related upwelling-favorable conditions near the Indigirka and Kolyma deltas. For all cold and turbid cases, the day of observation (filled red square) and/or the preceding day (empty red square) is located inside these dashed rectangles. On the other hand, the direction of the prevailing wind was different or its velocity was low during all periods when no cold and turbid areas were detected, i.e., all triangles in Figures 7 and 8 are located outside the dashed rectangles. Therefore, we presume that the cold and turbid areas observed to the north of the Indigirka and Kolyma deltas are surface manifestations of wind-driven upwelling events. Similarly to the Lena Delta region, we observe asymmetry in wind forcing conditions with almost absent wind forcing between 135° and 225°.

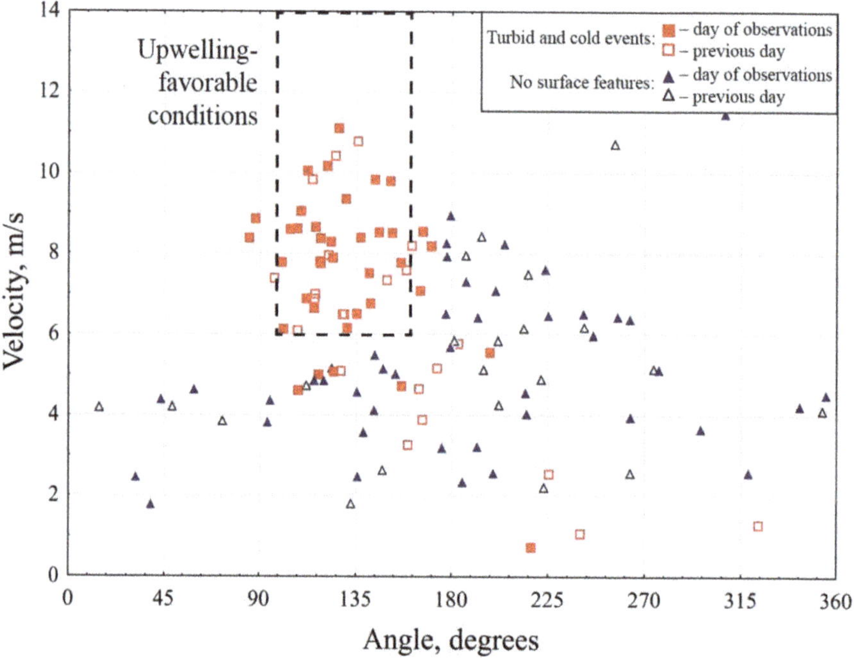

Figure 7. Wind forcing conditions at the Indigirka Delta region during periods of presence (red squares) and absence (blue triangles) of cold and turbid events detected on satellite imagery. For each satellite image, averaged wind forcing conditions are shown during the day of satellite observation (filled symbols) and during the preceding day (empty symbols). The black dashed rectangle indicates upwelling-favorable wind forcing conditions.

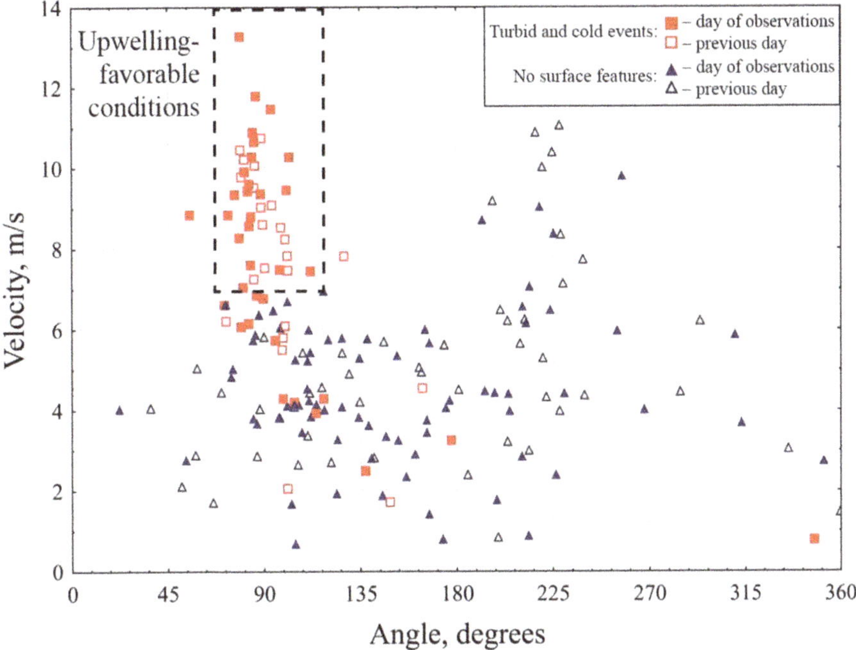

Figure 8. Wind forcing conditions at the Kolyma Delta region during periods of presence (red squares) and absence (blue triangles) of cold and turbid events detected on satellite imagery. For each satellite image, averaged wind forcing conditions are shown during the day of satellite observation (filled symbols) and during the preceding day (empty symbols). The black dashed rectangle indicates upwelling-favorable wind forcing conditions.

Upwelling areas in the East-Siberian Sea occupied a large part of the coastal sea adjacent to the Indigirka (3000–6000 km^2) and Kolyma (5000–9000 km^2) deltas (Figures 6, 9 and 10). Their southern borders are stretched along the northern coasts of the Indigirka and Kolyma deltas. Upwelling events were observed in 21 cases of the 40 considered periods near the Indigirka Delta and in 24 cases of the 62 considered periods near the Kolyma Delta. Similarly to upwelling events near the Lena Delta, we detected the process of development of upwelling events near the Indigirka and Kolyma deltas in response to changes of wind forcing regimes in August 2002 (Figures 9a and 10a), August 2010 (Figure 9b), and August 2014 (Figure 10b). Moderate (2–5 m/s) southeasterly wind forcing was prevailing in the study region on 3–5 August 2002 and its velocity increased to 7–8 m/s on 6 August 2002. No upwelling manifestations were observed on satellite images acquired on 4–5 August 2002 in the study area (Figures 9a and 10a). Then the areas of reduced surface temperature and increased surface turbidity were formed at the isobaths of 5–10 m near the Indigirka and Kolyma deltas on 6 August 2002. These areas increased on 7–10 August 2002, whereby their northern borders steadily propagated offshore, indicating the development of coastal upwelling events in response to strong southeasterly wind (9–10 m/s), which dominated in the study region till 11 August 2002. The upwelling area steadily increased to 6000 km^2 near the Indigirka Delta (Figure 9a) and to 9000 km^2 near the Kolyma Delta (Figure 10a).

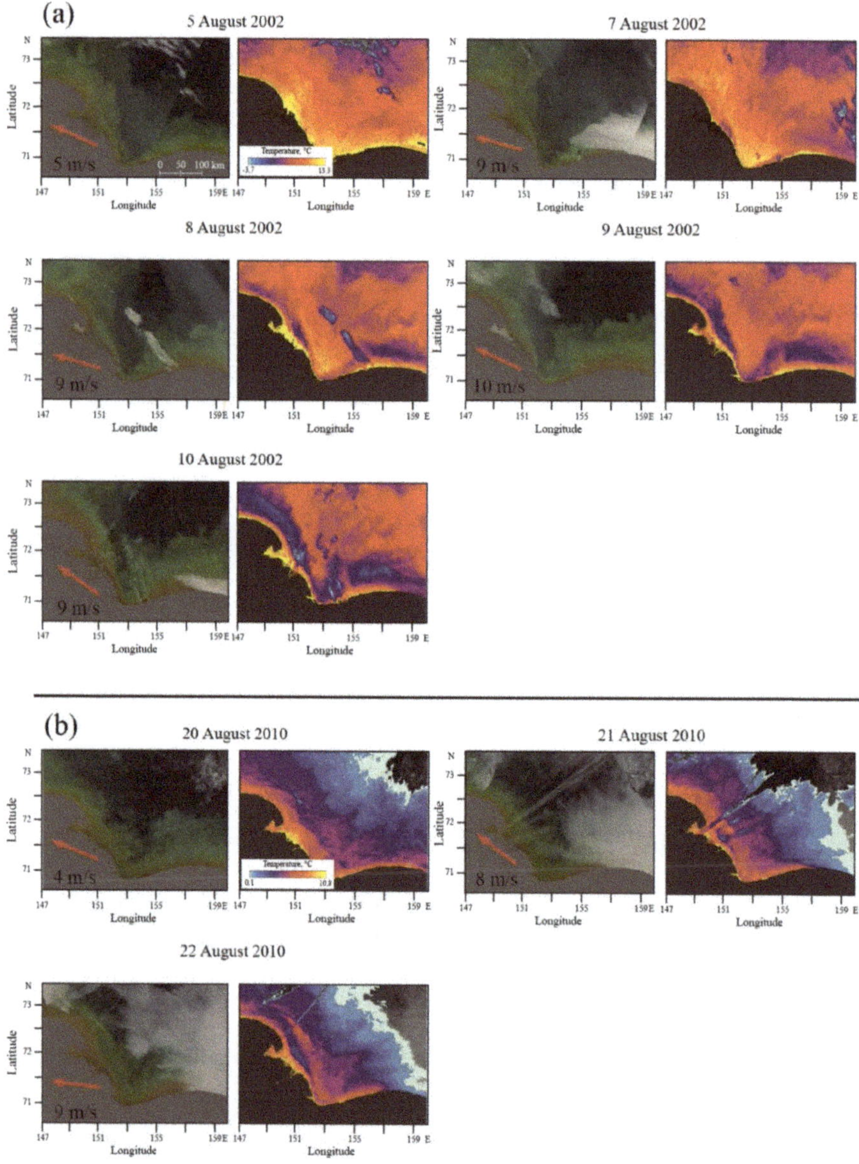

Figure 9. Corrected reflectance (left) and brightness temperature (right) from MODIS Terra and MODIS Aqua satellite images of the area adjacent to the Indigirka Delta acquired on (**a**) 5, 7–10 August 2002 and (**b**) 20–22 August 2010 illustrating formation of the upwelling event in response to wind forcing (arrows).

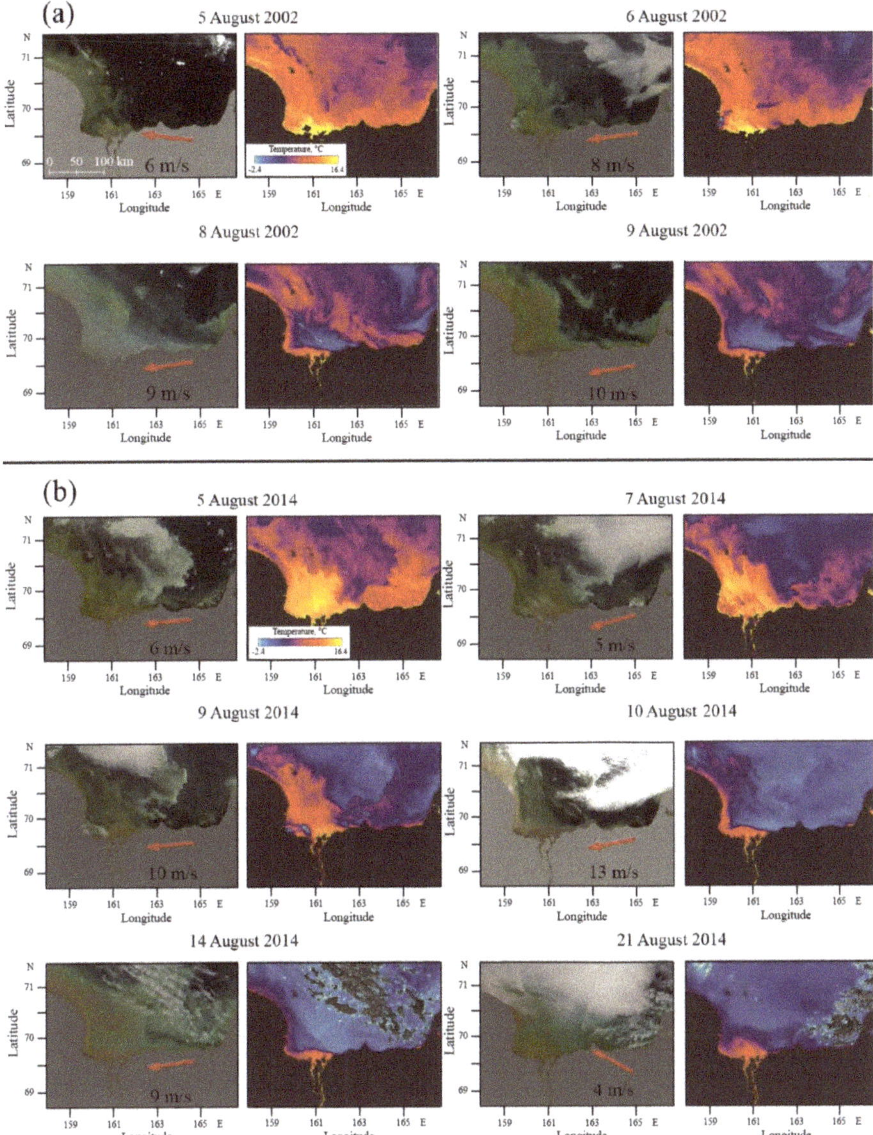

Figure 10. Corrected reflectance (left) and brightness temperature (right) from MODIS Terra and MODIS Aqua satellite images of the area adjacent to the Kolyma Delta acquired on (**a**) 5, 6, 8, 9 August 2002 and (**b**) 5, 7, 9, 10, 14, 21 August 2014 illustrating formation and dissipation of upwelling events in response to wind forcing (arrows).

Development of an upwelling event was also registered on 20–22 August 2010 (Figure 9b). No upwelling was observed on 20 August 2010 during moderate (4 m/s) southeasterly wind forcing. Then on 21 August 2010 upwelling wind increased to 8 m/s and formation of a cold and turbid zone started, which is visible on satellite image. The next day, 22 August 2010, a well-developed upwelling event was observed. Formation and dissipation of upwelling near the Kolyma Delta was observed on 5–24

August 2014 (Figure 10b). Satellite imagery show that warm river plume occupied the area adjacent to the Kolyma Delta on 5–8 August 2014 under moderate (4–6 m/s) wind forcing conditions. Coastal upwelling induced by easterly wind (9–13 m/s) on 9–14 August 2014 resulted in mixing of the Kolyma plume manifested by abrupt decrease of surface temperature at the upwelling area. The warm plume remained only in vicinity of the Kolyma Delta, and its area dramatically decreased from 13,000 to 1500 km^2. Relaxation of upwelling favorable wind (1–5 m/s) on 15–24 August 2014 was accompanied by steady increase of area of the Kolyma plume registered by satellite imagery on 21 and 24 August 2014.

4. Discussion

Upwelling winds near the Lena, Indigirka, and Kolyma deltas cause mixing and intense offshore transport of river plumes over sloping seafloor and upward penetration of cold subjacent sea water (Figure 11). The upwelling sea water induces resuspension of bottom sediments and transports them upward to the surface layer. This process strongly depends on local bathymetry, therefore it occurs only over certain zones of the coastal sea. As a result of detachment of river plumes from river delta and upwelling of subjacent sea water, large saline, cold, and turbid "holes" are formed within the Lena, Indigirka, and Kolyma plumes, which are detected on satellite imagery.

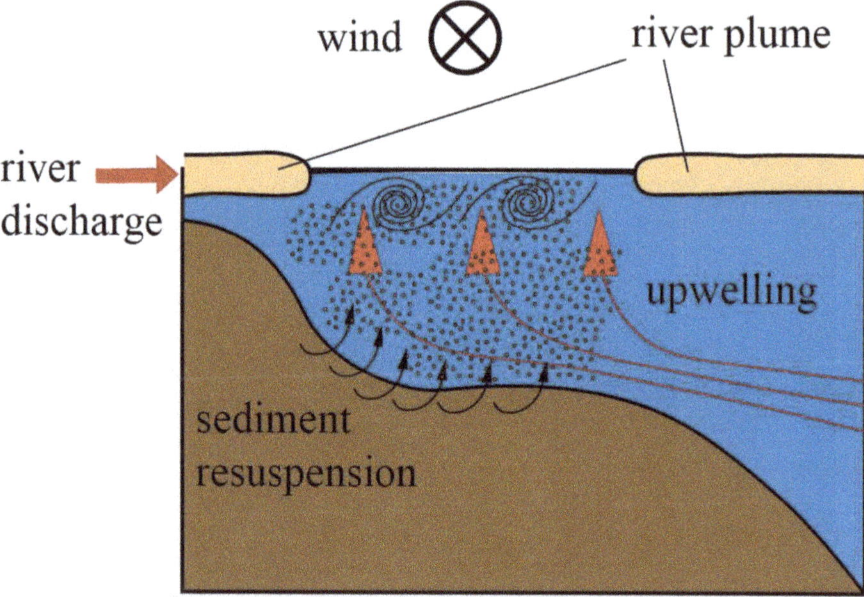

Figure 11. Schematic of formation of a saline, cold, and turbid "hole" within a plume as a result of advection and mixing of a river plume during wind-induced upwelling event.

Based on theory described by [39], we quantified the spatial and dynamical characteristics of the response of the Lena, Indigirka, and Kolyma plumes to upwelling-favorable winds. Given the speed of the upwelling wind, we can calculate three key parameters of this process, namely, the depth of the surface layer entrained by offshore displacement h_s, the time to separate the plume from the coast t_{sep}, and the time to halve the initial salinity anomaly of the plume t_s. The first parameter is determined by the equation

$$h_s = \sqrt{\frac{2Ri\rho_{sea}}{gh_p\Delta\rho_p}}U, \tag{1}$$

where $Ri = \frac{g\Delta\rho_p h_p^3}{\rho_{sea} U^2}$ is the Richardson number, ρ_{sea} is the ambient sea density, g is the gravity acceleration, h_p is the plume depth, $\Delta\rho_p$ is the plume salinity anomaly, $U = \frac{\tau}{\rho_{sea} f}$ is the Ekman transport, f is the Coriolis frequency, and τ is the wind stress. We obtain that if upwelling wind speed exceeds 9 m/s, i.e., U exceeds 1.46 m^2, for the Lena plume ($\rho_{sea} = 1016$ kg/m^3, $h_p = 5$ m, $\Delta\rho_p = 4$ kg/m^3, $f = 1.4 \times 10^{-4}$ 1/s, $Ri \sim 1$ according to [17,27]), then h_s is greater than the plume depth h_p, i.e., the whole depth of the Lena plumes is entrained into offshore displacement during an upwelling event. This theoretical estimation of the threshold value for wind speed (9 m/s) is in a good accordance with the threshold value (8 m/s) obtained from analysis of satellite imagery and wind reanalysis described in Section 3.1. Similar assessment of the upwelling wind threshold value for the Indigirka and Kolyma plumes ($h_p = 3$ m, $\Delta\rho_p = 4$ kg/m^3 according to [17,25]) is equal to 7 m/s, which is also consistent with the threshold values (6 m/s for the Indigirka plume and 7 m/s for the Kolyma plume) reconstructed from satellite imagery and wind reanalysis.

If wind speed exceeds the threshold value, a plume separates from the coast during several hours ($t_{sep} \sim 1/f = 1.4 \times 10^4$ s ~ 4 h) and halves its initial salinity anomaly during the time period quantified by the following equation:

$$t_s = \frac{2A_P}{\sqrt{Ri} U},$$

(2)

where $A_p = W_p \times h_p/2$ is the initial cross-sectional area of the plume and W_p is the initial cross-shore extent of the plume. For the considered plumes, we set $W_p \sim 5 \times 10^4$–10^5 m and obtain $t_s = 1.7 \times 10^5$–3.4×10^5 s ~ 2–4 days for the Lena plume and $t_s = 1.5 \times 10^5$–3×10^5 s ~ 1.5–3.5 days for the Indigirka and Kolyma plumes. As a result, several days of strong upwelling wind are estimated to induce northward offshore displacement of these plumes and halve their salinity anomalies due to intense mixing with subjacent sea. Several days of strong upwelling wind cause formation of $W_p/2 = 25$–50 km wide areas of saline ambient sea water between the plumes and the related deltas that is consistent with satellite observations of the study area (Figures 2 and 6). On the other hand, if wind speeds are smaller than the threshold values, separation of the plumes from the coast occurs after several days or more from the onset of upwelling wind. In this case, mixing of the plumes with ambient sea has low intensity; salinity anomalies of the plumes decrease slowly and halve only after several weeks of upwelling winds [39].

In Section 3 we described ranges of wind speed and wind direction that induce coastal upwelling events at the study regions visible on satellite imagery. We applied these ranges to wind reanalysis and identified periods of wind forcing favorable for formation of upwelling events near the Lena, Indigirka, and Kolyma deltas during the ice-free seasons of 1979–2019 (Figure 12). The average annual duration of upwelling events near the Lena, Indigirka, and Kolyma deltas during this period was equal to 5, 10, and 14 days per year, respectively. The frequencies of upwelling events detected on satellite imagery in the study areas are overestimated by several times, as compared to the obtained average frequencies reconstructed from wind reanalysis. This large bias is caused by, first, detection of residual upwelling events on satellite images that remain during several days after secession of upwelling wind and, second, by almost complete absence of cloud-free satellite imagery during non-upwelling northernly winds described in Section 3.

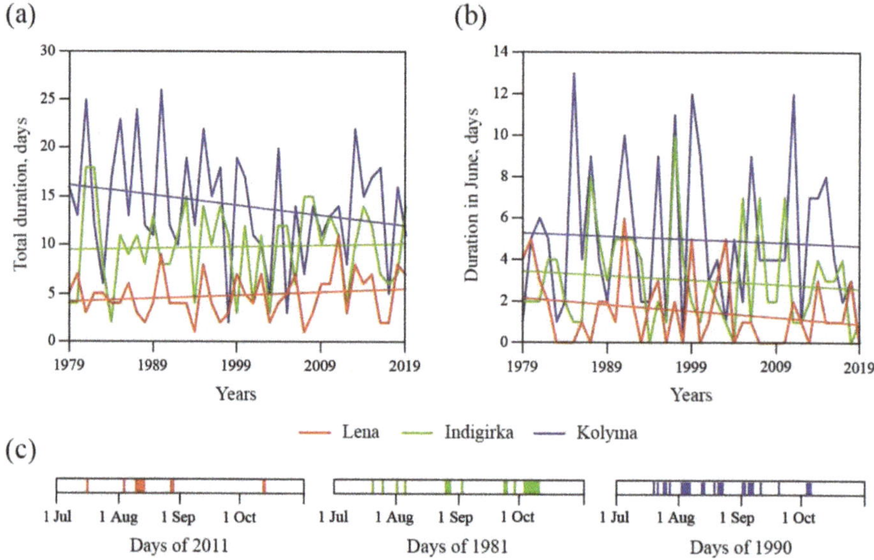

Figure 12. (**a**) The total annual duration and (**b**) duration in July of upwelling events near the Lena (red), Indigirka (green), and Kolyma (blue) deltas in 1979–2019. (**c**) Distributions of upwelling periods in July–October in 2011 near the Lena Delta (left), in 1981 near the Indigirka Delta (center), and in 1990 near the Kolyma Delta (right).

The Kolyma River discharge exhibits the longest upwelling-induced mixing, while influence of upwelling events on the Lena River discharge is the smallest among the considered rivers. However, the annual duration of upwelling events near the Kolyma Delta showed a strong negative trend decreasing by 25% from 1979 to 2019 (Figure 12a). The same characteristic for the Lena and Indigirka regions, in contrast, was increasing, albeit less dramatically than at the Kolyma region. The observed trends could be caused by the influence of the ongoing climate change on atmospheric circulation in the Arctic [100–102] and, therefore, on duration of upwelling winds in the study regions.

The annual duration of upwelling events showed substantial inter-annual variability caused by variability of local atmospheric circulation (Figure 12a). It varied from 2–3 days at all study regions to 11 days near the Lena Delta in 2011, 18 days near the Indigirka Delta in 1981 and 1982, and 26 days near the Kolyma Delta in 1990. Therefore, during certain years the wind-induced upwelling events and the related periods of intense mixing of the Lena, Indigirka, and Kolyma plumes account for up to 12%, 19%, and 28% of ice-free periods, respectively. As a result, the total duration of the upwelling periods, which is negligible on an annual scale, is much more significant during certain weeks and months. In particular, the longest registered durations of upwelling events during individual months are equal to 8, 10, and 14 days for the Lena, Indigirka, and Kolyma regions, respectively, i.e., upwelling events occurred during quarter to half of these months.

The diversity of duration of upwelling events in different years and months is illustrated by their durations in July in 1979–2019 (Figure 12b) and by uneven distributions of upwelling periods during the years with their maximal total duration, namely, 2011 for the Lena Delta region, 1981 for the Indigirka Delta region, and 1990 for the Kolyma Delta region (Figure 12c). In order to quantify the inter-annual variability of influence of upwelling events on mixing of freshwater discharge with sea water we analyzed the inter-annual variability of their durations in July (Figure 12b). Freshwater runoff from the Lena, Indigirka, and Kolyma rivers during the end of June and July provides approximately 60% of their total annual discharge and induces melting of sea ice at the areas adjacent to the river deltas [17]. As a result, long-term upwelling events in July can significantly increase mixing of the river plumes with the subjacent saline sea and strongly affect the structure and dynamics of the related river plumes. Indeed, in certain years upwelling events occurred over 5–6 days in July near the Lena Delta, 7–10 days near the Indigirka Delta, and 10–13 days near the Kolyma Delta. On the contrary, upwelling events and, therefore, upwelling-induced mixing, were completely absent in July during certain years for all three regions. In particular, upwelling events occurred during 0 or 1 day in July near the Lena Delta in 18 out of 41 considered years. As a result, the Lena discharge exhibited negligible upwelling-induced mixing near the delta in July in approximately half of the years during 1979–2019. The reconstructed durations of upwelling events in July showed similar slight negative trends in 1979–2019 at all three considered regions.

5. Conclusions

Satellite observations reveal upwelling events that regularly occur during ice-free seasons in the areas adjacent to the Lena, Indigirka, and Kolyma deltas in the Laptev and East-Siberian seas. These areas are manifested by decreased temperature and increased turbidity, as compared to the surrounding sea. Based on meteorological and satellite data, we estimated temporal characteristics of formation and dissipation of these upwelling events in response to variability of wind forcing. Surface manifestations of upwelling events occur after less than 1 day of strong upwelling winds at all three considered regions. Upwelling near the Lena Delta is fully developed and occupied an area of 15,000–17,000 km^2 after 4–5 days of strong upwelling-favorable southeasterly winds. Fully developed upwelling events near the Indigirka and Kolyma deltas are formed after 3–4 days of strong upwelling-favorable easterly and southeasterly winds; their areas are 5000–6000 and 8000–9000 km^2, respectively. Upwelling areas remain stable until secession of upwelling wind forcing and then steadily dissipate after several days of non-upwelling winds.

The importance of these upwelling events consists in their location near the large river deltas which provide the majority of freshwater discharge to the Laptev (70%) and East-Siberian (75%) seas. Upwelling events induce very intense advection and vertical mixing of freshened surface layer with subjacent saline sea near freshwater sources, as compared to mixing caused by wind-induced shear stress. As a result, the Lena, Indigirka, and Kolyma river plumes are significantly transformed and diluted near their sources during upwelling-favorable wind forcing periods. Frequency and duration of upwelling events govern the structure and dynamical characteristics of the large river plumes, which spread from the river deltas over the upwelling areas to the open sea. Therefore, despite their relatively small areas, upwellings can strongly influence transport and transformation of freshwater discharge over wide areas in the Laptev and East-Siberian seas.

Using NCEP/CFSR/CFSv2 wind reanalysis we reconstructed periods of upwelling events during ice-free seasons at the study areas in 1979–2019. Total annual duration of upwelling events shows large inter-annual variability from negligible (2–3 days) to significant, namely, 12% duration of ice-free periods near the Lena Delta, 19% for the Indigirka Delta, and 28% for the Kolyma Delta. Moreover, upwelling events are unevenly distributed within individual years. In particular, they can last for a quarter to a half of certain months followed by long periods of non-upwelling wind forcing. The most frequent upwelling events among the considered areas are observed near the Kolyma Delta, followed by the Indigirka Delta, and the Lena Delta. As a result, the Kolyma River discharge exhibits the strongest upwelling-induced mixing, however, with strong negative trend registered in 1979–2019. Durations of upwelling events near the Lena and Indigirka deltas, on the other hand, have slight positive trends that decrease their difference with duration of upwelling events near the Kolyma Delta. The revealed trends are presumably caused by long-term changes in atmospheric circulation in the study region induced by the ongoing climate change in the Arctic. Climate change also causes increase of river discharge and temperature of river water, as well as decrease of duration of ice coverage in the coastal areas. Influence of these complex processes on frequency and duration of upwelling events and intensity of upwelling-induced mixing near the Lena, Indigirka, and Kolyma deltas requires specific research and is a subject of future work.

Coastal upwelling events reported in this study can strongly affect salinity and stratification of the surface layer during ice-free periods and, therefore, influence variability of ice coverage in the Laptev and East-Siberian seas. In particular, enhanced duration and intensity of upwelling-induced mixing activity near the Lena, Indigirka, and Kolyma deltas can increase salinity of the related river plumes and, therefore, decelerate ice formation in the Laptev and East-Siberian seas, as was revealed for the other Arctic seas [103–105]. The considered upwelling events can also strongly influence primary productivity and local food webs. Upwelling causes upward penetration of nutrient-rich sea water [35,43–47,106], which is especially important for nutrient-poor areas at the shelf of the Laptev and East-Siberian seas where vertical convection is inhibited by strong stratification formed by large continental runoff. In particular, elevated concentrations of nitrates and increased biological productivity were reported in the vicinity of the upwelling area located near the Lena Delta shortly after an upwelling event [32]. Therefore, the results obtained in this study hold promise to provide improved assessments of the fate of freshwater discharge in the Laptev and East-Siberian seas, as well as its impact on local physical, biological, and geochemical processes. However, a detailed study of the influence of wind-driven coastal upwelling events on the structure and dynamics of the freshened surface layers in these seas requires specific in situ measurements during upwelling and non-upwelling events, as well as numerical modelling, and is within the scope of future work.

Author Contributions: Conceptualization, A.O.; formal analysis, A.O.; investigation, A.O., K.S. and S.M.; writing, A.O., K.S. and S.M. All authors have read and agreed to the published version of the manuscript.

Funding: This research was funded by the Ministry of Science and Higher Education of Russia, theme 0149-2019-0003 (collecting of wind reanalysis data); the Russian Foundation for Basic Research, research projects 18-05-60069 (collecting of satellite data), 20-35-70039 (analysis of satellite data), and 18-05-00019 (study of river plumes); the Russian Science Foundation, research project 18-17-00089 (analysis of wind reanalysis data); the Grant of the President of the Russian Federation for state support of young Russian scientists—candidates of science, research project MK-98.2020.5 (study of freshwater transport).

Acknowledgments: This research benefited substantially from fruitful discussions with Natalia Tilinina (Shirshov Institute of Oceanology). The authors are grateful to the editor Esther Liu and two anonymous reviewers for their comments and recommendations that served to improve the article.

Conflicts of Interest: The authors declare no conflicts of interest.

References

1. Carmack, E.C. The freshwater budget of the Arctic Ocean: Sources, storage and sinks. In *The Freshwater Budget of the Arctic Ocean*; Lewis, E.L., Jones, E.P., Eds.; Kluwer: Dordrecht, The Netherlands, 2000; pp. 91–126.
2. Dai, A.; Trenberth, K.E. Estimates of freshwater discharge from continents: Latitudinal and seasonal variations. *J. Hydrometeorol.* **2002**, *3*, 660–687. [CrossRef]
3. Aagaard, K.; Carmack, E.C. The Arctic Ocean and climate: A perspective. In *The Polar Oceans and Their Role in Shaping the Global Environment*; Johannessen, O.M., Muench, R.D., Overland, J.E., Eds.; AGU: Washington, DC, USA, 1994; Volume 85, pp. 5–20. [CrossRef]
4. Carmack, E.C. The alpha/beta ocean distinction: A perspective on freshwater fluxes, convection, nutrients and productivity in high-latitude seas. *Deep Sea Res. Part Ii* **2007**, *54*, 2578–2598. [CrossRef]
5. Carmack, E.C.; Winsor, P.; Williams, W. The contiguous panarctic Riverine Coastal DomaIn A unifying concept. *Prog. Oceanogr.* **2015**, *139*, 13–23. [CrossRef]
6. Yamamoto-Kawai, M.; McLaughlin, F.A.; Carmack, E.C.; Nishino, S.; Shimada, K.; Kurita, N. Surface freshening of the Canada Basin, 2003–2007: River runoff versus sea ice meltwater. *J. Geophys. Res.* **2009**, *114*, C00A05. [CrossRef]
7. Tremblay, J.E.; Gagnon, J. The effects of irradiance and nutrient supply on the productivity of Arctic waters: A perspective on climate change. In *Influence of Climate Change on the Changing Arctic and Subarctic Conditions*; Nihoul, C.J., Kostianoy, A.G., Eds.; Springer Science: Berlin, Germany, 2009; pp. 73–92. [CrossRef]

8. Li, W.K.W.; McLaughlin, F.A.; Lovejoy, C.; Carmack, E.C. Smallest algae thrive as the Arctic Ocean freshens. *Science* **2009**, *326*, 539. [CrossRef] [PubMed]

9. McLaughlin, F.A.; Carmack, E.C. Nutricline deepening in the Canada Basin, 2003–2009. *Geophys. Res. Lett.* **2010**, *37*, L24602. [CrossRef]

10. Zavialov, P.O.; Izhitskiy, A.S.; Osadchiev, A.A.; Pelevin, V.V.; Grabovskiy, A.B. The structure of thermohaline and bio-optical fields in the upper layer of the Kara Sea in September 2011. *Oceanology* **2015**, *55*, 461–471. [CrossRef]

11. Nummelin, A.; Ilicak, M.; Li, C.; Smedsrud, L.H. Consequences of future increased Arctic runoff on Arctic Ocean stratification, circulation, and sea ice cover. *J. Geophys. Res. Oceans* **2016**, *121*, 617–637. [CrossRef]

12. Carmack, E.C.M.; Yamamoto-Kawai, T.W.; Haine, S.; Bacon, B.A.; Bluhm, C.; Lique, H.; Melling, I.V.; Polyakov, F.; Straneo, M.L.; Williams, W.J. Freshwater and its role in the Arctic Marine System: Sources, disposition, storage, export, and physical and biogeochemical consequences in the Arctic and global oceans. *J. Geophys. Res. Biogeosci.* **2016**, *121*, 675–717. [CrossRef]

13. Osadchiev, A.A.; Izhitskiy, A.S.; Zavialov, P.O.; Kremenetskiy, V.V.; Polukhin, A.A.; Pelevin, V.V.; Toktamysova, Z.M. Structure of the buoyant plume formed by Ob and Yenisei river discharge in the southern part of the Kara Sea during summer and autumn. *J. Geophys. Res. Oceans* **2017**, *122*, 5916–5935. [CrossRef]

14. Osadchiev, A.A.; Asadulin, E.E.; Miroshnikov, A.Y.; Zavialov, I.B.; Dubinina, E.O.; Belyakova, P.A. Bottom sediments reveal inter-annual variability of interaction between the Ob and Yenisei plumes in the Kara Sea. *Sci. Rep.* **2019**, *9*, 18642. [CrossRef] [PubMed]

15. Gordeev, V.V.; Martin, J.M.; Sidorov, J.S.; Sidorova, M.V. A reassessment of the Eurasian river input of water, sediment, major elements, and nutrients to the Arctic Ocean. *Am. J. Sci.* **1996**, *296*, 664–691. [CrossRef]

16. Holmes, R.M.; McClelland, J.W.; Peterson, B.J.; Shiklomanov, I.A.; Shiklomanov, A.I.; Zhulidov, A.V.; Gorgeev, V.V.; Bobrovitskaya, N.N. A circumpolar perspective on fluvial sediment flux to the Arctic Ocean. *Glob. Biogeochem. Cycles* **2002**, *16*, 1098. [CrossRef]

17. Pavlov, V.K.; Timokhov, L.A.; Baskakov, G.A.; Kulakov, M.Y.; Kurazhov, V.K.; Pavlov, P.V.; Pivovarov, S.V.; Stanovoy, V.V. *Hydrometeorological Regime of the Kara, Laptev, and East-Siberian Seas*; University of Washington: Washington, DC, USA, 1996.

18. Aagaard, K.; Carmack, E.C. The role of sea ice and other fresh water in the Arctic circulation. *J. Geophys. Res.* **1989**, *94*, 14485–14498. [CrossRef]

19. Williams, W.J.; Carmack, E.C. The 'interior' shelves of the Arctic Ocean: Physical oceanographic setting, climatology and effects of sea-ice retreat on cross-shelf exchange. *Prog. Oceanogr.* **2015**, *139*, 24–31. [CrossRef]

20. Haine, T.W.N.; Curry, B.; Gerdes, R.; Hansen, E.; Karcher, M.; Lee, C.; Rudels, B.; Spreen, G.; Steur, L.; Stewart, K.D.; et al. Arctic freshwater export: Status, mechanisms, and prospects. *Glob. Planet. Chang.* **2015**, *125*, 13–35. [CrossRef]

21. Munchow, A.; Weingartner, T.J.; Cooper, L.W. The summer hydrography and surface circulation of the East Siberian Shelf Sea. *J. Phys. Oceanogr.* **1999**, *19*, 2167–2182. [CrossRef]

22. Semiletov, I.; Dudarev, O.; Luchin, V.; Charkin, A.; Shin, K.H.; Tanaka, N. The East Siberian Sea as a transition zone between Pacific-derived waters and Arctic shelf waters. *Geophys. Res. Lett.* **2005**, *32*, L10614. [CrossRef]

23. Dmitrenko, I.; Kirillov, S.; Eicken, H.; Markova, N. Wind-driven summer surface hydrography of the eastern Siberian shelf. *Geophys. Res. Lett.* **2005**, *32*, L14613. [CrossRef]

24. Dudarev, O.V.; Semiletov, I.P.; Charkin, A.N.; Botsul, A.I. Deposition settings on the continental shelf of the East Siberian Sea. *Dokl. Earth Sci.* **2005**, *409*, 1000–1005. [CrossRef]

25. Saveleva, N.I.; Semiletov, I.P.; Pipko, I.I. Impact of synoptic processes and river discharge on the thermohaline structure in the East-Siberian Sea shelf. *Russ. Meteorol. Hydrol.* **2008**, *33*, 240–246. [CrossRef]

26. Dmitrenko, I.A.; Kirillov, S.A.; Krumpen, T.; Makhotin, M.; Abrahamsen, E.P.; Willmes, S.; Bloshkina, E.; Holemann, J.A.; Kassens, H.; Wegner, C. Wind-driven diversion of summer river runoff preconditions the Laptev Sea coastal polynya hydrography: Evidence from summer-to-winter hydrographic records of 2007–2009. *Cont. Shelf Res.* **2010**, *30*, 1656–1664. [CrossRef]

27. Saveleva, N.I.; Salyuk, A.N.; Propp, L.N. Peculiar features of the thermohaline and hydrochemical water structure in the southeastern Laptev Sea. *Oceanology* **2010**, *50*, 869–876. [CrossRef]

28. Hölemann, J.A.; Kirillov, S.; Klagge, T.; Novikhin, A.; Kassens, H.; Timokhov, L. Near-bottom water warming in the Laptev Sea in response to atmospheric and sea-ice conditions in 2007. *Polar Res.* **2011**, *30*, 6425. [CrossRef]

29. Shakhova, N.; Semiletov, I.; Leifer, I.; Sergienko, V.; Salyuk, A.; Kosmach, D.; Chernykh, D.; Stubbs, C.; Nicolsky, D.; Tumskoy, V.; et al. Ebullition and storm-induced methane release from the East Siberian Arctic Shelf. *Nat. Geosci.* **2014**, *7*, 64. [CrossRef]

30. Dudarev, O.V.; Charkin, A.N.; Semiletov, I.P.; Pipko, I.I.; Pugach, S.P.; Chernykh, D.V.; Shakhova, N.E.; Sergienko, V.I. Peculiarities of the present-day morpholithogenesis on the Laptev Sea Shelf: Semenovskaya shoal (Vasema Land). *Dokl. Earth Sci.* **2015**, *462*, 510–516. [CrossRef]

31. Semiletov, I.; Pipko, I.; Gustafsson, O.; Anderson, L.G.; Sergienko, V.; Pugach, S.; Dudarev, O.V.; Charkin, A.N.; Gukov, A.; Broder, L.; et al. Acidification of East Siberian Arctic Shelf waters through addition of freshwater and terrestrial carbon. *Nat. Geosci.* **2016**, *9*, 361. [CrossRef]

32. Sukhanova, I.N.; Filint, M.V.; Georgieva, E.J.; Lange, E.K.; Kravchishina, M.D.; Demidov, A.B.; Nedospasov, A.A.; Polukhin, A.A. The structure of phytoplankton communities in the eastern part of the Laptev Sea. *Oceanology* **2017**, *57*, 75–90. [CrossRef]

33. Braga, F.; Zaggia, L.; Bellafiore, D.; Bresciani, M.; Giardino, C.; Lorenzetti, G.; Maicu, F.; Manzo, C.; Riminucci, F.; Ravaioli, M.; et al. Mapping turbidity patterns in the Po river prodelta using multi-temporal Landsat 8 imagery. *Estuar. Coast. Shelf Sci.* **2017**, *198*, 555–567. [CrossRef]

34. Semiletov, I.P.; Pipko, I.I.; Shakhova, N.E.; Dudarev, O.V.; Pugach, S.P.; Charkin, A.N.; Mcroy, C.P.; Kosmach, D.; Gustafsson, O. Carbon transport by the Lena River from its headwaters to the Arctic Ocean, with emphasis on fluvial input of terrestrial particulate organic carbon vs. carbon transport by coastal erosion. *Biogeoscience* **2011**, *8*, 2407–2426. [CrossRef]

35. Kosyan, R. *The Diversity of Russian Estuaries and Lagoons Exposed to Human Influence*; Springer: Basel, Switzerland, 2016. [CrossRef]

36. Sorokin, Y.I.; Sorokin, P.Y. Plankton and primary production in the Lena River estuary and in the south-eastern Laptev Sea. *Estuar. Coast. Shelf Sci.* **1996**, *43*, 399–418. [CrossRef]

37. Nikanorov, A.M.; Bryzgalo, V.A.; Kosmenko, S.; Reshetnyak, S. The Kolyma River mouth area under present conditions of anthropogenic impact. *Russ. Meteorol. Hydrol.* **2011**, *36*, 549–558. [CrossRef]

38. Kraberg, A.C.; Druzhkova, E.; Heim, B.; Loeder, M.J.G.; Wiltshire, K.H. Phytoplankton community structure in the Lena Delta (Siberia, Russia) in relation to hydrography. *Biogeoscience* **2013**, *10*, 7263–7277. [CrossRef]

39. Geyer, W.R.; Hill, P.S.; Kineke, G.C. The transport, transformation and dispersal of sediment by buoyant coastal flows. *Cont. Shelf Res.* **2004**, *24*, 927–949. [CrossRef]

40. Korotkina, O.A.; Zavialov, P.O.; Osadchiev, A.A. Submesoscale variability of the current and wind fields in the coastal region of Sochi. *Oceanology* **2011**, *51*, 745–754. [CrossRef]

41. Korotkina, O.A.; Zavialov, P.O.; Osadchiev, A.A. Synoptic variability of currents in the coastal waters of Soch. *Oceanology* **2014**, *54*, 545–556. [CrossRef]

42. Korotenko, K.A.; Osadchiev, A.A.; Zavialov, P.O.; Kao, R.C.; Ding, C.F. Effects of bottom topography on dynamics of river discharges in tidal regions: Case study of twin plumes in Taiwan Strait. *Ocean Sci.* **2014**, *10*, 865–879. [CrossRef]

43. Horner-Devine, A.R.; Hetland, R.D.; MacDonald, D.G. Mixing and transport in coastal river plumes. *Annu. Rev. Fluid Mech.* **2015**, *47*, 569–594. [CrossRef]

44. Lee, J.; Liu, J.T.; Hung, C.C.; Lin, S.; Du, X. River plume induced variability of suspended particle characteristics. *Mar. Geol.* **2016**, *380*, 219–230. [CrossRef]

45. Osadchiev, A.A.; Korotenko, K.A.; Zavialov, P.O.; Chiang, W.S.; Liu, C.C. Transport and bottom accumulation of fine river sediments under typhoon conditions and associated submarine landslides: Case study of the Peinan River, Taiwan. *Nat. Hazards Earth Syst. Sci.* **2016**, *16*, 41–54. [CrossRef]

46. Osadchiev, A.A.; Korshenko, E.A. Small river plumes off the north-eastern coast of the Black Sea under average climatic and flooding discharge conditions. *Ocean Sci.* **2017**, *13*, 465–482. [CrossRef]

47. Osadchiev, A.A.; Sedakov, R.O. Spreading dynamics of small river plumes off the northeastern coast of the Black Sea observed by Landsat 8 and Sentinel-2. *Rem. Sens. Environ.* **2019**, *221*, 522–533. [CrossRef]

48. Semiletov, I.P.; Shakhova, N.E.; Sergienko, V.I.; Pipko, I.I.; Dudarev, O.V. On carbon transport and fate in the East Siberian Arctic land–shelf–atmosphere system. *Environ. Res. Lett.* **2012**, *7*, 15201. [CrossRef]

49. Clarke, T.L.; Lesht, B.; Young, R.A.; Swift, D.J.P.; Freeland, G.L. Sediment resuspension by surface-wave action: An examination of possible mechanisms. *Mar. Geol.* **1982**, *49*, 43–59. [CrossRef]

50. De Jorge, V.N.; Van Beusekom, J.E.E. Wind-and tide-induced resuspension of sediment and microphytobenthos from tidal flats in the Ems estuary. *Limnol. Oceanogr.* **1995**, *40*, 776–778. [CrossRef]

51. Joordens, J.C.A.; Souza, A.J.; Visser, A. The influence of tidal straining and wind on suspended matter and phytoplankton distribution in the Rhine outflow region. *Cont. Shelf Res.* **2001**, *21*, 301–325. [CrossRef]

52. Kularatne, S.; Pattiaratchi, C. Turbulent kinetic energy and sediment resuspension due to wave groups. *Cont. Shelf Res.* **2008**, *28*, 726–736. [CrossRef]

53. Carlin, J.A.; Lee, G.H.; Dellapenna, T.M.; Laverty, P. Sediment resuspension by wind, waves, and currents during meteorological frontal passages in a micro-tidal lagoon. *Estuar. Coast. Shelf Sci.* **2016**, *172*, 24–33. [CrossRef]

54. Allen, J.S.; Newberger, P.A.; Federiuk, J. Upwelling circulation on the Oregon continental shelf. Part I: Response to idealized forcing. *J. Phys. Oceanogr.* **1995**, *25*, 1843–1866. [CrossRef]

55. Austin, J.A.; Lentz, S.J. The inner shelf response to wind-driven upwelling and downwelling. *J. Phys. Oceanogr.* **2002**, *32*, 2171–2193. [CrossRef]

56. Gan, J.; Lu, Z.; Dai, M.; Cheung, A.Y.; Liu, H.; Harrison, P. Biological response to intensified upwelling and to a river plume in the northeastern South China Sea: A modeling study. *J. Geophys. Res. Oceans* **2010**, *115*, C09001. [CrossRef]

57. Lentz, S.J.; Fewings, M.R. The wind- and wave-driven inner-shelf circulation. *Annu. Rev. Mar. Sci.* **2012**, *4*, 317–343. [CrossRef] [PubMed]

58. Fong, D.A.; Geyer, W.R. Response of a river plume during an upwelling favorable wind event. *J. Geophys. Res. Oceans* **2001**, *106*, 1067–1084. [CrossRef]

59. García Berdeal, I.; Hickey, B.M.; Kawase, M. Influence of wind stress and ambient flow on a high discharge river plume. *J. Geophys. Res. Oceans* **2002**, *107*, 3130. [CrossRef]

60. Lentz, S.J. The response of buoyant coastal plumes to upwelling-favorable winds. *J. Phys. Oceanogr.* **2004**, *34*, 2458–2469. [CrossRef]

61. Osadchiev, A.A.; Zavialov, P.O. Lagrangian model for surface-advected river plume. *Cont. Shelf Res.* **2013**, *58*, 96–106. [CrossRef]

62. Pimenta, F.M.; Kirwan, A.D., Jr. The response of large outflows to wind forcing. *Cont. Shelf Res.* **2014**, *89*, 24–37. [CrossRef]

63. Houghton, R.W.; Tilburg, C.E.; Garvine, R.W.; Fong, D.A. Delaware River plume response to a strong upwelling-favorable wind event. *Geophys. Res. Lett.* **2004**, *31*, L07302. [CrossRef]

64. Curtis Roegner, G.; Hickey, B.M.; Newton, J.A.; Shanks, A.L.; Armstrong, D.A. Wind-induced plume and bloom intrusions into Willapa Bay, Washington. *Limnol. Oceanogr.* **2002**, *47*, 1033–1042. [CrossRef]

65. Hill, J.K.; Wheeler, P.A. Organic carbon and nitrogen in the northern California current system: Comparison of offshore, river plume, and coastally upwelled waters. *Prog. Oceanogr.* **2002**, *53*, 369–387. [CrossRef]

66. Voss, M.D.; Bombar, N.; Loick, J.W. Dippner Riverine influence on nitrogen fixation in the upwelling region off Vietnam, South China Sea. *Geophys. Res. Lett.* **2006**, *33*, L07604. [CrossRef]

67. Han, A.; Dai, M.; Kao, S.J.; Gan, J.; Li, Q.; Wang, L.; Zhai, W.; Wang, L. Nutrient dynamics and biological consumption in a large continental shelf system under the influence of both a river plume and coastal upwelling. *Limnol. Oceanogr.* **2012**, *57*, 486–502. [CrossRef]

68. Tseng, Y.F.; Lin, J.; Dai, M.; Kao, S.J. Joint effect of freshwater plume and coastal upwelling on phytoplankton growth off the Changjiang River. *Biogeoscience* **2014**, *11*, 409–423. [CrossRef]

69. Pickart, R.S.; Moore, G.W.K.; Torres, D.J.; Fratantoni, P.S.; Goldsmith, R.A.; Yang, J. Upwelling on the continental slope of the Alaskan Beaufort Sea: Storms, ice, and oceanographic response. *J. Geophys. Res.* **2009**, *114*, C00A13. [CrossRef]

70. Tremblay, J.E.; Belanger, S.; Barber, D.G.; Asplin, M.; Martin, J.; Darnis, G.; Fortier, L.; Gratton, Y.; Link, H.; Archambault, P.; et al. Climate forcing multiplies biological productivity in the coastal Arctic Ocean. *Geophys. Res. Lett.* **2011**, *38*, L18604. [CrossRef]

71. Falk-Petersen, S.; Pavlov, V.; Berge, J.; Cottier, F.; Kovacs, K.M.; Lydersen, C. At the rainbow's end: High productivity fueled by winter upwelling along an Arctic shelf. *Polar Biol.* **2015**, *38*, 5–11. [CrossRef]

72. Sevigny, C.; Gratton, Y.; Galbraith, P.S. Frontal structures associated with coastal upwelling and ice-edge subduction events in southern Beaufort Sea during the Canadian Arctic Shelf Exchange Study. *J. Geophys. Res. Oceans* **2015**, *120*, 2523–2539. [CrossRef]

73. Dmitrenko, I.A.; Kirillov, S.A.; Rudels, B.; Babb, D.G.; Myers, P.G.; Stedmon, C.A.; Bendtsen, J.; Ehn, J.K.; Pedersen, L.T.; Rysgaard, S.; et al. Variability of the Pacific-Derived Arctic Water Over the Southeastern Wandel Sea Shelf (Northeast Greenland) in 2015–2016. *J. Geophys. Res. Oceans* **2019**, *124*, 349–373. [CrossRef]

74. Pisareva, M.N.; Pickart, R.S.; Lin, P.; Fratantoni, P.S.; Weingartner, T.J. On the nature of wind-forced upwelling in Barrow Canyon. *Deep Sea Res. Part Ii* **2019**, *162*, 63–78. [CrossRef]

75. Williams, W.J.; Carmack, E.C. Combined effect of wind-forcing and isobath divergence on upwelling at Cape Bathurst, Beaufort Sea. *J. Mar. Res.* **2008**, *66*, 645–663. [CrossRef]

76. Macdonald, R.W.; Yu, Y. The Mackenzie Estuary of the Arctic Ocean. In *Estuaries. The Handbook of Environmental Chemistry*; Wangersky, P.J., Ed.; Springer: Berlin, Germany, 2006; Volume 5H. [CrossRef]

77. Mulligan, R.P.; Perrie, W.; Solomon, S. Dynamics of the Mackenzie River plume on the inner Beaufort shelf during an open water period in summer. *Estuar. Coast. Shelf Sci.* **2010**, *89*, 214–220. [CrossRef]

78. Mulligan, R.P.; Perrie, W. Circulation and structure of the Mackenzie River plume in the coastal Arctic Ocean. *Cont. Shelf Res.* **2019**, *177*, 59–68. [CrossRef]

79. Janout, M.A.; Aksenov, Y.; Holemann, J.A.; Rabe, B.; Schauer, U.; Polyakov, I.V.; Bacon, S.; Coward, A.C.; Karcher, M.; Lenn, Y.D.; et al. Kara Sea freshwater transport through Vilkitsky Strait: Variability, forcing, and further pathways toward the western Arctic Ocean from a model and observations. *J. Geophys. Res. Oceans* **2015**, *120*, 4925–4944. [CrossRef]

80. Weingartner, T.J.; Danielson, S.; Sasaki, Y.; Pavlov, V.; Kulikov, M. The Siberian Coastal Current: A wind and buoyancy forced coastal current. *J. Geophys. Res.* **1999**, *104*, 29697–29713. [CrossRef]

81. Guay, C.K.; Falkner, K.K.; Muench, R.D.; Mensch, M.; Frank, M.; Bayer, R. Wind-driven transport pathways for Eurasian Arctic river discharge. *J. Geophys. Res.* **2001**, *106*, 11469–11480. [CrossRef]

82. Dmitrenko, I.; Kirillov, S.; Tremblay, L.B. The long-term and interannual variability of summer fresh water storage over the eastern Siberian shelf: Implication for climatic change. *J. Geophys. Res.* **2008**, *113*. [CrossRef]

83. Fofonova, V.; Danilov, S.; Androsov, A.; Janout, M.; Bauer, M.; Overduin, P.; Itkin, P.; Wiltshire, K.H. Impact of wind and tides on the Lena River freshwater plume dynamics in the summer sea. *Ocean Dyn.* **2015**, *65*, 951–968. [CrossRef]

84. Kowalik, Z.; Proshutinsky, A.Y. The Arctic Ocean tides. In *The Polar Oceans and Their Role in Shaping the Global Environment*; Johannessen, O.M., Muench, R.D., Overland, J.E., Eds.; AGU: Washington, DC, USA, 1994; Volume 85, pp. 137–158. [CrossRef]

85. Padman, L.; Erofeeva, S.A. Barotropic inverse tidal model for the Arctic Ocean. *Geophys. Res. Lett.* **2004**, *31*, L02303. [CrossRef]

86. Fofonova, V.; Androsov, A.; Danilov, S.; Janout, M.; Sofina, E.; Wiltshire, K. Semidiurnal tides in the Laptev Sea Shelf zone in the summer season. *Cont. Shelf Res.* **2014**, *73*, 119–132. [CrossRef]

87. Bareiss, J.; Gorgen, K. Spatial and temporal variability of sea ice in the Laptev Sea: Analyses and review of satellite passive-microwave data and model results, 1979 to 2002. *Glob. Planet. Chang.* **2005**, *48*, 28–54. [CrossRef]

88. Saha, S.; Moorthi, S.; Pan, H.; Wu, X.; Wang, J.; Nadiga, S.; Tripp, P.; Kistler, R.; Woollen, J.; Behringer, D.; et al. The NCEP Climate Forecast System Reanalysis. *Bull. Am. Meteorol. Soc.* **2010**, *91*, 1015–1057. [CrossRef]

89. Saha, S.; Moorthi, S.; Wu, X.; Wang, J.; Nadiga, S.; Tripp, P.; Behringer, D.; Hou, Y.; Chuang, H.; Iredell, M.; et al. The NCEP Climate Forecast System Version 2. *J. Clim.* **2014**, *27*, 2185–2208. [CrossRef]

90. Fedorova, I.; Chetverova, A.; Bolshiyanov, D.; Makarov, A.; Boike, J.; Heim, B.; Morgenstern, A.; Overduin, P.P.; Wegner, C.; Kashina, V.; et al. Lena Delta hydrology and geochemistry: Long-term hydrological data and recent field observations. *Biogeoscience* **2015**, *12*, 345–363. [CrossRef]

91. Klemas, V. Remote sensing of coastal and ocean currents: An overview. *J. Coast. Res.* **2012**, *28*, 576–586. [CrossRef]

92. Klemas, V. Airborne remote sensing of coastal features and processes: An overview. *J. Coast. Res.* **2013**, *29*, 239–255. [CrossRef]

93. Osadchiev, A.A. A method for quantifying freshwater discharge rates from satellite observations and Lagrangian numerical modeling of river plumes. *Environ. Res. Lett.* **2015**, *10*, 85009. [CrossRef]

94. Osadchiev, A.A. Estimation of river discharge based on remote sensing of a river plume. *Proc. SPIE* **2015**, *9638*, 96380H. [CrossRef]

95. Froidefond, J.M.; Castaing, P.; Jouannea, J.M. Distribution of suspended matter in a coastal upwelling area. Satellite data and in situ measurements. *J. Mar. Syst.* **1996**, *8*, 91–105. [CrossRef]

96. Dimarco, S.F.; Chapman, P.; Nowlin, W.D., Jr. Satellite observations of upwelling on the continental shelf south of Madagascar. *Geophys. Res. Lett.* **2000**, *27*, 3965–3968. [CrossRef]

97. Tang, D.L.; Kawamura, H.; Guan, L. Long-time observation of annual variation of Taiwan Strait upwelling in summer season. *Adv. Space Res.* **2004**, *33*, 307–312. [CrossRef]

98. Dabuleviciene, T.; Kozlov, I.E.; Vaiciute, D.; Dailidiene, I. Remote sensing of coastal upwelling in the south-eastern Baltic Sea: Statistical properties and implications for the coastal environment. *Remote Sens.* **2018**, *10*, 1752. [CrossRef]

99. Esiukova, E.E.; Chubarenko, I.P.; Stont, Z.I. Upwelling or differential cooling? Analysis of satellite SST images of the Southeastern Baltic Sea. *Water Resour.* **2015**, *44*, 69–77. [CrossRef]

100. Ding, Q.; Schweiger, A.; L'Heureux, M.; Battisti, D.S.; Po-Chedley, S.; Johnson, N.C.; Blanchard-Wrigglesworth, E.; Harnos, K.; Zhang, Q.; Eastman, R.; et al. Influence of high-latitude atmospheric circulation changes on summertime Arctic sea ice. *Nat. Clim. Chang.* **2017**, *7*, 289–295. [CrossRef]

101. Screen, J.A.; Bracegirdle, T.J.; Simmonds, I. Polar climate change as manifest in atmospheric circulation. *Curr. Clim. Chang. Rep.* **2018**, *4*, 383–395. [CrossRef] [PubMed]

102. Overland, J.E.; Wang, M.; Box, J.E. An integrated index of recent pan-Arctic climate change. *Environ. Res. Lett.* **2019**, *14*, 35006. [CrossRef]

103. Bauch, D.; Hölemann, J.A.; Nikulina, A.; Wegner, C.; Janout, M.A.; Timokhov, L.A.; Kassens, H. Correlation of river water and local sea-ice melting on the Laptev Sea shelf (Siberian Arctic). *J. Geophys. Res. Oceans* **2013**, *118*, 550–561. [CrossRef]

104. Nghiem, S.V.; Hall, D.K.; Rigor, I.G.; Li, P.; Neumann, G. Effects of Mackenzie River discharge and bathymetry on sea ice in the Beaufort Sea. *Geophys. Res. Lett.* **2014**, *41*, 873–879. [CrossRef]

105. Whitefield, J.; Winsor, P.; Mcclelland, J.; Menemenlis, D. A new river discharge and river temperature climatology data set for the pan-Arctic region. *Ocean Model.* **2015**, *88*, 1–15. [CrossRef]

106. Mathis, J.T.; Pickart, R.S.; Byrne, R.H.; McNeil, C.L.; Moore, G.W.K.; Juranek, L.W.; Liu, X.; Ma, J.; Easley, R.A.; Ellio, M.M.; et al. Storm-induced upwelling of high pCO_2 waters onto the continental shelf of the western Arctic Ocean and implications for carbonate mineral saturation states. *Geophys. Res. Lett.* **2012**, *39*, L07606. [CrossRef]

 *remote sensing*

Article

Towards Routine Mapping of Shallow Bathymetry in Environments with Variable Turbidity: Contribution of Sentinel-2A/B Satellites Mission

Isabel Caballero [1,2,*] and Richard P. Stumpf [1]

1 National Centers for Coastal Ocean Science, NOAA National Ocean Service, East West Highway, 1305, Silver Spring, MD 20910, USA; richard.stumpf@noaa.gov
2 Instituto de Ciencias Marinas de Andalucía (ICMAN), Consejo Superior de Investigaciones Científicas (CSIC), Avenida República Saharaui, 11510 Cadiz, Spain
* Correspondence: isabel.caballero@icman.csic.es

Received: 8 January 2020; Accepted: 30 January 2020; Published: 1 February 2020

Abstract: Satellite-Derived Bathymetry (SDB) has significant potential to enhance our knowledge of Earth's coastal regions. However, SDB still has limitations when applied to the turbid, but optically shallow, nearshore regions that encompass large areas of the world's coastal zone. Turbid water produces false shoaling in the imagery, constraining SDB for its routine application. This paper provides a framework that enables us to derive valid SDB over moderately turbid environments by using the high revisit time (5-day) of the Sentinel-2A/B twin mission from the Copernicus programme. The proposed methodology incorporates a robust atmospheric correction, a multi-scene compositing method to reduce the impact of turbidity, and a switching model to improve mapping in shallow water. Two study sites in United States are explored due to their varying water transparency conditions. Our results show that the approach yields accurate SDB, with median errors of under 0.5 m for depths 0–13 m when validated with lidar surveys, errors that favorably compare to uses of SDB in clear water. The approach allows for the semi-automated creation of bathymetric maps at 10 m spatial resolution, with manual intervention potentially limited only to the calibration to the absolute SDB. It also returns turbidity data to indicate areas that may still have residual shoaling bias. Because minimal in-situ information is required, this computationally-efficient technique has the potential for automated implementation, allowing rapid and repeated application in more environments than most existing methods, thereby helping with a range of issues in coastal research, management, and navigation.

Keywords: satellite-derived bathymetry; Copernicus programme; multi-temporal approach; atmospheric correction; lidar; turbidity

---

1. Introduction

Seafloor mapping plays a pivotal role in using and managing the world's oceans in a way that is in accordance with the United Nations Sustainable Development Goal 14 ("Life below water - conserve and sustainably use the oceans, seas and marine resources") that aims to achieve a better and more sustainable future by 2030 [1]. While bathymetric information is key to the world's management of coastal environments, we still have sparse, outdated, and spatially limited coverage. According to the International Hydrographic Organization (IHO) and the Intergovernmental Oceanographic Commission (IOC), the global bathymetry baseline available to date is surprisingly incomplete: an estimated 70% of the world's coastal seafloor remains unmapped, unobserved or inadequately surveyed to modern standards, and is, therefore, poorly understood [2–5]. The coastal shallow water zone can be a challenging environment in which to acquire water depth information using conventional methods, such as the vessel-based multi-beam sonar or the active non-imaging airborne lidar. These surveys are constrained by access, logistics and extremely high deployment cost. It is estimated that multibeam echo sounding (at best resolution) would take more than 200 ship-years and billions of dollars to complete a swath survey of the seafloor [6]. The problem is substantial; consider that some 50% of the USA territories—as an example—were surveyed by using old hydrographic methods that do not meet today's requirement [7]. In the opinion of the IHO, sea bottom information derived from satellite imagery, widely known as Satellite-Derived-Bathymetry (SDB), should be considered as a potential technology to improve the collection, timeliness, quality, and availability of bathymetric data worldwide. In this regard, IHO has started to evaluate SDB strategies and the IHO S-44 standards are currently under revision. Furthermore, SDB offers a low-cost and non-intrusive suitable solution because no mobilization is required, removing health and safety risk, and any environmental impact.

Approaches to SDB mapping vary on aim and rationale, spatial scale, and source of satellite system information. The concept is based on detection of sunlight reflected from the seafloor, and algorithms that use spectral information from this light to calculate water column depth. Several reviews of the methodologies are available in the literature [8,9]. Although the acquisition of imagery is not a problem today, image processing is more complex than data collected from conventional surveys, as it requires special treatment for a correction of the atmosphere, the air-water interface, and especially the local water characteristics [10]. Whereas SDB has been typically applied over environments with clear water (particularly coral reefs), its broader application has been constrained by water clarity, the most challenging constraint to routine satellite seabed mapping [8]. Even in optically shallow areas, where bottom features are visually detectable in imagery, a plume of suspended matter in the water has a significant and varying impact on the precision of SDB. There have been previous studies inspecting water quality issues on SDB [11–15]. Recent studies have already indicated the potential of multi-scene approaches in order to eliminate noise over clear waters [16–20]. The impact of turbidity on SDB is not random and often appears as a false shoal [21,22]. One solution created a global data set at 1 km for coral reef detection by statistical analysis of the entire Sea-Viewing Wide Field-of-View Sensor (SeaWiFS) ocean color data set [23,24]. However, reliable methodologies for evaluating or correcting the impact of turbidity on SDB are rare [21,22].

Likewise, given the large spatial extent and inaccessibility of many coastal regions, there is a pressing need for an Earth Observation program at appropriate spatial, spectral and temporal scales to fulfil the objective of operational cost-effective coastal monitoring [25,26]. The European Commission (EC) and the European Space Agency (ESA) have recently identified the need for improved Digital Elevation Models (DEMs) and bathymetry in order to develop the Copernicus-mission based coastal monitoring programme [27]. The provision of these services is based on the processing of environmental data collected from satellites called the Sentinels. As such, the potential to generate continuous bathymetry from the Sentinel Constellation has become a topic of increased interest worldwide, with an urgent demand of addressing both challenges and opportunities for implementing SDB within an operational production process [3–5]. During the first International Hydrographic Remote Sensing Workshop organized by the Canadian Hydrographic Service and the IHO in collaboration with the Service Hydrographique et Océanographique de la Marine (SHOM) and the National Oceanic and Atmospheric Administration (NOAA) in Ottawa, Canada, in September 2018 [28], the acceleration of SDB was identified as a core action in hydrography. Considering the advantages of SDB technology for aiding the blue economy, the scientific community, as well as the market and industry sectors, further research must be undertaken to implement strategies, especially in the challenging turbid environments, for adopting SDB as a future global product.

In the present study, we develop a framework for using satellite data to retrieve valid bathymetry in shallow water with moderate and varying turbidity. This approach uses the capabilities of the Sentinel-2 Multi Spectral Imagers (MSI) of the Copernicus programme, which include 10 m spatial resolution in the visible bands, five-day routine revisit, and freely available data. In this context, researchers have hypothesized that a higher temporal resolution would allow the development of guidelines for repeatable shallow water mapping approaches at regional to global scales [9,10,22].

This paper proposes the first multiple-image technique with Sentinel-2A and Sentinel-2B that allows semi-automated SDB estimation in areas of variable turbidity. Although some recent works have already obtained accurate results with Sentinel-2 and similar satellites, they were implemented in regions with transparent waters [18–20,29,30]. We focus on application of the log ratio model [31] because it was designed to support routine mapping in areas with extremely limited calibration data. Two different environments with varying water transparency conditions are examined and the main advantages and challenges of this new method are discussed.

2. Materials and Methods

2.1. Study Areas

The study areas were Saint Joseph Bay, Florida (29.75° N, 85.35° W, Figure 1a) and Cape Lookout, North Carolina (34.62° N, 76.54° W) (Figure 1b). Cape Lookout National Seashore preserves a 90-km-long section of the Southern Outer Banks of North Carolina with three undeveloped barrier islands. St. Joseph Bay is located in Gulf County between Apalachicola and Panama City. The north end of the bay is a relatively narrow opening to the Gulf of Mexico, and is approximately 24 km-long north to south and 10 km-wide at its widest point. The waters of St. Joseph Bay contain the St. Joseph Bay State Buffer Preserve and the St. Joseph Bay Aquatic Preserve.

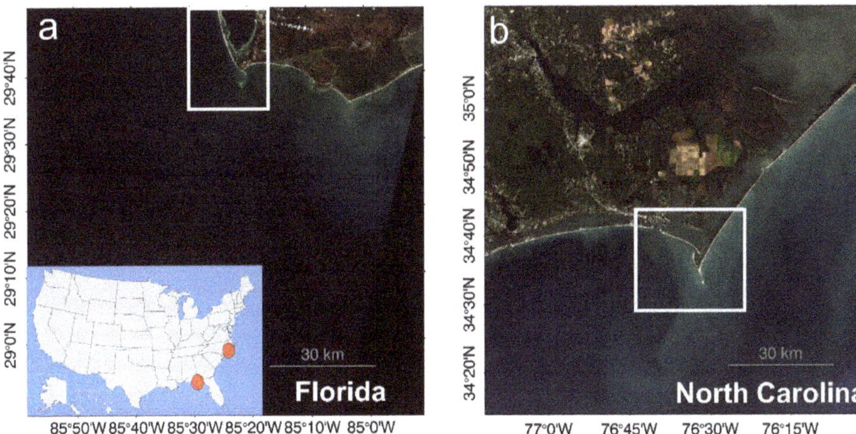

Figure 1. Location of the study regions in the east coast of United States. (**a**) RGB composite of Saint Joseph Bay in Florida from Sentinel-2A tile T16RFT (February 24, 2017), and (**b**) RGB composite of Cape Lookout in North Carolina from Sentinel-2B tile T18SUD (September 28, 2017). The rectangles indicate the Region of Interest (ROI) for bathymetric mapping.

These areas were selected because they have variable turbidity [32–35] and the availability of recent airborne lidar bathymetry (ALB) data for final validation and error analysis. In addition, the National Oceanic and Atmospheric Administration (NOAA) was interested in evaluating hurricane induced depth changes: after Hurricane Matthew (October 2016) in Cape Lookout and before Hurricane Irma (September 2017) in Joseph Bay. These study sites also represent complex microtidal environments with variable bottom types.

2.2. Lidar Data for Validation

NOAA's National Geodetic Survey (NGS) collected Airborne Lidar Bathymetry (hereinafter ALB) for Cape Lookout in December 2016 and for Joseph Bay in November 2016. These point cloud data sets were both collected using the Riegl VQ-880-G sensor, which provided high-resolution bathymetric data in nearshore waters. The Riegl VQ-880-G used a green laser that operates in a circular scan pattern, which could penetrate shallow clear water to the seafloor. The high-density point samples were combined with GPS and other positional data to create precise 3D topobathy elevation models. NGS used coastal elevation data to map the mean high-water shoreline, which is considered the nation's official shoreline. These high-resolution observations at 1 m were selected as reference data set in the two study sites and compared to SDB products. The Mean Lower Low Water (MLLW), standard chart datum was used as the reference. The range of depths within this data set is 0-13 m. ALB data referenced to the MLLW was gridded at the Sentinel-2 resolution (10 m) via arithmetic averaging.

2.3. Sentinel-2A/B Imagery

Sentinel-2A and 2B twin polar-orbiting satellites, developed by the ESA to meet the operational needs of the Copernicus programme, were used. Both Multispectral Instruments (MSI) on-board are now operational: Sentinel-2A was launched on 23 June 2015 and Sentinel-2B followed on 7 March 2017. The radiometric, spectral and spatial characteristics of the bands used in this study are specified in the User Handbook [36]. Sentinel-2 Level-1C products were downloaded from the Sentinel's Scientific Data Hub [37] and images of zone 18 in Cape Lookout (sub-tile SUD) and of zone 16 in Joseph Bay (sub-tile RFT) were used. A temporal examination provided Level-1C Sentinel-2 images were typically geo-located within two pixels of each other (20 m) which is within the stated quality requirements for absolute geo-location [38]. The study period was selected based on lidar collection and the passage of hurricanes given that intense resuspension and currents may have modified shallow seabed morphology, confounding comparison with the lidar survey. For Cape Lookout, a one-year data series was evaluated from January to December 2017. A total number of 59 Sentinel-2A and 2B scenes were available, but only 15 optimal final images (25%) were further processed due to intense cloud coverage and sunglint effects (Table 1). In St. Joseph Bay, only seven usable Sentinel-2A scenes were available owing to cloud cover and surface reflectance (glint) for a study period of December 2016–March 2017. St. Joseph Bay is on the east side of the swath, as a result, surface glint was severe during spring and summer. Furthermore, no images were considered after Hurricane Irma (September 2017). During that period, only the Sentinel-2A satellite was operational, with a less than 10-day revisit over the Florida zone. Matlab R.2016a software and QGIS (version 3.6.0) were used for visualization and processing of satellite data.

Table 1. List of Sentinel-2A and 2B images used in this study for the multi-temporal approach in Cape Lookout (15 scenes) and in Saint Joseph Bay (7 scenes). The percentage of pixels from each scene included in the final pSDBred and pSDBgreen (Equations (1) and (2), Section 2.5) after the multi-scene approach is indicated. The pixels evaluated (N) were only pixels with corresponding lidar data for final validation: N = 271635 in Cape Lookout and N = 678517 in Joseph Bay.

Number	Date	Sensor	pSDBred (%)	pSDBgreen (%)
1	01/23/2017	A	3.1	16.5
2	02/12/2017	A	9.46	0.5
3	02/25/2017	A	8.5	37.33
4	03/04/2017	A	5.28	25
5	03/17/2017	A	14.7	16.21
6	05/03/2017	A	2.9	0.045
7	07/22/2017	A	0.34	0.006
8	09/13/2017	A	5.35	0.005
9	09/20/2017	A	20.17	0.009
10	09/28/2017	B	5.1	0
11	10/18/2017	B	0.21	0.006
12	10/30/2017	A	0.08	0.029
13	11/17/2017	B	2.04	0.12
14	11/27/2017	B	22.7	4
15	12/14/2017	B	0.07	0.24

Table 1. Cont.

Number	Date	Sensor	pSDBred (%)	pSDBgreen (%)
1	12/13/2016	A	4.9	42.5
2	12/16/2016	A	21.8	0.17
3	12/23/2016	A	32.5	22.9
4	01/05/2017	A	17.58	1.9
5	02/14/2017	A	2.25	0.5
6	02/24/2017	A	0.87	2.03
7	03/16/2017	A	20.1	30

2.4. Atmospheric Correction

Sentinel-2 images were processed to Level-2A (L2A) with the ACOLITE processor developed by the Royal Belgian Institute of Natural Sciences (RBINS), which supports free processing, specifically for aquatic applications, of both Landsat-8 and Sentinel-2 [39–42]. ACOLITE products corresponded to Remote Sensing Reflectance (Rrs, 1/sr) in all visible and Near-Infrared (NIR) bands and chlorophyll concentration by the OC3 algorithm (Chl, mg/m^3) resampled to 10 m pixel size. We selected a combination of the NIR (865 nm) and Short Wave Infrared (SWIR) (1600 nm) bands in ACOLITE for the aerosol correction with a user defined epsilon value (maritime aerosol=1). This strategy has been shown to significantly improve the quality of the products by minimizing the influence of NIR/SWIR instrument noise [43]. A spatial filter (median filter 3x3) was conducted on the bands in order to remove some noise and inter-pixel variability. Recent experience with Sentinel-2 imagery indicated spatial filtering enhanced bathymetric products [10,22,44]. In this investigation, the spectral red-edge band at 704 nm (Rrs704) was used as a proxy for turbidity over optically shallow waters [22]. Several researchers have already indicated red-edge bands were appropriate for turbidity or suspended solids monitoring in optically shallow regions [43,45].

2.5. Satellite-Derived Bathymetry Model

In this study, the ratio model of log-transformed bands having different water absorption was applied (Equations (1) and (2)). The model was designed to support routine mapping in clear waters with extremely limited calibration data [31]. The model uses the Remote Sensing Reflectance (Rrs, units of sr^{-1}) of the blue (490 nm), green (560 nm) and red (664 nm) bands for each satellite image corrected for atmospheric effects (Equations (1) and (2)), and the log-transform addresses the exponential decrease in light with depth [31]. In this case, we used the ratio of blue (λ_i) to either green or red (λ_j) bands to produce the SDB (hereinafter called SDBgreen and SDBred, respectively). SDBgreen performs better in deep areas, and SDBred performs better in shallow water [22–24,44]:

$$SDB = m_1 pSDB - m_0 \tag{1}$$

where,

$$pSDB = \frac{\ln(n\,\pi Rrs(\lambda_i))}{\ln(n\,\pi Rrs(\lambda_j))} \tag{2}$$

pSDB is the relative or "pseudo" depth from satellite (dimensionless), SDB is the satellite-derived depth (meters), Rrs is the Remote Sensing Reflectance, m_1 and m_0 are tunable constants to linearly transform the model results to actual depth, π has units of sr, and n=1000 is a fixed constant to assure that both logarithms will be positive under any condition and that a residual non-linearity in the ratio is removed from depths that are retrievable from satellite [31].

2.6. Multi-Scene Approach

Thanks to the 5-day revisit of the Sentinel-2 twin satellites, the mission can offer imagery that would identify transient turbidity features that produce false shoals. Accordingly, in this study, we used a multi-scene approach (see Figure 2). The pSDBgreen and pSDBred were determined, per Equation (2), for all available scenes (Table 1). It was reasonable to assume that the contribution from the temporal variability in water-column turbidity to SDB models was greater than the temporal variability of the seafloor [46]. As turbidity produces a false shoaling of depth, all scenes were compared to identify the maximum pSDB at each pixel from all the scenes, and two resultant composite images were created: one of the maximum pSDBred, and one of the maximum pSDBgreen, with assumption being maximum pSDB values would most likely be without any shoal effect. Composite images of Rrs704 and chlorophyll from the OC3 algorithm were also returned, where the turbidity value at each pixel came from the same scene as the depth for that pixel. Both Rrs704 and chl were surrogates for addressing the false shallowing, not intended to describe actual turbidity or chl concentration.

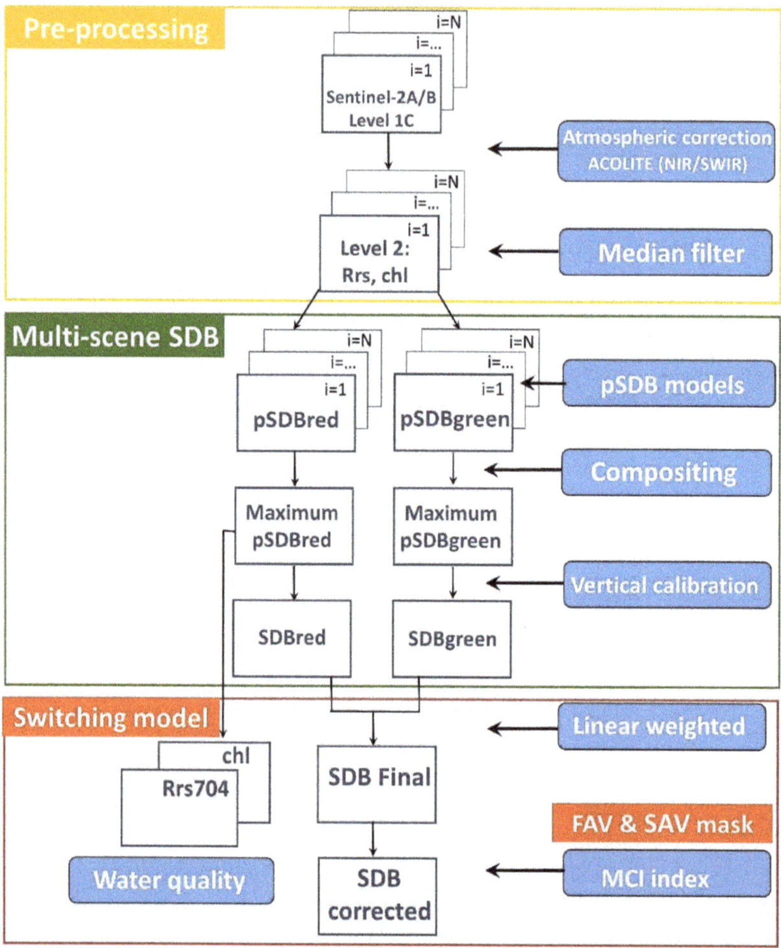

Figure 2. A schematic workflow of processing steps implemented for mapping Satellite-Derived Bathymetry with temporal imagery of Sentinel-2A and 2B satellites in Cape Lookout and Saint Joseph Bay: (1) Pre-processing, (2) Multi-scene SDB approach, (3) Switching model, and Submerged aquatic vegetation (SAV) and floating aquatic vegetation (FAV) masking (only in Joseph Bay).

2.7. Vertical Referencing with Chart Soundings

After generating the final pSDBred and pSDBgreen maps, the parameters m_1 and m_0 (Equation (1)) were tuned by linear regression with about ten control points from charts in each of the study regions and for pSDBred and pSDBgreen separately. Depth measurements (also known, as soundings) from NOAA charts [47] were used to select the control points over Cape Lookout (11545) and Joseph Bay (11389). Using chart points keeps the calibration independent of the lidar used for validation, and corresponds to methods that would be utilized in typical application [31]. The points were chosen as areas of uniform depth that were less likely to have changed over time (regions away from inlets and sand waves). The selected calibration points (pixels) for each of the four composite images were found to originate in different input scenes (3 and 4 different scenes respectively for SDBred and SDBgreen for St- Joseph Bay, and 5 for both Cape Lookout composites). The vertical calibration inherently corrects for the reference data by shifting the depths to the tidal datum, which in the case of USA is the MLLW, through the tuning of the coefficient m_0. Sea level depends on satellite observation time; in this case, both regions are microtidal environments. For this approach, our hypothesis was that the influence of the tide for a pSDB relative to each scene is small compared to the influence of the false shallowing generated by turbidity from multiple satellite images. The validation of this hypothesis and the improvement of accuracy by compensating for tide effects in meso and macrotidal regimes are subjects for future study.

In addition, to test the robustness of the multi-scene approach and the final SDB result in terms of a common calibration for remote unsurveyed locations, we interchanged the coefficients for pSDBred and pSDBgreen between the two regions, thus using the coefficients of Cape Lookout in St. Joseph Bay, and vice versa.

2.8. Switching Model

The final step for a corrected bathymetry mapping (SDB) corresponded to a switching model implementation between SDBred and SDBgreen due to the sensitivity of each model: while SDBred performs better over shallow regions [48], SDBgreen performs better over deeper regions [21,31,44,49]. Moreover, SDBgreen frequently yields severe overestimation in shallow regions [22,44,50], especially with dense seagrass, highly undesirable for navigational purposes. The switch between models used the conditions detailed below:

SDBred < 2 m, SDB=SDBred
SDBred > 2 m and SDBgreen > 3.5 m, SDB=SDBgreen
SDBred >= 2 m and SDBgreen <= 3.5 m, SDB=SDBweighted

SDBweighted was determined by a simple linear weighting calculation to account for a smooth transition (Equation (3)):

$$SDB = \alpha * SDBred + \beta * SDBgreen \tag{3}$$

where the depth weighting (for 3.5 m and 2 m) is determined by:

$$\alpha = \frac{3.5 - SDBred}{3.5 - 2} \text{ and } \beta = 1 - \alpha$$

The final SDB map was generated in each region then compared to the lidar surveys for validation. Assessment of the discrepancy between SDB and lidar used the mean difference as the bias metric, the Median Absolute Error (MedAE) providing the typical total error, and the interquartile range (IQR) as a measure of statistical dispersion. These are robust metrics that do not require an assumption of a Gaussian error distribution.

2.9. Submerged and Floating Aquatic Vegetation Masking

An additional step was carried out in St. Joseph Bay due to the existence of floating aquatic vegetation (FAV) and shallowed submerged aquatic vegetation (SAV). The complex coastal waters of Saint Joseph Bay are characterized by several benthic types, including high density submerged and floating aquatic vegetation at depths shallower than 2 m [51]. In order to correct this issue, we established a masking strategy for the FAV and SAV area. The MCI (Maximum Chlorophyll Index) first designed for the MEdium Resolution Imaging Spectrometer (MERIS), measuring the radiance peak at the red-edge band (709 nm) in water leaving radiance, indicates the presence of a high surface concentration of chlorophyll against a scattering background [52]. This index was formed with three MERIS narrow channels centered near 681, 709 and 754 nm used to define a linear baseline. The MCI has been used to map floating vegetation or benthic [53,54]. In this study, given that Sentinel-2 has three bands (665, 704, and 740 nm) similar to the MERIS bands used for MCI, we utilized Equation (4).

$$\mathrm{MCI} = Rrs704 - Rrs665 + (Rrs665 - Rrs740) * \frac{(Rrs704 - Rrs665)}{(Rrs740 - Rrs665)} \tag{4}$$

FAV and SAV areas were established for the pixels with positive values of MCI (MCI>0). This MCI mask was applied to the final SDB map after the multi-scene approach and switching model in St. Joseph Bay in order to remove possible overestimation for shallow depths where vegetation was present. Therefore, the MCI was not used to map seagrass, but to locate and remove an error source with SDB.

3. Results

3.1. Cape Lookout in North Carolina

Prior to implementing the multi-temporal approach with Sentinel-2A/B images in Cape Lookout (Figure 1b), we estimated bathymetry for several single scenes in order to evaluate the impact of varying turbidity. Figure 3 illustrates two different examples of SDBgreen model for Sentinel-2A on 23 January and 30 October 2017. The heterogeneous impact of water quality on satellite bathymetry is evident in the comparison between Airborne Lidar Bathymetry (hereinafter ALB) and satellite data (Figure 3a,b) and the residual errors (Figure 3c,d). The scatterplots reveal large spread and errors, particularly on 30 October, where an acute underestimation or false shoaling (Figure 3g) and large Median Absolute Error (MedAE = 2.95 m) resulted from the higher turbidity levels that are shown in the Remote Sensing Reflectance at 704 nm (Rrs704), a proxy for turbidity (Figure 3h). These outcomes demonstrated the impact of variable water quality conditions on the accuracy of the predicted SDB. Although the 23 January scene had lower turbidity overall—and might be selected as the optimal scene for analysis—there are still areas of elevated turbidity evident offshore (see the lower right of 3e and 3f). Over complex turbid environments such as Cape Lookout [32,34], using only a single scene for SDB has definite limitations.

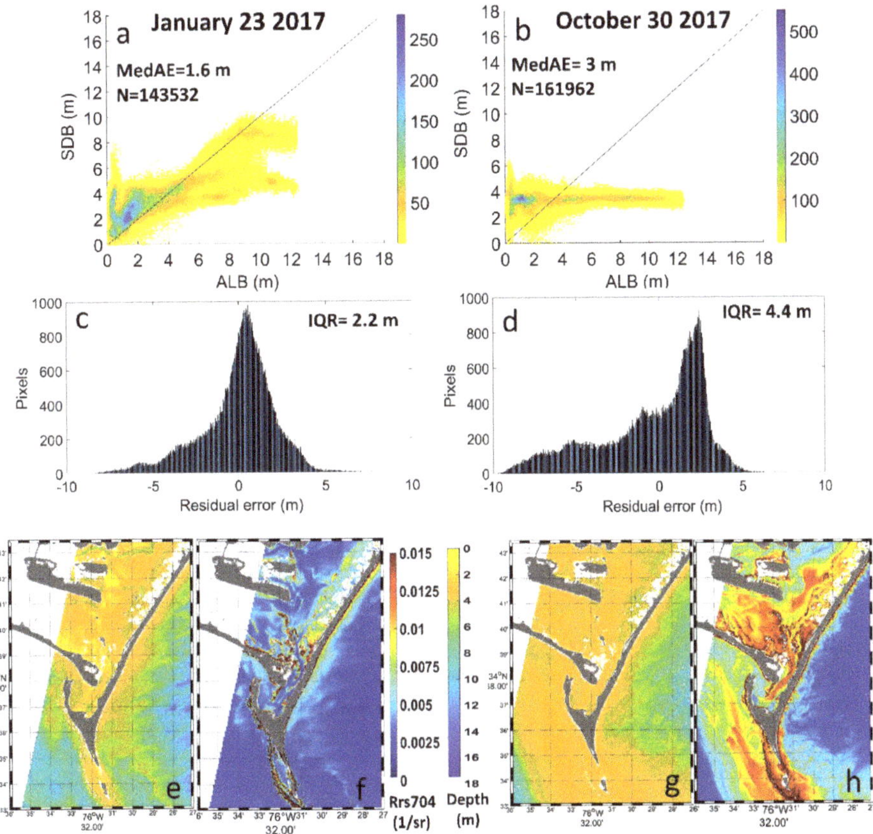

Figure 3. Comparison between lidar (ALB) and Sentinel-2 bathymetry (SDBgreen) for two individual scenes from the Cape Lookout, NC study area. (**a**) scatter frequency plots for 23 January and (**b**) 30 October 2017, the color bars indicate pixel frequency at the point; residual errors (SDBgreen—ALB) for (**c**) 23 January (standard deviation = 2.07 m, percentile 5%/95% = −4.06 m/2.93 m), and (**d**) 30 October (standard deviation = 3.17 m, percentile 5%/95% = −6.88 m/3.16 m), the bins of the histograms correspond to the number of pixels; maps of SDBgreen for (**e**) 23 January and (**g**) 30 October; and maps of the turbidity proxy used in this study, Remote Sensing Reflectance at 704 nm (Rrs704), for (**f**) 23 January and (**h**) 30 October. Gray color represents land mask and white color at the left is the limit of Sentinel-2 tile for each date.

The compositing selected the pixels of the multiple scenes that were slightly affected by turbidity. Compositing multiple scenes to identify the pixels in order to correct for water turbidity substantially improved the accuracy, as can be observed in Figures 4 and 5. The assumption of the approach is based on the submarine terrain remaining slightly unchanged during the study period (1 year) to accumulate a time series of scenes to use in the compositing, while turbidity (and noise: waves, cloud shadows, ships, and bubbles) affect the accuracy of bathymetric inversion. Using the multi-scene pixel compositing approach with the sixteen images (Table 1), and then applying the switching model between SDBred and SDBgreen, gave better performance from shallow to deeper regions. The composite solution scheme efficiently described the nearshore depth with minimum scatter and MedAE = 0.41 m over the range of ALB data (up to 12.5 m) compared to the individual scenes (Figure 4a,b for composite, and Figure 4c,d for best single scene).

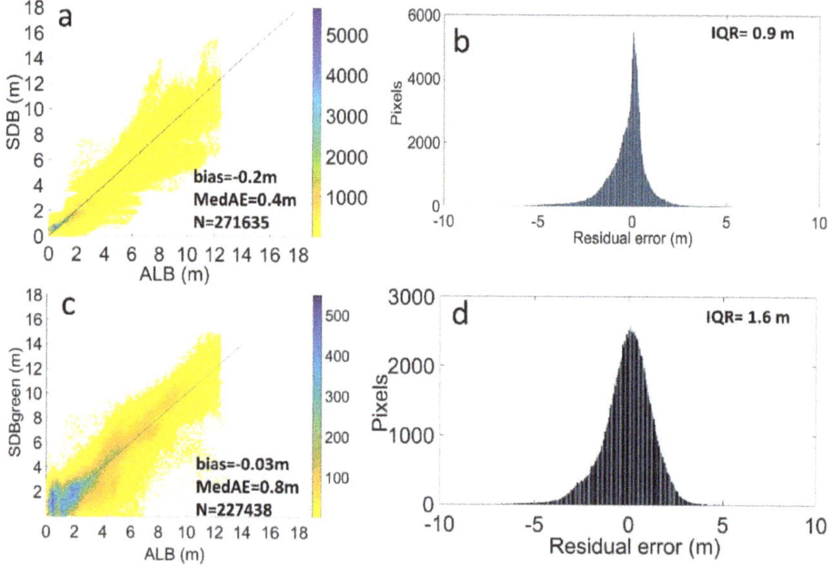

Figure 4. Comparison between lidar (ALB) and final SDB in Cape Lookout. (**a**) Scatter frequency of the final SDB after the multi-scene approach and switching model, the color bars indicate pixel frequency, (**b**) Associated histogram of residual errors with standard deviation=1.1m and percentile 5%/95%=-2.36m/0.93m, where the total number of pixels is N = 271635, divided in SDBred for depths <2 m with N = 148710 (54.7%), SDBweighted for depths 2-3.5 m with N = 30115 (11.1%), and SDBgreen for depths >3.5 m with N = 92810 (34.2%); (**c**), scatter frequency of the SDBgreen compared to ALB for the clearest scene acquired on 27 November 2017 (N = 227438), and (**f**) Associated histogram of residual errors and (**d**) with standard deviation=1.29m and percentile 5%/95%=-2.3m/1.89m, the bins of the histograms correspond to the number of pixels.

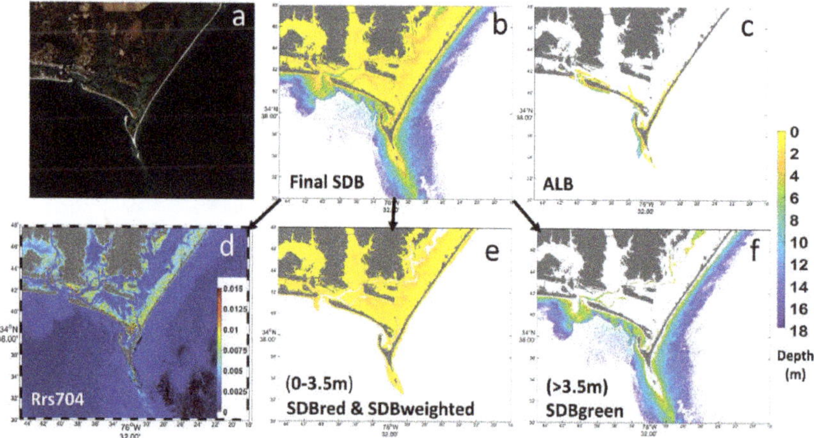

Figure 5. Satellite maps in Cape Lookout region. (**a**) Sentinel-2 RGB image composite, (**b**) Final SDB after the multi-scene approach and the switching model, (**c**) Lidar (ALB) data collected by NOAA, (**d**) Remote sensing reflectance at 704 nm (Rrs704) associated with the final SDB, segmentation of the final SDB with the switching model: (**e**) SDBred for depths <2 m and SDBweighted for depths 2–3.5 m, and (**f**) SDBgreen for depths >3.5 m. Data in the depth maps is presented as color-coded depths ranging from 0 to 18 m. Gray color represents land mask.

In addition, the switching strategy reduced error over using only SDBred (Figure 6a,c) or SDBgreen (Figure 6b,d). Recently, the switching algorithm has been suggested to be an opportune method for mapping regions from very shallow to deeper waters [22,44]. The quality of the combined models is shown in the histogram of the residual errors, which has a symmetric and narrow distribution (Figure 4b). The vertical calibration results are presented in Figure 7. Compared to the current standard approaches focused on the selection of the optimal scenes; in this case, the clearest image was acquired on 27 November 2017, compositing reduced the mean error by 50% and the IQR from 1.6 m to 0.9 m (Figure 4a,b vs. Figure 4c,d).

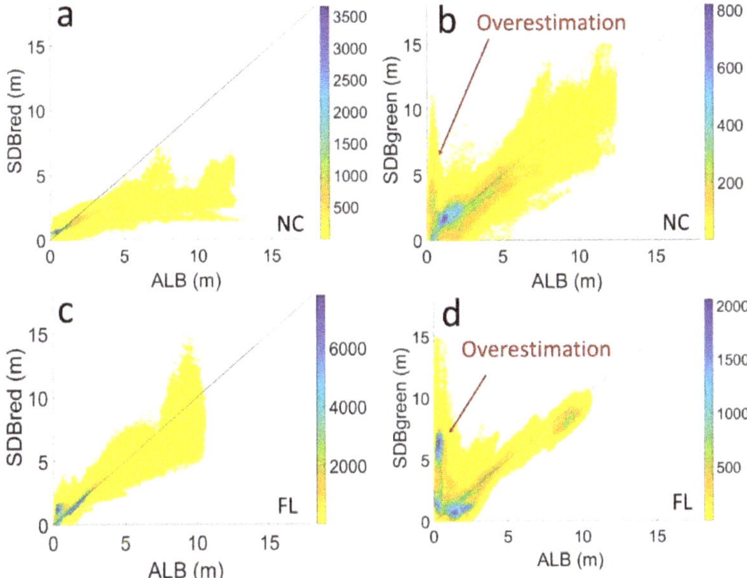

Figure 6. Validation of lidar (ALB) against final (**a**) SDBred (N = 269455) and (**b**) SDBgreen (N = 253220) after the multi-scene approach in Cape Lookout, North Carolina (NC); Validation of lidar against final (**c**) SDBred (N = 544793) and (**d**) final SDBgreen (N = 674659) after the multi-scene approach in Saint Joseph Bay, Florida (FL). Blue dotted lines indicate 1:1 line, the color bars indicate pixel frequency. The overestimation of SDBgreen over shallow areas (<2 m) is observed in both sites.

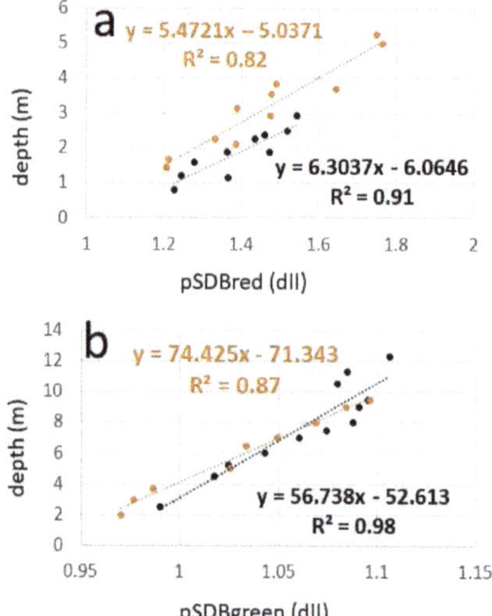

Figure 7. Vertical calibration for SDB models. (**a**) SDBred in Cape Lookout (orange, N = 11) and St. Joseph Bay (black, N = 10), (**b**) SDBgreen in Cape Lookout (orange, N = 9) and St. Joseph Bay (black, N = 12), where x is the pSDBred or pSDBgreen, y is the depth of soundings (chart), and R^2 corresponds to the coefficient of determination as a measurement of precision.

The method was evaluated by taking into account the selection of images, confirming that pixels from multiple scenes with high turbidity conditions (Rrs704) and chlorophyll (chl) were minimally incorporated into the model (<1%). Table 1 detailed the percentage of each scene incorporated into the final model. Turbidity (defined as either Rrs704 or chl) produces a false shoaling because the relative or "pseudo" depth pSDB (before the vertical calibration) decreases as turbidity increases (see an example for Cape Lookout in Figure 8). There is a relationship between the increase in the two water quality parameters and the decrease in relative depths for both models, pSDBred and pSDBgreen, thus confirming that intense false shoaling patterns were associated with highly turbid water. Similar results have been found in other sites along the Florida coastal waters using imagery from Sentinel-2A and Sentinel-3A [22]. Figure 5d represents the turbidity proxy (Rrs704) associated with the final SDB map, providing an indication of areas where residual turbidity may exhibit potential residual biases.

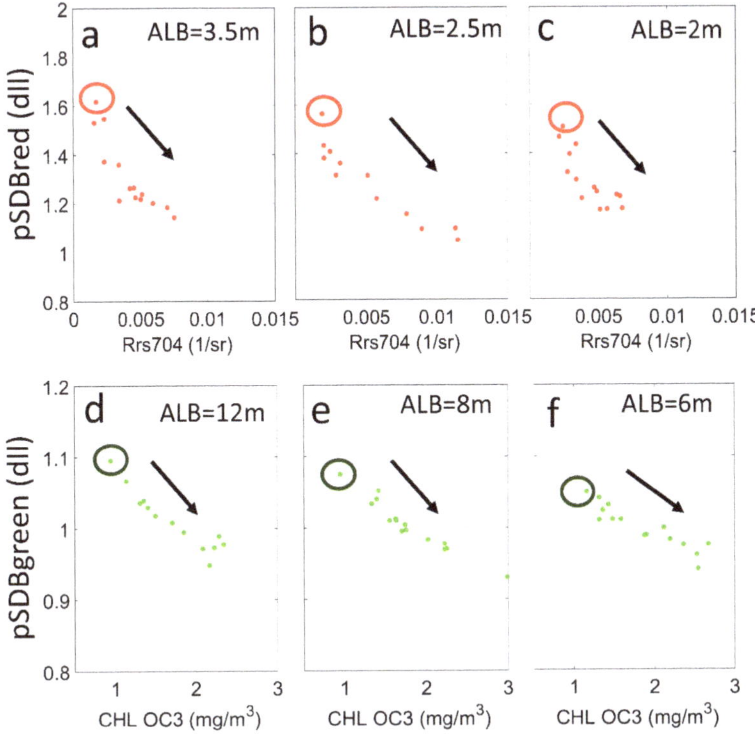

Figure 8. Temporal evaluation of the turbidity proxy Remote sensing reflectance at 704 nm (Rrs704) and chlorophyll (CHL OC3) against pSDBred and pSDBgreen, respectively. (**a**) pSDBred for lidar depths of 3.5 m (N = 14), (**b**) 2.5 m (N = 12), and (**c**) 2 m (N = 14), pSDBgreen for lidar depths of (**d**) 12 m (N = 13), (**e**) 8 m (N = 14), (**f**) 6 m (N = 15). The data corresponds to Cape Lookout and the study period from January to December 2017 (Table 1). The red and green circles indicate examples of the maximum pSDBred and pSDBgreen, respectively, selected for the multi-scene approach.

The method also exhibited some advantages over the in-situ data. The lidar survey was limited by water clarity as well, and some water areas nearshore were not retrieved (Figure 5c). In contrast, by using multiple dates of imagery, features could be identified by the composited satellite data, along the entire coastal fringe and within the inner banks (SDB on Figure 5b compared to ALB on Figure 5c). We also examined the switching model used in this study, so the final SDB map (Figure 5b) was split into the SDBred and SDBweighted (0-3.5 m, Figure 5e), and the SDBgreen (>3.5 m, Figure 5f). SDBred and SDBweighted mapped the shallow regions within the banks and barrier islands, while the SDBgreen mapped the offshore and the deep channels in the banks. In offshore areas where the bottom was not visible, the attributed depth was removed with a cut-off depth of 17 m. These results were support by visual comparison with the RGB composite (Figure 5a) and standard chart soundings collected by the National Oceanic and Atmospheric Administration [47], given that there was not high-resolution information available for validation of those deeper water areas.

3.2. Saint Joseph Bay in Florida

Previous studies have characterized the Saint Joseph Bay and the adjacent area as a moderately turbid region under influence of the Apalachicola Bay turbid waters [33,35]. The region also has areas of dense shallow aquatic vegetation [51], so a single scene was used to evaluate model performance over these areas. A preliminary inspection of a clear image acquired on 14 February 2017 for SDBred

(Figure 9e) and SDBgreen (Figure 9b) demonstrated the impact of FAV and SAV, showing overestimation at depths <1 m over the dense marine vegetation areas for both models. The overestimate was about 1 m for SDB red and severe at up to 6 m for SDBgreen. Inspection of other individual scenes also presented overall positive bias for depths <1 m, owing to an apparent higher absorption of the green band (560 nm) for SDBgreen, whereas lower relative absorption occurred over vegetation in the red band (664 nm) for SDBred. This was apparently due to the dark signatures of the floating aquatic vegetation (FAV) and submerged aquatic vegetation (SAV) that may cause the overestimation of depth by this model or other SDB approaches.

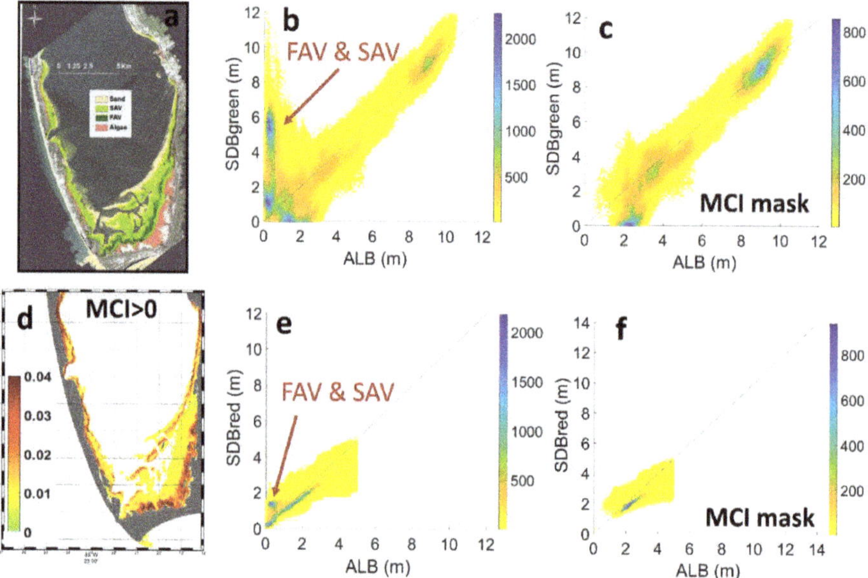

Figure 9. Saint Joseph Bay region. (**a**) Mapped distributions of red algae, submerged aquatic vegetation (SAV), and floating aquatic vegetation (FAV) [51], comparison of lidar (ALB) and SDBgreen (**b**) without MCI masking (N = 604736) and (**c**) with MCI masking (N = 323802), (**d**) MCI map used for the masking procedure (MCI>0), comparison of lidar (ALB) and SDBred (**e**) without MCI masking (N = 571199) and (**f**) with MCI masking (N = 140947), the color bars indicate pixel frequency. The Sentinel-2A scene was acquired on 14 February 2017.

In order to correct the overestimation issue, a masking procedure to locate and remove an error source based on a common algae index—the Maximum Chlorophyll Index MCI [53,54]—was developed. In [54], it is shown that slightly submerged vegetation will cause an MCI (false positive against phytoplankton). This approach allowed us to use the satellite without requiring any additional in-situ information. Pixels with positive MCI values (MCI>0) were identified as vegetation (Figure 5d) and then masked out. The scatterplots of ALB against SDBred and SDBgreen after masking vegetation (Figure 9c,f, respectively) show that the overestimation was eliminated, thus MCI can be applied to remove the overestimation. Very shallow SDBred pixels with good accuracy (lying on the 1:1 line) were also eliminated, because MCI does register on bright sand in very shallow water (owing to the signal from Rrs704 in <1 m). As a conservative solution, the FAV and SAV masking procedure would be appropriate owing to the severe consequences of overestimating very shallow bathymetry, as identified by the Chart Standards Groups (CSG) at NOAA's Office of Coast Survey/Marine Chart Division.

The same multi-scene approach and switching model as used for Cape Lookout was carried out in this region, although only seven cloud-free images were available during the study period (Table 1).

In this case, we tested the additional MCI masking to remove issues associated with FAV and SAV. The validation against lidar data confirmed the high accuracy obtained for the entire strategy with (Figure 10a) or without (Figure 10c) MCI masking. There was low scatter (MedAE=0.3 m) for depths up to 11 m (limit of ALB data set). The vertical calibration results are presented in Figure 7. Even without MCI correction, the switching model alone substantially reduced intense overestimation of SDBgreen for depths <1 m (Figure 10c). The importance of the switching strategy is evident in the validation of the final multi-scene SDBred and SDBgreen, where an acute overestimation occurred at depths <1 m for SDBgreen, similar to Cape Lookout (Figure 6a–d). In terms of error assessment, the distribution of the residuals indicated minimum discrepancies were achieved (Figure 10b,d). Furthermore, the compositing algorithm produced accurate SDB results compared to the single "optimal scene" approach (picking the scene with the lowest overall turbidity) combined with the most widely used model in the literature, SDBgreen (Figure 10e,f). The composite map of the proxy for turbidity (Rrs704) associated to the final SDB model shows minimum turbid conditions throughout (Figure 11d).

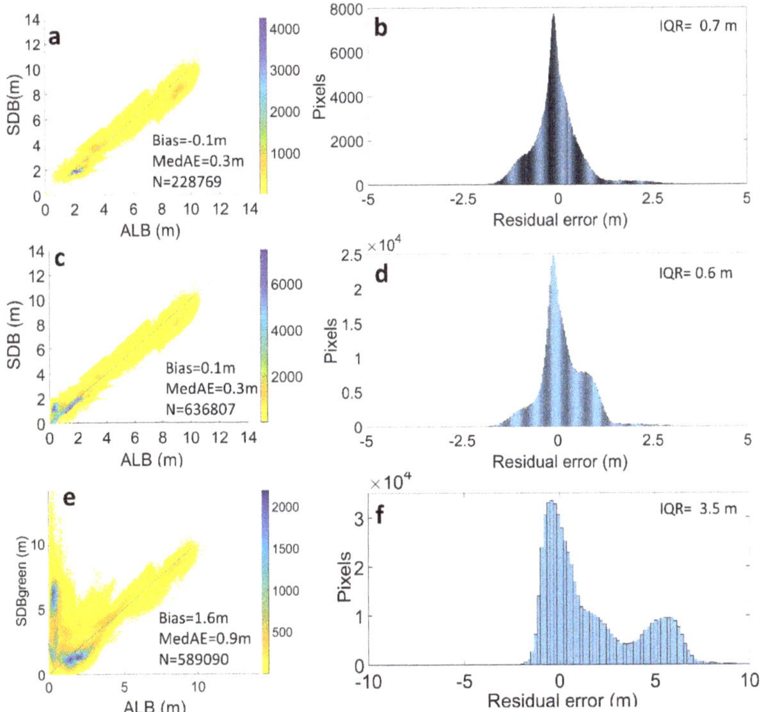

Figure 10. Comparison between lidar (ALB) and final SDB map in Saint Joseph Bay. (**a**) Scatter of ALB against SDB after the multi-scene approach, the switching model and the FAV and SAV masking procedures with MCI index, the color bars indicate pixel frequency, (**b**) Associated histogram of residual errors with standard deviation = 0.68m and percentile 5%/95% = −1.15 m/0.92 m, (**c**) The same as (**a**) without applying the MCI masking, (**d**) The same as (**b**) without applying the MCI masking with standard deviation = 0.69 m and percentile 5%/95% = −0.87 m/1.1 m, (**e**) scatter of the SDBgreen compared to ALB for the clearest scene acquired on 23 December 2016, and (**f**) Associated histogram of residual errors with standard deviation = 2.4m and percentile 5%/95% = −0.96 m/6.1 m (note different x-axis compared with b and d), the bins of the histograms correspond to the number of pixels. The blue dotted line indicates 1:1 line.

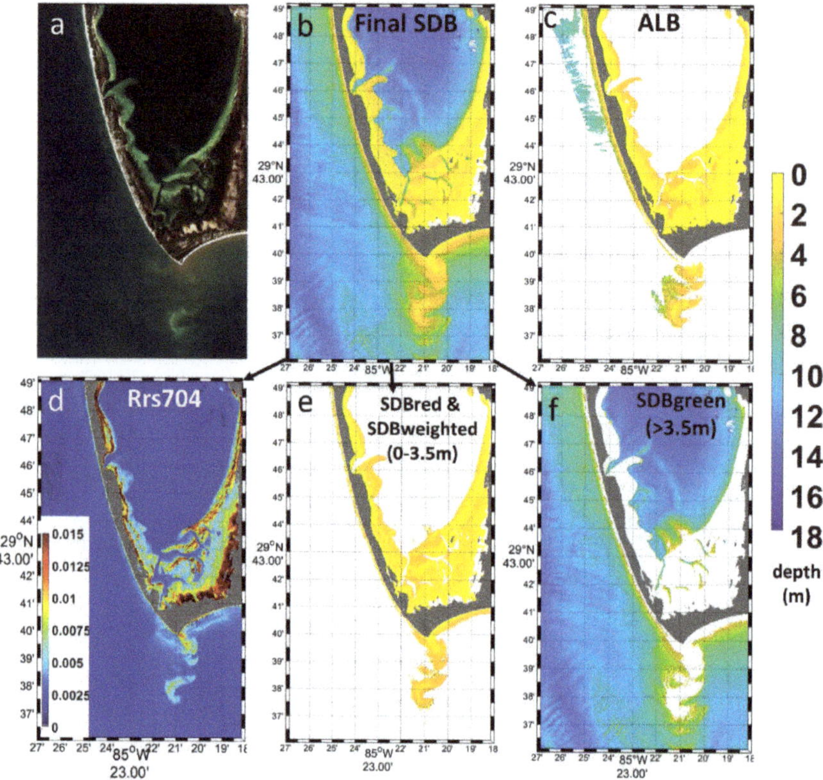

Figure 11. Satellite maps in Saint Joseph Bay region. (**a**) Sentinel-2 RGB image composite, (**b**) Final SDB after the multi-scene approach and the switching model, (**c**) Lidar (ALB) data collected by NOAA, (**d**) Remote sensing reflectance at 704 nm (Rrs704) associated with the final SDB, segmentation of the final SDB with the switching model: (**e**) SDBred for depths <2 m and SDBweighted for depths 2–3.5 m, and (**f**) SDBgreen for depths >3.5 m. Data in the depth maps is presented as color-coded depths ranging from 0 to 18 m. Gray color represents land mask.

The geographic distribution of features in the SDB (Figure 11b) corresponded to those identifiable in the lidar survey (Figure 11c), although much more spatial information is available in the SDB product. Similar to Cape Lookout region, the method offered a complete map of the St. Joseph coastal area compared to the restricted map provided by lidar surveys. In this case, we did not apply the MCI masking to add data over the vegetated area (depths <1 m), thereby allowing maximum coverage of the final SDB product. Shallow areas and shoals < 3.5 m were accurately described with SDBred and SDBweighted (Figure 11e) whereas depths >3.5 m were mapped with SDBgreen (Figure 11f).

4. Discussion

In the present investigation, we successfully applied both Sentinel-2A and 2B satellites to estimate water depth in two areas with complex bathymetry and water clarity, Cape Lookout, North Carolina, and Saint Joseph Bay, Florida. The multi-scene compositing approach addressed limitations inherent in conventional methods and reduced the impact of turbidity, performing better than the standard "pick the best scene" method that relies on a single image (Figures 3 and 4c vs. Figure 4a in Cape Lookout; and Figure 10e vs. Figure 10a,c in Saint Joseph Bay). The final corrected SDB produced robust depths up to the limit of the lidar surveys, with typical errors ≤0.4 m. These excellent results from Sentinel-2 compared favorably with those produced in relatively low turbidity water in south Florida [22,44], and in regions with transparent waters [10,18,20,30]. Whereas some researchers suggested there is still work to be performed regarding the identification of the optimal period throughout the year where bathymetric errors are minimized [18,29], others asked for novel strategies to allow seabed mapping without the laborious analysis per image and the visual inspection of the "clearest scene" [19]. Recent studies have already indicated the potential of multi-scene approaches in order to select the optimal scene or eliminate noise over clear waters [16,17,19,20]. However, our temporal compositing strategy successfully reduced the turbidity impact without requirement of visual inspection, thereby enhancing SDB performance in an easy way. The high temporal resolution and interchangeability of the Sentinel-2 twin mission may rapidly overcome SDB anomalies introduced by highly heterogeneous water transparency regimes [22].

The multi-scene strategy applied here did not require any screening or manual adjustment of the imagery prior to compositing. It automatically picked the pixels least impacted by turbidity (e.g., Figure 8) from the set of scenes provided (Table 1), substantially simplifying the effort compared with other studies where the selection of optimal images with variation in the water quality conditions were essential in the extraction of SDB [20,22,29,55]. Manual selection of the optimal scene is not only highly subjective, but requires considerable time and effort, and may still include regions having patches of turbidity. The consistency in the pSDB products, which were not yet calibrated to true depth, indicated that ACOLITE produced an effective and robust atmospheric correction across scenes (Figure 8), as previously demonstrated [22,44]. Inconsistencies or errors in ACOLITE (or any other atmospheric correction) would force manual intervention and make compositing impractical [10].

We chose chart values for the calibration to depth in order to demonstrate the benefit of the method for remote areas under likely applied conditions, rather than under the optimal (and extremely unlikely) condition of a contemporaneous lidar survey. Using limited chart soundings would have only introduced some of the error, most likely as bias in the depth. Some depth errors may also be due to the differences in resolution, i.e., 10 m from the satellite, and the 1 m spot size for lidar—a consideration in areas with steep gradients, such as channel edges. Using a median filter on the SDB reduced other artifacts that typically lead to random noise [49]. One potential error factor that needs to be considered in the future is tide range. The tide range in these areas is relatively small (<1 m). The influence of tide on the accuracy of this approach for areas with tidal ranges greater than 1 m requires further investigation. For meso-tidal areas, one option would be to constrain the input images to a common range of water level, such as 1 m maximum difference in water level. Macro-tidal areas are likely to be too turbid for optical SDB.

An interesting consideration is that the calibration to depth for this method may be dependent on water clarity rather than on geography. Applying the coefficients for composited pSDBred and pSDBgreen of Cape Lookout to St. Joseph Bay (Figure 7), and vice versa, generated precise SDB for both areas with errors ≤0.65 m (see results in Figure 12). This comparison suggests that an existing calibration after the multi-scene approach may be applicable to other remote and unsurveyed areas, assuming the compositing can retrieve the same turbidity. These results suggest the potential suitability of SDB estimates for charting applications using the IHO S-57 standard [56], which defines CATZOC levels containing depth accuracy specifications for depth ranges up to 30 m in order to allow for their incorporation into nautical products.

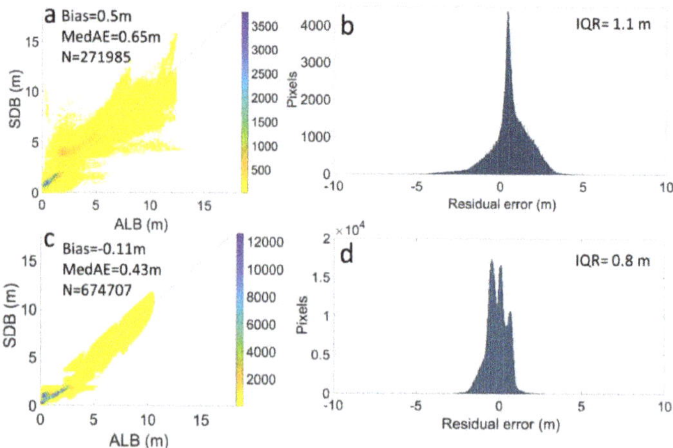

Figure 12. Validation of the final SDB. (**a**) Validation of lidar (ALB) against final SDB in Cape Lookout using the calibration coefficients of St. Joseph Bay (N = 271985), the color bars indicate pixel frequency and (**b**) residual errors (SDB - ALB), (**c**) Validation of lidar (ALB) against final SDB in St. Joseph Bay using the calibration coefficients of Cape Lookout (N = 674707) and (**d**) residual errors (SDB - ALB), the bins of the histograms correspond to the number of pixels. Blue dotted lines indicate 1:1 line. The vertical calibration for each study site is indicated in Figure 8.

The use of the SDBred for shallow water retrieval addressed the worst of the overestimation issues with the SDBgreen in both study regions. The use of different bands to treat overestimation in shallow water (or underestimation by SDBred in deep water) has been pointed out recently by some researchers [19,30,57]. Switching to SDBred has been previously used to provide better discrimination of depths in very shallow water over bright targets like carbonate sand [48], and other studies have applied each model over different depth ranges [22,44]. Here, we established an automated switching method between SDBred and SDBgreen that performed accurately over the two study sites. In addition, SDBgreen overestimated on dense seagrass in extremely shallow water using either the single scene (Figure 9b, Figure 10e) or the multi-scene approach (dense seagrass has a very low albedo, so in deeper water there is often no detectable signal to be retrieved from satellite). By switching to the SDBred, the combined product reduced most of the impact of shallow seagrass on SDBgreen (as seen in Figure 9b against Figure 9e or Figure 10e against Figure 10c). If more rigorous masking is required, the MCI can identify the presence of dense seagrass in extremely shallow water (<1–2 m, Figure 10a), without the need for external data sets on bottom type. Combined with calibration using only a few points (Figure 7), the approach detailed here offers an effective means of assessing bathymetry in regions lacking any data.

While the accuracy of SDB cannot match ALB, the spatial coverage of satellite imagery shown here surpasses ALB (Figures 5c and 11c vs. Figures 5b and 11b). This capability represents one of the great benefits of satellite monitoring; whereas in-situ surveys (lidar, sonar) are expensive and extremely limited due to technical and deployment cost, SDB can give us a broad picture of the entire study site with wide swaths, low-cost repeated coverage, and easy access to remote locations [8]. Lidar data is also constrained by availability and water quality, so information can be lost. We characterized post Hurricane Matthew bathymetry in Cape Lookout, extending the limited amount of lidar data to a much larger and more complete area.

Sentinel-2 may provide a way to use other satellites more effectively. Accordingly, Sentinel-2 might offer a low turbidity reference data set that can be used to inform mapping conducted with the various commercial very high-resolution sensors such as the World-View fleet [14,26,58] or the novel CubeSats from Planet [19,29]. On the other scale, Landsat has a four-decade history using Thematic

Mapper, and has been shown to be useful for some SDB in clear water [10,21,59,60], even with 30 m pixels. Comparisons of Sentinel-2 with Landsat-8 may ultimately lead to results that could expand the utility of the entire Landsat data record for change detection.

In this study, interpretation of error assessment and uncertainty can be achieved by means of the water quality composites of turbidity (Rrs704) associated with the final SDB maps (Figures 5d and 11d). These products may indicate areas that still have residual bias. They can also provide areas where the water is chronically optically deep due to turbidity, which can lead to better allocation of resources. Previous research has detailed the impact of water quality parameters on SDB results using the ratio model over several regions in Florida [22]. Recent studies using the ratio model showed that bottom reflectance has little influence on the accuracy of SDB estimates, while chlorophyll concentration has a strong influence [22,61]. This problem would also be resolved with the compositing method.

In accordance with our findings, the portability and reliability application of the proposed approach using minimum in-situ measurements is a distinct advantage in effectiveness and may especially benefit developing countries. In addition, using the semi-automated solution and a computer cloud-based system may allow exploitation of the Sentinel-2 data for regional to global scale coastal SDB [16]. The average time required for processing a Sentinel-2 image for bathymetry estimation is ~ 1 hour. For further avenues of research with Sentinel-2, we intend to upscale the study results to more environments with different water and atmospheric conditions in order to evaluate repeatability as well as implement the multi-temporal approach in cloud-based computing platforms such as the ESA Coastal Thematic Exploitation Platform—Coastal-TEP [62]. The promising transferability of this temporal technique exploiting the open and free archive of the Sentinel-2 mission will allow advancement of SDB applications and optimized scientific coastal mapping worldwide, especially in data-poor regions [63].

5. Concluding Remarks

This paper presents a framework to obtain SDB while potentially automated the image processing. Precise SDB was derived over moderately turbid environments by using the high revisit time (5-day) of the Sentinel-2A/B twin mission from the Copernicus programme. The proposed methodology incorporates a robust atmospheric correction with ACOLITE, a multi-scene compositing method to reduce the impact of turbidity, and a switching model to improve mapping in shallow water. Two study sites in United States were explored due to their varying water transparency conditions. The method establishes a plausible high-quality strategy for SDB with sufficient sensitivity to support interannual and pre-post hazards (e.g., hurricanes or tropical storms), which may benefit change detection analysis from multiple images to discriminate variability. The open data policy and long-term mission commitment of Sentinel-2 opens future promising time series evaluation over years and even decades that can be an important tool to provide crucial missing information on the bathymetry distribution, especially in data-poor or remote areas with large gaps in a retrospective, rapid and non-intrusive manner. It would enable immediate commencement of coastal mapping using archived and/or forthcoming imagery from the Copernicus programme at regional to global spatial scales, and any other source from which repeated scenes may be collected over the world's coastal systems. There is a requirement to promote the development of technology that provides low-cost solutions with enhanced data quality to address the needs of a wide range of user groups. The semi-automated framework demonstrated here may address that need, and help mitigate the long-standing gap of depth information for the majority of coastal regions on Earth. Such a capability can aid scientists, managers, and policymakers around the globe in assessing vulnerabilities to ecosystems, infrastructure, navigation and many other coastal concerns.

Author Contributions: I.C. and R.P.S. conceived and designed the research. I.C. collected the remote sensing data, processed the data, conducted the analysis, and prepared the figures. R.P.S. supervised the research. I.C. led the writing of the manuscript with contribution from R.P.S. All authors have read and agreed to the published version of the manuscript.

Funding: Isabel Caballero was funded by the National Oceanic and Atmospheric Administration and the National Academy of Science, Engineering and Medicine under the NRC postdoctoral Research Associateship Program (RAP) 2017-2019 with support from the NOAA-Korea Ministry of Oceans and Fisheries Joint Project Agreement. This research was also supported by the Sen2Coast Project (RTI2018-098784-J-I00) funded by the Ministry of Science, Innovation and Universities (MCIU), the State Research Agency (AEI) and the European Regional Development Fund (ERDF).

Acknowledgments: The authors thank the European Space Agency, the European Commission and the Copernicus programme for distributing Sentinel-2 imagery. Thanks to the National Oceanic and Atmospheric Administration and the National Geodetic Survey for providing the valuable high-resolution lidar data and the nautical charts used for this study. Thanks to Quinten Vanhellemont for his assistance in using ACOLITE processor. The authors acknowledge the three anonymous reviewers, whose comments helped to greatly improve an earlier version of this manuscript.

Conflicts of Interest: The authors declare no conflict of interest.

References

1. United Nations. Transforming Our World: The 2030 Agenda for Sustainable Development. 2015. Available online: https://sustainabledevelopment.un.org/post2015/transformingourworld/publication (accessed on 10 June 2019).

2. Charles, E.; Douvere, F. *Marine Spatial Planning: A Step-by-Step Approach toward Ecosystem-Based Management. Intergovernmental Oceanographic Commission and Man and the Biosphere Programme. IOC Manual and Guides*; ICAM Dossier No. 6; UNESCO: Paris, France, 2009; p. 99.

3. International Hydrographic Organization. International Hydrographic Review. 2017. Available online: https://www.iho.int/mtg_docs/IHReview/2017/IHR_May2017.pdf (accessed on 15 October 2019).

4. International Hydrographic Organization. International Hydrographic Review. 2017. Available online: https://www.iho.int/mtg_docs/IHReview/2017/IHR_November2017.pdf (accessed on 15 October 2019).

5. International Hydrographic Organization. International Hydrographic Publication C-55 Status of Hydrographic Surveying and Charting Worldwide. 2018. Available online: http://iho.int/mtg_docs/misc_docs/basic_docs/IHO_Work_Programme_for_2019_final.pdf (accessed on 15 October 2019).

6. Carron, M.J.; Vogt, P.R.; Jung, W.Y. A proposed international long-term project to systematically map the world's ocean floors from beach to trench: GOMaP (Global Ocean Mapping Program). *Int. Hydrogr. Rev.* **2001**, 2.

7. Marks, K.M. *International Hydrographic Organization, Intergovernmental Oceanographic Commission, The IHO-IOC GEBCO Cook Book*; IHO Publication B-11, IOC Manuals and Guides 63; IOC: Monaco, France, 2019; 493p.

8. Gao, J. Bathymetric mapping by means of remote sensing: Methods, accuracy and limitations. *Prog. Phys. Geogr.* **2009**, *33*, 103–116. [CrossRef]

9. Dekker, A.G.; Phinn, S.R.; Anstee, J.; Bissett, P.; Brando, V.E.; Casey, B.; Fearns, P.; Hedley, J.; Klonowski, W.; Lee, Z.P.; et al. Intercomparison of shallow water bathymetry, hydro-optics, and benthos mapping techniques in Australian and Caribbean coastal environments. *Limnol. Oceanogr. Methods* **2011**, *9*, 396–425. [CrossRef]

10. Hedley, J.D.; Roelfsema, C.; Brando, V.; Giardino, C.; Kutser, T.; Phinn, S.; Mumby, P.; Barrilero, O.; Laporte, J.; Koetzh, B. Coral reef applications of Sentinel-2: Coverage, characteristics, bathymetry and benthic mapping with comparison to Landsat 8. *Remote Sens. Environ.* **2018**, *216*, 598–614. [CrossRef]

11. Tripathi, N.K.; Rao, A.M. Bathymetric mapping in Kakinada Bay, India, using IRS-1D LISS-III data. *Int. J. Remote Sens.* **2002**, *23*, 1013–1025. [CrossRef]

12. Philpot, W.D.; Davis, C.O.; Bissett, W.P.; Mobley, C.D.; Kohler, D.C.D.; Lee, Z.; Bowles, J.; Steward, R.G.; Agrawal, Y.; Trowbridge, J.; et al. Bottom Characterization from Hyperspectral Image Data. *Oceanography* **2004**, *17*, 76–85. [CrossRef]

13. Minghelli-Roman, A.; Dupouy, C. Influence of water column chlorophyll concentration on bathymetric estimations in the lagoon of New Caledonia, using several MERIS images. *IEEE J. Sel. Top. Appl. Earth Obs. Remote Sens.* **2013**, *6*, 739–745. [CrossRef]

14. Hamylton, S.M.; Hedley, J.D.; Beaman, R.J. Derivation of high-resolution bathymetry from multispectral satellite imagery: A comparison of empirical and optimization methods through geographical error analysis. *Remote Sens.* **2015**, *7*, 16257–16273. [CrossRef]

15. Khondoker, M.S.I.; Siddiquee, M.Z.H.; Islam, M. The Challenges of River Bathymetry Survey Using Space Borne Remote Sensing in Bangladesh. *Atmos. Ocean. Sci.* **2016**, *1*, 7–13.

16. Traganos, D.; Poursanidis, D.; Aggarwal, B.; Chrysoulakis, N.; Reinartz, P. Estimating satellite-derived bathymetry (SDB) with the google earth engine and sentinel-2. *Remote Sens.* **2018**, *10*, 859. [CrossRef]

17. Chu, S.; Cheng, L.; Ruan, X.; Zhuang, Q.; Zhou, X.; Li, M.; Shi, Y. Technical Framework for Shallow-Water Bathymetry With High Reliability and No Missing Data Based on Time-Series Sentinel-2 Images. *IEEE Trans. Geosci. Remote Sens.* **2019**, *57*, 1–19. [CrossRef]

18. Evagorou, E.G.; Mettas, C.; Agapiou, A.; Themistocleous, K.; Hadjimitsis, D.G. Bathymetric maps from multi-temporal analysis of Sentinel-2 data: The case study of Limassol, Cyprus. *Adv. Geosci.* **2019**, *45*, 397–407. [CrossRef]

19. Poursanidis, D.; Traganos, D.; Chrysoulakis, N.; Reinartz, P. Cubesats Allow High Spatiotemporal Estimates of Satellite-Derived Bathymetry. *Remote Sens.* **2019**, *11*, 1299. [CrossRef]

20. Sagawa, T.; Yamashita, Y.; Okumura, T.; Yamanokuchi, T. Satellite Derived Bathymetry Using Machine Learning and Multi-Temporal Satellite Images. *Remote Sens.* **2019**, *11*, 1155. [CrossRef]

21. Pe'eri, S.; Madore, B.; Nyberg, J.; Snyder, L.; Parrish, C.; Smith, S. Identifying bathymetric differences over Alaska's North Slope using a satellite-derived bathymetry multi-temporal approach. *J. Coast. Res.* **2016**, *76*, 56–63. [CrossRef]

22. Caballero, I.; Stump, P.R.; Meredith, A. Preliminary assessment of turbidity and chlorophyll impact on bathymetry derived from Sentinel-2A and Sentinel-3A satellites in South Florida. *Remote Sens.* **2019**, *11*, 645. [CrossRef]

23. Stumpf, R.P.; Feldman, G.C.; Kuring, N.; Franz, B.; Green, E.; Robinson, J. SeaWiFS spies reefs. *Reef Encount.* **1990**, *26*, 29–30.

24. National Aeronautics and Space Administration. Remote Sensing of Coral Reefs: SeaWiFS Bathymetry and Data Archive Proof-of-Concept. 2003. Available online: https://oceancolor.gsfc.nasa.gov/cgi/reefs.pl (accessed on 15 September 2019).

25. Minghelli-Roman, A.; Goreac, A.; Mathieu, S.; Spigai, M.; Gouton, P. Comparison of bathymetric estimation using different satellite images in coastal sea waters. *Int. J. Remote Sens.* **2009**, *30*, 5737–5750. [CrossRef]

26. Kanno, A.; Tanaka, Y.; Shinohara, R.; Kurosawa, A.; Sekine, M. Which spectral bands of Worldview-2 are useful in remote sensing of water depth? A case study in coral reefs. *Mar. Geod.* **2014**, *37*, 283–292. [CrossRef]

27. European Commission. *Copernicus for Coastal Zone Monitoring and Management Workshop*; Technical Report; Copernicus Support Office, 2017; Available online: https://land.copernicus.eu/user-corner/technical-library/report-of-the-coastal-zone-monitoring-and-management-workshop (accessed on 1 October 2019).

28. Canadian Hydrographic Service. First international Hydrographic Remote Sensing (HRS) workshop, workshop report. In Proceedings of the First International Hydrographic Remote Sensing (HRS) Workshop, Ottawa, ON, Canada, 18–20 September 2018.

29. Li, J.; Schill, S.R.; Knapp, D.E.; Asner, G.P. Object-Based Mapping of Coral Reef Habitats Using Planet Dove Satellites. *Remote Sens.* **2019**, *11*, 1445. [CrossRef]

30. Poursanidis, D.; Traganos, D.; Reinartz, P.; Chrysoulakis, N. On the use of Sentinel-2 for coastal habitat mapping and satellite-derived bathymetry estimation using downscaled coastal aerosol band. *Int. J. Appl. Earth Obs. Geoinf.* **2019**, *80*, 58–70. [CrossRef]

31. Stumpf, R.P.; Holderied, K.; Sinclair, M. Determination of water depth with high-resolution satellite imagery over variable bottom types. *Limnol. Oceanogr.* **2003**, *48 Pt 2*, 547–556. [CrossRef]

32. Wells, J.T. Accumulation of fine-grained sediments in a periodically energetic clastic environment, Cape Lookout Bight, North Carolina. *J. Sediment. Res.* **1988**, *58*, 596–606.

33. Berndt, M.P.; Franklin, M.A. *Water-Quality and Discharge Data for St. Joseph Bay, Florida, 1997–1998*; Department of the Interior: Washington, DC, USA, 1999; Available online: https://apps.dtic.mil/dtic/tr/fulltext/u2/a442268.pdf (accessed on 1 October 2019).

34. Mienis, F.; Duineveld, G.C.A.; Davies, A.J.; Lavaleye, M.M.S.; Ross, S.W.; Seim, H.; Brooke, S. Cold-water coral growth under extreme environmental conditions, the Cape Lookout area, NW Atlantic. *Biogeosciences* **2014**, *11*, 2543–2560. [CrossRef]

35. Joshi, I.D.; D'Sa, E.J.; Osburn, C.L.; Bianchi, T.S. Turbidity in Apalachicola bay, Florida from Landsat 5 TM and field data: Seasonal patterns and response to extreme events. *Remote Sens.* **2017**, *9*, 367. [CrossRef]

36. European Space Agency. *Sentinel-2 User Handbook*; ESA Standard Document; ESA: Paris, France, 2015; Available online: https://sentinel.esa.int/documents/247904/685211/Sentinel-2_User_Handbook (accessed on 20 June 2019).

37. Sentinel's Scientific Data Hub. Available online: https://scihub.copernicus.eu/ (accessed on 15 May 2019).

38. European Space Agency. Sentinel-2 MSI Technical Guide. 2017. Available online: https://earth.esa.int/web/sentinel/technical-guides/sentinel-2-msi (accessed on 5 May 2019).

39. Vanhellemont, Q.; Ruddick, K. Turbid wakes associated with offshore wind turbines observed with Landsat 8. *Remote Sens. Environ.* **2014**, *145*, 105–115. [CrossRef]

40. Vanhellemont, Q.; Ruddick, K. Advantages of high quality SWIR bands for ocean colour processing: Examples from Landsat-8. *Remote Sens. Environ.* **2015**, *161*, 89–106. [CrossRef]

41. Vanhellemont, Q.; Ruddick, K. Acolite for Sentinel-2: Aquatic Applications of MSI Imagery. In Proceedings of the ESA Living Planet Symposium, Prague, Czech Republic, 9–13 May 2016.

42. Ruddick, K.; Vanhellemont, Q.; Dogliotti, A.; Nechad, B.; Pringle, N.; Van der Zande, D. New Opportunities and Challenges for High Resolution Remote Sensing of Water Colour. In Proceedings of the Ocean Optics XXIII, Victoria, BC, Canada, 7 October 2016.

43. Pahlevan, N.; Sarkar, S.; Franz, B.A.; Balasubramanian, S.V.; He, J. Sentinel-2 MultiSpectral Instrument (MSI) data processing for aquatic science applications: Demonstrations and validations. *Remote Sens. Environ.* **2017**, *201*, 47–56. [CrossRef]

44. Caballero, I.; Stumpf, R.P. Retrieval of nearshore bathymetry from Sentinel-2A and 2B satellites in South Florida coastal waters. *Estuar. Coast. Shelf Sci.* **2019**, *226*, 106277. [CrossRef]

45. Toming, K.; Kutser, T.; Laas, A.; Sepp, M.; Paavel, B.; Nõges, T. First experiences in mapping lake water quality parameters with Sentinel-2 MSI imagery. *Remote Sens.* **2016**, *8*, 640. [CrossRef]

46. Davis, R.A.; Fitzgerald, D.M. *Beaches and Coasts*; Blackwell Publishing Company: Malden, MA, USA, 2014; p. 419.

47. National Oceanic and Atmospheric Administration. NOAA Nautical Chart Catalog and Locator. 2019. Available online: http://www.charts.noaa.gov/ (accessed on 1 May 2019).

48. National Oceanic and Atmospheric Administration. Atlas of the Shallow-Water Benthic Habitats of the Northwestern Hawaiian Islands. 2003; p. 160. Available online: https://cdn.coastalscience.noaa.gov/datasets/e98/docs/NWHI_Atlas_sec1.pdf (accessed on 1 May 2019).

49. Linklater, M.; Hamylton, S.M.; Brooke, B.P.; Nichol, S.L.; Jordan, A.R.; Woodroffe, C.D. Development of a seamless, high-resolution bathymetric model to compare reef morphology around the subtropical island shelves of Lord Howe Island and Balls Pyramid, southwest Pacific Ocean. *Geosciences* **2018**, *8*, 11. [CrossRef]

50. Bramante, J.F.; Raju, D.K.; Sin, T.M. Multispectral derivation of bathymetry in Singapore's shallow, turbid waters. *Int. J. Remote Sens.* **2013**, *34*, 2070–2088. [CrossRef]

51. Hill, V.J.; Zimmerman, R.C.; Bissett, W.P.; Dierssen, H.; Kohler, D.D. Evaluating light availability, seagrass biomass, and productivity using hyperspectral airborne remote sensing in Saint Joseph's Bay, Florida. *Estuar. Coasts* **2014**, *37*, 1467–1489. [CrossRef]

52. Gitelson, A. The peak near 700 nm on radiance spectra of algae and water: Relationships of its magnitude and position with chlorophyll concentration. *Int. J. Remote Sens.* **1992**, *13*, 3367–3373. [CrossRef]

53. Gower, J.; King, S. Distribution of floating Sargassum in the Gulf of Mexico and the Atlantic Ocean mapped using MERIS. *Int. J. Remote Sens.* **2011**, *32*, 1917–1929. [CrossRef]

54. Gower, J.; King, S.; Borstad, G.; Brown, L. Detection of intense plankton blooms using the 709 nm band of the MERIS imaging spectrometer. *Int. J. Remote Sens.* **2005**, *26*, 2005–2012. [CrossRef]

55. Casal, G.; Monteys, X.; Hedley, J.; Harris, P.; Cahalane, C.; McCarthy, T. Assessment of empirical algorithms for bathymetry extraction using Sentinel-2 data. *Int. J. Remote Sens.* **2019**, *40*, 2855–2879. [CrossRef]

56. International Hydrographic Organization IHO. *S-57 Supplement No. 3-Supplementary Information for the Encoding of S-57 Edition 3.1 ENC Data*; International Hydrographic Organization: Monaco, France, 2014; Available online: https://www.iho.int/iho_pubs/standard/S-57Ed3.1/S-57_e3.1_Supp3_Jun14_EN.pdf (accessed on 5 September 2019).

57. Geyman, E.C.; Maloof, A.C. A simple method for extracting water depth from multispectral satellite imagery in regions of variable bottom type. *Earth Space Sci.* **2019**, *6*, 527–537. [CrossRef]

58. Eugenio, F.; Marcello, J.; Martin, J. High-resolution maps of bathymetry and benthic habitats in shallow-water environments using multispectral remote sensing imagery. *IEEE Trans. Geosci. Remote Sens.* **2015**, *53*, 3539–3549. [CrossRef]

59. Pacheco, A.; Horta, J.; Loureiro, C.; Ferreira, Ó. Retrieval of nearshore bathymetry from Landsat 8 images: A tool for coastal monitoring in shallow waters. *Remote Sens. Environ.* **2015**, *159*, 102–116. [CrossRef]

60. Kabiri, K. Accuracy assessment of near-shore bathymetry information retrieved from Landsat-8 imagery. *Earth Sci. Inform.* **2017**, *10*, 235–245.
61. Kerr, J.M.; Purkis, S. An algorithm for optically-deriving water depth from multispectral imagery in coral reef landscapes in the absence of ground-truth data. *Remote Sens. Environ.* **2018**, *210*, 307–324. [CrossRef]
62. Coastal Thematic Exploitation Platform—Coastal-TEP. Available online: https://www.coastal-tep.eu/geobrowser/ (accessed on 20 November 2019).
63. Wölfl, A.C.; Snaith, H.; Amirebrahimi, S.; Devey, C.W.; Dorschel, B.; Ferrini, V.; Lamarche, G. Seafloor Mapping–the challenge of a truly global ocean bathymetry. *Front. Mar. Sci.* **2019**, *6*, 283.

Article

Atmospheric Correction of GOCI Using Quasi-Synchronous VIIRS Data in Highly Turbid Coastal Waters

Jie Wu [1,2,3], Chuqun Chen [1,2,3,*] and Sravanthi Nukapothula [1,2]

1   State Key Laboratory of Tropical Oceanography, South China Sea Institute of Oceanology, Chinese Academy of Sciences, Guangzhou 510301, China; wujie@scsio.ac.cn (J.W.); sravantii73@gmail.com (S.N.)
2   Guangdong Key Laboratory of Ocean Remote Sensing, South China Sea Institute of Oceanology, Chinese Academy of Sciences, Guangzhou 510301, China
3   University of Chinese Academy of Sciences, Beijing 100049, China
*   Correspondence: cqchen@scsio.ac.cn

Received: 15 November 2019; Accepted: 24 December 2019; Published: 25 December 2019

Abstract: The Geostationary Ocean Color Imager (GOCI) sensor, with high temporal and spatial resolution (eight images per day at an interval of 1 hour, 500 m), is the world's first geostationary ocean color satellite sensor. GOCI provides good data for ocean color remote sensing in the Western Pacific, among the most turbid waters in the world. However, GOCI has no shortwave infrared (SWIR) bands making atmospheric correction (AC) challenging in highly turbid coastal regions. In this paper, we have developed a new AC algorithm for GOCI in turbid coastal waters by using quasi-synchronous Visible Infrared Imaging Radiometer Suite (VIIRS) data. This new algorithm estimates and removes the aerosol scattering reflectance according to the contributing aerosol models and the aerosol optical thickness estimated by VIIRS's near-infrared (NIR) and SWIR bands. Comparisons with other AC algorithms showed that the new algorithm provides a simple, effective, AC approach for GOCI to obtain reasonable results in highly turbid coastal waters.

Keywords: ocean color; GOCI; VIIRS; atmospheric correction; turbid waters

1. Introduction

The total radiance measured by an ocean color sensor is primarily composed of the water-leaving radiance, sea surface radiance, Rayleigh scattering radiance caused by air molecules, and aerosol scattering radiance (which includes aerosol single-scattering radiance and interactive scattering radiance between molecules and aerosols). In open waters, the atmospheric radiance can account for significant portions of the total satellite-measured radiance [1]. Therefore, to determine the properties of the upper ocean, such as colored dissolved organic matter, suspended particulate matter, and chlorophyll-a, accurate atmospheric correction (AC) is required. The Rayleigh scattering radiance can be theoretically computed accurately, owing to the stable distribution of the necessary atmospheric components [2,3]. Therefore, it is key to accurately estimate the aerosol scattering by aerosols and Rayleigh–aerosol interactions in order to determine the water-leaving radiance. For clear waters, the AC algorithm developed by Gordon and Wang [4] (herein named the GW94 algorithm) works quite well. It estimates the aerosol optical properties based on the black pixel assumption, according to which the water-leaving radiance at near-infrared (NIR) bands is assumed to be zero because the water can strongly absorb the light in these bands. However, in turbid coastal waters, the AC is more complicated because this assumption is rarely valid due to the significant suspended sediments backscattering in the NIR bands [5].

GOCI is the world's first geostationary ocean color satellite sensor with high spatial and very high temporal resolution (500 m and 1 h, respectively). It acquires eight images per day from 00:15 GMT to

07:15 GMT in hourly intervals [6]. GOCI exhibits considerable advantages when monitoring regional marine environment changes, and provides various useful products [7–9]. Proper AC of GOCI has been a challenge because it covers a 2500 km × 2500 km square, with the Korean Peninsula at the center, which is one of the most turbid areas in the world [10]. Various AC algorithms for turbid waters have been investigated to separate the water-leaving radiance and the aerosol scattering radiance at NIR bands, most of which are based on regional empirical models. The empirical AC algorithms are strongly dependent on the optical properties of water, which limits the applicability of these algorithms to other waters [11]. For several years, AC algorithms using shortwave infrared (SWIR) bands have been demonstrated, which neglects the water-leaving radiance in the SWIR bands as the water absorption in SWIR bands is much stronger than that in the NIR bands [12–14]. The SWIR-based algorithms are able to derive aerosol products in extremely turbid waters with higher accuracy [15] because they directly derive the aerosol scattering radiance at NIR bands, without any assumption of the marine radiance. However, there is a spectral limitation for GOCI, with only eight visible/NIR bands centered at 412, 443, 490, 555, 660, 680, 745, and 865 nm.

Including SWIR bands on a sensor is expensive but the SWIR bands of other sensors can be used to estimate the aerosol optical properties of the observing area, and then the aerosol radiance at GOCI observing geometries can be evaluated. The Visible Infrared Imaging Radiometer Suite (VIIRS) has two NIR bands with band centers 745 nm and 862 nm, and three SWIR bands with band centers 1238 nm, 1601 nm, and 2257 nm. It is on board the Suomi National Polar-Orbiting Partnership satellite, which was launched on 28 October, 2011, and is a new generation of polar-orbiting satellite [16]. VIIRS was intended to combine and improve the best characteristics of the Sea-Viewing Wide Field-of-View Sensor (SeaWiFS), the Moderate Resolution Imaging Spectroradiometer (MODIS), and other previous ocean color sensors. The spatial resolution of the moderate resolution imagery bands of VIIRS is 750 m at the viewing nadir. Previous research shows that the VIIRS can produce high-quality data for various applications [17–19].

Thus, the main objective of this study is to demonstrate and validate a new practical AC algorithm that estimates and removes the aerosol scattering radiance, according to the aerosol optical properties estimated by quasi-synchronous VIIRS's (QSV) NIR and SWIR bands for GOCI data.

2. Study Sites

The Bohai Sea, Yellow Sea, and East China Sea are parts of the Western Pacific marginal sea. Three highly turbid coastal regions are outlined in the boxes in this region in Figure 1a: (I) Bohai Sea, (II) the southwest coast of Korea, and (III) Changjiang Estuary.

The Bohai Sea is a shallow semi-enclosed sea with an average water depth of 18 m and a maximum depth of ~70 m. It deepens gradually from the coastal bays to the Central Bohai Sea and is characterized by a basin shape [20]. The dominant sediment source in this region is the sediment delivered by the Huanghe [21,22]. The high suspended sediment concentration is mainly distributed in the south of Bohai Sea, especially in the area around the Huanghe Delta [21].

The water depth of the area along the Southwestern Korean coast is less than 50 m. Numerous islands and vast tidal flats are located here and the coastlines in this area are complicated. The suspended sediment concentration of this area is relatively high (>20 g/m^3). The highest values (>100 g/m^3) occur in the winter owing to the stronger northwestern monsoon and shallow water depth, which can induce a resuspension of bottom sediments [23].

The Changjiang Estuary is located along the central-eastern coast of China, with the Hangzhou Bay to the south and the Subei Shallow to the north. The highly suspended sediment concentrations were observed here year-round [24]. The Changjiang River discharges about 390 × 10^6 tons of sediment into the East China Sea annually [25]. The sediment transportation inside the Hangzhou Bay is considerably affected by the secondary Changjiang plume [26]. Another major sediment source for the East China Sea is the resuspension of sediment at the Subei Shallow [27]. Wind-induced vertical mixing and bottom stress tend to resuspend a large amount of sediment, leading to high suspended sediment concentrations, especially in the winter [28,29].

One cloud-free example is chosen from each of these regions. The RGB pictures of the three examples are composed of the total radiance at 490 nm (B), 555 nm (G), and 680 nm (R) bands of GOCI and are shown in Figure 1b–d. More details of the three examples are listed in Table 1.

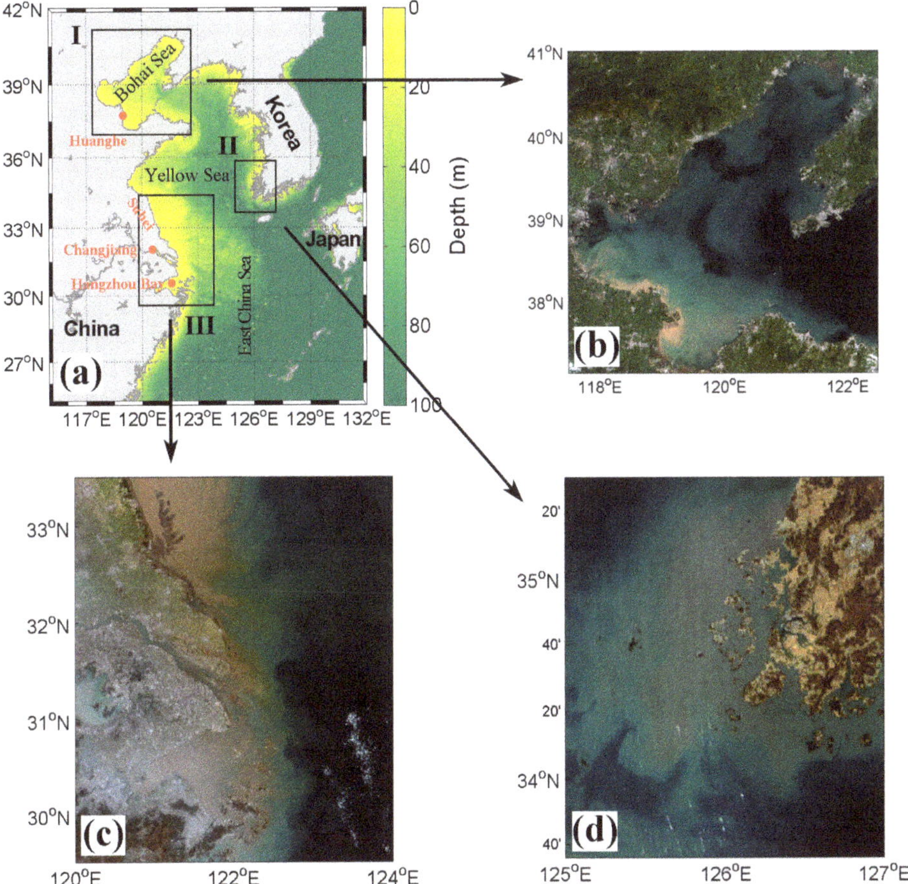

Figure 1. (a) Bathymetry for the Bohai Sea, Yellow Sea, and East China Sea. Three highly turbid water areas are outlined in three boxes (I, II, and III). Panels (**b–d**) are the composed RGB pictures of the three examples used for comparison (B: 490 nm, G: 555 nm, and R: 680 nm).

Table 1. Detail information on the processed images.

Area	Location	Date (MM-DD-YY)	Time Lag (min)
I	Bohai Sea	08-26-2016	9
II	Southwest coast of Korea	03-14-2017	21
III	Changjiang Estuary	03-01-2016	27

3. Method

Reflectance ρ is defined as:

$$\rho(\lambda) = \frac{\pi L(\lambda)}{F_0 \cos \theta_s} \tag{1}$$

where λ denotes the wavelength, F_0 is the extraterrestrial solar irradiance [30], θ_s is the solar zenith angle, and L is the radiance. It is more convenient to work with ρ because it is dimensionless. The total reflectance $\rho_t(\lambda)$ measured by a sensor can be written as:

$$\rho_t(\lambda) = \rho_r(\lambda) + \rho_a(\lambda) + t_v(\lambda)\rho_w(\lambda) \tag{2}$$

where ρ_r denotes Rayleigh scattering reflectance, ρ_a denotes aerosol multiple scattering reflectance, ρ_w denotes water-leaving reflectance, and t_v denotes the Rayleigh–aerosol diffuse transmittance from the sea surface to the satellite. It should be noted that the reflectance contributions from the sun glint and whitecaps are ignored [31,32].

Determining the remote sensing reflectance (R_{rs}, unit: sr^{-1}) and normalized water-leaving reflectance (ρ_{wn}) is the ultimate goal of AC because they are fundamental parameters that are widely used for deriving the water constituent concentrations and water quality parameters. These can be calculated by:

$$\rho_{wn}(\lambda) = \frac{\rho_w(\lambda)}{t_s(\lambda)} \tag{3}$$

$$R_{rs}(\lambda) = \frac{L_w(\lambda)}{F_0(\lambda)t_s(\lambda)\cos\theta_s} \tag{4}$$

where t_s is the diffuse transmittance from the sun to the sea surface.

In this section, we briefly review the previous AC algorithms that were adopted in the main GOCI processing software: SeaDAS (SeaWiFS Data Analysis System) and GDPS (GOCI data process system). Then, the development of the new algorithm is described.

3.1. Previous AC Algorithms

The key problem of AC over turbid waters is the removal of the water-leaving reflectance at NIR bands. Numerous red/NIR modeling approaches have been investigated to deal with non-zero ρ_w(red/NIR) values within the AC process. Three kinds of AC approaches are briefly reviewed herein—the B2010 AC algorithm proposed by Bailey et al. [33] is currently adopted in SeaDAS, while the A2012 algorithm proposed by Ahn et al. [34] and the L2013 algorithm proposed by Lee et al. [35] are implemented in GDPS.

The B2010 algorithm uses an iterative solution to separate ρ_w(NIR) and ρ_a(NIR). The ρ_w(NIR) is first assumed to be zero to complete the GW94 AC process. This gives an initial estimate of $R_{rs}(\lambda)$. Next, the concentration of chlorophyll-a is preliminarily estimated by the initial value of $R_{rs}(\lambda)$. Then, the absorption coefficient in the red band is obtained via a chlorophyll-a–based empirical relationship. The absorption coefficient and R_{rs} in the red band are used to solve the backscattering coefficient, which is used to extrapolate the ρ_w at the NIR band. Finally, the new values of ρ_w(NIR) are used to correct the non-zero water contribution, and the GW94 AC step is repeated for a new iteration.

The A2012 algorithm is the GOCI standard AC algorithm. It is theoretically based on the GW94 algorithm with partial modifications. The initial ρ_w(NIR) values are also assumed to be zero to execute the GW94 algorithm. The newly estimated ρ_{wn} in the red band is used to correct the ρ_{wn}(NIR) by an empirical polynomial relationship model [36]:

$$\rho_{wn}(745\mathrm{nm}) = \sum_{n=0}^{5} j_n[\rho_{wn}(660\mathrm{nm})]^n \tag{5}$$

$$\rho_{wn}(865\text{nm}) = \sum_{n=1}^{2} k_n [\rho_{wn}(745\text{nm})]^n \tag{6}$$

where j_n and k_n are known polynomial coefficients.

The L2013 algorithm is from the Management Unit of the North Sea Mathematical Models (MUMM) [37] algorithm, with some modified steps for extremely turbid water. In the original MUMM, the parameter α, representing the ratio of ρ_w at the NIR wavelengths, is assumed to have spatial homogeneity; however, the value of α changes with the concentration of suspended particles in extremely turbid waters, which would cause an underestimation of the water-leaving radiances [38]. The L2013 algorithm calculates the α value adaptively using an NIR water-leaving reflectance model:

$$\alpha = \sum_{n=0}^{3} c_n [\rho_w(745\text{nm})]^n \tag{7}$$

where the c_n values are known polynomial coefficients.

The algorithms mentioned above can improve the data quality in turbid waters, but the empirical relationships are highly reliant on in situ datasets; even the empirical relationship used in L2013 is derived from the nearest non-turbid AC algorithm [39]. These algorithms are, therefore, only regionally applicable [40]. For example, the B2010 method only works properly in low to moderately turbid waters [41,42], as the chlorophyll-a–based relationship used in the bio-optical model might not be appropriate to extrapolate the ρ_w(NIR) in waters whose optical properties are dominated by non-algal particles [41,42]. The A2012 method is suitable for sediment-dominated waters but also fails for extremely turbid waters because the ρ_{wn}(660 nm) can become optically saturated [43].

3.2. The Development of the New AC Algorithm

In order to distinguish the two sensors, the superscripts V and G represent VIIRS and GOCI, respectively. The procedure for the new algorithm can be divided into four parts:

- Extracting the $\rho_a(\text{NIR}^V)$ of VIIRS;
- Estimating the aerosol properties by $\rho_a(\text{NIR}^V)$;
- Calculating the $\rho_a(\lambda^G)$ at GOCI observing geometries according to the aerosol properties; and
- Removing the $\rho_a(\lambda^G)$ and completing the GOCI AC process.

The removal of the $\rho_w(\text{NIR}^V)$ of VIIRS follows the Shortwave Infrared Exponential (SWIRE) algorithm [12], in which the Rayleigh-corrected reflectance values ρ_{rc} ($\rho_{rc} = \rho_t - \rho_r$) of clear waters were assumed to be an exponential function of wavelength at the NIR/SWIR bands:

$$\rho_{ef}(\lambda) = ae^{b\lambda} \tag{8}$$

where ρ_{ef} is the extrapolated Rayleigh-corrected reflectance, and a and b are the fitting coefficients.

The ρ_{rc} values at the three VIIRS SWIR bands (1238, 1601, and 2257 nm) are used to calculate the coefficients a and b because the ρ_{rc} at the NIR bands is strongly influenced by suspended sediments scattering in turbid waters. The newly estimated ρ_{ef}(NIR) can be considered to be the ρ_{rc} of the optical equivalent clear water, influenced only by aerosol scattering, and can be used for the GW94 AC of the visible bands. The SWIR-based methods improve the ocean color products in the turbid coastal waters, but they require higher signal-to-noise ratios (SNR) for SWIR bands because of the longer extrapolated distances and lower signals [44]. Wang and Shi [14] used two SWIR bands to derive the aerosol scattering radiance, but their method did not achieve better results than the NIR-based methods in non-turbid waters because significant noise occurred in the derived products [45,46]. Generally, the more SWIR bands used, the more accurately can the aerosol scattering reflectance be estimated, and the requirements of SNR can probably be reduced. As a result, the SWIRE method represented an improvement over the SWIR-based methods [47].

The newly estimated $\rho_{ef}(NIR^V)$ can be considered as the $\rho_a(NIR^V)$ and can be applied to evaluate the aerosol properties of the observing area, including the contributing aerosol models and aerosol optical thickness τ_a. The contributing aerosol models are selected by the look-up-tables (LUTs) for 80 aerosol tables constructed by Ahmad et al. [48]. The parameters used to derive the aerosol single-scattering albedo (ω_a), aerosol scattering phase function (P_a), and extinction coefficient (β_a) are stored in the LUTs for each aerosol model and various geometries. The LUTs also contain the coefficients relating single to multiple scattering and the coefficients (A and B) used to calculate the Rayleigh–aerosol, diffuse transmittance in the following form [49]:

$$t = A \exp(-B\tau_a) \tag{9}$$

The aerosol optical thickness is the normalized extinction coefficient due to absorption and scattering from the direct beam and is a key parameter used to define the optical state of the atmosphere [50,51].

The estimation of the aerosol properties follows the GW94 method [4], which is one of the AC algorithms widely used for clear waters. In this algorithm, $\rho_a(NIR^V)$ is first converted into $\rho_{as}(NIR^V)$ [52] by solving the quadratic equation:

$$\ln(\rho_a) = \sum_{n=0}^{2} \ln[p_n(\rho_{as})^n] \tag{10}$$

where the p_n values are the quadratic coefficients stored in the LUTs. The AC parameter is defined as:

$$\varepsilon(NIR_S, NIR_L) = \frac{\rho_{as}(NIR_S)}{\rho_{as}(NIR_L)} \tag{11}$$

where NIR_S is the short-wavelength NIR band, and NIR_L is the long-wavelength NIR band. For the VIIRS sensor, the NIR_S and NIR_L are 745 nm and 862 nm, respectively. Theoretically, each candidate aerosol model has its own ε value (ε^M) for a specific geometry:

$$\varepsilon^M(NIR_S, NIR_L) = \frac{\omega_a(NIR_S)\beta_a(NIR_S)P_a(\theta_v, \theta_s, NIR_S)}{\omega_a(NIR_L)\beta_a(NIR_L)P_a(\theta_v, \theta_s, NIR_L)} \tag{12}$$

Two aerosol models with different weighting factors whose $\varepsilon(NIR_S, NIR_L)$ values are the closest to the $\varepsilon^M(NIR_S, NIR_L)$ averaged over all candidate models, are chosen to represent the aerosol type over the pixel. The selection of the closest aerosol models can dominate the estimation of the water-leaving reflectance, especially over turbid coastal waters [53]. The $\tau_a(745 \text{ nm})$ values of the two contributing models can be retrieved by [4]:

$$\tau_a(\lambda) = \frac{4\cos\theta_v \cos\theta_s \rho_{as}(\lambda)}{\omega_a(\lambda)P_a(\theta_v, \theta_s, \lambda)} \tag{13}$$

The VIIRS sensor is located in a sun-synchronous orbit and provides one image per day, whereas the GOCI sensor is located in a geostationary orbit and provides eight images per day. With the hourly observations from the GOCI, the VIIRS image has a corresponding GOCI image with a time difference of less than half an hour, which means the atmospheric conditions are nearly invariant. Therefore, the aerosol scattering reflectance of the GOCI observing geometries can be derived according to the known contributing aerosol models, weighting factor, and the $\tau_a(745 \text{ nm})$. First, calculate the $\tau_a(\lambda^G)$ of each contributing aerosol model by:

$$\frac{\tau_a(\lambda)}{\tau_a(745\text{nm})} = \frac{\beta_a(\lambda)}{\beta_a(745\text{nm})} \tag{14}$$

Second, calculate the $\rho_{as}(\lambda^G)$ by Equation (13). Next, convert $\rho_{as}(\lambda^G)$ to $\rho_{am}(\lambda^G)$ via Equation (10). Finally, the effects of the two contributing aerosol models are superimposed by weighting factors to estimate the total aerosol scattering reflectance of GOCI. In order to maintain the higher spatial resolution of GOCI, the aerosol parameters estimated from VIIRS are resampled to the GOCI data grid by the nearest neighbor method.

4. Results and Discussion

In the absence of in situ data, the results of the new algorithm (represented by QSV) are compared with those processed by the B2010 [33], A2012 [34], and L2013 [35] algorithms. VIIRS and GOCI Level-1 data were obtained online (https://oceancolor.gsfc.nasa.gov). For the QSV algorithm, the Rayleigh-corrected reflectance values were obtained by using the SeaDAS 7.4 l2gen processor, and these values were used as the inputs of the aerosol scattering correction procedure, which was performed by using Interactive Data Language (IDL). With default settings, the B2010 algorithm was realized by the SeaDAS 7.4 software, while the A2012 and L2013 algorithms were realized by the GDPS v1.4.1 software. The three examples listed in Table 1 are processed.

Figures 2–4 represent the R_{rs} values retrieved by four different algorithms centered at the Bohai Sea, the southwest coast of Korea, and the Changjiang Estuary, respectively. Panels (a–d), (e–h), and (i–l) are results at wavelengths of 490, 555, and 680 nm, respectively. The regions in which the algorithms fail to make an AC because of highly turbid waters or due to the presence of clouds and land, are shown with RGB pictures. It can be observed that the R_{rs} distributions are similar for all four algorithms, but the B2010, A2012, and L2013 algorithms result in varying degrees of failures, while the QSV algorithm can successfully execute the AC in the highly turbid coastal waters. The unmasked abnormal low estimations from L2013 also can be observed at 35°N, 126°E in Figure 3b,f,j, highlighted by the red circles.

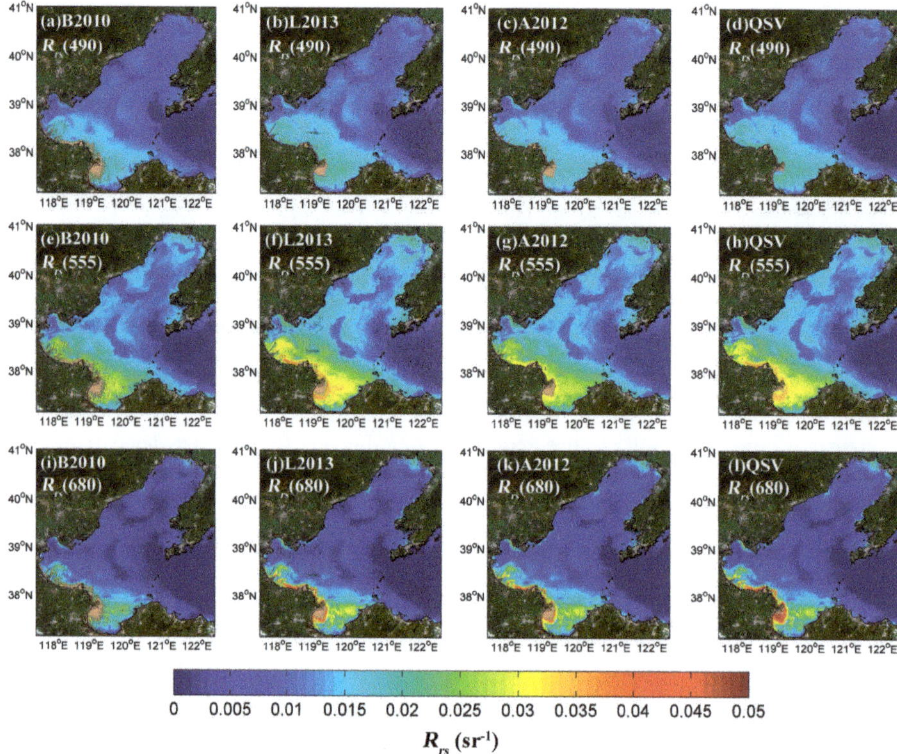

Figure 2. Comparisons of the R_{rs} distributions (unit: sr^{-1}) at (**a–d**) 490, (**e–h**) 555, and (**i–l**) 680 nm bands centered on the Bohai Sea on 26 August 2016 and processed by the B2010, L2013, A2012, and QSV algorithms. The regions in which the algorithms fail to make an AC because of highly turbid waters, or the presence of clouds and land are shown by the composed RGB pictures (B: 490 nm, G: 555 nm, R: 680 nm).

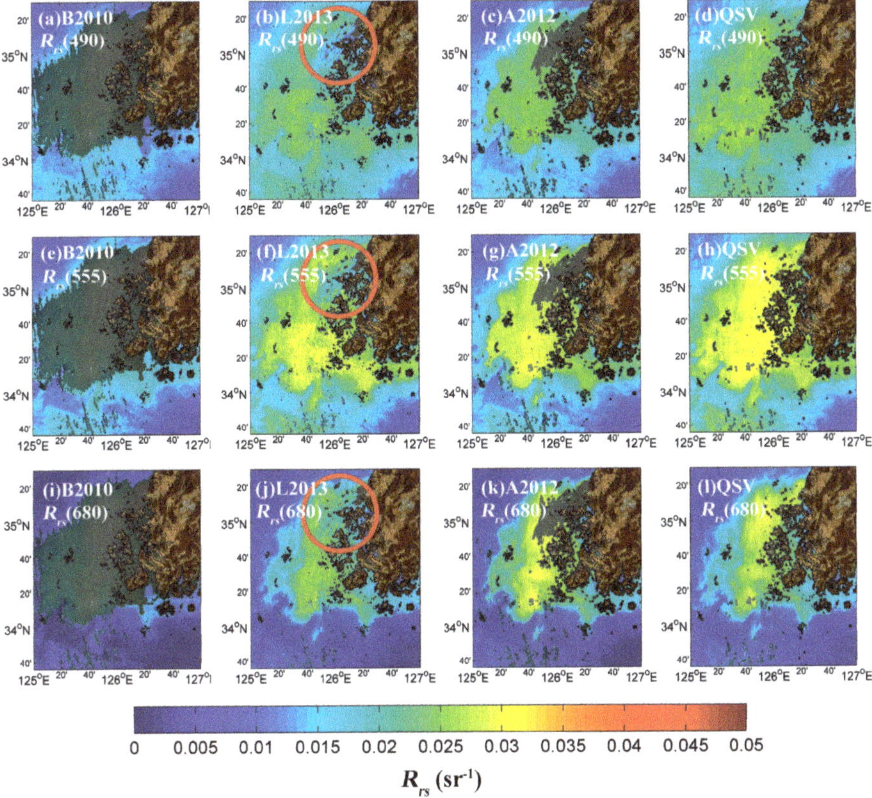

Figure 3. Comparisons of the R_{rs} distributions (unit: sr^{-1}) at the (**a–d**) 490, (**e–h**) 555, and (**i–l**) 680 nm bands centered on the southwest coast of Korea on 14 March 2017 and processed by the B2010, L2013, A2012, and QSV algorithms. The regions in which the algorithms fail to make an AC because of highly turbid waters, or the presence of clouds and land are shown by the composed RGB pictures (B: 490 nm, G: 555 nm, and R: 680 nm). The unmasked abnormal low estimations from L2013 are highlighted by the red circles in panels (**b,f,j**).

To further demonstrate the differences, we also extracted the $R_{rs}(\lambda)$ values for pixels along the red arrow in Figure 4a and plotted their profiles in Figure 5. A distance of 0 along the x-axis indicates the location closest to the coast. In general, the R_{rs} data are low in the open ocean and high near the coast. The mean absolute difference in the R_{rs} values between the A2012 and QSV algorithms is the smallest, with values of 0.00334, 0.00138, 0.00100, 0.00091, 0.00110, and 0.00182 sr^{-1} at wavelengths of 412, 443, 490, 555, 660, and 680 nm, respectively. At around 0 to 110 pixels along the transect, the B2010 algorithm is invalid. The results of the L2013 are relatively higher than those of the A2012 and QSV algorithms, at distances larger than 50 pixels along the transect. Abnormal fluctuations of the L2013 curves can be observed at around 25–50 pixels. The A2012 and L2013 methods both fail at around 0–25 pixels. The QSV is the only algorithm that can retrieve continuous and reasonable values of R_{rs} along the total transect.

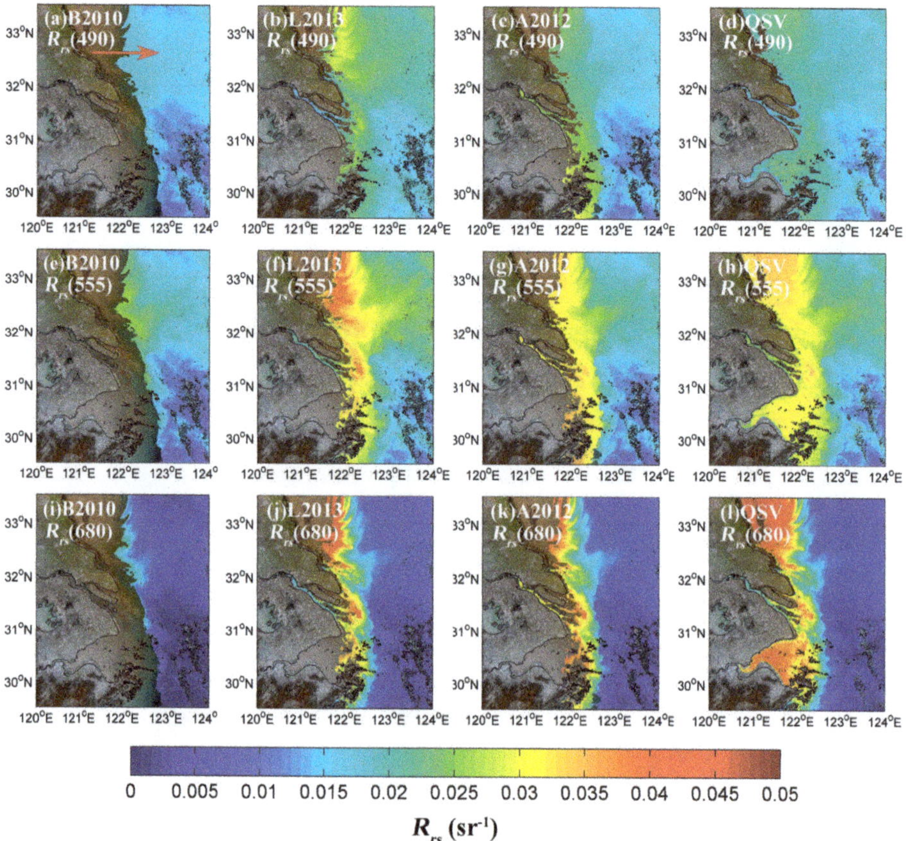

Figure 4. Comparisons of the R_{rs} distributions (unit: sr^{-1}) at (**a**–**d**) 490, (**e**–**h**) 555, and (**i**–**l**) 680 nm bands centered on the Changjiang Estuary on 1 March 2016 and processed by the B2010, L2013, A2012, and QSV algorithms. The red arrow in panel (**a**) refers to the location and direction of the transect used in Figure 5. The regions in which the algorithms fail to make an AC because of highly turbid waters, or the presence of clouds and land are shown by the composed RGB pictures (B: 490 nm, G: 555 nm, and R: 680 nm).

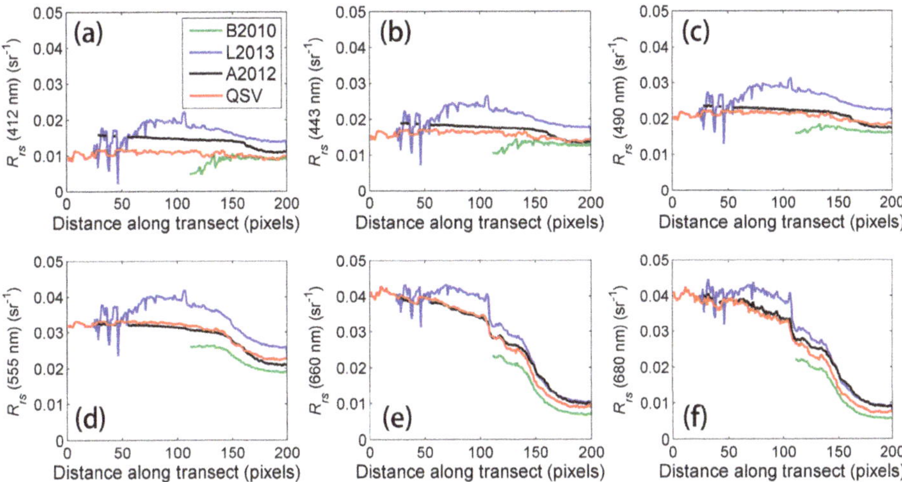

Figure 5. Comparison of the R_{rs} values extracted from the pixels along the red arrow in Figure 4a at wavelengths of (**a**) 412, (**b**) 443, (**c**) 490, (**d**) 555, (**e**) 660, and (**f**) 680 nm. A distance of 0 in the x-axis indicates the location closest to the coast.

Three sets of R_{rs} spectra derived from four different AC algorithms in highly turbid waters at station (a), moderately turbid water at station (b), and clear water at station (c) are compared and shown in Figure 6. At station (c), the curves of the B2010, A2012, and QSV algorithms tend to be coincident, while the R_{rs} values retrieved by L2013 are relatively high. At station (b), the B2010 algorithm shows slightly lower estimations. At station (a), the B2010 algorithm has no valid results, and the difference between L2013 and A2012/QSV is more pronounced.

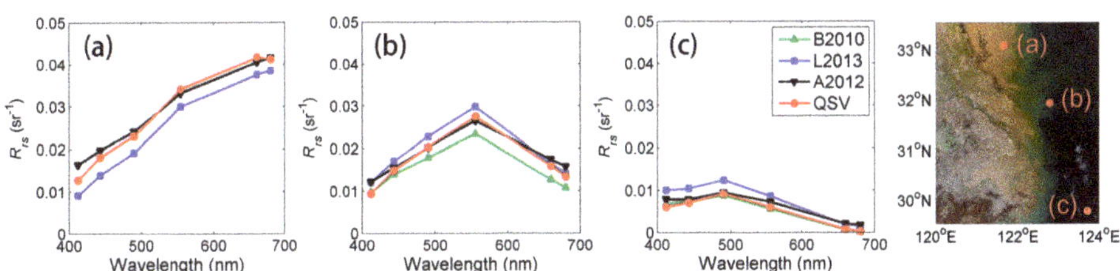

Figure 6. Comparisons of the R_{rs} spectra derived from four different algorithms in highly turbid waters at station (**a**), moderately turbid waters at station (**b**), and clear waters at station (**c**). The RGB image shows the locations of the three stations.

Since these four algorithms are developed for turbid waters, they can all improve the data quality in turbid waters. The B2010, A2012, and L2013 algorithms are based on empirical models and are highly regionally dependent. Significant errors or failures can be induced when these methods are applied to the ocean regions they are not suitable for. Even though the QSV algorithm can be applied to only one scene of the GOCI data, it can obtain reasonable results in extremely turbid coastal waters. The use of three SWIR bands can improve the data quality of the AC retrievals in clear waters without obvious noise. In general, the QSV algorithm provides a simple, effective, new AC approach for GOCI to obtain reasonable results in highly turbid waters and provides an alternative method to validating other AC methods.

5. Conclusions

An alternative AC algorithm using quasi-synchronous VIIRS data for GOCI in highly turbid waters is presented in this paper. GOCI covers highly turbid waters in the Western Pacific region; AC has been a challenge in these turbid ocean regions because of the lack of SWIR bands. The new algorithm shows its superiority in highly turbid coastal waters, and the results could be at least be used for the validation of other AC methods. However, there are still some limitations that should be noted. The major disadvantage of this algorithm is that only one VIIRS image per day is not sufficient for GOCI, and more in situ data are needed for the validation. In the future, GOCI-2 will have full-disk coverage with higher resolution and five more bands than GOCI does, but it will still not have any SWIR bands [54]. Including SWIR bands on a sensor is expensive, therefore, further research is necessary on how to best take advantage of SWIR bands from other ocean color sensors.

Author Contributions: Conceptualization, C.C. and J.W.; implementation, C.C. and J.W.; validation, J.W. and S.N.; writing-original draft preparation, J.W. and S.N.; writing-review and editing, C.C., J.W., and S.N.; funding acquisition: C.C. All authors have read and agreed to the published version of the manuscript.

Funding: This research was funded by the National Natural Science Foundation of China [grant number U1901215], the National Key Research and Development Projects [grant number 2018YFC1406604], the Scientific Research Key Project from Guangzhou City [grant number 201707020031], the National Natural Science Foundation of China [grant number 41276182], and the Project of State Key Laboratory of Tropical Oceanography [grant number LTOZZ1705].

Acknowledgments: We wish to thank NASA's Ocean Biology Processing Group (OBPG) for providing the SeaDAS software package and VIIRS/SNPP Level 1-A data. We also wish to thank the Korea Ocean Satellite Center for providing the GDPS software package and GOCI Level-1 data.

Conflicts of Interest: The authors declare no conflict of interest.

References

1. Wang, M.H. A sensitivity study of the SeaWiFS atmospheric correction algorithm: Effects of spectral band variations. *Remote Sens. Environ.* **1999**, *67*, 348–359. [CrossRef]
2. Wang, M.H. A refinement for the Rayleigh radiance computation with variation of the atmospheric pressure. *Int. J. Remote Sens.* **2005**, *26*, 5651–5663. [CrossRef]
3. Wang, M.H. Rayleigh radiance computations for satellite remote sensing: Accounting for the effect of sensor spectral response function. *Opt. Express* **2016**, *24*, 2414–2429. [CrossRef] [PubMed]
4. Gordon, H.R.; Wang, M. Retrieval of water-leaving radiance and aerosol optical thickness over the oceans with SeaWiFS: A preliminary algorithm. *Appl. Opt.* **1994**, *33*, 443–452. [CrossRef] [PubMed]
5. Emberton, S.; Chittka, L.; Cavallaro, A.; Wang, M.H. Sensor Capability and Atmospheric Correction in Ocean Colour Remote Sensing. *Remote Sens.* **2016**, *8*, 1. [CrossRef]
6. Ryu, J.H.; Han, H.J.; Cho, S.; Park, Y.J.; Ahn, Y.H. Overview of Geostationary Ocean Color Imager (GOCI) and GOCI Data Processing System (GDPS). *Ocean Sci. J.* **2012**, *47*, 223–233. [CrossRef]
7. Hu, Z.F.; Pan, D.L.; He, X.Q.; Bai, Y. Diurnal Variability of Turbidity Fronts Observed by Geostationary Satellite Ocean Color Remote Sensing. *Remote Sens.* **2016**, *8*, 147. [CrossRef]
8. Choi, J.K.; Min, J.E.; Noh, J.H.; Han, T.H.; Yoon, S.; Park, Y.J.; Moon, J.E.; Ahn, J.H.; Ahn, S.M.; Park, J.H. Harmful algal bloom (HAB) in the East Sea identified by the Geostationary Ocean Color Imager (GOCI). *Harmful Algae* **2014**, *39*, 295–302. [CrossRef]
9. Liu, R.J.; Zhang, J.; Yao, H.Y.; Cui, T.W.; Wang, N.; Zhang, Y.; Wu, L.J.; An, J.B. Hourly changes in sea surface salinity in coastal waters recorded by Geostationary Ocean Color Imager. *Estuar Coast Shelf Sci.* **2017**, *196*, 227–236. [CrossRef]
10. Choi, J.K.; Park, Y.J.; Ahn, J.H.; Lim, H.S.; Eom, J.; Ryu, J.H. GOCI, the world's first geostationary ocean color observation satellite, for the monitoring of temporal variability in coastal water turbidity. *J. Geophys. Res. Ocean.* **2012**, *117*. [CrossRef]
11. Vanhellemont, Q.; Ruddick, K. Advantages of high quality SWIR bands for ocean colour processing: Examples from Landsat-8. *Remote Sens. Environ.* **2015**, *161*, 89–106. [CrossRef]

12. He, Q.J.; Chen, C.Q. A new approach for atmospheric correction of MODIS imagery in turbid coastal waters: A case study for the Pearl River Estuary. *Remote Sens. Lett.* **2014**, *5*, 249–257. [CrossRef]

13. Wang, M.H.; Shi, W. Estimation of ocean contribution at the MODIS near-infrared wavelengths along the east coast of the US: Two case studies. *Geophys. Res. Lett.* **2005**, *32*. [CrossRef]

14. Wang, M.H.; Shi, W. The NIR-SWIR combined atmospheric correction approach for MODIS ocean color data processing. *Opt. Express* **2007**, *15*, 15722–15733. [CrossRef] [PubMed]

15. Jiang, L.D.; Wang, M.H. Improved near-infrared ocean reflectance correction algorithm for satellite ocean color data processing. *Opt. Express* **2014**, *22*, 21657–21678. [CrossRef]

16. Cao, C.Y.; de Luccia, F.J.; Xiong, X.X.; Wolfe, R.; Weng, F.Z. Early On-Orbit Performance of the Visible Infrared Imaging Radiometer Suite Onboard the Suomi National Polar-Orbiting Partnership (S-NPP) Satellite. *IEEE Trans. Geosci. Remote* **2014**, *52*, 1142–1156. [CrossRef]

17. Shi, W.; Zhang, Y.L.; Wang, M.H. Deriving Total Suspended Matter Concentration from the Near-Infrared-Based Inherent Optical Properties over Turbid Waters: A Case Study in Lake Taihu. *Remote Sens.* **2018**, *10*, 333. [CrossRef]

18. Cao, C.Y.; Xiong, J.; Blonski, S.; Liu, Q.H.; Uprety, S.; Shao, X.; Bai, Y.; Weng, F.Z. Suomi NPP VIIRS sensor data record verification, validation, and long-term performance monitoring. *J. Geophys. Res. Atmos.* **2013**, *118*, 11664–11678. [CrossRef]

19. Wang, M.H.; Son, S. VIIRS-derived chlorophyll-a using the ocean color index method. *Remote Sens. Environ.* **2016**, *182*, 141–149. [CrossRef]

20. Bian, C.W.; Jiang, W.S.; Pohlmann, T.; Sundermann, J. Hydrography-Physical Description of the Bohai Sea. *J. Coast. Res.* **2016**, *74*, 1–12. [CrossRef]

21. Wang, H.J.; Wang, A.M.; Bi, N.H.; Zeng, X.M.; Xiao, H.H. Seasonal distribution of suspended sediment in the Bohai Sea, China. *Cont. Shelf Res.* **2014**, *90*, 17–32. [CrossRef]

22. Milliman, J.D.; Li, F.; Zhao, Y.Y.; Zheng, T.M.; Limeburner, R. Suspended Matter Regime in the Yellow Sea. *Prog. Oceanogr.* **1986**, *17*, 215–227. [CrossRef]

23. Min, J.E.; Choi, J.K.; Yang, H.; Lee, S.; Ryu, J.H. Monitoring changes in suspended sediment concentration on the southwestern coast of Korea. *J. Coast. Res.* **2014**, 133–138. [CrossRef]

24. Dong, L.X.; Guan, W.B.; Chen, Q.; Li, X.H.; Liu, X.H.; Zeng, X.M. Sediment transport in the Yellow Sea and East China Sea. *Estuar. Coast. Shelf Sci.* **2011**, *93*, 248–258. [CrossRef]

25. Wang, Y.; Shen, J.; He, Q.; Zhu, L.; Zhang, D. Seasonal variations of transport time of freshwater exchanges between Changjiang Estuary and its adjacent regions. *Estuar. Coast. Shelf Sci.* **2015**, *157*, 109–119. [CrossRef]

26. Su, J.L.; Wang, K.S. Changjiang River Plume and Suspended Sediment Transport in Hangzhou Bay. *Cont. Shelf Res.* **1989**, *9*, 93–111.

27. Saito, Y.; Yang, Z.S. Historical change of the Huanghe (Yellow River) and its impact on the sediment budget of the East China Sea. In *Global Fluxs of Carbon and Its Related Substances in the Coastal Sea-Ocean Atmosphere System*; M & J International: Yokohama, Japan, 1995.

28. Bian, C.W.; Jiang, W.S.; Quan, Q.; Wang, T.; Greatbatch, R.J.; Li, W. Distributions of suspended sediment concentration in the Yellow Sea and the East China Sea based on field surveys during the four seasons of 2011. *J. Mar. Syst.* **2013**, *121*, 24–35. [CrossRef]

29. Bian, C.W.; Jiang, W.S.; Greatbatch, R.J. An exploratory model study of sediment transport sources and deposits in the Bohai Sea, Yellow Sea, and East China Sea. *J. Geophys. Res. Ocean.* **2013**, *118*, 5908–5923. [CrossRef]

30. Thuillier, G.; Herse, M.; Simon, P.C.; Labs, D.; Mandel, H.; Gillotay, D.; Foujols, T. The visible solar spectral irradiance from 350 to 850 nm as measured by the SOLSPEC spectrometer during the ATLAS I mission. *Sol. Phys.* **1998**, *177*, 41–61. [CrossRef]

31. Gordon, H.R.; Wang, M.H. Influence of Oceanic Whitecaps on Atmospheric Correction of Ocean-Color Sensors. *Appl. Opt.* **1994**, *33*, 7754–7763. [CrossRef]

32. Wang, M.H.; Bailey, S.W. Correction of sun glint contamination on the SeaWiFS ocean and atmosphere products. *Appl. Opt.* **2001**, *40*, 4790–4798. [CrossRef]

33. Bailey, S.W.; Franz, B.A.; Werdell, P.J. Estimation of near-infrared water-leaving reflectance for satellite ocean color data processing. *Opt. Express* **2010**, *18*, 7521–7527. [CrossRef] [PubMed]

34. Ahn, J.H.; Park, Y.J.; Ryu, J.H.; Lee, B.; Oh, I.S. Development of Atmospheric Correction Algorithm for Geostationary Ocean Color Imager (GOCI). *Ocean Sci. J.* **2012**, *47*, 247–259. [CrossRef]

35. Lee, B.; Ahn, J.H.; Park, Y.J.; Kim, S.W. Turbid water atmospheric correction for GOCI: Modification of MUMM algorithm. *Korean J. Remote Sens.* **2013**, *29*, 173–182. [CrossRef]

36. Park, Y.J.; Ahn, Y.H.; Han, H.J.; Yang, H.; Moon, J.E.; Ahn, J.H.; Lee, B.R.; Min, J.E.; Lee, S.J.; Kim, K.S.; et al. *GOCI Level 2 Ocean Color Products (GDPS 1.3) Brief Algorithm Description*; Korea Ocean Satellite Center, Korea Institute of Ocean Science and Technology: Ansan, Korea, 2014.

37. Ruddick, K.G.; Ovidio, F.; Rijkeboer, M. Atmospheric correction of SeaWiFS imagery for turbid coastal and inland waters. *Appl. Opt.* **2000**, *39*, 897–912. [CrossRef] [PubMed]

38. Goyens, C.; Jamet, C.; Schroeder, T. Evaluation of four atmospheric correction algorithms for MODIS-Aqua images over contrasted coastal waters. *Remote Sens. Environ.* **2013**, *131*, 63–75. [CrossRef]

39. Hu, C.M.; Carder, K.L.; Muller-Karger, F.E. Atmospheric correction of SeaWiFS imagery over turbid coastal waters: A practical method. *Remote Sens. Environ.* **2000**, *74*, 195–206. [CrossRef]

40. Huang, X.C.; Zhu, J.H.; Han, B.; Jamet, C.; Tian, Z.; Zhao, Y.L.; Li, J.; Li, T.J. Evaluation of Four Atmospheric Correction Algorithms for GOCI Images over the Yellow Sea. *Remote Sens.* **2019**, *11*, 1631. [CrossRef]

41. Goyens, C.; Jamet, C.; Ruddick, K.G. Spectral relationships for atmospheric correction. II. Improving NASA's standard and MUMM near infra-red modeling schemes. *Opt. Express* **2013**, *21*, 21176–21187. [CrossRef]

42. Goyens, C.; Jamet, C.; Ruddick, K.G. Spectral relationships for atmospheric correction. I. Validation of red and near infra-red marine reflectance relationships. *Opt. Express* **2013**, *21*, 21162–21175. [CrossRef]

43. Doxaran, D.; Froidefond, J.M.; Lavender, S.; Castaing, P. Spectral signature of highly turbid waters—Application with SPOT data to quantify suspended particulate matter concentrations. *Remote Sens. Environ.* **2002**, *81*, 149–161. [CrossRef]

44. Wang, M.H.; Son, S.; Shi, W. Evaluation of MODIS SWIR and NIR-SWIR atmospheric correction algorithms using SeaBASS data. *Remote Sens. Environ.* **2009**, *113*, 635–644. [CrossRef]

45. Wang, M.H.; Shi, W. Sensor Noise Effects of the SWIR Bands on MODIS-Derived Ocean Color Products. *IEEE Trans. Geosci. Remote* **2012**, *50*, 3280–3292. [CrossRef]

46. Carswell, T.; Costa, M.; Young, E.; Komick, N.; Gower, J.; Sweeting, R. Evaluation of MODIS-Aqua Atmospheric Correction and Chlorophyll Products of Western North American Coastal Waters Based on 13 Years of Data. *Remote Sens.* **2017**, *9*, 1063. [CrossRef]

47. Ye, H.B.; Chen, C.Q.; Yang, C.Y. Atmospheric Correction of Landsat-8/OLI Imagery in Turbid Estuarine Waters: A Case Study for the Pearl River Estuary. *IEEE J. STARS* **2017**, *10*, 252–261. [CrossRef]

48. Ahmad, Z.; Franz, B.A.; McClain, C.R.; Kwiatkowska, E.J.; Werdell, J.; Shettle, E.P.; Holben, B.N. New aerosol models for the retrieval of aerosol optical thickness and normalized water-leaving radiances from the SeaWiFS and MODIS sensors over coastal regions and open oceans. *Appl. Opt.* **2010**, *49*, 5545–5560. [CrossRef]

49. Yang, H.Y.; Gordon, H.R. Remote sensing of ocean color: Assessment of water-leaving radiance bidirectional effects on atmospheric diffuse transmittance. *Appl. Opt.* **1997**, *36*, 7887–7897. [CrossRef]

50. Masmoudi, M.; Chaabane, M.; Tanre, D.; Gouloup, P.; Blarel, L.; Elleuch, F. Spatial and temporal variability of aerosol: Size distribution and optical properties. *Atmos. Res.* **2003**, *66*, 1–19. [CrossRef]

51. Queface, A.J.; Piketh, S.J.; Annegarn, H.J.; Holben, B.N.; Uthui, R.J. Retrieval of aerosol optical thickness and size distribution from the CIMEL Sun photometer over Inhaca Island, Mozambique. *J. Geophys. Res. Atmos.* **2003**, *108*. [CrossRef]

52. Franz, B.A. rhoa_to_rhoas—MS aerosol reflectance to SS aerosol reflectance. In *Aerosol.c in SeaDAS Code*; NASA's Ocean Biology Processing Group (OBPG), 2004. Available online: http://seadas.gsfc.nasa.gov (accessed on 8 January 2016).

53. McCarthy, S.C.; Gould, R.W.; Richman, J.; Kearney, C.; Lawson, A. Impact of Aerosol Model Selection on Water-Leaving Radiance Retrievals from Satellite Ocean Color Imagery. *Remote Sens.* **2012**, *4*, 3638–3665. [CrossRef]

54. Cho, S. Introduction of GOCI and GOCI-II Mission with Lunar Calibration. In *Lunar Calibration Workshop*; EUMETSAT: Darmstadt, Germany; Available online: http://gsics.atmos.umd.edu/pub/Development/LunarCalibrationWorkshop/4b_Cho_GOCI2.pdf (accessed on 10 August 2018).

Article

Synergy of Satellite Remote Sensing and Numerical Ocean Modelling for Coastal Geomorphology Diagnosis

Mario Benincasa [1,2,*]**, Federico Falcini** [1]**, Claudia Adduce** [2] **and Gianmaria Sannino** [3] **and Rosalia Santoleri** [1]

[1]   CNR-ISMAR, Institute of Marine Sciences, National Research Council, 00133 Rome, Italy; federico.falcini@cnr.it (F.F.); rosalia.santoleri@cnr.it (R.S.)

[2]   Department of Engineering, Roma Tre University, 00146 Rome, Italy; claudia.adduce@uniroma3.it

[3]   ENEA—Climate Modelling and Impacts Laboratory, 00060 Rome, Italy; gianmaria.sannino@enea.it

*   Correspondence: mario.benincasa@artov.ismar.cnr.it

Received: 28 August 2019; Accepted: 1 November 2019; Published: 12 November 2019

Abstract: Sediment dynamics is the primary driver of the evolution of the coastal geomorphology and of the underwater shelf clinoforms. In this paper, we focus on mesoscale and sub-mesoscale processes, such as coastal currents and river plumes, and how they shape the sediment dynamics at regional or basin spatial scales. A new methodology is developed that combines observational data with numerical modelling: the aim is to pair satellite measurements of suspended sediment with velocity fields from numerical oceanographic models, to obtain an estimation of the sediment flux. A numerical divergence of this flux is then computed. The divergence field thus obtained shows how the aforementioned mesoscale processes distribute the sediments. The approach was applied and discussed on the Adriatic Sea, for the winter of 2012, using data provided by the ESA Coastcolour project and the output of a run of the MIT General Circulation Model.

Keywords: remote sensing; satellite; sediment transport; coastal geomorphology

---

1. Introduction

The morphological evolution of shoreline, coastal features, lagoons, deltas and clinoforms on continental shelves, in absence of relative sea level changes, due to either eustacy or local tectonism, is primarily driven by the sediment dynamics and the sediment transport [1–5]. Seminal studies, indeed, highlight the role of coastal plumes or river runoff, and their sediment loads, in contributing to sediment availability for coastal and continental shelf morphodynamics [5,6].

In particular, in the Adriatic sub-basin (Mediterranean Sea; Figure 1), Bignami et al. [7] investigated the variability of the Western Adriatic Coastal Current (WACC), and its turbid, Case 2 waters, in relation to circulation patterns and to wind regimes. Brando et al. [8] investigated river plumes at mesoscale and submesoscale in the North Adriatic Sea (NAS), by means of coupled satellite data and numerical modelling. On the other hand, Cattaneo et al. [9] and Sherwood et al. [10] focused on the sediment erosion, transport and deposition in the Adriatic Sea. The Adriatic Sea, especially its western side, the Italian east coast, is a case study for this kind of phenomena, because of its characteristic coastal flow dynamics and the significant role of river sediment inputs on the northwest side of the basin [7,9–11].

Indeed, the circulatory regime of this basin, as described in [12], is dominated by:

- a cyclonic gyre with a component that flows parallel to the western shoreline (Figure 1), mainly generated by the wind patterns;

- the North Adriatic Dense Water (NAdDW; Figure 1) that forms in the northern Adriatic through winter cooling and then, once a density threshold is reached, flows southward until cascading toward the deep southern Adriatic Basin; and
- the Levantine Intermediate Water (LIW), a salty water that forms in the Eastern Mediterranean Sea and intrudes in the Adriatic basin flowing at depths of 200–600 m.

An intensified western boundary current, the Western Adriatic Coastal Current (WACC), flows southward with long-term average speeds that reach 0.20 m s^{-1} at some locations [13]. The overall thermohaline circulation runs along the Italian coast, constraining the main sediment flux to deposit in a prism parallel to the coast. The investigation of sediment dynamics along the coastal Adriatic currents, besides the use of in-situ measurements and numerical modelling, is largely improved by remote sensing observations and analyses [4,14,15].

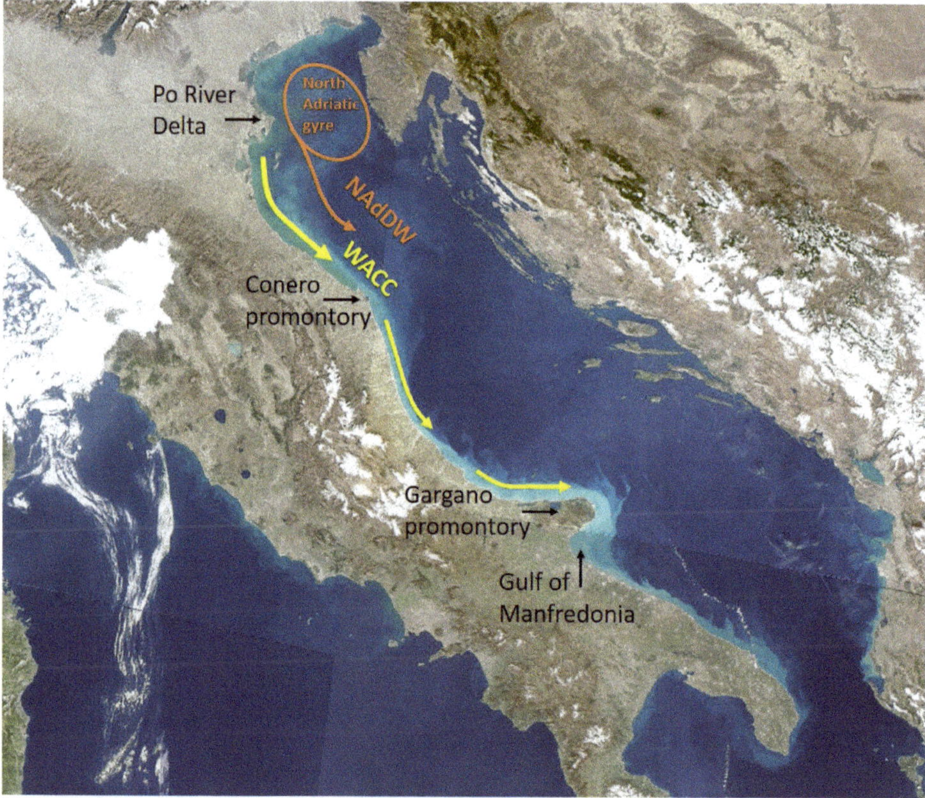

Figure 1. True Colour (bands 9,6,4) image of the Adriatic Sea, 28 February 2019, by the OLCI sensors on board of the Sentinel-3A and Sentinel-3B spacecrafts. The Western Adriatic Coastal Current (yellow arrows) is clearly visible, due to its suspended sediment load, transported along the coast [13]. The North Adriatic Gyre and the subsequent path of the North Adriatic Dense Water are highlighted in orange [13]. The black arrows indicate the geographic locations mentioned in the text.

Figure 1 shows a combination of the views acquired on 28 February 2019 by the OLCI sensors on board of the Sentinel-3A and Sentinel-3B spacecrafts. In this true colour picture the Western Adriatic Coastal Current (WACC) is clearly visible, by means of its suspended sediment load, which is transported along the coast. The sediment laden plume originates from river inputs at the northwest sector of the sub-basin [8] and receives the contribution of other small rivers in the central part.

Sediment dynamics does not reduce to the settling of the sediment suspended in the coastal plume [16]. Longshore sediment transport in the littoral and coastal plume systems are the main cause of the sediment load distribution along the coast, without, or with a limited, net sediment offshore export. The redistribution of the sediment load over the entire inner shelf is instead caused by the action of meteorological events, varying currents and wind waves, which resuspend and transport sediments in deeper water.

The link between the sediment dynamics and the most general model for morphodynamics is provided by the Exner equation. Exner [17,18] proposed that the change in time of the local elevation $\eta(x,t)$ of the bed of a river or channel is proportional to the spatial rate of change of the average flow velocity:

$$\frac{\partial \eta}{\partial t} = -A\frac{\partial v}{\partial x} \tag{1}$$

In the text describing the equation, Exner explained that he intended the flow velocity v as a proxy for the sediment flux [19].

Thus, the Exner equation, in its general formulation, is now written as a classical conservation law:

$$\frac{\partial \eta}{\partial t} = -\frac{1}{\epsilon_0}\nabla \cdot \vec{q}_s \tag{2}$$

that relates the seabed elevation (η), relative to a fixed datum, with the sediment that is transported by water ($\vec{q}_s$ is the sediment flux). The bed elevation increases ($\frac{\partial \eta}{\partial t} > 0$) proportionally to the sediment that is dropping out of the transport (negative sediment flux divergence, $\nabla \cdot \vec{q}_s < 0$). Conversely, the bed elevation decreases proportionally to the sediment that becomes entrained by the flow (positive sediment flux divergence, $\nabla \cdot \vec{q}_s > 0$). The proportionality factor is the inverse of the grain packing density, ϵ_0.

Paola and Voller [19] provided a complete generalisation of the Exner equation to address several different processes over a wide array of temporal scales.

Equation (2) can be related to coastal geomorphology and, in particular, to sediment availability along the inner shelf, since a positive divergence of sediment flux suggests sediment erosion, while a negative divergence suggests sediment deposition. Figure 2 sketches a simplified geometry that can be adopted to represent the seabed near the shoreline. The x-axis is usually taken parallel to the shoreline itself, and it is possibly a curvilinear coordinate, while the y-axis is usually taken seaward, orthogonal to the shoreline. The z-axis is usually the vertical, with the positive direction pointing up and the zero value at the sea surface. The seabed height η is clearly indicated. ρ represents the sediment concentration in the water.

The practical application of the Exener equation (Equation (2)) relies on a proper estimation of the sediment flux $\vec{q}_s$. At local scale, sediment dynamics is primarily dominated by waves and wave-induced currents, especially during strong meteorological events. Several empirical or semi-empirical relations have been proposed to correlate the sediment flux to the wave field parameters and to the sediment grain size [20–23]. All these approaches, however, when considering time scales longer than the single meteorological event, lead to considering only a prevalent wave field, or a small set of prevalent wave fields that could happen in the time frame under investigation. While they are able to produce good estimations of erosion vs. deposition in local spatial and temporal scales, or to stable vs. unstable shoreline profiles, they are usually not applied to wider scales.

The scope of our work, instead, is to investigate the sediment dynamics at larger spatial and temporal scales, by analysing coastal sediment plumes and coastal currents with a synergic approach, combining observational data with numerical modelling. Marine coastal currents are strongly influenced by the input of fresh waters, tides, topographic features and winds. Moreover, hydrodynamics typically observed in coastal areas involves processes interacting on a wide range of spatial and temporal scales. In our work, we make use of a suitable numerical run that is able to capture the coastal processes, such as buoyant plume formation and propagation, as well as associated coastal upwelling and downwelling [24] .

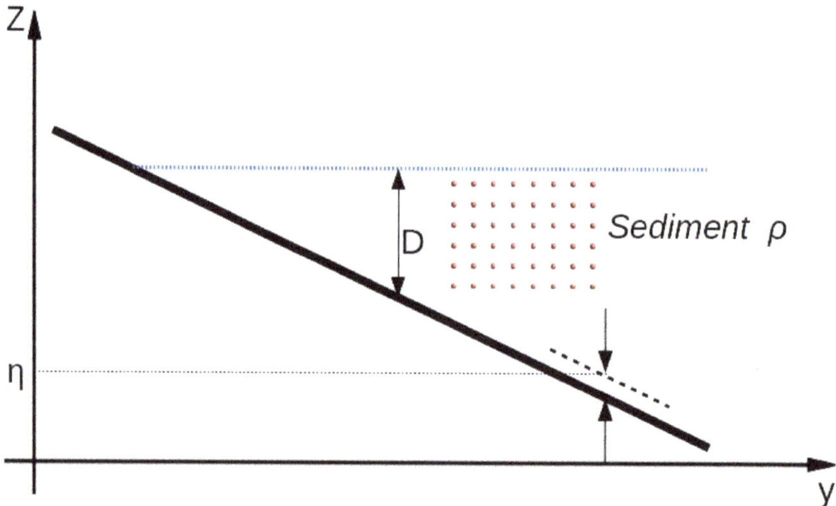

Figure 2. Schematic representation of the seabed geometry: ρ represents the sediment concentration in water; D represents the "closure depth", i.e., the maximum depth at which the waves are able to generate sediment resuspension from the seabed; and η represents the seabed height, relative to a fixed datum.

We envisage the need to move from an empirical to an observational approach, and thus we look for alternative estimations of the sediment flux. As a general statement, the flux of sediment in the water, can be expressed in terms of the suspended sediment concentration in the water ρ and the water velocity $\vec{u}$:

$$\vec{q}_{3D} = \rho\vec{u} \tag{3}$$

As better explained in Section 2, optical observations are able to estimate physical, chemical and biological properties of the water. IOPs (Inherent Optical Properties), AOPs (Apparent Optical Properties) and water constituents can be retrieved by means of proven algorithms from remote sensing observations in the visible and in the near-infrared spectrum. The concentration of suspended matter is one the water constituent that can be retrieved from such observations [25–27]. Spaceborne sensors are one of the most important sources of optical observations, due to their great temporal and spatial coverage.

It is then natural to plug the remotely-sensed TSM (Total Suspended Matter) as the sediment concentration ρ in Equation (3). In the same way, a water velocity field derived from an oceanographic model can be plugged as the velocity $\vec{u}$ in Equation (3).

2. Materials and Methods

To show the feasibility of our approach, we need to combine a remotely sensed sediment concentration field with a velocity field from an oceanographic model to obtain a sediment flux to feed in the Exner equation (Equation (2)) and thus estimate the sediment erosion and deposition processes (i.e., the temporal evolution of the seabed height η). The data sources we chose for our first experiments are:

- for the TSM fields, the ESA Coastcolour project [28]; and
- for the velocity field, a MITgcm (Massachusetts Institute of Technology General Circulation Model [29])-based model for the Adriatic Sea, run by ENEA-ISMAR

The reasons for such a choice were the quality and the ready availability of these two datasets. As better explained in the following paragraphs, these two datasets overlap for a small time window. Actually the model availability time frame (just four months) almost entirely lies within the ESA Coastcolour project time frame (nine years).

2.1. Remote Sensing Data

The Coastcolour project [28], launched by ESA to fully exploit the potential of the MERIS instrument, provides us a complete (from 4 January 2003 to 7 April 2012, when the mission ended) series of ocean optics observation of a set of basins where the presence of Case 2, optically complex waters is important. The whole Mediterranean, and thus the Adriatic Sea, is among these basins. The Coastcolour project provides three levels of products:

- The Level 1P product (L1P) provides top of atmosphere radiance, with geolocation, equalisation to reduce coherent noise, smile correction, pixel characterisation information (cloud, snow, etc.) and a precise coastline.
- The Level 2R (L2R) product is the result of a neural network based atmospheric correction, which is applicable for a large range of water type, from clear to extreme scattering waters; it contains water leaving reflectance, normalised water leaving reflectance and different information about atmospheric properties.
- The Level 2W (L2W) product provides information about water properties such as IOPs (Inherent Optical Properties) and water constituent concentrations.

Among the water constituent concentrations provided by L2W product, there is the concentration of the Total Suspended Matter variable (conc_tsm) that we use as an estimation of the sediment concentration in water, ρ. As per the "Validation Report" from the "Publication" section of the Coastcolour website [28], the correlation coefficient and the coefficient of determination (R^2) of the (conc_tsm) versus the in-situ validation dataset ([30]) are, respectively, 0.831 and 0.691. Refer to the "Publication" section of the Coastcolour website [28] for the complete documentation on the algorithms and on the validation processes that have been used to create and validate all the Coastcolour products.

2.2. Numerical Model Outputs

In this work we make use of a MITgcm run that has been setup to investigate coastal upwelling and downwelling processes in the Adriatic Sea during a strong dense water event that occurred in winter 2012 [24]. The model domain, which covers the entire Adriatic Sea, is discretised by a non-uniform curvilinear orthogonal grid of 432 × 1296 points, with 100 vertical levels. This grid has a variable resolution, ranging from less than 500 m in the nearshore area up to 1000–2000 m offshore. Furthermore, it is orthogonal, or almost orthogonal, to most of the coastline. Regarding the vertical resolution, the grid has 100 vertical z levels with a thickness of 1 m in the upper 23 m gradually increasing to a maximum of 17 m for the remaining 64 levels.

The model simulations started at the beginning of December 2011 and finished at the end of April 2012. Simulated fields and diagnostic were produced every three simulated hours. The bathymetry used by MITgcm is provided by the National Group of Operational Oceanography (GNOO; http: //gnoo.bo.ingv.it/bathymetry/). As in [31,32], an implicit linear formulation of the free surface is used. The river runoff was considered explicitly and modelled as a lateral open-boundary condition. Rivers were included by introducing small channels in the bathymetry that simulate the river bed close to the coast. Velocity was imposed at the upstream end of each channel, with the prescribed discharge rate being obtained by multiplying the velocity by the cross sectional area of the channel. No flux conditions for either momentum or tracers and no slip conditions for momentum were imposed at the solid boundaries. Bottom drag was expressed as a quadratic function of the mean flow in the bottom layer. The net transport through the southern open boundary was corrected during run-time at each time step to balance the effects of river discharge and of the evaporation minus precipitation budget on the surface level. This solution prevented any unrealistic drift in the sea surface elevation. Tides were imposed as a barotropic velocity at the southern boundary. At the surface, the wind drag coefficient was computed following the default MITgcm formulation:

$$C_d = \frac{0.0027}{U_{10}} + 0.000142 + 0.0000764 \cdot U_{10} \tag{4}$$

where U_{10} is the wind speed at 10 m. The wind speed and direction, together with the other surface forcings (air temperature, relative humidity, and cloud cover), were provided by means of hourly meteorological forecasts from the MOLOCH (MOdello LOCale in H coordinates) model, developed and run at the ISAC (Institute of Atmospheric Sciences and Climate—National Research Council) CNR, Bologna, Italy [33–35]. The numerical run was finally validated by using time series of surface and bottom temperature as well as surface salinity from the VIDA buoy [24]. Refer to McKiver et al. [24] for the full description of the model set-up, the experiments and their validation. The model has been geared to investigate coastal processes, and the MITgcm capabilities in capturing them. This feature and its horizontal resolution that, near the shore, is comparable with the Coastcolour TSM field, are the key advantage of this model in our study. Other models and other velocity fields (e.g., from reanalysis products) we tested did not provide meaningful divergence fields. Hereafter, we call Winter2012 this model run as well as its grid.

2.3. Combining Marine Currents and TSM Upstream Data

As aforesaid, these two datasets (i.e., Coastcolour TSM satellite product and the Winter2012 MITgcm run for marine currents) overlap for only a four-month time window; however we considered this sufficient to test our approach. The TSM concentration field ρ from the satellite data is inherently bidimensional, while the velocity field $\vec{u}$ from the model is tridimensional. We have to consider this aspect when formulating our sediment flux: $\vec{q}_s = \rho \vec{u}$. Furthermore, the two datasets are referred to two different spatial grids as well as to two different temporal grids.

Time-wise, the model data are available every 3 h while the satellite data are available with a variable period, roughly close to 24 h, depending on the satellite overflying times. We chose to pick one single model result per day, i.e., the closest to the satellite observation. In this way, we used the coarsest temporal resolution between the two datasets, thus we could avoid temporal interpolations of any sort. We therefore remapped the model time-grid to the satellite time-grid.

Space-wise, on the other hand, we remapped the satellite data on the model grid.

In both cases, the remapping of the data from one grid to another was performed with a "nearest-point" algorithm. "Nearest-point" algorithms are computationally efficient and correctly handle the quantities that need to be conserved. To remap the data from one spatial grid to another we used the pyresample package, which is part of the PyTroll suite [36].

Dimension-wise, we wanted to converge to a 2D approach: the formulation of our problem is indeed bi-dimensional. We therefore assumed that the sediment concentration is constant along the water column, and identical to the value provided by the Coastcolour product, i.e.,

$$\rho(x,y,z) = \rho(x,y) = \text{conc\_tsm}(x,y) \tag{5}$$

This is justified by the fact that the north- and central-west Adriatic shelf is very shallow (i.e., 5–20 m depth) and the water column depth is comparable with optical penetration depth. Thus, we define the 3D sediment flux vector as:

$$\vec{q}_{3D} = \rho(x,y,z)\vec{u} \tag{6}$$

with $\vec{u}$:

$$\vec{u} = (u,v,w)$$

A 2D sediment flux can, therefore, be defined as:

$$\vec{q}_s = \int_{b(x,y)}^{0} \vec{q}_{3D} \, dz \tag{7}$$

where $b(x,y)$ is the bathymetry of point (x,y). The bathymetry used in this study is the EMODnet Digital Bathymetry (DTM) [37], which provides the water depth (referring to the Lowest Astronomical Tide Datum, LAT) in gridded form on a DTM grid of $1/8 \times 1/8$ arc minute of longitude and latitude (ca. 230×230 m).

From now on, our $\vec{q}_s$ is a bi-dimensional sediment flux and all the considerations and computations we produced are on the bi-dimensional space of the sea (earth) surface. The horizontal divergence of the sediment flux is simply defined as:

$$\nabla_H \cdot \vec{q}_s = \left(\frac{\partial}{\partial x}, \frac{\partial}{\partial y} \right) \cdot (q_{sx}, q_{sy}) = \frac{\partial q_{sx}}{\partial x} + \frac{\partial q_{sy}}{\partial y} \tag{8}$$

Its vertical component is zero:

$$q_{sz} = \int_{b(x,y)}^{0} \rho w \, dz = \rho \int_{b(x,y)}^{0} w \, dz \approx 0 \tag{9}$$

or negligible with respect to the other components because, for the timescale we dealt with (3 or 24 h time-steps), we can safely assume that the sea surface displacement ($\int_b^0 w \, dz$) averages to zero in the time step. Of course, we also neglected the vertical currents and the vertical transport of the sediment. Our horizontal divergence term $\nabla_H \cdot \vec{q}_s$ actually takes into account the settling and re-suspension of the sediment besides its transport by the water velocity. The approximation in Equation (9) we made on vertical transport refers only to sediment transport by water velocity, not to sediment motion due to gravity and waves. As we have density measurements only from the satellite, thus inherently bidimensional, and we applied the approximation in Equation (9), the divergence $\nabla_H \cdot \vec{q}_s$ we estimated is actually the sum of the seventh and the eighth terms of Equation 17b of Paola and Voller [19]:

$$\nabla_H \cdot \vec{q}_s = \frac{\partial}{\partial t} \int_{\eta}^{\eta + h_f} \alpha_f \, dz + \nabla_H \vec{\Phi}_f \tag{10}$$

where η is the seabed height, h_f is the sea bathymetry, α_f is the density of the sediment laden seawater, and Φ_f is the line flux, i.e., the bi-dimensional, vertically integrated, flux.

The model data are available in an Arakawa-C grid, with vector quantities, i.e., meridional and zonal components of the velocity, located on the edges of the cell, while the scalar quantities and the vertical component of the velocity are located inside the cell (Figure 3).

On the other hand, the Level-2 MERIS data are available on a swath-based grid. To harmonise all the quantities, we projected the ρ concentration field on all three different grids: the grid of the u-points, on the "west" and "east" edges of the cells; the grid of the v-points, on the "north" and "south" edges of the cells; and the grid of the w-points, in the centres of the cells. The vertical component of the velocity (w) is considered by the model as one of the scalar quantities, such as temperature, salinity and elevation.

The computation of the x and y components of the 2D sediment flux, i.e., the vertical integrals

$$q_{sx} = \int_{b(x,y)}^{0} \rho u \, dz$$

$$q_{sy} = \int_{b(x,y)}^{0} \rho v \, dz$$

were performed on the cells' edges, by means of a simple trapezoidal rule, on every point (x, y) of the horizontal grid:

$$q_{sx} = \int_{b(x,y)}^{0} \rho u \, dz = \rho \int_{b(x,y)}^{0} u \, dz \approx -\rho \sum_{k=0}^{N-1} \frac{u(z_{k+1}) + u(z_k)}{2} (z_{k+1} - z_k)$$

$$q_{sy} = \int_{b(x,y)}^{0} \rho v \, dz = \rho \int_{b(x,y)}^{0} v \, dz \approx -\rho \sum_{k=0}^{N-1} \frac{v(z_{k+1}) + v(z_k)}{2} (z_{k+1} - z_k)$$

$$(11)$$

where $N = 100$ is the number of vertical levels of the grid. The minus sign is because $k = 0$ means the surface, thus we integrated from surface downwards.

We now have a 2D sediment flux, whose components were computed on the cells' edges: the zonal component on the u-points and the meridional component on the v-points.

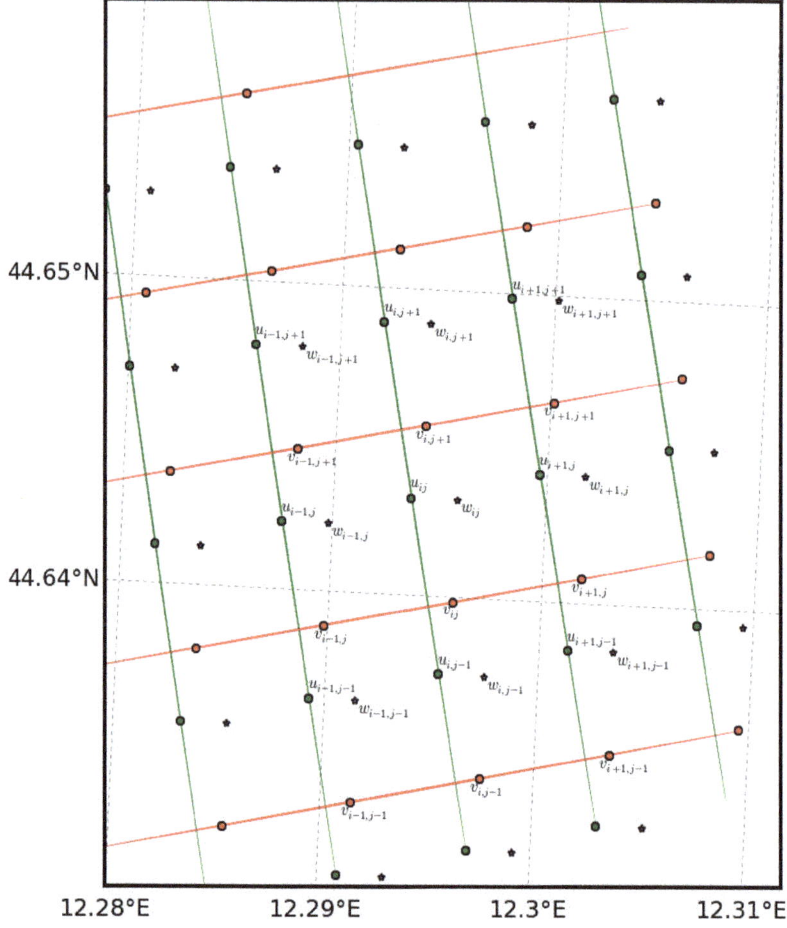

Figure 3. Grid detail: cell and edge points.

According to Hyman et al. [38], to approximate the divergence of the sediment flux, we used a local version of the divergence theorem, applied at every single cell of the grid:

$$\left(\overline{\nabla_H \cdot \vec{q}_s}\right)_{ij} = \frac{1}{A_{ij}} \int_{\Omega_{ij}} \nabla_H \cdot \vec{q}_s \, d\Omega = \frac{1}{A_{ij}} \int_{\partial\Omega_{ij}} \vec{q}_s \cdot \vec{n} \, dl \tag{12}$$

where Ω_{ij} is the ij 2D cell, A_{ij} is its area, $\partial\Omega_{ij}$ is its frontier, and $\vec{q}_s$ is the 2D sediment flux.

According to Figure 4, we call $\mathbf{P}_{ij} = (\xi_{ij}, \psi_{ij})$ the southern-most vertex of the ij cell, and ξ_{ij} and ψ_{ij} its longitude and its latitude.

Let us call also

$$\Delta\xi_{ij} = \left\|\mathbf{P}_{i+1,j} - \mathbf{P}_{ij}\right\|$$
$$\Delta\psi_{ij} = \left\|\mathbf{P}_{i,j+1} - \mathbf{P}_{ij}\right\| \tag{13}$$

the length of the edges of cell ij, again as depicted in Figure 4.

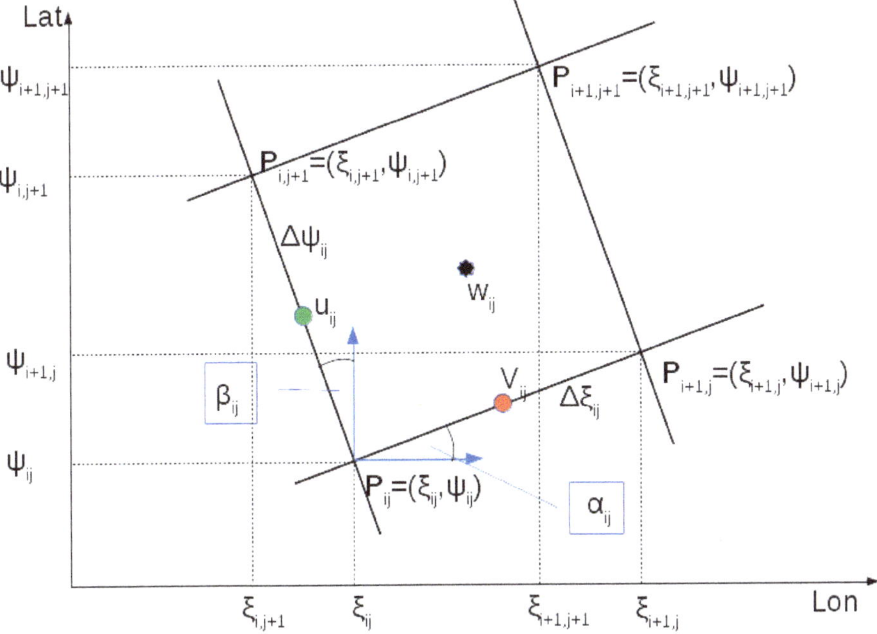

Figure 4. Geometry of ij cell.

Let us call α_{ij} the angle between the edge $\overline{\mathbf{P}_{ij}\mathbf{P}_{i+1,j}}$ and the parallel passing by $\mathbf{P}_{ij}$ and β_{ij} the angle between the edge $\overline{\mathbf{P}_{ij}\mathbf{P}_{i,j+1}}$ and the meridian passing by $\mathbf{P}_{ij}$

The area of the cell A_{ij} can be calculated by:

$$A_{ij} = \Delta\xi_{ij}\Delta\psi_{ij}\sin(\hat{\mathbf{P}}_{ij}) = \Delta\xi_{ij}\Delta\psi_{ij}\cos(\alpha_{ij} - \beta_{ij}) \tag{14}$$

where $\hat{\mathbf{P}}_{ij}$ is the angle in $\mathbf{P}_{ij}$ between the segments $\overline{\mathbf{P}_{ij}\mathbf{P}_{i+1,j}}$ and $\overline{\mathbf{P}_{ij}\mathbf{P}_{i,j+1}}$. The Winter2012 grid is non-uniform and curvilinear but locally orthogonal, thus $\alpha_{ij} = \beta_{ij}$, but in general this may not be the case.

The flux across the border of the ij cell can be decomposed in the four fluxes across the four different edges of the cell. As the edges of the cells are not parallel to meridians and parallels, we have to take into account both components of the velocity field on any edge. Thus, to compute the divergence we rewrite Equation (12) as:

$$A_{ij} \cdot \left(\overline{\overline{\nabla_H \cdot \vec{q}_s}} \right)_{ij} = \Delta\xi_{ij}\Delta\psi_{ij}\cos(\alpha_{ij} - \beta_{ij}) \left(\overline{\overline{\nabla_H \cdot \vec{q}_s}} \right)_{ij} =$$

$$= \int_{\partial\Omega_{ij}} \vec{q}_s \cdot \vec{n} \, dl =$$

$$= + (-u_{ij}\cos\beta_{ij} - v_{ij}\sin\beta_{ij})\rho_{ij}^{(u)}\Delta\psi_{ij} +$$

$$+ (-v_{ij}\cos\alpha_{ij} + u_{ij}\sin\alpha_{ij})\rho_{ij}^{(v)}\Delta\xi_{ij} +$$

$$+ (u_{i+1,j}\cos\beta_{i+1,j} + v_{i+1,j}\sin\beta_{i+1,j})\rho_{i+1,j}^{(u)}\Delta\psi_{i+1,j} +$$

$$+ (v_{i,j+1}\cos\alpha_{i,j+1} - u_{i,j+1}\sin\alpha_{i,j+1})\rho_{i,j+1}^{(v)}\Delta\xi_{i+1,j} \tag{15}$$

where $\rho_{ij}^{(u)}$ and $\rho_{ij}^{(v)}$ are the projection of the tsm_conc field on the u-point and on the v-point, respectively, of cell ij; and and u_{ij} and v_{ij} are the bi-dimensional velocities, i.e., the vertically integrated velocities computed in Equation (11), again for cell ij:

$$u_{ij} = -\sum_{k=0}^{N-1} \frac{u_{i,j,k+1} + u_{ijk}}{2}(z_{k+1} - z_k)$$

$$v_{ij} = -\sum_{k=0}^{N-1} \frac{v_{i,j,k+1} + v_{ijk}}{2}(z_{k+1} - z_k) \tag{16}$$

The whole computation is than arranged as per the following flow:

- Vompute ξ_{ij} and ψ_{ij}, coordinates of the vertexes $\mathbf{P}_{ij}$ of the cells.
- Compute angles α_{ij} and β_{ij}.
- Start a DAILY cycle: at each day compute u_{ij} and v_{ij} according to Equation (16) and then compute the flux divergence $\left(\overline{\overline{\nabla_H \cdot \vec{q}_s}} \right)_{ij}$ according to Equation (15).
- Compute temporal average of the flux divergence $\left(\overline{\overline{\nabla_H \cdot \vec{q}_s}} \right)_{ij}$.

3. Results and Discussion

By using a sediment mass balance approach (i.e. the Exner equation, Equation (2)), and thus, by mapping the divergence of the sediment flux, we intend to recognise those zones that are characterised by sediment deposition or erosion. All this does not depend on the specific grain-size (i.e., settling velocity) of the suspended sediment. That is, if a sediment laden pixel does not conserve its sediment concentration along its motion (regardless its grain-size), this necessarily means that some sediment has been lost or gained. In Figure 5, we show the average, over the whole time frame under investigation (four months, from 6 December 2011 to 7 April 2012), of our input fields: the satellite TSM (Total Suspended Matter) concentration field and the surface velocity field. The Western Adriatic Coastal Current (WACC) is clearly visible along the Italian shoreline, from both the TSM and the velocity patterns.

We notice that the TSM satellite product, for the study period, highlights the riverine input along the northern Adriatic coast (see Figure 5a), which significantly contributes in driving a southeastward current along the Italian coast. In particular, Figure 5a shows that the Po, along with all the other North Adriatic river inputs, produces an almost single river plume, contributing 84% of the total freshwater discharge delivered to the basin [39,40]. We recognise the presence of mid-to-high TSM concentration waters, in the entire northern basin and, partly, in the middle part of the Adriatic Sea. For the cold period we analysed, high TSM values are also recorded at the exit to the Ionian Sea, through the Otranto Strait [7]. Finally, we remark that the spatial distribution of optically complex waters, marked by high turbidity parameters such as TSM or a diffusion attenuation coefficient, closely matches slow-settling particle deposition patterns of the Adriatic Sea [7], and thus highlights zones that are affected by sedimentary processes.

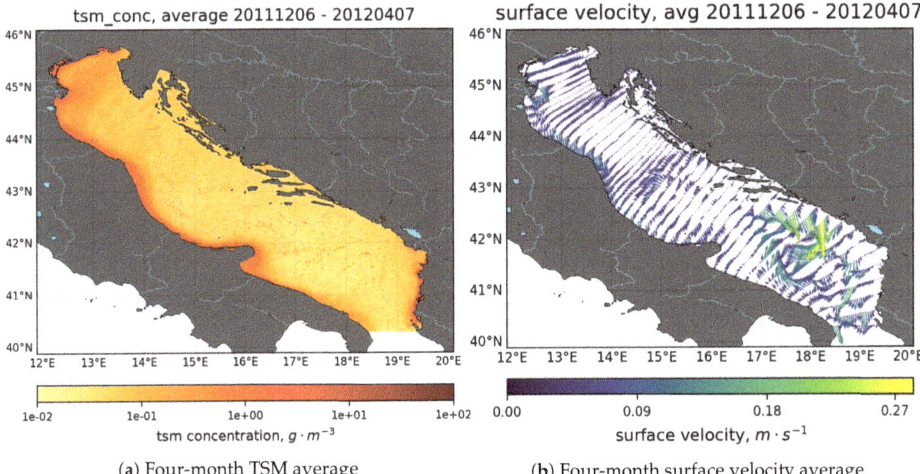

(a) Four-month TSM average

(b) Four-month surface velocity average

Figure 5. Source data: Four-month averages (from 6 December 2011 to 7 April 2012) of surface TSM concentration (**a**) and surface velocity field (**b**).

Accordingly, the highly resolved marine currents, provided by the MITgcm numerical outputs, show the correct alongshore momentum (Figure 5b), where the inclusion of lateral freshwater inputs affects the capability to reproduce buoyant processes in the coastal area. The model output confirms, for the winter season, a basin-wide cyclonic circulation with a strong southward currents along the Italian Peninsula on the western side (i.e., the WACC). Such a geostrophic pattern is known to be due to the estuarine circulation [41], forced by river inputs, mainly the Po River, and by strong air–sea fluxes. In particular, the Po River input results in a wide extension of surface freshwater, particularly evident along the northern littoral of the basin [24]. Between the Po River mouth and the Conero Promontory (Figure 5b), the WACC width is about 40 km and its maximum speed reaches $0.25 \, \mathrm{m \, s^{-1}}$. This well-recognisable, southeastward current along the Italian coast (i.e., the WACC), overlays the TSM pattern, providing a comforting agreement between the satellite and the numerical upstream data. Moreover, for the study period, the effect of surface wind stress (i.e., the Bora event recorded from 25 January to 14 February 2012) led to a more confined strip of freshwater [24].

On the western Adriatic coast, Bora is a downwelling favourable wind and can generate large waves with significant wave heights of 1 m, and period of 5 s [42]. Wave-driven sediment resuspension is an important resuspension mechanism in the shallow coastal areas of the NAS [43], and contributes significantly to the complexity of the sediment distribution and flux features in the region. Therefore, waves should not be neglected in the study of dynamics of sediment transport and resuspension in the shallow coastal seas. However, the fact that the hydrologic model we used does not include waves did not affect our ability to estimate suspended sediment concentration, which is directly retrieved from satellite, regardless of the mechanism that keep the sediment in suspension. Moreover, sediment transport is known to be insensitive to the angle between directions of wave propagation and current in the NAS, where large waves were generated by the Bora storm for strong wave conditions [44,45]. This suggests that wave-induced currents, not included in the MITgcm runs, have no significant effect on the southward sediment flux along the western Adriatic coast. Finally, it is worth mentioning that the effect of mixing due to wave breaking on sediment distribution and fluxes in the NAS is not significant since the water column is well mixed due to strong current shear driven by the Bora winds and shallow water depth [45].

In Figure 6, from Figure 6a to Figure 6d, on the other hand, we show the sediment flux divergence fields obtained with the methodology introduced in Section 2.3. For an easier readability of the computed maps, monthly averages are shown, i.e., one map for each of the months under investigation: December 2011, January 2012, February 2012 and March 2012. From Equation (2), a pixel (a cell) with positive divergence (a yellow or a red pixel) means a sediment erosion point, or, more correctly, a point where we have sediment from the bottom layer entrained into the water flow. Conversely a pixel (a cell) with negative divergence (a blue or light blue pixel) means a sediment deposition point, or a point where suspended sediment is precipitating on the bottom layer.

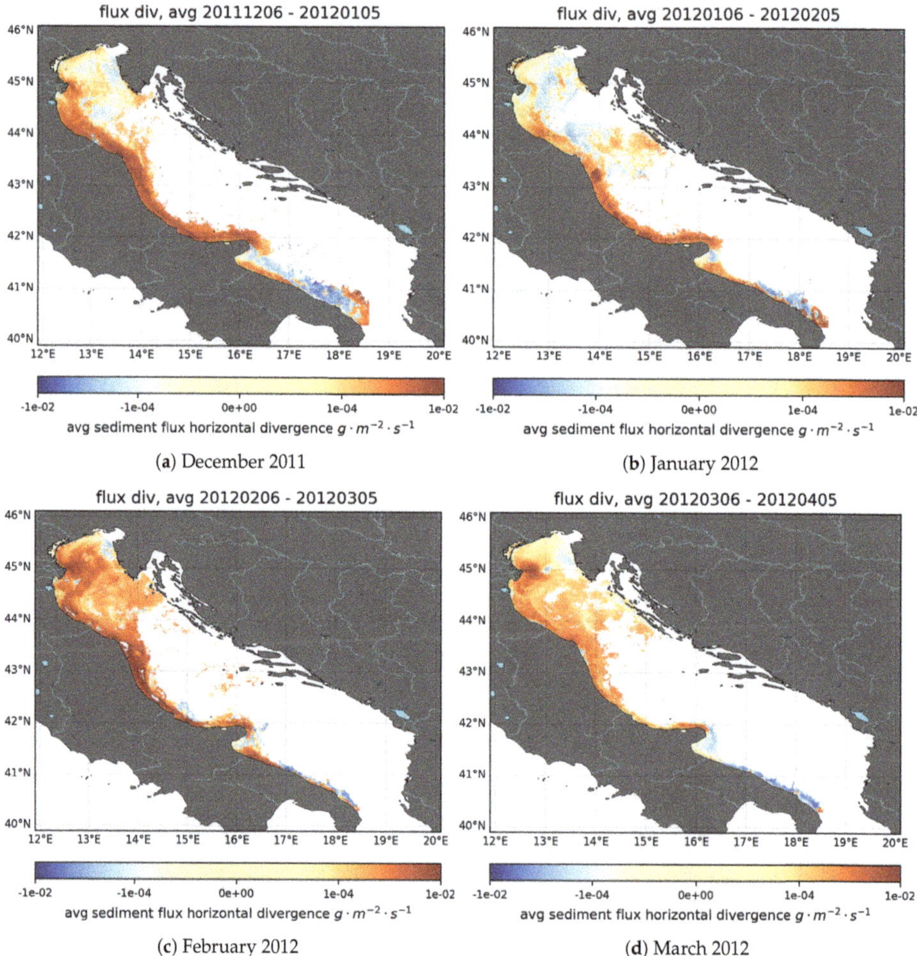

Figure 6. Monthly averages of sediment flux divergence: (**a**) December 2011; (**b**) January 2012; (**c**) February 2012; and (**d**) March 2012.

A long-time scale persistence of erosive or depositional pattern along the coastal plume can be related to sediment starvation or sediment availability over the inner shelf (Figure 6). Indeed, for the time frame under observation, we found a persistent sediment erosion condition along the coast traits between the Po river delta and the Conero promontory and between the Conero and Gargano rocky promontories (Figure 6). This, in general, is in agreement with the overall shoreline retrograding shape reported by ISPRA [46] and ISPRA [47], found in [48]. On the other hand, there are depositional patterns in the Gulf of Manfredonia, likely due to recirculation of the southern Adriatic currents around the Gargano promontory, as depicted in [49,50]. The Po river delta shows depositional and erosive behaviours, depending on the season. In this area, the availability of long time series of both observation and model data would help identify the predominant long-term behaviour.

A further confirmation of our results can be obtained from the knowledge of the subaqueous clinoform anatomy [51]. In particular, we can use our computed divergence patterns as a proxy for the thickness of subaqueous clinoform: clinoform thicknesses are expected to decrease within our computed erosional zones and to increase where we estimate sediment deposition. According to Correggiari et al. [51] (see Figure 2 in [51], reported also in [12,45]), we infer that our depositional areas, e.g., south of the Po River delta and south of the Gargano promontory, correspond to those areas where the highest clinoform thickness is observed to be attached to the shoreline.

Finally, we remark that, in the Adriatic Sea, the majority of horizontal structures that are observed in the coastal zone are characterised by Rossby and Richardson numbers of around 1 (sub-mesoscale), representing areas where vertical fluxes and buoyancy are enhanced [52]. In such a complex environment, sudden changes in the wind forcing can trigger strong hydrodynamic events, such as the formation of dense water (DW) and wind driven upwelling. In the model we used, the direct effect of wind forcing is assimilated in order to reproduce DW formation events [24] as well as a number of small scale features that show high horizontal variability of vertical processes. This proves that the high resolution of MITgcm allows for the reproduction of more small scale vortical structures, identifying a wider spatial range of processes. Therefore, pairing high resolution TSM fields from remote sensing with the MITgcm outputs, which well represent submesoscale processes, made us confident in capturing coastal sediment transport dynamics.

As already pointed out, waves play a major role in sediment suspension and resuspension (see [45] for a study on the Adriatic sea). While MITgcm based simulations do not include waves, their effect in sediment suspension is taken into account in the observation dataset, i.e., in the satellite provided TSM maps. The availability of such dataset, furthermore, allows us to avoid the execution of a sediment dynamics model. The goal of our work is to investigate the feasibility of an operative product that, with a low computational effort, can represent the sediment dynamics at basin or regional spatial scales, and at decadal time scales. The observational-numerical approach we propose, at these spatial and temporal scales, with these spatial and temporal resolutions, appears quite adequate to fulfil this objective.

4. Conclusions

This work proposes a new approach to investigate the sediment erosive and depositional patterns that occur at regional or basin scale, and that are driven by mesoscale and sub-mesoscale processes, such as coastal currents and fluvial plumes. Our approach is based on a synergy between satellite observations and numerical simulations. The computed maps of the sediment flux divergence show patterns of sediment erosion and deposition that are in agreement with the general knowledge of the sediment dynamics and coastal geomorphology in the Adriatic Sea.

The main agents in the sediment distribution on the shore, and the subsequent morphodynamics, are waves and wave-induced currents. However, the daily Total Suspended Matter as retrieved from Ocean Colour remote sensing is able to capture the material resuspended by a large variety of phenomena (including wave-induced resuspension). We therefore believe that the divergence of the sediment flux, estimated from the synergy of remote sensing and high resolution horizontal marine currents, is able to highlight both depositional and erosive areas resulting from mass conservation, i.e., those pixel where we can expect sediment deposition/erosion over the continental shelf, regardless of what caused its deposition or entrainment.

Our investigation was limited by the short intersection of the two time domains: the Coastcolour project time domain (from 4 January 2003 to 7 April 2012) and the Winter2012 model run time domain (from 6 December 2011 to 30 April 2012). The short duration of this intersection (i.e., four months) did not allow us to provide statistical analyses and further validation of our results. However, the use of the two upstream datasets, as well as the remapping and interpolation schemes, resulted to be suitable tools for the detection of realistic depositional or erosive spatial patterns, which were confirmed by the authors of [12,49,50]. We finally remark that the goal of our work is to provide a potential operative product that might be suitable for long-term analyses and monitoring programmes.

A longer, multi-year duration would allow one to extract trend analysis that could be matched with the long time series of in-situ information on the Adriatic shoreline. This time series of in-situ information is the "ground truth" that can validate and tell the accuracy of the methodology that has been introduced.

In envisioning a longer time frame, a suitable observational part (i.e., spaceborne remotely sensed data) can be represented by the 2002–2012 and 2016-onward TSM time series from either Coastcolour [28] or OLCI [53], with daily time resolution and 300-m spatial resolution. A further application of our approach could also be the design and use of ad hoc models, which could assimilate the sediment concentration from the satellite measurements, as done by Stroud et al. [15] for Lake Michigan, and that directly output the divergence field.

Finally, it would be promising to consider the assimilation of two other kinds of spaceborne sensors: geostationary satellite and hyperspectral missions. Geostationary sensors such as SEVIRI–Meteosat [54] and their successors show a very high temporal resolution (15 min–1 h), at the expenses of a very coarse spatial resolution, e.g., 4 km × 6 km at our latitudes). On the contrary, hyper-spectral missions (PRISMA [55] and EnMAP [56]) show coarser temporal resolution, but much finer spatial and spectral resolutions. The high temporal resolution of geostationary satellites can enhance the near daily temporal resolution of Low Earth Orbit satellites and mitigate the presence of cloud coverage, while the high spectral resolution of hyperspectral sensors can give insight on the chemical composition of the sediment as well as on its size distribution.

Author Contributions: F.F. proposed the idea. M.B. and G.S. developed the method. M.B. and F.F. analyzed the results and wrote the text. F.F., R.S. and C.A. supervised the work. All authors commented the paper.

Funding: This research received no external funding.

Conflicts of Interest: The authors declare no conflict of interest.

Abbreviations

η	sea-bed, river-bed or channel-bed elevation (m)
ϵ_0	grain packing density of the sediment
D	depth of closure (m)
$\vec{q}_s$	Sediment flux, be-dimensional ($\mathrm{g\,m^{-1}\,s^{-1}}$)
$\vec{q}_{3D}$	Sediment flux, three-dimensional ($\mathrm{g\,m^{-2}\,s^{-1}}$)
∇_H	bi-dimensional version of the Nabla operator, $\left(\frac{\partial}{\partial x}, \frac{\partial}{\partial y}\right)$
$\vec{u}$	velocity field ($\mathrm{m\,s^{-1}}$)
u	zonal velocity component ($\mathrm{m\,s^{-1}}$)
v	meridional velocity component ($\mathrm{m\,s^{-1}}$)
w	vertical velocity component ($\mathrm{m\,s^{-1}}$)
x	alongshore coordinate, or longitude (m)
y	cross-shore coordinate, or latitude (m)
z	vertical coordinate (m)
ρ	sediment concentration in water ($\mathrm{g\,m^{-3}}$)
b	bathymetry, negative under sea surface (m)
ξ_{ij}	longitude of gridpoint ij (m)
ψ_{ij}	latitude of gridpoint ij (m)
α_{ij}	angle between south edge of cell ij and parallel
β_{ij}	angle between west edge of cell ij and meridian
$\Delta\xi_{ij}$	length of the south edge of cell ij (m)
$\Delta\psi_{ij}$	length of the west edge of cell ij (m)

References

1. Dyer, K. *Coastal and Estuarine Sediment Dynamics*; Wiley: New York, NY, USA, 1986.
2. Masselink, G.; Hughes, M.G. *An Introduction to Coastal Processes and Geomorphology*; Routledge: Abingdon, UK, 2014.
3. Syvitski, J.P.M.; Harvey, N.; Wolanski, E.; Burnett, W.C.; Perillo, G.M.E.; Gornitz, V. Dynamics of the coastal zone. In *Coastal Change and the Anthropocene: The Land-Ocean Interactions in the Coastal Zone Project of the International Geosphere-Biosphere Programme*; Crossland, C.J., Kremer, H.H., Lindeboom, H.J., Crossland, J.I.M., Le Tissier, M.D.A., Eds.; Global Change—The IGBP Series; Springer: Texel, The Netherlands, 2005; pp. 39–94.
4. Carniello, L.; Silvestri, S.; Marani, M.; D'Alpaos, A.; Volpe, V.; Defina, A. Sediment dynamics in shallow tidal basins: In situ observations, satellite retrievals, and numerical modeling in the Venice Lagoon. *J. Geophys. Res. Earth Surf.* **2014**, *119*, 802–815. [CrossRef]
5. Parker, G.; Garcia, M. Theory for a clinoform of permanent form on a continental margin emplaced by weak, dilute muddy turbidity currents. In Proceedings of the 4th IAHR Symposium on River, Coastal and Estuarine Morphodynamics (RCEM 2005), Urbana, IL, USA, 4–7 October 2005; Taylor and Frances: Philadelphia, PA, USA, 2006; pp. 553–561.
6. Fagherazzi, S.; Overeem, I. Models of deltaic and inner continental shelf landform evolution. *Annu. Rev. Earth Planet. Sci.* **2007**, *35*, 685–715. [CrossRef]
7. Bignami, F.; Sciarra, R.; Carniel, S.; Santoleri, R. Variability of Adriatic Sea coastal turbid waters from SeaWiFS imagery. *J. Geophys. Res. Oceans* **2007**, *112*. [CrossRef]
8. Brando, V.; Braga, F.; Zaggia, L.; Giardino, C.; Bresciani, M.; Matta, E.; Bellafiore, D.; Ferrarin, C.; Maicu, F.; Benetazzo, A.; et al. High-resolution satellite turbidity and sea surface temperature observations of river plume interactions during a significant flood event. *Ocean Sci.* **2015**, *11*, 909–920. [CrossRef]
9. Cattaneo, A.; Trincardi, F.; Asioli, A.; Correggiari, A. The Western Adriatic shelf clinoform: energy-limited bottomset. *Cont. Shelf Res.* **2007**, *27*, 506–525. [CrossRef]
10. Sherwood, C.R.; Book, J.W.; Carniel, S.; Cavaleri, L.; Chiggiato, J.; Das, H.; Doyle, J.D.; Harris, C.K.; Niedoroda, A.W.; Perkins, H. Sediment Dynamics in the Adriatic Sea Investigated with Coupled Models. *Oceanography* **2004**, *17*, 58–69. [CrossRef]
11. Harris, C.K.; Sherwood, C.R.; Signell, R.P.; Bever, A.J.; Warner, J.C. Sediment dispersal in the northwestern Adriatic Sea. *J. Geophys. Res. Oceans* **2008**, *113*. [CrossRef]
12. Pellegrini, C.; Maselli, V.; Cattaneo, A.; Piva, A.; Ceregato, A.; Trincardi, F. Anatomy of a compound delta from the post-glacial transgressive record in the Adriatic Sea. *Mar. Geol.* **2015**, *362*, 43–59. [CrossRef]
13. Poulain, P.M. Adriatic Sea surface circulation as derived from drifter data between 1990 and 1999. *J. Mar. Syst.* **2001**, *29*, 3–32. [CrossRef]
14. Ouillon, S.; Douillet, P.; Andréfouët, S. Coupling satellite data with in situ measurements and numerical modeling to study fine suspended-sediment transport: a study for the lagoon of New Caledonia. *Coral Reefs* **2004**, *23*, 109–122.
15. Stroud, J.R.; Lesht, B.M.; Schwab, D.J.; Beletsky, D.; Stein, M.L. Assimilation of satellite images into a sediment transport model of Lake Michigan. *Water Resour. Res.* **2009**, *45*. [CrossRef]
16. Niedoroda, A.W.; Reed, C.W.; Das, H.; Fagherazzi, S.; Donoghue, J.F.; Cattaneo, A. Analyses of a large-scale depositional clinoformal wedge along the Italian Adriatic coast. *Mar. Geol.* **2005**, *222*, 179–192. [CrossRef]
17. Exner, F.M. Uber die wechselwirkung zwischen wasser und geschiebe in flussen. *Akad. Wiss. Wien Math. Naturwiss. Klasse* **1925**, *134*, 165–204.
18. Exner, F.M. *Zur Physik der Dünen*; Hölder: Wien, Austria, 1920.
19. Paola, C.; Voller, V. A generalized Exner equation for sediment mass balance. *J. Geophys. Res. Earth Surf.* **2005**, *110*. [CrossRef]
20. Komar, P.D. The mechanics of sand transport on beaches. *J. Geophys. Res.* **1971**, *76*, 713–721. [CrossRef]
21. Rosati, J.; Walton, T.; Bodge, K. Longshore sediment transport. In *Coastal Engineering Manual*; USACE: Washington DC, USA, 2002.
22. Ashton, A.; Murray, A.B.; Arnoult, O. Formation of coastline features by large-scale instabilities induced by high-angle waves. *Nature* **2001**, *414*, 296–300. [CrossRef] [PubMed]
23. Ashton, A.D.; Murray, A.B. High-angle wave instability and emergent shoreline shapes: 1. Modeling of sand waves, flying spits, and capes. *J. Geophys. Res. Earth Surf. (2003–2012)* **2006**, *111*. [CrossRef]

24. McKiver, W.; Sannino, G.; Braga, F.; Bellafiore, D. Investigation of model capability in capturing vertical hydrodynamic coastal processes: A case study in the north Adriatic Sea. *Ocean Sci.* **2016**, *12*, 51. [CrossRef]

25. D'Sa, E.J.; Miller, R.L.; McKee, B.A. Suspended particulate matter dynamics in coastal waters from ocean color: Application to the northern Gulf of Mexico. *Geophys. Res. Lett.* **2007**, *34*. [CrossRef]

26. Dogliotti, A.I.; Ruddick, K.; Nechad, B.; Doxaran, D.; Knaeps, E. A single algorithm to retrieve turbidity from remotely-sensed data in all coastal and estuarine waters. *Remote Sens. Environ.* **2015**, *156*, 157–168. [CrossRef]

27. Nechad, B.; Ruddick, K.; Park, Y. Calibration and validation of a generic multisensor algorithm for mapping of total suspended matter in turbid waters. *Remote Sens. Environ.* **2010**, *114*, 854–866. [CrossRef]

28. ESA. MERIS Instrument for Remote Sensing of the Coastal Zone. Available online: http://www.coastcolour.info (accessed on 18 June 2019).

29. MITgcm Group. MITgcm Website. Available online: http://mitgcm.org (accessed on 19 July 2019).

30. ESA. MERMAID—Meris Matchup In-Situ Database. Available online: http://mermaid.acri.fr/ (accessed on 15 July 2019).

31. Sánchez-Garrido, J.C.; Sannino, G.; Liberti, L.; García Lafuente, J.; Pratt, L. Numerical modeling of three-dimensional stratified tidal flow over Camarinal Sill, Strait of Gibraltar. *J. Geophys. Res. Oceans* **2011**, *116*. [CrossRef]

32. Sannino, G.; Sanchez Garrido, J.; Liberti, L.; Pratt, L. Exchange Flow through the Strait of Gibraltar as Simulated by a σ-Coordinate Hydrostatic Model and az-Coordinate Nonhydrostatic Model. In *The Mediterranean Sea: Temporal Variability and Spatial Patterns*; John Wiley & Sons: New York, NY, USA; pp. 25–50.

33. Drofa, O.; Malguzzi, P. In Proceedings of the 14th International Conference on Clouds and Precipitation, Bologna, Italy, 19–23 July 2004. Available online: https://www.sciencedirect.com/journal/atmospheric-research/vol/82/issue/1 (accessed on 13 May 2019).

34. Malguzzi, P.; Grossi, G.; Buzzi, A.; Ranzi, R.; Buizza, R. The 1966 "century" flood in Italy: A meteorological and hydrological revisitation. *J. Geophys. Res. Atmos.* **2006**, *111*. [CrossRef]

35. Ferrarin, C.; Roland, A.; Bajo, M.; Umgiesser, G.; Cucco, A.; Davolio, S.; Buzzi, A.; Malguzzi, P.; Drofa, O. Tide-surge-wave modelling and forecasting in the Mediterranean Sea with focus on the Italian coast. *Ocean Model.* **2013**, *61*, 38–48. [CrossRef]

36. Raspaud, M.; Hoese, D.; Dybbroe, A.; Lahtinen, P.; Devasthale, A.; Itkin, M.; Hamann, U.; Ørum Rasmussen, L.; Nielsen, E.S.; Leppelt, T.; et al. PyTroll: An open source, community driven Python framework to process Earth Observation satellite data. *Bull. Am. Meteorol. Soc.* **2018**, *99*, 1329–1336. [CrossRef]

37. EMODnet Bathymetry Consortium. EMODnet Digital Bathymetry (DTM). EMODnet Bathymetry 2016. 2016. Available online: https://doi.org/10.12770/c7b53704-999d-4721-b1a3-04ec60c87238 (accessed on 19 July 2019).

38. Hyman, J.M.; Knapp, R.J.; Scovel, J.C. High order finite volume approximations of differential operators on nonuniform grids. *Phys. D Nmiscar Phenomena* **1992**, *60*, 112–138. [CrossRef]

39. Cozzi, S.; Giani, M. River water and nutrient discharges in the Northern Adriatic Sea: Current importance and long term changes. *Cont. Shelf Res.* **2011**, *31*, 1881–1893. [CrossRef]

40. Falcieri, F.M.; Benetazzo, A.; Sclavo, M.; Russo, A.; Carniel, S. Po River plume pattern variability investigated from model data. *Cont. Shelf Res.* **2014**, *87*, 84–95. [CrossRef]

41. Hopkins, T.; Artegiani, A.; Kinder, C.; Pariante, R. A discussion of the northern Adriatic circulation and flushing as determined from the ELNA hydrography. In *The Adriatic Sea*; European Commission: Brussels, Belgium, 1999; Volume 32, pp. 85–106.

42. Cavaleri, L.; Bertotti, L. In search of the correct wind and wave fields in a minor basin. *Mon. Weather Rev.* **1997**, *125*, 1964–1975. [CrossRef]

43. Wang, X.; Pinardi, N. Modeling the dynamics of sediment transport and resuspension in the northern Adriatic Sea. *J. Geophys. Res. Oceans* **2002**, *107*. [CrossRef]

44. Grant, W.D.; Madsen, O.S. Combined wave and current interaction with a rough bottom. *J. Geophys. Res. Oceans* **1979**, *84*, 1797–1808. [CrossRef]

45. Wang, X.; Pinardi, N.; Malacic, V. Sediment transport and resuspension due to combined motion of wave and current in the northern Adriatic Sea during a Bora event in January 2001: A numerical modelling study. *Cont. Shelf Res.* **2007**, *27*, 613–633. [CrossRef]

46. ISPRA. MARE E AMBIENTE COSTIERO. Available online: http://www.isprambiente.gov.it/files/pubblicazioni/statoambiente/tematiche2011/05_%20Mare_e_ambiente_costiero_2011.pdf (accessed on 13 May 2019).

47. ISPRA. MARE E AMBIENTE COSTIERO. Available online: http://www.isprambiente.gov.it/files/pubblicazioni/statoambiente/tematiche-2012/Cap.5_Mare_ambiente_costiero.pdf (accessed on 13 May 2019).

48. ISPRA. Pubblicazioni ISPRA—Stato Ambiente. Available online: http://www.isprambiente.gov.it/files/pubblicazioni/statoambiente (accessed on 13 May 2019).

49. Artegiani, A.; Paschini, E.; Russo, A.; Bregant, D.; Raicich, F.; Pinardi, N. The Adriatic Sea general circulation. Part II: baroclinic circulation structure. *J. Phys. Oceanogr.* **1997**, *27*, 1515–1532. [CrossRef]

50. Limic, N.; Orlic, M. Objective analysis of geostrophic currents in the Adriatic Sea. *Geofizika* **1986**, *3*, 75–84.

51. Correggiari, A.; Trincardi, F.; Langone, L.; Roveri, M. Styles of failure in late Holocene highstand prodelta wedges on the Adriatic shelf. *J. Sediment. Res.* **2001**, *71*, 218–236. [CrossRef]

52. Thomas, L.N.; Tandon, A.; Mahadevan, A. Submesoscale processes and dynamics. *Ocean Model. Eddying Regime* **2008**, *177*, 17–38.

53. ESA. OLCI Instrument. 2016. Available online: https://sentinel.esa.int/web/sentinel/technical-guides/sentinel-3-olci/olci-instrument (accessed on 13 June 2019).

54. Aminou, D. MSG's SEVIRI instrument. *ESA Bull. (0376-4265)* **2002**, *111*, 15–17.

55. Lopinto, E.; Ananasso, C. The Prisma Hyperspectral Mission. In Proceedings of the 33rd EARSeL Symposium Towards Horizon 2020, Matera, Italy, 3–6 June 2013.

56. Stuffler, T.; Kaufmann, C.; Hofer, S.; Förster, K.; Schreier, G.; Mueller, A.; Eckardt, A.; Bach, H.; Penne, B.; Benz, U.; et al. The EnMAP hyperspectral imager—An advanced optical payload for future applications in Earth observation programmes. *Acta Astronaut.* **2007**, *61*, 115–120. [CrossRef]

Article

Ocean Color Quality Control Masks Contain the High Phytoplankton Fraction of Coastal Ocean Observations

Henry F. Houskeeper * and Raphael M. Kudela

Ocean Sciences Department, University of California Santa Cruz, Santa Cruz, CA 95064, USA; kudela@ucsc.edu
* Correspondence: hhouskee@ucsc.edu

Received: 2 August 2019; Accepted: 16 September 2019; Published: 18 September 2019

Abstract: Satellite estimation of oceanic chlorophyll-a content has enabled characterization of global phytoplankton stocks, but the quality of retrieval for many ocean color products (including chlorophyll-a) degrades with increasing phytoplankton biomass in eutrophic waters. Quality control of ocean color products is achieved primarily through the application of masks based on standard thresholds designed to identify suspect or low-quality retrievals. This study compares the masked and unmasked fractions of ocean color datasets from two Eastern Boundary Current upwelling ecosystems (the California and Benguela Current Systems) using satellite proxies for phytoplankton biomass that are applicable to satellite imagery without correction for atmospheric aerosols. Evaluation of the differences between the masked and unmasked fractions indicates that high biomass observations are preferentially masked in National Aeronautics and Space Administration (NASA) ocean color datasets as a result of decreased retrieval quality for waters with high concentrations of phytoplankton. This study tests whether dataset modification persists into the default composite data tier commonly disseminated to science end users. Further, this study suggests that statistics describing a dataset's masked fraction can be helpful in assessing the quality of a composite dataset and in determining the extent to which retrieval quality is linked to biological processes in a given study region.

Keywords: phytoplankton remote sensing; coastal ocean; red tides; black pixel assumption; atmospheric correction

---

1. Introduction

Ocean color remote sensing has greatly improved our ability to monitor global scale biological processes of ocean systems [1,2] but the potential for conventional satellite ocean color tools to characterize coastal ecosystems is limited by the assumptions used in various algorithms, for example that diverse phytoplankton communities match global bio-optical relationships [3] or that backscattered light from particles does not interfere with atmospheric correction [4]. Although these assumptions are often not valid for coastal waters [5], satellite assessment of coastal marine ecosystems is an area of intense focus in part because of reported increases in the frequency of coastal phytoplankton blooms considered harmful to humans and wildlife [6–10]. Fundamental challenges for ocean color remote sensing of coastal marine ecosystems arise from the increased complexity of water constituents, as well as the entanglement of atmospheric and oceanic signals. Overcoming these difficulties motivates next-generation ocean color satellite missions, with the aim to characterize oceanic ecosystems spanning oligotrophic to eutrophic waters, for example through increasing spectral resolution to resolve variability in phytoplankton pigmentation [11] and through increasing spectral range to discern the effects of absorbing aerosols and colored dissolved organic matter and to improve aerosol characterization [12]. Nonetheless, data from existing satellite platforms is presently required to assess coastal marine ecosystems. When next-generation sensing platforms become operational, interpreting legacy measurements of coastal waters will still be necessary in order to construct and interpret climate-quality data records and will require methods to detect regional bias of legacy retrievals [13–15]. For coastal targets, constructing climate-quality datasets will require an approach to maintain atmospheric correction efficacy across variable phytoplankton concentrations.

Conventional approaches for atmospheric correction of ocean color satellite imagery take advantage of the strong light absorption by water at longer wavelengths, for example, in the near-infrared (NIR), to estimate that the water-leaving radiance (L_W) in the NIR (L_W(NIR)) is negligible [16]. Thus, after removal of glint and white capping effects, the derived top-of-atmosphere (TOA) radiance in the NIR (L_{TOA}(NIR)) is attributed to the atmospheric contributions by Rayleigh scattering, aerosol scattering and multiple interactions between Rayleigh and aerosol scattering [17], allowing a solution to discern aerosol thickness.

The approximation that L_W(NIR) is zero, termed the "Black Pixel Assumption" [4], is often incorrect, frequently so in coastal waters where high near-surface particle loads (organic or inorganic) can strongly backscatter light, such that the L_W(NIR) domain is appreciably non-zero. Because L_{TOA}(NIR) is attributed to atmospheric constituents, L_W(NIR) contributions cause overestimation of backscattering by atmospheric aerosols and thus result in incorrect (often negative) derivation of L_W in the visible domain (L_W(VIS)), particularly in the blue bands used for, among others, chlorophyll-a derivation [18]. As a result, atmospheric correction is more problematic for water masses with high particle loads, including of phytoplankton cells and high biomass pixels are frequently removed during quality control processing of satellite datasets (Figure 1).

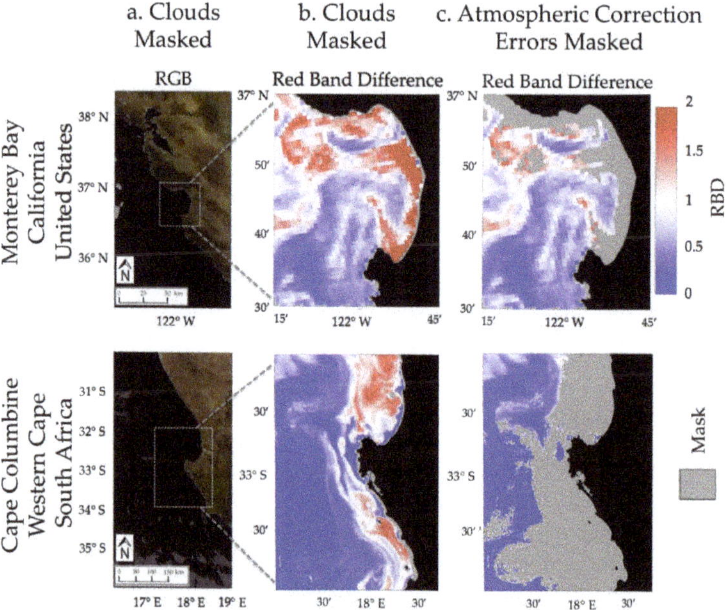

Figure 1. MODIS Aqua imagery of phytoplankton blooms obtained on October 26th, 2016 in Monterey Bay, California (upper) and September 27th, 2011 near Cape Columbine, Western Cape (lower). (**a**) Pseudo-true color images with clouds masked in grey; (**b**) Red band difference (RBD) algorithm (a proxy for phytoplankton biomass) with clouds masked in grey; (**c**) RBD algorithm with clouds and suspect atmospheric correction (defined as maximum aerosol iterations reached and low water-leaving radiances) masked in grey.

Alternate atmospheric correction methods have been developed with the goal of improving L_W(VIS) retrieval in coastal waters, for example by estimating aerosol contributions from longer (short-wave infrared) bands [19], by assuming stable NIR reflectance ratios within a scene [20] or by neural network determination of atmospheric contribution [21]. Alternate methods for atmospheric correction of coastal imagery have generally improved performance of nearshore ocean color retrievals compared with the conventional (NIR-based) methodology [22], although users still must decide when use of these methods is preferable given tradeoffs of noisy wavebands, non-analytical solutions and increased difficulty in obtaining alternate processing for National Aeronautics and Space Administration (NASA) imagery. Another potential reason for usage of default NIR aerosol corrected imagery, evaluated in detail within this work, is that the alteration of satellite dataset distributions by atmospheric correction errors may be obscured by the default quality control masks applied to composited imagery.

Quality control of NASA ocean color imagery is achieved in part through flag assignments that trigger masking or removal of individual pixels which do not satisfy pre-defined thresholds. Increasingly rigorous flag criteria are applied to mask observations from sequential data tiers, based on the quality requirements of the tier's expected end-users. Pixels within atmospherically-corrected imagery (level 2 data tier, L2) are masked by default when the derivation of meaningful products is severely inhibited, for example when the sensor is viewing land or clouds or when the sensor saturates. L2 datasets contain shifting pixel coordinates, frequent data gaps (i.e., from clouds) and large file sizes inconvenient for users requiring continuous or less computationally expensive products [23]. To satisfy these user needs, statistical composites of geophysical variables binned in space and time (level 3 data tier, L3) are provided by the NASA Ocean Biology Processing Group (OBPG) and are valuable to users beyond the ocean color community, for example as inputs into biogeochemical models. In order to provide higher quality composites for a larger end-user community, the default masks applied during L3 processing are more rigorous than during L2 processing, for example removing observations flagged for likely or known errors in atmospheric correction. Spatial distortions may also arise during compositing and although not evaluated here, are likewise relevant to L3 end-users [24].

In this study, we compare estimates of phytoplankton biomass obtained without aerosol correction for observations that satisfy (versus fail) standard quality control thresholds for two Eastern Boundary Current (EBC) ecosystems. Characterization of the changes that occur from removing portions of satellite datasets enables assessment of whether quality control methodology alters satellite perspectives of biology in coastal ecosystems. We assess whether observations that satisfy quality control methods—hereafter the masked fraction—provide an unbiased perspective of phytoplankton biomass in coastal ecosystems and we provide examples for L3 end-users to consider when determining whether use of standard composite products may be reasonable for a specific study region.

2. Materials and Methods

2.1. Site Selection:

The mid-latitude eastern margins (or EBCs) of the world's oceans are regions of heightened biological primary production due to coastal upwelling or the wind-driven transport of nutrient-rich subsurface waters to the illuminated surface layer. Heightened nutrient availability, coupled with the persistence of seed stocks from the shelf and from retention in the lee of headlands, support high phytoplankton concentrations that periodically form blooms, some of which may be harmful to humans and wildlife [7,25].

Here we consider EBC ecosystems of Monterey Bay (MB), California, USA and St. Helena Bay (SHB), Western Cape, South Africa. MB is within a marine sanctuary and is partially sheltered from the predominant alongshore winds of the central California Current System (CCS) by coastline geometry. Phytoplankton in the region follow a distinct climatology, with spring onset of upwelling-favorable winds supporting diatom-rich phytoplankton blooms, followed by a mid-summer reduction in phytoplankton associated with rapid advection to offshore waters [26]. Fall relaxation of upwelling-favorable winds and the resulting increased vertical stratification of the surface layer facilitates a community shift towards dinoflagellates, which periodically form dense red tides with concentrations that may reach or exceed those of spring diatom blooms [25].

SHB is an upwelling ecosystem in the lee of Cape Columbine within the southern Benguela Current System (BCS). The region's proximity to the Cape Peninsula upwelling cell, coupled with shelter from the lee and a widened shelf, provides high nutrient loads in a relatively stable environment, which support persistently high phytoplankton production [27]. In addition, sea surface temperature is relatively high within SHB compared with other EBCs, allowing elevated phytoplankton populations to persist throughout all seasons [28]. As with MB, phytoplankton succession in the southern BCS, including within SHB, is dictated by the intensification and relaxation of alongshore winds, with characteristic diatom and dinoflagellate regimes dominating in the spring and fall, respectively [29].

2.2. Atmospheric Dataset:

Climatological datasets for Ångstrom exponent and aerosol optical depth (500 nm) were obtained from the Aerosol Robotic Network (AERONET; aeronet.gsfc.nasa.gov) for Monterey, California, USA (36.59°N, 121.85°W) and Simonstown, Western Cape, South Africa (34.18°S, 18.43°E; Figure 2). The Monterey AERONET site is located to the southeast of MB and separated from a nearby agricultural region by a coastal mountain range, although diurnal sea breeze north of this range may increase mixing between terrestrial and marine airmasses. The Simonstown AERONET site is located on the eastern slope of the Cape Peninsula, roughly 150 km south of Cape Columbine. Predominant windstress is equatorward (towards SHB) with summertime intensification [30].

Matchup Sites

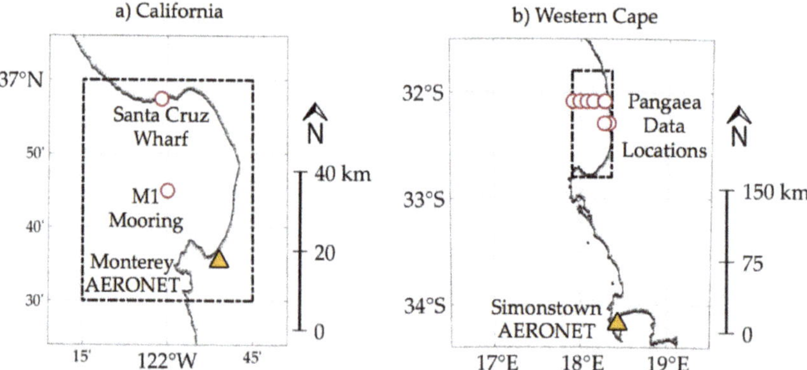

Figure 2. Location of biological and atmospheric measurements used in this study for (**a**) Monterey Bay, California and (**b**) St. Helena Bay, Western Cape. Chlorophyll-a and fluorometer measurement sites denoted with red circles, atmospheric measurement sites denoted with orange triangles. Regions of satellite L2 datasets used for analysis and matchups indicated by dashed black lines.

2.3. Biological Field Data

Weekly fluorometric Chlorophyll-a (Chla) measurements (*in vitro*) were obtained from the Southern California Coastal Ocean Observing System portal (sccoos.org/data/habs/) for the Santa Cruz Wharf (SCW; 36.958°N, 122.017°W; Figure 2) in northern MB. Daily mean in situ surface (1 m) fluorescence measurements were obtained from a HydroScat-2 fluorometer (HOBI Labs) mounted on an oceanographic mooring in central MB (M1; 36.750°N, −122.000°E), maintained by the Monterey Bay Aquarium Research Institute (MBARI; mbari.org). Measurements of Chla within SHB were obtained from the European Space Agency (ESA) Ocean Color Climate Change Initiative (OC-CCI) dataset, available on the Pangaea portal (doi.pangaea.de). Only fluorometric measurements (*in vitro*) were used for match-ups with satellite products for SHB since the majority of Chla estimates in the database were based on that methodology.

2.4. Satellite Data

MODIS Aqua (MODISA) calibrated, non-atmospherically corrected imagery (L1A) was obtained from the NASA Ocean Color website (oceancolor.gsfc.nasa.gov) for dates spanning July 2002 to September 2018 within MB (36.50–37.00°N; 121.75–122.25°W) and SHB (31.80–32.80°S; 17.90–18.35°E; Figure 2) for matchup validation and for analysis of L2 datasets. Surface reflectances (ρ_S) were obtained for both regions from geo-referenced and atmospherically corrected imagery produced using NASA OBPG software SeaDAS (version 7.5) with observations removed if viewing land or clouds or for non-physical retrievals (i.e., ρ_S outside of the range 0–1). Spatial subsets used for comparison with AERONET results were selected from the MB and SHB domains based on the AERONET site location and the local topography (MB latitude < 36.65°N; longitude < 121.92°W; SHB latitude > 32.10°S; longitude > 18.10°E). Data for L3 analysis was obtained for the same timeframe and for similar regions in MB (36.50–37.00°N; 121.75–122.25°W) and SHB (32.20–32.80°S; 17.90–18.35°E), as well as for two nearby transects placed in regions with relatively north-south coastline of the CCS (37.10–37.50°N, 122.40–123.40°W) and BCS (31.80–32.20°S, 17.30–18.35°E). Processing for the L3 products applied additional SeaDAS software *l2bin* and *l2mapgen* to form daily, 4km standard map grid composites.

Quality control flags were assigned to all pixels during the L2 processing chain according to standard OBPG L2 flag thresholds. Masks were applied to one identical set of the L2 data if flag assignments indicated likely errors in atmospheric correction. Flags chosen included warnings for low water-leaving radiance (LOWLW), maximum iterations exceeded during atmospheric correction processing (MAXAERITER) and out-of-range spectral slope of derived aerosol radiances (ATMWARN). This combination of flags will be hereafter referred to as AC flags. More detailed information on thresholds and applications of default flags can be found in SeaWiFS postlaunch documentation [31].

Neural network Chla estimates were included for evaluating satellite products using processing tools provided by the Coast Color project (coastcolour.org). In short, MODISA imagery was georeferenced and calibrated using SeaDAS. Atmospheric correction and derivation of Chla was then performed using the Sentinel Application Platform (SNAP; step.esa.int) with the Case-2 Regional Coast Color (C2RCC; brockmann-consult.de) plugin.

2.5. Remote Estimation of Phytoplankton Biomass:

Remote measurements of the spectral radiance anomaly generated by sun-induced fluorescence of the Chla molecule have been applied as a proxy for phytoplankton biomass for over four decades [32,33]. The fluorescence line height approach (FLH), which subtracts a red to NIR baseline from the Chla fluorescence peak to correct for brightness effects, is the most widely used of these satellite tools. FLH has been proposed to complement traditional Chla satellite algorithms in high–biomass coastal waters [34,35] and is disseminated in standard L2 and L3 OBPG data derived from normalized L_W (nFLH). Another FLH-type method, the red band difference algorithm (RBD) [36], subtracts the signal derived at the nearest shorter wavelength from the signal measured at the Chla fluorescence peak. RBD was chosen for this work because of its relative robustness to sediment effects [36].

Fluorescence line height products, including RBD, may be derived at TOA, thus bypassing the potential errors arising during the atmospheric correction procedure. Here we use a partial atmospheric correction product that accounts for Rayleigh but not aerosol effects, termed the surface reflectance ($\rho_S(\lambda)$), which may be defined as

$$\rho_S(\lambda) = \left(\frac{\pi}{F_0\mu_0}\right)\left(\frac{L_{TOA}}{t_{solar}t_{sensor}} - L_r\right)\left(t'_{solar}t'_{sensor}t'_{O_2}t'_{H_2O}\right)^{-1} \tag{1}$$

where F_0 is the solar downward irradiance, μ_0 is the cosine of the solar zenith angle, t and t' are the direct and diffuse atmospheric transmittances, respectively, for the sun-surface and surface-sensor path lengths and for the atmospheric effects of oxygen and water vapor. RBD is thus derived as the difference between MODISA surface reflectances:

$$RBD = \rho_S(678nm) - \rho_S(667nm) \tag{2}$$

where $\rho_S(678nm)$ corresponds to the height of the Chla fluorescence maximum (approximately $683nm$) and $\rho_S(667nm)$ provides an adjacent baseline to account for overall spectral brightness.

2.6. Match-up Procedure, Derivation of Climatological Averages and Error Statistics

Validation statistics for all satellite products were derived using only same-day, 3×3 pixel median match-ups centered on the in-water samples due to high spatial and temporal variability at the match-up sites. OC3M and C2RCC, as well as fluorescence and Chla measured in situ were $\log_{10}$-transformed prior to derivation of matchups. RBD and flag climatologies were derived as the mean monthly values for each region. Composite datasets were compared using standardized bias (SB), derived as the absolute bias due to masking of the composite data normalized by the standard deviation:

$$SB(i) = \left(\frac{\overline{X}_i - \overline{Y}_i}{\sigma_Y} \right) \tag{3}$$

where $\overline{X}_i$ and $\overline{Y}_i$ correspond to the mean composite RBD values for datasets with AC flagged pixels omitted and retained, respectively, normalized by the standard deviation of the dataset with AC flagged pixels retained (σ_Y). *SB* was partitioned (i) by longitude or by the fraction of underlying pixels (L2) which were assigned AC flags before spatial binning.

3. Results

3.1. Association Between Red Band Difference and Phytoplankton Biomarkers

Satellite match-ups at both MB and SHB indicate that the RBD algorithm associated more strongly (based on a Pearson test) with in situ proxies for phytoplankton concentrations than either a standard NASA blue-green Chla algorithm (OC3M) or a neural-network-based Chla algorithm (Coast Color), although the comparison presented here is not intended as a rigorous inter-comparison of Chla algorithms. An attempt to model surface Chla from satellite measurements (using a linear, least squares approach) resulted in generally higher (never lower) accuracy of the RBD method versus the other remote products, suggesting that RBD is a useful proxy for Chla within our study regions. Greater frequency of valid match-ups were possible at the M1 buoy (MB) location because of increased distance from land and because of the greater number of in situ records. More valid match-ups were also possible for RBD versus OC3M because the ρ_S derivation (which does not account for aerosols) avoided retrieval failures. Visual inspection of Coast Colour match-up scenes suggested that common culprits for reduction of valid match-ups were both the out-of-range inputs to the atmospheric neural network as well as incorrect cloud mask assignment (Table 1).

The lowest Pearson coefficients for all remote products occur at the Santa Cruz Wharf (MB), where matchups are more difficult because of the increased spatial and temporal heterogeneity of the near-shore environment and where fewer adjacent pixels are available due to blockage by the shoreline. The Pearson coefficient for the RBD product is highest relative to the other products at this site, suggesting that the nearshore match-ups also were strongly affected by resuspended sediment, riverine discharge or terrestrial aerosols given the relative robustness of RBD to signal brightening effects and to absorption by riverine constituents, such as colored dissolved organic material. Error between modeled and in situ Chla for the OC3M algorithm was lowest at the M1 buoy (MB), the site with the greatest prevalence of optically simple (case-1) water types among our validation sets. The highest

Pearson coefficient for each product was derived from SHB matchups, with RBD showing the strongest association with in situ Chla among the evaluated remote products. The SHB matchups were unique from the two MB sites in that the in situ measurements were obtained by ship at various distances from shore, allowing the SHB matchup dataset to include a wider diversity of water types than either the wharf or mooring datasets in MB.

Table 1. Match-up statistics for 3 MODIS Aqua phytoplankton biomass proxies in Monterey Bay (MB), California, USA and St. Helena Bay (SHB), Western Cape, South Africa where n is the number of valid match-ups, $P(r)$ is the Pearson coefficient and $nRMSE$ is the root mean square error of the linear fit of the satellite data to the in situ data, normalized by the in situ data range.

Location	Product	n	$P(r)$	$nRMSE$
Monterey Bay,	Red Band Difference	1012	0.4190	14.4%
California	Blue-Green Chla Algorithm[1]	773	0.3344	14.8%
M1 Buoy	Neural Network Chla Algorithm[2]	840	0.2179	17.0%
Monterey Bay,	Red Band Difference	361	0.2055	14.1%
California	Blue-Green Chla Algorithm[1]	8	0.0359	28.8%
Santa Cruz Wharf	Neural Network Chla Algorithm[2]	132	0.0488	17.3%
St. Helena Bay,	Red Band Difference	90	0.5493	19.2%
South Africa	Blue-Green Chla Algorithm[1]	74	0.4283	20.8%
Various Locations	Neural Network Chla Algorithm[2]	75	0.3973	19.9%

[1] NASA OBPG standard Chla product (OC3M algorithm). [2] Coast Color standard Chla product.

3.2. Climatology of Atmospheric Correction Flags

Comparison of AERONET and satellite (MODISA) climatologies did not reveal similarities between atmospheric constituents and satellite flags. In particular, results from a Pearson's correlation test were not significant between AC flags and either aerosol optical depth (p = 0.25, p = 0.29) or Ångstrom exponent (p = 0.67, p = 0.84) for the Monterey or Simonstown sites, respectively. AC flag assignments correlated positively with RBD at both sites, with correlation significant for MB (p < 0.01) but not for SHB (p = 0.43). Both AERONET sites revealed local maxima of both aerosol optical depth and Ångstrom exponent during summer that did not correspond to a spike in AC flag assignments during the same month. For the Monterey site, the summertime peak in atmospheric complexity coincided in local minima in AC flag assignments suggesting that pollution or aerosol loading during summer months are not dominant mechanisms for low atmospheric correction efficacy in this sample. The seasonality of AC flags in the Simonstown region was more uniform than the Monterey region but was similarly incongruous with the AERONET results (Figure 3).

Although not elucidated by this analysis, the relatively higher Ångstrom exponent and aerosol optical depth measured from spring through fall at the Monterey AERONET station may in part be responsible for the decreased performance in the OC3M match-ups within MB compared with SHB. Other differences between the Monterey and Simonstown results may be due to the AERONET locations, with the Monterey site nearer the sheltered retentive zone in southern MB and the Simonstown site located near a headland with more exposure to wind and currents and further from the SHB subset.

3.3. Impact of Atmospheric Correction Masks on Level 2 RBD datasets

Satellite retrievals with higher RBD values were more frequently assigned AC flags, with 33.8% and 33.1% of observations assigned AC flags within the upper quartile of RBD data and only 5.3% and 8.2% within the lower quartile at MB and SHB, respectively. As a result, masking of AC flagged retrievals decreased the right-hand tails of the RBD dataset distributions at both sites (Figure 4). The resultant masked fraction describes lower average RBD values (mean: −18.1% and −11.0%; median: −14.7% and −8.0%) with less variance (standard deviation: −13.3% and −13.0%) compared with the initial (AC flagged pixels not masked) datasets for MB and SHB, respectively.

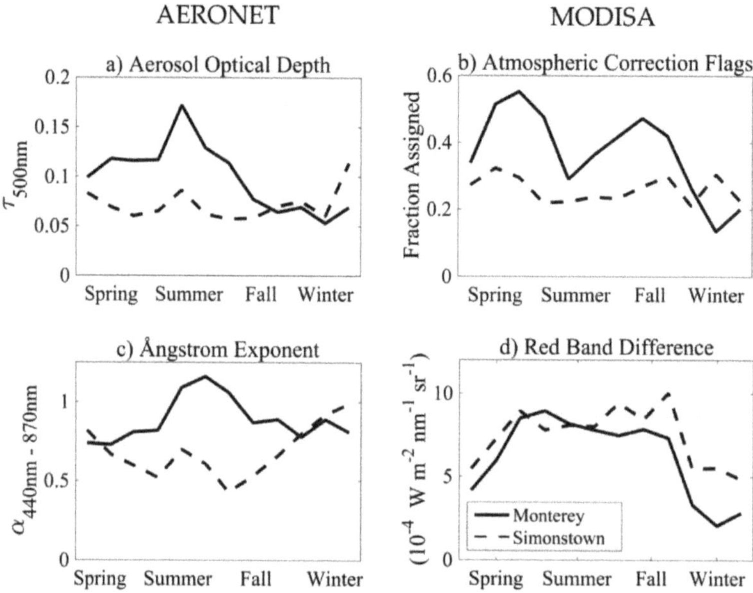

Figure 3. Climatology of atmospheric and satellite products near Monterey Bay (MB) and SHB, shifted for the phase timing of northern and southern hemispheres (Monterey: solid black line, March–February; Simonstown: dashed black line, September–August): (**a**) Aerosol optical depth, (**b**) assignment frequency of AC flags (**c**) Ångstrom exponent and (**d**) Red Band Difference (RBD).

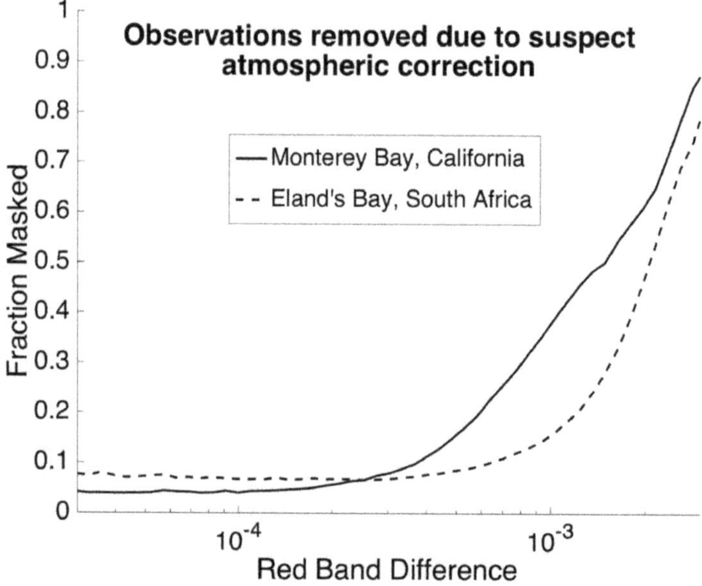

Figure 4. Fraction of pixels assigned atmospheric correction (AC) flags as a function of RBD, with MB and SHB shown with solid and dashed black lines, respectively. The range shown encompasses greater than 98% of the data for both regions.

3.4. Impact of Atmospheric Correction Masks on Level 3 RBD Datasets

L3 spatial composites (4 km, 1 day) were compared between the masked and unmasked RBD datasets. Maximum negative SB occurred in the lee of retentive features that outline MB and SHB, regions prone to frequent phytoplankton blooms due to recirculation of water-masses and protection from offshore advection during upwelling pulses (Figure 5). SB was more negative in near-shore composite grids within the BCS compared with the CCS, with near-shore SB approximately a fifth of a standard deviation lower in the masked versus relaxed dataset. Amplitudes in SHB and MB were comparable, with SB of near-shore composite grids negative by approximately one quarter of a standard deviation.

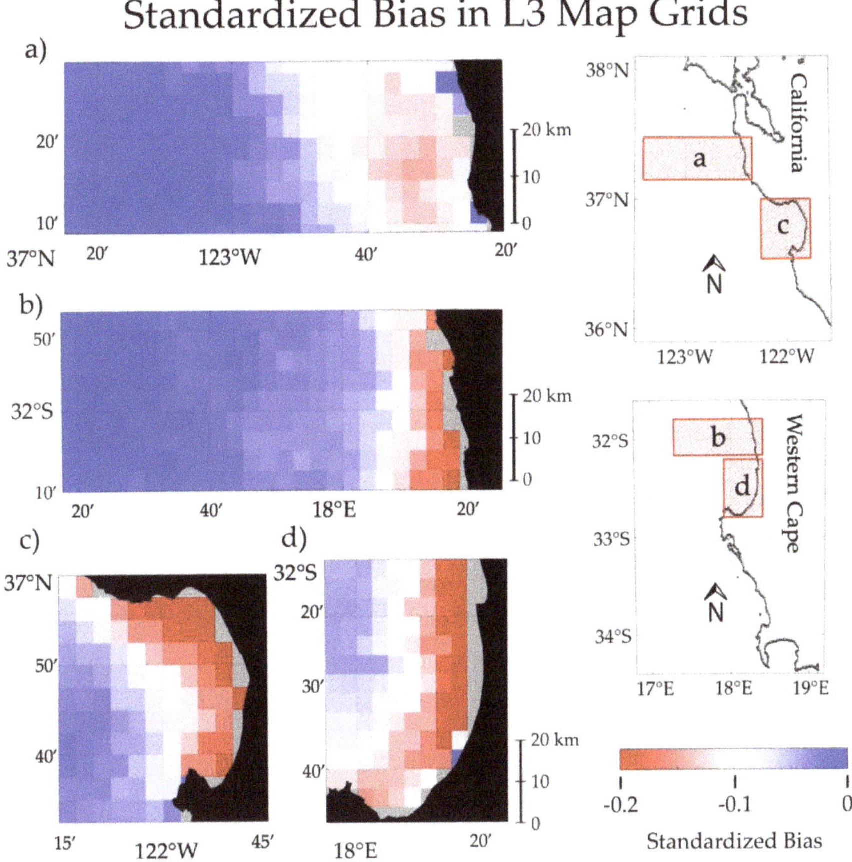

Figure 5. Standardized bias in MODISA RBD composites (L3) due to the removal of underlying (L2) pixels, shown for standard map grids within the central California Current System (CCS) (**a**) and southern Benguela Current System (BCS) (**b**) and within MB (**c**) and SHB (**d**).

Composites within the CCS and BCS transects (Figure 5a,b) were partitioned by longitude to derive SB as a function of distance from shore. Transect regions were adjacent to relatively north-south coastlines and were each greater than 50 km equatorward of the largest nearby coastline points (e.g., Point Reyes and Cape Columbine). The BCS transect showed more rapid improvement of data quality as a function of distance from shore, with SB less than a tenth of one standard deviation negative beyond approximately 30 km and 15 km within the CCS and BCS respectively (Figure 6). The greater offshore persistence of the negative SB for the CCS transect may be due in part to regional circulation differences such as proximity to upwelling hotspots and retentions zones. For example, high RBD amplitudes persist at all latitudes within the nearby SHB domain (Figure 5d), indicating that the BCS transect is in close proximity to a phytoplankton-rich, retentive zone. In comparison, the CCS transect lies poleward of MB and intersects a relatively unprotected stretch of coastline, favoring upwelling dynamics that generate offshore flow. Beyond approximately 50 km from shore, where phytoplankton concentrations are lower, both transects indicate convergence to a small SB, although the signs remain negative across our dataset.

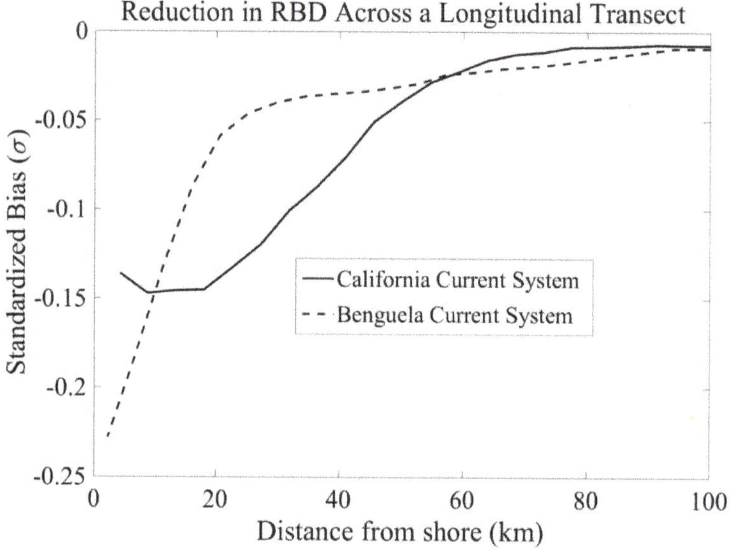

Figure 6. Standardized bias in RBD composites (L3) across longitudinal transects due to the removal of underlying (L2) pixels assigned AC flags within the CCS (solid line) and BCS (dashed line).

From all four masked and unmasked L3 datasets, composites were partitioned by the fraction of AC flag assignments within each composite's underlying (L2) pixels, which was recorded during L3 processing. The data products show near zero SB for composites with few AC flag assignments but the degradation in data quality increases for composites with greater fractions of L2 pixels masked by AC flags (Figure 7). The change in SB is strongly negative as the composites contain increasing pixels masked by AC flags, with the reduction approaching approximately four tenths of one standard deviation for heavily masked composites.

Despite the atmospheric, ecological and topographic differences between the BCS and CCS regions, the slope of the composite reduction is similar, implying consistency in the sites' sensitivity to bias from the association between biology and retrieval quality. The slopes for all sites also imply a linear relationship due to the mixing of two distributions (masked and unmasked) within the composites. L3 transects used for this comparison include a broader range in water masses (onshore and offshore) and 4% of the composites in both the CCS and BCS contained a quarter or more L2 pixels assigned AC flags. Within the more productive regions of MB and SHB, 8% and 10% of the composites generated, respectively, contained a quarter or more L2 pixels assigned AC flags.

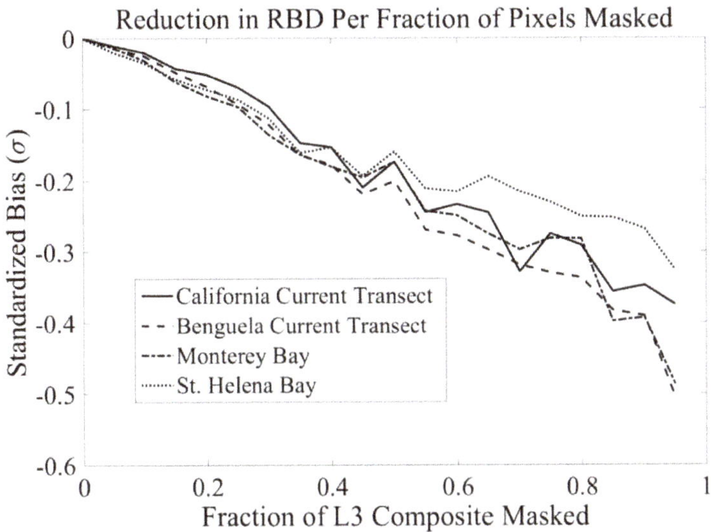

Figure 7. Standardized bias in RBD composites (L3) due to the removal of the underlying (L2) pixels assigned AC flags, as a function of L2 AC flag assignment, for L3 transects within the CCS (solid line) and BCS (dashed line) transects and within the MB (dashed-dotted) and SHB (dotted) composited regions.

4. Discussion

4.1. Performance of Satellite Products at Match-up Sites

Based on the match-up results, RBD is a reasonable proxy to describe relative changes in Chla within the study regions, although this comparison is not intended as a validation activity to assess OC3M, C2RCC or other alternate processing methods. Indeed, for the OC3M products, atmospheric correction quality was a fundamental problem for the match-up regions but rigorous quality screening of the match-ups would have been counter to the goals targeted by this study, namely, to characterize the observations that fail such screenings. C2RCC performed marginally worse than RBD in the comparisons but it should be noted that although the C2RCC network is compatible with MODISA, development was not primarily targeted towards NASA products. Moreover, neural network algorithms require training sets representative of the regions assessed and our results are in no way intended to suggest that C2RCC would not outperform RBD under a different match-up set, or after addition of a larger training set. Indeed, in a recent intercomparison of atmospheric correction methods for coastal waters, strong improvements were shown for a C2RCC model after the inclusion of an expanded training set [22].

FLH algorithms such as RBD are useful for scene comparisons and as general Chla proxies but are not a satisfactory full solution to remote sensing challenges in coastal waters. Although Chla fluorescence and concentration generally covary, their relationship is inconsistent. Factors that may alter the relationship between Chla fluorescence and concentration include phytoplankton species composition, pigment packaging effects, physiology, limitation of nutrients or light or solar-induced fluorescence quenching [37]. The ability to measure fluorescence is also strongly affected by attenuation from water and its constituents, particularly by non-algal particles [38] and from sensor-specific response functions, for example, if the fluorescence peak shifts between response bands [39,40].

Despite the inherent difficulties in quantifying phytoplankton concentrations with Chla fluorescence products, such proxies are reasonable for the analysis shown here because of their relative robustness to atmospheric correction errors and because the RBD biomass comparison is not used across large spatial domains. Comparisons here are assumed to be relative within the region and as a result, this work did not focus on modeling RBD onto in situ Chla.

Products that use inputs from blue wavelengths (such as OC3M) were considered the most sensitive to the decrease in atmospheric correction efficacy addressed by the AC flag assignments and were not reasonable options for the comparison described here. Similarly, products such as OC3M often cannot be derived for AC flagged pixels, for example, when overestimation of aerosol thickness causes derivation of negative radiances at blue wavelengths. Finally, while RBD was preferred here over the OBPG default nFLH product because of its reported robustness to sediment effects, nFLH is expected to be a reasonable alternative for users who intend to perform a similar analysis of their study region but who require direct downloads of default L2 products.

4.2. Variability of Atmospheric Constituents and Efficacy of Ocean Color AC Flags

Atmospheric correction errors arising from elevated L_W(NIR) have been a focus for improving satellite retrievals over sediment- or phytoplankton-loaded waters. The associations shown here between AC flags and phytoplankton concentrations within MB and SHB are intended to demonstrate the frequency in which retrieval quality is linked to biology and to assess whether the set of observations which satisfy quality control thresholds are able to accurately characterize coastal marine ecosystems. In our study areas, the dataset fractions that satisfy default retrieval criteria (i.e., are not assigned AC flags) describe ecosystems which are generally reduced in biomass and have lower variability than described by the parent dataset. The results of this study, however, do not suggest that users relax AC flag criteria, because the flag assignments are in most cases reasonable indicators of degraded data quality, particularly in the portion of the spectrum relevant to blue-green band ratio algorithms (i.e., OC3M).

The removal of high biomass observations from the satellite record additionally screens out important biological processes, such as the formation of phytoplankton dense fronts and removes regions that may be disproportionately important to the ecosystem dynamics and species succession. For example, a northern MB retentive zone, which maintains dinoflagellate stocks that play an important role in species succession by seeding the surrounding waters [41], was frequently masked from the MB satellite record. We also note anecdotally that during red tide events within MB, ocean color retrievals on clear-sky days are often fully masked, with the satellite record resuming upon bloom termination.

4.3. Potential for User Evaluation of L2 and L3 Datasets

Defining the transition zone between regions where default NIR-based atmospheric correction methods can and cannot be used is challenging, as evidenced by the different transect results for the masked and unmasked composites for the two EBC ecosystems. The severity of the elevated L_W(NIR) effects may extend farther offshore than anticipated given the physics of the region (e.g., advection offshore by mesoscale eddies or jets). For regions where the removal of ephemeral high biomass events may be more infrequent, research targeting ecosystem processes may suffer from the loss of rare but high-impact events.

How can L2 and L3 end-users test whether satellite datasets contain a bias from the removal of high phytoplankton observations? L2 users with a priori knowledge of a region can compare flag assignments with expected phytoplankton dynamics to determine whether flags covary with target environmental parameters. For some flags, seasonality due to the Earth-sun geometry or cloud dynamics may resemble biological parameters without tracking phytoplankton biomass within an individual image. Interpretation of the flag assignment frequency should be considered cautiously, because infrequent flag assignments can have outsized effects in regions with high environmental variability. In the absence of a priori knowledge of a region, nFLH is anticipated to provide useful comparisons of the masked and unmasked data fractions, as shown herein. L3 end-users could make use of compositing statistics in order to assess the representativeness of spatial or temporal, quality-controlled averages. The similarity between sites in the relationship between composite SB and L2 flag assignments (Figure 7) suggests that L2 flag assignments are useful parameters for interpreting

L3 composites. As such, metadata that includes flag assignment statistics may be a beneficial addition for disseminated L3 products, particularly for users requiring inputs for coastal ocean models.

5. Conclusions

Wide recognition of decreased performance of ocean color products in coastal waters has encouraged development of a variety of alternative methods aimed to overcome difficulties such as high organic and inorganic particle concentrations. However, no direct comparison of the masked versus unmasked fraction of satellite datasets or suggestions for determining a dataset's sensitivity to the loss of high biomass retrievals, has been presented. This work compares non-aerosol-corrected biomass proxies to test whether high quality satellite retrievals are representative of initial datasets in coastal regions. Key findings shown are that, for productive ecosystems like MB and SHB, the changes to dataset distributions reflect a bias towards decreased biomass due to difficulty in removing the satellite signal's atmospheric component over phytoplankton-rich waters. The changes in the biomass of L2 datasets are apparent in L3 composites and the distance that the changes extend offshore is variable even among broadly similar systems (i.e., EBCs). Finally, users may assess the sensitivity of their study site using a similar approach by comparing FLH products between the masked and unmasked fractions of their dataset or by deriving compositing statistics when generating L3 products. In cases where the masked fraction is dissimilar to the unmasked fraction, users may prefer to use alternative atmospheric correction methods regardless of the strength of validation results obtained from the masked fraction only. Research directed towards coastal ocean ecosystems should evaluate whether quality-controlled satellite estimates of phytoplankton concentrations are representative compared to the statistics of the parent dataset. When possible, TOA proxies are useful tools for such comparisons.

Author Contributions: Conceptualization, H.F.H. and R.M.K.; methodology, H.F.H. and R.M.K.; formal analysis, H.F.H.; writing—original draft preparation, H.F.H.; writing—review and editing, H.F.H. and R.M.K. Both authors made substantial contributions to this paper.

Funding: Support for H.F.H. was partially provided by NASA C-HARRIER Campaign grant number NNX17AK89G, and the NOAA Central and Northern California Ocean Observing System (CeNCOOS) grant number NA16NOS0120021.

Acknowledgments: We are grateful to Christopher Edwards (University of California, Santa Cruz), Stan Hooker (NASA Goddard Space Flight Center), John Ryan (Monterey Bay Aquarium Research Institute; MBARI) and David Draper (University of California, Santa Cruz) for advising on this work. The Pangaea dataset used is available through (doi.pangaea.de), the Santa Cruz Wharf Chla data through the Southern California Coastal Ocean Observing System portal (sccoos.org/data/habs/), M1 mooring fluorometry data through the MBARI repository (dods.mbari.org), the AERONET climatologies through (aeronet.gsfc.nasa.gov), the MODIS Aqua data and SeaDAS processing software through (oceancolor.gsfc.nasa.gov), the SNAP software through (step.esa.int) and the C2RCC plugin from Brockmann consulting (brockmann-consult.de). The final version of the manuscript was improved by feedback from anonymous reviewers.

Conflicts of Interest: The authors declare no conflict of interest.

References

1. Gregg, W.W.; Conkright, M.E. Decadal changes in global ocean chlorophyll. *Geophys. Res. Lett.* **2002**, *29*, 1–4. [CrossRef]
2. McClain, C.R. A Decade of Satellite Ocean Color Observations. *Annu. Rev. Mar. Sci.* **2009**, *1*, 19–42. [CrossRef] [PubMed]
3. Dierssen, H.M. Perspectives on empirical approaches for ocean color remote sensing of chlorophyll in a changing climate. *Proc. Natl. Acad. Sci. USA* **2010**, *107*, 17073–17078. [CrossRef] [PubMed]
4. Siegel, D.A.; Wang, M.; Maritorena, S.; Robinson, W. Atmospheric correction of satellite imagery: The black pixel assumption. *Appl. Opt.* **2000**, *39*, 3582–3591. [CrossRef] [PubMed]

5. Sathyendranath, S.; Bukata, R.P.; Arnone, R.; Dowell, M.D.; Davis, C.O.; Babin, M.; Berthon, J.F.; Kopelevich, J.; Cmpbell, J.W. Color of Case 2 Waters. In *Remote Sensing of Ocean Colour in Coastal and Other Optically-Complex, Waters*; Sathyendranath, S., Ed.; Reports of the International Ocean-Color Coordinating Group: Dartmouth, NS, Canada, 2010; Volume 3, pp. 23–46.

6. Kahru, M.; Mitchell, B.G. Ocean color reveals increased blooms in various parts of the world. *EOS Trans. Am. Geophys. Union* **2008**, *89*, 170–172. [CrossRef]

7. Jessup, D.A.; Miller, M.A.; Ryan, J.P.; Nevins, H.M.; Kerkering, H.A.; Mekebri, A.; Crane, D.B.; Johnson, T.A.; Kudela, R.M. Mass stranding of marine birds caused by a surfactant-producing red tide. *PLoS ONE* **2009**, *4*, e4550. [CrossRef]

8. Lewitus, A.J.; Horner, R.A.; Carn, D.A.; Garcia-Mendoza, E.; Hickey, B.M.; Hunter, M.; Huppert, D.D.; Kudela, R.M.; Langlois, G.W.; Largier, J.L.; et al. Harmful algal blooms along the North American west coast region: History, trends, causes and impacts. *Harmful Algae* **2012**, *19*, 133–159. [CrossRef]

9. McCabe, R.M.; Hickey, B.M.; Kudela, R.M.; Lefebvre, K.A.; Adams, N.G.; Bill, B.D.; Gulland, F.M.D.; Thomson, R.E.; Cochlan, W.P.; Trainer, V.L. An unprecedented coastwide toxic algal bloom linked to anomalous ocean conditions. *Geophys. Res. Lett.* **2016**, *43*, 10366–10376. [CrossRef]

10. Pitcher, G.C.; Figueiras, F.G.; Kudela, R.M.; Moita, T.; Reguera, B.; Ruiz-Villareal, M. Key questions and recent research advances on harmful algal blooms in eastern boundary upwelling systems. In *Global Ecology and Oceanography of Harmful Algal Blooms*, 1st ed.; Gilbert, P.M., Berdalet, E., Burford, M.A., Pitcher, G.C., Zhou, M., Eds.; Springer: Cham, Switzerland, 2018; Volume 232, pp. 205–227.

11. Bracher, A.; Bouman, H.A.; Brewin, R.J.; Bricaud, A.; Brotas, V.; Ciotti, A.M.; Clementson, L.; Devred, E.; Di Cicco, A.; Dutkiewicz, S.; et al. Obtaining phytoplankton diversity from ocean color: A scientific roadmap for future development. *Front. Mar. Sci.* **2017**, *4*, 1–15. [CrossRef]

12. Frouin, R.J.; Franz, B.A.; Ibrahim, A.; Knobelspiesse, K.; Ahmad, Z.; Cairns, B.; Chowdhary, J.; Dierssen, H.M.; Tan, J.; Dubovik, O.; et al. Atmospheric correction of satellite ocean-color imagery during the PACE era. *Front. Earth Sci.* **2019**, *7*, 1–43. [CrossRef]

13. Barnes, R.A.; Clark, D.K.; Esaias, W.E.; Fargion, G.S.; Feldman, C.R.; McClain, C.R. Development of a consistent multi-sensor global ocean colour time series. *Int. J. Remote Sens.* **2003**, *24*, 4047–4064. [CrossRef]

14. Gregg, W.W.; Casey, N.W. Improving the consistency of ocean color data: A step toward climate data records. *Geophys. Res. Lett.* **2010**, *37*, 1–5. [CrossRef]

15. Kahru, M.; Kudela, R.M.; Manzano-Sarabia, M.; Mitchell, B.G. Trends in the surface chlorophyll of the California Current: Merging data from multiple ocean color satellites. *Deep Sea Res. Part II Top. Stud. Oceanogr.* **2012**, *77*, 89–98. [CrossRef]

16. Gordon, H.R.; Wang, M. Retrieval of water-leaving radiance and aerosol optical thickness over the oceans with SeaWiFS: A preliminary algorithm. *Appl. Opt.* **1994**, *33*, 443–452. [CrossRef] [PubMed]

17. Mobley, C.D.; Werdell, J.; Franz, B.; Ahmad, Z.; Bailey, S. Atmospheric correction for satellite ocean color radiometry. *NASA Tech. Rep. Serv.* **2016**, *1*, 1–85.

18. Wang, M.; Antoine, D.; Fruoin, R.; Gordon, H.R.; Fukushima, H.; Morel, A.; Nicolas, J.; Deschamps, P. Comparison Results. In *Atmospheric Correction for Remotely-Sensed Ocean-Colour Products*; Wang, A., Ed.; Reports of the International Ocean-Color Coordinating Group: Dartmouth, NS, Canada, 2010; Volume 10, pp. 23–38.

19. Wang, M. Remote sensing of the ocean contributions from ultraviolet to near-infrared using the shortwave infrared bands: Simulations. *Appl. Opt.* **2007**, *46*, 1535–1547. [CrossRef] [PubMed]

20. Ruddick, K.G.; Ovidio, F.; Rijkeboer, M. Atmospheric correction of SeaWiFS imagery for turbid coastal and inland waters. *Appl. Opt.* **2000**, *39*, 897–912. [CrossRef] [PubMed]

21. Brockmann, C.; Doerffer, R.; Peters, M.; Stelzer, K.; Embacher, S.; Ruescas, A. Evolution of the C2RCC neural network for Sentinel 2 and 3 for the retrieval of ocean colour products in normal and extreme optically complex waters. In Proceedings of the Living Planet Symposium, Prague, Czech Republic, 9–13 May 2016; Volume 740, p. 54.

22. Mograne, M.A.; Jamet, C.; Loisel, H.; Vantrepotte, V.; Mériaux, X.; Cauvin, A. Evaluation of five atmospheric correction algorithms over French optically-complex waters for the Sentinel-3A OLCI ocean color sensor. *Remote Sens.* **2019**, *11*, 668. [CrossRef]

23. Campbell, J.W.; Blaisdell, J.M.; Darzi, M. Level-3 Sea WiFS data products: Spatial and temporal binning algorithms. *Oceano. Lit. Rev.* **1996**, *9*, 952.

24. Scott, J.P.; Werdell, P.J. Comparing level-2 and level-3 satellite ocean color retrieval validation methodologies. Accepted Optics Express. 2019. [CrossRef]

25. Ryan, J.P.; Fischer, A.M.; Kudela, R.M.; Gower, J.F.R.; King, S.A.; Marin III, R.; Chavez, F.P. Influences of upwelling and downwelling winds on red tide bloom dynamics in Monterey Bay, California. *Cont. Shelf Res.* **2009**, *29*, 785–795. [CrossRef]

26. Pennington, J.T.; Chavez, F.P. Seasonal fluctuations of temperature, salinity, nitrate, chlorophyll and primary production at station H3/M1 over 1989-1996 in Monterey Bay, California. *Deep Sea Res. Part II* **2016**, *47*, 947–973. [CrossRef]

27. Pitcher, G.C.; Brown, P.C.; Mitchell-Innes, B.A. Spatio-temporal variability of phytoplankton in the southern Benguela upwelling system. *S. Afr. J. Mar. Sci.* **1992**, *12*, 439–456. [CrossRef]

28. Barlow, R.; Sessions, H.; Balarin, M.; Weeks, S.; Whittle, C.; Hutchings, L. Seasonal variation in phytoplankton in the southern Benguela: Pigment indices and ocean colour. *Afr. J. Mar. Sci.* **2005**, *27*, 275–287. [CrossRef]

29. Fawcett, A.; Pitcher, G.C.; Bernard, S.; Cembella, A.D.; Kudela, R.M. Contrasting wind patterns and toxigenic phytoplankton in the southern Benguela upwelling system. *Mar. Ecol. Prog. Ser.* **2007**, *348*, 19–31. [CrossRef]

30. Schumann, E.H.; Martin, J.A. Climatological aspects of the coastal wind field at Cape Town, Port Elizabeth and Durban. *S. Afr. Geogr. J.* **1991**, *73*, 48–51. [CrossRef]

31. Patt, F.S.; Barnes, R.A.; Eplee, R.E., Jr.; Franz, B.A.; Robinson, W.D.; Feldman, G.C.; Bailey, S.W.; Gales, J.; Werdell, P.J.; Wang, M.; et al. Algorithm updates for the fourth SeaWiFS Data Reprocessing. In *SeaWiFS Postlaunch Technical Report Series*; Hooker, S.B., Firestone, E.R., Eds.; NASA Technical Reports; NASA: Greenbelt, MD, USA, 2002; Volume 22, pp. 34–40.

32. Neville, R.A.; Gower, J.F.R. Passive remote sensing of phytoplankton via chlorophyll a fluorescence. *J. Geophys. Res.* **1977**, *82*, 3487–3493. [CrossRef]

33. Gordon, H.R. Diffuse reflectance of the ocean: The theory of its augmentation by chlorophyll *a* fluorescence at 685 nm. *Appl. Opt.* **1979**, *18*, 1161–1166. [CrossRef] [PubMed]

34. Letelier, R.M.; Abbott, M.R. An analysis of chlorophyll fluorescence algorithms for the Moderate Resolution Imaging Spectrometer (MODIS). *Remote Sens. Environ.* **1996**, *58*, 215–223. [CrossRef]

35. Gower, J.F.R. On the use of satellite-measured chlorophyll fluorescence for monitoring coastal waters. *Int. J. Remote Sens.* **2016**, *37*, 2077–2086. [CrossRef]

36. Amin, R.; Zhou, J.; Gilerson, A.; Gross, B.; Moshary, F.; Ahmed, S. Novel optical techniques for detecting and classifying toxic dinoflagellate Karenia brevis blooms using satellite imagery. *Opt. Express* **2009**, *17*, 9126–9144. [CrossRef] [PubMed]

37. Roesler, C.S.; Perry, M.J. In situ phytoplankton absorption, fluorescence emission and particulate backscattering spectra determined from reflectance. *J. Geophys. Res.* **1995**, *100*, 13279–13294. [CrossRef]

38. Gilerson, A.; Zhou, J.; Hlaing, S.; Ioannou, I.; Schalles, J.; Gross, B.; Moshary, F.; Ahmed, S. Fluorescence component in the reflectance spectra from coastal waters. Dependence on water composition. *Opt. Express* **2007**, *15*, 15702–15721. [CrossRef] [PubMed]

39. Gower, J.F.R.; Borstad, G.A. On the potential of MODIS and MERIS for imaging chlorophyll fluorescence from space. *Int. J. Remote Sens.* **2004**, *25*, 1459–1464. [CrossRef]

40. Ryan, J.P.; Davis, C.O.; Tufillaro, N.B.; Kudela, R.M.; Gao, B.C. Application of the Hyperspectral Imager for the Coastal Ocean to phytoplankton ecology studies in Monterey Bay, CA, USA. *Remote Sens.* **2014**, *6*, 1007–1025. [CrossRef]

41. Ryan, J.P.; Gower, J.F.R.; King, S.A.; Bissett, W.P.; Fischer, A.M.; Kudela, R.M.; Kolber, Z.; Mazzillo, F.; Rienecker, E.V.; Chavez, F.P. A coastal ocean extreme bloom incubator. *Geophys. Res. Lett.* **2008**, *35*, 1–5. [CrossRef]

Article

Radon-Augmented Sentinel-2 Satellite Imagery to Derive Wave-Patterns and Regional Bathymetry

Erwin W. J. Bergsma [1,*], Rafael Almar [2] and Philippe Maisongrande [3]

[1] CNES-LEGOS, UMR-5566, 14 Avenue Edouard Belin, 31400 Toulouse, France
[2] IRD-LEGOS, UMR-5566, 14 Avenue Edouard Belin, 31400 Toulouse, France
[3] CNES, 18 Avenue Edouard Belin, 31400 Toulouse, France
*   Correspondence: erwin.bergsma@legos.obs-mip.fr

Received: 5 July 2019; Accepted: 13 August 2019; Published: 16 August 2019

Abstract: Climatological changes occur globally but have local impacts. Increased storminess, sea level rise and more powerful waves are expected to batter the coastal zone more often and more intense. To understand climate change impacts, regional bathymetry information is paramount. A major issue is that the bathymetries are often non-existent or if they do exist, outdated. This sparsity can be overcome by space-borne satellite techniques to derive bathymetry. Sentinel-2 optical imagery is collected continuously and has a revisit-time around a few days depending on the orbital-position around the world. In this work, Sentinel-2 imagery derived wave patterns are extracted using a localized radon transform. A discrete fast-Fourier (DFT) procedure per direction in Radon space (sinogram) is then applied to derive wave spectra. Sentinel-2 time-lag between detector bands is employed to compute the spectral wave-phase shift and depth using the gravity wave linear dispersion. With this novel technique, regional bathymetries are derived at the test-site of Capbreton, France with an root mean squared (RMS)-error of 2.58 m and a correlation coefficient of 0.82 when compared to the survey for depths until 30 m. With the proposed method, the 10 m Sentinel-2 resolution is sufficient to adequately estimate bathymetries for a wave period of 6.5 s or greater. For shorter periods, the pixel resolution does not allow to detect a stable celerity. In addition to the wave-signature enhancement, the capability of the Radon Transform to augment Sentinel-2 20 m resolution imagery to 10 m is demonstrated, increasing the number of suitable bands for the depth inversion.

Keywords: Sentinel-2; radon transform; remote sensing; bathymetry inversion; multi-scale monitoring; image augmentation

---

1. Introduction

Climatological extremes are occurring more frequently and with greater intensity, in particular, coastal flooding and higher intensity storms [1,2]. These environmental changes and associated risks are often described in terms of sea level at the coast. Sea level at the coast can be broken down into several contributors such as regional sea level and wave contribution. The latter is dictated by the underlying bathymetry. To effectively mitigate environmental changes, in other words, to manage coastal environments, one requires to know bathymetry and its evolution. Coastal bathymetries are far from static and change continuously depending on incident waves, currents and sea level. As a matter of concept, powerful conditions such as storms can be considered large sediment transport drivers in the near-shore. These events often lead to abrupt and large erosion (e.g., net offshore sediment transport) while recovery during calm conditions (after initial immediate storm-recovery) is a slower process (net onshore sediment transport). Considering the dynamic coastal bathymetry to better predict its action on waves is thus paramount.

Major current issues to enhance this knowledge are spatial data-concentration and temporal data-sparsity. Field campaigns often sample locally, with an insufficient temporal resolution to observe climatic modes, storm impact and recovery/resilience. For example, on the one hand, specific sites are well sampled with real-time kinematic GPS (RTK-GPS) survey-campaigns (relatively often in time (every month) and dense in space) but on a local domain ($O(\text{km}^2)$). This also holds for remote-sensing techniques such as shore-based video cameras that have a high temporal frequency but a local domain. On the other hand, traditional large echo-sounding campaigns for bathymetries cover km^2's but are often sporadic. An alternative is the use of space-borne observations [3] such as optical image products and radar that now sample the entire globe regularly (at best every few days—ESA's Sentinel constellation [4]).

In overview, two approaches to obtain nearshore depths are most widely applied: (1) depth through color absorption depending on the height of the water column (rule-of-thumb: the darker, the deeper) [5] and (2) depth through the physical relation between wave celerity and depth (rule-of-thumb: the faster, the deeper) [6–8]. The first method is uniquely applicable to optical imagery and it is capable of estimating reasonable depth until 10 s of meters but the method is sensitive to turbid or aerated waters. The wave-based method is applicable to both radar and optical imagery. Its limitation is mainly the observability of waves: as long as waves are visible, depth can theoretically be found from shore to intermediate water depth (0 to depths: $h_{int} = \frac{\lambda}{2}$, further details in the methodology section). The relation between wave celerity has been applied to numerous observational techniques from shore-based systems [6,7,9–12], airborne [13] to space-borne [14,15]. To find depth related to a certain wave celerity through the linear dispersion relation, one needs to obtain any pair from celerity c, wavenumber k, wave frequency ω, wavelength λ ($1/k$) or wave period T ($1/\omega$), either in the spectral domain (ω, k) or temporal/spatial (T, λ) domain. Compared to space-borne imagery (often one snapshot), shore-based or airborne camera systems have significantly more temporal resolution. The lack of temporal information limits depth estimations from space-borne imagery to the determination of spatial wave characteristics (k or L). Most commonly applied 2D-discrete fast-Fourier transform (DFT) or wavelet analyses require sub-domains with the size of a few wavelengths to overcome wave-stochasticity issues [15–18], which on its own leads to significant spatial smoothing. To a certain degree, these mathematical applications depend on image resolution and visibility of the wave pattern. In this work, we extract the wave pattern using a Radon transform and obtain physical wave characteristics using a 1D-DFT for the most energetic incident wave direction in Radon space (sinogram).

The article has the following structure: the next section describes the proposed methodology mathematically and which applicability is thereafter demonstrated using a synthetic deep-water case in the subsequent section. Wave pattern extraction and depth estimation from or to a Sentinel-2 image are shown in Section 4. The discussion section focusses on the sensitivity to image resolution for wave-number, celerity and depth-sensing using the proposed method. The final part of the discussion elaborates on Sentinel-2 image augmentation for wave patterns allowing for the use of additional low-resolution bands to estimate wave characteristics.

2. Methodology

Underlying bathymetry dictates wave-celerity in case the wave is propagating through intermediate to shallow water. Intermediate and shallow water limits are wavelength dependent and typically $h_{int} = \frac{L}{2}$ for intermediate water depth and $h_{sh} = \frac{L}{20}$ for shallow. Hence, from intermediate water until shore the linear dispersion relation (1) can be used to estimate a local depth.

$$c^2 = \frac{g}{k} \tanh(kh) \Leftrightarrow h = \frac{\tanh^{-1}\left(\frac{c^2 k}{g}\right)}{k} \tag{1}$$

in which c is wave celerity, g represents the gravitational acceleration, k is the wavenumber. It requires knowledge about two of the celerity c, wavenumber k, wave frequency ω, wavelength λ ($1/k$) or wave period T $(1/\omega)$ set to solve (1). Between the temporal and spatial domain, it is common practice to use the highest resolution of both to find the other. For example, in the case of video products or X-band radar, in which the temporal resolution is often the highest, one isolates wave frequencies ω to find the wavenumber (k) [15,16]. Sentinel-2 imagery does not allow for such an approach due to the lack of temporal information.

Although the Sentinel-2 products seem just a snapshot, there is underlying temporal information. The sensors collect imagery one wavelength specific-band at the time and hence, a time-lag between the different image bands exists. Such time-lag is common in optical spatial imagery and can also be found in, for example, the SPOT (max 2.04 s) and Pleiades (0.165 s between the detector bands: max 0.66 s) constellations. If we consider the bands with 10-meter resolution (red, green, blue and near-infrared), a maximum time lag of 1.005 s can be found between the blue and red band, illustrated in Figure 1. This time-lag information of Sentinel-2 has been used to determine the direction of ocean waves (a 2D-DFT has duplicate quadrants) in [19] and earlier with the SPOT constellation to estimate wave propagation through cross-correlation [16].

Figure 1. Illustration of the time-lag between detector bands B02 (blue), B03 (green) and B04 (red) by the different colors of the flying airplane at Capbreton, France. The time-lag between each band is 0.527 s.

2.1. Radon Transform-Derived Wave Signal and Spectra

Waves can be observed from space with satellite imagery due to sunlight reflected from the sea surface (sun glitter). Satellite Sun Glitter Imagery is shown to contain valuable information to obtain wave statistics such as wave height, period and direction and even a reconstruction of the 3D surface [19]. Wave-visibility depends on several factors such as cloud coverage, satellite incident angles and wave conditions (height, period and direction). The quality/visibility of the wave pattern, and thus the ability to invert depth, differs per satellite image even for similar wave conditions. A common technique to enhance linear patterns in imagery is the Radon transform [20] (RT). Even if wave patterns are not obviously apparent and/or contain a great amount of noise, the RT distils linear features. This particular feature makes the RT a powerful tool to process imagery. RT are extensively used for tomography (for example in CT-scans) in order to reconstruct finite projections. In addition, [21,22] have shown the RT's power in separating incident from outgoing reflected waves in the nearshore coastal zone. Here, we use the RT to extract wave signal after which we apply a 1D-DFT to find the spectral phase of the wave (per band).

In principle, the RT accentuates linear features in an image by integrating image intensity ($I(x,y)$) along lines defined by angle θ and offset ρ following (2):

$$R_I(\theta,\rho) = \oiint_D I(x,y)\, \delta(\rho - x\cos(\theta) - y\sin(\theta))\, dy\, dx \tag{2}$$

wherein δ represents the Dirac delta function. R_I represents a sinogram: the signal per direction (θ) over the associated beam with length (ρ). The angular limits of (2) are commonly set to $0° > \theta > 180°$. Likewise to, for example, an inverse DFT, the original input signal can be reconstructed applying an inverse RT. RT filtering means that only a limited number of angles are used for the inverse reconstruction. To isolate wave patterns, only the most energetic wave-related θ-ρ-pair can be used for the inverse RT.

Wave-like patterns are observed in the RT-sinogram with wave patterns in the original image. This allows for the calculation of the wave amplitude and phase per direction θ through a DFT (here in a discrete form). This results in a wave-number spectrum.

2.2. Waves Phase-Shift, Celerity and Depth

Imagery collected by the Sentinel-2 constellation consists of multiple bands with their specific sensing wavelength and resolution. Bands are collected one at the time with a fractional time-lag in between them. The time-lag combined with the RT-based wave number spectrum and its phase shift between bands results in the wave celerity. For each point in space, a sinogram (2) is calculated over the sub-domain. The maximum variance for all angles θ over ρ is considered the propagative direction of incident waves. Over the beam ρ with maximum variance, a DFT ($\mathscr{G}_{R_I} = DFT(RT_{\theta_{mv}})$) is used to obtain the phase per band following:

$$\Phi\left(x, y, \theta, \rho\right) = \tan^{-1}\left(\frac{\Im\left(\mathscr{G}_{R_I}\right)}{\Re\left(\mathscr{G}_{R_I}\right)}\right) \tag{3}$$

wherein $\mathscr{G}_{R_I}$ represents the DFT over the sinogram $R_I(\theta, \rho)$ in polar space over the sub-domain around point x, y. $\Im$ and $\Re$ respectively denote the imaginary and real part of complex numbers. For each x, y location we can then calculate the spectral phase-shift ($\Delta\Phi$ in rad) between two (or more) detector bands at different times (t). Since the wave number (k) is kept fixed, a shift in spectral phase-shift represents $\omega(t)$ and the celerity can be calculated:

$$c = \frac{\Delta\Phi}{2\pi k \Delta t} = \frac{\Delta\Phi\lambda}{2\pi\Delta t}. \tag{4}$$

With the derived wave celerity c and wave number k (or wavelength λ) depth is found solving (1). The method as presented here selects a single, most energetic, peak in the Radon-DFT (a single wave number k) to compute the phase-shift and hence the water depth.

3. A Synthetic Deep-Water Case

The process from a satellite image to bathymetry can be split into two main components; (1) sensing wave parameters and (2) depth-inversion. The two parts have their own associated error [12] and limits. In this section, the method steps are illustrated and the sensing capabilities of the method are scrutinized with synthetic data. Since the wave sensing principally does not depend on the relative water depth (deep, intermediate or shallow waters), it makes sense to test the sensing-method in deep water and consider pure sinusoidal waveforms. In an attempt to introduce some reality to the synthetic dataset, input-parameters are set to represent Sentinel-2 settings such as 10 m resolution and 1.005 s time-lag between two snapshots. The chosen sinusoidal has 1 m amplitude and a 9-second period which roughly correspond to the annual mean at Capbreton, France (see Section 4), resulting in two-wave patterns as shown in Figure 2. Given the wave period and considering the deep water assumption, a wavelength (λ) of 126.36 m ($k_x = 0.008$ 1/m) and wave celerity (c) of 14.04 m/s can be found. This wave propagates purely in x-direction ($k_y = 0$). In addition, a second case is considered in which the wave is oblique incident $45°$ ($k_x = k_y$) and results in a wavenumber k of 0.006 1/m and a celerity of 19.85 m/s. Here we attempt to find the appropriate wavenumber k and celerity c.

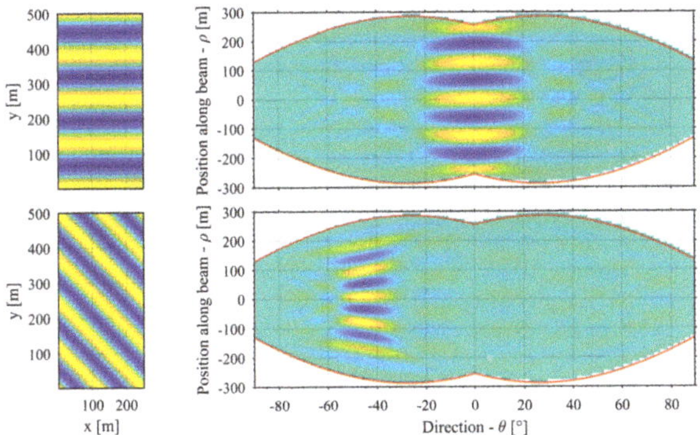

Figure 2. Radon transform sinograms for the synthetic datasets containing (1) a shore normal wave (0 degrees) and (2) an oblique example where $k_x = k_y$. The red lines indicate the sinograms limits along the beam length.

For this synthetic case, the sub-sample domain was set to $2L$ by $4L$ (252 m width $\times$ 505 m length). This is further elaborated in the discussion. Conceivably, the subsample domain size limits the appearance of the number of full wavelengths and hence the performance of the DFT (not the RT). The application of (2) to the subsample results in a sinogram, as presented in Figure 2. The upper and lower limits (red lines in Figure 2) of the beam length are computed from the center of the subsample domain. These limits were determined exclusively by the object the RT is analyzing, in this case, a rectangular sub-domain. Note that at zero degrees the limits were -252.5 m to 252.5 m (total length of 505 m) and at -90 or 90 degrees the beam stretches between -126 m and 126 m, representing the total width of 252 m. In between -90, 0 and 90 degrees the sinogram limits are determined by the furthest perpendicular line of sight along the beam that senses the object (here the sub-domain). Hence, the slight increase in total beam length from zero degrees (both ways) before decreasing to the smallest beam length on the shortest side of the subdomain.

The sinograms in Figure 2 have a radial increment ($\Delta\theta$) of half a degree, resulting in 361 beams. A DFT is applied to every beam individually. For each rotational angle θ, the resolution $\Delta\rho$ along the RT-beam (ρ—crossing data-points) changes and follows in case of $\Delta x = \Delta y$:

$$\Delta\rho(\theta) = 2\left(\sqrt{2\Delta x^2} - \Delta x\right)|\sin(\theta)\cos(\theta)| + \Delta x \tag{5}$$

Whilst the sinogram' resolution changes per beam, the number of sample points remains constant. Resulting RT-spectra as shown in Figure 3 have a typical cone shape: higher resolution at $\theta = -90$, 0 or 90 degrees, results in the ability to resolve higher frequencies to a greater extent and vice versa for the diagonal. Figure 3 shows the wave-number spectra derived from the sinograms in Figure 2.

A first inspection of the energy peak in Figure 3 reveals that the incident directions correspond to the synthetic input: 0 degrees for the left plot and -45 degrees in the right plot. The energy peaks also correspond to the appropriate wave number, respectively 0.008 [1/m] for 0 degrees case and 0.006 [1/m] for the oblique case. In both spectra, but in particular for the left plot, energy spreading is visible (in the shape of $\wedge$). Along the wave track, the Radon integration solves the wave signal. As the beam rotates and the relative angle between the propagative wave and the beam direction is small, a wave signal with a longer wavelength is found (smaller wavenumber k). As this relative angle increases the energy fades out, hence the $\wedge$ shaped energy distribution. In the case of oblique waves, the integrated energy over the wave crest perpendicular to the RT-beam (with angle θ) is higher closer to the center of the sub-domain and lower towards the edges, compared to a constant crest length for 0 degrees. This effect is visible in the sinogram in Figure 2. In the lower plot ($\theta = -45$ degrees) most of the energy is concentrated (around $\rho = 0$ m, between -100 m $> \rho > 100$ m) while in the top plot ($\theta = 0$ degrees) the energy is more spread, between -250 m $> \rho > 250$ m.

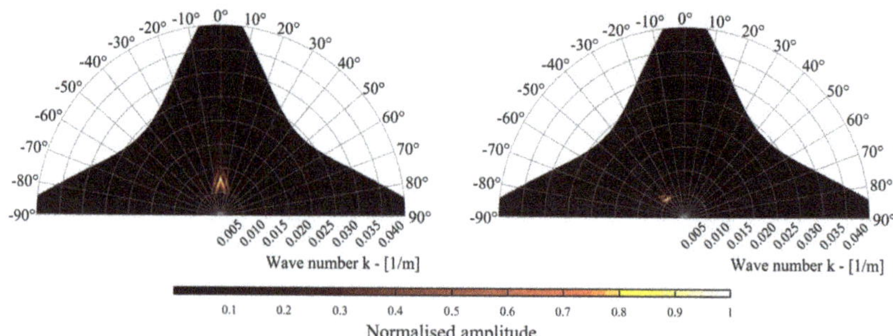

Figure 3. Normalized Radon-discrete fast-Fourier transform (DFT) wave number spectra for a wave in shore normal direction at the top and oblique (−45 deg) incident wave in the bottom. The wave number range is between 0 and 0.04 [1/m] with 0.001 [1/m] resolution, the circular bands indicate 0.005 [1/m] intervals. The direction ranges from −90 to 90 with a $\Delta\theta$ of 0.5 degree, the spokes represent a 10-degree interval.

The RT-DFT allows for the calculation of a spectral-phase per position on the beam, per direction. Let us impose a time lag of 1.005 s between two snapshots to the sinusoidal above and compare the phase-shift at the energy peak. For the zero degrees incident wave with a period of 9 s, the theoretical phase shift, $(2\pi\Delta T/T_p)$ is 0.701 rad. The RT-DFT for the peak gives 0.701 rad shift in phase, a 0.016% offset. Given the phase shift and time-lag (ΔT), a celerity of 14.04 m/s is obtained. For the oblique waves, the estimated phase shift is 0.701 rad compared to 0.701 rad for the theoretical phase shift (0.039% offset). From the synthetic case, we can say that it is possible to estimate wave celerity accurately, in an ideal-case setting.

4. Regional Wave-Pattern and Bathymetry

To scrutinize the method's performance in a realistic case, it is applied to a real-world configuration. The area surrounding Capbreton in France (Figure 4) was selected considering the availability of an in-situ dataset. A field campaign was conducted from 5–18 November 2017 which was initially designed to accommodate in-situ validation of spaceborne-derived bathymetries and Digital Elevation Models (topography) using CNES' Pleiades constellation [23,24]. The coastal zone around Capbreton is particularly suited to method-validation considering that within several kilometers one finds a variety of coastal features such as a deep-water canyon, a port entrance, hard (walls) and soft (dunes) coastal-defense structures.

During the field campaign, hydrodynamics, topography and bathymetry were measured in various ways. The hydrodynamics are captured by an acoustic doppler current profiler (ADCP) likewise waves and currents are modeled, all executed by BRGM (Pessac). The topography was measured using real-time-kinematic GPS, structure for motion (SfM) from a drone and airborne LiDAR (BRGM). The bathymetry was measured with an echo-sounder mounted on a boat and the shore-based video camera systems that deploy cBathy and the temporal method [12]. For this work, the echo-sounding measured bathymetry was considered the ground-truth. Wave conditions were of special interest as incident waves might influence the space-borne bathymetry inversion quality [12]. Here, the two closest nearshore wave buoys in the Bay of Biscay were used to obtain wave data (Wave buoy number 62066 – Anglet and number 62064 – Arcachon). During the field campaign, the dominant wave conditions were quite energetic. Wave height conditions ranged between 0.7 and 4.4 m with a mean significant wave height of 2.22 m. The maximum wave height had an associated period of 15.4 s. The maximum wave period went up to 18.4 s with an associated wave height of 1.2 m. The wave direction was relatively constant considering a mean direction 307.2 degrees ± 7.56 degrees.

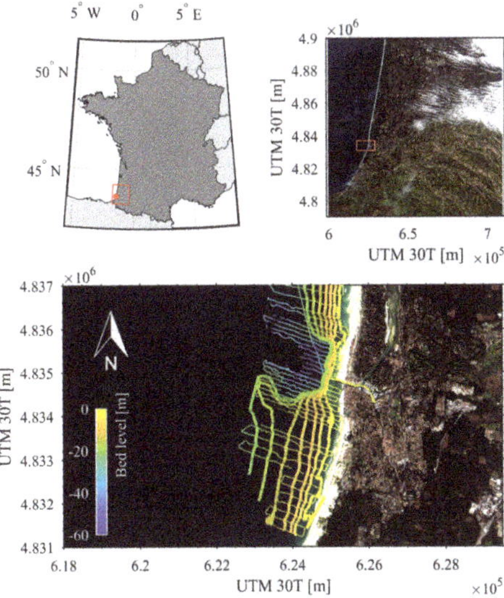

Figure 4. Location of Capbreton in South–West France. The top-left overview shows part of Western Europe in which France is highlighted as the darker grey area (coordinate system WGS84). The red dot represents Capbreton and the box around the dot represents Sentinel-2 tile 30TXP projected on the UTM-zone 30T (Universal Transverse Mercator) . The tiled Sentinel-2 image is presented at the top-right (image taken on 20 November 2017). The red box in the top-right figure highlights the zoomed-in area of the bottom plot. The small dots (mainly on cross shore arrays) represent the echo-sounder measured depths. The thicker predominantly alongshore lines represent the depth isocontours. The deep-water canyon (dark blue dots) is evident close to shore, West–Northwest of the harbour entrance. The echo-sounder was realistically capable of measuring from the shallowest as the boat would go until 60–70 m water depth.

In this paper we test RT-based wave pattern extraction on 10×10 m resolution Sentinel-2 imagery [25] covering the region surrounding Capbreton—Sentinel-2 relative orbit 94, tile 30TXP—on two dates: (1) 2 days after the end field campaign on 20 November 2017 and (2) preferable wave conditions (30 March 2018). The bathymetry is only derived for the latter to demonstrate the methods' performance.

4.1. Parameter Settings for the rt Method

Few parameter settings were required to apply the RT for wave pattern extraction and wave-number spectra, and these were only limited to the spatial domain. An RT-filter was applied every 50 m in x and y direction, (Δx and $\Delta y = 50$ m), in other words, the depth estimation results have a horizontal resolution of 50 m. The windowing sub-domain for this real-world case is currently set to 30×20 pixels which practically relates to 300×200 m and an amplification factor (κ) is applied as a function of the distance (D in km) from the coastline $\kappa = 1 + 0.3D$. The size of the sub-sample domain is in a similar order compared to [15,17,18], and relates mostly to the stochasticity of the sub-sampled wave pattern. In other words, to apply typical methods such a wavelet or DFT analysis, more than a single wavelength should be sub-sampled, the same holds here.

For the RT-based phase-shift derived celerity, limits are imposed. The estimated celerity should be greater than zero and cannot exceed the deep water limit related to a user-defined maximum wave period. Here the cut-off wave period is 18 s, which relates to 28.1 m/s. This should be considered quite a generous upper limit, as these waves are not expected to travel with these celerities in the coastal zone. In addition to the minimum wave celerity threshold, sensed depth can only be positive (before it is referenced to a vertical datum).

4.2. Wave Extraction: Illustration

To show the power of RT-filtering we apply it on Sentinel-2 imagery collected at Capbreton on 20 November 2017. A wave pattern is not apparent in the raw level-2 Sentinel-2 imagery. The two closest wave buoys in the Bay of Biscay (buoy-ID 62066 and 62064—Global GlobWave database) measured an average significant wave height of 0.58 to 0.66 m, an average peak period of 11.72 to 11.76 s and a peak-associated wave direction of 274 to 326 degrees. These relative calm wave conditions result in an imperceptible wave pattern in the satellite imagery. Often these images are discarded for further analysis but let's consider this image as a nice example-case for RT filtering. The RT filtering as in phase I of the methodology is applied per sub-domain and normalized over the full domain. Figure 5 shows the filtered result over approximately 30 km^2.

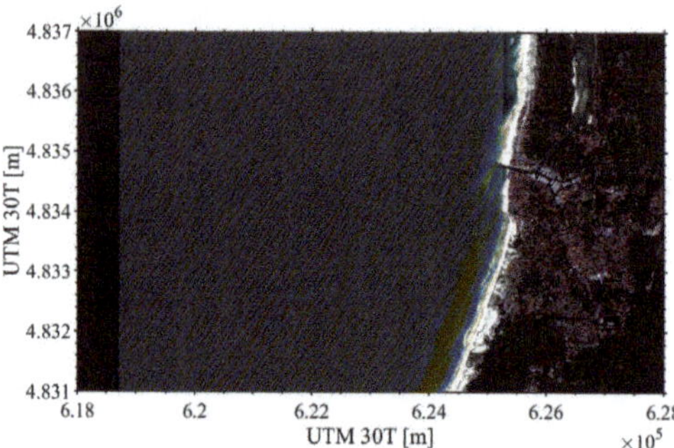

Figure 5. Radon-filtered wave pattern (normalized) from Sentinel-2 imagery collected on 20 November 2017.

The filtered image clearly shows the incident wave pattern in the appropriate direction. In addition, secondary wave-like patterns are apparent such as the satellite sensor direction (lines with 290 degrees angle). This seems to be linked to the signal to noise ratio, as the wave signal is imperceptible and the sensing leaves a trace of a similar order of magnitude the RT also amplifies this artifact. At the coast, particularly South of the harbor entrance, larger, non-physical, intensities are observed (wide yellow band). This is due to the fact that the RT sees the coast as the most energetic linear feature in that sub-domain and the results are maxed-out.

In case incident waves are steeper, greater wave height and similar period, wave patterns are easier to observe from the satellite imagery. Figure 6 shows RT results for the wave pattern extraction on 30 March 2018 under wave conditions of 3.26 to 3.3 m average significant wave height, an average peak wave period between 12.69 and 12.79 s and an incident wave-angle between 280 to 309 degrees. Also, the incident swell was cleaner and had less directional spread (10–20 degrees) on 30 March 2018 in comparison to 20 November 2017 (13–30 degrees). The wave pattern was significantly more distinctive in comparison to Figure 5. For example, refraction patterns around the harbor entrance were visible.

For the wave pattern in Figure 6, unfiltered (all directions), a RT-based spectrum can be derived for every location (x_p, y_p). Figure 7 shows an example of such spectrum for point $x = 621{,}000$ m and $y = 4{,}834{,}000$ m, roughly in the middle of the domain as indicated by the red circle in Figure 6. The spectral coloring in Figure 7 relates to the normalized amplitude. This RT wavenumber spectrum confirms shows similar wave direction and spreading as the offshore wave buoy. Figure 7 also indicates the RT-related energy spreading, likewise to the synthetic spectra.

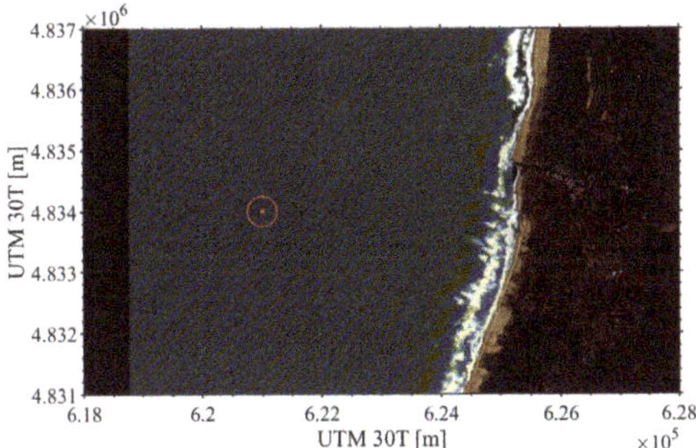

Figure 6. Radon-filtered wave pattern (normalized) from Sentinel-2 imagery collected on 30 March 2018. The red circle indicates the position related to the Radon-derived wave spectrum.

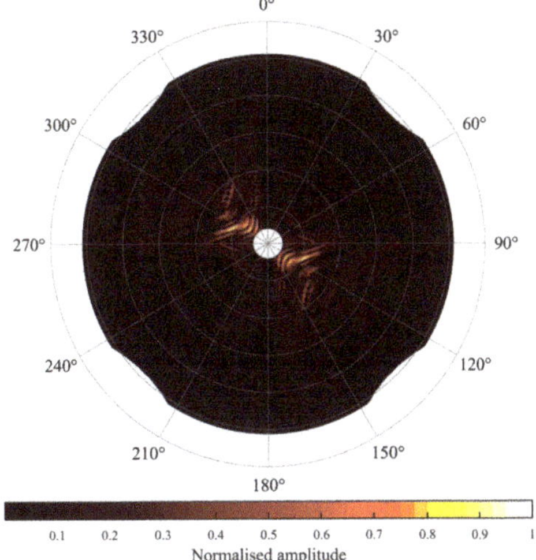

Figure 7. Full Radon spectrum derived from Sentinel-2 imagery that was sensed on 30 March 2018. The circular bands follow the same increment (0.005 [1/m]) as Figure 3 but the range is limited to 0.03 [1/m]. The off-shore wave-period range in this spectrum was 5 s (0.0256 [1/m]) to 18 s (0.002 [1/m]).

4.3. Wave Number and Depth Estimation

Several distinct energy concentrations are apparent in the wave-number spectrum around the 300 degrees spoke for wave numbers around 0.005 m^{-1}. The derived spectrum confirms what the wave buoy data told us in the sections above. Which energy peak to use for the depth inversion for each spectrum is not trivial. Here, the current set-up takes the direction from the variance peak in the sinogram. For the wave number the 99% interval is calculated, so the 1% most energetic points, from the normalized amplitude spectrum. For the most energetic 1% wavenumber k is computed and used to calculate the phase-shift. Figure 8 shows the measured depth and depth derived from Sentinel-2 imagery acquired on 30 March 2018. Bear in mind, these are tide-corrected depths (tidal elevation = −0.81 m). Here, depths are estimated over the same domain as the field-campaign covered.

The measured and estimated depth shows comparable morphological features. The deep-water canyon (300 m depth close to shore) in front of the harbor is recognizable. Features like these are often lacking in state-of-the-art applications because they are relatively hard to resolve. Between the survey and estimation, a correlation coefficient of 0.82 and RMS difference of 2.58 m is found for measured depths up to 30 m. The correlation coefficient shows that the estimated morphological features have a physical meaning and represent reality. Figure 9 presents sectioned profiles: North, central (with the deep-water canyon) and the South section. In the North, one can see that the depth is well estimated until the sub-tiles hit the white water due to wave breaking. In the canyon section, we do estimate the canyon outline but we underestimate the canyon' slope nearshore. This offset in the vicinity of the canyon can be explained by complex wave attenuation on the canyon banks as the waves make a quick transition between deep-water to shallow water: multiple spectral peaks disperse the computed phase shift. This could be solved by introducing depth estimation over multiple wave numbers (not included in the current version). In the Southern section, the waves were less affected by the deepwater canyon and the wave breaking zone is less wide, the depth estimation technique does quite well. These results, the presence of the canyon and adequate estimation of the shallowest depths, show the potential of this novel estimation of the wave-phase shift, and celerity in polar (RT) space.

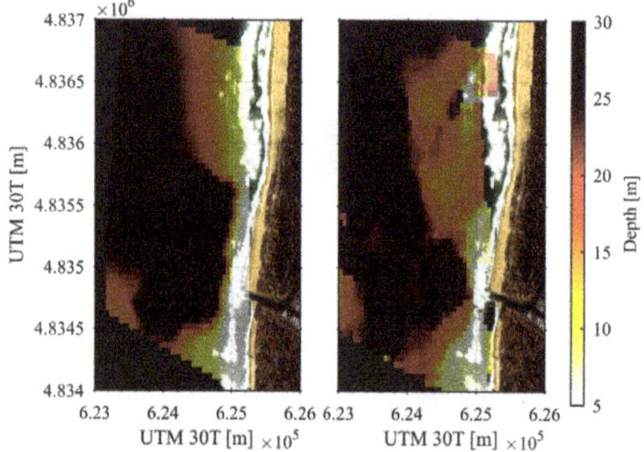

Figure 8. Measured (**left**) and Sentinel-2 estimated (**right**) water depths in the vicinity of the Capbreton harbour. The measured bathymetry is interpolated on the depth estimation locations. This highly complex bathymetry, in particular the contours of the deep-water channel and nearshore zone South of the harbour entrance show similarities.

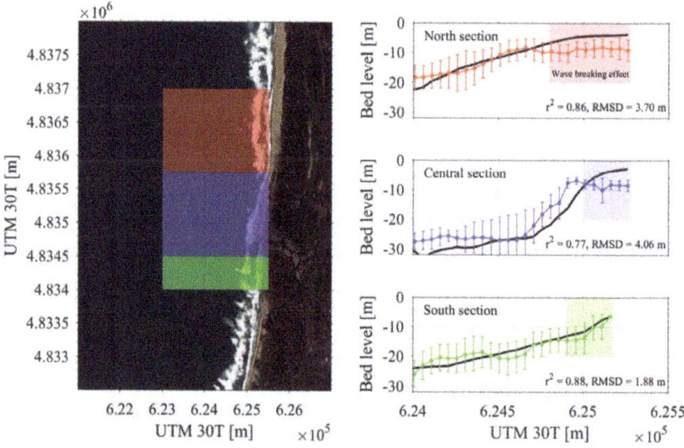

Figure 9. Sectioned profiles considering North of the canyon (red), a central part including the deep water canyon (blue) and South of the canyon (green). The black line illustrates the measured profile while the colors are the estimated profiles using the RT-based depth estimation.

5. Discussion

The results show that wave patterns can be extracted by a local application of the RT for two different wave conditions and depth can be estimated. Wave-enhancement is an intrinsic characteristic of the RT while depth inversion from the RT spectra depends on the physical and sensing conditions. What are the limitations, in particular at other different coastal wave environments? To a certain degree, all coastal zones host waves passing through. Depth inversion depends, in the first instance, on the wave period/length and image resolution while the wave observability greatly depends on the relative angles between incident waves, satellite view angle and position of the sun [26] but also wave characteristics such as wave steepness (H/λ). Here, we focus on the methodological aspect imposed limits by the image resolution.

5.1. Required Sensing Resolution

The image resolution predominantly determines whether one can see waves or not; this holds for all imagery from shore-based video, airborne systems to space-borne optical and radar imagery. One could argue that at the very least 2 points over a full wavelength (Nyquist criterion) are required to start recognizing a waveform. At the same time, waveforms are asymptotically better visible than the more points per wavelength. In order to isolate and scrutinize the methodology's limits, synthetic data is used. Let's consider a synthetic dataset representing an image-resolution range between 0.5 m (Pleiades) and 100 m (THIR bands of NASA' Landsat-8) in combination with a set of wavelengths related to their offshore period (from 4 to 20 s with half a second interval). This results in 33 different wave periods and 29 resolutions, hence a total of 957 configurations. Perhaps it is stating the obvious, but the most coarse resolution will not be sufficient to resolve the shortest waves and are then neglected (resolution $> 0.5\lambda$). For each wave and resolution pair, the phase-shift difference between the theoretical and estimated shift is determined. Figure 10 shows the percent error as a function of the resolution/λ.

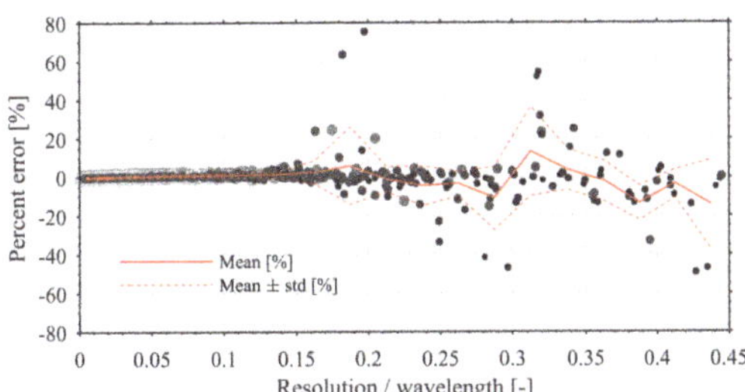

Figure 10. Percent error per image resolution (Δx) over wavelength (λ). The solid red line represents the mean per bin 0.025 [-] (Res/λ) and the dotted red lines represent the associated standard deviation ($+/-$). The color and size of the scatter relates to the wave period: the larger the period, the bigger and lighter the dots.

Figure 10 shows that independently of the wave period/length the percent error starts to increase and vary significantly after an image resolution over the wavelength-ratio of 0.15. A careful look at the data reveals that for ratios near zero to 0.13 the percent error is smaller than 1% and the standard deviation is smaller than 1.5%. Beyond the 0.17 ratio, the mean percent error varies between 4.25% and -14% with standard deviations up to 23.3%. The 0.13 and 0.15 ratios mean that 6.67 (0.15) to 7.69 (0.13) points are required on the full wavelength to resolve the wave phase shift appropriately. For Sentinel-2's best resolution (10 m) this means 7–8 pixels, and thus, wavelengths smaller than 70–80 m are not sufficiently resolved to estimate the phase shift. From this analysis, a required sensing resolution can be determined per wave period or wavelength, as shown in Figure 11.

The blue solid line in Figure 11 indicates the 0.15 ratio between resolution and wavelength. The right of this line means that the resolution is sufficient to resolve the wave phase shift with the current method (green area), and the vice versa for the left-hand side of the blue line (red area). With the Sentinel-2 constellation, one should typically be able to resolve waves with a period equal to or greater than 7 s. This represents the offshore period here, in the coastal zone, nearshore hydrodynamic processes, such as shoaling, shift this boundary upwards. In addition to the Sentinel-2 constellation, the Pleiades (0.5 m) and SPOT 6/7 (1.5 m) are represented by a horizontal line in Figure 11. Both constellations do not have limitation to observe the smallest wind-wavelengths and periods. For example, for the lower wind-wave period, let's say 3 s, the Pleiades constellation has already 14 points over a full offshore wavelength. However, the limitation for the Pleiades constellation is the interval between images, either 0.146 s between image-bands (as used here) or 8 s between individual snapshots. Hence waves with a period smaller than 8 s cannot be resolved [26].

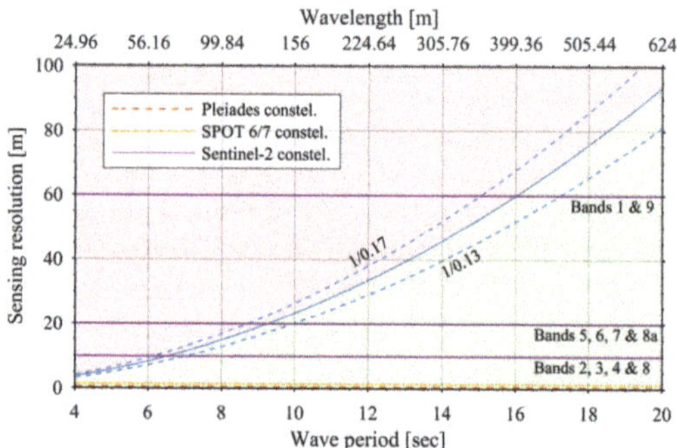

Figure 11. Sensing resolution as a function of wave period/length. The solid blue line represents the resolution's limit to resolve an associated deep-water wave period/length. The green area indicates sufficient resolution while the opposite is true for the red area. The purple lines (Sentinel-2 constellation) are plotted for bands 1 to 9 with 10, 20 and 60 m resolution.

5.2. Sentinel-2 Bands Resolution Augmentation

Besides Sentinel-2' highest 10 m resolution bands (RGB+NIR), lower resolution (20 m and 60 m) visible bands are acquired [27]. Presuming that all these bands observe a similar wave pattern (which depends on the detector-wavelength). The time-lag between the detector-bands and the first band (Blue–band-2) increases from 0.264 to 2.586 s for the last band (band-9 with 60 m resolution) as shown in Table 1. To observe wave propagation one could argue that the resolution of the imagery should be smaller than the distance covered by the wave over Δt. The larger the Δt, the better observable wave propagation should be. For example, considering the 60 m Sentinel-2 resolution, waves have to travel 60 m in 2.586 s requiring a celerity of 23.2 m/s, which is quite fast for free surface waves in the nearshore. For the 10 m (1.005 s) and 20 m (2.055 s) resolution bands, the optimum is respectively 9.95 m/s and 9.73 m/s. Furthermore, according to Figure 11, lower resolution images with 60 m would only be sufficient to resolve celerity for waves with a period of 16 s or longer. Stepping up to the higher-resolution 20 m Sentinel-2 bands waves with a 9.25 s period can be resolved.

A multi-band/multi-resolution combination to compute the celerity is a possibility. [16] uses two bands with different resolution from SPOT-5 for a sub-pixel cross-correlation. However, this approach requires significant computational power. As mentioned before, the RT is used in image processing to augment image resolution. The RT sinogram is interpolated and subsequently used in the inverse RT, and so, augmenting the image resolution. Let us focus on Sentinel-2 band 8a which has 20 m resolution and the largest time-lag with respect to band 2 (Table 1). The time-lag between Band-2 (the first band—Figure 6) and the 20 m resolution bands range between 1.269 s to 2.055 s making them very useful in the phase-shift analysis if augmented to 10 m resolution.

Table 1. Detector bands Sentinel-2: visible and near-infrared [27].

Band	Δt [Sec]	Av. Wavelength (S2A/S2B) [nm]	Resolution [m]
2	0	492.3	10
8	0.264	832.9	10
3	0.527	559.4	10
4	1.005	664.8	10
5	1.269	704.0	20
6	1.525	739.8	20
7	1.790	781.3	20
8a	2.055	864.4	20
1	2.314	442.5	60
9	2.586	944.1	60

From Figure 12 one can see that the 20 m resolution resembles the incident wave pattern at large, for the longer wavelengths. Smaller wave features, present in Figure 6 are not visible and particularly at the coast the wave pattern is not well resolved. The RT-augmented wave pattern in Figure 12b contains smaller wave patterns, like the original 10 m resolution wave pattern in Figure 6. At the coast, the refraction pattern around the harbor is clearly visible now. Focussing on the sub-sampled wave patterns in Figure 12b–d, the effect of the augmentation is even more apparent. The ocean wave pattern in Figure 12c is incomparable to that in Figure 12b. Using correlation methods to find the celerity would be a challenge here. However, as in Figure 12d in case the Radon-based augmentation is applied, similar (shifted) wave patterns become visible. Figure 12f highlights this in even greater detail. For example, between 150–300, hardly any wave signal is visible in the blue line (B05-original) while the red line (B05-augmented), represents a similar wave pattern as found with the 10 m resolution band B02.

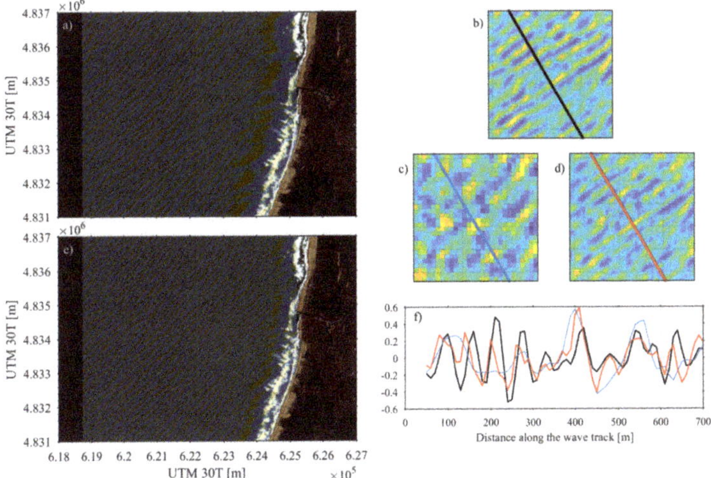

Figure 12. Augmentation of Sentinel-2' mid-resolution bands (20 m) to the high-resolution (10 m). (**a**) shows the sensed 20 m resolution wave pattern (normalized) derived from detector-band 5 and (**e**) shows the augmented wave pattern (normalized) using a localized Radon transform and interpolation of the local Radon-sinogram. (**b–d,f**) show the effect on the obtained wave signal. (**b**) is a subsample of the B02 band (10 m resolution), (**c**) represents the sensed subsample of band B05 (20 m resolution) and (**d**) is the radon-augmented version (10 m resolution). (**f**) presents the spatial wave-signal along wave-track following the color-coding as in (**b–d**).

The RT-augmented wave pattern augmentation allows the 20 m resolution bands to be used alongside the 10 m bands, so that wave phase can be computed over bands 2 to 8a. Discarding band 3 and 8 (time-lag < 0.55 s), a wave phase-shift can then be computed between detector band-2 and the five other bands (B4, B5, B6, B7, B8a). Conceivably, more usable frames will improve the method' robustness. Noteworthy, this technique can also be applied to augment 10 m imagery to higher resolutions. If so, to a certain degree this augmentation diminishes the minimally required wave period to sense the wave celerity. This first application shows promising but the limits of Radon-based augmentation must be explored as well as the use of multiple bands to determine celerity, and a large variety of wave conditions (likewise for the depth estimation).

6. Conclusions

In this work, an RT-based wave-pattern extraction, augmentation and depth inversion method are presented. Wave patterns are extracted by applying an RT and subsequent angle filtering to Sentinel-2 imagery so that the wave signal contains the most-dominant wave directions. Depth is derived exploiting the time-lag between detector-bands. The wave-phase per band, and after the phase shift, is obtained by applying a DFT to the RT sinogram over a local sub-domain. Depths are derived with a good correlation of 0.82 and 2.58 m RMS error over the surveyed domain. These results are beyond the expectations considering the challenging environment including a deep-water canyon and its effect on surrounding wave patterns. In terms of international admiralty standard [28] the measurements could be qualified as level C (up to 30 m). Besides physical limits of the linear dispersion relation, image resolution limits adequate observation of the wave celerity. This method is shown to work stably for waves with a period larger than 6.5 s. In addition to the wave pattern extraction and depth derivation, the RT can be used to augment image resolution. 20 m resolution image bands, from Sentinel-2, are augmented to match the 10 m resolution bands allowing those four extra bands to be used in the depth estimation method, with time-lags ranging between 1.005 s to 2.055 s.

Author Contributions: Conceptualization, E.W.J.B. and R.A.; methodology, E.W.J.B.; validation, E.W.J.B.; resources, E.W.J.B., R.A. and P.M.; data curation, E.W.J.B. and R.A.; writing—original draft preparation, E.W.J.B.; writing—review and editing, E.W.J.B., R.A. and P.M.; visualization, E.W.J.B.; supervision, R.A. and P.M.

Funding: E. Bergsma was partly funded by IRD (Institut de Recherche pour le Développement) and is currently part of the CNES post-doctoral fellowship-program to develop this work. The authors would like to thank CNRS/LEGOS for funding the COMBI-project that initiated this work and provided the in-situ data.

Acknowledgments: We would like to thank the French National Space Agency (CNES) for facilitating fruitful discussions between their Sentinel team, support and scientists. Evidently, this work would not be possible without ESA's Earth Observation and EU's Copernicus programme. We are grateful to ESA and EU for their efforts to make satellite imagery (optical and radar) more accessible to the public. The authors would to thank Luis Pedro Almeida and Paulo Baganha Baptista for the execution of the bathymetry survey and in cooperation with, and amazing support from D. Legros.

Conflicts of Interest: The authors declare no conflict of interest. The funders had no role in the design of the study; in the collection, analyses, or interpretation of data; in the writing of the manuscript, or in the decision to publish the results.

References

1. Vitousek, S.; Barnard, P.L.; Fletcher, C.H.; Frazer, N.; Curt, D.; Storlazzi, L.E. Doubling of coastal flooding frequency within decades due to sea-level rise. *Sci. Rep.* **2017**, *7*, 1399. [CrossRef] [PubMed]
2. Vousdoukas, M.; Mentaschi, L.; Voukouvalas, E.; Feyen, L. Climatic and socioeconomic controls on coastal flooding impacts in Europe. *Nat. Clim. Chang.* **2018**, *8*, 776–780. [CrossRef]
3. Cazenave, A.; Le Cozannet, G.; Benveniste, J.; Woodworth, P.; Champollion, N. Monitoring coastal zone changes from space. *EOS* **2017**, *98*. Available online: https://eos.org/opinions/monitoring-coastal-zone-changes-from-space (accessed on 5 July 2019). [CrossRef]
4. Li, J.; Roy, D.P. A Global Analysis of Sentinel-2A, Sentinel-2B and Landsat-8 Data Revisit Intervals and Implications for Terrestrial Monitoring. *Remote Sens.* **2017**, *9*, 902–919.

5.   Lyzenga, D. Passive remote sensing techniques for mapping water depth and bottom features. *Appl. Opt.* **2018**, *17*, 379–383. [CrossRef] [PubMed]

6.   Holman, R.A.; Plant, N.; Holland, T. cBathy: A Robust Algorithm For Estimating Nearshore Bathymetry. *J. Geophys. Res. Oceans* **2013**, *118*, 2595–2609. [CrossRef]

7.   Bergsma, E.W.J.; Conley, D.C.; Davidson, M.A.; O'Hare, T.J. Video-Based Nearshore Bathymetry Estimation in Macro-Tidal Environments. *Mar. Geol.* **2016**, *374*, 31–41. [CrossRef]

8.   Brodie, K.; Palmsten, M.; Hesser, T.; Dickhudt, P.; Raubenheimer, B.; Ladner, H.; Elgar, S. Evaluation of video-based linear depth inversion performance and applications using altimeters and hydrographic surveys in a wide range of environmental conditions. *Coast. Eng.* **2018**, *136*, 147–160. [CrossRef]

9.   Stockdon, H.F.; Holman, R.A. Estimation of wave phase speed and nearshore bathymetry from video imagery. *J. Geophys. Res.* **2000**, *105*, 22015–22033. [CrossRef]

10.   Plant, N.G.; Holland, K.T.; Haller, M.C. Ocean Wavenumber Estimation from Wave-Resolving Time Series Imagery. *IEEE Trans. Geosci. Remote Sens.* **2008**, *46*, 2644–2658. [CrossRef]

11.   Almar, R.; Bonneton, P.; Senechal, N.; Roelvink, D. Wave Celerity from Video Imaging: A new method. In Proceedings of the 31st International Conference Coastal Engineering, Hamburg, Germany, 31 August–5 September 2008; pp. 1–14.

12.   Bergsma, E.W.J.; Almar, R. Video-based depth inversion techniques, a method comparison with synthetic cases. *Coast. Eng.* **2018**, *138*, 199–209. [CrossRef]

13.   Williams, W.W. The determineation of Gradients on enemy-held beaches. *Geogr. J.* **1946**, *109*, 76–90. [CrossRef]

14.   Abileah, R. Mapping shallow water depth from satellite. In Proceedings of the ASPRS Annual Conference, Reno, NV, USA, 1 May 2006.

15.   Poupardin, A.; Idier, D.; de Michele, M.; Raucoules, D. Water Depth Inversion From a Single SPOT-5 Dataset. *IEEE Trans. Geosci. Remote Sens.* **2016**, *119*, 2329–2342. [CrossRef]

16.   de Michele, M.; Leprince, S.; Thiébot, J.; Raucoules, D.; Binet, R. Direct Measurement of Ocean Waves Velocity Field from a Single SPOT-5 Dataset. *Remote Sens. Environ.* **2012**, *119*, 266–271. [CrossRef]

17.   Poupardin, A.; de Michele, M.; Raucoules, D.; Idier, D. Water depth inversion from satellite dataset. In Proceedings of the 2014 IEEE Geoscience and Remote Sensing Symposium, Quebec City, QC, Canada, 13–18 July 2014; pp. 2277–2280.

18.   Danilo, C.; Farid, M. Wave period and coastal bathymetry using wave propagation on optical images. *IEEE Trans. Geosci. Remote Sens.* **2016**, *54*, 6307–6319. [CrossRef]

19.   Kudryavtsev, V.; Yurovskaya, M.; Chapron, B.; Collard, F.; Donlon, C. Sun glitter imagery of ocean surface waves. Part 1: Directional spectrum retrieval and validation. *J. Geophys. Res. Oceans* **2017**, *122*, 1369–1383. [CrossRef]

20.   Radon, J. Ufiber die bestimmung von funktionen durch ihre integral-werte langs gewisser mannigfaltigkeiten. *Sachs. Akad. Der Wiss. Leipz. Math-Phys* **1917**, *62*, 262–267.

21.   Almar, R.; Michallet, H.; Cienfuegos, R.; Bonneton, P.; Tissier, M.; Ruessink, G. On the use of the Radon Transform in studying nearshore wave dynamics. *Coast. Eng.* **2014**, *92*, 24–30. [CrossRef]

22.   Martins, K.; Blenkinsopp, C.E.; Almar, R.; Zang, J. The influence of swash-based reflection on surf zone hydrodynamics: A wave-by-wave approach. *Coast. Eng.* **2017**, *122*, 27–43. [CrossRef]

23.   Berthier, E.; Vincent, C.; Magnússon, E.; Gunnlaugsson, A.; Pitte, P.; Meur, E.L.; Masiokas, M.; Ruiz, L.; Pálsson, F.; Belart, J.M.C.; Wagnon, P. Glacier topography and elevation changes derived from Pléiades sub-meter stereo images. *Cryosphere* **2014**, *8*, 2275–2291. [CrossRef]

24.   Almeida, L.P.; Almar, R.; Bergsma, E.W.J.; Berthier, E.; Baptista, P.; Garel, E.; Dada, O.; Alves, B. High resolution coastal topography from satellite stereo imagery. *Remote Sens.* **2019**, *11*, 590. [CrossRef]

25.   Drusch, M.; Bello, U.D.; Carlier, S.; Colin, O.; Fernandez, V.; Gascon, F.; Hoersch, B.; Isola, C.; Laberinti, P.; Martimort, P.; et al. Sentinel-2: ESA's Optical High-Resolution Mission for GMES Operational Services. *Remote Sens. Environ.* **2012**, *120*, 25–36. [CrossRef]

26.   Almar, R.; Bergsma, E.W.J.; Maisongrande, P.; Almeida, L.P. Coastal bathymetry from optical submetric satellite video sequence: A showcase with Pleiades persistent mode. *Remote Sens. Environ.* **2019**, *231*, 111263. [CrossRef]

27. SUHET. *Sentinel-2 User Handbook*; ESA-ESTEC: Noordwijk, The Netherlands, 2015.
28. Admiralty Charts: Zones of Confidence (ZOC) Table. Available online: https://www.admiralty.co.uk/AdmiraltyDownloadMedia/Blog/CATZOC%20Table.pdf (accessed on 5 August 2019).

Article

First Results of Phytoplankton Spatial Dynamics in Two NW-Mediterranean Bays from Chlorophyll-*a* Estimates Using Sentinel 2: Potential Implications for Aquaculture

Jesús Soriano-González [1,2], Eduard Angelats [1,*], Margarita Fernández-Tejedor [2], Jorge Diogene [2] and Carles Alcaraz [2]

[1] Centre Tecnològic de Telecomunicacions de Catalunya (CTTC/CERCA), Av. Carl Friedrich Gauss, 7. Building B4, 08860 Castelldefels, Spain
[2] Institut de Recerca Tecnològica Agroalimentària (IRTA/CERCA), Carretera de Poblenou, km 5.5, 43540 Sant Carles de la Ràpita, Spain
* Correspondence: eduard.angelats@cttc.es; Tel.: +34-936-452-900 (ext. 2243)

Received: 19 June 2019; Accepted: 23 July 2019; Published: 25 July 2019

Abstract: Shellfish aquaculture has a major socioeconomic impact on coastal areas, thus it is necessary to develop support tools for its management. In this sense, phytoplankton monitoring is crucial, as it is the main source of food for shellfish farming. The aim of this study was to assess the applicability of Sentinel 2 multispectral imagery (MSI) to monitor the phytoplankton biomass at Ebro Delta bays and to assess its potential as a tool for shellfish management. In situ chlorophyll-*a* data from Ebro Delta bays (NE Spain) were coupled with several band combination and band ratio spectral indices derived from Sentinel 2A levels 1C and 2A for time-series mapping. The best results (AIC = 72.17, APD < 10%, and MAE < 0.7 mg/m^3) were obtained with a simple blue-to-green ratio applied over Rayleigh corrected images. Sentinel 2–derived maps provided coverage of the farm sites at both bays allowing relating the spatiotemporal distribution of phytoplankton with the environmental forcing under different states of the bays. The applied methodology will be further improved but the results show the potential of using Sentinel 2 MSI imagery as a tool for assessing phytoplankton spatiotemporal dynamics and to encourage better future practices in the management of the aquaculture in Ebro Delta bays.

Keywords: ACOLITE; coastal waters; atmospheric correction; time-series; management

1. Introduction

Shellfish are filter-feeding organisms that feed on different types of suspended particles in the water column, thus their production is mainly related to phytoplankton availability [1]. Spain is the leading producer and consumer of bivalves in Europe, Catalonia being the most important producer area in the Spanish Mediterranean, with most of the production concentrated in the Ebro Delta (Figure 1). The most important species for aquaculture are the Mediterranean mussel (*Mytilus galloprovincialis*) and the Pacific oyster (*Crassostrea gigas*), but other bivalves such as clams (e.g., *Ruditapes philippinarum*) and cockles (e.g., *Venus verrucosa)* are also harvested. Bivalve culture is mainly developed inside its two bays, Alfacs and Fangar, representing 1.8% and 6.5% of their respective surfaces [2]. Since 1990, an official monitoring program carried out by the Regional Government of Catalonia establishes a weekly analysis of the phytoplankton community and water physicochemical parameters at different locations of both bays (12 samples per week). However, the sampling procedure is temporally and spatially limited, so global extrapolations are subject to large uncertainties.

Temporal phytoplankton dynamics are highly influenced by the nutrient input from rice fields trough the irrigation network [3]. Furthermore, freshwater inputs have a great physicochemical impact in both bays, increasing water column stratification and dominating over wind on a seasonal scale [4]. Therefore, freshwater input imposes a double layer circulation system like typical estuarine circulation patterns. However, when channels are closed (from October to April), the water renewal time of the bays increases, forming retention areas that can become accumulation zones. Both scenarios may favor phytoplankton growth. On shorter time scales (days to weeks), the wind is the main controlling factor of water mixing [5] by breaking the vertical stratification. Therefore, water circulation patterns, and hence phytoplankton temporal and spatial variability, depend on freshwater inputs, meteorology, and coastal geomorphology [6]. Remote sensing allows obtaining information of marine and continental processes at different spatiotemporal scales [7]. Chlorophyll-*a* (chl-*a*) is the main photosynthetic pigment present in algae and an optically active seawater constituent; thus, it is commonly used as indicator of phytoplankton biomass and has significant implications on remote sensing [8–11]. The estimation of chl-*a* concentration from remotely sensed data requires the development of algorithms with a maximal sensitivity to chl-*a* and minimal to the rest of constituents present in the water [12]. Different authors have proposed several methodologies to estimate chl-*a* from satellite remote sensing imagery (see a review in [13–15]); for instance, a classical approach is developing relationships between band-ratios (namely color indices) or their combinations [14]. Several ratio-based and 3-band combination algorithms have been proposed, including the common Blue to Green ratios, the Ocean Color-based algorithms [16,17], and those including the Red edge [18–21], which take advantage of pigment's absorption maxima (i.e., at 665 nm) [22,23]. Other approximations are based on spectral band difference by using band triplets from the Red and Near Infrared (NIR), such as the Fluorescence Line Height (FLH) [24], the Maximum Chlorophyll Index (MCI) [24], and the Maximum Peak Height (MPH) [25]. The properties of coastal waters, however, are controlled by complex interactions and fluxes of material between land, ocean, and atmosphere, which makes challenging to achieve reasonable estimates of water-leaving radiance (removing atmospheric contributions from a signal received at the TOA), and to obtain a robust relationship between water quality and satellite-based parameters [26] (integrating the remote sensing and in situ measurements). Although a large amount of satellite data is available for remote sensing of chl-*a* (e.g., SeaWiFS, MODIS, MERIS), the fast dynamics of phytoplankton in coastal areas, both temporally and spatially, cannot be fully resolved because of either their coarse spectral, spatial and/or temporal resolution. Currently, the increased frequency (up to five days under ideal conditions) and higher spatial resolution (10 to 60 m^2) of Sentinel 2 together with its spectral band configuration has opened a new potential to remote sensing of chl-*a* in coastal zones of small geographical extension, and hence as an alternative for phytoplankton monitoring in coastal areas.

The overall purpose of this study was to analyze the potential of Sentinel 2 multispectral imagery (MSI) data as a support tool for the future management of shellfish cultures through the monitoring of phytoplankton biomass in the Ebro Delta bays. Thus, this paper is a first attempt to assess chl-*a* in a shallow coastal environment with Sentinel 2 imagery data, a free public resource. The objectives of this study were to

1. Generate 20 m^2 resolution chl-*a* maps from Sentinel-2 MSI imagery covering the whole system;
2. Understand the spatiotemporal phytoplankton biomass dynamics by using the derived chl-*a* maps and to relate them to environmental variables and the rice farming year;
3. Assess the applicability of the results to shellfish aquaculture management in the area.

2. Materials and Methods

2.1. Study Sites

The Ebro Delta is one of the largest (320 km^2) deltas in the northwestern Mediterranean Basin. The climate is Mediterranean temperate with warm dry summers and cool wet winters, annual mean temperature ranges between 5 and 33 °C, and annual precipitation from 500 to 600 mm, being maximum in autumn and minimum in summer. The study area was located in the two bays of the Ebro Delta (Figure 1). Fangar Bay, with an area of 12 km^2, is connected to the sea by a narrow mouth (*ca.* 1 km wide) that is currently closing. Maximum depth is ca. 4 m, which makes it very sensitive to environmental variations. Water temperature ranges between 6.5 and 32 °C, salinity varies from 9 to 37 PSU (Practical Salinity Unit) and renewal time is about four days when channels are open [6]. Alfacs Bay, covering an area of 56 km^2 and connected to the sea by a channel of 2.5 km wide, has an average depth of 3.13 m (maximum depth is 7 m). Water temperature ranges between 8 and 32 °C, the salinity varies from 26 to 37 PSU, and the renewal time is about 15 days with open channels [6]. The hydrology of both bays is highly influenced by freshwater inputs from the irrigation network (Figure 1). Freshwater and nutrient inputs from the river allowed the development of prosperous fishery and farming activities. The production of bivalves in Ebro Delta bays constitutes a major economic activity in the area (Figure 1), together with agriculture, since 210 km^2 of the delta plain are devoted to rice production (Figure 1).

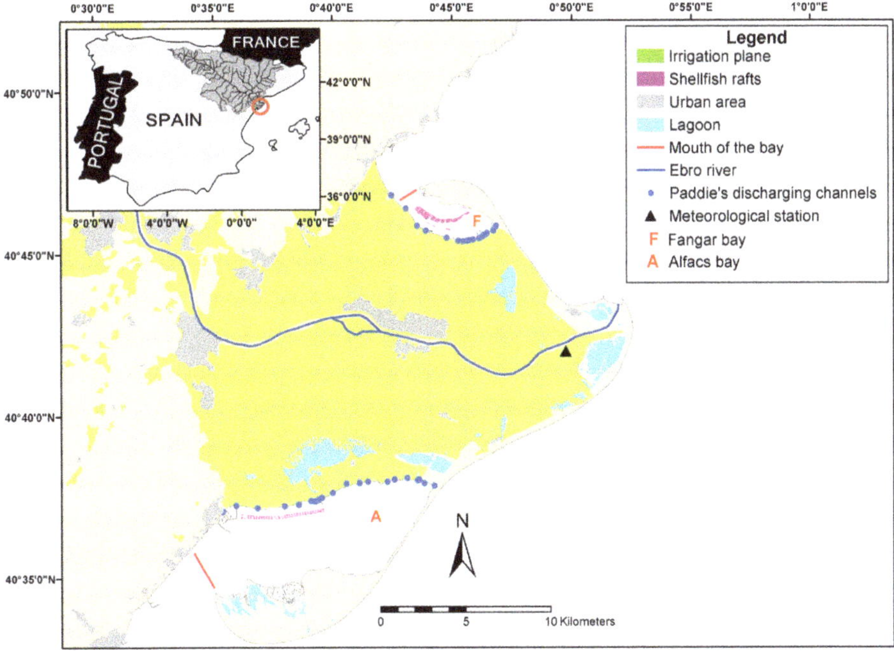

Figure 1. Location of the study bays, meteorological station, mussel rafts, coastal lagoons, irrigation fields, and the discharging channels in Ebro Delta.

2.2. *In situ Data:* chl-*a*

Eight water samplings campaigns were carried out from April 2016 to August 2017 coinciding with the Sentinel 2A satellite pass (Table 1). Different sampling grids were used (see Table 1) for different days, and not both of the bays were sampled every day. Integrated water samples were collected using the Lindahl methodology [27] ($N = 106$). In addition, on 25 July 2017 and 4 August 2017, surface water samples were collected ($N = 40$) with polypropylene bottles. Seawater samples were kept in a portable cool box until arrival to the laboratory. In the laboratory, three different methods were used to measure chl-*a* concentration, in vivo fluorimetry [28] (hereafter in vivo), and after acetone extraction both in a fluorometer (corrected chl-*a*; hereafter FL) [29], and in a spectrophotometer (chl-*a*; hereafter SP) [30]. For all samples ($N = 106$), chl-*a* was estimated in vivo, and in 58 of them, chl-*a* was measured after acetone extraction. Briefly, water samples (550–1000 mL) were filtered using fiberglass filters (GF/F), and filters were submerged in 10 mL of acetone inside 15 mL labelled conical centrifuge tubes. After 24 h in the fridge (4 °C), they were sonicated for 5 min (ultrasonic processor) and centrifuged for 10 min at 4000 rpm at 4 °C. The chl-*a* concentration was then measured in a SHIMADZU UV-1800 spectrophotometer (Shimadzu Corporation, Kyoto, Japan) and/or in a TURNER Trilogy ® fluorometer (Turner Designs, San Jose, CA, USA) (Table 1). The datasets generated during the current study are available from margarita.fernandez@irta.cat on reasonable request.

Table 1. Summary of chl-*a* samples coinciding with Sentinel 2A pass. Grid 1: routine sampling for the official water-quality monitoring program (see orange dots in Figure 2a,b). Grid 2: specific sampling grid designed for ground truth of Sentinel images (green dots in Figure 2a,b). Grid 3: sampling grid of the project "Model of water circulation in Fangar Bay from the European Maritime and Fisheries Fund (EMFF)" (white dots in Figure 2b).

Date	Sampling Grid	Bay	Number of Samples per Method		
			in vivo	Fluorimeter	Spectrophotometer
11 April 2016	1	Fangar	5	1	0
20 June 2016	1	Alfacs	7	1	0
		Fangar	5	1	0
16 January 2017	1	Alfacs	7	1	1
17 March 2017	1	Alfacs	7	7	7
		Fangar	5	5	5
6 April 2017	1	Alfacs	7	7	7
		Fangar	5	5	5
26 May 2017	2	Alfacs	6	6	6
		Fangar	6	6	6
15 June 2017	2	Alfacs	6	6	6
25 July 2017	3	Fangar	40 [a]	6	0
4 August 2017	3	Fangar	40 [a]	6	0

a: 20 integrated water column samples, and 20 surface water samples at same locations.

2.3. Sentinel 2 Data

A set of 47 Sentinel 2A L1C images (i.e., not cloud covered) were downloaded from Copernicus Open Acces Hub (https://scihub.copernicus.eu/). Thirteen of them (six from Alfacs and seven from Fangar) within the period April 2016–August 2017 were selected for calibration and validation purposes (Figure 3). The remaining images between January 2017 and January 2018 (Table 1), 18 from Alfacs and 16 from Fangar, were used for time-series estimation (Figure 3). Although the calibration/validation (CalVal) image sets covered mainly spring and summer, the time-series was estimated for a full year in order to include the full rice growing season. Additional meteorological data, including daily air temperature (°C), wind direction (°), wind speed (m/s), and precipitation (mm), were obtained from the Illa de Buda meteorological station (Station Id. 11043, located at 1 m above sea level) of the Catalan Meteorological Service, http://www.meteo.cat (Figure 1).

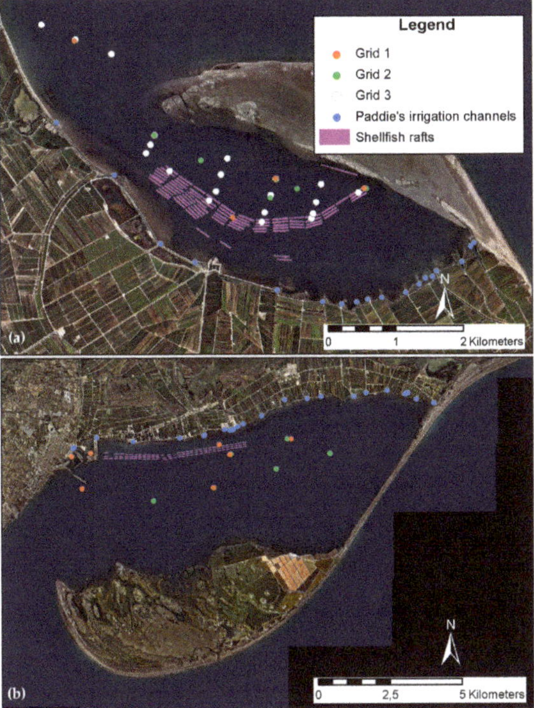

Figure 2. Official water-quality monitoring program sampling grid (Grid 1), specific sampling grid (Grid 2), and European Maritime and Fisheries Fund (EMFF) project sampling grid (Grid 3) at Fangar (**a**) and Alfacs (**b**) bays.

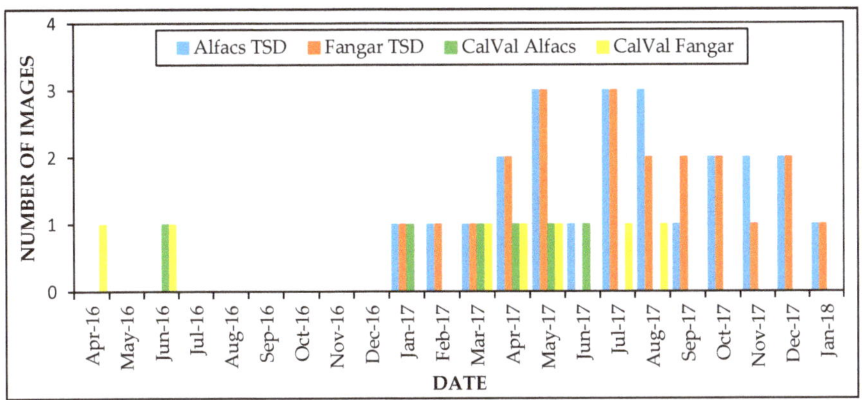

Figure 3. Temporal distribution of Sentinel 2 images used in this study for calibration and validation and time-series development (TSD) at Alfacs and Fangar bays.

2.4. Atmospheric Correction: ACOLITE

Sentinel 2A L1C imagery were atmospherically corrected with ACOLITE processor. It bundles the atmospheric correction algorithms and processing software developed by the Royal Belgian Institute of Natural Sciences (RBINS) for aquatic applications of Landsat (5/7/8) and Sentinel 2 (A/B) satellite data. The Dark Spectrum Fitting (DSF) [31], used here, computes the atmospheric path radiance based on multiple targets in the scene or sub-scene, with no a priori dark band, allowing an aerosol correction. ACOLITE includes a sun glint correction, which uses the short-wave infra-red (SWIR) bands to estimate a glint signal [32] and to establish the threshold to determine which pixels need to be corrected (0.05 by default). Sentinel 2A B11 and B12 bands (SWIR at 1604 nm and 2202 nm) were used for sun glint correction. The thresholds were set manually image-by-image after a SWIR analysis that was carried out considering the response of Sentinel 2A B11 over water pixels compared to non-water pixels. For each day and bay land/water mask and sun-glint correction thresholds were defined, ranging between 0.0215 and 0.1. Therefore, the atmospheric correction procedure output included, for each image, uncorrected (a), partially corrected (b), and fully corrected atmosphere (c and d) reflectance data.

(a)   Rhot: top of atmosphere reflectance (TOA) derived from the original input file.
(b)   Rhorc: Rayleigh corrected reflectance. This is the Rhot with removed and corrected reflectance for two-way Rayleigh transmittance. An additional pre-processing step was made to avoid high reflectance pixels by fixing a maximum threshold (Rhorc reflectance at 492 nm or 560 nm $\geq$ 0.11) above which pixels were assigned as invalids.
(c)   Rrs: remote sensing reflectance (sr^{-1}) for water pixels (Rrs = Rhow/π).
(d)   Rhow: surface reflectance for water pixels.

2.5. Chlorophyll-a Estimation Algorithms

Seven different spectral algorithms band-ratio and band-combination based, were applied to each product resulting of ACOLITE processing (Rhot, Rhorc, Rrs and Rhow). Briefly,

I.   BG: The Ratio between Blue and Green spectra uses the reflectance at 490 nm (blue) and 560 nm (green). At 490 nm carotenoids absorb light strongly, while at 560 nm the absorption of all photosynthetic pigments is minimal (i.e., green reflection). This algorithm was initially proposed by [33]. *R* stands for Rhot, Rhorc, Rrs, or Rhow reflectance.

$$[\text{chl–}a] \propto \frac{R(490)}{R(560)} \tag{1}$$

II.   BR: The Blue–Red ratio is based on the two chl-*a* maximal absorption peaks.

$$[\text{chl–}a] \propto \frac{R(490)}{R(665)} \tag{2}$$

III.   RG: The Green–Red ratio is based on the minimal and maximal absorption peaks of chl-*a*, thus avoiding the use of the blue bands [23,34].

$$[\text{chl–}a] \propto \frac{R(665)}{R(560)} \tag{3}$$

IV.   NR: The ratio between Red and NIR assumes that the absorption by non-algal particles, yellow substances and the backscattering are insignificant when compared to chl-*a* absorption at red wavelengths (665 nm) [35]. Between 700 and 720 nm, the absorption due to water constituents is minimal.

$$[\text{chl–}a] \propto \frac{R(705)}{R(665)} \tag{4}$$

V.   NDCI: The Normal Difference Chlorophyll Index developed by Mishra et al. [36] for turbid productive waters uses the information of the reflectance peak centered at 700 nm, which is maximally sensitive to variations in chl-*a* concentration in water. Furthermore, a wide spectral absorption peak between 665 nm and 675 nm is generally associated to the absorption by chl-*a* pigments. The normalization through the NDCI eliminates uncertainties in the estimation of the remote sensing reflectance, seasonal solar azimuth differences, and atmospheric contributions.

$$[\text{chl-}a] \propto \frac{[R(705) - R(665)]}{[R(705) + R(665)]} \tag{5}$$

VI.   DO5: Dall'Olmo and Gitelson [37] presented a three-band model using Red and NIR bands. It assumes that (i) the absorption by coloured dissolved organic matter (CDOM) and total suspended matter (TSM) beyond 700 nm is approximately equal to that at 665–675 nm and the difference between them can be neglected; (ii) total chlorophyll, CDOM, and TSM absorption beyond 730 nm is almost 0; and (iii) backscattering coefficient of chl-*a* is spectrally invariant [36,37].

$$[\text{chl-}a] \propto \left[\frac{1}{R(665)} - \frac{1}{R(705)}\right] \times R(740) \tag{6}$$

VII.   MCI: The Maximum Chlorophyll Index allows the detection of red tides and aquatic vegetation [24]. For Sentinel 2, it uses the band 5 (705 nm), perfectly located to detect high biomass water bodies against the baselines of the bands 4 and 6 (665 and 740 nm). In Equation (7), k is the thin cloud correction factor fixed at 1.005 for thin clouds.

$$[\text{chl-}a] \propto R(705) - k \times \left(R(665) + (R(740) - R(665)) \times \frac{705 - 665}{740 - 665}\right) \tag{7}$$

2.6. Model Calibration and Validation

Sentinel 2A (Level 1C and 2) images and all in situ chl-*a* of coinciding days were used for model calibration and validation. In order to reduce the effect of noise from the sensor and the time-difference between the image (20 m^2 resolution) and water samples acquisition, reflectance was averaged over a 3×3 pixel-box centered at the in situ measurements. However, not all of the nine pixels per in situ sampling location could be used as there might be outliers coming from different sources such as bottom contamination, different affection of sun glint and adjacency or infrastructures as rafts or harbor jetties interfering in some pixels. For this reason, a pre-processing step was carried out on each spectral band used and for all atmospheric correction levels. For each day and bay, considering together all in situ sampling locations, outliers were detected and removed by Tukey's fences method (Boxplot). The criteria flagged as invalid a pixel if in one of the five spectral bands (see Equations (1)–(7)) the reflectance value was classified as outlier. A second step to clean the remaining outliers was carried out applying the same methodology to each 3×3 pixel-box centered at in situ sampling sites, individually. To ensure the possibility of using the averaged reflectance of 2–9 pixels, without corrupting the methodology, standard deviation (SD) of the average at each sampling location was computed against the number of pixels used for the average.

After outlier deletion, the seven algorithms were computed using the averaged reflectance at each chl-*a* sampling location. Model calibration was done with 70% of the data (with ordinary least of squares fitting, OLS) and the remaining 30% was used for model validation. Models were calibrated and validated in two different ways: (i) using only those samples where in situ chl-*a* was measured by the three methodologies (i.e., in vivo, FL, SP) and (ii) for each methodology including all the available data. In both cases, model development was carried out considering all possible combinations of ACOLITE-derived imagery together with two different scenarios (individually or both bays together).

Model performance was assessed graphically by plotting observed and predicted values, and efficiency was measured with the Akaike Information Criterion (AIC), the Averaged Percentage Difference (APD), and the Mean Absolut Error (MAE). AIC combines fit and parsimony (number of parameters) of models, with the best fitting model having the lowest AIC. MAE and APD were applied following the criteria of [38], who suggested that these metrics account better for accuracy of the models over non-Gaussian distributions by not amplifying outliers and precisely reflecting the error magnitude. Models with lowest AIC, MAE, and APD, in this order, were considered better. Although the coefficient of determination (i.e., R^2) and Normalized Root Mean Squared Error (RMSE) are widely used goodness-of-fit measures, they are not recommended for non-Gaussian distributions [38]. Thus, both measures were only included to allow the comparison with previous works.

2.7. Time-Series Estimation

The best model was selected to construct chl-*a* time-series maps with the available Sentinel 2A images in 2017. Pixel-stability was assessed by using an unsupervised classification cluster analysis (2 classes) based on the inter-pixel slope of the averaged time-series chl-*a* and the coefficient of variation (CV; Equation (8)) of chl-*a* of the same set of images.

$$CV = \frac{\sigma}{\overline{X}} \tag{8}$$

where σ stands for standard deviation (SD) and $\overline{X}$ for the average.

2.8. Workflow

The proposed workflow (Figure 4) started with the selection and download of Sentinel 2A L1C images. The images were processed with ACOLITE after the SWIR analysis, including a resampling of all bands to 20 m^2, image cropping to the region of interest (Ebro Delta bays), and the atmospheric correction to derive Rhorc, Rrs, and Rhow reflectance (see Section 2.4). After ACOLITE processing, for the spectral bands of interest, outliers were detected and removed. Then, for each image of the calibration set, the spectral algorithms were computed, and the resulting values were extracted at each chl-*a* sampling location. Models were calibrated and validated with Rhot, Rhorc, Rrs, and Rhow imagery together with ground truth data. The best algorithm and methodology were selected, applied to all the available images in 2017, and the pixel-stability analysis was carried out. Finally, the resulting time series was then used to analyze spatiotemporal patterns of chl-*a* (as indicator of phytoplankton biomass dynamics), thus covering different seasons and the full rice farming cycle.

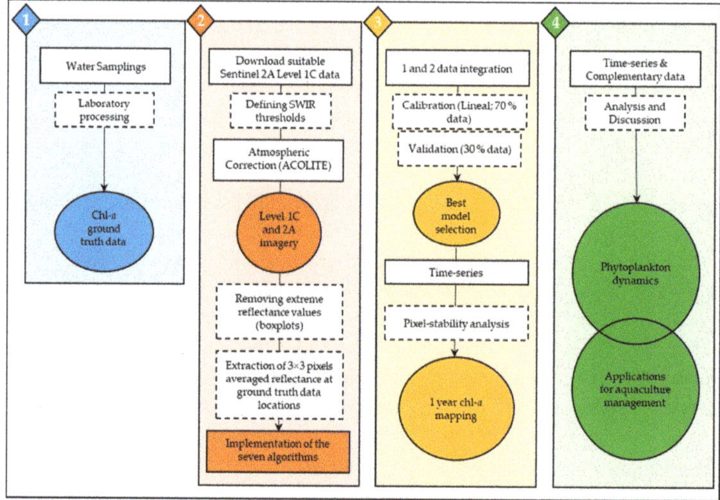

Figure 4. Workflow to derive chl-a time-series from Sentinel 2A multispectral imagery (MSI) data at Ebro Delta bays for aquaculture management purposes.

All statistical analyses were performed with R version 3.5.2; the packages foreign 0.8.71, xlsx 0.6.1, xlsxjars 0.6.1, ncdf 1.16.1, and raster 2.8.19 were used to load external data with different formats. Packages rgdal 1.4.3, spatstat 1.59.0, and maptools 0.9.5 were used to work with geospatial data (create masks, band math calculator, and pixel extraction). Packages FSA 0.8.24, NCStats 0.4.7, nlstools 1.0.2, and minipack.lm 1.2.1 were used to evaluate the model performance (ROC curves and associated statistical parameters).

3. Results

3.1. In situ Data: chl-a

Overall, chlorophyll-*a* concentration varied among seasons and sites, with different spatial distribution patterns in both bays. In Alfacs Bay, chl-*a* showed a spatial gradient trend defined generally by higher concentrations from the central zone with higher concentration values, to the inner area, with minimum chl-*a* concentrations in the shellfish rafts (Figure 1). In Fangar Bay, maximum chl-*a* concentrations were found in the mouth and minimum concentrations in the inner part of the bay, which showed similar values to those in the shellfish rafts. Table 2 summarize chl-*a* results per bay. In general, Alfacs Bay showed higher chl-*a* concentrations.

Among the different laboratory methodologies used to measure chl-*a* concentration, in vivo results showed moderate correlation values with both FL (Pearson's $r = 0.60$, $N = 55$, $P < 0.001$) and SP ($r = 0.62$, $N = 43$, $P < 0.001$), while these two methods (FL and SP) were highly correlated ($r = 0.93$, $N = 43$, $P < 0.001$). The average percentage difference (APD) between methodologies was 98% between in vivo and FL, 56% between in vivo and SP and 20% between FL and SP. Surface and integrated water column (sampled in Fangar Bay on both 25 July and 17 August) in vivo chl-*a* concentrations were strongly correlated ($r = 0.80$, $N = 40$, $P < 0.001$), with an APD of 7.6%.

Table 2. Descriptive statistics of chl-*a* concentration (mg/m^3) per bay and measuring method, during the study period. FL = Fluorometer; SP = Spectrophotometer. N is the number of samples, SD is the Standard Deviation, and CV is the Coefficient of Variation.

Bay	Method	N	Minimum	Maximum	Median	Mean	SD	CV
Fangar	In vivo	66	0.512	6.553	2.719	2.716	1.497	0.551
	FL	30	0.170	4.992	1.365	1.836	1.278	0.696
	SP	16	0.222	2.597	1.604	1.501	0.732	0.487
Alfacs	In vivo	40	0.774	8.880	2.807	3.197	1.867	0.584
	FL	28	1.010	4.750	1.705	2.206	1.131	0.513
	SP	27	1.373	5.596	2.613	2.988	1.291	0.432

3.2. Atmospheric Correction and Outlier Removal

The averaged reflectance at the sampling locations for the different atmospheric correction products (i.e., Rhot, Rhorc, Rhow, and Rrs) for each CalVal date and bay (Figure 5), and at each Sentinel 2A band, showed less reflectance from uncorrected to full corrected levels, this being more pronounced for shorter wavelengths. Fangar Bay showed higher averaged reflectance than Alfacs, when comparing the same day, and for all different Level products.

Averaged reflectance of a 3×3–pixel box centered at the in situ sampling points was used as the reflectance at each location; however, outlier pixels were removed. After outlier detection, 18 sampling points were completely removed and were not used in the CalVal process. Sixteen of the 18 removed points corresponded to Fangar Bay and were mostly located within the shellfish rafts, the mouth of the bay, and the inner area. Two points were removed from Alfacs Bay, both located in the harbor on 20 June 2016. Final available chl-*a* data are summarized in Table 3. In order to evaluate the impact of outlier pixels on the reflectance estimation, it was assessed the reflectance SD relative to the number of valid pixels (2 to 9), at each sampling site, for all type of Sentinel 2A products and all the bands used in algorithm calculation. Pearson's correlation coefficient, in absolute value, ranged between 6.11×10^{-3} and 0.23, thus, reflectance values were similar, independent of the size of the pixel-box around the sampling point (from 2 to 9), and outliers can be removed without introducing significant errors.

Table 3. Summary of in situ chl-*a* data set used in the calibration and validation process of the Sentinel 2 derived data.

Date	Sampling Grid	Bay	Number of Samples per Method		
			In Vivo	FL	SP
11 April 2016	1	Fangar	5	1	0
20 June 2016	1	Alfacs	5	1	0
		Fangar	4	1	0
16 January 2017	1	Alfacs	7	1	1
17 March 2017	1	Alfacs	7	7	7
		Fangar	4	4	4
6 April 2017	1	Alfacs	7	7	7
		Fangar	3	3	3
26 May 2017	2	Alfacs	6	6	6
		Fangar	6	6	6
15 June 2017	2	Alfacs	6	6	6
25 July 2017	3	Fangar	12	5	0
4 August 2017	3	Fangar	16	5	0

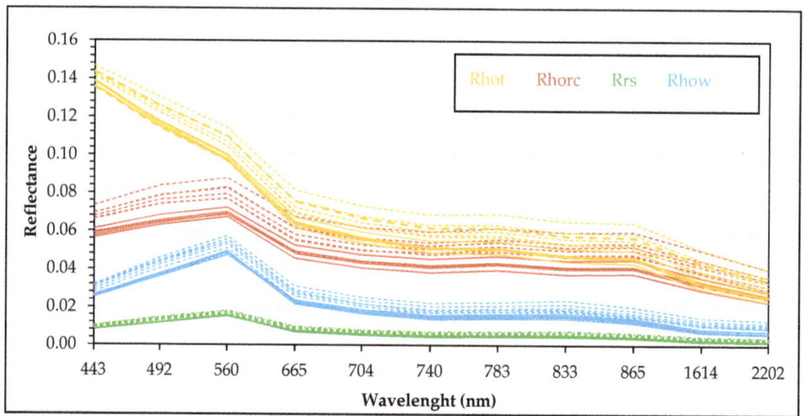

Figure 5. Daily averaged reflectance spectra per bay for each band of Sentinel 1C and 2A products on Calibration and Validation dates. Alfacs Bay: solid line. Fangar Bay: dashed line.

3.3. Model Calibration and Validation

All variable combinations resulted in 252 models (see Table S1); chl-*a* methodology (in vivo, FL and SP) × bay (Alfacs, Fangar, and both bays together) × Sentinel 2A images (Rhot, Rhorc, Rrs and Rhow) × spectral algorithm (BG, BR, RG, NR, NDCI, DO5, MCI) (see Table S1). Overall, considering all the possible models, the algorithms performed better when applied to Rhorc images, although Red-to-Green (RG) and, especially MCI, showed less sensibility to the atmospheric correction and similar results were achieved with Rhot, Rhorc, Rrs, or Rhow reflectance. The best results were obtained combining Rhorc images with spectrophotometer chl-*a* measures (SP). Within the "Rhorc_SP" models, the best performing algorithms were BG (Blue-to-Green ratio) for Alfacs Bay and for both bays together, and the NDCI (Normal Difference Chlorophyll Index) algorithm returned the best results for Fangar Bay (Table 4). Close results to BG were achieved in Alfacs Bay and both bays together with RG (Red-to-Green) band ratio, while worse results in both cases were obtained with Maximum Chlorophyll Index (MCI). In Fangar Bay, despite differences among the performance of the different algorithms were smaller than in Alfacs Bay (Table S1), NIR-to-Red (NR) band ratio and MCI performed similar to NDCI, while BG performed worse.

Chlorophyll-*a* was not measured by the three methodologies (in vivo, FL, and SP) in all the samples; thus, models had different sample size. In order to avoid the influence of sample size on results, the models were also fitted using only those chl-*a* samples measured by the three methodologies (Table S2). There were not significant changes associated to *N*, but changes on model performance were more related to the range of chl-*a* covered by the samples (e.g., the lower variability of chl-*a* in Fangar Bay).

Different algorithms performed better in Alfacs and Fangar Bay. The low number of available SP data and the good results obtained calibrating and validating the model including both bays together suggest the use of "Rhorc_SP" configuration (Figure 6) until more data are available. Despite BG performance in Fangar Bay was worse than the achieved with other algorithms (i.e., NDCI, NR, and MCI), probably it was due to the lack of variability towards higher concentrations and the weight of few extreme values over a small dataset. In fact, the linear distribution of chl-*a* SP in Fangar fit with the trend of Alfacs (Figure 7). Also, the trend line using data of both bays or using data only from Alfacs Bay was almost equal, denoting that Fangar samples were in agreement with the global trend described (Figure 7). These results support the idea of using both bays together and reinforce the assumptions for applying the same model to both bays.

Table 4. Summary of the best performing models per bay for the calibration dataset. "Algorithm" refers to the spectral algorithm applied to Rhorc images and calibrated with chl-*a* spectrophotometer (SP).

Bay	Algorithm	Min chl-*a*	Max chl-*a*	Intercept	Slope	*N*	MAE	APD	R^2	AIC
Fangar	NDCI	0.46	2.39	0.56	−12.80	9	0.41	0.27	0.43	17.12
Alfacs	BG	1.83	5.12	13.88	−12.24	18	0.71	3.30	0.61	49.54
Both	BG	1.37	5.60	13.93	−12.50	27	0.63	5.58	0.58	72.17

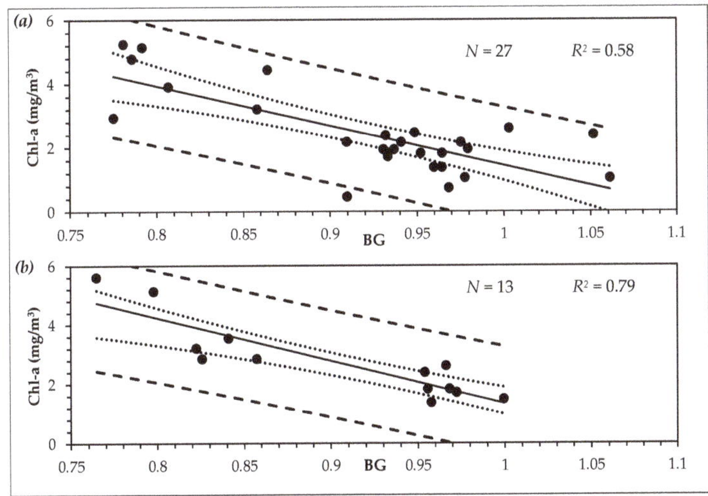

Figure 6. Calibration (**a**) and Validation (**b**) results of the Blue-to-Green ratio (BG) algorithm, chl-*a* SP, and both bays together over the set of Rhorc images. The 95% prediction (dashed line) and the 95% confidence interval (dotted line) are also shown.

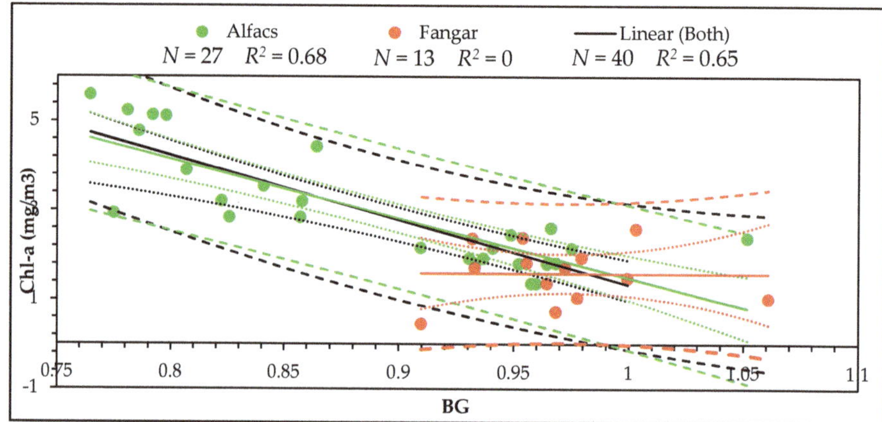

Figure 7. Linear relationship between all available chl-a SP and BG per bay. The 95% prediction (dashed line) and the 95% confidence interval (dotted line) are also shown.3.4 Chlorophyll-a Time-Series.

According to previous results, chl-*a* time-series was generated from the Blue-to-Green ratio (BG) model using partially corrected images (Rhorc) for both bays together, and chl-*a* measured by spectrophotometer (SP). Two pre-processing steps were applied to reduce the sources of error on the bottom of Rayleigh corrected reflectance. First, although images were selected according to cloud absence, in two of them, small areas at the extremes of Alfacs Bay were contaminated by clouds. There, the threshold applied to the Rhorc images removed pixels associated with thick clouds (Figure 8), but the thinnest clouds were not successfully detected, and the ground information was not restored in either case. The second pre-processing step consisted in the generation of a mask to avoid areas where BG did not responded only to chl-*a*, but to other sources such as bottom reflectance or macrophytes (Figure 9). The clustering used to make the mask highlighted the boundaries where maximum changes occurred, namely, shallow waters with bottom or seagrass contribution, hard structures such as rafts, and semi-static objects like the ships in the Alfacs Bay harbor. Finally, before chl-*a* time-series estimation, a 20 m buffer (i.e., 1 pixel) was applied, around each raft, created in order to delete mixed border pixels.

Figure 8. Masking Rhorc high reflectance over blue and green bands (threshold = 0.11). Exempla of cloud presence in Alfacs Bay on 22 November.

Both for Fangar and Alfacs bays, one-year chl-*a* time-series were processed (Figure S1). Overall, during winter and early spring, higher concentrations of chl-*a* were observed in Alfacs Bay. From April to October, chl-*a* concentrations were comparable between bays; after, in Fangar Bay, chl-*a* concentrations decreased more sharply. Despite the differences in chl-*a* concentration, the general trend was similar in both bays almost all the year, but chl-*a* peaks differed. In Alfacs Bay, chl-*a* peaked during February–April and October–November, achieving maximum concentrations in March. In Fangar Bay, chl-*a* peaked on May and September–October, being the most productive along the year in this last period. Minimum chl-*a* concentrations were found in winter in both bays, January and November for Alfacs and Fangar Bay, respectively. The coefficient of variation (CV) of chl-*a* (Figure 10a,b) along the year showed, in general, higher CV in Fangar than Alfacs Bay. In Fangar, higher variability was observed in the mouth of the bay, associated to higher chl-*a* concentrations, while lower CV values were found at the inner area and within the shellfish rafts, where lowest values of chl-*a* were found. Conversely, in Alfacs Bay, higher CV was observed in the eastern half of the bay, especially in the inner area and the eastern half of the rafts polygon, with lower averaged chl-*a* concentration. The harbor area and its surroundings, including the western half of rafts and the mouth of the bay, showed lower values of CV, related to higher concentrations of chl-*a*.

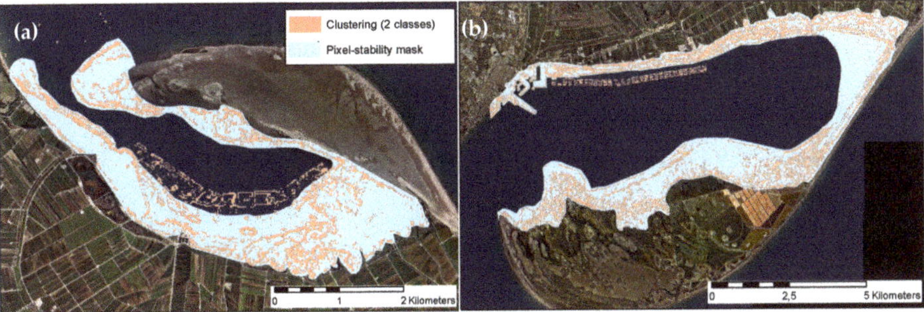

Figure 9. Time-series pixel-stability mask. (**a**) Fangar Bay; (**b**) Alfacs Bay.

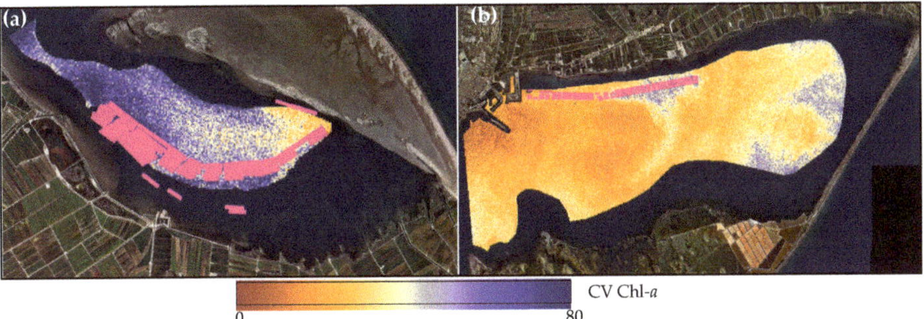

Figure 10. Time-series coefficient of variation (CV) of chl-*a*. (**a**) Fangar (**b**) Alfacs.

The time-series (Figure S1) was revised according to the four different rice-paddies irrigation network scenarios (i.e., Closed channels in winter, semi-closed channels in spring, opened channels in summer and semi-opened channels in autumn)and aquaculture production. The closure of the discharging channels (closed, semi-closed, and semi-opened) propitiated a more eutrophic environment, reaching higher chl-*a* concentrations than during the opened channels stage, this phenomenon being more evident in Alfacs Bay. During the closed and semi-closed stage (from January to April), chl-*a* tended to increase in both bays, but the increment was much more pronounced and long-lasting in Alfacs. During these months rice paddies are dry and so, the supply of water from the channels is minimum. Regarding the chl-*a* within the shellfish rafts, while Fangar Bay showed similar chl-*a* concentrations inside and outside the rafts (more homogeneous bay), in Alfacs Bay, lower concentrations of chl-*a* were observed inside the rafts' polygon. During the opened channels stage (from April to September), chl-*a* concentration decreased in Alfacs and remained the same in Fangar Bay, compared with the prior period. However, from late July, both bays showed an increase of chl-*a* concentration that lasted until the end of September, when chl-*a* dropped sharply, achieving values close to 0 mg/m^3 in both bays. The opened channels stage is characterized by high freshwater inputs with the maximum occurring in September–October. Despite shellfish filter more actively during the warm months, no significant differences in chl-*a* were observed between the rafts and their surroundings in neither of the two bays. Finally, the semi-opened channels stage (from October to December) started with a strong increase of chl-*a* at both bays in October, which lasted until late November, when chl-*a* concentrations dropped close to 0 at both bays. During December, Alfacs Bay recovered chl-*a* concentrations similar to those of the opened channels stage, but Fangar Bay kept low chl-*a* values. The semi-opened channels stage implies that water is still being discharged in the bay, but contributions decrease with time. Most of the shellfish are harvested during summer, so the bivalve grazing pressure is reduced the last months of the year. Although similar chl-*a* concentrations were found between the rafts and the rest of the bays, lower values of chl-*a* tended to aggregate in the middle area of the Alfacs rafts' polygon, and the Northern rafts of Fangar Bay.

4. Discussion

4.1. In situ chl-a Data

Three different laboratory methods for chl-*a* quantification from water samples were compared. Chlorophyll-*a* concentration measured by spectrophotometer (SP) after acetone extraction was better correlated with satellite data. The in vivo method is only used as a fast qualitative proxy of chl-*a* due to its sensibility to errors with unknown uncertainty (i.e., overestimation due to non-phytoplanktonic contribution), while extracting the pigment with a solvent (i.e., alcohol-based or acetone) and measuring it with the fluorometer or spectrophotometer is the common procedure in remote sensing of chl-*a* [21,39,40]. Regarding the use of surface or integrated water samples for ground truth chl-*a* quantification, the vertical distribution of the phytoplankton biomass might have a significant impact on the remote sensing signal. In Fangar Bay, significant differences were not found between surface and integrated water column chl-*a* concentrations. This finding is in agreement with Ramón et al. [41], who found homogeneous chl-*a* concentration by depth in a 10 month study (1 sample per month) in Fangar. These results suggest the use of an integrated water column chl-*a* for remote sensing model calibration and validation in coastal shallow waters, but further research should include data of both bays under different scenarios to prove the validity of this assumption during the year.

4.2. Atmospheric Correction and chl-a Estimation Algorithms

ACOLITE was used for atmospheric correction of Sentinel 2A L1C images using the Dark Spectrum Fitting (DSF) based on the SWIR bands. The results showed that TOA contributed over 50% for all MSI over surface reflectance of water pixels (Rhow). This might be related with non-negligible water reflectance in the SWIR band. According to [42], the invalidity of the SWIR black pixel assumption could lead to an overcorrection of the reflectance (SWIR reflectance for water pixels was up to 10 times larger when solar zenith < 42°; i.e., spring and summer). In this study, it was not possible to validate the atmospheric correction with field radiometric measurements, but the drop of reflectance of Rhow images compared to Rhot in the blue bands was noticeable. However, a strong reflectance peak was observed in the green part, independent of the level of atmospheric correction applied. Similar results were obtained by [43] using ACOLITE without sun glint correction in an estuarine area, and they also found higher water reflectance in all bands in areas with higher concentration of total suspended matter. In the Ebro Delta, Fangar Bay always showed larger reflectance at all spectrum compared to Alfacs Bay. Fangar is shallower and thus is more susceptible to wind-driven mixing and sediment resuspension. However, the increased reflectance of Fangar Bay might be also related to bottom reflectance or contamination due to adjacency effects. As suggested by other authors [44,45], adjacency effects have significant impact in coastal waters due to typical lower reflectance in relation with their neighbourhood surfaces (i.e., sand beaches and rice paddies), increasing the apparent brightness. This effect might be more pronounced in Fangar Bay due to its geomorphological characteristics (smaller, shallower, and more closed).

The simplified atmospheric correction procedure, which normalizes the TOA signal for Rayleigh effects, was preferred in favor of a full aerosol atmospheric correction given the large uncertainties associated with water leaving reflectance over turbid waters. Ref. [31] found a similar performance between the median spectra derived for full atmospheric correction and only Rayleigh correction. Our models showed better performance using Rhorc images instead of Rhow. For both bays together, the best performing algorithm was the BG ratio. Common ocean color algorithms based on the ratio of blue and green bands do not perform well in optically complex coastal waters (less sensivity to chl-*a* concentration changes) [20,23,36,46]. However, Ref. [46] suggested that, in case 2 oligotrophic waters ([chl-*a*] < 4 mg/m^3), the use of blue and green wavelengths is more appropriate. In the present study, averaged chl-*a* concentrations (measured in situ) were 1.50 and 2.99 mg/m^3, with maximums of 2.60 and 5.60 mg/m^3, for Fangar and Alfacs Bays, respectively, and the results achieved were in agreement with [47,48]. The chl-*a* estimates were reasonably well derived (MAE = 0.63 and APD < 10%) using the BG ratio.

4.3. Model Calibration and Validation

In order to reduce noise and minimize temporal-gap effects, ground truth chl-*a* data were averaged over a 3 × 3–pixel box, centered on the sampling point. Despite the averaged reflectance is commonly used, it might be a poor measure of central tendency if the set of pixels used for its calculation contains outliers. Here, outliers were removed before averaging the 3 × 3–pixel box. Although the number of valid pixels differed among locations and dates, the number of valid pixels (between 2 and 9) was not correlated with the mean standard deviation, thus demonstrating the suitability of the results. After model selection, the BG applied over Rhorc images was preferred and used to make a "pixel-stability mask" to identify and reject those areas where the values obtained with the integration of the remote sensing and the model were not responding to the changes of chl-*a*. Based on k-means clustering, the applied methodology allowed us to distinguish the boundaries where maximum changes occurred, thus defining the edges for the delimitation of the mask. In Alfacs Bay, better results were achieved that were able to differentiate each raft individually (rafts more separated than in Fangar) and masking the shallow waters (confined only to the margins of the bay).

The applied model was based on algorithms specifically tuned for Alfacs and Fangar bays. Despite the good results achieved for the CalVal dates, the suitability of the model depends on the ratio between the range of remote sensing and the range of the available ground data and its representativeness along different seasons or scenarios. In our study, not many samples were available, but their spatial distribution covered a wide range of in-day scenarios at each bay. However, most of the samples for CalVal purposes included only the seasons of summer and spring so the application of the models over winter and autumn was subjected to higher uncertainty. Indeed, the range of chl-*a* measured in situ included low number of samples with concentrations under 1 mg/m^3, which are highly representative in the winter season. This fact coupled with the linearity of the developed models, increase the error related to low chl-*a* concentrations, tending more rapidly to negative numbers (e.g., Figure S1b).

4.4. Spatiotemporal chl-*a* Dynamics

The time-series of chl-*a* covered all year 2017, including the different channel stages at both Fangar and Alfacs bays. Overall, the temperature increased from winter (T_{mean} ~ 14.4 °C) to summer (T_{mean} ~ 27.7 °C), and in autumn, the temperature (T_{mean} ~ 22.5 °C) was similar, even higher, than in spring (T_{mean} ~ 20.1 °C). The most frequent winds during the year came from the NW sector, predominantly in the morning, with strong influence of southern winds (spring and summer), switching to SSE (winter and summer) and to SSW (spring and autumn) from noon onward. Within the dates included in this study, highest intensities were registered in March and December, both related to direction of 300–360°

(NW and NNW). The rainiest month along year was January (72.4 mm in seven days), followed by March (36.5 mm in 14 days), and within the days included in the time-series, it rained on 4 August between 4:30 and 5:00 am (accumulated precipitation of 0.1 mm) and on 23 September from 4:30–6:30 am (accumulated precipitation of 0.3 mm).

Regarding the variability of chl-*a* inside each bay (in terms of CV), Fangar Bay showed higher heterogeneity along time, but more homogeneity along space than Alfacs Bay. Fangar Bay is smaller and shallower, which makes it more susceptible to environmental variations, making the changes faster and affecting more of the bay's area. In this study, it has been observed that in Fangar maximum variations were associated to more energetic areas with more chl-*a* (mouth), while in Alfacs, higher variabilities were associated to less energetic areas where chl-*a* dynamics depend largely on the wind-driven mixing (inner NE area). These findings are related with the renewal time of the bays (higher in Alfacs) and linked to the capacity for developing larger phytoplankton populations (higher chl-*a* concentrations). In this sense, Alfacs Bay characteristics (larger and deeper bay more perpendicular to NW and N winds with higher water residence times) allow nutrients to sink and get stored in the sediment of the bay and, at the same time, allow them to be released and suspended during more time (increased nutrient availability for phytoplankton). Conversely, quicker changes in Fangar Bay make chl-*a* to be diluted faster by the Mediterranean water inputs (less productive waters).

In relation to chl-*a* concentrations, besides the seasonal temperature-driven dynamics, wind was the environmental parameter more related to the maximum variations of chl-*a* inside the bays. In terms of temporal dynamics, overall, chl-*a* increased more with prolonged NW and N strong winds episodes occurred in 17 March both bays (Figure S1e,f), 25 July both bays (Figure S1u,v), 13 September Fangar Bay (Figure S1ab), 23 October both bays (Figure S1ag,ah), and 12 November Alfacs Bay (Figure S1ai). The highest accumulations of chl-*a* at both bays occurred on March (Figure S1e,f), October (Figure S1ae–ah) and early November (Figure S1ai), when channels were closed or semi-closed. Conversely, weaker winds from southern components were related to decreases in chl-*a* concentrations as happened in 15 July both bays (Figure S1s,t), 23 September both bays (Figure S1ac,ad) and 22 November both bays (Figure S1aj,ak). Reduction of chl-*a* concentration in both bays was enhanced after rainy events as in 23 September (Figure S1ac,ad) and 22 November (Figure S1aj,ak). These results suggest that wind plays a major role in the nutrient load of the water column. On one hand, mixing the water re-suspending the sediment, thus making the nutrients available (wind-induced turbulence). On the other, enhancing water renewal which increases flushing cells ratio and does not allow time enough for development of large phytoplankton populations [49] (wind-enhanced circulation). Therefore, higher chl-*a* concentrations are expected to occur when the estuarine circulation is weakened and the turbulence increases. This effect might be enhanced at the end or after drainage of the irrigation channels stage (August–November) which increase the nutrients stored in the bays. In the time-series presented in this study, this occurred in August at both bays (Figure S1w–aa) and 23 October both bays (Figure S1ag, ah), especially when winds blow from land (NW and N). In general, the observed trend was in agreement with previous studies [49], which found high chl-*a* concentrations of chl-*a* in October in Alfacs Bay and high concentrations from July to November in Fangar Bay.

In terms of spatial dynamics (in-day scenarios), high chl-*a* concentrations were related more frequently to the mouth area of both bays. There, the exchange with the Mediterranean Sea leads to a more instable scenario in which, despite water renewal might be higher, increased turbulence favors phytoplankton growth prevailing over the wind's regime. High concentrations in the mouth of Fangar Bay were related with more chl-*a* within the central channel of the bay (northern face of rafts), while in Alfacs Bay there was not so clear relation. In this bay, highest chl-*a* concentrations were also found in the inner area (NE), which is more retentive and concentrate more nutrients [5].

Because of all the aforementioned, shellfish aquaculture in the bays benefits from increased chl-*a* concentrations compared to the open sea. However, the retentiveness that characterizes the bays become a double-edged sword due to the high temperatures that water reaches during summer (>30 °C), which negatively affects the feeding rate of shellfish, becoming lethal when temperatures above 28 °C are maintained for days [41]. In order to develop a feasible method for aquaculture management by means of remote sensing monitoring, temperature must be included as one of the main factors, together with chl-*a*, controlling shellfish growth and conditioning the sustainability of the cultures.

In this article, the first results have been presented, and measures to enhance aquaculture can be proposed. However, the feasibility of implementing them is subjected to the availability of bio-geophysical models considering longer time-series, which would allow to make a more integrated and robust approach. Including more data (parameters considered, increased number of data, wider dynamic range) and integrating them into the models would lead to carry out studies in line with [27,30], which coupled remote sensed chl-*a* with other environmental parameters to establish shellfish farming suitability index, to determinate the load capacity of the bays, and to rezoning the rafts' locations.

5. Conclusions

Moderate spatial resolution (10–60 m^2) Sentinel 2 imagery offers a new opportunity for remote sensing of water quality at small coastal geographic areas. In the Ebro Delta bays, the main Spanish Mediterranean shellfish production site, Sentinel 2 imagery has demonstrated the potential to become a suitable tool for resolving the fast dynamics of phytoplankton in the area (in terms of chl-*a* concentration), within short space and time-frames. The monitoring using satellite remote sensing improves the standard in situ sampling-based methodology, allowing moving from punctual to full coverage, thus enabling holistic analyses (time-series) to enhace coastal management (e.g., aquaculture).

After testing different levels of atmospheric correction, it is not feasable to use uncorrected atmosphere images (TOA), but the full correction of the atmosphere is still highly uncertain. The results obtained suggest the use of Rayleight corrected Sentinel 2 imagery together with a simple Blue-to-Green ratio for chl-*a* estimation, until full correction is completely solved/validated. With this configuration, APD < 10% and MAE < 0.7 mg/m^3 were achieved, being able to derive credible chl-*a* maps of the bays, including the preservation of some information within the rafts polygons.

Despite the aforementioned success, remote sensing of small complex coastal geographic areas still faces several challenges. The main limitations found in this study were (i) full atmospheric correction accounting for aerosol, Rayleigh, sun glint, and adjacency effects and (ii) uncertainties associated to shallower areas contaminated by bottom reflectance, contributions of seagrasses to the total chl-*a* concentration, and validity of the results out the range of derivation of the model (location of ground truth data, wider range of chl-*a* concentrations, and seasonality). Further research should be directed to solve these shortcommings by improving the atmospheric correction and gathering more field data covering higher number of scenarios. With these, a sensivity test should be conducted for algorithm bounding, and, ideally, specific tunned models should be developed for each scenario (bay/season/water optical properties).

Supplementary Materials: The following are available online at http://www.mdpi.com/2072-4292/11/15/1756/s1, Table S1: Model performance (in situ chl-a range and *N*, spectral algorithms' range, intercept, slope, RMSE, APD, Pearson's *r*, R^2, AIC and BIC) with all variable combinations (chl-*a* method, bay, atmospheric correction level and spectral algorithm) using all available in situ chl-*a* data. Table S2: Model performance (in situ chl-a range and *N*, spectral algorithms' range, intercept, slope, RMSE, APD, Pearson's *r*, R^2, AIC and BIC) with all variable combinations (chl-*a* method, bay, atmospheric correction level and spectral algorithm) using only the samples for which chl-*a* was measured by the three methodologies (in vivo, FL and SP). Figure S1: One year (2017) chl-*a* time-series at Fangar and Alfacs bay generated with the model with which better performance was achieved (Blue-to-Green ratio on Rayleigh corrected S2A images calibrated with chl-*a* measured with spectrophotometer after acetone extraction).

Author Contributions: Conceptualization, J.S.-G. and E.A.; Data curation, M.F.-T.; Formal analysis, J.S.-G. and C.A.; Methodology, E.A. and C.A.; Supervision, E.A. and J.D.; Validation, C.A.; Writing—original draft, J.S.-G.; Writing—review & editing, E.A., M.F.-T. and C.A.

Funding: This research was funded by Agència de Gestió d'Ajuts Universitaris i de Recerca (grant number 2018 FI_B00705), the European Maritime and Fisheries Fund (ARP005/17/00076) and Fisheries Directorate from Generalitat de Catalunya.

Acknowledgments: The authors thank the IRTA staff for providing support during the water sampling and chlorophyll analysis.

Conflicts of Interest: The authors declare no conflict of interest.

References

1. Duarte, P.; Fernández-Reiriz, M.J.; Labarta, U. Modelling Mussel Growth in Ecosystems with Low Suspended Matter Loads Using a Dynamic Energy Budget Approach. *J. Sea Res.* **2012**, *67*, 44–57. [CrossRef]

2. Ramón, M.; Cano, J.; Peña, J.B.; Campos, M.J. Current Status and Perspectives of Mollusc (Bivalves and Gastropods) Culture in the Spanish Mediterranean. *Boletín Inst. Español Oceanogr.* **2005**, *21*, 361–373.

3. Prat, N.; Muñoz, I.; Camp, J.; Comin, F.A.; Lucena, J.R.; Romero, J.; Vidal, M. Seasonal Changes in Particulate Organic Carbon and Nitrogen in the River and Drainage Channels of the Ebro Delta (N.E. Spain). *SIL Proc.* **1988**, *23*, 1344–1349. [CrossRef]

4. Llebot, C.; Rueda, F.J.; Solé, J.; Artigas, M.L.; Estrada, M. Hydrodynamic States in a Wind-Driven Microtidal Estuary (Alfacs Bay). *J. Sea Res.* **2014**, *85*, 263–276. [CrossRef]

5. Artigas, M.L.; Llebot, C.; Ross, O.N.; Neszi, N.Z.; Rodellas, V.; Garcia-Orellana, J.; Masqué, P.; Piera, J.; Estrada, M.; Berdalet, E. Understanding the Spatio-Temporal Variability of Phytoplankton Biomass Distribution in a Microtidal Mediterranean Estuary. *Deep Res. Part II Top. Stud. Oceanogr.* **2014**, *101*, 180–192. [CrossRef]

6. Camp, J.; Delgado, M. Hidrografía de Las Bahías Del Delta Del Ebro. In *Investigación Pesquera*; Instituto de Ciencias del Mar: Barcelona, Spain, 1987; pp. 351–369.

7. Forget, M.-H.; Stuart, V.; Platt, T. Reports and Monographs of the International Ocean-Colour Coordinating Group Remote Sensing in Fisheries and Aquaculture. *Aquaculture* **2009**, *1998*, 1–128.

8. Garcia, L.E.; Rodriguez, D.J.; Wijen, M.; Pakulski, I. (Eds.) *Earth Observation for Water Resources Management: Current Use and Future Opportunities for the Water Sector*; World Bank Group: Washington, DC, USA, 2016.

9. Gregor, J.; Maršálek, B. Freshwater Phytoplankton Quantification by Chlorophyll a: A Comparative Study of in Vitro, in Vivo and in Situ Methods. *Water Res.* **2004**, *38*, 517–522. [CrossRef] [PubMed]

10. Del López-Rodríguez, M.C.; Leira, M.; Valle, R.; Moyà-Niell, G. El Fitoplancton Como Indicador de Calidad de Masas de Agua Muy Modificadas En La DMA. El Lago Artificial de As Pontes (A Coruña. España). *Nov. Acta Cient. Compostel.* **2016**, *23*, 85–97.

11. Kutser, T.; Paavel, B.; Verpoorter, C.; Ligi, M.; Soomets, T.; Toming, K.; Casal, G. Remote Sensing of Black Lakes and Using 810 Nm Reflectance Peak for Retrieving Water Quality Parameters of Optically Complex Waters. *Remote Sens.* **2016**, *8*, 497. [CrossRef]

12. Gurlin, D.; Gitelson, A.A.; Moses, W.J. Remote Estimation of Chl-a Concentration in Turbid Productive Waters-Return to a Simple Two-Band NIR-Red Model? *Remote Sens. Environ.* **2011**, *115*, 3479–3490. [CrossRef]

13. Blondeau-Patissier, D.; Gower, J.F.R.; Dekker, A.G.; Phinn, S.R.; Brando, V.E. A Review of Ocean Color Remote Sensing Methods and Statistical Techniques for the Detection, Mapping and Analysis of Phytoplankton Blooms in Coastal and Open Oceans. *Prog. Oceanogr.* **2014**, *123*, 23–144. [CrossRef]

14. Matthews, M.W. A Current Review of Empirical Procedures of Remote Sensing in Inland and Near-Coastal Transitional Waters. *Int. J. Remote Sens.* **2011**, *32*, 6855–6899. [CrossRef]

15. Gholizadeh, M.; Melesse, A.; Reddi, L. A Comprehensive Review on Water Quality Parameters Estimation Using Remote Sensing Techniques. *Sensors* **2016**, *16*, 1298. [CrossRef] [PubMed]

16. Volpe, G.; Santoleri, R.; Vellucci, V.; Ribera d'Alcalà, M.; Marullo, S.; D'Ortenzio, F. The Colour of the Mediterranean Sea: Global versus Regional Bio-Optical Algorithms Evaluation and Implication for Satellite Chlorophyll Estimates. *Remote Sens. Environ.* **2007**, *107*, 625–638. [CrossRef]

17. Campbell, J.W.; O'Reilly, J.E. Metrics for Quantifying the Uncertainty in a Chlorophyll Algorithm: Explicit Equations and Examples Using the OC4.v4 Algorithm and NOMAD Data. *Ocean Color Bio-Opt. Algorithm Mini-Workshop* **2006**, *4*, 1–15.

18. Gitelson, A.A.; Yacobi, Y.Z.; Karnieli, A.; Nurit, K. Reflectance Spectra of Polluted Marine Waters in Haifa Bay, Southeastern Mediterranean:Features and Application for Remote Estimation of Chlorophyll Concentration. *J. Earth Sci.* **1996**, *45*, 127–136.

19. Odermatt, D.; Gitelson, A.; Brando, V.E.; Schaepman, M. Review of Constituent Retrieval in Optically Deep and Complex Waters from Satellite Imagery. *Remote Sens. Environ.* **2012**, *118*, 116–126. [CrossRef]

20. Le, C.; Hu, C.; Cannizzaro, J.; English, D.; Muller-Karger, F.; Lee, Z. Evaluation of Chlorophyll-a Remote Sensing Algorithms for an Optically Complex Estuary. *Remote Sens. Environ.* **2013**, *129*, 75–89. [CrossRef]
21. Gitelson, A.A.; Dall'Olmo, G.; Moses, W.; Rundquist, D.C.; Barrow, T.; Fisher, T.R.; Gurlin, D.; Holz, J. A Simple Semi-Analytical Model for Remote Estimation of Chlorophyll-a in Turbid Waters: Validation. *Remote Sens. Environ.* **2008**, *112*, 3582–3593. [CrossRef]
22. Gitelson, A.A.; Kondratyev, K.Y. Optical Models of Mesotrophic and Eutrophic Water Bodies. *Int. J. Remote Sens.* **1991**, *12*, 373–385. [CrossRef]
23. Oliveira, E.N.; Fernandes, A.M.; Kampel, M.; Cordeiro, R.C.; Brandini, N.; Vinzon, S.B.; Grassi, R.M.; Pinto, F.N.; Fillipo, A.M.; Paranhos, R. Assessment of Remotely Sensed Chlorophyll- a Concentration in Guanabara Bay, Brazil. *J. Appl. Remote Sens.* **2016**, *10*, 026003. [CrossRef]
24. Gower, J.F.R.; Doerffer, R.; Borstad, G.A. Interpretation of the 685nm Peak in Water-Leaving Radiance Spectra in Terms of Fluorescence, Absorption and Scattering, and Its Observation by MERIS. *Int. J. Remote Sens.* **1999**, *20*, 1771–1786. [CrossRef]
25. Matthews, M.W.; Bernard, S.; Robertson, L. An Algorithm for Detecting Trophic Status (Chlorophyll-a), Cyanobacterial-Dominance, Surface Scums and Floating Vegetation in Inland and Coastal Waters. *Remote Sens. Environ.* **2012**, *124*, 637–652. [CrossRef]
26. Joshi, I.D.; D'Sa, E.J.; Osburn, C.L.; Bianchi, T.S. Turbidity in Apalachicola Bay, Florida from Landsat 5 TM and Field Data: Seasonal Patterns and Response to Extreme Events. *Remote Sens.* **2017**, *9*, 367. [CrossRef]
27. Sutherland, T.F.; Leonard, C.; Taylor, F.J.R. A Segmented Pipe Sampler for Integrated Profiling of the Upper Water Column. *J. Plankton Res.* **1992**, *14*, 915–923. [CrossRef]
28. Lorenzen, C.J. A Method for the Continuous Measurement of in Vivo Chlorophyll Concentration. *Deep Res.* **1996**, *13*, 223–227. [CrossRef]
29. Yentsch, C.S.; Menzel, D.W. A Method for the Determination of Phytoplankton Chlorophyll and Phaeophytin by Fluorescence. *Deep Res. Oceanogr. Abstr.* **1963**, *10*, 221–231. [CrossRef]
30. Jeffrey, S.W.; Humphrey, G.F. New Spectrophotometric Equations for Determining Chlorophylls a, b, C1 and C2 in Higher Plants, Algae and Natural Phytoplankton. *Biochem. Physiol. Pflanz.* **1975**, *167*, 191–194. [CrossRef]
31. Vanhellemont, Q.; Ruddick, K. Atmospheric Correction of Metre-Scale Optical Satellite Data for Inland and Coastal Water Applications. *Remote Sens. Environ.* **2018**, *216*, 586–597. [CrossRef]
32. Harmel, T.; Chami, M.; Tormos, T.; Reynaud, N.; Danis, P.A. Sunglint Correction of the Multi-Spectral Instrument (MSI)-SENTINEL-2 Imagery over Inland and Sea Waters from SWIR Bands. *Remote Sens. Environ.* **2018**, *204*, 308–321. [CrossRef]
33. Morel, A.; Prieur, L. Analysis of Variations in Ocean Color. *Limnol. Oceanogr.* **1977**, *22*, 709–722. [CrossRef]
34. Gitelson, A.; Nikanorov, A.; Szabo, G.; Szilagyi, F. Etude de La Qualite Des Eaux de Surface Télédétéction. In *Monitoring to Detect Chamges in Water Quality Series*; IAHS Publication: Budapest, Hungary, 1986; Volume 157, pp. 111–121.
35. Lins, R.; Martinez, J.M.; Motta Marques, D.; Cirilo, J.; Fragoso, C. Assessment of Chlorophyll-a Remote Sensing Algorithms in a Productive Tropical Estuarine-Lagoon System. *Remote Sens.* **2017**, *9*, 516. [CrossRef]
36. Mishra, S.; Mishra, D.R. Normalized Difference Chlorophyll Index: A Novel Model for Remote Estimation of Chlorophyll-a Concentration in Turbid Productive Waters. *Remote Sens. Environ.* **2012**, *117*, 394–406. [CrossRef]
37. Dall'Olmo, G.; Gitelson, A.A. Effect of Bio-Optical Parameter Variability on the Remote Estimation of Chlorophyll-a Concentration in Turbid Productive Waters: Experimental Results. *Appl. Opt.* **2005**, *44*, 412. [CrossRef]
38. Seegers, B.N.; Stumpf, R.P.; Schaeffer, B.A.; Loftin, K.A.; Jeremy Werdell, P. Performance Metrics for the Assessment of Satellite Data Products: An Ocean Color Case Study. *Opt. Express* **2018**, *26*, 7404–7422. [CrossRef]
39. Cannizzaro, J.P.; Carder, K.L. Estimating Chlorophyll a Concentrations from Remote-Sensing Reflectance in Optically Shallow Waters. *Remote Sens. Environ.* **2006**, *101*, 13–24. [CrossRef]
40. Gons, H.J.; Auer, M.T.; Effler, S.W. MERIS Satellite Chlorophyll Mapping of Oligotrophic and Eutrophic Waters in the Laurentian Great Lakes. *Remote Sens. Environ.* **2008**, *112*, 4098–4106. [CrossRef]

41. Ramón, M.; Fernández, M.; Galimany, E. Development of Mussel (Mytilus Galloprovincialis) Seed from Two Different Origins in a Semi-Enclosed Mediterranean Bay (N.E. Spain). *Aquaculture* **2007**, *264*, 148–159. [CrossRef]

42. Dogliotti, A.; Ruddick, K. Improving Water Reflectance Retrieval from MODIS Imagery in the Highly Turbid Waters of La Plata River. In Proceedings of the VI International Conference in Current Problems in Optics of Natural Waters, St. Petersburg, Russia, 10 September 2011.

43. Caballero, I.; Steinmetz, F.; Navarro, G. Evaluation of the First Year of Operational Sentinel-2A Data for Retrieval of Suspended Solids in Medium- to High-Turbiditywaters. *Remote Sens.* **2018**, *10*, 982. [CrossRef]

44. Novoa, S.; Doxaran, D.; Ody, A.; Vanhellemont, Q.; Lafon, V.; Lubac, B.; Gernez, P. Atmospheric Corrections and Multi-Conditional Algorithm for Multi-Sensor Remote Sensing of Suspended Particulate Matter in Low-to-High Turbidity Levels Coastal Waters. *Remote Sens.* **2017**, *9*, 61. [CrossRef]

45. Sei, A. Efficient Correction of Adjacency Effects for High-Resolution Imagery: Integral Equations, Analytic Continuation, and Padé Approximants. *Appl. Opt.* **2015**, *54*, 3748. [CrossRef]

46. Gons, H.J.; Rijkeboer, M.; Ruddick, K.G. A Chlorophyll-Retrieval Algorithm for Satellite Imagery (Medium Resolution Imaging Spectrometer) of Inland and Coastal Waters. *J. Planktont Res.* **2002**, *24*, 947–951. [CrossRef]

47. Gernez, P.; Doxaran, D.; Barillé, L. Shellfish Aquaculture from Space: Potential of Sentinel2 to Monitor Tide-Driven Changes in Turbidity, Chlorophyll Concentration and Oyster Physiological Response at the Scale of an Oyster Farm. *Front. Mar. Sci.* **2017**, *4*, 1–15. [CrossRef]

48. Busch, J.A. Phytoplankton Dynamics and Bio-Optical Variables Associated with Harmful Algal Blooms in Aquaculture Zones. Ph.D. Thesis, Universität Bremen, Bremen, Germany, 2013.

49. Llebot, C.; Solé, J.; Delgado, M.; Fernández-Tejedor, M.; Camp, J.; Estrada, M. Hydrographical Forcing and Phytoplankton Variability in Two Semi-Enclosed Estuarine Bays. *J. Mar. Syst.* **2011**, *86*, 69–86. [CrossRef]

Technical Note

A Preliminary Study of Wave Energy Resource Using an HF Marine Radar, Application to an Eastern Southern Pacific Location: Advantages and Opportunities

Valeria Mundaca-Moraga [1,*], Rodrigo Abarca-del-Rio [2,3], Dante Figueroa [2,3] and James Morales [1]

[1] Graduate Program in Energy, Faculty of Engineering, Universidad de Concepción, Concepción 4030000, Chile; jmoralesl@udec.cl
[2] Geophysics Department, Universidad de Concepción, Casilla 160-C, Concepción 4030000, Chile; roabarca@udec.cl (R.A.-d.-R.); dantefigueroa@udec.cl (D.F.)
[3] Chilean Integrated Ocean Observing System, Universidad de Concepción, Concepción 4030000, Chile
* Correspondence: valemundaca@udec.cl

Abstract: As climate change is of global concern, the electric generation through fossil fuel is progressively shifted to renewable energies. Among the renewables, the most common solar and wind, the wave energy stands for its high-power density. Studies about wave energy resource have been increasing over the years, especially in coastal countries. Several research investigations have assessed the global wave power, with higher values at high latitudes. However, to have a precise assessment of this resource, the measurement systems need to provide a high temporal and spatial resolution, and due to the lack of in-situ measurements, the way to estimate this value is numerical. Here, we use a high-frequency radar to estimate the wave energy resource in a nearshore central Chile at a high resolution. The study focuses near Concepción city (36.5° S), using a WERA (WavE RAdar) high frequency (HF) radar. The amount of annual energy collected is calculated. Analysis of coefficient of variation (COV), seasonal variability (SV), and monthly variability (MV) shows the area's suitability for installing a wave energy converter device due to a relatively low variability and the high concentration of wave power obtained. The utility of HF radars in energy terms relies on its high resolution, both temporal and spatial. It can then compare the location of interest within small areas and use them as a complement to satellite measurements or numerical models, demonstrating its versatility.

Keywords: remote sensing; HF marine radars; wave energy

Citation: Mundaca-Moraga, V.; Abarca-del-Rio, R.; Figueroa, D.; Morales, J. A Preliminary Study of Wave Energy Resource Using an HF Marine Radar, Application to an Eastern Southern Pacific Location: Advantages and Opportunities. *Remote Sens.* **2021**, *13*, 203. https://doi.org/10.3390/rs13020203

Received: 21 November 2020
Accepted: 5 January 2021
Published: 8 January 2021

Publisher's Note: MDPI stays neutral with regard to jurisdictional claims in published maps and institutional affiliations.

1. Introduction

Worldwide marine energies have been widely studied. Recent work [1–5] establishes a more significant wave potential at higher latitudes. The most favored coasts being, to name only a few, Australia, New Zealand, South Africa, Chile, the western U.K., and Canada. Wave devices are currently lagging tidal in terms of technological development. As many as 170 types of wave energy converter have been designed, fewer than 20% are at the full-scale prototype stage [6]. By 2017, the world's installed capacity for marine energy was 536 MW, compared to 267 MW in 2007 [7]. However, Ocean Energy Europe has projected at a low grown scenario that 1300 MW of tidal energy and 170 MW of wave energy could be installed by 2030 [8].

Chile is one of the countries with the most extensive maritime territory, with a coastline that extends from 19° S to 56° S, positioning it as one of the best countries to develop wave energy. Marine gravity waves are considered one of the coastal ocean's essential features due to their impact on security issues such as harbor activities, traffic planning, beach management, and energy potential. It is vital to have constant monitoring of the ocean, especially when a strong presence of coastal industrial and residential activity is present.

445

Several methodologies, numerical or instrumental, provide relevant information of waves, either in-situ (buoys), remotely (radars, satellites), or through computational models—the set of methodologies and instruments constituting an integrated observation system [9]. Remote instruments deliver information from the study area on a large scale through data acquired by a device that is not in contact with what is being investigated [10]. It is of great importance for ocean data since the environment can be quite hostile and ample to be covered with in-situ instruments. The most used are satellites that cover a large part of the Earth, and for which information is available since the early 90s [11].

Other instruments of remote sensing are the high-frequency marine radars. They are located on the coastline and study oceanographic variables providing high spatial and temporal coverage. These systems' applicability includes port security, tsunami detection, renewable energy, storm surges, and flooding, among others [12–17]. The advantage of using HF radars instead of instruments such as buoys or satellites is their high spatial coverage. It is also possible to complement satellites' information with the radar information (which does not provide quality measurements near the coastal area [18]). The HF radar range can be up to 100 km from the coast.

The best option to measure wave by HF radars is through a dual system. Two radars sweep a shared area, illuminating each surface water parcel from two directions, thus avoiding ambiguities when measuring the wave height [19]. The same work compares data from single and dual HF radar systems: It shows that the dual radar algorithm is significantly more accurate and reliable. One consideration to have in mind is that single or dual radar measurements tend to be less accurate in high sea states because of limitations in the backscatter model in these conditions [19]. More precisely, Ref. [20] shows that the HF radar wave height's accuracy is around 12.6%. However, one radar does obtain significant wave height with a range of approximately 30 km from the coast [12,13,21]. A single radar can be used in particular applications to look directly into the operationally-important wave propagation [20] and always as a complement to other instruments.

The use of HF marine radars is not uncommon in first-world countries (United States, Europe, Australia, and Japan [22]. In other countries, however, their use is somewhat limited. For example, Mexico and Chile use these systems in the rest of the American continent [23] (see Section 2.1).

Considering the wave energy opportunities in Chile, this work aims to present a preliminary study of an HF marine radar application. Our goal is to assess the wave energy potential and prove the usefulness of this instrument to establish the wave energy resource of a specific location. This way, we can provide a high-resolution grid to determine specific hotspots that could optimize the available energy harness.

2. Materials and Methods

2.1. Study Area and Data

2.1.1. Study Area

The study site is located in south-central Chile's coastal area facing the Pacific Ocean (Figure 1). The continental shelf is approximately 100 m deep and extends about 60 km west to 73.5° W. The red dot in Figure 1 shows the location of the radar used in the present study. Information from a global climatology [24], as specific data from the zone [25], indicates that the wave height has a uniform spatial pattern with average values close to 3 m and lowest variability [26] and typical periods between 9 and 12 s [25]. The winds in this area come predominantly from the southwest [25], impacting the wave's direction.

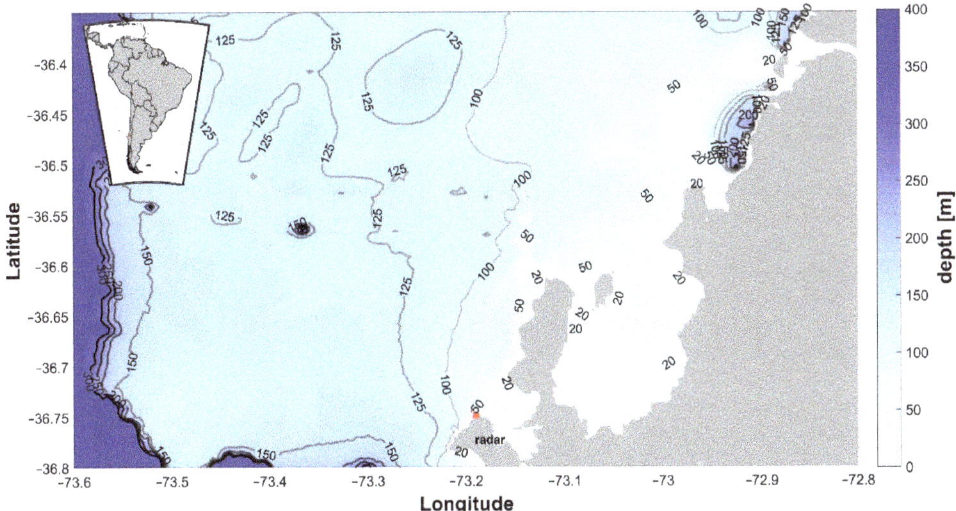

Figure 1. Area of interest; bathymetry and HF marine radar location.

2.1.2. HF Marine Radar

Wave measurements in front of the Bío-Bío region were made by a WERA HF radar [12], working at a frequency of 16.15 MHz. It is located in the "Faro Hualpén" Bío-Bío region (36.74° S–73.19° W). The system has four (4) transmitting and eight (8) receiving antennas. It has a sampling frequency of 33 s. The maximum range is 100 km, with an aperture of +/−50° spatial resolution of 2 km and an azimuthal resolution of 4°.

The waves were measured in a regular grid of 2 km resolution and a temporal interval of 1 h. For this study, we used approximately one year of measurements between December 2017 and January 2019.

This radar is part of the ocean observation system CHIOOS (www.chioos.cl) in charge of the Department of Geophysics (DGEO) of the University of Concepción (UDEC). It is the only HF radar in the entire southeast Pacific coast [23] and, together with those located in Australia [22], to the best of our knowledge, the only ones in the South Pacific.

2.2. Methodology

2.2.1. Data Processing

The radar sweep area is not constant throughout the study period. There were times when no measurement took place, having more of fewer grid cell to be studied. Although the same area is always covered, the amount of spatial data is not the same. Moreover, the further away from the radar, the lower the spatial coverage and the quality of the data. Although the radar area varies at each measurement time, the measured points are grouped in the same area and not scattered throughout the sweep. It generates both spatial and temporal discontinuity in the data. Figure 2 display a flow chart that resumes the data treatment process. It is divided into 3 main parts. The 1st represents the initial dataset state. The 2nd division is the data processing performed; this involves 3 stages (A, B, and C). The first part was the suppression of extreme elements (A), two interpolation parts (B: 2D and C: 1D), and the 3rd the dataset obtained.

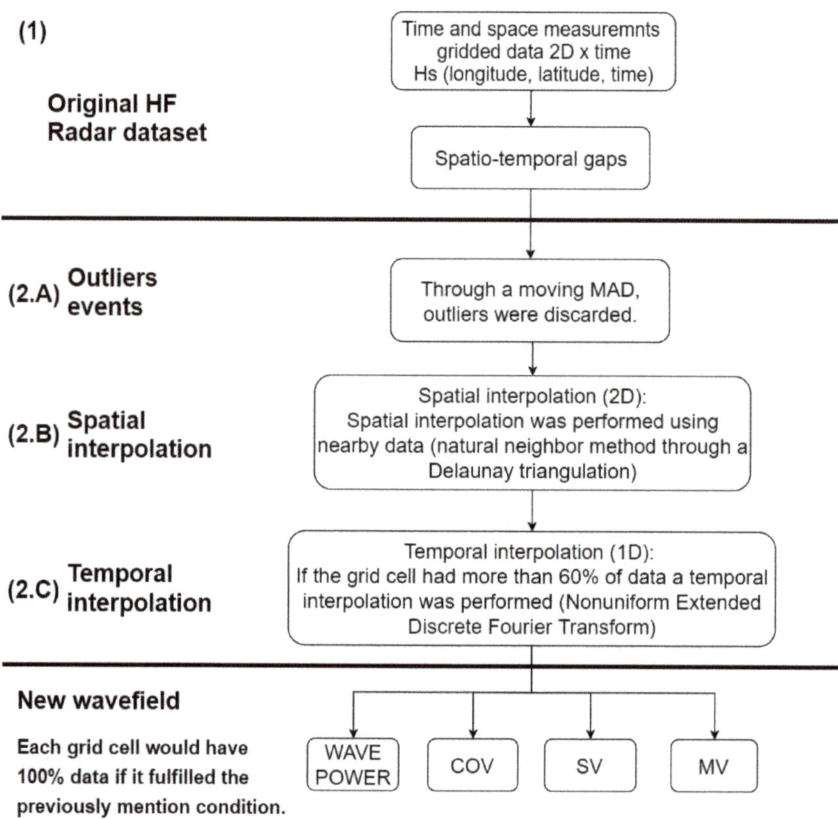

Figure 2. Flow chart with the stages to obtain the wave fields.

Radar measurements showed noise and outliers, with extreme values exceeding 3 or even 5 standard deviations (see Figure 5), exceptionally with hourly waves as high as 10–13 m, even close to the coast. The region's wave height for those dates was verified by checking with the only buoy in the area and analyzing the Chilean navy's information for those times. In the end, for those epochs, the radar reported erroneous peaks. Therefore, to eliminate outliers, the gridded time series were first filtered by a 3-day moving median [27]. All the data above or below the 1.5 MAD were discarded. It generated a cleaner data field and time series without those extreme values (2A stage in Figure 2). Then, in order to have a coherent space-time data field, we performed a spatial-temporal interpolation (see stages 2B and 2C in flow diagram, Figure 2). When the grid point met the natural neighbor 2D interpolation specific applicability conditions (employing a natural neighbor method through a Delaunay triangulation [28]), spatial interpolation was executed. It allowed filling empty grid points with surrounding radar measurements (stage 2B in Figure 2). However, despite the spatial interpolation, there were still grid points with temporal gaps. On average, only 15% of the data was filled because of these restrictions. Then, in those places where there were still missing data, we performed a temporal interpolation applying a Nonuniform Extended Discrete Fourier Transform [29,30]. It is applied only if the grid complies with the condition that there were more than 60% of the data (see stage 2C in Figure 2).

Note that, interpolating solely by 1D (time), crude discontinuities may appear in both time and space. Thus, if each grid cell time series is treated as an independent time series, it creates many spatial discontinuities, which may well not represent the reality of the spatial variability within the area interpolated. Inversely, spatial interpolation followed by 1D (time) interpolation ensures the field variability's spatial continuity is preserved.

The interpolation method, as the methodology of temporal interpolation through different ways, was verified. For example, comparing the interpolation made in the buoy or the radar series that had a greater number of original points. The interpolation technique for missing data that most preserves the data's reality is the Nonuniform Extended Discrete Fourier Transform [29,30] method. Thus, the grid points that did not have data before and after the spatial interpolation were subtracted. Then, the correlation between the original buoy series and the temporal interpolated buoy series was calculated. On the buoy, the correlation was 0.81 and 0.86, respectively. Next, the same test was performed with the radar series that was initially the most complete; 88% of the data and 15 kilometers off the coast (location 73.1665° W, 36.6441° S). In this case, the correlation was 0.864 and 0.858. In summary, this shows that the method of temporal interpolation used is sufficiently robust.

Figure 3 shows the radar's sweep coverage area, between 36.2° S–36.7° S and 73.0° W–73.6° W. The percentage of the radar's original coverage is also shown as a data field. The final percentage after spatial interpolation is shown in contour lines. Blue/red color shows the places with less/more than average data. From the coast (oblique) to 30 km offshore (i.e., the equivalent of 0.27 degrees), it has 60 to 80% spatial coverage (red to orange colors) of the data for one year of study. Further away from this area, it has the worst coverage, between 40 and 10% of data (blueish colors). So we will only focus on the first 30 km.

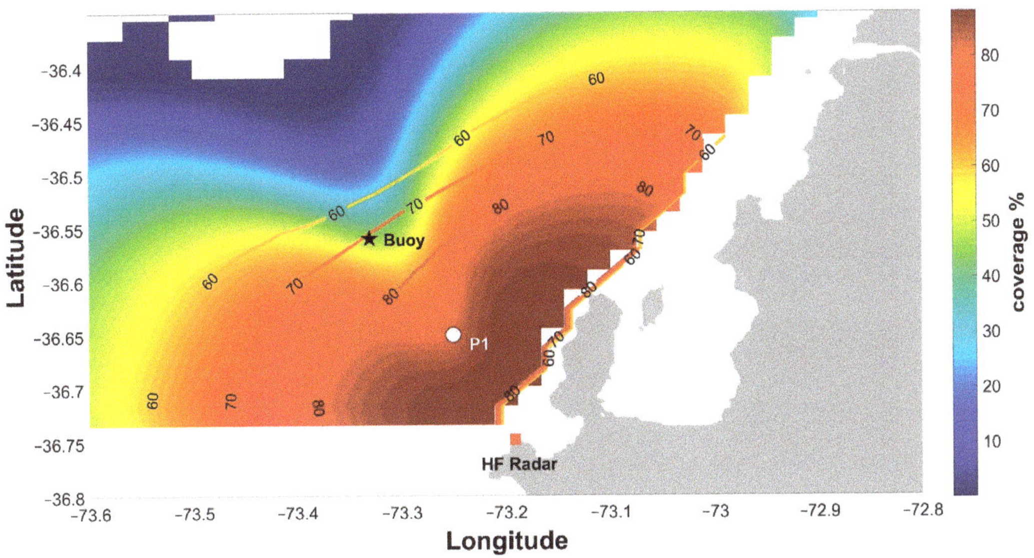

Figure 3. HF radar coverage field for the duration of the study. The contour lines show the coverage percentage after spatial interpolation of the data. The star represents the buoy's location. The white dot represents an analyzed time series location (P1).

The treatment previously mentioned allowed the construction of a regular spatial-temporal grid data set. That is, 20 latitude × 28 longitude points (corresponding to 560 grid points), with a spatial resolution of 2 km and 9444 hourly time points from December 2017 to January 2019). Statistics in Figure 2 are derived from the interpolated fields.

2.2.2. Wave Potential

Once the data set is complete, the wave potential, energy period, coefficient of variation, and seasonal and monthly variability coefficients are calculated [1]. These parameters are essential to establish the locations where the extraction of energy is favorable either by having a high potential or a low variability, or a combination of both [31].

For the calculation of the potential, the following equation in W/m was used, which is valid for deep waters, $h > \frac{\lambda}{2}$ [32,33]

$$P = \rho g \frac{H_s^2 T_e}{64\pi} \tag{1}$$

where ρ is the density of seawater (1027 kg/m^3), g is the acceleration due to gravity (9.81 m/s^2), H_s is the significant wave height in meters, and T_e being the energy period in seconds. It is the period of a monochromatic wave that contains the same mean energy as the rough sea [34]. The HF radar gives the wave height. However, the radar's measurements do not provide all the information required to obtain the energy period.

The wave energy period can be calculated from the wave spectra ($S(f)$) as [34]

$$Te = \frac{m_{-1}}{m_0} \tag{2}$$

where the nth spectral moment is defined as

$$m_n = \int_0^\infty f^n S(f) \, df \tag{3}$$

Ref. [1] specifies that T_e should be estimated from other variables when the spectral form is unknown. Therefore, we will use the peak period (T_p) estimated in previous works [24,25,35], as

$$T_e = 0.9 \, T_p \tag{4}$$

The use of a bibliographic value lies in the lack of correlation between the HF radar and in-situ information (see below Section 3). This way, the wave power calculation would be more representative and robust (this topic is discussed in more detail in Section 4). For this area, the peak period is around 10 s (e.g., Chilean wave atlas [35]). Although the value of T_e is constant and does not vary throughout the year of study, it is expected that the errors associated with the period are not significant compared to the wave height because $P \propto T_e H_s^2$ [1,36]

The coefficient of variation (COV) [1,37] allows investigating the temporal energy variability. It is obtained from the quotient between the standard deviation of the power and the average power

$$COV = \frac{\sigma(P(t))}{\overline{P}} \tag{5}$$

This index gives a guideline on how reliable the study site is. Having a low COV means that the waves' energy density remains constant over time, so the energy extraction will not significantly alter the system [37]. Conversely, high values indicate that the energy is very variable, having sudden changes in a few periods, putting at risk the supply's reliability [1].

To determine how the potential changes during the year, the indices of monthly (MV) and seasonal (SV) variability used [1,26] are defined as

$$MV = \frac{P_{M1} - P_{M2}}{P_{annual}} \tag{6}$$

where P_{M1} and P_{M2} are the average power for the most and least energetic month, respectively, P_{annual} being the annual power average. For the seasonal variability [1], we have

$$SV = \frac{P_{S1} - P_{S2}}{P_{annual}} \tag{7}$$

where P_{S1} is the average power for the more energetic season while P_{S2} for the less energetic season.

As it compares the most energetic months or seasons with the less energetic ones, these indicators also provide information about the energy stability (as performed in [1,26,38]). The monthly and seasonal scales are also essential to study the viability of an energy project (installation, right weather windows for maintenance, etc.). The lower these indexes are, the more stable the studied area is [31].

2.3. Layout

From a data field comprising a total area of 800 km^2 with a very high spatial resolution of 2 km, and temporal resolution of 1 h, we estimate the wave potential in front of the Biobío region. Simultaneously, we perform the following studies to understand wave power performance: Mean, median, NRMSE, NBIAS, COV, SV, MV.

For example, the most energetic location off the coast (73.25° W 36.65° S) is collected and analyzed (see Figure 3). The monthly boxplots (Figure 3) show the average, extreme values, and outliers per month from this time series. Next, the climatology field allows determining maximum and minimum wave height values, seasons, and months with the most significant resource. In that way, it will allow detecting regions close to the coast, where this type of energy is more favorable and viable to harness. Monthly average spatial fields are presented and their respective variability indexes, which relate the maximum and minimum values for each grid point (Figures 4 and 5, respectively).

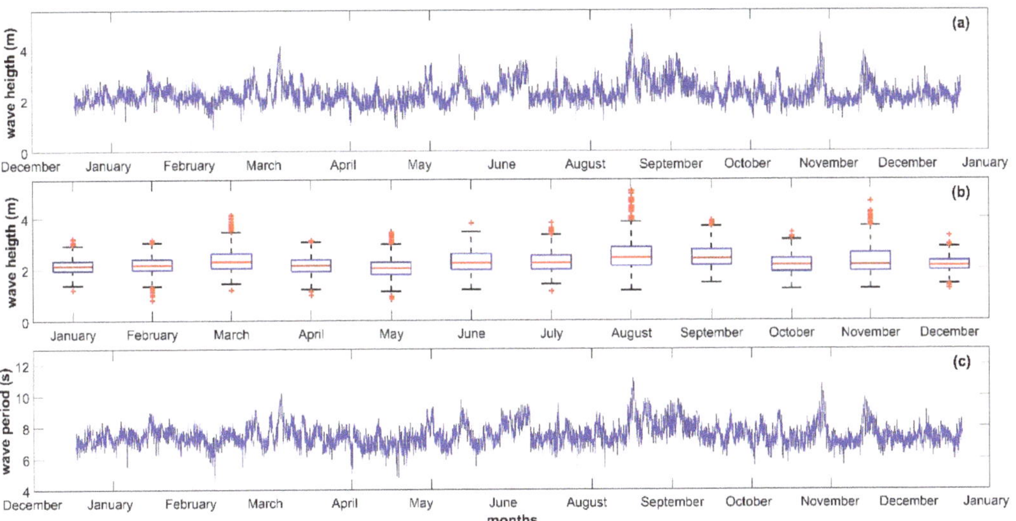

Figure 4. Time series (panel **a**), whisker boxplot (panel **b**) of monthly wave height and wave period (panel **c**) for a reference hotspot location, white dot (P1) in Figure 1. (**b**) The red line in each box is the 50th percentile (median). The blue box represents the interquartile range (IQR) (25th percentile-bottom blue line and 75th percentile-top blue line).

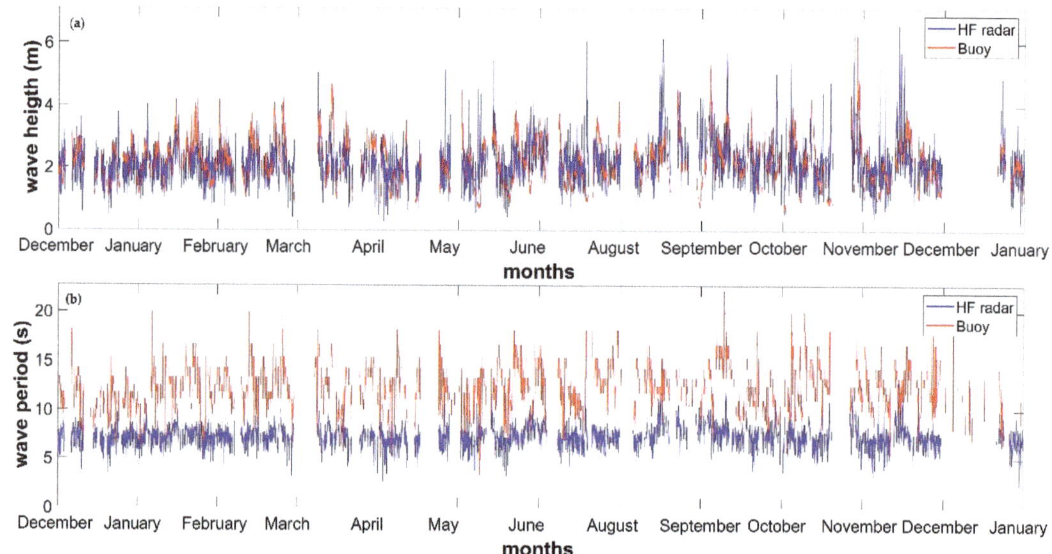

Figure 5. Comparison of HF radar (blue) and buoy (red) time series, (**a**) represents wave height and (**b**) wave period.

3. Results

Figure 4a shows the time series of the wave height for the point P1 of Figure 3, located at 73.25° W and 36.65° S (see Section 2.3). This series' hourly annual mean is 2.5 m, with a standard deviation of 0.5 m (energy of 27.6 kW/m and 1.0 kW/m, respectively). While the annual median is 2.45 m (26.47 kW/m), and 75 and 25% are in the values 2.18 m (21 kW/m) and 2.79 (34 kW/m), respectively. The maximum height for the whole period was 4.96 m (11 August 2018), which corresponds to an energy of 108 kW/m.

The boxplot (Figure 4b) shows that the median varies with the seasons, having its maximum and minimum values in winter and summer, respectively. Summer presents less variability and with more stable values, with minima and maxima of 1.33 m (7.5 kW/m) and 3.54 m (55 kW/m), respectively. Simultaneously for this season, the percentiles show that 75% of summertime wave reaches heights greater than 2.1 m, corresponding to 20 kW/m. During winter, extreme values frequency increases to wave heights higher than 4 m (70 kW/m). The maxima and minima being 4.96 m (108 kW/m) and 1.04 m (5 kW/m), respectively. One can also note that, 75% of the time, it reaches values greater than 2.31 m (23 kW/m) during this season. The wave peak period (Figure 4c) has a mean of 7.8 s and presents its maximum and minimum values in winter and summer.

The amount of energy collected throughout the year for the most energetic sweep point (presented in Figure 1) is 270 MWh/m/year. The lowest as perceived by the radar is 160 MWh/m/year. Finally, a periodogram (not shown) showed amplitudes of the annual and semi-annual cycles being 0.2 and 0.1 m, respectively.

The HF radar raw data (before the outliers' detection method was applied) is compared with the buoy (36.56° S–73.33° W) in Figure 5.

The buoy is quite far from the coast and is not precisely in a radar grid point. It is noticeable that, in that location (see Figure 3), the HF radar presents initially 50% fewer data than the buoy. In Figure 5, gaps are shown in both time series to demonstrate how damaging data loss can be. Thus, finally, our study only investigated grid cells where raw gaps do not exceed 40%. However, even if that place is out of the correct radar cover zone, although some extreme variation in the buoy data not captured in the HF radar, the main frequencies (semi-annual, seasonal) amplitudes agree remarkably. Figure 5b shows the time series of the period measured by radar and buoy. While the values show a similar trend, the magnitude of both does not match. Despite the previous mention, the buoy's statistics are still equivalent to that of the chosen grid analyzed in the field. The annual mean is 2.26 m for radar and 2.12 m for buoy, with a standard deviation of 0.71 and 0.66 m, respectively. The index results for this grid cell are COV 0.9 for HF radar and 0.76 for buoy; SV is 0.1 for HF radar and 0.05 for buoy; MV is 0.16 for HF radar and 0.12 for the buoy. The mean wave power was also estimated, resulting from 23.8 kW/m for the radar data and 31.6 kW/ for the buoy. The normalized RMS is 0.27, and the normalized BIAS is 1.3.

The spatial analysis shows that the monthly averages shown in Figure 6 present a wave power variation between 10 and 40 kW per meter of the coast for the less and more energetic months. The maximum and minimum values, as expected, are obtained for the winter and summer months, respectively. In summer (December–February), the field values vary between 15 and 25 kW/m. The least energetic month is January, with a constant 20 kW/m. In winter (June–August), the values are between 20 and 40 KW/m, approximately twice as much as summer. August and September are the months in which more energy can be obtained, having values over 28 kW/m for approximately all the radar sweep.

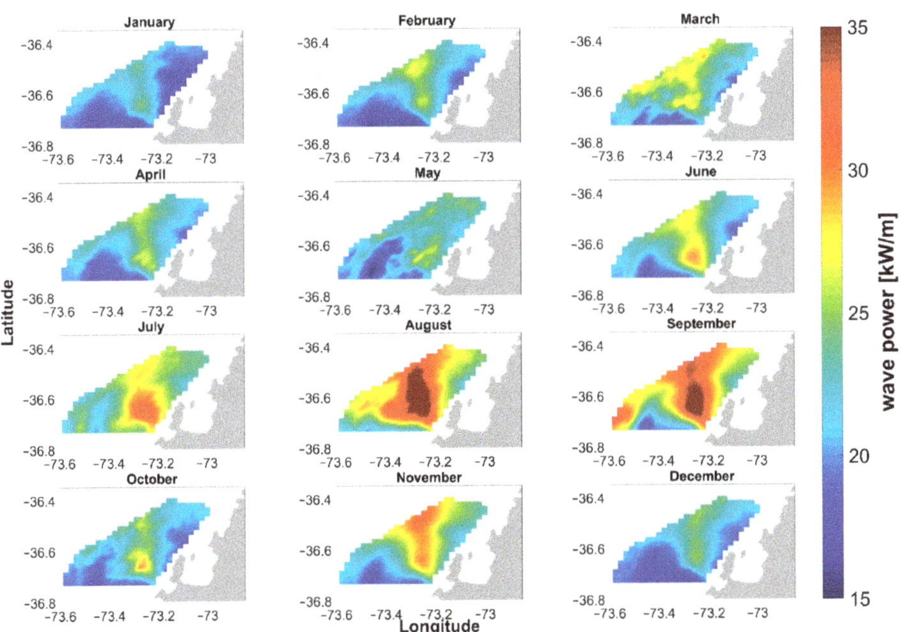

Figure 6. Wave potential monthly mean fields.

Figure 7 shows both the annual accumulation and the coefficient of variation (COV, see Equation (5)) for the study area. The COV presents values between 0.3 and 0.5, with the highest values along the latitude 36.7°S, which coincides with one of the radar sweeps' edges. In turn, the indices of seasonal and monthly variability were also calculated (see Equations (6) and (7)), which presented values between 0.06–0.4 and 0.1–0.6. The COV values are relatively small compared to the other indices (SV and MV), which indicates the wave field has a small variability, and the values are close to the average.

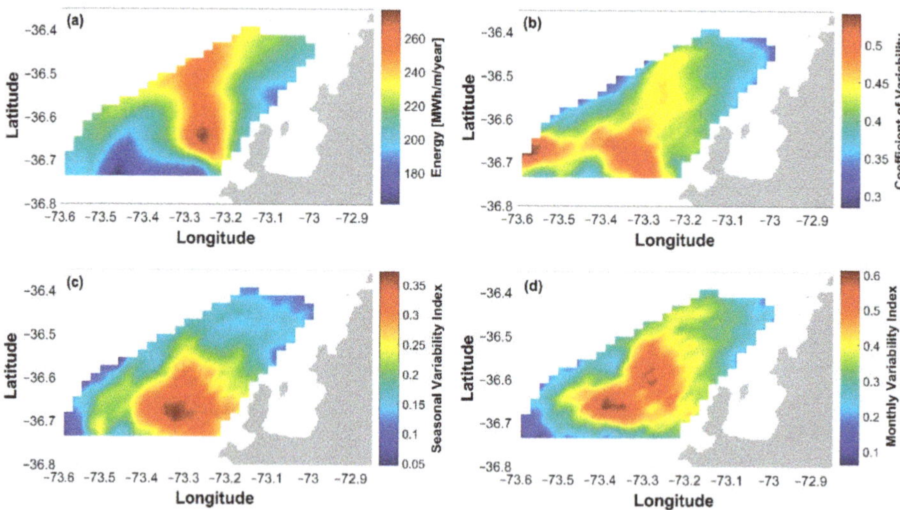

Figure 7. (**a**) Annual accumulated energy field, (**b**) coefficient of variation (COV), (**c**) Seasonal Variability index, and (**d**) Monthly Variability index.

From the MV index analysis, it is clear that the values are higher than SV since the months compared (August and January) present significant differences, especially in the radar sweep's central sector, where the energy presents its highest values.

In the location exemplified in Figure 4, the coefficients mentioned above obtained from the HF radar have the following values COV = 0.42, SV = 0.3, MV = 0.47. The site shows the highest wave power compared to the surrounding area and presents the highest variability indexes. Although wave power value assesses the location's suitability, sites with constant energy flows are more attractive than those with high variability [31]. They are more reliable and allow constant energy injection into the power grid. The site selection must combine both characteristics, high power with low variability.

4. Discussion and Conclusions

Chile's energy goals are clear; by 2050, 70% of the national energy matrix must be covered by renewable energy [39]. According to reports [40] on energy projects under construction, most of them are wind and solar energy since northern Chile has one of the highest solar radiation levels. On the other hand, wind energy has sectors south of the country where the potential is quite favorable. However, due to the long Chilean coastline, wave energy is also considered a potential source to supply the energy demand [41,42]. Notably, the work of [26,43–46] has already estimated the potential using numerical models compared to buoys and satellites. However, when it comes to obtaining information in high spatial resolution, there is no comparable method. It is the main reason for our study.

As mentioned throughout this paper, the advantage of HF radars is that they can provide high spatial and temporal resolution with scales from 0.3 to 50 km and a minimum of 30 min, respectively [12,23,47,48]. Additionally, they allow for sweeping a large area of the ocean. Being remote instruments, they are not affected by the hostile state that it can present. Therefore, despite some inaccuracies that HF radars can present [19], this study is valuable to understand their usefulness. In addition, the measurement they provide is also a resource to validate and complement information from other sources.

We worked with an HF radar to obtain the wave power fields of approximately 800 km^2 in front of Concepción. The treated data's spatial resolution is 2 km, and temporal resolution is 1 h, from December 2017 to January 2019. It is necessary to highlight that a one-year period is not long enough to understand the wavefield, limiting the work thoroughly. However, it can be significant if used as a complement for other observations, either remote or in situ, to understand the area's wave climate.

As the first interest of the installation of HF radars by CHIOOS was the detection of a tsunami, for the moment, this work only uses data from a single radar. While single radar estimates of significant wave height do not provide as accurate a measurement as dual radars, let us note that under homogeneous wave conditions, individual radars have a slight advantage over a conventional buoy [19]. Similarly, the local values collected from the radar were close to those found in other works [24,25,35].

The HF radar data were contrasted with in-situ data (buoy). However, given the buoy location at 36.56°S 73.33°W, where the HF radar's temporal coverage is less than 50%, it provides very low representativeness. Nevertheless, the buoy's statistics are still equivalent to that of the chosen grid analyzed in the field. The annual mean is 2.26 m for radar and 2.12 m for buoy, with a standard deviation of 0.71 m and 0.66 m. The annual and semi-annual cycles were compared with data from a wave-atlas produced by Universidad de Valparaiso [35] through numerical models (WW3). Thus, although the wave-atlas epoch (1980–2015) does not coincide with ours, the results show similar annual and semi-annual cycle characteristics; 0.25 m and 0.11 m for the radar, the atlas showing 0.3 m and 0.1 m, respectively.

The potential annual mean average power is 23 kW/m. However, this value depends on the used peak period. For the peak period, as mentioned in Section 2.2.2, we used one obtained from the official bibliography; the Chilean marine atlas [35,49]. However, while we could have used the buoy peak period instead, the variations of these appear relatively discretized (see Figure 5b), showing a considerable variability around 12.4 s. The majority of works investigated, including the atlas, showed values around 10 s without presenting many variations throughout the year. In addition, the comparison between the radar and buoy period data series also shown in Figure 5b reveals, due to the absence of data of the radar sweep around this location, the lack of correlation between what the buoy measures and what the radar provides there. It is also interesting to know how much the final energy would have varied if we had taken one period value or the other; HF radar's or the buoy's. The average mean peak period for the entire radar sweep (not shown) is 7.48 s with a standard deviation of 0.19 but varies between 7.1 and 7.9 s over the entire zone, which is less than 1 s, decreasing while approaching the coast. The radar mean peak period closest to the buoy is 7.6 s, with a standard deviation of 0.7 s (Figure 5b). The buoy mean peak period, calculated only for the overlapping radar times, is 12.4 s with a standard deviation of 2.6 s (Figure 5b). Using the buoy data as reality, assuming that the radar nearest grid is correct, we computed a calibrated period gridded field. Thus, using the calibrated mean peak period the wave mean power in that location would have increased by 20%, 4.6 kW/m. On the contrary, having used the radar's mean peak period, 24% less, that is 4.4 kW/m.

Nevertheless, the potential annual average powers fluctuate between 15 and 30 kW/m, depending on the sector within the sweep. As stated in [50], approximately 20 kW/m will make wave energy economically viable. As wave energy is not as mature as other renewable energies (like solar or wind), the economic viability not only depends on the amount of power a location can provide. The cost of the WECs is an essential part of the cost of a wave farm. The cost of operation and maintenance is also high, as corresponds to a facility in the sea [51]. The maximum power occurs during the winter season at 73.25 W and 36.65 S, approximately 15 km from the coast. The seasonal average is 37 kW/m, corresponding to a wave height of approximately 2.9 m, with a standard deviation of 1.6 kW/m. The maximum height at this point was 4.96 m (108 kW/m). In summer (December-February), the values vary between 15 and 25 kW/m. The least energetic month is January, presenting a constant 20 kW/m, half the one obtained in winter. One can note that throughout the year, 75% of the time, the wave height is above 2.2 m, which means that 75% of the time, the power will be a minimum of 22 kW/m. If this is carried out in terms of energy, it corresponds to a minimum of 193 MWh/m/year.

The SV and MV indices calculated presented values between 0.06–0.4 and 0.1–0.6, similar to those presented by [26]. These indices are relatively low compared to other studies that present similar potential magnitudes, such as [31,52]. Additionally, as mentioned before, the site selection must consider the wave power and variability of the area; here, the COV presents a lower variability than the considerable wave potential found. As a whole, the different indices make the area a more than convenient place for energy extraction.

We investigated the energy consumption of the areas close by. On average, a household consumes approximately 200 kWh per month, presenting a maximum peak in winter with values close to 230 kWh/month and a minimum in summer of 150 kWh/month [53]. Our work shows that this value is less than what can be extracted from a single grid cell, which is 2 × 2 km, since 75% of the time, the energy extracted would be larger than 16.6 MWh/month. Annually the maximum that could be collected is 270 MWh/m/year, and the lowest as perceived by the radar is 160 MWh/m/year. Considering a simple device, for example, an OWC with an efficiency of 40% [54,55], a single wave energy collector could have an approximate annual savings of USD 13,500, which would supply the average annual energy demand of approximately 33 houses, equivalent to avoiding 32.31 tonCO2eq/year [56]. The most significant advantage of wave energy is its high-power density, meaning that it can harness more energy than other renewable energies. Due to the novelty and newness of wave energy converters, it is impossible to provide a specific number of devices that can attain powers as big as wind or solar energy. For example, Ref. [51] shows several devices' performance, and the installed power varies between 6 kW and 15 MW. Thus, to reach 1 GW of installed power, the quantity of WEC can vary between 160,000 and 66. Likewise, the amount of area used to display the WECs also will change. Noteworthy, there are so many WECs types and locations where this can be installed that each case has to be evaluated case by case. Let us note something interesting to consider; hybrid systems, coupling waves and wind energy in the same facility [57]. It will provide an even more stable supply using the same area.

This work made it possible to estimate the wave power in front of the Biobío region, determining the area's suitability for its low variability and high energy concentration. With the high-resolution information provided by the HF radar, the best location for energy extraction can be identified. The utility of HF radars in energy terms relies on its high resolution, both temporal and spatial, given the ability to compare locations of interest within the same area. The possibility to implement a transportable radar to measure the high frequencies waves rank this instrument as an excellent asset to establish the wave power of a location.

As established throughout this work, even when a single radar can provide wave information, it can present ambiguities. This cannot be solved using only the radar's information. Of course, eventually, two radar systems would be ideal. However, not being cheap systems, we must enhance what we already have. Thus, one of our research's next stages could be to move the transportable HF radar close to the buoy to calibrate it correctly. Instead, we can use coastal video imagery for doing so [58,59]. We should also investigate an up-and-coming smartphone-based camera system recently assessed for coastal image classification [60].

Another great applicability of an HF marine radar system is its complementary use to satellite measurements or numerical models. In turn, radar data can be a valuable tool for the calibration of satellite data such as CFOSAT [61] and SWOT [62]. For example, SWOT's main objective is to measure ocean topography with centimeter-scale accuracy over kilometer-scale spatial resolution [63]. Therefore, HF radar measurements that also have a high resolution can be an accurate validation system. It shows the usefulness of HF marine radar as an instrument of remote sensing to study wave patterns and wave energy potential of vast areas at a high resolution.

Author Contributions: Conceptualization, V.M.-M. and R.A.-d.-R.; methodology, V.M.-M. and R.A.-d.-R.; software, V.M.-M. and R.A.-d.-R.; validation, V.M.-M. and R.A.-d.-R.; formal analysis, V.M.-M. and R.A.-d.-R.; investigation, V.M.-M. and R.A.-d.-R.; data curation, V.M.-M. and R.A.-d.-R.; visualization, V.M.-M.; supervision, R.A.-d.-R.; project administration, R.A.-d.-R.; V.M.-M.; writing—original draft: V.M.-M. and R.A.-d.-R.; writing—review and editing, V.M.-M., R.A.-d.-R., D.F., and J.M. All authors have read and agreed to the published version of the manuscript.

Funding: The National Agency for Research and Development (ANID)/Scholarship Program funded this research through grant DOCTORADO BECAS CHILE/2018–21181031.

Data Availability Statement: The HF radar fields resulting from this research work are available upon request to the first author.

Acknowledgments: We appreciated the reviewers' expert work that allowed this research, both in form and substance, to improve more than substantially. Buoy data was provided by the CENDHOC, Chilean Navy's Hydrographic and Oceanographic Service (SHOA).

Conflicts of Interest: The authors declare no conflict of interest.

References

1. Cornett, A. A Global Wave Energy Resource Assessment. In Proceedings of the Eighteenth International Offshore and Polar Conference, Vancouver, BC, Canada, 6–11 July 2008.
2. Arinaga, R.A.; Cheung, K.F. Atlas of global wave energy from 10 years of reanalysis and hindcast data. *Renew. Energy* **2012**, *39*, 49–64. [CrossRef]
3. Reeve, D.E.; Chen, Y.; Pan, S.; Magar, V.; Simmonds, D.J.; Zacharioudaki, A. An investigation of the impacts of climate change on wave energy generation: The Wave Hub, Cornwall, UK. *Renew. Energy* **2011**, *36*, 2404–2413. [CrossRef]
4. Gunn, K.; Stock-Williams, C. Quantifying the global wave power resource. *Renew. Energy* **2012**, *44*, 296–304. [CrossRef]
5. Mork, G.; Barstow, S.; Kabuth, A.; Pontes, M.T. Assessing the Global Wave Energy Potential. In Proceedings of the 29th International Conference on Ocean, Offsh-ore and Arctic Engineering, Shanghai, China, 6–11 June 2010; Volume 3, pp. 447–454.
6. European Commission. *Technology Information: Ocean Energy*; European Commission: Brussels, Belgium, 2014.
7. IRENA. *Renewable Capacity Statistics 2017*; International Renewable Energy Agency (IRENA): Abu Dhabi, UAE, 2017.
8. Europe Ocean Energy. *2030 Ocean Energy Vision*; Europe Ocean Energy: Brussels, Belgium, 2019.
9. Trowbridge, J.; Weller, R.; Kelley, D.; Dever, E.; Plueddemann, A.; Barth, J.A.; Kawka, O. The Ocean Observatories Initiative. *Front. Mar. Sci.* **2019**, *6*. [CrossRef]
10. Lillesand, T.; Kiefer, R.; Chipman, J. *Remote Sensing and Image Interpretation*, 7th ed.; Wiley: Hoboken, NJ, USA, 2015.
11. Ribal, A.; Young, I.R. 33 years of globally calibrated wave height and wind speed data based on altimeter observations. *Sci. Data* **2019**, *6*. [CrossRef]
12. Gurgel, K.W.; Antonischiki, G. Remote Sensing of Surface Currents and Waves by the Hf Radar Wera. In Proceedings of the International Conference on Electronic Engineering in Oceanography, Southampton, UK, 23–25 June 1997.
13. Essen, H.H.; Gurgel, K.W.; Schlick, T. Measurement of ocean surface waves by HF radars using a direction finding receive antenna. In Proceedings of the IGARSS 2000. IEEE 2000 International Geoscience and Remote Sensing Symposium. Taking the Pulse of the Planet: The Role of Remote Sensing in Managing the Environment. Proceedings (Cat. No.00CH37120), Honolulu, HI, USA, 24–28 July 2000.
14. Lipa, B.; Barrick, D.; Broug, J.; Nyden, B. HF Radar Detection of Tsunamis. *J. Oceanogr.* **2006**, *62*, 705–716. [CrossRef]
15. Lopez, G.; Conley, D.C. Comparison of HF Radar Fields of Directional Wave Spectra Against In Situ Measurements at Multiple Locations. *J. Mar. Sci. Eng.* **2019**, *7*, 271. [CrossRef]
16. Saviano, S.; Kalampokis, A.; Zambianchi, E.; Uttieri, M. A year-long assessment of wave measurements retrieved from an HF radar network in the Gulf of Naples (Tyrrhenian Sea, Western Mediterranean Sea). *J. Oper. Oceanogr.* **2019**, *12*, 1–15. [CrossRef]
17. Helzel, T.; Lopez, O.; Wyatt, L.R. *Ocean Radar for the Planning and Operational Phase of Off-Shore Renewable Energy System*; IEEE: Piscataway, NJ, USA, 2011.
18. Patel, R.P.; Nagababu, G.; Kumar, S.V.A.; Seemanth, M.; Kachhwaha, S.S. Wave resource assessment and wave energy exploitation along the Indian coast. *Ocean Eng.* **2020**, *217*, 107834. [CrossRef]
19. Wyatt, L.R. An evaluation of wave parameters measured using a single HF radar system. *Can. J. Remote Sens.* **2002**, *28*, 205–218. [CrossRef]
20. Wyatt, L.R. Significant waveheight measurement with h.f. radar. *Int. J. Remote Sens.* **1988**, *9*, 1087–1095. [CrossRef]
21. Heron, M.L.; Atwater, D.P. Temporal and spatial resolution of HF ocean radars. *Ocean Sci. J.* **2013**, *48*, 99–103. [CrossRef]
22. Fujii, S.; Heron, M.L.; Kim, K.; Lai, J.-W.; Lee, S.-H.; Wu, X.; Wu, X.; Wyatt, L.R.; Yang, W.-C. An overview of developments and applications of oceanographic radar networks in Asia and Oceania countries. *Ocean Sci. J.* **2013**, *48*, 69–97. [CrossRef]
23. Roarty, H.; Cook, T.; Hazard, L.; George, D.; Harlan, J.; Cosoli, S.; Wyatt, L.; Alvarez Fanjul, E.; Terrill, E.; Otero, M.; et al. The Global High Frequency Radar Network. *Front. Mar. Sci.* **2019**, *6*. [CrossRef]
24. Young, I.R. Seasonal Variability Of The Global Ocean Wind And Wave Climate. *Int. J. Climatol.* **1999**, *19*, 931–950. [CrossRef]

25. Aguirre, C.; Rutllant, J.A.; Falvey, M. Wind waves climatology of the Southeast Pacific Ocean. *Int. J. Climatol.* **2017**, *37*, 4288–4301. [CrossRef]
26. Lucero, F.; Catalán, P.A.; Ossandón, Á.; Beyá, J.; Puelma, A.; Zamorano, L. Wave energy assessment in the central-south coast of Chile. *Renew. Energy* **2017**, *114*, 120–131. [CrossRef]
27. Leys, C.; Ley, C.; Klein, O.; Bernard, P.; Licata, L. Detecting outliers: Do not use standard deviation around the mean, use absolute deviation around the median. *J. Exp. Soc. Psychol.* **2013**, *49*, 764–766. [CrossRef]
28. Amidror, I. Scattered data interpolation methods for electronic imaging systems: A survey. *J. Electron. Imaging* **2002**, 11. [CrossRef]
29. Liepins, V. An algorithm for evaluating a discrete Fourier transform for incomplete data. *Autom. Control Comput. Sci.* **1996**, *30*, 20–29.
30. Liepins, V. Extended Fourier analysis of signals. *arXiv* **2013**, arXiv:1303.2033.
31. Bingölbali, B.; Jafali, H.; Akpınar, A.; Bekiroğlu, S. Wave energy potential and variability for the south west coasts of the Black Sea: The WEB-based wave energy atlas. *Renew. Energy* **2020**, *154*, 136–150. [CrossRef]
32. Holthuijsen, L.H. *Waves in Oceanic and Coastal Waters*; Cambridge University Press: Cambridge, UK, 2007; Volume 1.
33. Soares, C.G. (Ed.) *Advances in Renewable Energies Offshore*; Taylor & Francis: London, UK, 2018.
34. Ingram, D.M.; Smith, G.; Bittencourt Ferreira, C.; Smith, H. *Protocols for the Equitable Assessment of Marine Energy Converters*; The Institute for Energy Systems: Edinburgh, UK, 2011.
35. Beyá, J.; Álvarez, M.; Gallardo, A.; Hidalgo, H.; Aguirre, C.; Valdivia, J.; Parra, C.; Méndez, L.; Contreras, F.; Winckler, P.; et al. *Atlas de Oleaje de Chile*; Escuela de Ingeniería Civil Oceánica—Universidad de Valparaiso: Valparaiso, Chile, 2016.
36. Reguero, B.G.; Losada, I.J.; Méndez, F.J. A global wave power resource and its seasonal, interannual and long-term variability. *Appl. Energy* **2015**, *148*, 366–380. [CrossRef]
37. Silva, D.; Martinho, P.; Guedes Soares, C. Wave energy distribution along the Portuguese continental coast based on a thirty three years hindcast. *Renew. Energy* **2018**, *127*, 1064–1075. [CrossRef]
38. Alizadeh, M.J.; Alinejad-Tabrizi, T.; Kavianpour, M.R.; Shamshirband, S. Projection of spatiotemporal variability of wave power in the Persian Gulf by the end of 21st century: GCM and CORDEX ensemble. *J. Clean. Prod.* **2020**, *256*, 120400. [CrossRef]
39. de Energía, C.C. *Energía 2050, Política Energética de Chile*; Ministerio de Energía: Santiago, Chile, 2017.
40. Comisión Nacional de Energía. *Anuario Estadístico de Energía*; Ministerio de Energía: Santiago, Chile, 2018; pp. 1–162.
41. Villavicencio, P. Chile Cuenta Con Nuevo Centro de Investigación, Pionero en Energía Marina. Available online: https://www.elciudadano.com/ciencia-tecnologia/centro-de-energia-marina/06/18/ (accessed on 10 September 2020).
42. MERIC. Ministro de Energía Inaugura Pionero Centro de Investigación en Energía Marina. Available online: https://www.meric.cl/en/ministro-de-energia-inaugura-pionero-centro-de-investigacion-en-energia-marina-2/ (accessed on 10 September 2020).
43. Mediavilla, D.G.; Figueroa, D. Assessment, sources and predictability of the swell wave power arriving to Chile. *Renew. Energy* **2017**, *114*, 108–119. [CrossRef]
44. Mediavilla, D.G.; Sepúlveda, H.H. Nearshore assessment of wave energy resources in central Chile (2009–2010). *Renew. Energy* **2016**, *90*, 136–144. [CrossRef]
45. Mazzaretto, O.M.; Lucero, F.; Besio, G.; Cienfuegos, R. Perspectives for harnessing the energetic persistent high swells reaching the coast of Chile. *Renew. Energy* **2020**, *159*, 494–505. [CrossRef]
46. Monárdez, P.; Acuña, H.; Scott, D. Evaluation of the Potential of Wave Energy in Chile. In Proceedings of the ASME 2008 27th International Conference on Offshore Mechanics and Arctic Engineering, Estoril, Portugal, 15–20 June 2008; Volume 6, pp. 801–809.
47. Wyatt, L.R. Wave and Tidal Power measurement using HF radar. *IMEJ* **2018**, 1. [CrossRef]
48. Wyatt, L.R. High frequency radar applications in coastal monitoring, planning and engineering. *Aust. J. Civ. Eng.* **2014**, *12*, 1–15. [CrossRef]
49. Beyá, J.; Álvarez, M.; Gallardo, A.; Hidalgo, H.; Winckler, P. Generation and validation of the Chilean Wave Atlas database. *Ocean Model.* **2017**, *116*, 16–32. [CrossRef]
50. Sandberg, A.; Klementsen, E.; Muller, G.; de Andres, A.; Maillet, J. Critical Factors Influencing Viability of Wave Energy Converters in Off-Grid Luxury Resorts and Small Utilities. *Sustainability* **2016**, *8*, 1274. [CrossRef]
51. Astariz, S.; Iglesias, G. The economics of wave energy: A review. *Renew. Sustain. Energy Rev.* **2015**, *45*, 397–408. [CrossRef]
52. Amrutha, M.M.; Sanil Kumar, V. Spatial and temporal variations of wave energy in the nearshore waters of the central west coast of India. *Ann. Geophys.* **2016**, *34*, 1197–1208. [CrossRef]
53. Comisión Nacional de Energía. *Consumo Eléctrico Anual por Comuna y Tipo de Cliente*; Energía Abierta; Comisión Nacional de Energía: Santiago, Chile, 2019.
54. Aderinto, T.; Li, H. Review on Power Performance and Efficiency of Wave Energy Converters. *Energies* **2019**, *12*, 4329. [CrossRef]
55. Maldonado, C. *Stochastic Modelling of Owc Device and Power Production*; Pontificia Universidad Catolica De Chile: Santiago, Chile, 2017.
56. Comisión Nacional de Energía. *Factor de Emisión, Promedio Anual*; Energía Abierta; Comisión Nacional de Energía: Santiago, Chile, 2019.
57. Astariz, S.; Iglesias, G. Co-located wind and wave energy farms: Uniformly distributed arrays. *Energy* **2016**, *113*, 497–508. [CrossRef]

58. Cienfuegos, R.; Villagran, M.; Aguilera, J.C.; Catalán, P.; Castelle, B.; Almar, R. Video monitoring and field measurements of a rapidly evolving coastal system: The river mouth and sand spit of the Mataquito River in Chile. *J. Coast. Res.* **2014**, *70*, 639–644. [CrossRef]
59. Cang, Y.; He, H.; Qiao, Y. Measuring the Wave Height Based on Binocular Cameras. *Sensors* **2019**, *19*, 1338. [CrossRef]
60. Valentini, N.; Balouin, Y. Assessment of a Smartphone-Based Camera System for Coastal Image Segmentation and Sargassum monitoring. *J. Mar. Sci. Eng.* **2020**, *8*, 23. [CrossRef]
61. Hauser, D.; Tourain, C.; Lachiver, J.M. CFOSAT: A New Mission in Orbit to Observe Simultaneously Wind and Waves at the Ocean Surface. *Space Res. Today* **2019**, *206*, 15–21. [CrossRef]
62. Aouf, L.; Hauser, D.; Dalphinet, A.; Giordani, H. *SWOT-Waves: For the Improvement of Operational Wave Forecasting*; NASA: Washington, DC, USA, 2020.
63. Peral, E.; Rodríguez, E.; Esteban-Fernández, D. Impact of Surface Waves on SWOT's Projected Ocean Accuracy. *Remote Sens.* **2015**, *7*, 14509–14529. [CrossRef]

Letter

Seasonal Variability of SST Fronts in the Inner Sea of Chiloé and Its Adjacent Coastal Ocean, Northern Patagonia

Gonzalo S. Saldías [1,2,*], Wilber Hernández [3], Carlos Lara [4,5], Richard Muñoz [1], Cristian Rojas [3], Sebastián Vásquez [6], Iván Pérez-Santos [7,8] and Luis Soto-Mardones [1]

[1] Departamento de Física, Facultad de Ciencias, Universidad del Bío-Bío, Concepción 4051381, Chile; richmunoz@udec.cl (R.M.); lsoto@ubiobio.cl (L.S.-M.)

[2] Centro FONDAP de Investigación en Dinámica de Ecosistemas Marinos de Altas Latitudes (IDEAL), Valdivia 5090000, Chile

[3] Programa de Magíster en Ciencias Físicas, Facultad de Ciencias, Universidad del Bío-Bío, Concepción 4051381, Chile; wilber.hernandez1701@alumnos.ubiobio.cl (W.H.); cristian.rojas2001@alumnos.ubiobio.cl (C.R.)

[4] Departamento de Ecología, Facultad de Ciencias, Universidad Católica de la Santísima Concepción, Concepción 4090541, Chile; carlos.lara@ucsc.cl

[5] Centro de Investigación en Recursos Naturales y Sustentabilidad (CIRENYS), Universidad Bernardo O'Higgins, Santiago 8370993, Chile

[6] Instituto de Investigación Pesquera, Talcahuano 4260000, Chile; svasquez@inpesca.cl

[7] Centro i-mar, Universidad de los Lagos, Puerto Mont 5480000, Chile; ivan.perez@ulagos.cl

[8] Centro de Investigación Oceanográfica COPAS Sur-Austral, Concepción 4030000, Chile

* Correspondence: gsaldias@ubiobio.cl

Citation: Saldías, G.S.; Hernandez, W.; Lara, C.; Muñoz, R.; Rojas, C.; Vásquez, S.; Pérez-Santos, I.; Soto-Mardones, L. Seasonal Variability of SST Fronts in the Inner Sea of Chiloé and Its Adjacent Coastal Ocean, Northern Patagonia. *Remote Sens.* **2021**, *13*, 181. https://doi.org/10.3390/rs13020181

Received: 30 November 2020
Accepted: 4 January 2021
Published: 7 January 2021

Publisher's Note: MDPI stays neutral with regard to jurisdictional claims in published maps and institutional affiliations.

Abstract: Surface oceanic fronts are regions characterized by high biological activity. Here, Sea Surface Temperature (SST) fronts are analyzed for the period 2003–2019 using the Multi-scale Ultra-high Resolution (MUR) SST product in northern Patagonia, a coastal region with high environmental variability through river discharges and coastal upwelling events. SST gradient magnitudes were maximum off Chiloé Island in summer and fall, coherent with the highest frontal probability in the coastal oceanic area, which would correspond to the formation of a coastal upwelling front in the meridional direction. Increased gradient magnitudes in the Inner Sea of Chiloé (ISC) were found primarily in spring and summer. The frontal probability analysis revealed the highest occurrences were confined to the northern area (north of Desertores Islands) and around the southern border of Boca del Guafo. An Empirical Orthogonal Function analysis was performed to clarify the dominant modes of variability in SST gradient magnitudes. The meridional coastal fronts explained the dominant mode (78% of the variance) off Chiloé Island, which dominates in summer, whereas the SST fronts inside the ISC (second mode; 15.8%) were found to dominate in spring and early summer (October–January). Future efforts are suggested focusing on high frontal probability areas to study the vertical structure and variability of the coastal fronts in the ISC and its adjacent coastal ocean.

Keywords: MUR SST; SST fronts; Inner Sea of Chiloé; northern Patagonia

1. Introduction

Oceanic fronts are relatively narrow regions with high gradients of physical, chemical, biological, and optical properties. They are generally associated with convergence at the surface [1] and high aggregation of organisms and biological activity e.g., [2,3]. Lately, enhanced submesoscale activity has been identified around fronts [4,5], which involves ageostrophic vertical circulation with increased vertical fluxes of tracers and momentum. In general, frontal features and currents such as jets and meanders e.g., [6,7], filaments e.g., [8,9], and river discharges e.g., [10,11] present distinct patterns of variation in strength and duration over multiple temporal scales and are identified from satellite sea surface temperature (SST) fields.

Remotely sensed data have been crucial in the study of the evolution of ocean fronts at multiple spatio-temporal scales [9,12–16]. The improved accuracy and spatial resolution of SST sensors over time have allowed the quantification and spatial distribution of sharp fronts in coastal regions [11,15,17–20]. Here, high-resolution (1 km) SST images are used to analyze the variability of SST fronts in the Inner Sea of Chiloé (ISC) and adjacent coastal ocean (northern Chilean Patagonia). The ISC is a long inner sea (about 260 km) with high heterogeneity of hydrographic conditions due to the influence of freshwater from several rivers and the marine influence by the active water exchange through Boca del Guafo and Chacao channel [21,22]. Several islands (Desertores Islands) promote contrasting hydrographic and SST characteristics between the northern and southern areas. In general, northern Patagonia is characterized by elevated surface chlorophyll-a concentration and primary productivity during spring-summer-fall [23]. Recently, harmful algal blooms events were described in the region in association with anomalous oceanographic conditions during summer 2016 [24]. Although the development of chlorophyll patches and blooms have been suggested to develop around frontal regions [25], there are no major insights into the evolution and distribution of surface fronts along the ISC.

This study aims to characterize the seasonal and interannual variability of surface thermal fronts along the Inner Sea of Chiloé and its adjacent coastal ocean. For the first time, high-resolution satellite images were used to achieve a frontal analysis of this region. Section 2 describes the data and methods. Section 3 resumes the principal results and discusses the seasonal and interannual evolution of SST fronts. Finally, a summary is presented in Section 4.

2. Satellite Data and Methods

We used daily SST data from the Multi-scale Ultra-high Resolution (MUR) product [18] obtained from the Physical Oceanography Distributed Active Archive Center (PODAAC; see Data Availability Statement below). MUR data were chosen over other SST products due to the increased data cover and improved performance detecting SST gradients in the coastal region [18]. The data have a spatial resolution of 0.01 × 0.01 degrees. A comparison with a near-surface temperature time series from a buoy located in Seno Reloncaví (see Figure 1) validates the performance of MUR SST in the ISC (Figure 1b,c). The buoy has been maintained by the research center i-mar since 2017 (see Data Availability Statement below).

The gradient magnitude (GM) and the Canny edge-detection algorithm [26] were used to detect and quantify the frontal regions and their evolution. The GM was computed following other studies e.g., [11,18]:

$$\nabla_x TSM_i = (TSM_{i-1} - TSM_{i+1})/(X_{i-1} - X_{i+1}) \tag{1}$$

$$\nabla_y TSM_j = (TSM_{j-1} - TSM_{j+1})/(Y_{j-1} - Y_{j+1}) \tag{2}$$

$$| \nabla TSM | = \{(\nabla_x TSM_i)^2 + (\nabla_y TSM_j)^2\}^{1/2} \tag{3}$$

where $\nabla_x SST_i$ and $\nabla_y SST_j$ are the zonal and meridional components of the SST gradient, respectively, and $| \nabla SST|$ is the gradient magnitude. The Canny edge-detection algorithm [26] was applied on daily SST fields to identify coherent frontal segments and compare them with the regions of increased gradient magnitudes. The Canny method tracks the direction of SST gradients using a threshold value. The gradients are calculated using the derivative of a Gaussian filter. After studying the range of gradient magnitudes for all images (Figure 1d), we used a high threshold value of 0.1 °C/km, which separates most values in the low range of the PDF from the higher values associated with the formation of SST fronts. The probability of finding a front was calculated as the number of times a pixel is classified as a front divided by the total images considered in a time window (i.e., seasonal aggregates). Further details of the application of the Canny method can be found elsewhere [15,19].

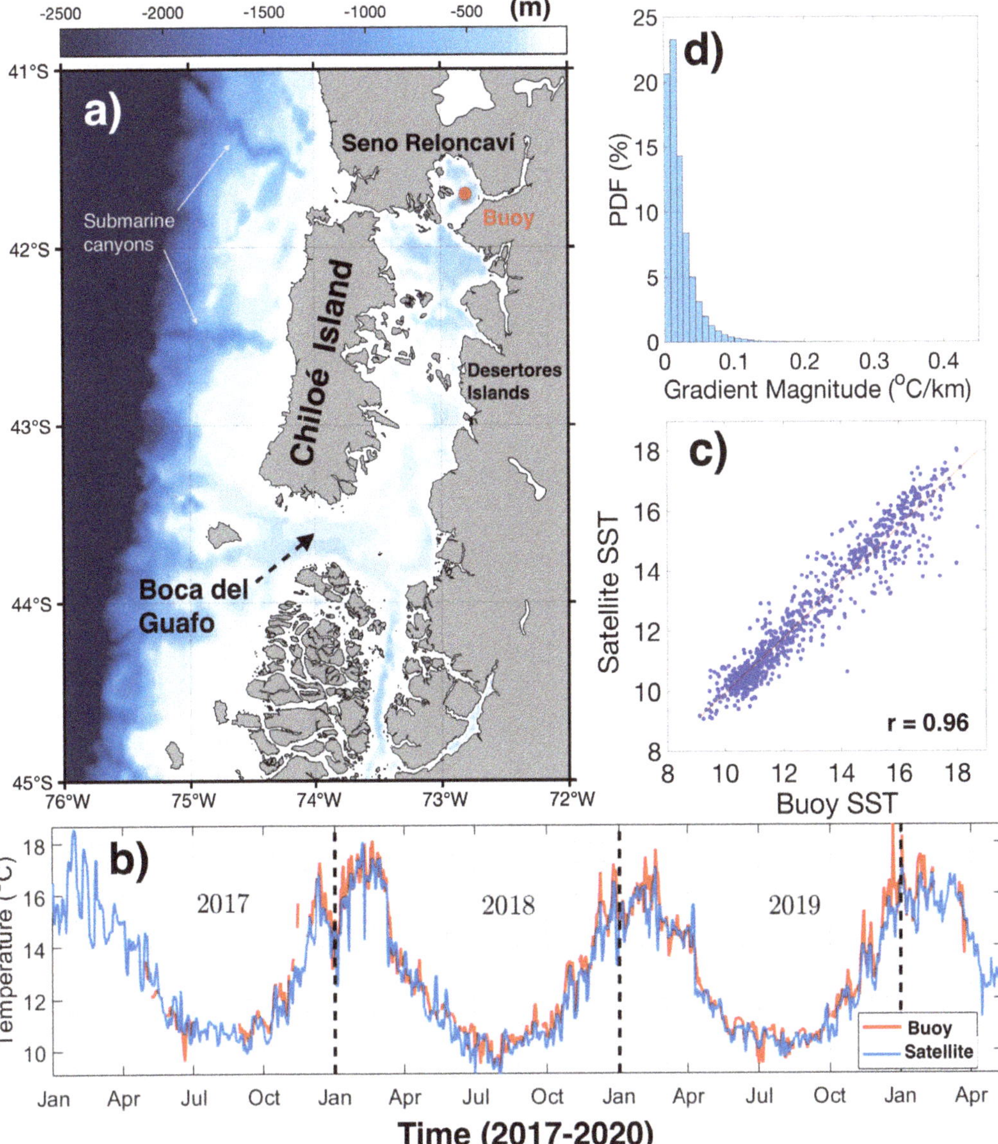

Figure 1. (**a**) Map of the Inner Sea of Chiloé and its adjacent coastal ocean in northern Patagonia. The bathymetry is shown in a blue-white color scale. The position of the oceanographic buoy located in Seno Reloncaví is denoted by a red dot. The comparison between satellite SST and near-surface (1 m) temperature at the buoy is presented in (**b**,**c**). In situ temperature has been daily averaged to match satellite SST data. (**d**) Probability Density Function (in %) of the distribution of all SST gradient magnitudes for the period 2003–2019.

Satellite chlorophyll fluorescence (Fluorescence Line Height, nFLH) data from the Moderate Resolution Imaging Spectroradiometer (MODIS, on-board Aqua) and for the period 2003–2019 were obtained from the ocean color website (see Data Availability Statement below). These data were used to compute the seasonal climatology. We chose to use the chlorophyll fluorescence over the chlorophyll product because of the characteristic estuarine turbid conditions of the ISC and the relatively poor performance of default chlorophyll algorithms [27].

An Empirical Orthogonal Function (EOF) analysis was performed on the gradient magnitudes to separate the main modes of variability. The EOF was computed following the Singular Value Decomposition (SVD) approach to avoid a large covariance matrix associated with the high resolution of the images [28]. Please note that each time series is demeaned and detrended in the process.

3. Results and Discussion

The temporal pattern of SST variability is well-captured by MUR SST compared to the buoy measurements at Seno Reloncaví for the period 2017–2020, as shown in Figure 1b. However, some events with peaks in *in situ* temperature were not well recorded by the satellite data, especially during spring-summer (Figure 1b). There is a slight underestimation by MUR SST (Figure 1c). In general, MUR data reproduced the temporal SST variability with high correlation (r = 0.96), which gave us confidence that the MUR product is reasonably accurate in these coastal waters. Given the recent deployment of this buoy in Seno Reloncaví (starting in 2017), this is the first comparison of satellite SST and *in situ* temperature in the ISC.

The seasonal climatology of SST fields revealed a typical pattern for temperate ecosystems, i.e., marked spatial and seasonal variability. During austral spring and summer, the higher SST is observed in the northern area (north of Desertores Islands) and adjacent coastal ocean (Figure 2a,d). In fall, The adjacent coastal ocean showed the highest values compared to the ISC (Figure 2b), whereas the entire coastal ocean shows the lowest temperatures (<11 °C) in winter (Figure 2c). The persistence of these mean fields is variable depending on the location and season (Figure 2e–h). The mean warm pattern observed in summer in the northern ISC (Seno Reloncaví) is also highly variable (Figure 2e). This strong variability is also presented in the fall and spring (Figure 2f,h). In general, the ICS presents the largest SST variability (greater than 1.2 °C) in spring (Figure 2h). Notice that low variability south of Desertores Islands is associated with the lowest averaged SST fields in connection with the coastal ocean through Boca del Guafo (Figure 2b,d), characterized by intrusions of Sub-Antarctic waters (SAAW) [29,30]. Finally, winter represents the coldest season with the annual cycle's greatest spatial homogeneity (Figure 2c,g). The annual cycle of SST is coherent with previous studies in northern Patagonia [22,31,32].

The quantification of the climatological SST gradient magnitudes for the entire 17 years of study is shown in Figure 3. During spring and summer, the SST gradient magnitude fields suggest a high frontal activity in the northern area of the ISC (Figure 3a,d). Summer also represents a period with increased SST fronts in the coastal ocean with an extended meridional band of high gradients (Figure 3a). This would be linked with the development of a coastal upwelling front off Chiloé Island in summer due to predominant northward winds [30,33]. The extended band with medium values of gradient magnitude in fall would represent the weakening of upwelling-favorable winds, and consequently, a weaker upwelling front (Figure 3b). Medium values in the standard deviation corresponding to the upwelling front would explain a synoptic variability with pulses of upwelling events through summer and fall (Figure 3e,f). High values in SST gradient (>0.05 °C/km) and variability (>0.03 °C/km) on the southern area of the ISC in summer could be associated with the generation of SST fronts from the intrusions of oceanic waters through Boca del Guafo (~44°S) (Figure 3a,e). In spring, the highest gradients are presented in the northern ISC with a band of increased variability near Desertores Islands (Figure 3d,h). Finally, SST's gradient magnitudes are less pronounced in winter (Figure 3c,g) which is coherent with a more homogeneous and cold temperature field (Figure 2c,g). Water column stratification in Seno Reloncaví (not shown) presents a coherent annual cycle with maximum (minimum) stratification in spring-summer (fall-winter).

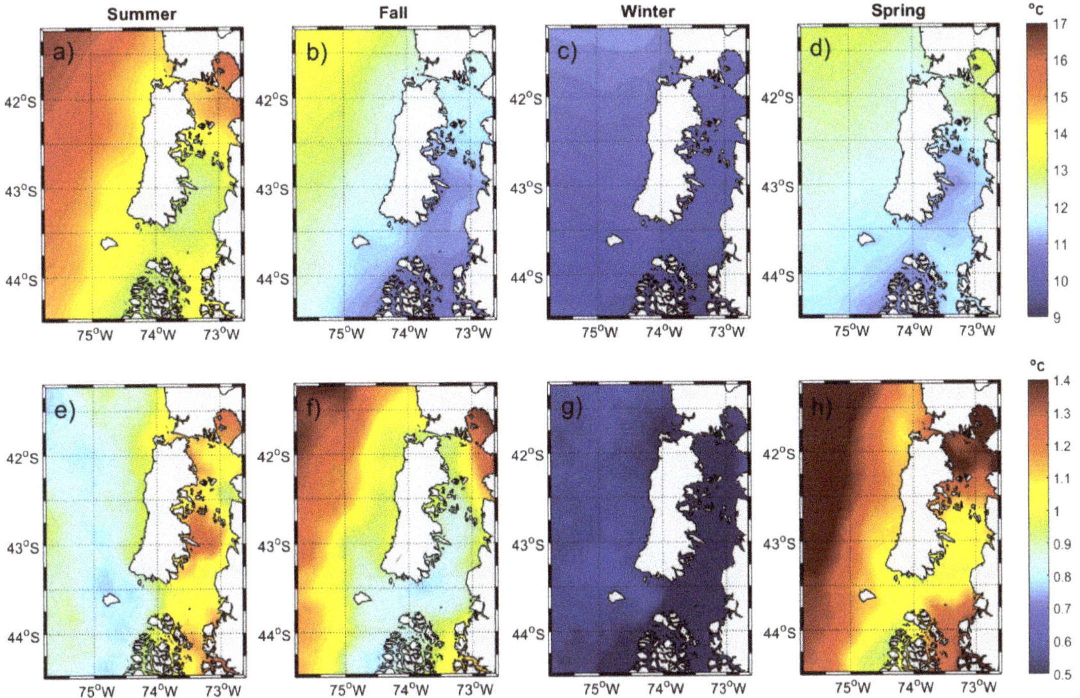

Figure 2. Seasonal climatology (2003–2019) of SST in the Inner Sea of Chiloé and its adjacent coastal ocean. (upper panels) averages and (lower panels) standard deviations for (**a,e**) summer (January, February, March), (**b,f**) fall (April, May, June), (**c,g**) winter (July, August, September), and (**d,h**) spring (October, November, December). For reference of locations see Figure 1a.

To further understand the generation and evolution of fronts along the ISC and its adjacent coastal ocean, the seasonal variability of frontal probability (FP) is shown in Figure 4. The use of an edge-detection algorithm is crucial for calculating the FP since fronts tend to be narrow and coherent bands of increased gradients of ocean properties [14], which can be overlooked and or not correctly identified through the gradient magnitude. The FP maps reveal that the formation of SST fronts off Chiloé Island in summer-fall (Figure 4a,b), potentially associated with the coastal upwelling, is consistent with the increased gradient magnitudes shown in Figure 3. The FP reaches its largest values (>9%) in the northern ISC in spring-summer (Figure 4a,d). Also, the lowest SST frontal activity in spring-summer is shown on the western section of the southern ISC (Figure 4a,d), which might be associated with a greater oceanic influence on the eastern side. The eastern side also has several river outflows, which could influence the temperature field, creating increased gradients. It is interesting to note that the FP in fall is <3% along most of the ISC, representing the season with the lowest SST frontal activity (Figure 4b). In contrast, the frontal probability values during winter indicate higher PF as compared to fall (Figure 4c vs. Figure 4b), especially at Boca del Guafo (values up to 7%), which was not demonstrated through the gradient magnitude (Figure 3).

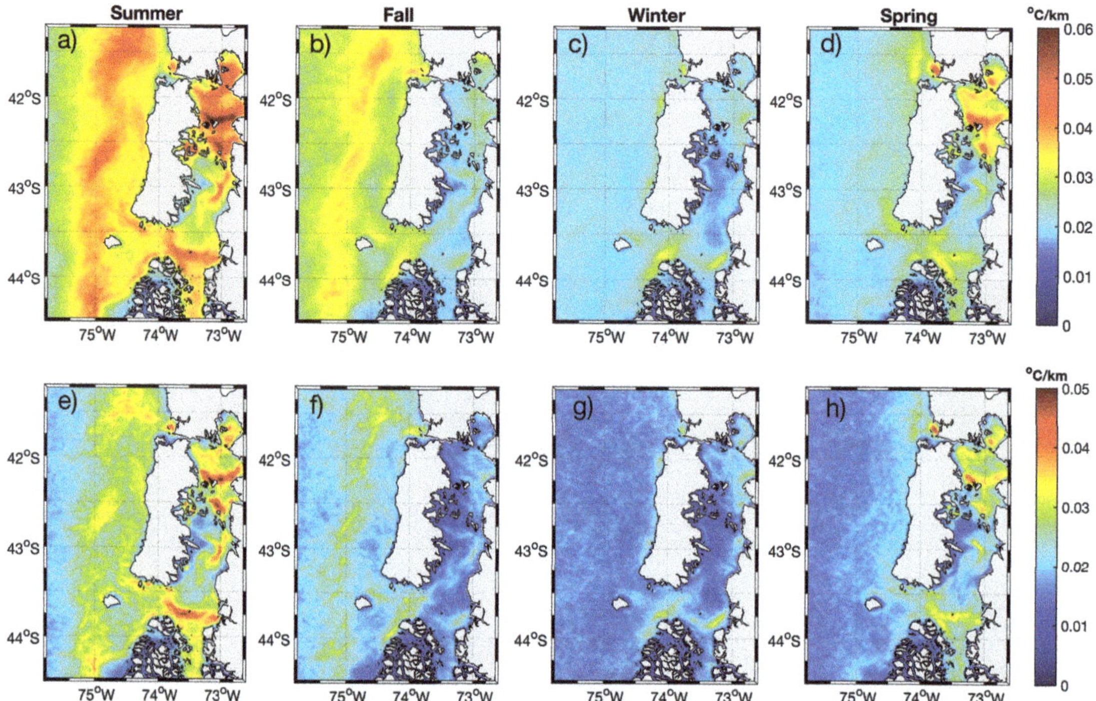

Figure 3. Seasonal climatology (2003–2019) of gradient magnitude of SST in the Inner Sea of Chiloé and its adjacent coastal ocean. (upper panels) averages and (lower panels) standard deviations for (**a,e**) summer (January, February, March), (**b,f**) fall (April, May, June), (**c,g**) winter (July, August, September), and (**d,h**) spring (October, November, December). For reference of locations see Figure 1a.

We performed an EOF analysis to separate the main modes of variability of the SST gradient magnitude and evidence the periods when the development of SST fronts in the coastal ocean and the ISC is more likely to occur. The EOFs show that most of the variance is explained in the meridional band of high SST gradients off Chiloé (Figure 5a). These features, highly associated with the upwelling front [22,32], occurred most of the years during December-April (Figure 5c,d). The second EOF isolates the enhanced SST frontal activity in the ISC (Figure 5b), predominantly in spring and early summer (October–February; Figure 5c,d). The temporal oscillations of the EOFs reveal a persistent annual cycle, especially after the end of 2005 (Figure 5d). A maximum SST gradient magnitude in the coastal ocean occurred in early 2008 as seen through the peak of mode 1 (Figure 5d). In northern Patagonia, large-scale climatic influence has been suggested to produce changes in the coastal oceanography, and concomitantly, in ecological patterns around the Inner Sea of Chiloé [21,34]. Future studies focused on the interannual variability of SST fronts should consider the potential impact of climate variability e.g., [21] on the generation or blocking of SST fronts in the ISC.

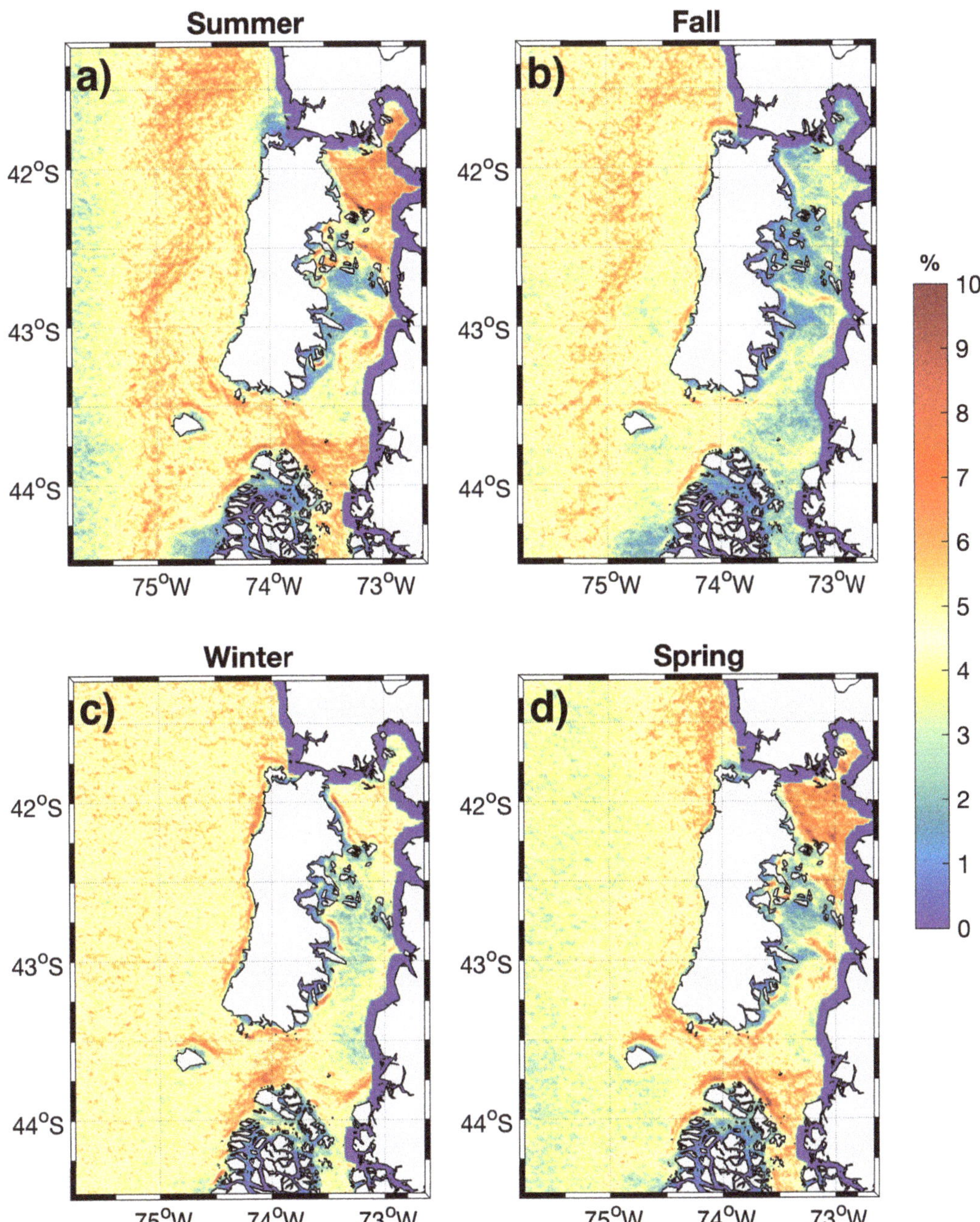

Figure 4. Seasonal climatology (2003–2019) of SST frontal probability (FP; %), based on the Canny edge-detection algorithm, in the Inner Sea of Chiloé and its adjacent coastal ocean for (**a**) summer (January, February, March), (**b**) fall (April, May, June), (**c**) winter (July, August, September), and (**d**) spring (October, November, December). For reference of locations see Figure 1a.

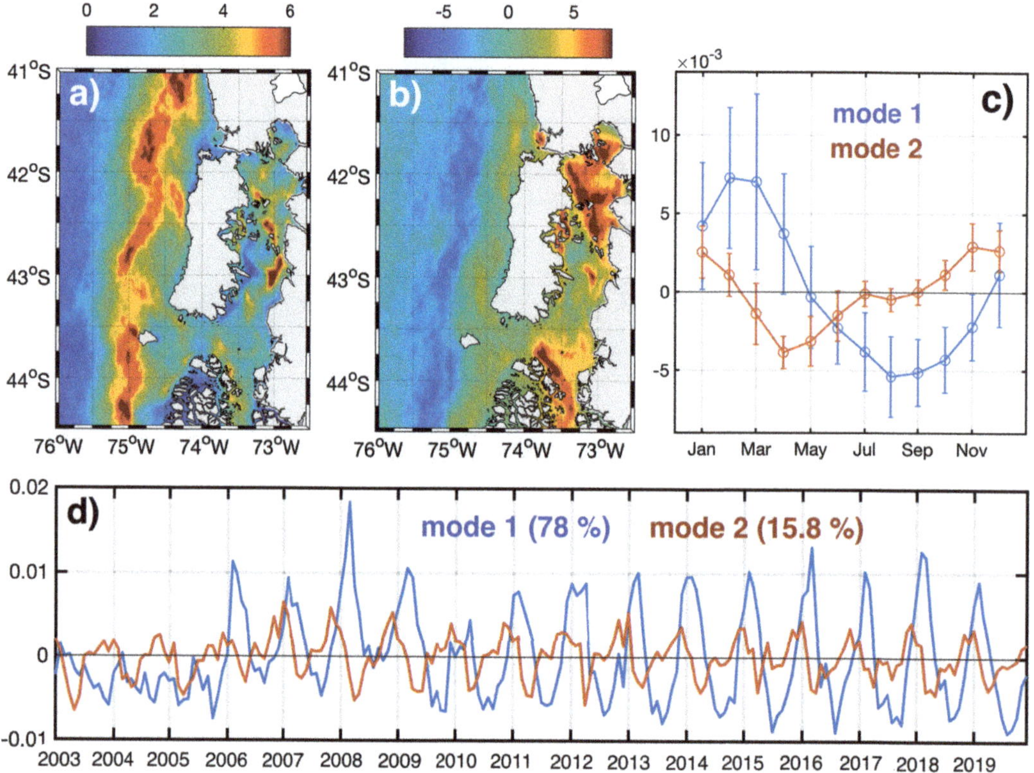

Figure 5. EOF analysis of SST gradient magnitudes. (**a**) First and (**b**) second EOF modes. (**c**) Mean annual cycle of EOF time series shown in (**d**). The error bars in (**c**) correspond to the monthly standard deviations.

The use of satellite platforms to monitor oceanographic properties provides important insights into oceanographic fronts' spatial and temporal variability. An important feature of the northern Patagonian shelf is its significant freshwater inputs [35,36] which could also generate thermal variability and fronts. The river discharges in the northern ISC have been associated with phytoplankton blooms [37]. Freshwater discharge from large rivers and or glacial melting into the Inner Sea of Chiloé (41–45°S) creates a freshwater plume with high levels of biological activity which extends into the coastal ocean [32,38]. While the local impacts of those freshwater discharges remain unknown, it is expected that they would favor the formation of oceanographic fronts and the aggregation of large organisms, such as whales, in the region [39]. Ocean fronts are hotspots of high biological activity [19,40–42] by which increased primary productivity is expected during spring-summer-fall and linked to the areas of high frontal activity (i.e., northern ISC, coastal band off Chiloé Island, southern border of Boca del Guafo). The seasonal climatology of chlorophyll fluorescence (Figure 6) suggests there is enhanced phytoplankton activity around the SST fronts in northern ISC and a maximum fluorescence in a meridional band off Chiloé in fall (Figure 6b), which would agree with the presence of a coastal upwelling frontal band in the coastal ocean (Figure 3b). Thus, northern Patagonia is a highly dynamic region where biophysical interactions over coastal waters remain largely unstudied. The use of satellite products across multiple spatial and temporal scales provides fundamental insights into the oceanographic processes around coastal fronts [43]. Future studies assessing the variability of coastal chlorophyll and productivity in northern Patagonia should focus on regions with enhanced frontal activity to further understand the biophysical coupling at ocean fronts.

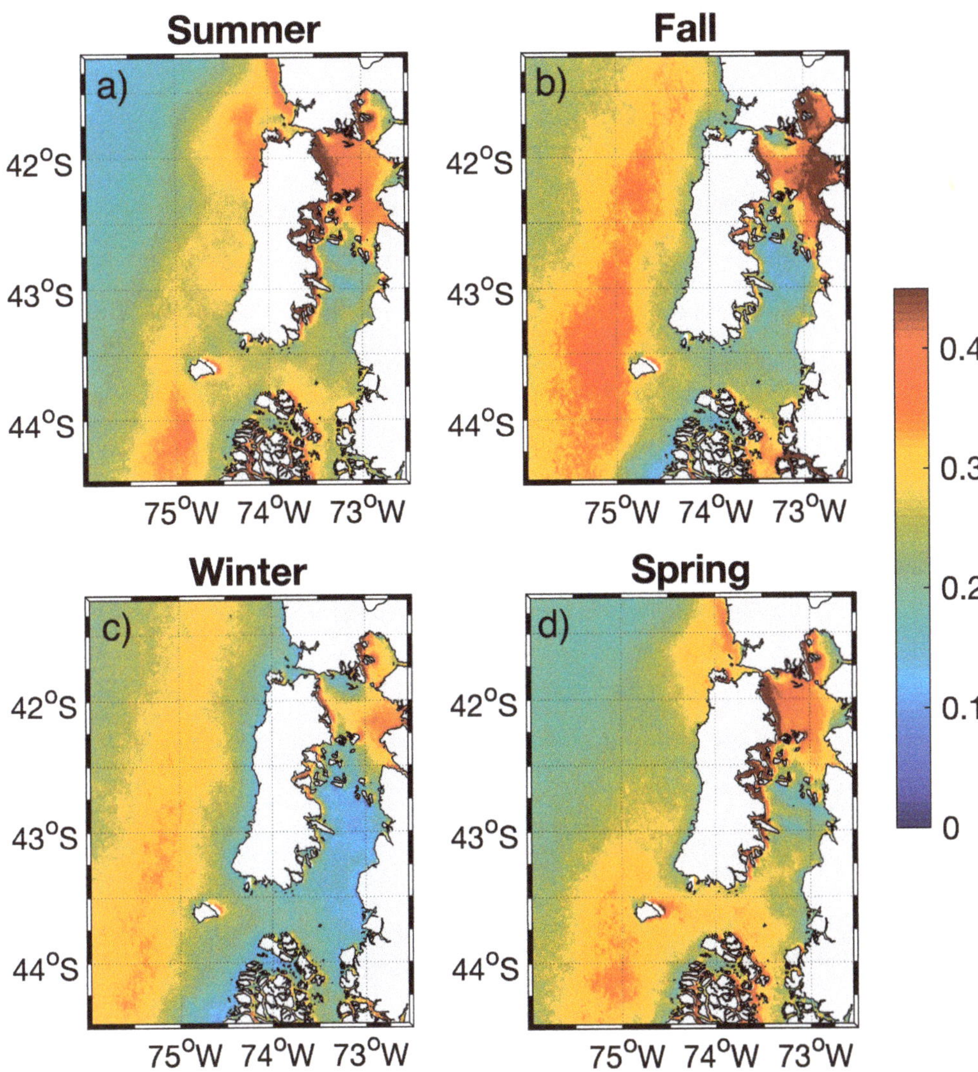

Figure 6. Seasonal climatology (2003–2019) of chlorophyll fluorescence (nFLH; W m^{-2} µm^{-1} sr^{-1}) in the Inner Sea of Chiloé and its adjacent coastal ocean for (**a**) summer (January, February, March), (**b**) fall (April, May, June), (**c**) winter (July, August, September), and (**d**) spring (October, November, December). For reference of locations see Figure 1a.

Potential mechanisms leading to the generation of these fronts could also be related to the bathymetry along the ISC. The northern basin (north of Desertores Islands) has average depths around 300–400 m, whereas the southern ISC is considerably shallower (Figure 1a) [44]. The presence of Desertores Islands limits the water exchange and the circulation between these two sub-regions with distinct regimes of environmental variability [21], which would explain the sharp contrast in SST and frontal activity between these areas (e.g., Figures 2 and 3). The dynamics at Boca del Guafo is likely a major factor influencing the intrusion of oceanic waters and generation of fronts in the southern ISC. Future studies, including field measurements, should focus on sampling the locations with high FP (Figure 4) to understand better the vertical structure and variability of fronts in northern Patagonia.

4. Summary

This study presents the first analyses of SST frontal variability in the Inner Sea of Chiloé and its adjacent coastal ocean. A high correlation with an *in situ* time series in Seno Reloncaví, a region with enhanced frontal variability and high seasonal fluctuations, validates the use of MUR SST fields. The annual cycle of SST gradient magnitudes suggested enhanced frontal activity in northern ISC (north of Desertores Islands) in spring-summer, whereas the coastal ocean off Chiloé presented the highest average gradients in summer and fall. These seasonal patterns are, in general, confirmed by the quantification of the SST frontal probability. Maximum probabilities reached about 10% in northern ISC and off Chiloé. Overall, the southern ISC presented low SST gradient magnitudes and frontal probability yearlong, except for the southern side of Boca del Guafo. An EOF analysis clarified the dominant modes of variability of SST gradient magnitude, highlighting (i) a coastal band of enhanced SST gradients off Chiloé in summer-fall which is coherent with coastal upwelling events and fronts, and (ii) maximum SST fronts in northern ISC and around the southern side of Boca del Guafo in spring-summer. A preliminary inspection of the annual cycle of chlorophyll fluorescence suggests an increased physical-biological coupling around ocean fronts since the highest fluorescence is found in northern ISC and off Chiloé Island, where SST gradient magnitudes and frontal probabilities are maximum. Future studies are suggested to occur in the regions with high frontal activity and considering high-frequency field observations in the surface mixed layer to understand further the dynamics of ocean fronts and their biological implications in northern Patagonia.

Author Contributions: Conceptualization: G.S.S. and W.H.; Methodology: G.S.S., W.H., R.M., C.R.; Investigation: G.S.S., W.H. and L.S.-M.; Resources: G.S.S., C.L. and I.P.-S.; Data curation: G.S.S., S.V., I.P.-S. and L.S.-M.; Writing—original draft preparation: G.S.S. with contribution from all coauthors. All authors have read and agreed to the published version of the manuscript.

Funding: This research was funded by FONDECYT grant number 11190209 and 1190805. GSS was partially funded by the Millennium Nucleus Center for the Study of Multiple Drivers on Marine Socio-Ecological Systems (MUSELS) funded by MINECON NC120028. WH has been funded by MUSELS and Universidad del Bío-Bío. CL has been funded by MUSELS and UPWELL Millenium Nucleus (grant ICM NCN 19153). RM has been funded by Universidad del Bío-Bío. The oceanographic buoy at Seno Reloncaví was funded by FONDEQUIP Grant EQM160167.

Data Availability Statement: MUR data are available at https://podaac.jpl.nasa.gov/dataset/MUR-JPL-L4-GLOB-v4.1). Buoy data at Seno Reloncaví are available at http://www.cdom.cl/estaciones.php?estacion=RLCV&seccion=Oceano. Chlorophyll fluorescence data are available at NASA's ocean color website http://oceancolor.gsfc.nasa.gov.

Conflicts of Interest: The authors declare no conflict of interest. The funders had no role in the design of the study; in the collection, analyses, or interpretation of data; in the writing of the manuscript, or in the decision to publish the results.

References

1. Franks, P.J. Sink or swim: Accumulation of biomass at fronts. *Mar. Ecol. Prog. Ser.* **1992**, *82*, 1–12. [CrossRef]
2. Reese, D.C.; O'Malley, R.T.; Brodeur, R.D.; Churnside, J.H. Epipelagic fish distributions in relation to thermal fronts in a coastal upwelling system using high-resolution remote-sensing techniques. *ICES J. Mar. Sci.* **2011**, *68*, 1865–1874. [CrossRef]
3. Danell-Jiménez, A.; Sanchez-Velasco, L.; Lavín, M.; Marinone, S. Three-dimensional distribution of larval fish assemblages across a surface thermal/chlorophyll front in a semienclosed sea. *Estuar. Coast. Shelf Sci.* **2009**, *85*, 487–496. [CrossRef]
4. Mahadevan, A. Modeling vertical motion at ocean fronts: Are nonhydrostatic effects relevant at submesoscales? *Ocean Model.* **2006**, *14*, 222–240. [CrossRef]
5. Thomas, L.N.; Tandon, A.; Mahadevan, A. Submesoscale processes and dynamics. *Ocean. Model. Eddying Regime* **2008**, *177*, 17–38.
6. Castelao, R.M.; Barth, J.A.; Mavor, T.P. Flow-topography interactions in the northern California Current System observed from geostationary satellite data. *Geophys. Res. Lett.* **2005**, *32*, L24612. [CrossRef]

7. Saldías, G.S.; Allen, S.E. The Influence of a Submarine Canyon on the Circulation and Cross-Shore Exchanges around an Upwelling Front. *J. Phys. Oceanogr.* **2020**, *50*, 1677–1698. [CrossRef]
8. Strub, P.T.; Kosro, P.M.; Huyer, A. The nature of the cold filaments in the California Current System. *J. Geophys. Res.* **1991**, *96*, 14743–14768. [CrossRef]
9. Nieto, K.; Demarcq, H.; McClatchie, S. Mesoscale frontal structures in the Canary Upwelling System: New front and filament detection algorithms applied to spatial and temporal patterns. *Remote Sens. Environ.* **2012**, *123*, 339–346. [CrossRef]
10. Otero, P.; Ruiz-Villarreal, M.; Peliz, A. River plume fronts off NW Iberia from satellite observations and model data. *ICES J. Mar. Sci.* **2009**, *66*, 1853–1864. [CrossRef]
11. Saldías, G.S.; Lara, C. Satellite-derived sea surface temperature fronts in a river-influenced coastal upwelling area off central-southern Chile. *Reg. Stud. Mar. Sci.* **2020**, *37*, 101322. [CrossRef]
12. Ullman, D.S.; Cornillon, P.C. Satellite-derived sea surface temperature fronts on the continental shelf off the northeast US coast. *J. Geophys. Res.* **1999**, *104*, 23459–23478. [CrossRef]
13. Belkin, I. New challenge: Ocean fronts. *J. Mar. Syst.* **2002**, *1*, 1–2. [CrossRef]
14. Belkin, I.M.; Cornillon, P.C.; Sherman, K. Fronts in large marine ecosystems. *Prog. Oceanogr.* **2009**, *81*, 223–236. [CrossRef]
15. Castelao, R.M.; Wang, Y. Wind-driven variability in sea surface temperature front distribution in the California Current System. *J. Geophys. Res.* **2014**, *119*, 1861–1875. [CrossRef]
16. Vazquez-Cuervo, J.; Torres, H.S.; Menemenlis, D.; Chin, T.; Armstrong, E.M. Relationship between SST gradients and upwelling off Peru and Chile: Model/satellite data analysis. *Int. J. Remote Sens.* **2017**, *38*, 6599–6622. [CrossRef]
17. Castelao, R.M.; Mavor, T.P.; Barth, J.A.; Breaker, L.C. Sea surface temperature fronts in the California Current System from geostationary satellite observations. *J. Geophys. Res.* **2006**, *111*. [CrossRef]
18. Vazquez-Cuervo, J.; Dewitte, B.; Chin, T.M.; Armstrong, E.M.; Purca, S.; Alburqueque, E. An analysis of SST gradients off the Peruvian Coast: The impact of going to higher resolution. *Remote Sens. Environ.* **2013**, *131*, 76–84. [CrossRef]
19. Wang, Y.; Castelao, R.M.; Yuan, Y. Seasonal variability of alongshore winds and sea surface temperature fronts in Eastern Boundary Current Systems. *J. Geophys. Res.* **2015**, *120*, 2385–2400. [CrossRef]
20. Wang, Y.; Yu, Y.; Zhang, Y.; Zhang, H.R.; Chai, F. Distribution and variability of sea surface temperature fronts in the south China sea. *Estuar. Coast. Shelf Sci.* **2020**, *240*, 106793. [CrossRef]
21. Lara, C.; Saldías, G.S.; Tapia, F.J.; Iriarte, J.L.; Broitman, B.R. Interannual variability in temporal patterns of Chlorophyll–a and their potential influence on the supply of mussel larvae to inner waters in northern Patagonia (41–44 S). *J. Mar. Syst.* **2016**, *155*, 11–18. [CrossRef]
22. Narváez, D.A.; Vargas, C.A.; Cuevas, L.A.; García-Loyola, S.A.; Lara, C.; Segura, C.; Tapia, F.J.; Broitman, B.R. Dominant scales of subtidal variability in coastal hydrography of the Northern Chilean Patagonia. *J. Mar. Syst.* **2019**, *193*, 59–73. [CrossRef]
23. Iriarte, J.; González, H.; Liu, K.; Rivas, C.; Valenzuela, C. Spatial and temporal variability of chlorophyll and primary productivity in surface waters of southern Chile (41.5–43 S). *Estuar. Coast. Shelf Sci.* **2007**, *74*, 471–480. [CrossRef]
24. León-Muñoz, J.; Urbina, M.A.; Garreaud, R.; Iriarte, J.L. Hydroclimatic conditions trigger record harmful algal bloom in western Patagonia (summer 2016). *Sci. Rep.* **2018**, *8*, 1–10. [CrossRef] [PubMed]
25. Lara, C.; Miranda, M.; Montecino, V.; Iriarte, J.L. Chlorophyll-a MODIS mesoscale variability in the Inner Sea of Chiloé, Patagonia, Chile (41–43° S): Patches and gradients? *Rev. Biol. Mar. Oceanogr.* **2010**, *45*, 217–225. [CrossRef]
26. Canny, J. A computational approach to edge detection. *IEEE Trans. Pattern Anal.* **1986**, *8*, 679–698. [CrossRef]
27. Lara, C.; Saldías, G.S.; Westberry, T.K.; Behrenfeld, M.J.; Broitman, B.R. First assessment of MODIS satellite ocean color products (OC3 and nFLH) in the Inner Sea of Chiloé, northern Patagonia. *Latin Am. J. Aquat. Res.* **2017**, *45*, 822–827. [CrossRef]
28. Emery, W.J.; Thomson, R.E. *Data Analysis Methods in Physical Oceanography*, 2nd ed.; Elsevier: Amsterdam, The Netherlands, 2004.
29. Silva, N.; Calvete, C.; Sievers, H. Masas de agua y circulación general para algunos canales australes entre Puerto Montt y Laguna San Rafael, Chile (Crucero Cimar-Fiordo 1). *Cien. Tecnol. Mar.* **1998**, *21*, 17–48.
30. Acha, E.M.; Mianzan, H.W.; Guerrero, R.A.; Favero, M.; Bava, J. Marine fronts at the continental shelves of austral South America: physical and ecological processes. *J. Mar. Syst.* **2004**, *44*, 83–105. [CrossRef]
31. Tello, A.; Rodriguez-Benito, C. Characterization of mesoscale spatio-temporal patterns and variability of remotely sensed Chl a and SST in the Interior Sea of Chiloe (41.4–43.5°S). *Int. J. Remote Sens.* **2009**, *30*, 1521–1536.
32. Strub, P.T.; James, C.; Montecino, V.; Rutllant, J.A.; Blanco, J.L. Ocean circulation along the southern Chile transition region (38°–46°S): Mean, seasonal and interannual variability, with a focus on 2014–2016. *Prog. Oceanogr.* **2019**, *172*, 159–198. [CrossRef] [PubMed]
33. Pérez-Santos, I.; Seguel, R.; Schneider, W.; Linford, P.; Donoso, D.; Navarro, E.; Amaya-Cárcamo, C.; Pinilla, E.; Daneri, G. Synoptic-scale variability of surface winds and ocean response to atmospheric forcing in the eastern austral Pacific Ocean. *Ocean Sci.* **2019**, *15*, 1247–1266. [CrossRef]
34. Giesecke, R.; Clement, A.; Garcés-Vargas, J.; Mardones, J.I.; González, H.E.; Caputo, L.; Castro, L. Proliferaciones masivas de salpas en el mar interior de la isla de Chiloé (sur de Chile): Posibles causas y consecuencias ecológicas. *Lat. Am. J. Aquat. Res.* **2014**, *42*, 604–621. [CrossRef]
35. Dávila, P.M.; Figueroa, D.; Müller, E. Freshwater input into the coastal ocean and its relation with the salinity distribution off austral Chile (35–55 S). *Cont. Shelf Res.* **2002**, *22*, 521–534. [CrossRef]

36. Calvete, C.; Sobarzo, M. Quantification of the surface brackish water layer and frontal zones in southern Chilean fjords between Boca del Guafo (43 30′ S) and Estero Elefantes (46 30′ S). *Cont. Shelf Res.* **2011**, *31*, 162–171. [CrossRef]
37. Iriarte, J.; León-Muñoz, J.; Marcé, R.; Clément, A.; Lara, C. Influence of seasonal freshwater streamflow regimes on phytoplankton blooms in a Patagonian fjord. *N. Z. J. Mar. Freshw. Res.* **2017**, *51*, 304–315. [CrossRef]
38. Saldías, G.S.; Sobarzo, M.; Quiñones, R. Freshwater structure and its seasonal variability off western Patagonia. *Prog. Oceanogr.* **2019**, *174*, 143–153. [CrossRef]
39. Buchan, S.J.; Quiñones, R.A. First insights into the oceanographic characteristics of a blue whale feeding ground in northern Patagonia, Chile. *Mar. Ecol. Prog. Ser.* **2016**, *554*, 183–199. [CrossRef]
40. Bost, C.A.; Cotté, C.; Bailleul, F.; Cherel, Y.; Charrassin, J.B.; Guinet, C.; Ainley, D.G.; Weimerskirch, H. The importance of oceanographic fronts to marine birds and mammals of the southern oceans. *J. Mar. Syst.* **2009**, *78*, 363–376. [CrossRef]
41. Taylor, J.R.; Ferrari, R. Ocean fronts trigger high latitude phytoplankton blooms. *Geophys. Res. Lett.* **2011**, *38*. [CrossRef]
42. Nieto, K.; Xu, Y.; Teo, S.L.; McClatchie, S.; Holmes, J. How important are coastal fronts to albacore tuna (Thunnus alalunga) habitat in the Northeast Pacific Ocean? *Prog. Oceanogr.* **2017**, *150*, 62–71. [CrossRef]
43. Belkin, I.M.; Cornillon, P.C. Fronts in the world ocean's large marine ecosystems. *ICES CM* **2007**, *500*, 21.
44. González, H.; Calderón, M.; Castro, L.; Clement, A.; Cuevas, L.; Daneri, G.; Iriarte, J.; Lizárraga, L.; Martínez, R.; Menschel, E.; et al. Primary production and plankton dynamics in the Reloncaví Fjord and the Interior Sea of Chiloé, Northern Patagonia, Chile. *Mar. Ecol. Prog. Ser.* **2010**, *402*, 13–30. [CrossRef]

 *remote sensing*

Letter

Comparison of Measured Surface Currents from High Frequency (HF) and X-Band Radar in a Marine Protected Coastal Area of the Ligurian Sea: Toward an Integrated Monitoring System

Lyuba Novi [1], Francesco Raffa [1,*] and Francesco Serafino [2]

[1] Institute of Geosciences and Earth Resources (IGG) National Research Council (CNR) Via Moruzzi, 1-56124 Pisa, Italy; ljuba.novi@igg.cnr.it

[2] Institute of Bioeconomy (IBE) National Research Council (CNR) Via Madonna del Piano, 10-50019 Sesto Fiorentino (FI), Italy; francesco.serafino@cnr.it

* Correspondence: francesco.raffa@igg.cnr.it

Received: 31 July 2020; Accepted: 17 September 2020; Published: 19 September 2020

Abstract: Two different ground-based remote sensing instruments can be used for the near-real-time monitoring of surface waves and currents, namely the high frequency HF radar and the microwave X-band radar. The HF system reaches larger offshore distances at lower spatial resolutions and provides a poorer measurement of the wave-induced currents in very shallow waters. On the other hand, the X-band system achieves significantly higher spatial resolutions with a smaller offshore coverage. This study provides a preliminary comparison of the measured surface currents, obtained by the two different tools where they overlap. The comparison showed a good agreement between the measures with some discrepancies ascribable to the difference in the characteristics of the two radar technologies.

Keywords: wave radar; sea waves; model data; Mediterranean sea

---

1. Introduction

The observation and monitoring of marine coastal currents is an important task for coastal protection, erosion control, and flood mitigation as well as near-shore fishing management and marine operations such as installations of offshore wind farms or oil and gas plants [1].

In recent years, the monitoring of surface currents with remote sensing techniques has greatly improved, making it possible to even perform real-time observations over sea surface areas of different extension. Among these techniques, two different ground-based remote sensing instruments can be deployed for the near-real-time monitoring of surface waves and currents, namely the high frequency HF radar and the microwave X-band radar. They directly measure the directional wave spectra at a spatial resolution from 250 m to 15 km, which depends on the specific allocated bandwidth and antenna design.

The overall spatial coverage of these tools significantly differs, as well does their spatial resolution. The HF system reaches larger offshore distances at lower spatial resolutions and provides a poorer measurement of the wave-induced currents in very shallow waters. On the other hand, the X-band system achieves significantly higher spatial resolutions with a smaller offshore coverage. The inherent differences of HF and X-band radars open new routes toward an integrated monitoring technique, which exploits the complementary nature of the output provided separately by the two systems [2,3].

High frequency (HF) coastal radars are very powerful instruments, providing information on surface velocity in terms of hourly maps over extended regions (range up to 100 km) and with high spatial resolution (order of 1–3 km). This information can be used to address several societal needs such as navigation safety, search and rescue, oil spill or other pollutant tracking, marine protected areas and fishery management [4,5].

The X-band radar represents a practical remote sensing system for sea waves and current monitoring in coastal and shallow waters. It is used for the acquisition and the analysis of consecutive sea surface images [6,7]. The surface current is retrieved from a sequence of these radar images by an inversion procedure, that accounts for the modulation effects that depend on both the sea state and the radar parameters as well as on the acquisition geometry [7–11].

This work is devoted to explore whether an integrated monitoring system can be successfully employed to measure surface currents in near real time across a variety of spatial scales. By blending HF and X-band radar data such an integrated system aims at reaching a high near-shore spatial resolution still covering a large off-shore area. This study provides a preliminary comparison of the measured surface currents, obtained by the two different tools where they overlap. Measurements taken at a selected study site located in the Ligurian Sea were analyzed.

As the present work focused on comparing two different measuring tools, rather than to study the local marine dynamics, an analysis of the surface circulation of the Ligurian Sea was beyond our aim and is already quite a well-covered topic in literature [12].

2. Materials and Methods

2.1. Study Site and Analyzed Sea Conditions

The study site is located within the Eastern Ligurian Sea, in the North West Mediterranean Sea as depicted in Figure 1, along a 15 km-long coastline in an area situated in front of which bounds an important Marine Protected Area (Cinque Terre).

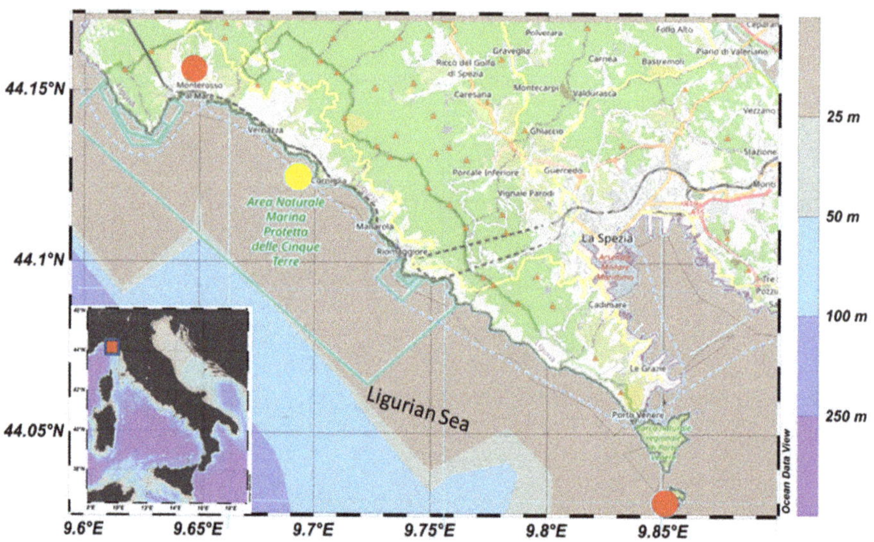

Figure 1. Study area in the Ligurian Sea: red circles indicate the High Frequency (HF) radar stations. The yellow circle indicate the X-band radar location.

A CNR-ISMAR HF Radar Network has been installed along the coast of Eastern Liguria, near La Spezia and Cinque Terre, in year 2016 and is composed by two CODAR SeaSonde HF radar stations operating in the frequency band of 25 MHz. A CONSILIUM/SELESMAR X-band radar was installed at Corniglia (SP) about 60 meters above sea level. The radar locations are shown in Figure 1.

At the installation site, the HF radar and the X-band radar both worked from 12 September 2017 to 1 April 2018.

The analysis was carried out as follows:

- As a first preliminary step, a qualitative snapshot comparison of the spatially-varying time-averaged surface velocity fields (horizontal components) derived by HF and X-Band is shown as a time average over a reduced time range. Despite this part has no quantitative aims, it allows us to show the overlapping points of the two instruments. Due to the different spatial resolutions involved, a linear interpolation in space was carried out to have measurements on matching grids. HF outputs were evaluated on the X-Band grid before qualitative comparison of the surface velocity time averaged field. However, due to the small overlap among them, only a few HF grid points resided within the X-Band grid. This likely makes the interpolated HF field oversmoothed, and a significant quantitative comparison at these scales is therefore not significant.

- A quantitative comparison at overlapping points was carried out for the measured time-series sampled from 12 September 2017 to 1 April 2018. The overlapping points between the HF and X-band grids, without any spatial interpolation, were identified and selected as comparison sites, namely A and B. The time-varying zonal (U) and meridional (V) surface velocity components, independently derived by the HF and X-band, were analyzed and compared at these locations. A comparison between the HF and X-Band time signatures, means, and standard deviations is given. Root mean square errors between X-band velocities and HF velocities at A and B were also computed.

2.2. HF Radar Data Collection and Analysis

The HF radar network was designed, implemented, and managed through the efforts of Institute of Marine Sciences - National Research Council (ISMAR-CNR La Spezia) [13]. HF radar data were collected and processed by ISMAR-CNR within the Ritmare and Jerico–Next projects [14]. The datasets hereinafter considered were downloaded from the website http://ritmare.artov.isac.cnr.it/thredds/catalog.html. Depending on the sea state, estimated errors ranged from 3 to 10 cm·s^{-1} and explained only part of the rms difference of 10–20 cm·s^{-1} found between HF and the in situ current measurements. The rest was assumed to be due the differences of the quantities measured (e.g., the spatial averaging [15]).

The acquisition settings are listed in Table 1.

Table 1. HF Radar system parameters.

Frequency Band	Radial Coverage	Radial Range Cutoff	Radial Resolution	Angular Resolution
(MHz)	(km)	(km)	(km)	(deg)
26	35–45	45	1	5

HF radar is appropriate to detect surface ocean currents due to the diffraction grating effects of the rough sea surface [16,17]. Just when the radar signal scatters off a wave that is exactly half the transmitted signal wavelength, and that wave is traveling in a radial path either directly away from or toward the radar, the radar signal will return directly to its source. The scattered radar electromagnetic waves coherently add up, resulting in a strong energy return at a certain specific wavelength. The returning signal exhibits a Doppler-frequency shift that would always turn up at a known position in the frequency spectrum in the absence of ocean currents. Nevertheless, the observed Doppler-frequency shift does not match up exactly with the theoretical wave speed. The Doppler-frequency shift includes the information of the principal ocean current on the wave velocity in a radial pathway, jointly with the theoretical wave speed. Total velocities are derived using least square fit, which maps radial velocities measured from individual sites on a Cartesian grid. The final result is a map of the horizontal components of the ocean currents, on a regular grid, in the area covered by two or more radar stations [14].

2.3. X-Band Radar Data Collection and Analysis

A CONSILIUM SELESMAR marine X-band radar was installed on the roof of the sewage treatment plant at Corniglia (SP) about 60 meters above sea level. The radar antenna was located at the coordinates 44°07′10″ N, 9°42′20″ E.

The radar system radiates a maximum power of 25 KW, operates in the short pulse mode (i.e., pulse duration of about 90 ns), and is equipped with an 9-ft (2.7 m) long antenna with horizontal polarization (HH). These features enable reaching a spatial resolution of about 9m and an angular resolution of approximately 0.9°. The signal received by the antenna was converted through an analog–digital converter and interpolated on a Cartesian grid with a regular spacing of about 10 m to obtain two-dimensional (2D) sea surface images. The image sequence acquired by the X-band radar was stored and processed, and each raw data sequence consisted of 64 individual images stored every 2.4 s. The accuracy of the X-band radar in terms of measured velocities was of the order 10 cm s^{-1} [18].

The acquisition settings are listed in Table 2.

Table 2. X-band radar system parameters.

Frequency Band	Radial Coverage	Time Range	Spatial Resolution	Angular Resolution
(MHz)	(km)	(s)	(m)	(deg)
9200	5.55	2.4	9	0.9

The image processing to extract the inhomogeneous surface current fields from the X-band radar data were based on the so called "Local Method", proposed in [19,20] and can be applied to data acquired in coastal areas, where the presence of coastlines and varying bathymetry cause a spatial inhomogeneity of the wave motion [3,21–23].

A block diagram of the inversion procedure is presented in Figure 2.

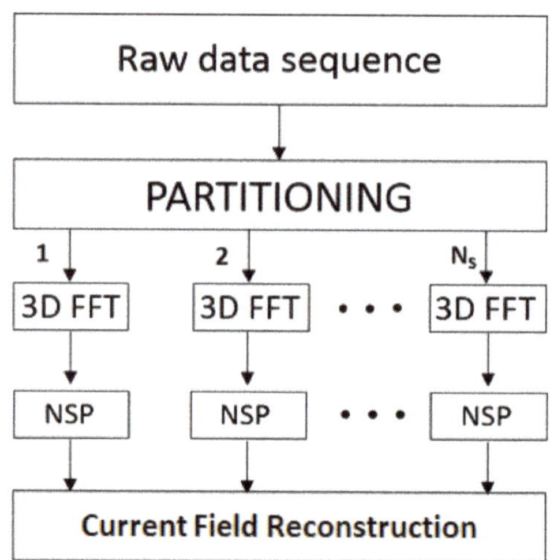

Figure 2. Block diagram of the current field reconstruction procedure, where 3D FFT is the fast Fourier transformto obtain the 3D radar sub-spectra and NSP is the normalized scalar product technique.

The partitioning procedure is needed to extract N_s spatially overlapping sub-areas, so it is possible to assume the waves' homogeneity and uniformity from the analyzed radar data temporal sequence. After that, the fast Fourier transform (FFT) is applied to the N_s temporal sub-sequences to obtain the 3D radar sub-spectra.

Each spectrum is expressed as $\left\{F^j\big(\vec{k},\omega\big)\right\}_{j=1,\dots,N_s}$, where $\vec{k} = \big(k_x, k_y\big)$ is the wave-number vectorand ω the angular frequency; spectra are then analyzedby applying the normalized scalar product (NSP) technique [7], in order to retrieve the local surface current vector through the following estimator:

$$
V^j\big(\vec{U}\big) = \underset{h}{argmax}\,\frac{\big\langle \big|F^j\big(\vec{k},\omega\big)\big|, G\big(\vec{k},\omega,\vec{U}\big)\big\rangle}{\sqrt{P_F P_G}}
\tag{1}
$$

where $G\big(\vec{k},\omega,\vec{U}\big) = \delta\big(\omega - \sqrt{gk} - \vec{k}\cdot\vec{U}\big)$ is the characteristic function based on the dispersion relation; $\delta(\cdot)$ is the Dirac delta distribution; $\langle|F|, G\rangle$ represents the scalar product between the functions $|F|$ and G; and P_F and P_G are the powers associated with $|F|$ and G, respectively.

Once the local (sub-areas) current vectors have been estimated, it is possible to define the 'global' (applied to the whole radar spectrum) band-pass (BP) filter [3,20,21].

3. Results and Discussion

Figure 3 shows a qualitative snapshot of the HF and X-band surface velocity fields, time averaged over a sample period on the original spatial grids, with the purpose to qualitatively show the coverage overlaps and the overlapping points. More in detail, red (blue) arrows are located at the HF (X-band) grid points, whereas colored dots indicate the instruments overlaps.

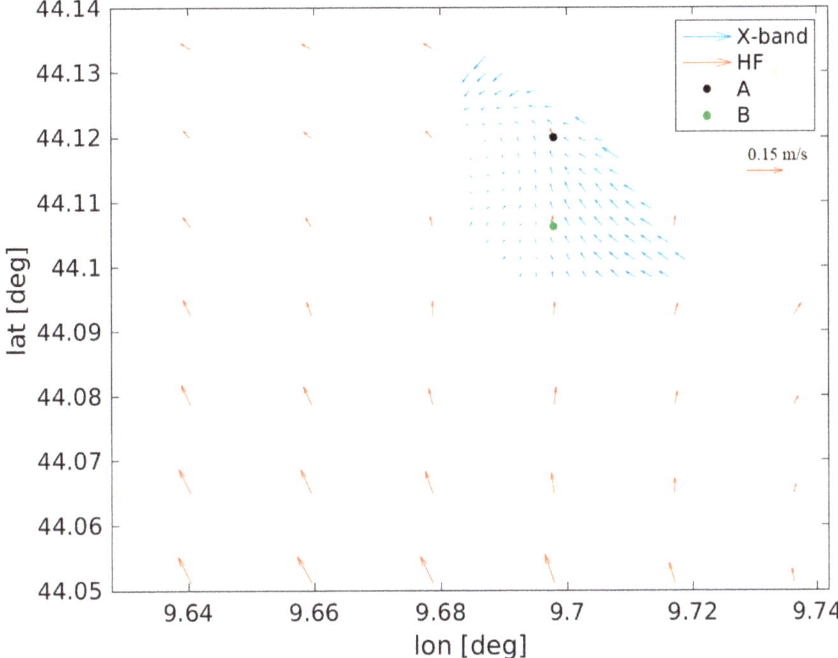

Figure 3. Qualitative snapshot of surface currents from X-band (blue arrows) and HF (red arrows), time averaged from 12 September 2017 to 18 September 2017 on original grids. Overlapping points A and B are indicated by the black and greed dots respectively.

Due to the very different spatial resolutions, some differences in the spatial velocity patterns may arise between the HF and X-band, which capture different spatial scales. Strong spatial variability at the HF sub-grid level may not be completely captured by the low resolution radar, resulting in an over-smoothed surface circulation, especially in the coastal zone. In order to cover coastal waters with

HF measurements at a higher spatial resolution, a rather trivial option is to linearly interpolate the HF-derived currents on a finer grid to compensate for the missing locations. However, it of course does not improve the quality of data, as the sub-grid processes still remain unresolved. Although the main large-scale current direction is consistently measured by the two instruments, the HF-derived circulation pattern does not capture the details of the near-shore spatial variability, especially in the west–northwest portion of the domain. Here, the X-band measurements revealed the existence of a cyclonic branch at the western edge of the grid, which was instead missed by the HF-derived data at the same location.

As clearly visible in Figure 3, at intermediate off-shore distances, an overlapping zone exists between the HF and X-band grids, where the time-series of surface velocities can be directly compared without any additional interpolation in space. In such an intermediate zone, X-band and HF derived data without spatial interpolation are expected to give similar results over time if the X-band surface currents are correctly derived. Seaward of these overlapping locations, the HF radar has the advantage of a long distance coverage suitable to capture larger scale circulation structures, whereas the X-band becomes advantageous shoreward of the overlapping areas, where smaller scale dynamics needs to be resolved.

Time series of the northward and eastward surface velocity components, derived by HF and X-band radar at overlapping points A and B, are reported in Figures 4 and 5, respectively, from 12 September 2017 to 1 April 2018. In each panel of Figures 4 and 5, the green line refers to the HF measurements, whereas the black line shows the X-band ones. Missing data in the time series corresponds to periods where the X-band radar system did not work or the surface dynamics in the near shore area, covered by the X-band radar, cannot be measured with enough accuracy due to low sea state or rain that affect the current field estimation.

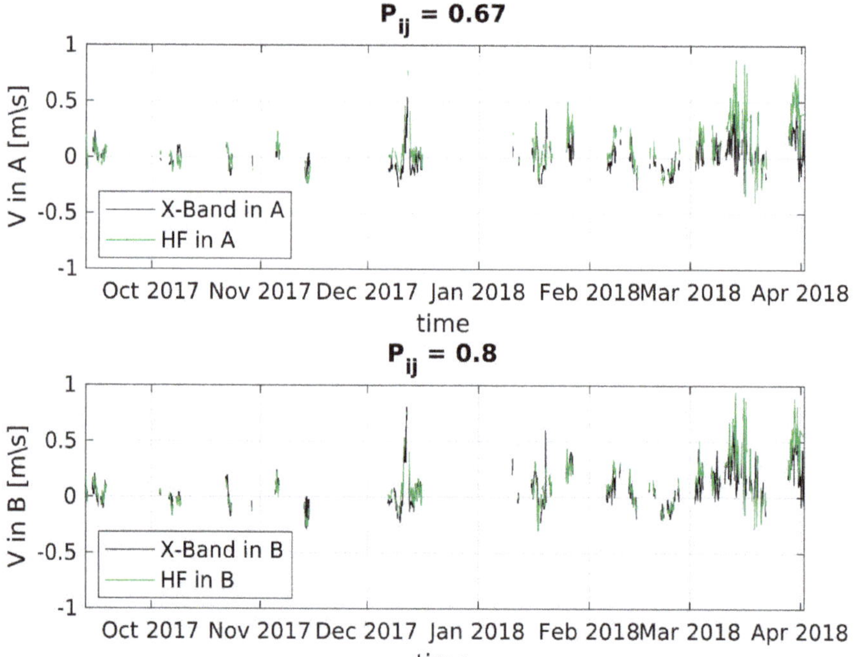

Figure 4. Time series of the northward surface velocity components, derived by HF and X-band radar at overlapping points A and B. The values on top (Pij) denote the Pearson's correlation coefficient.

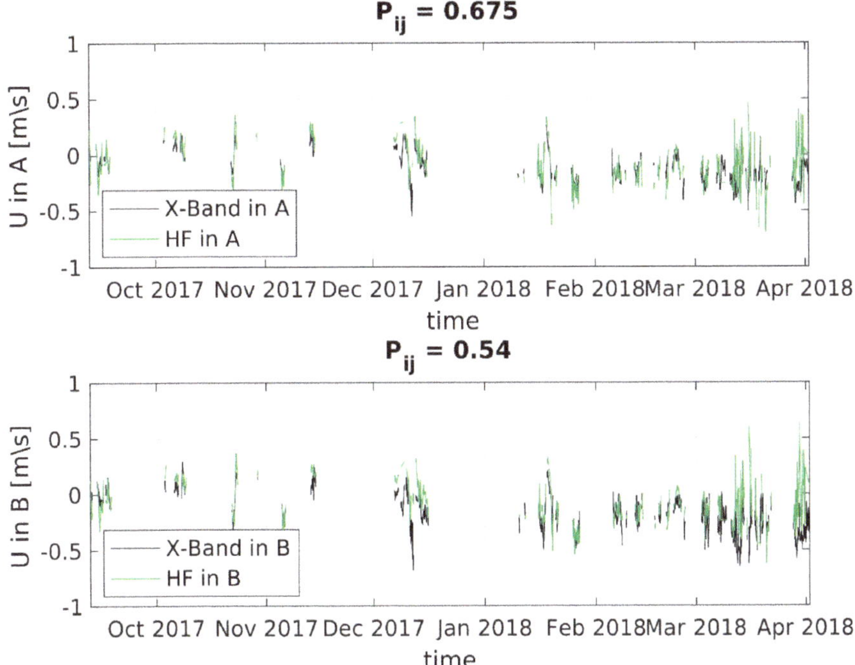

Figure 5. Time series of the eastward surface velocity components, derived by HF and X-band radar at overlapping points A and B. The values on top (Pij) denote the Pearson's correlation coefficient.

HF derived currents are provided as hourly means, whereas X-band measurements are obtained as instantaneous values at irregular time steps (multiple time steps per hour). In order to get a clearer comparison, the X-band data were therefore averaged over time to get hourly means. The HF values were then linearly interpolated in time in order to match the X-band hourly mean time spacing.

X-band derived velocity components display a good agreement with the HF counterpart throughout the sampling. A significant (p value $\ll 0.01$) positive correlation among the data was also indicated by the Pearson's linear correlation coefficients Pij, here computed. Pij was 0.675 and 0.54 for the U components in A and B, respectively, whereas it had a value of 0.67 and 0.8 for V in A and B, respectively. The root mean square errors of U at point A and B were 0.14 m/s 0.17 m/s, respectively, while for the northward components, it assumed the values of 0.14 m/s and 0.13 m/s in A and B, respectively. Figure 6 shows the resulting time signature of the velocity intensity at overlapping points A and B as derived by the two instruments; a close up of a shorter timeslot is shown in Figure 7 only for clearer visualization purposes. Figure 8 finally reports a scatter plot of the HF and X-band surface velocity components in A and B, separately.

As a final step, we show in Figures 9 and 10 a close up on the measured components over a reduced time range (12 September 2017 to 18 September 2017) characterized by the time-average condition reported in Figure 3 for that time range. The reported time-series are shown at regular hourly time spacing. The corresponding root mean square errors, for the shorter set, were 0.1 m/s and 0.12 m/s for U in A and B, respectively, and 0.07 m/s and 0.05 m/s for V in A and B, respectively. It is interesting to note the substantial disagreement in northward components at location A, occurring at 15 September and neighboring times (upper panel of Figure 9). Here, the X-band measurements showed a positive peak as opposed to the local decrease captured by the HF measurements. The computed spatial standard deviation of the northward component, at this time (and neighboring times), exceeded the 90% of its maximum values, revealing the existence of a high spatial variability that might be not completely captured by the low resolution grid of the HF radar.

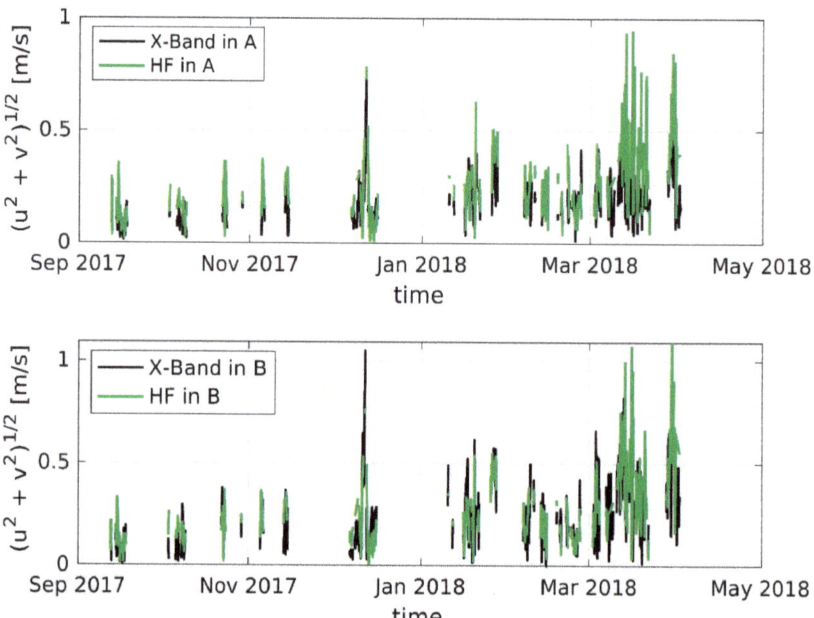

Figure 6. Velocity intensities derived by HF and X-band radar at overlapping points A and B.

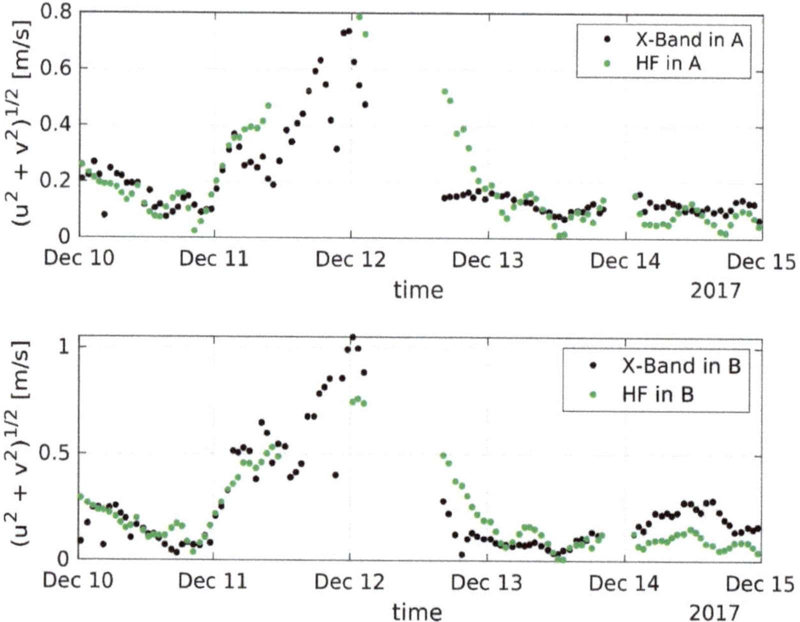

Figure 7. Velocity intensities derived by HF and X-band radar at overlapping points A and B. Close up of Figure 6 over a reduced time slot (10 December 2017 at 00:24 to 14 December 2017 at 23:23).

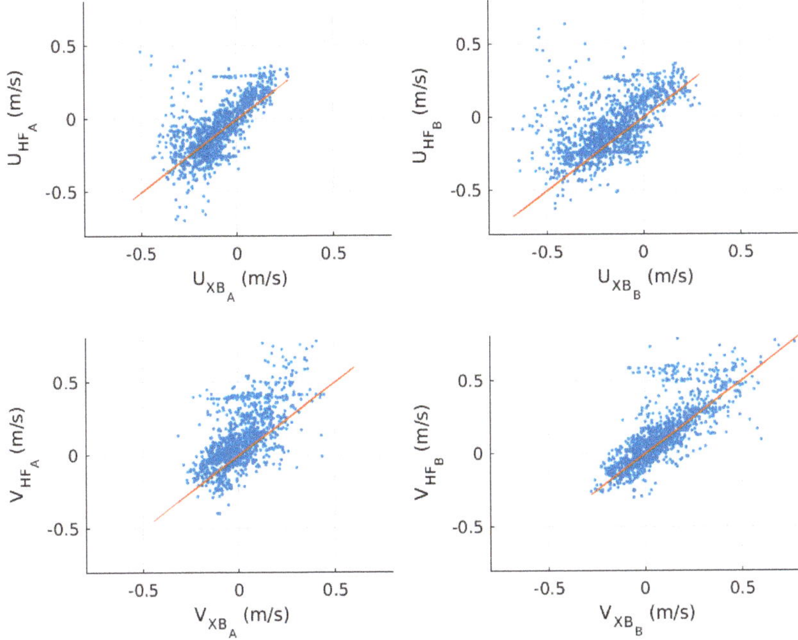

Figure 8. Scatter plot of the HF and X-band surface velocity components in A (**left**) and B (**right**).

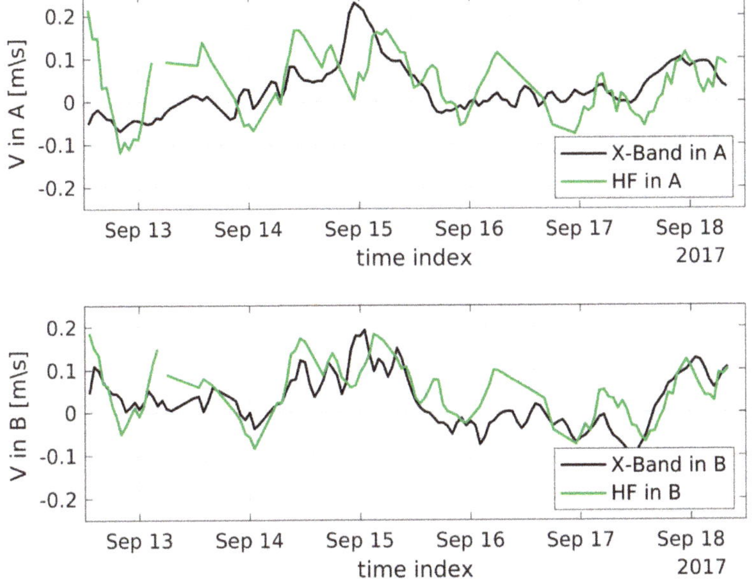

Figure 9. Time-series of the northward surface velocity components, derived by HF and X-band radar at overlapping points A and B for a reduced time period (12–18 September 2017).

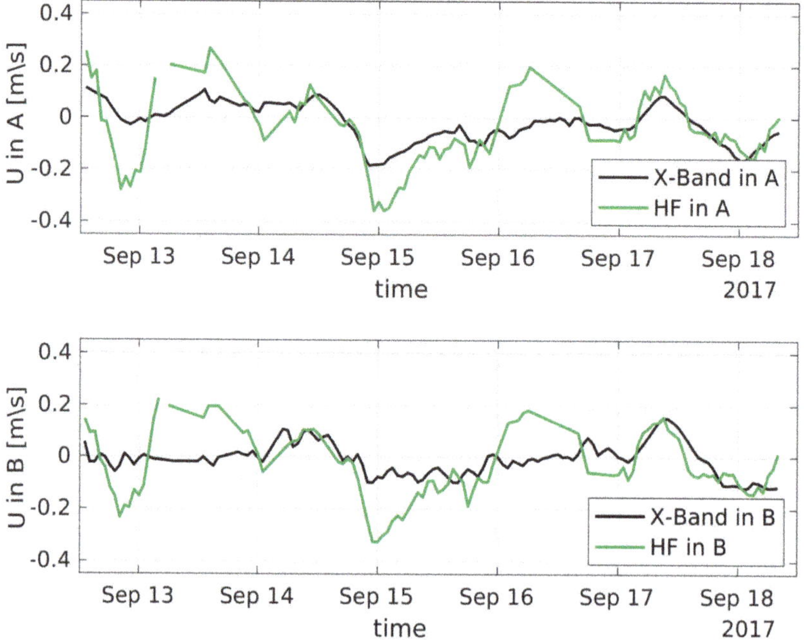

Figure 10. Time series of the eastward surface velocity components, derived by HF and X-band radar at overlapping points A and B for a reduced time period (12–18 September 2017).

Similarly, the eastward components at the same time locations were also characterized by a local overestimation of the velocity intensity by the HF radar compared to the X-band one, at location A (upper panel of Figure 10). Additionally, in this case, a field spatial standard deviation above 85% of the maximum level was found. At early times (i.e., between 12–13 September (Figure 9 and upper panel of Figure 10), discrepancies between HF and X-band data were associated with values of the spatial standard deviations ranging between 50% and 70% of the maximum value for the northward component, and around 50% of the maximum value for the eastward component, revealing a quite significant spatial variability that may affect the HF derived values in the analyzed sea condition.

4. Conclusions

In this work, the surface current fields measured by an X-band radar were compared with those provided by a HF-band radar. The comparison showed good agreement between the measures, although some discrepancies were also detected. At this stage, we cannot rigorously explain the nature of such differences among the data. A possible explanation resides in the different spatial scales that can be captured by the two tools. The HF radar is likely to over-smooth the small-scale features typically occurring in coastal waters or complex nearshore bathymetries. The high spatial variability of such features might not be completely captured by the HF coarse resolution, largely remaining a not-resolved sub grid process. On the other hand, the higher spatial resolution of the X-band radar allows for the capture of finer processes that are likely to carry higher sources of field local variance. The second possible source of differences between the HF and X-band derived currents is the inherent limitation of the latter, which loses accuracy when estimating surface currents in under-developed sea conditions. An important aspect that emerges from this work is the possible integrated usage of the two instruments, which exploits the advantages provided by the individual parts (wider spatial coverage for the HF band and higher spatial resolution for the X-band). Consequently, more accurate measurements of surface currents can be carried out by the combined system on large spatial domains still capturing smaller scale effects in the nearshore area. The analysis carried out in this study represents a preliminary assessment for the system performance. Its effective usage aimed at unveiling the physical processes of oceanographic relevance still remains beyond the target of this work. A physically-oriented application aiming at enlarging the background of physical oceanography is clearly needed to fully exploit the system potential, and will be explored in a future study.

Author Contributions: F.S. proposed the idea. L.N. and F.S. developed the method. L.N. and F.R. analyzed the results and wrote the text. F.S. supervised the work. All authors commented on the paper. All authors have read and agreed to the published version of the manuscript.

Funding: This research received no external funding.

Acknowledgments: For the HF radar data considered in the present letter, the authors are very grateful to the manager of the datasets of the HF radar data, collected and processed by ISMAR-CNR within the Ritmare and Jerico-Next projects.

Conflicts of Interest: The authors declare no conflict of interest.

References

1. Klemas, V. Remote Sensing of Coastal and Ocean Currents: An Overview. *J. Coast. Res.* **2012**, *28*, 576–586. [CrossRef]
2. Capodici, F.; Ciraolo, G.; Cosoli, S.; Maltese, A.; Mangano, M.C.; Sarà, G. Downscaling Hydrodynamics Features to Depict Causes of Major Productivity of Sicilian—Maltese Area and Implications for Resource Management. *Sci. Total Environ.* **2018**, *628*, 815–825. [CrossRef] [PubMed]
3. Serafino, F.; Lugni, C.; Ludeno, G.; Arturi, D.; Uttieri, M.; Buonocore, B.; Zambianchi, E.; Budillon, G.; Soldovieri, F. REMOCEAN: A flexible X-band radar system for sea-state monitoring and surface current estimation. *IEEE Geosci. Remote Sens. Lett.* **2012**, *9*, 822–826. [CrossRef]
4. Sciascia, R.; Berta, M.; Carlson, D.F.; Griffa, A.; Panfili, M.; La Mesa, M.; Corgnati, L.; Mantovani, C.; Domenella, E.; Fredj, E.; et al. Linking Sardine Recruitment in Coastal Areas to Ocean Currents Using Surface Drifters and HF Radar: A Case Study in the Gulf of Manfredonia, Adriatic Sea. *Ocean Sci.* **2018**, *14*, 1461. [CrossRef]
5. Abascal, A.J.; Sanchez, J.; Chiri, H.; Ferrer, M.I.; Cárdenas, M.; Gallego, A.; Castanedo, S.; Medina, R.; Alonso-Martirena, A.; Berx, B.; et al. Operational Oil Spill Trajectory Modelling Using HF Radar Currents: A Northwest European Continental Shelf Case Study. *Mar. Pollut. Bull.* **2017**, *119*, 336–350. [CrossRef] [PubMed]
6. Ludeno, G.; Raffa, F.; Soldovieri, F.; Serafino, F. X-Band Radar for the Monitoring of Sea Waves and Currents: A Comparison between Medium and Short Radar Pulses. *Geosci. Instrum. Methods Data Syst. Discuss.* **2017**, *2017*, 1–11. [CrossRef]
7. Serafino, F.; Lugni, C.; Soldovieri, F. A novel strategy for the surface current determination from marine X-Band radar data. *IEEE Geosci. Remote Sens. Lett.* **2010**, *7*, 231–235. [CrossRef]
8. Plant, W.J.; Keller, W.C. Evidence of Bragg Scattering in Microwave Doppler Spectra of Sea Return. *J. Geophys. Res. Ocean.* **1990**, *95*, 16299–16310. [CrossRef]
9. Lee, P.H.Y.; Barter, J.D.; Beach, K.L.; Hindman, C.L.; Lake, B.M.; Rungaldier, H.; Shelton, J.C.; Williams, A.B.; Yee, R.; Yuen, H.C. X-band microwave backscattering from ocean waves. *J. Geophys. Res. Ocean.* **1995**, *100*, 2591–2611. [CrossRef]
10. Wenzel, L.B. Electromagnetic scattering from the sea at low grazing angles. In *Suface Waves and Fluxes*; Geernaert, G.L., Plant, W.J., Eds.; Kluwer Academic: Norwell, MA, USA, 1990; pp. 41–108.
11. Borge, J.C.N.; Rodríquez, G.R.; Hessner, K.; González, P.I. Inversion of marine radar images for surface wave analysis. *J. Atmos. Ocean. Technol.* **2004**, *21*, 1291–1300. [CrossRef]
12. Astraldi, M.; Gasparini, G.P.; Manzella, G.M.R.; Hopkins, T.S. Temporal variability of currents in the eastern Ligurian Sea. *J. Geophys. Res.* **1990**, *95*, 1515–1522. [CrossRef]
13. Carrara, P.; Corgnati, L.; Griffa, A.; Oggioni, A.; Pepe, M.; Kalampokis, A.; Zambianchi, E.; Mantovani, C.; Cosoli, S.; Raffa, F.; et al. The RITMARE coastal radar network and applications to monitor marine transport infrastructures. In Proceedings of the European Geosciences Union General Assembly, Vienna, Austria, 27 April–2 May 2014.

14. Corgnati, L.; Mantovani, C.; Griffa, A.; Bellomo, L.; Carlson, D.F.; Magaldi, M.G.; Berta, M.; Pazienza, G.; D'Adamo, R. The ISMAR high frequency coastal radar network: Monitoring surface currents for management of marine resources. In Proceedings of the MTS/IEEE OCEANS 2015—Genova: Discovering Sustainable Ocean Energy for a New World, Genova, Italy, 18–21 May 2015.

15. Essen, H.-H.; Gurgel, K.-W.; Schlick, T. On the accuracy of current measurements by means of HF radar. *IEEE J. Ocean. Eng.* **2000**, *25*, 472–480. [CrossRef]

16. Barrick, D.E.; Evans, M.W.; Weber, B.L. Ocean surface currents mapped by radar. *Science* **1977**, *198*, 138–144. [CrossRef] [PubMed]

17. Barrick, D.E.; Lipa, B.J. An Evaluation of Least-Squares and Closed-Form Dual-Angle Methods for CODAR Surface-Current Applications. *IEEE J. Ocean. Eng.* **1986**, *11*, 322–326. [CrossRef]

18. Nasello, C.; Ciraolo, G.; Serafino, F.; Ludeno, G.; Soldovieri, F.; Raffa, F. A comparison between drifter and X-band wave radar for sea surface current estimation. *Remote Sens.* **2016**, *8*, 696–706.

19. Ludeno, G.; Reale, F.; Dentale, F.; Carratelli, E.P.; Natale, A.; Soldovieri, F.; Serafino, F. An X-band radar system for bathymetry and wave field analysis in a harbour area. *Sensors* **2015**, *15*, 1691–1707. [CrossRef] [PubMed]

20. Ludeno, G.; Flampouris, S.; Lugni, C.; Soldovieri, F.; Serafino, F. A novel approach based on marine radar data analysis for high-resolution bathymetry map generation. *IEEE Geosci. Remote Sens. Lett.* **2014**, *11*, 234–238. [CrossRef]

21. Ludeno, G.; Brandini, C.; Lugni, C.; Arturi, D.; Natale, A.; Soldovieri, F.; Gozzini, B.; Serafino, F. Remocean system for the detection of the reflected waves from the costa concordia ship wreck. *IEEE J. Sel. Top. Appl. Earth Obs. Remote Sens.* **2014**, *7*, 3011–3018. [CrossRef]

22. Bell, P.S. Mapping Shallow Water Coastal Areas Using a Standard Marine X-Band Radar. In Proceedings of the Hydro8, Liverpool, UK, 4–6 November 2008; pp. 1–9.

23. Senet, C.M.; Seemann, J.; Flampouris, S.; Ziemer, F. Determination of Bathymetric and Current Maps by the Method DiSC Based on the Analysis of Nautical X-Band Radar Image Sequences of the Sea Surface (November 2007). *IEEE Trans. Geosci. Remote Sens.* **2008**, *46*, 2267–2279. [CrossRef]

MDPI
St. Alban-Anlage 66
4052 Basel
Switzerland
Tel. +41 61 683 77 34
Fax +41 61 302 89 18
www.mdpi.com

Remote Sensing Editorial Office
E-mail: remotesensing@mdpi.com
www.mdpi.com/journal/remotesensing